PLUMBING

Harold E. Babbitt

Professor Emeritus of Sanitary Engineering
University of Illinois
Visiting Professor of Civil Engineering
Iowa State University

THIRD EDITION

McGRAW-HILL BOOK COMPANY

New York Toronto London

1960

PLUMBING

 Library of Congress Catalog Card Number: 59-11919

18 19 VBVB 89876

ISBN 07-002688 2

PREFACE

The preface of the second edition of this book called attention to the increase of information, of research, and of publications relating to plumbing subsequent to the appearance of the first edition in 1922. During the period until 1950, when the second edition was published, such notable reports appeared as: "Recommended Minimum Requirements for Plumbing in Dwellings and Similar Buildings," published by the National Bureau of Standards in 1923 and widely known as the Hoover Report; the "Plumbing Manual," Report BMS66, published by the National Bureau of Standards in 1940; "The Uniform Plumbing Code for Housing," Technical Paper 6, published by the Housing and Home Finance Agency in 1948; "Report of the Uniform Plumbing Code Committee," published by the Housing and Home Finance Agency in 1949; and the "American Standard Plumbing Code," American Standards Association A 40.7–1949, published by the American Society of Mechanical Engineers in 1949.

The need for codifying, assembling, and adjusting such a wealth of information resulted, in part, in the publication, in 1955, of the "National Plumbing Code," ASA A 40.8–1955, by the American Society of Mechanical Engineers. The Code, published with but three illustrations and couched in technical language, has been interpreted by illustrations and explanatory remarks by Vincent T. Manas in "National Plumbing Code Handbook," published by the McGraw-Hill Book Company in 1957.

Quotations from the "National Plumbing Code," ASA A 40.8–1955, included in this third edition have been "extracted from the 'American Standard National Plumbing Code' (ASA A 40.8–1955) with the permission of the publisher, The American Society of Mechanical Engineers, 19 West 39th Street, New York 18, N.Y." The quotation is from the permission to publish which appears on the copyright sheet of the Code. It is to be noted that the Code does not profess to be mandatory. Although the requirements of the Code are stated in detail, a "saving" clause is inserted frequently which permits deviations from the Code "when approved by the Administrative Authority" having local jurisdiction.

This third edition embodies changes from the second edition based on the "National Plumbing Code" of 1955 whenever the Code is applicable. Where the NPC makes no change in the intent or wording of earlier authoritative reports, references continue to be made to the original source. Partly because of the demands for space in this book and partly because of the large amount of material available in the literature, the former chapter on Heating and the Appendix of Problems have been eliminated. In their place there have been added a chapter on Hot-water Supplies, a short chapter on Storage Tanks, and an extension of all but one or two other chapters and appendixes. An appendix has been added showing incorrect plumbing installations for correction, together with a statement of how each correction should be made.

This book continues to aim to be serviceable to engineers, architects, owners, designers, and builders of buildings involving the installation of plumbing.

Harold E. Babbitt

CONTENTS

CHAPTER 1

ELEMENTS OF PLUMBING

1-1. Plumbing. The plumbing[1] of a building includes the pipes, fixtures, equipment, and accessories that convey water or other fluids from the public water supply or other source to and through the building to points of use. The plumbing conveys also the used fluids to the nearest sewer or other place of disposal. The fluids most commonly conveyed in household plumbing are cold water, hot water, and gas for heating, commonly called illuminating gas. The business of house heating is frequently combined with that of plumbing, but it is not strictly included therein. Fluids conveyed in heating are steam, hot water, and warm air. Chilled water may be conveyed in hotel plumbing, and a wide variety of liquids and gases may be conveyed in industrial plants. Supply or waste pipes with one or more branches outside a building are commonly considered to be water-distribution systems or sewer systems rather than plumbing. Plumbing is also the art, trade, or business involved in the design, installation, and maintenance of the plumbing of a building.

1-2. The Plumber. A plumber[1] is a person engaged in the art of plumbing. Three grades of experience are recognized in the trade: the apprentice, the journeyman, and the master. An apprentice is a beginner who usually serves 3 to 5 years as a helper to a journeyman. A journeyman has served his apprenticeship and is competent to perform the task of installing and repairing plumbing. A master is skilled in the art of plumbing and either is in the business of plumbing or employs one or more journeymen, or both. A journeyman should be skilled in his trade. The master should be as competent as a journeyman, and he should have, in addition, knowledge of the physical laws affecting the things with which he deals and the installations he makes, of the laws and ordinances affecting plumbing, and of business methods and procedures.

1-3. Purpose of Plumbing. Plumbing is installed in a building for convenience and comfort, as well as for sanitation and health. Supply pipes bring water or gas, and drainage pipes carry off the used water. Wholesome water may be supplied by a publicly or privately owned

[1] Definitions are given in Appendix I.

public service enterprise. The quality of the water is under the supervision of the local and state health authorities. The used water is generally discharged into a common sewer where its disposal is the responsibility of governmental agencies. Where no public water supply is available, where the quality of the water supply is unsatisfactory, or where no common sewer is available, obtaining water, treating it, and attending to sewerage and to sewage disposal become necessary. Under such conditions the plumber may be called on for information, equipment, and service.

The householder at home and the occupants of public buildings are accustomed to and demand types of plumbing fixtures and installations unknown a generation ago. The conditions of plumbing that existed

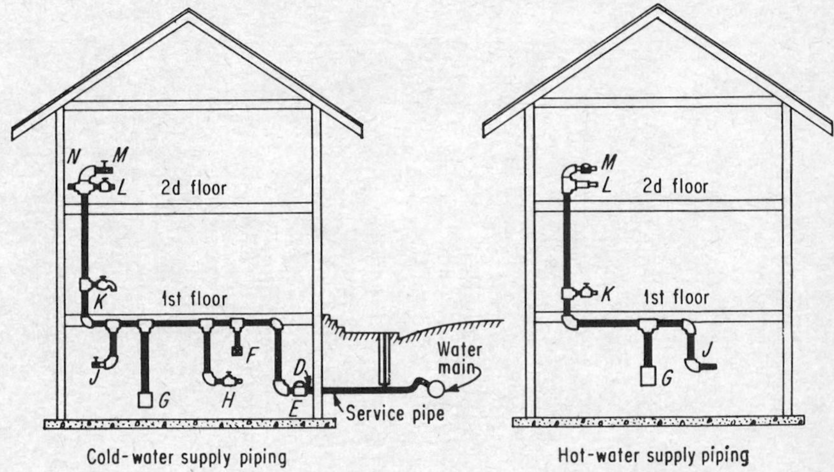

FIG. 1-1. Water-supply pipes in a small, single-family dwelling.

even so short a time ago as the span of one human life would not now be tolerated. The scale of living in both urban and rural communities is measured, in part, by the completeness of the plumbing in the home. Good plumbing is a mark of good taste and of good living. It is a source of pride to the householder, and it is necessary in apartment, commercial, and other large buildings. Good plumbing is prized and appreciated around the world.

1-4. Simplicity of Plumbing. A plumbing system, reduced to its simplest terms, consists of a supply pipe leading to a fixture and a drain pipe taking the used water away from that fixture. As the number of fixtures is increased, the branching of the supply and drainage pipes is increased and venting is required. Typical water-supply piping for a small dwelling is shown in Figs. 1-1 and 1-2*a* to *c*. As the kinds of fluid supplied are increased, the complication of the piping is increased.

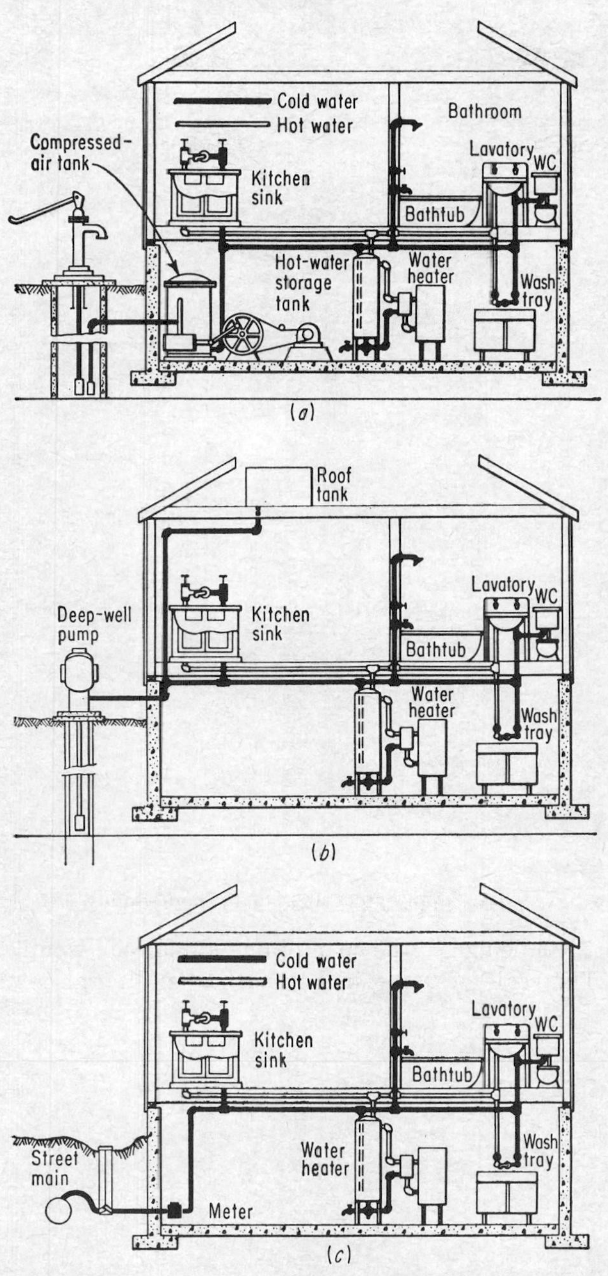

FIG. 1-2. (*a*) Hydropneumatic water-supply system for a rural residence. (*b*) Deep-well water supply, with force pump and roof tank, for a rural residence. (*c*) Water-supply piping in an urban residence.

Most plumbing fixtures have two supply pipes, one bringing cold water and the other bringing hot water. In some buildings hard-water, soft-water, chilled-water, and other supply pipes may be installed. When each supply system is considered independently, the piping arrangements can be more easily understood.

Drainage pipes are trapped to prevent the passage of sewer air into the building, and they are vented to the air outside the building to prevent the loss of seal in the traps. When the purpose of these supply, drainage, and vent pipes is understood and each system is considered independently, a plumbing system is not too complicated to be comprehended.

1-5. Objections to Plumbing. Undesirable features of plumbing installations include dangers to health and discomforts that may result from the pollution of water supplies through cross connections or from the escape of odors, gases, vapors, insects, and rodents from drainage pipes, and noises resulting from various causes. The escape of explosive, poisonous, or asphyxiating gases into a building has caused injury and death to human beings, animals, and plants in the building and has resulted in damage to the structure and its contents. Cross connections offer an equally serious but more subtle danger to health.

1-6. Sewer Air. Sewer air is a mixture of vapors, odors, and gases found in a sewer. It is not a specific substance with a definite chemical composition. Among the gases present in sewer air may be included methane (CH_4), known also as *natural gas, fire damp*, and by other names; gasoline vapor; illuminating gas; hydrogen sulfide (H_2S); carbon monoxide (CO); acetylene (C_2H_2); hydrogen (H_2); ammonia (NH_3); carbon dioxide (CO_2); oxygen (O_2); and nitrogen (N). The first seven are

TABLE 1-1. CHARACTERISTICS OF GASES FOUND IN SEWER AIR

Name of gas	Lower explosive limit*	Max. safe conc.*	Physiological effect†	Sp gr ref. to air	Btu per cu ft
Carbon monoxide	12.5	0.01	1, 2		
Methane (natural gas)	5.6	See note‡	1	0.63	1,000
Hydrogen sulfide	4.3	0.02–0.002	3		
Carbon dioxide	Not explosive	2–3	1		
Gasoline	1.4	1.0	4		
Ammonia	16	0.03	5		
Sulfur dioxide	Not explosive	0.005	5		
Illuminating gas	5.0	0.01	1	0.6	500

* Percentage, by weight, in mixture with air.

† Classification: (1) asphyxiating; (2) extremely dangerous—odorless, colorless, and subtle, lighter than air; (3) irritant, systemic poisoning; (4) anesthetic—produces headache, nausea, "jag"; (5) respiratory, eye, and mucus irritant.

‡ Dangerous when oxygen is displaced more than 8 per cent.

combustible and explosive in the presence of oxygen. Hydrogen sulfide and carbon monoxide are highly poisonous, even in relatively small concentrations. All may be asphyxiating when the concentration of oxygen in the atmosphere is reduced. Hydrogen sulfide and other gases and volatile matter may be present in low concentration, but nevertheless sufficient to create highly offensive odors and to cause corrosion. Information concerning the characteristics of gases sometimes found in sewers is given in Table 1-1.

The possibility that sewer air may be dangerous to health has resulted in controversy. Men have worked for years in sewers without apparent bad effect on their health, and men have died from asphyxiation or gas poisoning on entering a manhole. No scientific data can be found to prove or to disprove to the satisfaction of all the claims that any specific

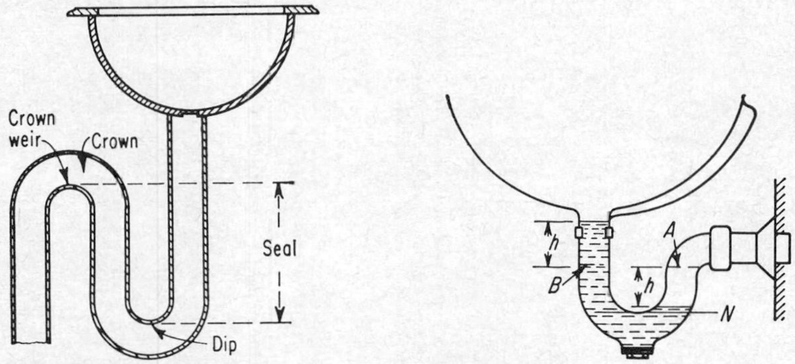

FIG. 1-3. Parts of a trap, showing also the effect of back pressure on the trap seal.

disease results from the presence of so-called "sewer gas" in a building. Nevertheless, in view of the uncertainties concerning specific diseases and dangers resulting from the admission of gases from sewers and drainage pipes into buildings, the greatest effort should be made to confine sewer air to sewers.

1-7. Plumbing Traps. A plumbing trap is a fitting placed in the drain pipe from a plumbing fixture for the purpose of holding water or other fluid to form a seal that will prevent the passage of gases and odors from the drain pipes into the air within a building. The word *trap* is defined in Appendix I. The parts of a trap are illustrated in Fig. 1-3. Its hydraulics are discussed in Sec. 2-24, and its performance is discussed in Sec. 11-34.

The maintenance of the water seal in a trap offers difficulties that add greatly to the cost of plumbing. The seal may be destroyed when water is blown or sucked out of the trap as a result of variations in pressure in the drainage pipes, when water discharging through a trap at a high velocity does not fall back sufficiently to maintain or restore the seal, or when water evaporates from the trap. The difficulties, except for evaporation, can be overcome or controlled by venting.

1-8. Purposes of Venting. Vent pipes are used to prevent the loss of water seals in traps due to causes other than evaporation. Unfortunately, venting accelerates evaporation of the water in a trap because it increases the rate of change of air over the trap seal. Since the loss of water seal by evaporation is slow, possibly requiring weeks or months to become dangerous, the normal use of a plumbing fixture restores the water seal in the trap. In cities where main or house traps are not required, vent stacks aid in the ventilation of the common sewers.

No safe and certain method has been devised to prevent the evaporation of water from a trap when the fixture is not used. In unoccupied premises

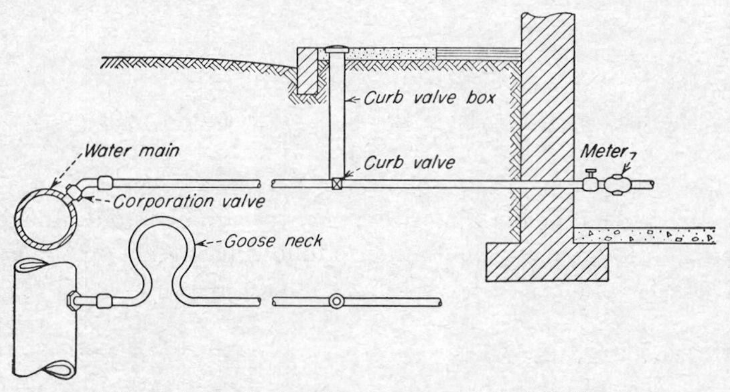

FIG. 1-4. Typical water-service pipe installation.

water should be removed from the traps and the seal restored with kerosene or other fluid that evaporates more slowly than water.

1-9. Water and Gas Pipes. The water supply and the gas supply for a city building are distributed in separate pipes that are sometimes called the *water main* and the *gas main*, respectively. These pipes are commonly buried under the street near the building served. A smaller pipe, called the *service pipe*, is connected to the supply main to convey the water or the gas into the building.

A typical water service pipe is shown in Fig. 1-4. The water-supply pipes in the building start from the main valve or stop-and-waste valve, shown at *D* in Fig. 1-1. Layouts of cold-water supply pipes for small dwellings are shown in Figs. 1-1 and 1-2, and gas supplies are discussed in Chap. 9.

1-10. Drainage and Vent Pipes. Drainage pipes are installed to remove used water, rainfall, and other waste liquids from a building. The design and layout of the drainage pipes are often the most difficult to understand because of the complications involved in maintaining slopes and scouring velocities, proper pipe sizes, and proper capacities and arrangements of vent pipes to prevent the development of abnormal air pressures in the drainage pipes.

Water falling through vertical drainage pipes, called *stacks*, and flowing through horizontal drainage pipes, called *branches*, will entrap, compress, and rarefy air. A result may be the breaking of trap seals unless adequate vent pipes have been installed. Simple installations of drainage and vent pipes in small dwellings are shown in Figs. 11-3 and 11-11.

1-11. Installation of Plumbing. The installation of that part of the plumbing pipes known as the *roughing-in* proceeds simultaneously with the erection of the structural members of a building, since the pipes are often concealed in the walls and floors. The pipes must, therefore, be installed before the walls and floors are completed. Because of the necessity for correct installation the plumber must be able to read architectural drawings, to make sketches, and sometimes to make drawings of his own to supplement those of the architect.

Usually the first step in the installation of plumbing is the connection of the service pipe with the water main and of the house sewer with the public sewer. Supports for the plumbing stacks are then placed, and the stacks are erected as the building rises. Water-supply and drainage pipes are placed simultaneously, the branches following closely on the erection of riser pipes and stacks. All roughing-in should be completed before the walls are finished or the flooring laid. The prefabrication of plumbing,[1] consisting of special assemblies suitable for mass designs, is said to be reducing the cost of plumbing. Plumbing fixtures, except built-in bathtubs and some special fixtures, are installed after the completion of the flooring and plastering. Their installation, called *finishing*, is among the last things done before the completion of a building.

1-12. Plumbing Facilities Required. In estimating the plumbing facilities required in a building, recommendations in the "National Plumbing Code"[2] and in the "Uniform Plumbing Code for Housing," shown in Table 1-2, may be useful.

1-13. Plumbing Codes and Standards. The improper installation of plumbing may affect the health of the occupants of a building and create a focal center for the spread of disease. The possibility is so real and of sufficient public interest as to require the regulation of plumbing by law. The police power of the state is invoked, and the right of the government

[1] See *Eng. News-Record*, May 24, 1956, p. 49.

[2] The "National Plumbing Code," ASA A 40.8–1955. Published jointly by the American Society of Mechanical Engineers and the American Public Health Association. Quotations in this book have been extracted from "American Standard Plumbing Code" (ASA A 40.8–1955) with the permission of the publisher, the American Society of Mechanical Engineers, 29 West 39th St., New York, 18, N.Y.

TABLE 1-2a. MINIMUM PLUMBING FACILITIES RECOMMENDED BY "UNIFORM PLUMBING CODE FOR HOUSING"*

Type of building or occupancy†	Water closets			Urinals		Lavatories		Bathtubs or showers	Drinking fountains‡
	No. of persons	Male	Female	No. of persons	Male	No. of persons	No. of fixtures		
Schools	1–15	1	1	1–30	1	1–30	1	None required except for gymnasium or special purpose	One for each 75 pupils
	16–30	1	2	31–55	2	31–55	2		
	31–55	2	3	56–80	2	56–80	2		
	56–80	3	4	81–110	3	81–110	3		
	81–110	4	5	111–150	4	111–150	3		
	111–150	6	7	151–190	4	151–190	4		
	151–190	7	8	191–240	6	191–240	4		
	191–240	8	10	241–300	6	241–300	5		
	241–300	9	12	Over 300	One for each 50 additional persons	Over 300	One for each 50 additional persons		
	Over 300	One for each 30 additional persons							
Office buildings or public buildings§	1–15	1	1	1–15	0	1–15	1		
	16–35	2	2	16–35	1	16–35	2		
	36–60	3	4	36–60	1	36–60	3		
	61–90	4	5	61–90	1	61–90	4		
	91–120	5	7	91–120	2	91–120	4		
	121–150	6	8	121–150	2	121–150	5		
	151–190	6	9	151–190	3	151–190	5		
	191–240	8	11	191–240	4	191–240	6		
	241–300	9	12	241–300	4	241–300	7		
	Over 300	One for each additional 35	25	Over 300	One for each 75 additional persons	Over 300	One for each 45 additional persons		
Public assembly auditoriums, etc.	1–100	1	1	1–200	1	1–100	1	....	One for each 100 persons
	101–200	2	2	201–750	2	101–200	1		
	201–400	3	3	Over 750	One for each additional 300	201–400	2		
	401–750	3	4			401–750	3		
	Over 750	One for each additional 500	300			Over 750	One for each additional 500		

* The figures as shown are based upon one fixture being the minimum required for the number of persons indicated or any fraction thereof.
† No requirements have been given for hospitals, sanitariums, or hotels, and each will have to be considered individually by the Administrative Authority.
‡ Drinking fountains shall not be installed in toilet rooms.
§ No laundry tubs or kitchen sinks required.

TABLE 1-2*b*. MINIMUM PLUMBING FACILITIES RECOMMENDED BY "UNIFORM PLUMBING CODE FOR HOUSING"

Type of building or occupancy	Water closets			Urinals		Lavatories			Bathtubs or showers			Drinking fountains	Laundry tubs		Kitchen sinks
	No. of persons	Male	Female	No. of persons	Male	No. of persons†	No. of fixtures								
Manufacturing, warehouses, workshops, loft buildings, foundry, mine, and similar types of building*	1–9 10–24 25–40 41–60 61–90 91–120 121–150 Over 150	1 2 3 4 4 5 5 One for each additional 30 persons	1 2 3 4 4 5 5	Whenever urinals are provided one water closet less than the number specified herein may be provided for each urinal except that water closets shall not be reduced to less than ⅔ of the number specified		1–15 16–35 36–60 61–100 101–150 Over 150‡	1 2 4 7 10 One for each additional 25 persons		One for each person who may be exposed to excessive heat or to skin contamination with poisonous or irritating material			One for each 75 persons			
							Male	Female	No. of persons	Bath	Shower		No. of persons	No. of fixtures	
Dormitories	1–15 16–30 31–50 51–75 76–100 101–150 Over 150	1 2 3 4 6 8 One for each additional 25	1 2 4 6 8 10 20	1–30 31–50 51–100 101–150 One for each additional 50 persons over 150	1 2 3 4	1–15 16–30 31–50 51–75 76–100 101–150 Over 150	1 2 3 4 6 8 One for each additional 20	2 3 4 6 8 10 15	1–15 16–30 31–45 46–60 61–100 101–150 Over 150	1 1 2 2 3 4 One for each additional 50	1 2 3 6 10 12 20	One for each 75 persons	1–30 31–75 76–125 126–200 Over 200	1 2 3 4 One for each 50 additional persons	One for each utility kitchen provided
Residence dwelling or apartment building	One for each dwelling or apartment unit			. . . .		One for each dwelling or apartment unit			One for each dwelling or apartment unit			. . . .	One for each dwelling or apartment unit or 2 for each 10 apartments		One for each dwelling or apartment unit

NOTE: Footnotes to Table 1-2*a* are applicable also to data given here.

* See "American Standard Safety Code for Industrial Sanitation in Manufacturing Establishments" (ASA Z 4.1–1935).

† Where there is exposure to skin contamination with poisonous or irritating materials, then provide one lavatory for each 5 persons.

‡ Eighteen inches of wash sink circumference (circular type) shall be equal to one lavatory and straight type of wash sinks in 18-in. spaces with equivalent spouts shall be equal to one lavatory.

Temporary Workingmen Sanitary Facilities: One water closet and one urinal for each 30 workers; 24-in. urinal trough—one urinal; 36-in.—two urinals; 60-in.—three urinals; 72-in.—four urinals.

TABLE 1-2c. MINIMUM PLUMBING FACILITIES RECOMMENDED BY "NATIONAL PLUMBING CODE"[a,b]

Type of building or occupancy	Water closets	Urinals	Lavatories	Bathtubs or showers	Drinking fountains
Dwelling or apartment house[c]	One for each dwelling or apartment unit	....	One for each dwelling or apartment unit	One for each dwelling or apartment unit	
Office or public buildings	No. of persons / No. of fixtures: 1– 15 / 1 16– 35 / 2 36– 55 / 3 56– 80 / 4 81–110 / 5 111–150 / 6 One fixture for each additional 40 persons	[d]	No. of persons / No. of fixtures: 1– 15 / 1 16– 35 / 2 36– 60 / 3 61– 90 / 4 91–125 / 5 One fixture for each 45 additional persons	....	1 for each 75 persons
Schools:[e] Elementary	Male: One per 100 Female: One per 35	One per 30 male	One per 60 persons	....	One per 75 persons
Secondary	Male: One per 100 Female: One per 45	One per 30 male	One per 100 persons	....	One per 75 persons
Manufacturing, warehouses, workshops, loft buildings, foundries, and similar establishments[f]	No. of persons / No. of fixtures: 1– 9 / 1 10– 24 / 2 25– 49 / 3 50– 74 / 4 75–100 / 5 One fixture for each additional 30 employees	[d]	1–100 persons, one fixture for each 10 persons. Over 100, one for each 15 persons[g,h]	[i]	One for each 75 persons
Dormitories[j]	Male: one for each 10 persons Female: One for each 8 persons[k]	One for each 25 men[l]	One for each 12 persons[m]	One for each 8 persons[n]	One for each 75 persons
Theatres, auditoriums	No. of persons[o] / No. of fixtures Male / Female: 1–100 / 1 / 1 101–200 / 2 / 2 201–400 / 3 / 3	No. of persons, male[p] / No. of fixtures: 1–200 / 1 201–400 / 2 401–600 / 3	No. of persons[q] / No. of fixtures: 1–200 / 1 201–400 / 2 401–750 / 3	....	1 for each 100 persons

[a] "National Plumbing Code," Sec. 7.21.2. NPC is also ASA A 40.8–1955.

[b] The figures shown are based on one fixture being a minimum for the number of persons indicated or any fraction thereof.

[c] Laundry trays—one single-compartment tray for each dwelling unit or two 2-compartment trays for each 10 apartments. Kitchen sinks—one for each dwelling or apartment unit.

[d] Wherever urinals are provided for men, one water closet less than the number specified may be

to regulate the details of plumbing is based on the principle of the protection of public health and life. Municipal ordinances regulating plumbing are commonly called *plumbing codes.* Some states and most cities with a population above about 10,000 have a plumbing code and an organization to enforce its provisions.

The aim of a plumbing code should be to cover every contingency that may arise in the installation of plumbing. A complete plumbing code is necessarily long. Every builder should assure himself that his plans are in accord with the requirements of the code of the city in which he is building. Plumbing codes are being improved. Impetus to their improvement was given through the publication, in 1923, of "Recommended Minimum Requirements for Dwelling and Similar Buildings" by the National Bureau of Standards. It is widely known as the "Hoover Report." Four codes or standards subsequently published are the "Plumbing Manual," published as Report BMS66 by the National Bureau of Standards in 1940; "The Uniform Plumbing Code for Housing," published as Technical Paper 6, in February, 1948, by the Housing and Home Finance Agency, hereinafter referred to as the "Housing Code"; "Report of the Uniform Plumbing Code Committee," published by the Housing and Home Finance Agency, Government Printing Office, in 1940; and "American Standard Plumbing Code" referred to in Sec. 1-12.

---

provided for each urinal installed except that the number of water closets in such cases shall not be reduced to less than two-thirds of the minimum specified. Building category not shown on this table will be considered separately by the Administrative Authority.

[e] This schedule has been adopted (1945) by the National Council on Schoolhouse Construction.

[f] As required by the "American Standard Safety Code for Industrial Sanitation in Manufacturing Establishments" (ASA Z 4.1–1935).

[g] Where there is exposure to skin contamination with poisonous, infectious, or irritating materials, provide one lavatory for each five persons.

[h] 24 lin in. of wash sink or 18 in. of a circular basin, when provided with water outlets for such space, shall be considered equivalent to one lavatory.

[i] One shower for each 15 persons exposed to excessive heat or to skin contamination with poisonous, infectious, or irritating material.

[j] Laundry trays, one for each 50 persons. Slop sinks, one for each 100 persons.

[k] Over 10 persons, add one fixture for each 25 additional males and one for each 20 additional females.

[l] Over 150 persons, add one fixture for each additional 50 men.

[m] Separate dental lavatories should be provided in community toilet rooms. Ration of dental lavatories for each 50 persons is recommended. Add one lavatory for each 20 males and one for each 15 females [*sic*].

[n] In the case of women's dormitories, additional bathtubs should be installed at the ratio of one for each 30 females. Over 150 persons add one fixture for each 20 persons.

[o] Over 400, add one fixture for each additional 500 males and one for each 300 females.

[p] Over 600, one for each additional 300 males.

[q] Over 750, one for each additional 500 persons.

General. In applying this schedule of facilities, consideration must be given to the accessibility of the fixtures. Confirmation purely on a numerical basis may not result in an installation suited to the need of the individual establishment. For example, schools should be provided with toilet facilities for each floor having classrooms.

Temporary Workingmen Facilities: One water closet and one urinal for each 30 workers; 24-in. urinal trough—one urinal; 36-in.—two urinals; 48-in.—two urinals; 60-in.—three urinals; 72-in.—four urinals.

1-14. Cross Connections.[1] A cross connection is any physical connection, either direct or indirect, that will permit or may possibly permit the flow of undrinkable water into a conduit or receptacle containing or intended to contain water that is fit to drink. A direct connection consists of a continuous conduit leading undrinkable water into the drinking-water supply. An indirect connection consists of a gap or space across which undrinkable water can be sucked, blown, or otherwise made to enter the drinking-water supply.

The existence of permanent or temporary, active or potential cross connections is more extensive than is generally realized. Thousands of cases of water-borne disease resulting from unsuspected cross connections are reported in the literature. The plumber may be both criminally and financially liable for damages resulting from careless or ignorant installation or creation of cross connections.

Types of cross connection are almost countless. Some of the more common are listed in Table 1-3. Cross connections are made for many reasons, among the principal of which may be included: (1) to supplement an inadequate private supply with the public supply or vice versa, (2) to furnish water from or to the public water supply for fire protection, and (3) to permit the flushing, cleaning, cooling, or operation of plumbing fixtures with nonpotable water. Cross connections may be created inadvertently during the installation of plumbing, during repairs and renovations, or by "do-it-yourself" building occupants. The plumber should be alert to recognize a cross connection, to warn the building occupants of its existence, and to see that the cross connection is removed if it is possible for him to do so.

Some confusion exists among the terms *cross connection*, *crossover*, and *interconnection*. Definitions of these terms are given in Appendix I.

1-15. Prevention of Cross Connections and Backflow. The installation of cross connections is prevented, in part, by the enforcement of requirements such as the following, taken from the "Housing Code."

8.2.5. Backflow. The water distribution system shall be protected against backflow. Every fixture supply pipe shall be protected from backflow preferably by having the outlet end from which the water flows spaced a distance above the flood-level rim of the receptacle into which the water flows sufficient to provide a "minimum required air gap" as defined in ASA A 40.4–1912. Where it is not possible to provide a minimum air gap the fixture shall be equipped with an accessibly located backflow preventer complying with ASA A 40.6–1943 installed beyond the manual control valve.

[1] See also R. F. Goudy, Practical Aspects of Cross Connections, Interconnections, and Backflow, *J. Am. Water Works Assoc.*, March, 1941, p. 391.

TABLE 1-3. WHERE TO LOOK FOR CROSS CONNECTIONS*,†

The following conditions offer hazards that may result in a cross connection:

1. Water-closet bowls equipped with flush valves, or flush tanks having submerged float-operated ball cocks
2. Frostproof closets
3. Hopper closets with hand-operated flush valves
4. Seat-acting water closets with flush valve in or attached to the bowl
5. Bedpan washers
6. Bidets
7. Sterilizers with water inlets subject to pollution by either gravity or siphonic action
8. Therapeutic baths with inlets below the rim of the fixture
9. Water-operated waste ejectors, such as are used by dentists, undertakers, and those who practice colonic irrigation
10. Cellar drainers of the water-jet type
11. All fixtures with inlets below the rim of the fixture such as bathtubs, lavatories, dishwashers, and washing machines
12. Cuspidors with water-supply connections
13. Dental cuspidors with water-supply connections
14. Hospital appliances generally, such as sterilizers, condensers, filters, stills, aspirators, and operating tables, with underrim outlets
15. Mixing or combination faucets with one supply safe and the other unsafe
16. Floor drains with water-flush connections
17. Swimming pools with water-supply inlets below the overflow line or having a physical connection between potable water and the circulating mains
18. Yard hydrants arranged so that polluted ground water can drain into the water-supply line
19. Automatic, water-supplied siphon flush tanks with inlet below the water line. This includes sewer flush tanks
20. Dual water pumps operated from an unsafe water supply and pumping a safe supply
21. Industrial vats, tanks, or other equipment of any description that has a water-supply connection below the top of the spill rim or is so constructed that polluted water can splash back or vapors can pollute the water-supply inlet; or devices without air gaps
22. A flexible hose attached to a water-supply pipe, the hose being of sufficient length to dangle into a water container
23. Leaking water pipes near sewers
24. Any physical connection between safe and unsafe water supplies
25. Hydrant drains leading to sewers
26. Water supplies to (*a*) sealing rings on pumps handling unsafe fluids such as sewage, (*b*) float valves in sewage plants, (*c*) priming pumps handling unsafe fluids, (*d*) sludge-pipe blowouts, (*e*) laundry equipment, (*f*) water-jacketed grease-intercepting traps with cast partitions
27. Drains from fire sprinklers to sewers or wastes
28. Filters with the backwash directly connected to the water supply
29. Check-valves to prevent access of unsafe water to safe water-supply pipes

* A cross connection exists in any plumbing fixture which can hold water so that the supply pipe opening to the fixture is below the flood-level rim shown in Fig. 1-5 and the air gap is less than the distance *A* shown in the figure and given in Table 1-4.

† From A. R. McGonegal, Discussion of Plumbing Hazards, *J. Am. Water Works Assoc.*, vol. 31, p. 987, 1939.

Where it is not possible to provide a minimum air gap or backflow preventer, as may be the case in connection with cooling jackets, condensers, or other industrial or special appliances, then the Administrative Authority shall require other approved means of protection.

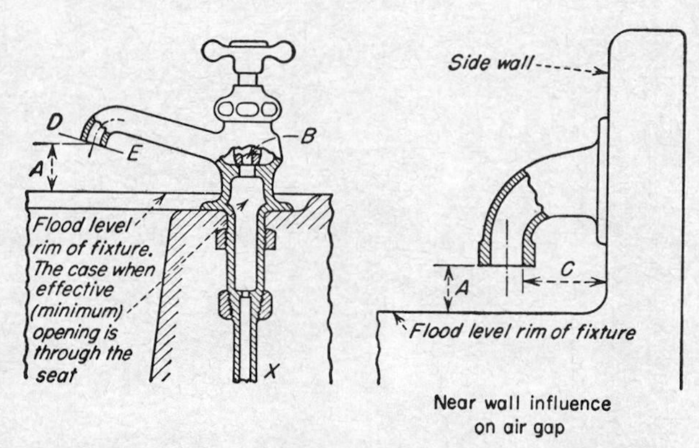

FIG. 1-5. Air gap and effective opening (see Table 1-4).

Only an adequate air gap can be depended on to prevent backflow under all conditions. Standard minimum air gaps are illustrated in Fig. 1-5 and are listed in Table 1-4. Other devices that are used include (1) air valves, usually good on 2-in. and larger pipes; (2) mechanical vacuum

TABLE 1-4. MINIMUM AIR GAPS FOR GENERALLY USED FIXTURES*

Fixture	Minimum air gap, in.	
	When not affected by near wall†	When affected by near wall‡
Lavatories with effective openings not greater than ½ in. diameter	1.0	1.50
Sink, laundry-tray, and gooseneck bath faucets with effective openings not greater than ¾ in. diameter	1.5	2.25
Overrim bath fillers with effective openings not greater than 1 in. diameter	2.0	3.00
Effective openings greater than 1 in.	2× effective opening	3× effective opening

* From "The Uniform Plumbing Code for Housing." Technical Paper 6, p. 67, Housing and Home Finance Agency, February, 1948, and "National Plumbing Code," Table C.2.4.

† Side walls, ribs, or similar obstructions do not affect the air gaps when spaced from inside edge of spout opening a distance greater than three times the diameter of the effective opening for a single wall, or a distance greater than four times the diameter of the effective opening for two intersecting walls.

‡ Vertical walls, ribs, or similar constructions extending from the water surface to or above the horizontal plane of the spout opening require greater air gap when spaced closer to the nearest inside edge of the spout opening than specified in the table. The effect of three or more such vertical walls or ribs has not been determined. In such cases, the air gap shall be measured from the top of the wall.

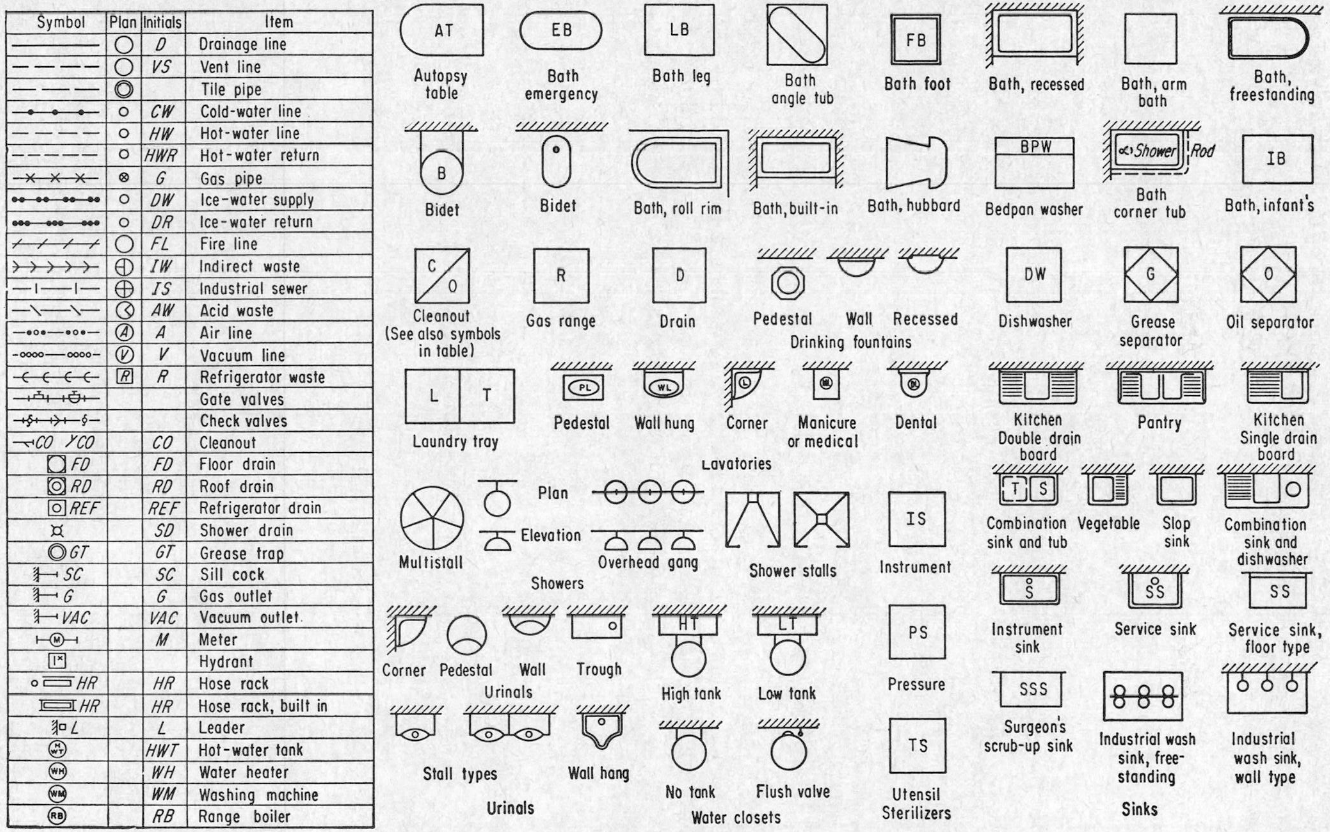

Symbol	Plan	Initials	Item
		D	Drainage line
		VS	Vent line
			Tile pipe
		CW	Cold-water line
		HW	Hot-water line
		HWR	Hot-water return
		G	Gas pipe
		DW	Ice-water supply
		DR	Ice-water return
		FL	Fire line
		IW	Indirect waste
		IS	Industrial sewer
		AW	Acid waste
	A	A	Air line
	V	V	Vacuum line
	R	R	Refrigerator waste
			Gate valves
			Check valves
CO YCO		CO	Cleanout
FD		FD	Floor drain
RD		RD	Roof drain
REF		REF	Refrigerator drain
		SD	Shower drain
GT		GT	Grease trap
SC		SC	Sill cock
G		G	Gas outlet
VAC		VAC	Vacuum outlet
M		M	Meter
			Hydrant
HR		HR	Hose rack
HR		HR	Hose rack, built in
L		L	Leader
HWT		HWT	Hot-water tank
WH		WH	Water heater
WM		WM	Washing machine
RB		RB	Range boiler

AT Autopsy table
EB Bath emergency
LB Bath leg
Bath angle tub
FB Bath foot
Bath, recessed
Bath, arm bath
Bath, freestanding

B Bidet
Bidet
Bath, roll rim
Bath, built-in
Bath, hubbard
BPW Bedpan washer
Shower Rod Bath corner tub
IB Bath, infant's

C O Cleanout (See also symbols in table)
R Gas range
D Drain
Pedestal Wall Recessed
Drinking fountains
DW Dishwasher
G Grease separator
O Oil separator

L T Laundry tray
PL Pedestal
WL Wall hung
L Corner
M Manicure or medical
D Dental
Lavatories
Kitchen Double drain board
Pantry
Kitchen Single drain board

Multistall
Plan
Elevation
Overhead gang
Showers
Shower stalls
IS Instrument
T S Combination sink and tub
Vegetable
Slop sink
Combination sink and dishwasher

Corner Pedestal Wall Trough
Urinals
HT High tank
LT Low tank
PS Pressure
S Instrument sink
SS Service sink
SS Service sink, floor type

Stall types
Wall hang
Urinals
No tank
Flush valve
Water closets
TS Utensil Sterilizers
SSS Surgeon's scrub-up sink
Industrial wash sink, free-standing
Industrial wash sink, wall type
Sinks

FIG. 1-6. Standard plumbing symbols.

Table 1-5. Graphical Symbols for Use on Drawings
(American Standard, ASA 32.2.3–1949)

Plumbing:	
Soil, waste or leader (above grade)	
Soil, waste or leader (below grade)	
Vent	
Cold water	
Hot water	
Hot-water return	
Fire line	–F—F–
Gas	–G—G–
Acid waste	Acid
Drinking-water flow	
Drinking-water return	
Vacuum cleaning	–V—V–
Compressed air	—A—
Sprinklers:	
Main supplies	—S—
Branch and head	
Drain	–S––S–
Pneumatic tubes	
Tube runs	
Fittings:	
Elbow, 90 deg	
Fittings (contd.):	
Elbow, 45 deg	
Elbow, turned up	
Elbow, turned down	
Elbow, long radius	LR
Side outlet elbow, outlet down	
Side outlet elbow, outlet up	
Base elbow	
Double branch elbow	
Single sweep tee	
Double sweep tee	
Reducing elbow	
Tee	
Tee, outlet up	
Tee, outlet down	
Side outlet tee, outlet up	
Side outlet tee, outlet down	
Cross	
Reducer	
Eccentric reducer	
Fittings (contd.):	
Lateral	
Crossover	
Orifice and flange	
Plug, pipe	
Sleeve	
Joints or ends:	
Joint	
Flanged	
Screwed	
Bell and spigot	
Bell and spigot	
Welded	
Soldered	
Cap	
Expansion-joint flange	
Reducing flange	
Union	
Bushing	
Valves:	
Gate valve	
Valves (contd.):	
Globe valve	
Angle gate elevation	
plan	
Angle globe elevation	
plan	
Reducing	
Diaphragm	
Check valve	
Angle check valve	
Stopcock	
Safety valve	
Quick-opening valve	
Float-operating valve	
Motor-operated gate valve	
Globe, motor-operated	M
	M
Hose, angle	
Hose, gate	
Hose, globe	

breakers, illustrated in Sec. 14-13; (3) a "broken-connection" valve that closes automatically when the inlet pressure is zero or less; (4) check valves, usually considered to be unreliable; (5) swing connections, consisting of a piece of pipe joined to a private distribution system, that can be connected to only one of two or more different supply pipes; (6) nipples that are inserted as a temporary connection between two water systems during an emergency, as a fire; (7) a vertical water-supply pipe in the form of a vertical bend, thus ∩, more than 34 ft high, leading down to a fixture; and (8) double check valves. Double check valves in series on a pipe, with a blowoff or drain valve between them, are highly reliable in preventing backflow. Mechanical backflow preventers are discussed in Sec. 14-13.

1-16. Control of Pests. Improperly designed, installed, or maintained plumbing makes possible infestations by rats, cockroaches, water bugs, mosquitoes, silverfish, and other pests. Their presence caused by the plumbing in a building can be discouraged or prevented by proper precautions in the installation and maintenance of the plumbing. Precautions to be taken include: the making of tight joints in drainage pipes; the use of adequate seals in traps, although rats and water bugs have been known to pass through trap seals; the avoidance of standing water in the presence of light to discourage egg-laying by flying insects; the use of tight escutcheons and collars on pipes passing through walls and floors to prevent travel of pests along the pipes; the use of toxic substances in pastes and glues on pipe coverings to kill silverfish; and other expedients.

1-17. Plumbing Plans and Symbols. Plans for the roughing-in of plumbing are drawn with conventional projections, somewhat as indicated in Fig. 1-6, using standard symbols, such as those shown in the figure and in Table 1-5. Piping dimensions are shown to the center line of pipes.

CHAPTER 2

HYDRAULICS AND PNEUMATICS

2-1. Problems Involved. Among the hydraulic and pneumatic problems involved in plumbing may be included water pressures; gas, air, and steam pressures; the flow of fluids under pressure, as in supply pipes to fixtures; the flow of liquids at or near atmospheric pressure under conditions of open-channel flow, as in drainage pipes; the measurement of rates of flow of fluids; and the characteristics of pressures resulting from the movement of air, water, and solids in drainage pipes and of the flow of air in vent pipes.

The selection of the proper sizes of pipe for plumbing installations is an important problem, the correct solution of which involves the principles of hydraulics and pneumatics of the flow of fluids in closed conduits, i.e., under a pressure other than atmospheric pressure. In practice, "rules of thumb," authoritative publications, and code requirements are commonly used. Such practical aids to the selection of pipe sizes are usually based on experience rather than on theory, and properly so. However, the plumber depending on them alone is restricted in his practice. Knowledge of the principles involved in the flow of fluids under pressure in supply pipes is of value in checking practice and in supplementing empirical rules. Knowledge of the conditions of flow in drainage pipes is of value in understanding their design and installation.

2-2. Static Water Pressures. Water is a liquid that, at any point, exerts equal intensity of pressure in all directions. The intensity of pressure at any point depends on the depth of submergence of the point below the free surface and on the unit weight of the liquid. *Free surface* means a surface of water exposed to atmospheric pressure.

The unit weight of water, or its specific gravity, varies with the temperature, as shown in Table 2-1. Since water is practically incompressible within the limits of pressure encountered in plumbing, the unit weight of water is unaffected by such pressures.

The intensity of pressure on a submerged surface is usually expressed in pounds per square inch (psi) or in feet of water above the pressure of the atmosphere at the point where the pressure is being read. It can be

TABLE 2-1. TEMPERATURE, VOLUME, AND BOILING POINT OF WATER

Temperature, °F	Relative volume	Relative density	Lb per cu ft	Temperature, °F	Relative volume	Relative density	Lb per cu ft	Gage pressure, approx psi	Boiling point, °F	Gage pressure, approx psi	Boiling point, °F
32	1.00000	1.00000	62.418	140	1.01690	0.98339	61.381	−14	102.018	17	254.002
35	0.99993	1.00007	62.422	145	1.01839	0.98194	61.291	−13	126.302	19	257.523
39.1[a]	0.99989	1.00011	62.425	150	1.01889	0.98050	61.201	−12	141.654	21	260.883
40	0.99989	1.00011	62.425	155	1.02164	0.97882	61.096	−11	153.122	23	264.093
45	0.99993	1.00007	62.422	160	1.02340	0.97714	60.991	−10	162.370	25	267.168
46	1.0000	1.00000	62.418	165	1.02589	0.97477	60.983	− 9	170.173	27	270.122
50	1.00015	0.99985	62.409	170	1.02690	0.97380	60.783	− 8	176.945	29	272.965
52.3	1.00029	0.99971	62.400	175	1.02906	0.97193	60.665	− 7	182.952	31	275.704
55	1.00038	0.99961	62.394	180	1.03100	0.97006	60.548	− 6	188.357	33	278.348
60	1.00074	0.99926	62.372	185	1.03300	0.96828	60.430	− 5	193.284	35	280.904
62	1.00101	0.99899	62.355	190	1.03500	0.96632	60.314	− 4	197.814	37	283.381
65	1.00119	0.99881	62.344	195	1.03700	0.96440	60.198	− 3	202.012	39	285.781
70	1.00160	0.99832	62.313	200	1.03889	0.96256	60.081	− 2	205.929	41	288.111
75	1.00239	0.99771	62.275	205	1.04140	0.96020	59.930	− 1	205.604	45	292.575
80	1.00299	0.99702	62.240	210	1.0434	0.95840	59.820	0[g]	212.00	51	298.842
85	1.00379	0.99622	62.182	212[b]	1.0444	0.95750	59.760	0	213.670	55	302.744
90	1.00454	0.99543	62.133	230	1.0529	0.94990	59.260	1	216.347	59	306.526
95	1.00554	0.99449	62.074	250	1.0628	0.94110	58.750	2	219.452	65	311.866
100	1.00639	0.99365	62.022	270	1.0727	0.93230	58.18	3	222.424	71	316.893
105	1.00739	0.99260	61.960	290	1.0838	0.92270	57.59	4	225.255	75	320.094
110	1.00889	0.99119	61.868	298[c]	1.0899	0.91750	57.27	5	227.964	81	324.688
115	1.00989	0.99021	61.807	338[d]	1.1118	0.89940	56.14	7	233.069	85	327.625
120	1.01139	0.98874	61.715	366[e]	1.1301	0.88500	55.29	9	237.803	90	331.169
125	1.01239	0.98808	61.654	390[f]	1.1444	0.87380	54.54	11	242.225	95	334.582
130	1.01390	0.98630	61.563	....	....	....	.....	13	246.376	100	337.874
135	1.01539	0.98484	61.472	....	....	....	.....	15	250.293	105	341.058

[a] Point of greatest density.
[b] Water boils at this temperature when under atmospheric pressure. Its temperature can be increased above 212°F only when the water is under a pressure higher than atmospheric.
[c] Steam at 50 psi.
[d] Steam at 100 psi.
[e] Steam at 150 psi.
[f] Steam at 205 psi.
[g] Absolute pressure exactly 14.69 psi.

formulated as

$$h = 2.31p \tag{2-1}$$

where h = depth of submergence, ft
p = density of pressure, psi[1]

The value of p in this expression is known as the gage pressure, since it is the pressure that would be shown on a pressure gage. Absolute pressure

[1] Where not otherwise stated, psi refers to pounds per square inch, gage.

is equal to the sum of atmospheric pressure and gage pressure. It may be expressed as psia to distinguish it from gage pressure, sometimes written as psig. It is evident from Eq. (2-1) that a pressure of 1 psi corresponds to a column of water 2.31 ft high and that a column of water 1 ft high corresponds to a pressure of 0.43 psi.

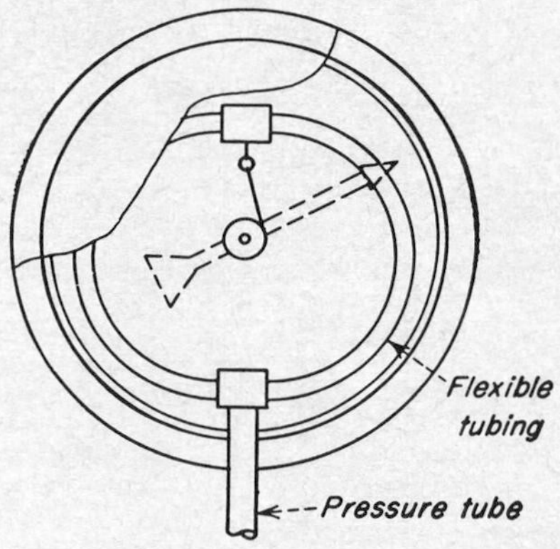

FIG. 2-1. Cutaway rear view of simple Bourdon gage.

2-3. Measurement of Pressures. Fluid pressures are measured by pressure gages, piezometers, and manometers. A common type of spring actuated gage, known as a Bourdon gage, is shown in Fig. 2-1. This gage consists of a bent, flattened, flexible, springlike metallic tube that tends to straighten as fluid pressure increases in it and to return to normal position when the pressure is decreased. The movement of the tube is transmitted to the calibrated indicating arm by means of a gear train.

Pressures above atmospheric are usually shown on pressure gages in pounds per square inch or in feet of water. Pressures below atmospheric are shown in pounds per square inch below atmospheric, in pounds per

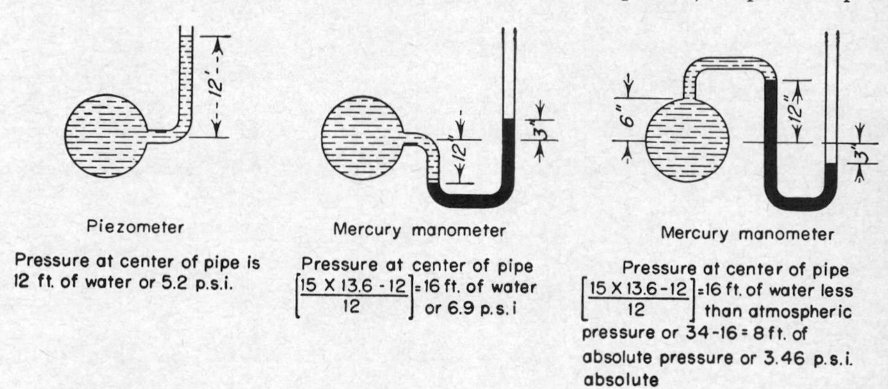

FIG. 2-2. Piezometer and manometers.

square inch absolute, in feet of water, and in inches of mercury below atmospheric pressure.

Piezometers and manometers are open-end tubes connected to the conduit or vessel under pressure. The intensity of pressure is indicated by the height of the column of liquid in the tube, as shown in Fig. 2-2. A piezometer shows only pressures above atmospheric. A manometer may show pressures above or below atmospheric. When mercury is used in the arms of a manometer, the difference in the elevations of the two columns of mercury should be multiplied by 13.6 to convert to equivalent height of water, since the specific gravity of mercury is 13.6.

2-4. Pressures on a Submerged Surface. The total pressure against a submerged surface can be expressed as

$$P = k_1 A p = k_2 A \bar{h} \tag{2-2}$$

where P = total pressure in a direction normal to the surface
A = area of submerged surface
p = intensity of pressure at center of gravity of area
$\bar{h}$ = vertical distance from water surface to center of gravity of area
k_1 and k_2 = coefficients depending on the units used. If P is in pounds, p in psi, A in square inches, and h in feet, then $k_1 = 1$ and $k_2 = 0.434$

The total pressure in a horizontal direction against a submerged surface, whether plane or otherwise, is equal to the product of the area of the horizontal projection of the area against a vertical plane and the intensity of pressure at the centroid of the submerged area of projection. That is,

$$P = k_1 A_h p \tag{2-3}$$

where A_h = area of horizontal projection of the surface against a vertical plane
k_1 = a constant depending on the units used. When the units are pounds and feet, $k_1 = 1.0$

The following example illustrates an application of the preceding principles to the solution of a problem:

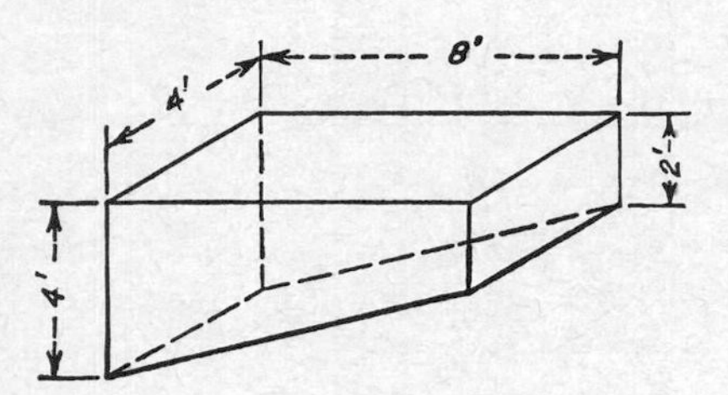

Example. Find the total pressure normal to each side and to the bottom of the tank shown in the accompanying figure, and find the total pressure on the bottom of the tank in both a horizontal and a vertical direction. Find also the total weight of water in the tank. It is to be assumed that the tank is full of water that weighs 62.5 lb per cu ft.

Solution: First find the pressure against the 8-ft side.

$$P = k_1 A\bar{h} = 0.434A\bar{h} \text{ (where } \bar{h} \text{ is distance to centroid)}$$

$$A = 144 \times \frac{4+2}{2} 8 = 24 \times 144 = 3{,}460 \text{ sq in.}$$

$$\bar{h} = \frac{(8 \times 2 \times 1) + (8 \times 1 \times 2.33)}{(8 \times 2) + (8 \times 1)} = 1.44 \text{ ft}$$

Therefore

$$P = 0.434(24 \times 144)1.44 = 2{,}164 \text{ lb}$$

The total pressure against the 4- by 4-ft end is

$$0.434(4 \times 4 \times 144)2 = 2{,}000 \text{ lb}$$

The total pressure against the 2- by 4-ft end is

$$0.434(2 \times 4 \times 144)1 = 500 \text{ lb}$$

The total pressure against the bottom is

$$0.434 \times 4 \times \sqrt{4 + 64} \times 3 \times 144 = 6{,}190 \text{ lb}$$

The pressure against the bottom, in a horizontal direction, is

$$0.434(4 \times 2)144 \times 3 = 1{,}500 \text{ lb}$$

The pressure against the bottom, in a vertical direction, is

$$62.5[(2 \times 4 \times 8) + (2 \times 4 \times 8 \times 0.5)] = 6{,}000 \text{ lb}$$

The weight of water in the tank = the total pressure in a vertical direction = 6,000 lb

2-5. Bursting Pressures. The tension in the walls of a circular pipe or a tank of unit length due to internal bursting pressure can be formulated as

$$T = \frac{kpd}{2} \tag{2-4}$$

where d = diameter of pipe or tank
k = a factor depending on the units used
p = intensity of internal pressure
T = total tensile stress in tank or pipe walls, per unit length

The intensity of stress in the tank or pipe walls is

$$i = \frac{T}{t} \tag{2-5}$$

where i = intensity of stress
t = thickness of pipe or tank wall

Example. Find the intensity of stress in the walls of a 1-in.-nominal-diameter pipe, with an actual diameter of 1.049 in., when the internal bursting pressure is 100 psi and the wall thickness is 0.133 in.

Solution:

$$T = \frac{100 \times 1.049}{2 \times 1} = 52.45$$

$$i = \frac{52.45}{0\ 133 \times 1} = 394 \text{ psi}$$

2-6. Measurement of Rate of Flow. Methods for the measurement of the rate of flow of water that may be applicable in plumbing include volumetric tanks, gravimetric tanks, orifices, short tubes, weirs, displacement meters, and velocity meters.

When a volumetric tank is used, the rate of flow is measured by observing the time in which the known volume of the tank is filled. When a gravimetric tank is used, the weight of water caught in the tank in a known period of time is converted to volume by dividing the weight of water caught in the tank by the weight of a unit volume of water. The unit weights of water at various temperatures are shown in Table 2-1.

Various forms of orifices, weirs, and short tubes for the measurement of the rate of flow of water are described in Table 2-2. The table shows also formulas applicable to the flow of water through orifices and short tubes and over weirs. Some solutions of these formulas for conditions within the range of convenient use in plumbing are given in Table 2-3.

TABLE 2-2. FORMULAS FOR FLOW THROUGH ORIFICES, THROUGH SHORT TUBES, AND OVER WEIRS

Type of orifice, tube, or weir	Description	Formula
Standard circular	Sharp edges flush with inside vertical wall or horizontal bottom of orifice tank. Orifice removed from disturbance caused by sides or bottom of tank. Head not less than diameter of orifice	$Q = 0.61a\sqrt{2gh}$*
Standard short tube	Cylindrical short tube of length 2 to 2½ diameters, with outlet flush with plane of inside vertical wall of tank, and removed from wall or bottom disturbances	$Q = 0.82a\sqrt{2gh}$*
Standard weir	With no end contractions and no velocity of approach	$Q = 3.33lh^{3/2}$
	With no end contractions but with velocity of approach	$Q = 3.33l[(h + h_1)^{3/2} - h_1^{3/2}]$
	With end contractions and no velocity of approach	$Q = 3.33(l - 0.2h)h^{3/2}$
	With end contractions and with velocity of approach	$Q = 3.33(l - 0.2h)[(h + h_1)^{3/2} - h_1^{3/2}]$
Triangular notch	Sharp-edged notch with edge in plane of vertical wall of tank and with 90° angle of notch	$Q = 2.53h_2^{5/2}$

* a is the cross-sectional area of the orifice; d is the diameter of the orifice; h is the distance from the center of the circle or from the edge of the weir to the surface of water above the orifice or weir; h_1 is the velocity head of the approach velocity; h_2 is the vertical distance from the bottom of the notch (angular vertex) to the free surface of the water above the notch; l is the length of the crest of the weir; g is the acceleration due to gravity.

TABLE 2-3. EXAMPLES OF SOLUTION OF FORMULAS FOR FLOW THROUGH ORIFICES, THROUGH SHORT TUBES, AND OVER WEIRS

Type of orifice, tube, or weir	Data	Solution
Standard circular orifice	$d = 2$ in. $h = 2$ ft	$a = \frac{\pi(d^2)}{4} = \frac{\pi 4}{4} \div 144 = 0.0218$ sq ft $Q = 0.61 \times 0.0218 \times 8 \times 1.414 \times 60 \times 7.5 = 67.7$ gpm
Standard short tube	$d = 2$ in. $h = 2$ ft	$Q = 0.82 \times 0.0218 \times 8 \times 1.414 \times 60 \times 7.5 = 91.0$ gpm
Standard weir	Length = 10 ft h = 4 ft	No contractions or velocity of approach: $Q = 3.33 \times 10 \times 8 = 266$ cfs No contractions but velocity of approach is 3 fps: $h_1 = 9/64 = 0.141$ Therefore, $Q = 3.33 \times 10[(4.141)^{3/2} - (0.141)^{3/2}]$ $= 33.3(8.42 - 0.05) = 279$ cfs Contractions, but no velocity of approach: $Q = 3.33(10 - 0.8)8 = 245$ cfs Contractions and velocity of approach: $Q = 3.33 \times 9.2 \times 8.37 = 256$ cfs
Triangular weir	Head = 2 ft	$Q = 2.53(2)^{5/2} = 2.53 \times 5.65 = 14.3$ cfs

A velocity meter, as used in plumbing, is a device for observing the velocity of flow of water through a transverse cross section of known area. The rate of flow is the product of the velocity and the area of the cross section. A displacement meter is a device which measures the rate of flow of water by filling and emptying a chamber of known volume an observed number of times. The rate of flow is the product of the volume of the chamber and the number of times it has been filled. Displacement and velocity meters are discussed in Sec. 14-30.

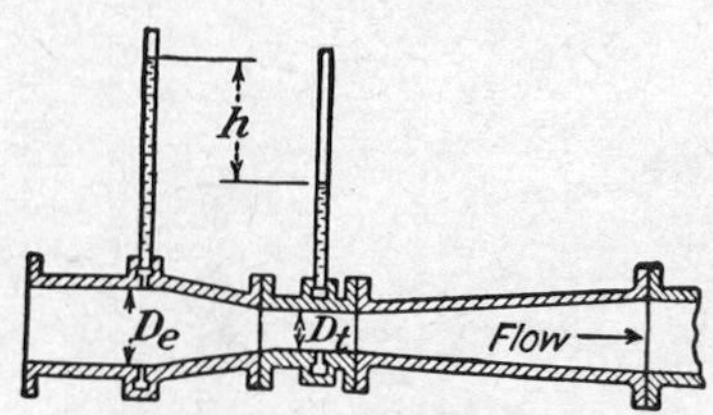

FIG. 2-3. Venturi meter.

A venturi meter is a special form of velocity meter in which the pressure at the throat, shown in Fig. 2-3, can be converted into rate of flow by means of the formula

$$Q = \frac{19.7 d_e^2 d_t^2 h^{1/2}}{(d_e^4 - d_t^4)^{1/2}} C \tag{2-6}$$

where Q = flow, gpm

d_e = diameter of entrance to meter, in.

d_t = diameter of throat of meter, in.

h = difference between pressure at throat and at entrance, ft of water

C = a coefficient depending on the construction of the meter

The value of C can be determined by calibration of the meter. It is usually within 1 or 2 per cent of unity. Venturi meters are used for the measurement of high rates of flow, and with the aid of recording and integrating devices, they will measure large volumes of flow.

Elbow meters[1] and orifice plates may be useful in measuring rates of flow with relatively simple equipment. An elbow meter is shown in Fig. 2-4. Almost any constriction in a circular conduit conveying a fluid that results in a difference of pressure above and below the constriction can be converted, by calibration, into a velocity meter through which the rate of flow can be found by the following equation:

$$Q = \frac{kd^2}{4}\sqrt{h} \tag{2-7}$$

where Q = rate of flow

h = difference in pressure above and below the elbow or other constriction

d = diameter of conduit

k = a coefficient depending on the units used and on the form of the elbow or other constriction. The value of k must be determined by calibration

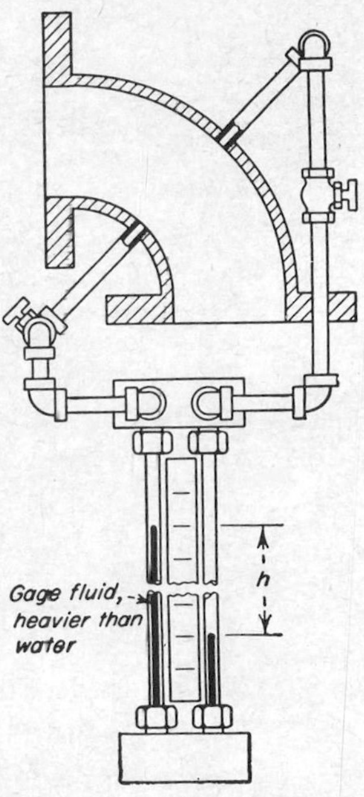

FIG. 2-4. Diagrammatic sketch of elbow meter. (*Univ. Illinois, Eng. Expt. Sta. Bull.* 289, 1936.)

Lansford[2] indicated that where the units are feet and seconds, values of k vary from 7 for 1-in. pipe to about 5 for 24-in. pipe.

A plate placed in a straight run of pipe, with a hole of any shape in the plate, can be used for measuring the rate of flow by observing the difference in pressures above and below the plate and substituting in Eq. (2-7), the value of k having been determined by calibration.

When observations are made on orifice-plate meters, the pressure gages must be placed far enough above and below the plane of the plate to avoid the turbulence in flow caused by the constriction. Generally, a

[1] See also D. C. Taylor and M. B. McPherson, Elbow Meter Performance, *J. Am. Water Works Assoc.*, November, 1954, p. 1087.

[2] W. M. Lansford, *Univ. Illinois Eng. Expt. Sta. Bull.* 289, 1936.

distance of at least one diameter of pipe upstream and three to four diameters downstream is sufficient.

2-7. Flow of Fluids. If two vessels containing liquids at different elevations are connected by a tube, as shown in Fig. 2-5, the liquid in the higher container will flow toward the lower container. The total energy at any point in the tube, such as point A, is equal to the total energy at any other point in the tube, such as point B, plus the energy lost by friction due to flow between the two points. The total energy at any point is equal to the sum of th...

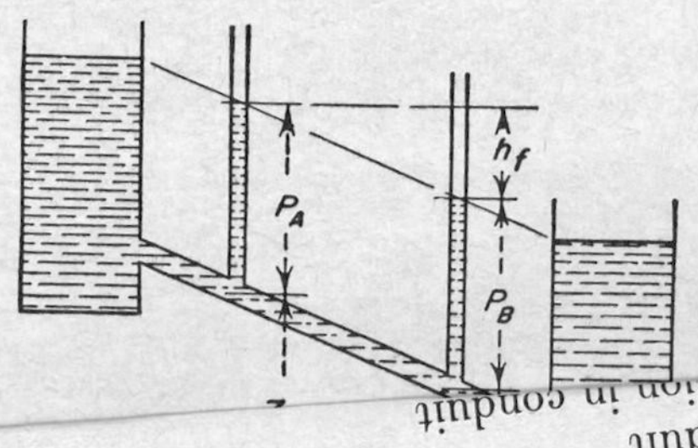

(2-10)

d^2g

$$Q = AC\sqrt{RS}$$

or

in which f = friction factor, depending on character of wall of conduit

g = acceleration due to gravity

l = length of pipe or conduit

d = diameter of pipe or conduit

... due to friction in conduit

					eter, in.		eter, in.	gpm
3/8	0.488	2.45	0.438	1.75	0.338	0.97	0.188	0.26
1/2	0.618	4.55	0.568	3.30	0.468	2.02	0.318	0.84
3/4	0.820	9.60	0.770	7.20	0.670	4.65	0.520	2.60
1	1.04	18.0	0.99	13.6	0.89	8.75	0.74	5.85
1 1/4	1.37	37.6	1.32	28.5	1.22	19.1	1.07	13.7
1 1/2	1.60	56.0	1.55	42.5	1.45	29.8	1.30	22.6
2	2.06	108.0	2.01	82.4	1.91	59.1	1.76	47.7
2 1/2	2.46	173.0	2.41	134.0	2.31	94.5	2.16	79.0
3	3.06	310	3.01	238	2.91	170	2.76	147
4	4.02	630	3.97	485	3.87	345	3.72	310
5	5.04	1125	4.99	880	4.89	617	4.74	575
6	6.05	1805	6.00	1420	5.90	993	5.75	930

* Based mainly on information given in "Plumbing Manual," p. 42, National Bureau of Standards, 1940.

† Flow is approximate, because it is based on a head loss of 10 psi per 100 ft of pipe.

TABLE 2-6. VISCOSITIES OF WATER AT VARIOUS TEMPERATURES

Temperature, °F	32	50	68.4	100	160
Absolute viscosity, μ poises	0.0179	0.0131	0.0100	0.0069	0.0040

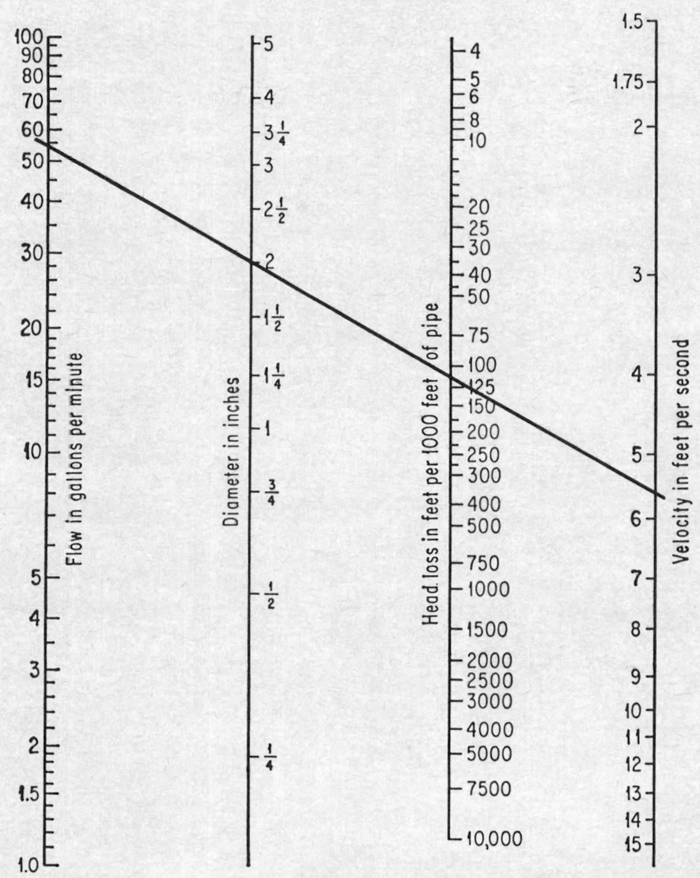

FIG. 2-6. Nomograph for the determination of pressure losses in small pipes by the Hazen and Williams formula, with $C = 100$.

2-9. Flow of Hot Water.[1] The effect of temperature on the head losses due to the flow of water in pipes, particularly in relatively small pipes, may be large. Poiseuille's formula states

$$\frac{h_f}{l} = 0.0326\left(\frac{\mu}{S}\right)\left(\frac{V}{d^2}\right) \quad \text{in cgs units} \tag{2-12}$$

where h_f = head loss due to friction, cm of water
l = length of pipe, cm
μ = absolute viscosity, poises
S = specific gravity of fluid
V = average velocity of flow, cm per sec
d = diameter of pipe, cm

Values of μ for water are shown at various temperatures in Table 2-6.

[1] See also Chap. 8.

The specific gravities of water at various temperatures are shown in Table 2-1. Poiseuille's equation shows that the head loss varies directly as the viscosity and inversely as the specific gravity. Tests by Saph and Schoder[1] show that loss of head decreases about 0.4 per cent for each degree Fahrenheit of rise in temperature from 40 to 70°F but that near the boiling point the loss of head increases. Tests by Giesecke[2] led to the conclusion that for new, clean, black standard pipe

$$\frac{h_f}{l} = \frac{0.01533V^{1.77}}{t^{0.19}d^{1.275}} \tag{2-13}$$

where h_f = head loss due to friction, ft of water of the same density as that flowing in the pipe
l = length of pipe, ft
V = velocity of flow, fps
t = temperature, °F
d = diameter of pipe, in.

It follows from the Giesecke formula that the friction loss at 160°F is about 87.5 per cent of that at 50°F.

2-10. Flow of Gas under Pressure. The friction loss due to the flow of a gas, such as illuminating gas or air, under such slight differences of pressure that the density of the gas remains approximately constant can be expressed as

$$\frac{h_f}{l} = \frac{4fV^2}{2gD} \tag{2-14}$$

where h_f = head loss, ft of water
l = length of pipe, ft
V = velocity of flow of gas, fps
D = pipe diameter, ft
g = acceleration due to gravity, fps²
$f = 0.0044\left(\frac{W_g}{W}\right)\left(1 + \frac{1}{7D}\right)$
W_g = weight of gas, lb per cu ft
W = weight of water, lb per cu ft

It has been found by test that for air

$$f = 0.0027\left(\frac{W_a}{W}\right)\left(1 + \frac{3}{10D}\right)$$

[1] A. V. Saph and E. H. Schoder, An Experimental Study of the Resistance to the Flow of Water in Pipes, *Trans. Am. Soc. Civil Engrs.*, vol. 91, p. 253, 1903.

[2] F. E. Giesecke, "The Friction of Water in Pipes and Fittings," *Univ. Texas Bull.* 1759, Oct. 20, 1917.

Where the flow of gas is under such pressures that the density changes appreciably, then

$$P_B = P_A \sqrt{1 - \frac{f \dfrac{W}{W_g} V^2 l}{229{,}000D}} \tag{2-15}$$

where P_A and P_B are the pressures, respectively, at points A and B, in pounds per square inch.

The volume of a gas varies directly with the absolute temperature and inversely with the absolute pressure. The absolute temperature on the Fahrenheit scale is the Fahrenheit reading minus 459.6°, and on the centigrade scale it is the centigrade reading minus 273°. The absolute pressure is equal to gage pressure plus the atmospheric pressure, as indicated by a barometric reading.

Some values of friction losses due to the flow of illuminating gas in pipes are shown in Sec. 9-6.

2-11. Hydraulic Gradient. A line drawn through the tops of the liquid columns in a row of piezometers connected to a pipe through which water is flowing under pressure, as shown in Fig. 2-7, is known as the *hydraulic grade line.* Its slope is the *hydraulic gradient.* The slope of the hydraulic grade line is downward in the direction of flow if the velocity is constant, regardless of the direction of flow in the pipe. Although the diagram shows that water can flow uphill, the hydraulic gradient is downward in the direction of flow when the velocity of flow is constant. The intensity of pressure in a water main in the street is greater than that in any pipe in a building at or above the elevation of the pipe in the street as long as the direction of flow is from the street toward the building. The difference in pressures between those in the street main and in the pipes in the building is the sum of the friction loss and the difference in elevation between the two points at which pressures are observed. When conditions are static, that is, when there is no flow, there is no loss due to friction and the pressures in the pipes within the building are equal to the pressures in the street main less the differences in elevations at which the pressures are observed. When the rate of drawing water on lower floors in a building is high, causing high friction losses in an inadequate service pipe, the pressures in the pipes on upper floors of the building may be seriously reduced, even to the extent of causing siphonage, as indicated by the hydraulic grade line in Fig. 2-8.

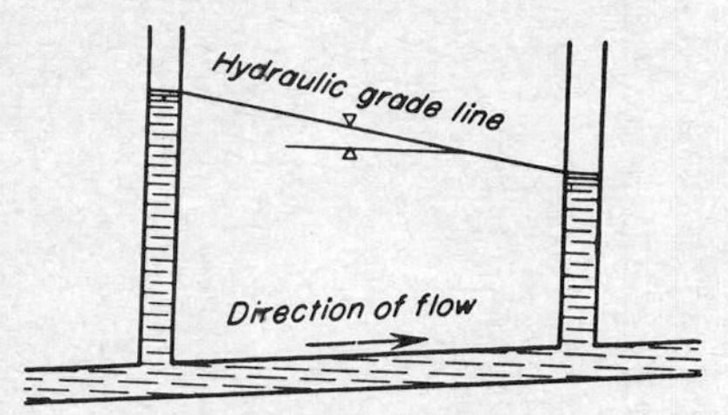

FIG. 2-7. Hydraulic gradient.

2-12. Siphons and Siphonage. A true siphon consists of a bent tube, one leg of which is vertically longer than the other, as shown in Fig. 2-9. When flow is taking place in the siphon the shorter leg is upstream. The

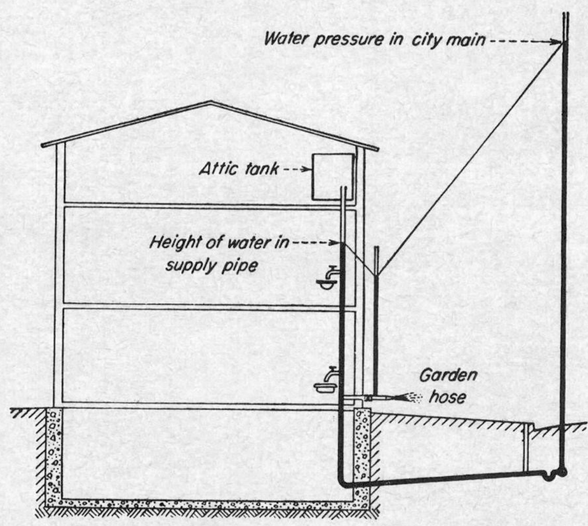

FIG. 2-8. Diagram showing siphonage in attic tank resulting from the use of garden hose.

flow through a siphon is due to the difference in elevation of the free surfaces above and below the siphon. However, since the absolute pressures in the shorter, or rising, leg of the siphon vary inversely with the height above the upper free surface, the greatest height over which a siphon can apparently lift water is equal to the intensity of the atmospheric pressure on the upper free surface. For example, the pressure of the atmosphere at sea level is equal, approximately, to a column of water 34 ft high. Normal atmospheric pressures at various elevations above sea level are shown in Table 2-7. Hence, a siphon at sea level should, apparently, lift water to a height of 34 ft. It will not do so, however, because of the effect of vapor pressure.

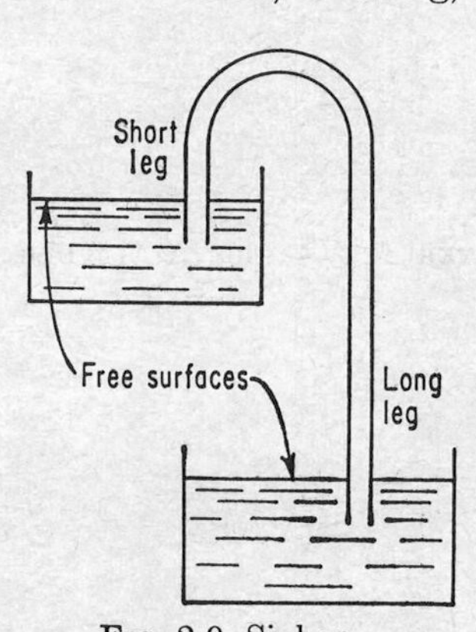

FIG. 2-9. Siphon.

Siphonage is highly undesirable in plumbing. In supply pipes it may cause damage to appurtenances, such as the collapse of hot-water tanks, as indicated in Fig. 2-10, or pollution of the water supply. In drainage pipes it may cause the loss of seal in traps, admitting sewer air to the building.

Vapor pressure is the pressure exerted by the tendency of a liquid to vaporize. The tendency varies with the temperature of the liquid, as

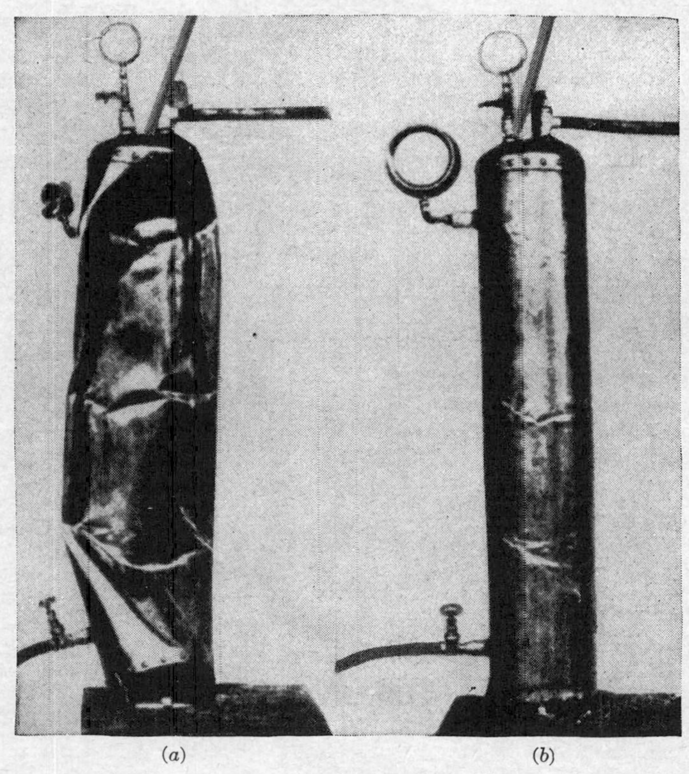

(*a*) (*b*)

FIG. 2-10. Collapsed hot-water storage tank: (*a*) after collapse; (*b*) after restoration of pressure. (*From Water Works Eng.*, *July* 12, 1944.)

TABLE 2-7. ATMOSPHERIC PRESSURE IN FEET OF WATER AT VARIOUS ALTITUDES ABOVE SEA LEVEL
(Temperature 60°F)

Elevation above sea level, ft	Atmospheric pressure, ft of water	Elevation above sea level, ft	Atmospheric pressure, ft of water	Elevation above sea level, ft	Atmospheric pressure, ft of water	Elevation above sea level, ft	Atmospheric pressure, ft of water
0	34.0	2,000	31.5	4,000	29.4	6,000	27.4
500	33.3	2,500	31.0	4,500	28.9	7,000	26.4
1,000	32.7	3,000	30.4	5,000	28.4	8,000	25.5
1,500	32.1	3,500	29.9	5,500	27.9	10,000	23.7

shown in Table 2-8. The height to which a column of liquid will rise under a vacuum, i.e., the height to which it can be lifted by suction, is equal to the atmospheric pressure less the vapor pressure. *Vacuum, suction*, and *siphonage* are synonymous.

TABLE 2-8. VAPOR TENSION IN FEET OF WATER

Temperature, °F	Pressure, psi	Pressure, ft of water	Temperature, °F	Pressure, psi	Pressure, ft of water	Temperature, °F	Pressure, psi	Pressure, ft of water
32	0.09	0.21	100	0.95	2.19	180	7.53	17.4
40	0.12	0.28	120	1.69	3.91	200	11.56	26.7
60	0.26	0.60	140	2.89	6.68	212	14.70	34.0
80	0.50	1.15	160	4.74	11.0			

2-13. Suction Lift. The height to which a pump can lift water by suction is limited by atmospheric pressure, vapor tension, friction losses in the suction pipe, and velocity head in the suction pipe. Some limitations on suction lift are shown in Table 2-9. The height of suction lift can be formulated as

TABLE 2-9. MAXIMUM THEORETICAL SUCTION LIFT OF PUMPS, FEET

Temperature, °F	Elevation above sea level, ft										
	0	500	1,000	1,500	2,000	2,500	3,000	4,000	5,000	7,000	10,000
40	33.7	33.0	32.4	31.8	31.2	30.7	30.1	29.1	28.1	26.1	23.4
60	33.4	32.7	32.1	31.5	30.9	30.4	29.8	28.8	27.8	25.8	23.1
80	32.8	32.1	31.5	30.9	30.3	29.8	29.2	28.2	27.2	25.2	22.5
100	31.8	31.1	30.5	29.9	29.3	28.8	28.2	27.2	26.2	24.2	21.5
120	25.0	24.3	23.7	23.1	22.5	22.0	21.4	20.4	19.4	17.4	14.7
160	23.0	22.3	21.7	21.1	20.5	20.0	19.4	18.4	17.4	15.4	12.7
180	16.6	15.9	15.3	14.7	14.1	13.6	13.0	12.0	11.0	9.0	6.3

$$H = A - V_p - H_f - \frac{V^2}{2g} \tag{2-16}$$

in which A = atmospheric pressure (values are shown in Table 2-7)
V_p = vapor tension (values for water are shown in Table 2-8)
H_f = head loss due to friction caused by flow in suction line
$V^2/2g$ = one velocity head, where V is velocity of flow in suction pipe and g is acceleration due to gravity

For most practical purposes the suction lift of pumps lifting cold water at elevations less than 1,000 ft above sea level is limited to a maximum of 22 to 25 ft.

Example. What is the maximum suction lift of a pump at 5,000 ft above sea level, where the temperature of the water to be lifted is 60°F and the friction and velocity heads are 2.61 ft of water?

Solution: A, from Table 2-7, is 28.4 ft of water; V_p, from Table 2-8, is 0.26; hence, $H = 28.4 - 0.26 - 2.61 = 25.53$ ft.

Another problem in suction lift is solved in Sec. 5-4.

2-14. Air and Gas Locks. A bend or hump extending upward above the regular line of the run of a pipe, as shown in Fig. 2-11, or extending

FIG. 2-11. Air trap in pipeline.

above the hydraulic grade line, as in a siphon, may permit the accumulation of air or gas in the bend. The effect may be either to reduce or to cut off flow in the pipe or to require pressure to force the trapped air through the pipe. The trapped air will diminish or stop flow through the pipe by reducing the cross-sectional area available for flow. It will act as a stoppage that no amount of "rodding" will remove and that is not to be found when the pipe is opened for examination. Examples of

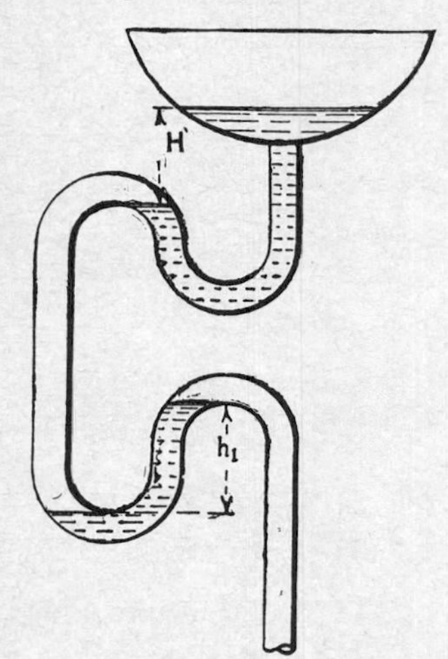

FIG. 2-12. Diagram of double-trapped fixture.

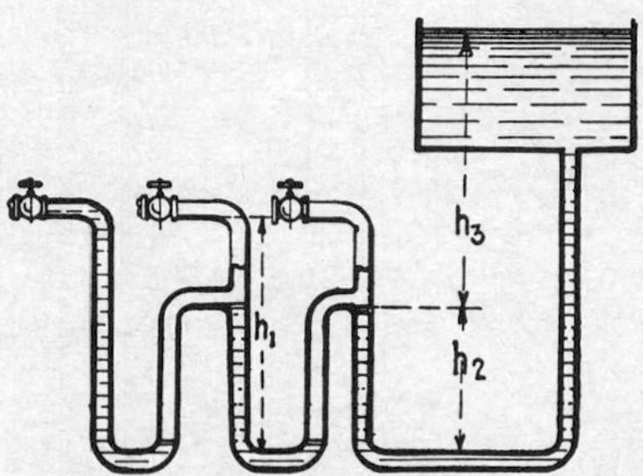

FIG. 2-13. Diagram illustrating air binding.

such stoppages, sometimes called *air binding*, are illustrated in the double-trapped fixture shown in Figs. 2-12 and 2-13. In the condition shown in Fig. 2-13, flow will be stopped if $h_1 + h_2$ is materially greater than h_3. If $h_1 + h_2$ is approximately equal to h_3, a small amount of water will flow over the top of the loops and from faucet 1 when it is opened without affecting the existence of the air-lock.

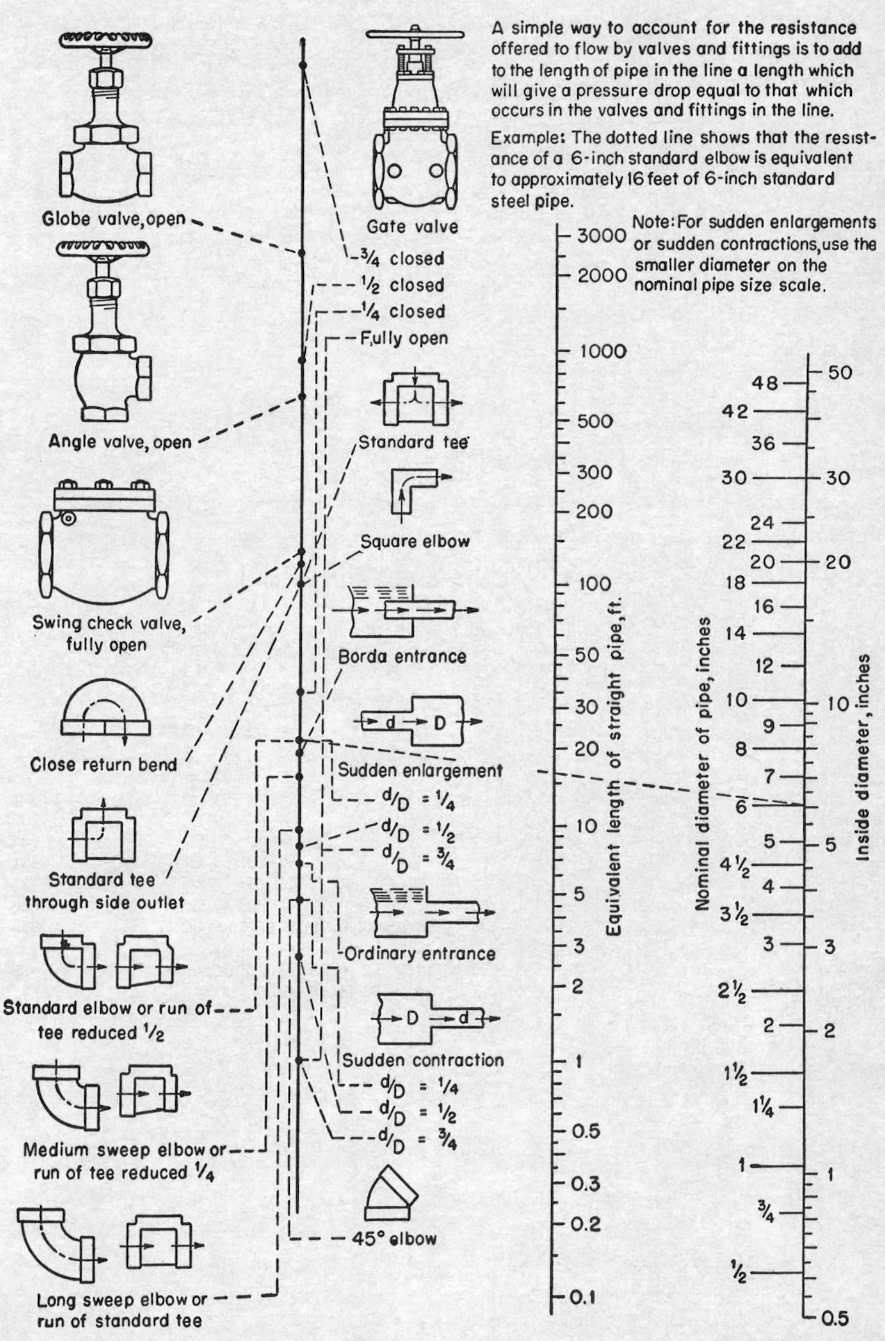

FIG. 2-14. Resistance of valves and fittings to flow of fluids. (*Crane Co.*)

Air-locks are likely to give trouble in pipes under low pressure and in siphons because of inadequate force to push the air along the pipe. The formation of an air-lock can be prevented by avoiding the creation of upward humps in a pipeline or by the installation of an air-release valve at the high point where gas is likely to accumulate.

TABLE 2-10. LOSS OF PRESSURE IN 90° ELBOWS
(psi*)
(See also Tables 2-11 to 2-14)

Flow, gpm	Diameter of pipe, in.										Flow, gpm	Diameter pipe, in.		
	½	¾	1	1¼	1½	2	2½	3	4	6		8	10	12
5	0.40	0.080	0.025	0.013	.....	0.002	.....	.....	.....	.....	100	0.003		
10	1.60	0.318	0.100	0.041	0.020	0.006	0.003	.....	.....	.....	125	0.004	0.002	
15	3.60	0.715	0.225	0.092	0.045	0.014	0.005	.....	.....	.....	150	0.006	0.003	
20	6.40	1.27	0.400	0.164	0.079	0.025	0.010	0.005	.....	.....	175	0.009	0.004	
25	10.0	1.99	0.625	0.256	0.124	0.039	0.016	0.008	.....	.....	200	0.011	0.005	
30	.....	2.86	0.900	0.369	0.178	0.056	0.023	0.011	.....	.....	250	0.017	0.007	
35	.....	3.90	1.230	0.502	0.243	0.077	0.031	0.015	.....	.....	300	0.025	0.010	
40	.....	5.09	1.60	0.656	0.317	0.100	0.041	0.020	0.007	.....	350	0.034	0.014	
45	.....	6.45	2.03	0.830	0.402	0.127	0.052	0.026	0.009	.....	400	0.044	0.018	
50	.....	7.95	2.50	1.250	0.495	0.156	0.064	0.032	0.010	.....	450	0.057	0.023	
60	.....	.....	3.60	1.48	0.714	0.225	0.092	0.044	0.015	0.003	500	0.068	0.028	0.017
70	.....	.....	4.90	2.01	0.971	0.306	0.125	0.060	0.021	0.004	750	0.156	0.063	0.031
80	.....	.....	6.40	2.62	1.27	0.400	0.164	0.080	0.027	0.005	1,000	0.272	0.112	0.062
90	.....	.....	8.10	3.32	1.60	0.506	0.207	0.104	0.035	0.007	1,250	0.435	0.175	0.108
100	.....	.....	10.0	4.10	1.98	0.625	0.255	0.128	0.043	0.008	1,500	0.624	0.252	0.128
125	.....	.....	.....	6.42	3.10	0.976	0.400	0.200	0.067	0.013	2,000	1.08	0.446	0.207
150	.....	.....	.....	9.22	4.45	1.41	0.575	0.286	0.096	0.019	2,500	1.63	0.693	0.331
175	.....	.....	.....	.....	6.08	1.92	0.782	0.390	0.132	0.026	3,000	.....	0.995	0.476
200	.....	.....	.....	.....	7.93	2.50	1.02	0.512	0.172	0.032	3,500	.....	1.35	0.650
250	.....	.....	.....	.....	.....	3.86	1.60	0.800	0.268	0.055	4,000	.....	.....	0.849
300	.....	.....	.....	.....	.....	5.63	2.30	1.14	0.384	0.076	4,200	.....	.....	0.968
350	.....	.....	.....	.....	.....	.....	4.09	1.58	0.530	0.103	4,500	.....	.....	1.07
400	.....	.....	.....	.....	.....	10.0	5.12	2.05	0.688	0.128	5,000	.....	.....	1.33
450	.....	.....	.....	.....	.....	.....	6.20	2.58	0.870	0.170				
500	.....	.....	.....	.....	.....	.....	7.64	3.20	1.11	0.280				
750	.....	.....	.....	.....	.....	.....	.....	.....	2.42	0.470				
1,000	.....	.....	.....	.....	.....	.....	.....	.....	4.28	0.832				
1,250	.....	.....	.....	.....	.....	.....	.....	.....	6.70	1.31				
1,500	.....	.....	.....	.....	.....	.....	.....	.....	9.68	1.88				

* Computed from Weisbach formula $H_e = (0.131 + 1.847)\left(\frac{r}{R}\right)^{3.5} \times \frac{V^2}{2g} \times \frac{a}{180}$.

where H_e = head loss, psi
r = internal radius
R = radius of axis of pipe
V = velocity, fps
a = central angle or angle subtended by the bend, deg

These figures are approximations of the actual quantities.

TABLE 2-11. EQUIVALENT LENGTHS OF IRON PIPE TO GIVE SAME LOSS OF PRESSURE DUE TO FRICTION AS IS GIVEN BY THE FITTINGS AND EQUIPMENT SHOWN
(Lengths given in feet)

Fitting or equipment[a]	Pipe, ft[b]	Fitting or equipment[c]	Pipe length, ft, for diameter, in.			
			½	¾	1	1¼
6-in. swing check valve	50	30-gal vertical hot-water tank. ¾-in. pipe	4	17	56	
6-in. lift check valve	200	30-gal horizontal hot-water tank. ¾-in. pipe	1.2	5	16	
4-in. swing check valve	25	⅝-in. meter with ½-in. connections[d]	6.7	28	90	
2½- to 8-in. long-turn ells	130	⅝-in. meter with ¾-in. connections[d]	4.8	20	64	
2½- to 8-in. short-turn ells	4	¾-in. meter with ¾-in. connections[d]	3.4	14	45	
3- to 8-in. long-turn tees	9	1-in. meter with 1-in. connections[d]	....	9	30	115
3- to 8-in. short-turn tees	9	1¼-in. meter with 1-in. connections[d]	....	4.4	14	54
One-eighth bend 45°	17	Water softener	50 to 200			
4-in. Grinnel dry-pipe valves	47	6-in. Grinnel dry-pipe valves[e]	....	......	...	89[f]
4-in. Grinnel alarm check valves	47	6-in. Grinnel alarm check valves[e]	....	......	...	100[f]

[a] From John R. Freeman.
[b] Same diameter as fitting or equipment.
[c] "National Plumbing Code," Sec. D.7.10.
[d] Valves not included.
[e] From John R. Freeman.
[f] Diameter 6 in.

2-15. Losses of Pressure Due to Fittings and Appurtenances.[1] Losses of pressure due to flow through a valve or fitting can be expressed in terms of the length of straight pipe of the same diameter that will cause the same loss of head. Losses of pressure or energy in orifices, short tubes, changes in pipe cross section, and similar circumstances can be shown in terms of one velocity head, that is, $V^2/(2g)$. Results of determinations of such losses are shown in Tables 2-10 to 2-15, inclusive, and in Fig. 2-14, p. 36.

Example. Let it be desired to determine the pressure at the meter in Fig. 1-4, under the following conditions: pressure in the street main is 40 psi; rate of flow is 20 gpm; diameter of service pipe is ¾ in.; fittings and valves are one corporation cock, one 45° ell, one coupling, one curb valve, one gate valve, and 70 ft of ¾-in. galvanized-iron pipe.

Solution: Equivalent lengths, in feet: corporation cock or globe valve, from Tabl. 2-14, 20 ft; 45° ell, 1.5 ft; coupling, 0.8 ft; one curb valve, 20 ft; one gate valve, 0.5 fte

[1] See also E. P. De Craene, *Heating, Piping, Air Conditioning*, October, 1955, p. 90.

Total equivalent pipe is

$$70 + 20 + 1.5 + 0.8 + 20 + 0.5 = 112.8 \text{ ft of pipe}$$

The rate of head loss in feet of water per foot of pipe, from Table 2-12, is 0.776. Hence, total head loss is 0.776 × 112.8 or 87.5 ft of water, equivalent to 38.0 psi. The pressure at the meter is, therefore, 40 − 38 = 2 psi.

A hydraulic grade line showing the results of sudden changes in pipe sizes is shown in Fig. 2-15. The rise of the grade line at the sudden enlargement is due to the recovery of pressure energy when velocity energy is diminished.

TABLE 2-12. LOSS OF PRESSURE DUE TO FRICTION IN IRON PIPE AND FITTINGS*

Description of pipe or fitting, in.	Loss of head in feet of water per foot or per fitting for flow, gpm†							
	5	10	15	20	25	50	75	100
½ black pipe	0.29	0.95	1.89	3.2	4.74			
¾ black pipe	0.063	0.22	0.45	0.776	1.13	4.02	8.2	
1 black pipe	0.019	0.065	0.13	0.22	0.33	1.12	2.23	3.8
1¼ black pipe	....	0.018	0.047	0.061	0.09	0.30	0.61	1.011
1½ black pipe	0.0021	0.00074	0.015	0.025	0.038	0.13	0.27	0.46
2 black pipe	....	0.00085	0.002	0.0036	0.0059	0.025	0.061	0.11
2½ black pipe	....	0.00016	0.00030	0.00047	0.00067	0.0020	0.0037	0.0068
3 black pipe	....	....	....	0.00013	0.00019	0.00060	0.0012	0.0019
1¼ galvanized pipe	....	0.020	0.041	0.070	0.10	0.36	0.74	1.23
1½ galvanized pipe, new	....	0.010	0.020	0.032	0.047	0.15	0.30	0.50
1½ galvanized pipe, old	....	0.0084	0.015	0.036	0.055	0.21	0.46	0.80
¾ unreamed ends	0.048	0.12	0.21	0.32	0.43	1.08		
1 unreamed ends	0.016	0.068	0.16	0.283	0.46	1.95	4.6	8.50
¾ coupling	0.0047	0.0105	0.017	0.024	0.031	0.070		
1 coupling	....	0.0022	0.0039	0.0059	0.0083	0.016	0.041	0.063
½ elbow	0.51	1.73	3.52	5.80	8.60	29.0		
¾ elbow	0.016	0.38	0.80	1.35	2.05	7.35	15.4	26.2
1 elbow	0.045	0.166	0.36	0.611	0.94	3.45	7.50	
1½ elbow	0.0090	0.036	0.081	0.144	0.224	0.89	2.00	3.57
2 elbow	....	0.0098	0.0215	0.0378	0.0585	0.223	0.50	0.83
2½ elbow	0.00142	0.0056	0.0125	0.0200	0.0340	0.132	0.33	0.52
1¼ elbow, short radius	0.0127	0.055	0.127	0.231	0.370	1.53	3.54	6.51
1½ elbow, long radius	0.043	0.029	0.088	0.182	0.353	2.32	7.15	16.65
1¼ drainage elbow‡	....	0.086	0.189	0.331	0.510	1.98	4.41	7.85
1¼ drainage elbow§	0.0105	0.044	0.102	0.185	0.291	1.21	2.80	5.11
	0.0098	0.036	0.076	0.131	0.200	0.722	1.54	2.64

* Computed from F. E. Giesecke, *Univ. Texas Bull.*, Oct. 20, 1917.

† Head losses in pipes with other friction coefficients and at other rates of flow can be read from Table 2-4 and Fig. 2-6.

‡ Short radius.

§ Long radius.

TABLE 2-13. LOSSES OF HEAD IN VALVES
(Feet of water)

Discharge, gpm	1-in. gate, University of Illinois*	1-in. globe, University of Illinois*	1-in. angle, University of Illinois*	1½-in. gate, Purdue†	1½-in. globe, Purdue†	1½-in. check, Purdue†	Discharge, gpm	2-in. gate, University of Illinois*	2-in. globe, University of Illinois*	2-in. angle, University of Illinois*
5	...	....	....	...	0.2	0.5	30	...	0.7	0.3
10	0.2	1.5	0.6	...	0.7	0.9	40	...	1.2	0.5
15	0.3	3.2	1.6	0.8	1.4	1.7	50	...	1.8	0.8
20	0.6	6.1	2.9	1.2	2.5	3.0	75	0.1	4.0	1.2
30	1.4	13.9	6.1	2.5	4.3	6.7	100	0.4	6.6	2.0
40	2.6	25	11.8	4.1	10.8	13	150	0.9	16	4.8
50	4.0	38	17.5	6.5	13	19	200	1.8	28	8.6
75	8.7	....	36	...	32	39	250	2.6	....	12.7
100	...	....	....	...	....	....	300	3.6	....	17

* *Eng. Expt. Sta. Bull.* 105, 1918.
† *Eng. Expt. Sta. Bull.* 1, 1918.

TABLE 2-14. ALLOWANCE IN EQUIVALENT LENGTH OF PIPE FOR FRICTION LOSS IN VALVES AND THREADED FITTINGS*

Diameter of fitting, in.	Equivalent length of pipe for various fittings, ft						
	90° standard ell	45° standard ell	90° side tee	Coupling or straight run of tee	Gate valve	Globe valve	Angle valve
⅜	1	0.6	1.5	0.3	0.2	8	4
½	2	1.2	3	0.6	0.4	15	8
¾	2.5	1.5	4	0.8	0.5	20	12
1	3	1.8	5	0.9	0.6	25	15
1¼	4	2.4	6	1.2	0.8	35	18
1½	5	3	7	1.5	1.0	45	22
2	7	4	10	2.0	1.3	55	28
2½	8	5	12	2.5	1.6	65	34
3	10	6	15	3.0	2.0	80	40
3½	12	7	18	3.6	2.4	100	50
4	14	8	21	4.0	2.7	125	55
5	17	10	25	5	3.3	140	70
6	20	12	30	6	4	165	80

* From "Plumbing Manual," p. 41, National Bureau of Standards, 1940, and "National Plumbing Code," Sec. D.7.8.

TABLE 2-15. HEAD LOSSES IN COPPER PIPE AND FITTINGS APPLICABLE ESPECIALLY TO SERVICE PIPES*

Flow, gpm	Head losses, psi									
	Corporation stop†		Curb stop†		Globe valve†		Disk meter‡		Copper pipe per 100 ft§	
	¾ in.	1 in.	¾ in.	1 in.	¾ in.	1 in.	⅝ in.	1 in.	¾ in.	1 in.
1	0.015	......	0.011	....	0.060	.....	0.05	....	0.26	
2	0.050	......	0.034	....	0.195	.....	0.10	....	0.85	
5	0.500	0.070	0.180	0.04	1.00	0.29	0.63	0.1	4.3	1.0
10	0.879	0.23	0.606	0.13	3.44	1.0	2.8	0.3	15.0	3.5
20	2.75	0.80	1.90	0.46	10.76	3.5	10.8	1.3	47.0	12.0
40	.....	2.7	6.3	1.5	35	11.6	40	5.1	160	40
50	.....	4.0	9.7	2.3	90	17.5	90	8.3	235	60
100	.....	12.5	.....	7.5	......	60	.....	34.0	......	220

* From James G. Carns, Jr., Service Line and Water Meter Requirements, *J. Am. Water Works Assoc.*, October, 1956, p. 1243.

† These figures are based on the equivalent number of feet of ¾-in. type K copper tubing as follows: corporation stop, 5.86; curb stop, 4.04; and globe valve, 22.9.

‡ Based on ⅝-in. meter manufactured by Neptune Mfg. Co.

§ Head loss varies directly with length of pipe; for example, loss for 1 ft of pipe is ¹⁄₁₀₀ of the value in this table.

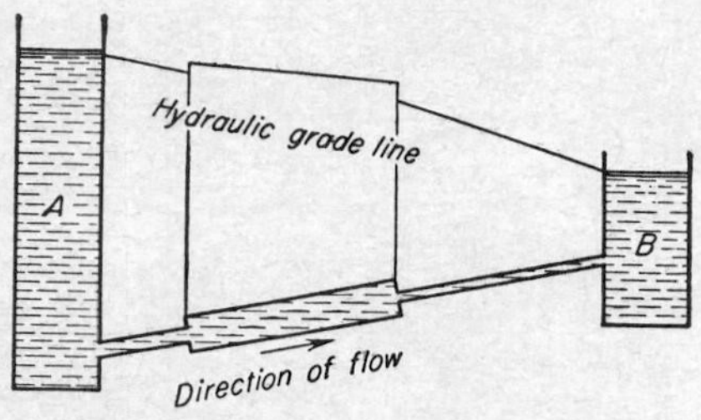

FIG. 2-15. Hydraulic grade line showing effect of change in size of pipe.

2-16. Equivalent Pipes. Pipes are said to be *equivalent* when they will carry the same rate of flow with the same loss of head.

Example. Find the length of a ¾-in. pipe that is equivalent to 100 ft of 1-in. pipe, where C in the Hazen and Williams formula is 100.

Solution: Assume any convenient rate of flow through the given pipe and the equivalent pipe, such as 10 gpm. The head loss, from Fig. 2-6, in the 1-in. pipe is 150 ft per 1,000 ft, or 15 ft in the length given. The head loss in the ¾-in. pipe, with the same rate of flow, is 525 ft per 1,000 ft, or 0.525 ft per ft. Hence, the length of ¾-in. pipe required to give a head loss of 15 ft is $15/0.525 = 28.5$ ft.

Some lengths and diameters of equivalent pipes are given in Table 2-16.

2-17. Siphonage. Siphonage, aspiration, suction, negative pressure, partial vacuum or vacuum, and other terms are used synonymously to indicate a pressure below atmospheric or below gage pressure. The condition is illustrated in Fig. 2-16. In Fig. 2-16*A* the velocity of flow and the rate of head loss are such that *P* has a positive value at points *A* and *B*. In Fig. 2-16*B*, however, the velocity head is so great at *B* that in order to maintain the same total energy the pressure head P_B must be negative, that is, less than atmospheric. Siphonage is discussed also in Sec. 2-12.

TABLE 2-16. SIZES AND LENGTHS OF EQUIVALENT PIPES

Diameter of pipe, in.		Capacity as percentage of 2-in. pipe with $C = 100$	Number of feet of pipe required to give the same head loss with the same rate of flow as 1,000 ft of 2-in. pipe with $C = 100$			
Nominal	Actual		$C = 75$	$C = 100$	$C = 125$	$C = 150$
¼	0.364	1.15	0.137	0.232	0.35	0.49
½	0.622	4.67	1.73	2.94	4.4	6.2
¾	0.824	9.80	7.4	12.5	18.9	26.5
1	1.049	18.6	21	37	56	78
1¼	1.380	30.0	100	170	271	382
1½	1.610	57.5	191	324	490	666
2	2.067	100	590	1,000	1,510	2,120
2½	2.469	175	1,518	2,570	3,880	5,450
3	3.068	312	4,360	7,400	11,180	15,700
4	4.026	636	17,070	27,900	42,100	59,100

(The number of ½-in. pipes that will discharge as much as a single pipe of any other size for the same pressure loss)

Size of pipe, in.	½	⅝	¾	1	1¼	1½	2	2½	3	4	6	8	10
Number of ½-in. pipes with same capacity	1	1.7	2.9	6.2	10.9	17.4	37.8	65.5	110.5	189	527	1,200	2,090

Example. What size of main pipe will supply two 1-in. and one ¾-in. branch pipes? Two 1-in. pipes = 12.4, ½-in. pipes, and one ¾-in. pipe = 2.9, ½-in. pipes, or a total of 15.3, ½-in. pipes. The nearest (next largest) size of pipe is a 1½-in. pipe that should be used for the main.

2-18. Cavitation. Water vaporizes or boils when its temperature is raised to 212°F at atmospheric pressure. Water will vaporize or boil at a lower temperature if the pressure is reduced. In fact, if the pressure is reduced sufficiently, water will boil at room temperature or lower. This phenomenon may occur in plumbing pipes, equipment, and pumps. It is called *cavitation*. It may be defined as a rupture of the continuity of a liquid as it turns to vapor owing to a sudden reduction of pressure. Low pressures are produced in conduits by a sudden increase of velocity. They are produced in equipment as, for example, in pumps when a moving object such as an impeller passes rapidly through the water. In other words, the pressure reduces as the velocity head increases in order that their sum may remain constant.

The rapidity with which a high vacuum is made and broken and water changes from a liquid to a vapor and back to liquid again may be so great as to create sounds varying from a rattle to a loudroar. A corrosive

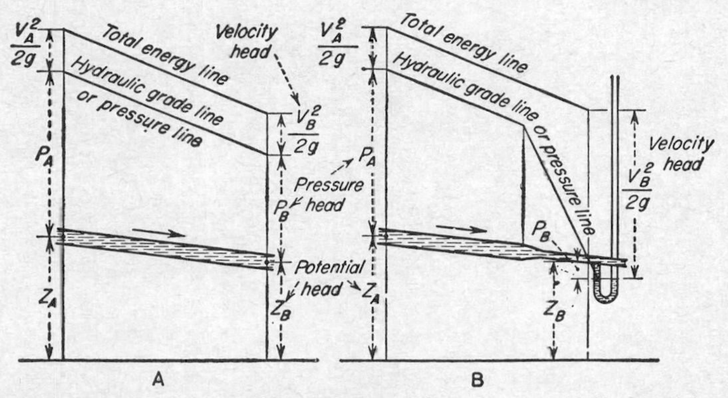

Fig. 2-16. Diagram showing production of negative pressure (siphonage) by increase of velocity head without change of total energy.

effect may appear on the surfaces of metal exposed to cavitation. The phenomenon is avoided by maintaining low velocities between liquids and surfaces in contact and by avoiding sudden accelerations in velocity of flow in closed conduits.

2-19. Water Hammer. *Water hammer* is a pressure that results from a sudden deceleration or stoppage of the velocity of flow of water in a closed conduit. An effect is the propagation of waves of pressure alternately above and below normal in the conduit. The duration of each intensity of pressure wave is the time required for the wave to travel from the point of velocity deceleration to a point of relief and back again.

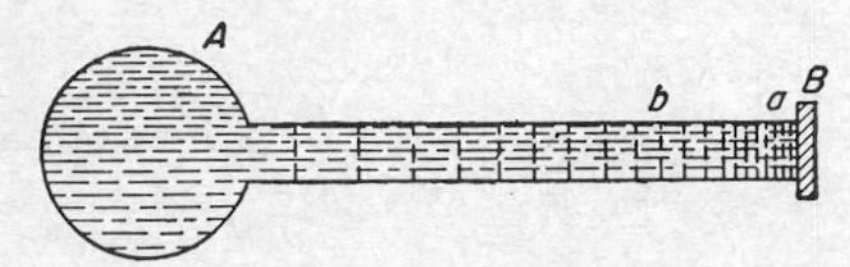

Fig. 2-17. Diagram showing water hammer resulting from sudden stoppage of flow.

Water flowing through a long, closed conduit may be compared with a heavy, incompressible battering ram. When the ram is moving and is suddenly stopped by the closing of a valve, the condition illustrated in Fig. 2-17 is created. When the valve at B is closed, the deceleration of the particle of water at b increases the pressure on the particle a. Since water is practically incompressible, the increased intensity of pressure forces the walls of the pipe to stretch (expand) and to permit particles at b to move at a diminishing velocity toward the valve. The wave of increased pressure moves upstream in the pipe with the speed of sound in water. At point A, which may be called a point of relief, the motion of the pressure wave is stopped. Pipe AB is now expanded, and the water in it is under a pressure higher than normal. The pipe now contracts, ejecting water from it, and the direction of the pressure wave is reversed as a subnormal pressure wave. This subnormal pressure wave travels with the speed of sound back to point B. Abnormal and subnormal pressure waves continue to travel back and forth through the pipe, the intensities of pressure being diminished by the absorption of energy in the expansion and contraction of the pipe walls. If a pressure gage were placed in the pipe near B, the fluctuations of pressure might be indicated somewhat as shown in Fig. 2-18.

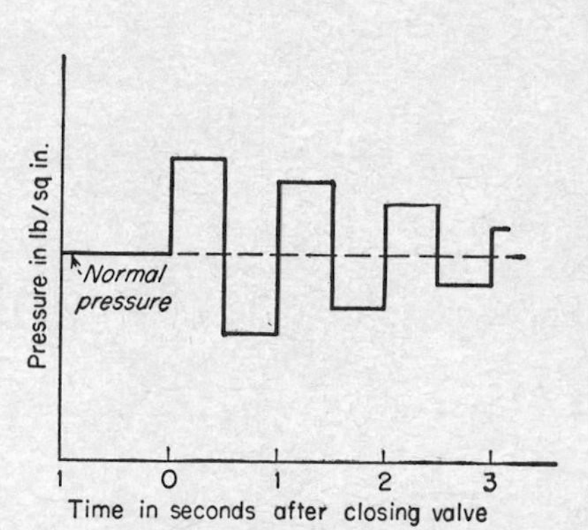

FIG. 2-18. Diagram showing changes of pressure due to water hammer.

Water hammer is silent, but the rapid fluctuations of pressure may set up vibrations or movements of the conduit and its supports, resulting in alarming noises; the pressures created may burst the conduit; or other difficulties may result. The bursting of a pipe due to water hammer has been legally termed an explosion, and insurance covering damage due to explosion has been collected. Water hammer may result in plumbing from such causes as the sudden closing of a valve, the sudden stopping or reversal of flow through a centrifugal pump, and the too rapid operation of a reciprocating pump and from admitting water too rapidly into a closed tank filled with air.

Water-hammer pressures can be approximated from the "rule-of-thumb" expression

$$P = 64V \tag{2-17}$$

where P = increased pressure due to water hammer, psi

V = velocity of flow of water, psi

Two formulas suitable for use in long, straight runs of pipe are

$$P = \frac{4{,}665V_w}{144g}\sqrt{\frac{1}{1 + \dfrac{Kd}{Et}}} \tag{2-18}$$

$$V_w = 4{,}665\sqrt{\frac{1}{1 + \dfrac{Kd}{Et}}} \tag{2-19}$$

where P = excess pressure due to water hammer, psi
V = velocity of flow of liquid, fps
w = weight per unit volume of liquid, lb per cu ft
E = modulus of elasticity of pipe material, psi
K = bulk modulus of elasticity of the fluid (294,000 psi for water)
V_w = velocity of travel of pressure wave, fps
d = diameter of pipe, in.
t = thickness of pipe shell, in.
g = acceleration due to gravity, 32.2 fps^2

TABLE 2-17. RATIOS OF K/E*

Plate steel	Cast iron	Concrete	Wood
0.01	0.02–0.022	0.10	0.20

* R. W. Angus, Water Hammer Pressures in Compound and Branched Pipes, *Proc. Am. Soc. Civil Engrs.*, January, 1938, p. 133.

Values of K/E for different pipe materials are given in Table 2-17. If L is the length of pipe from valve to point of relief, the time of closure must be less than or equal to $2L/V_w$ to develop maximum pressure. An approximate value of the pressure rise that occurs when the time of closure T_a is greater than the critical time $T_c = 2L/V_w$ can be obtained by multiplying P in Eq. (2-18) by T_c/T_a.

The advantage of the slow closing of a valve in controlling the intensity of pressure due to water hammer is indicated by the solution of the following problem:

Example. Water is flowing with a velocity of 10 fps in a 2-in. steel service pipe that is 200 ft long. (1) A quick-closing valve at the building end of the pipe is slammed shut in 1 sec. What would be the excess pressure due to water hammer? (2) What would be the excess pressure if the valve were closed in 10 sec?

Solution: (1) Using formula (2-19), t from Table II-49 in Appendix II, and K/E from Table 2-17 we have

$$V_w = 4{,}665 \sqrt{\frac{1}{1 + [(0.01)(2)/0.154]}}$$
$$= 4{,}385$$

and from Eq. (2-18) we have

$$P = \frac{(4{,}655)(4{,}385)}{(144)(32)}(0.94) = 4{,}180 \text{ psi}$$

$$T_c = \frac{2L}{V_w} = \frac{400}{4{,}385} = 0.091$$

Hence, when $T_a = 1$, $P = \dfrac{(0.091)(4{,}180)}{1} = 380 \text{ psi} = \text{answer 1}$

When $T_a = 10$, $P = \dfrac{(0.091)(4{,}180)}{10} = 38 \text{ psi} = \text{answer 2}$

The preceding formulas are not widely used in the solution of water-hammer problems in plumbing because of the short runs and many branches in plumbing pipes. The formulas do, however, give some indication of the magnitude of the pressures involved. Water-hammer pressures in mains cause high shocks in plumbing because of the greater volumes of water in the mains than in the plumbing pipes. Approximate maximum water-hammer pressures that may be produced in iron pipes are shown in Table 2-18.

TABLE 2-18. APPROXIMATE MAXIMUM WATER PRESSURES, ABOVE STATIC PRESSURE, THAT MAY BE CAUSED BY WATER HAMMER IN IRON PIPE

Velocity of flow in pipe, fps	1	2	3	5	8	10	15	20
Pressure, psi	59	118	177	354	472	590	885	1,180

2-20. Conditions of Flow in Stacks. The flow through drainage pipes is intermittent, and the pipes do not always flow full of water. The flow down a soil stack or a downspout will consist of alternate slugs of air and of water mixed with air or, momentarily, of an unbroken column of water. In the case of the discharge of water from a water closet, the first part of the discharge to reach the soil stack from the sloping branch pipe will have fallen some distance and will have gained appreciable velocity before the last part of the discharge from the fixture reaches the soil stack. As a result, air from the upper part of the soil stack or other connected pipes is drawn in to relieve the partial vacuum formed by the separating slugs of water. The entrained air mixes with the falling water, the mixture occupying a greater space than that occupied by the water alone.

The falling water and air occupy the full cross section of the stack and fall downward together, acting somewhat as a long, elastic piston expelling air ahead of it and drawing air in behind it. The air mixed with the water allows parts of the piston to expand and contract as the water at the lower end is accelerated or retarded. Siphonage and back pressure are thus created to be transmitted throughout the piping system or relieved by venting.

Siphonage may be expected to develop when the first of the falling water has acquired a velocity greater than that of the water following it, and back pressure to develop when the velocity of the leading water is retarded, as at the time it reaches the bottom of the stack. If the curvature of the footpiece at the base of the stack is gentle and the house drain and house sewer are unobstructed by a main trap so that entrained air can be discharged freely into the common sewer, or if some other satisfactory vent is provided, the back pressure formed at the head of the falling column of water will be small, generally of the order of 0.01 ft of water. Similarly, the siphonage at the top of the stack will be small if there is adequate venting. If, however, there is inadequate venting at either end of the stack, the back pressures and siphonages produced may be measured in many feet of water. Cases have been observed in which alternation of back pressures and siphonages were so rapid that water in traps was churned without being expelled.

Air pressures may be transmitted through pipes by changing the volume of air in the pipe by compression or expansion or by concussion, that is, the transmittal of a pressure wave with the speed of sound in air by successive collisions of particles of air. For pipes relatively long with respect to their diameter, the sudden transmission of such pressures may be sufficient to rupture the seal of a trap. Only a small change in the volume of the air confined in a pipe is required to force the seal of a trap. If normal atmospheric pressure is assumed to be 34 ft of water, the 2-in. seal of a trap corresponds to about $\frac{1}{200}$ of the absolute air pressure. Hence, a change of $\frac{1}{200}$ of the absolute air pressure in the stack should break a 2-in. trap seal.

2-21. Effect of Solids on Pressures in Stacks. The effect on the intensities of back pressures and siphonages resulting from the mixture of solids with water discharged from plumbing fixtures, particularly from water closets, may be appreciable when the rate of flow down the stack is low, but it is negligible when the rate of discharge approaches the capacity of the drainage pipes. Under this condition pressures and siphonages are greatest. Hence, for practical purposes in determining the pressures in drainage and vent pipes, the effect of solids in the discharges can be disregarded. However, it is important to consider slopes, angles of turns, types of connections, smoothness of passages, cleanouts, and similar features in order to avoid clogging of the pipes by solids.

2-22. Pressures in Horizontal Drainage Pipes. The characteristics of flow in horizontal drainage pipes differ from those in stacks, since only a mixture of water and solids, without slugs of air, is usually involved in the former. Horizontal pipes should be designed to flow less than full, with vents provided to permit egress and ingress of air, since the pipes are alternately partly filled with and emptied of water, which causes air movements. Pressures may be created in horizontal pipes when two or more fixtures are discharged simultaneously at a rate sufficient to fill the pipe between them without adequate venting between them. Such a condition would exist in Figs. 11-18 and 11-19 if there were no vents. Methods of controlling these pressures are discussed in Chap. 11.

Pressures created in stacks may be transmitted undiminished to and through connected horizontal pipes. This fact is of importance in considering the distance between traps, stacks, and vent pipes. There is no practicable limit at which a trap can be placed along a horizontal pipe connected to a stack to avoid pressures created in the stack without an interposed vent. The trap close to the stack may be affected in the same manner and intensity as the trap some distance from it.

2-23. Characteristics of Plumbing Traps. The principal parts of a plumbing trap are illustrated in Fig. 1-3. The *seal* is defined as the vertical distance between the crown weir and the *dip*, as shown in Fig. 1-3. The *dip* of a trap is the lowest point in the trap to which the liquid surface can sink before air or gas can pass through the trap. The *crown weir* is the lowest point in the trap over which the liquid must flow to leave the trap. The depth of seal used in ordinary plumbing traps is about 2 in.

Maintenance of the liquid seal in a trap requires venting and other expedients that increase the cost of plumbing installations. The seal of a trap may be destroyed by direct air pressure or back pressure; by the capillary action of an absorbent material, such as a string or cloth, lying across the crown weir; by inertia of water passing rapidly through the trap; by evaporation, which is discussed in Sec. 11-37; and by siphonage and self-siphonage, discussed in Sec. 2-24.

2-24. Hydraulics of Traps. The strength or ability of a trap seal to resist the passage of air or gas through it is determined by the vertical height to which water in the trap can rise above the dip to resist this passage. For practical purposes the strength of a trap seal is equal to its depth. However, under circumstances depending on the configuration of the trap and the volume of water in it, the strength may be greater or less than the seal.

Back pressure on a trap is pressure applied against the downstream free surface of water in the trap, *A* in Fig. 1-3. This pressure tends to force the water up the drain pipe toward or into the fixture. *Siphonage* is the reduction of the pressure against the downstream free surface of water in the trap so that the greater air pressure on the upstream free surface forces water to flow from the trap into the drain pipe. *Self-siphonage* is the reduction of pressure against the downstream free surface of water in the trap by the creation of an unvented column of water in the drain pipe continuous with and downstream from the water in the trap. The intensity of the strength of siphonage and of self-siphonage is equal to the difference between the pressures on the free surfaces of water upstream and downstream from the trap.

Siphonage and self-siphonage are more undesirable than back pressure because the former remove water from the trap so that the strength of the seal is weakened or destroyed, whereas after back pressure is removed, water may flow from the fixture to restore the seal.

The strength of a simple form of trap to prevent the breaking of its seal should be equal to the depth of the seal, provided the pressure is developed slowly and is applied steadily. The strength of a trap seal may be weakened if the rhythm of impulses of back pressure or siphonage corresponds to the swinging period of water in the trap. Back pressure and siphonage are seldom slowly and steadily developed and applied. They often consist of sudden impulses of short duration, alternating between pressure and siphonage.

Conditions affecting self-siphonage include seal, rate of discharge, momentum of mass of water causing self-siphonage, type of trap, and vertical length of the downstream leg of the drain pipe which forms the lower leg of a siphon. Conditions that will prevent or diminish self-siphonage are a short, downstream length of drain pipe below the crown weir; a low rate of discharge, particularly one that is insufficient to fill the downstream drain pipe; a nonsiphon trap; a fixture that will create a vortex as it discharges, thus admitting air to break the siphonage; a fixture with a flat bottom that will cause a low rate of flow toward the end of its discharge, thus refilling the trap seal; or venting.

Traps in plumbing fixtures do not always stand full. Some water is drawn from the trap by the discharge of an adjacent fixture. However, as long as the volume of water held within a trap is sufficient to fill either leg of the trap on either side of the dip of the trap for a vertical height equal to the seal height, the strength of the trap to resist the passage of air is equal to the height of its seal.

CHAPTER 3

WATER SUPPLIES

3-1. The Water Supply. Many cities have a public water supply that provides water of safe quality under adequate pressure to meet normal requirements of rate of flow into buildings of moderate height. It is a matter of pride that the quality of the local water supply meets the standards of the U.S. Public Health Service[1] for water that may be delivered to the public by interstate carriers. In locations remote from a public water supply the builder must seek a source of water of good quality and adequate quantity.[2] The pressure required and the rate of use must then be considered.

3-2. Quality of Water. A decision concerning the potability, or fitness to drink, of a water supply should be based on a study of the source, the surroundings, and chemical and bacteriological analyses of the supply. A physical analysis showing satisfactory tint (color), taste, temperature, and turbidity is insufficient evidence on which to risk the health of the water users. Lives have been lost as a result of such reliance.[3]

The sanitary quality of a water supply is not necessarily a permanent condition. The purity of the supply today is no assurance that it will be pure tomorrow unless the source is adequately protected. The sanitary quality of a water supply can be compared with the health of an individual. Both may be excellent for years, but without adequate protection, either may suddenly become unsatisfactory. Nor can the quality of either be expressed as a percentage of anything. No water supply is "99 per cent pure." In fact, "pure" water may be unhygienic. For example, dissolved oxygen and dissolved carbon dioxide are necessary to make water palatable, and certain dissolved minerals are required for good hygiene.

Physical, chemical, bacteriological, microscopic, and possibly mineral analyses must be made to determine the sanitary and industrial qualities

[1] Standards of the U.S. Public Health Service for Potable Water in Interstate Traffic, *J. Am. Water Works Assoc.*, March, 1946, p. 361.

[2] See also J. Horsford, How Big Should Water Systems Be? *Domestic Eng.*, May, 1957, p. 136.

[3] "Report on Manteno State Hospital Epidemic, 1939," Illinois State Department of Public Health, 1945.

of a sample of water. The making of these analyses requires the services of technical specialists. No conclusive opinion on the potability and hygienic safety of a proposed source for water supply can be reached without a sanitary survey.

A sanitary survey consists of a study of the surroundings of the source and the catchment area on which the water is collected or through which it runs. In general, no surface water can be expected to be safe at all times unless continuously patrolled and police-protected. Natural springs should be viewed with suspicion as a source of water unless it is known that the water has flowed an appreciable distance through fine, porous material. Shallow dug wells, that is, wells less than 40 to 60 ft in depth, of the "old oaken bucket" type are commonly dangerous unless specially protected. A deep well in fine, porous material, removed 100 ft or more from the nearest source of pollution and adequately protected against surface contamination, can generally be relied on to provide safe drinking water.

Where the sanitary quality of a water that must be used is unsatisfactory—where it is too hard or where it stains fixtures, corrodes pipes, or is otherwise unsatisfactory—it must be purified. Methods for the purification of water for domestic purposes are discussed in Sec. 17-1.

3-3. Water-borne Diseases. Infectious diseases that may result from the drinking of polluted water include typhoid fever, paratyphoid fever, cholera, bacillary and amoebic dysentery, hepatitis or infectious jaundice, and possibly poliomyelitis. That poliomyelitis can be conveyed by drinking water has not been proved to the satisfaction of all health authorities. Swimming or bathing in polluted water may result in any of the aforementioned diseases and, in addition, may result in infections of schistosomiasis, sometimes called "swimmer's itch," or other parasitic infections of the skin and of the eye, ear, nose, and throat. Athlete's foot, a fungus growth on the skin between the toes, is sometimes contracted in swimming pools. Water has sometimes been suspected of conveying infected organic matter containing organisms of such diseases as anthrax and tapeworm.

Natural water supplies contain minerals that are conducive to good health. For example, the absence of dissolved iodides from water is usually accompanied by a relatively high endemic index of goiter in the locality; a content of fluorides of less than 1.0 ppm, by weight, may be desirable to prevent dental caries,[1] but larger concentrations may cause mottling of tooth enamel in infants.

Nonspecific physiological derangements are evidenced mainly by diarrhea, nausea, or even constipation from some forms of intestinal

[1] Standards of the U.S. Public Health Service for Potable Water in Interstate Traffic, *J. Am. Water Works Assoc.*, March, 1946, p. 361.

bacteria or from a high concentration of such materials as calcium sulfate and magnesium sulfate. On the other hand there is a minimum desirable concentration of certain minerals in water.

It is rare that toxic substances are present in natural waters in sufficient concentration to be dangerous to man. The limiting concentrations of some substances are shown in Table 3-1. The presence of copper sulfate and of free chlorine, even in low concentration, may be fatal to fish and to plants. Cattle and other large animals can safely drink water

TABLE 3-1. MAXIMUM LIMITS OF SOME SUBSTANCES PERMITTED IN POTABLE WATER*

Substance or characteristic	Concentration, ppm† unless noted	Substance or characteristic	Concentration, ppm† unless noted
Turbidity	10	Copper	3.0
Color	20‡	Iron and manganese	0.3§
Lead	0.1	Magnesium	125
Fluoride	1.5	Zinc	15
Arsenic	0.05	Chloride	250
Selenium	0.05	Sulfate	250
Chromium¶	0.05	Total solids	500–1,000

* From Standards of the United States Public Health Service for Potable Water in Interstate Traffic (see *J. Am. Water Works Assoc.*, March, 1946, p. 361).

† ppm means parts per million by weight.

‡ Cobalt scale.

§ Either one alone or both combined.

¶ Hexavalent.

containing some organisms pathogenic to man or containing somewhat larger concentrations of minerals than are suitable in drinking water for man.

3-4. Quantity of Water Required. Personal daily needs of water for drinking only are small, measurable in terms of a few pints or less. The requirements for all purposes, such as bathing, washing, cleaning, and cooking, are greater. The average per capita daily use of water for all purposes varies among cities, following no rule of location, climate, scale of living, or otherwise. Figures showing water use for various purposes and the amounts recommended for building supplies and fixtures are listed in Tables 3-2 to 3-4. The "National Plumbing Code"[1] states that 50 gal per capita per day may be considered a safe figure for an apartment building and 40 gal per capita per day for dwellings in a housing project. The "Housing Code"[2] recommends (p. 56) an allowance of a minimum

[1] Secs. D.2.5 and D.2.6.

[2] "The Uniform Plumbing Code for Housing," Housing and Home Finance Agency, February, 1948.

TABLE 3-2. RATES OF WATER USAGE FOR VARIOUS PURPOSES[a]

Purpose	Amount, gal per capita per day, unless otherwise stated	Purpose	Amount, gal per capita per day, unless otherwise stated
Domestic only	10	Office buildings[b]	27–45
Cities in New York, New Jersey, and New England	26–44.5	Hospitals[b]	125–350[c]
		Hospitals[d]	66–1,144[e]
Single-family houses in 1927:		Hotels[b]	306–525[e]
West Haven, Conn	36.8	Laundries[b]	3–5[f]
New Haven, Conn	59.4	Restaurants[b]	0.5–4[g]

[a] From M. A. Pond, *J. Am. Water Works Assoc.*, December, 1939, p. 2,003.
[b] From H. E. Jordan, *ibid.*, January, 1946, p. 65.
[c] Gallons per bed per day.
[d] From G. C. St. Laurent, *Hotel Eng.*, 1940.
[e] Gallons per day per occupied room, with average of 400 for hospitals.
[f] Gallons per pound.
[g] Gallons per meal.

TABLE 3-3. RATE OF FLOW AND REQUIRED PRESSURE DURING FLOW FOR DIFFERENT FIXTURES*

Fixture	Flow pressure, psi†	Flow rate, gpm
Ordinary basin faucet	8	3.0
Self-closing basin faucet	12	2.5
Sink faucet, 3/8 in	10	4.5
Sink faucet, 1/2 in	5	4.5
Bathtub faucet	5	6.0
Laundry tub cock, 1/2 in	5	5.0
Shower	12	5.0
Ball cock for closet	15	3.0
Flush valve‡ for closet	10–20	15–40§
Flush valve‡ for urinal	15	15.0
Garden hose, 50 ft, and sill cock	30	5.0

* From "National Plumbing Code," Sec. D.3.2.
† Flow pressure is the pressure in the pipe at the entrance to the particular fixture considered.
‡ Flushometer type.
§ Wide range due to variation in design and type of flush-valve‡ closets.

of 50 gal per capita per day, to be supplied at a minimum rate of 5 gpm per capita.

The quantity of water is a controlling factor in the selection of the source and, possibly, in determining the volume of storage to be provided. The rate of use is important in determining the capacity of pumps and the sizes of pipe, since it is the peak load that must usually be cared for

TABLE 3-4. RATES OF WATER SUPPLY TO AND DRAINAGE FROM PLUMBING FIXTURES, EXPRESSED IN FIXTURE UNITS*

Fixture	Units
Bathroom group:†	
With flushometer (*H*, *N*, and *P*)	8
With flush valve (*H*, *N*, and *P*)	6
Bathtub:	
Private (*H*, *N*, and *P*)	2
Public	4
Bedpan washer, public (*H*)	10
Bidet (*B*)	1
Private (*H*)	3
Public (*H*)	4
Combination fixture (*H*, *N*, and *P*)	3
Dental cuspidor (*H*)	1
Dental lavatory:	
Private (*H*)	1
Public (*H*)	2
Dishwashing machine‡	1
Drinking fountain:	
Private (*H*)	½
Public (*H*)	1
Electric water cooler (*H*)	1
Kitchen sink:	
Private (*H*, *N*, and *P*)	2
Public (*H*, *N*, and *P*)	4
Garden hose (*B*)	10
Lavatory:	
Private (*H*, *N*, and *P*)	1
Public (*H*, *N*, and *P*)	2
Barber, beauty parlor (*H*)	3
Surgeon's (*H*)	3
Laundry tray, 1 or 2 tubs:	
Private (*H*)	2
Public (*H*)	4
1 to 3 tubs (*P* and *N*)	3
Shower, separate head:	
Private (*H*, *N*, and *P*)	2
Public (*H*, *N*, and *P*)	4
Sinks:	
Surgeon's (*H*)	3
Soda fountain (*H*)	2
Flushing rim, flushometer (*H*)	10
Service (*H*, *N*, and *P*)§	3
Pot or scullery (*H*)	5
Pantry (*B*)	1
Slop (*B*)	1
Urinal:	
Pedestal, with flushometer (*H*, *N*, and *P*)	10
Wall or stall (*H*, *N*, and *P*)	3–5
With flush tank (*H*)	3
Trough (every 2 ft) (*H*)	2
Wash sink, circular or multiple, each set of faucets (*H*)	2
Water closet:	
Flushometer:	
Private (*H*, *N*, and *P*)	6
Public (*H*, *N*, and *P*)	10
With flush tank:	
Private (*H*, *N*, and *P*)	3
Public (*H*, *N*, and *P*)	5

* One fixture unit is equivalent to a flow of 7.5 gpm. For continuous outlets likely to impose continuous demands, estimate continuous supply separately and add to the total demand of fixtures (see Table 11-1). For fixtures not listed, weights can be assigned by comparing the fixture to a listed one using water in similar quantities and at similar rates. The given weights are for total demand. For fixtures with both hot- and cold-water supplies the weights for maximum separate demands may be taken as three-quarters of the listed demand for supply.

† A bathroom group consists of one bathtub, one water closet or shower head, and one lavatory.

‡ From *Domestic Eng.*, May, 1957, p. 136.

§ "Plumbing Manual" (note *P*) states: For supply outlets likely to impose a continuous demand when other fixtures are in extensive use, add the estimated continuous demand to the total demand for fixtures, *e.g.*, 5 gpm for a sill cock or hose connection is a liberal but not excessive allowance.

Fixtures not listed in the table, if installed in relatively small numbers compared to the rated fixtures, may usually be ignored in estimating for the building main and large distributing branches. If installed in sufficiently large numbers to justify their consideration, they may be assigned fixture unit ratings on the basis of a comparison with a rated fixture that uses water in similar quantities and at similar rates, *e.g.*, if wash sinks and wash troughs with multiple supply outlets are to be installed, each supply outlet may be considered as comparable in demand to that of a wash basin in public service.

The ratings given in the table are for the total hot- and cold-water demand. The following are suggested as ample allowances under favorable conditions: for main hot-water branches allow three-fourths of the total fixture units as given by the table for all fixtures using hot water; for main cold-water branches compute the total fixture units separately for fixtures that are and are not supplied with hot water and add three-fourths of the total for fixtures that are supplied with hot water to the total for fixtures that are supplied with cold water only. Ignore demands for service sinks except for hot-water supply and that for the cold-water branch to the fixture itself. Other fixtures used similarly out of hours may be treated similarly.

B. From H. E. Babbitt, "Plumbing," 2d ed., McGraw-Hill Book Company, Inc., New York, 1950.

H. From "The Uniform Plumbing Code for Housing," Technical Paper 6, Housing and Home Finance Agency, 1948.

P. From "Plumbing Manual," Report BMS66, National Bureau of Standards, 1940.

N. From "National Plumbing Code," ASA A 40.8–1955.

by pumps. Peak rates of supply to be provided for some plumbing fixtures are stated in Table 3-4.

3-5. Probability of the Simultaneous Use of Fixtures. The maximum rate of water use to be provided for in a building is not the sum of the maximum rates of supply to each fixture, since the probability is remote that all fixtures will be using water at the maximum rate at the same

TABLE 3-5. RECOMMENDED SIZES OF WATER-SUPPLY PIPES TO FIXTURES*
(Standard wrought pipe)

Fixture	Number of fixtures								
	1	2	4	8	12	16	24	32	40
Water closet:									
Tank: gpm	8	16	24	48	60	80	96	128	150
Tank: pipe size, in.	½	¾	1	1¼	1½	1½	2	2	2
Flush valve: gpm	30	50	80	120	140	160	200	250	300
Flush valve: pipe size, in.	1	1¼	1½	2	2	2	2½	2½	2½
Urinal:									
Tank: gpm	6	12	20	32	42	56	72	90	120
Tank: pipe size, in.	½	¾	1	1¼	1¼	1¼	1½	2	2
Flush valve: gpm	25	37	45	75	85	100	125	150	175
Flush valve: pipe size, in.	1	1¼	1¼	1½	1½	2	2	2	2
Wash basin†: gpm	4	8	12	24	30	40	48	64	75
Wash basin†: pipe size, in.	½	½	¾	1	1	1¼	1¼	1½	1½
Bathtub: gpm	15	30	40	80	96	112	144	192	240
Bathtub: pipe size, in.	¾	1	1¼	1½	2	2	2	2½	2½
Shower bath: gpm	8	16	32	64	96	128	192	256	320
Shower bath: pipe size, in.	½	¾	1¼	1½	2	2	2½	2½	3
Sinks† slop, kitchen: gpm	15	25	40	64	84	96	120	150	200
Sinks† slop, kitchen: pipe size, in.	¾	1	1¼	1½	1½	2	2	2	2½

* W. S. Timmis, *J. Am. Soc. Heating Ventilating Engrs.*, vol. 28, p. 397, 1922.

† Each faucet. Sizes based on pressure drop of 30 lb per 100 ft. Hot-water faucets to be disregarded when estimating sizes of risers and mains.

instant. This fact has been allowed for in Table 3-5. Kessler[1] suggests the effect of simultaneous use of fixtures on the rate of use in residences to be in accordance with the figures shown in Table 3-6. An application of the mathematical theory of probability to the simultaneous use of plumbing fixtures taken from page 76 of the "Housing Code"[2] is shown in Fig. 3-1. Cleverdon[3] recommends provision for a peak load sixteen

[1] L. H. Kessler, Some Points on Hydraulics in the Program for Adequate Piping, *Domestic Eng.*, June, 1946, p. 110.

[2] "The Uniform Plumbing Code for Housing," Housing and Home Finance Agency, February, 1948.

[3] W. S. L. Cleverdon, "Plumbing Engineering," p. 79, Pitman Publishing Corporation, New York, 1937.

times the average daily consumption for small dwellings and thirty times the average for apartment buildings.

TABLE 3-6. DEMAND FOR COLD WATER AND FOR HOT WATER

Small residences and farms[a]					Hot water for restaurants and hotels[b]					
Number and type of fixture	Small single-family house		Large single-family house		Establishment	Hours of max. hot-water demand			IV	V
						Use periods		III		
	I	II	I	II		No.	Hr each			
2 sill cocks	10	5	10	5	Hotels	2	6	4	1:15	10
2 laundry trays	20	10	20	10	All-day restaurant	2	6	4	1:15	10
1 kitchen sink	7.5	...	7.5		Two-meal restaurant	1	4	0	1:10	20
1 lavatory	5	5	15[e]	5[e]	One meal restaurant	1	3	0	1:15	45
Water closet[c]	2.5[d]	...	7.5[e]	2.5[e]	Luncheonettes	2	6	4	1:10	30
Bathtub	10[d]	...	20[f]	10[e]						
Total	55	20	80	32.5						

I. Total flow in all fixtures, gpm.
II. Actual flow from fixtures in simultaneous use, gpm.
III. Intervals between usage periods, hr.
IV. Ratio of maximum hourly demand to 24-hr use.
V. Heater capacity. Per cent of maximum daily.

[a] From L. H. Kessler, Some Points on Hydraulics in the Program for Adequate Piping, *Domestic Eng.*, June, 1946, p. 110.

[b] From F. M. Reiter, *Air Conditioning, Heating, and Ventilating*, February, 1954, p. 90.

[c] Water closets are flush-tank style

[d] One fixture.

[e] Two fixtures.

[f] Three fixtures.

The following rates of consumption are fair for metered residences: 1 adult, per day, 50 gal; filling lavatory, 1.5 gal; bathtub, 30 gal; flushing water closet, 6 to 8 gal; shower bath, 6 to 8 gpm; sprinkling lawns, 2 to 3 gpm; garden house, nozzle full open, 5 to 8 gpm; for sprinkling 100 sq ft of lawn, 8 to 10 gal; horse, per day, 5 to 10 gal; cow, 7 to 12 gal; hog, 12 gal; sheep, 1 gal; 100 chickens, 4 gal.

3-6. Water Pressures Required. Pressures maintained in the water mains in most cities that are not excessively hilly range between 30 and 80 psi. Such pressures are unsatisfactory for fire protection and must be supplemented by mobile pumping engines, or they must be increased throughout the distribution system by raising the pressure at the pumping station Fire pressures up to 100 psi or more may be needed at the hydrant or at the mobile pumping engine to provide the needed minimum of 40 to 50 psi at the base of the nozzle on a fire hose.

A pressure of 30 psi in the street main is somewhat low, but it may be sufficient to supply water to a bungalow or a two-story building. It is usually insufficient, however, in any district where the buildings are more than 40 ft high. In rough topography it may be necessary to maintain pressures up to 100 psi in lower districts in order to supply water to the upper regions of a city, or it may be necessary to divide the distribution

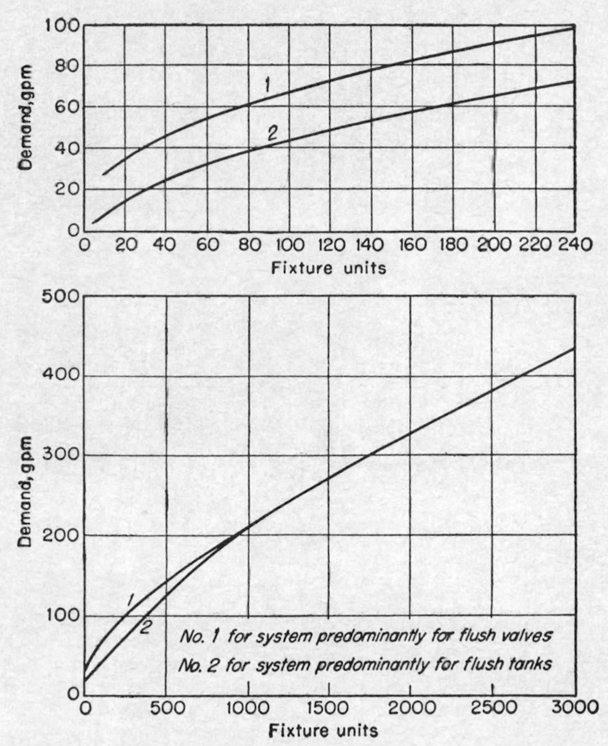

FIG. 3-1. Flow corresponding to various fixture units. (*From "The Uniform Plumbing Code for Housing," February,* 1948.)

system into pressure zones depending on the elevations on which buildings have been constructed. Where pressures are above 70 to 100 psi, care should be taken in the installation of plumbing to allow for the additional pressure. Water must be pumped to the top of tall buildings by pumps located either in the basement of the building or on different floors.

Before a building is erected, the plumber should be certain that the water pressure in the street main will be sufficient to supply water to the highest fixture or roof tank or that pumps therefor have been provided in the plumbing design. The required minimum pressure in the street main is equal to the vertical height of the highest fixture or the roof tank above the street main added to the head losses in the plumbing system when the maximum probable number of fixtures are being used simultaneously. Because of uncertainties in the computation of such head losses, it may be sufficient to provide for a pressure of 25 psi in addition to the height of the highest fixture or roof tank. Pressures at fixtures recommended by the "National Plumbing Code"[1] are shown in Table 3-3. If the pressure in the water main or other source of supply is less than the recommended intensity it is necessary to pump. The "Housing Code"[2] recommends (p. 36) a minimum pressure of 8 psi for each fixture, except for flushometers, which require 15 psi. If the pressure is unsteady, dropping below the minimum required pressure for short periods of time, an elevated storage tank or a pneumatic storage tank may supply water during the period of low pressure in the supply main.

3-7. Effect of Meters on Water Use and Required Pressure. The installation of a meter on a service not previously metered may be expected to reduce the total amount of water used, but it will probably not affect the maximum rate of use. The loss of pressure through a water meter may be appreciable. The presence of a meter on a water-supply pipe may be a large item of head loss between the water main and the plumbing fixture. Characteristics of water meters are discussed in Secs. 14-30 to 14-38, inclusive.

3-8. The Service Pipe. The public water supply is commonly delivered to the consumer through a pipe lying in a trench, extending from the street main to and through the wall of the building. This pipe is known as the *water service pipe.* Typical service-pipe installations for pipes 2 in. in diameter and smaller are shown in Figs. 1-4, 3-2, and 3-3. Larger service pipes may be connected by a tee, somewhat as shown in Fig. 3-4. It is undesirable to tap a 2-in. hole in a cast-iron pipe less than 8 in. in diameter. The depth of the trench in which the service pipe is laid should be sufficient to protect the service pipe from traffic loads and from frost. The service pipe should be laid on a straight line and a continuous slope to prevent air binding, to permit drainage, and to provide ease in rodding. Materials used for service pipes are discussed in Sec. 3-9.

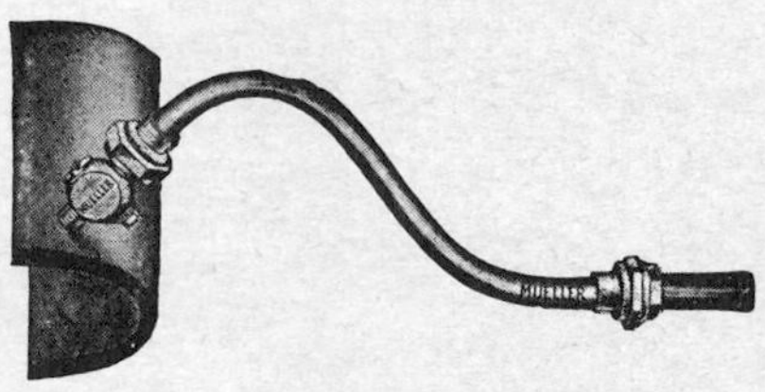

Fig. 3-2. Lead gooseneck with brass fittings. (*Mueller Co.*)

Large buildings are sometimes provided with two service pipes con-

[1] Table D.3.2.

[2] "The Uniform Plumbing Code for Housing," Housing and Home Finance Agency, February, 1948.

nected to different water mains or to different points on the same main either as a factor of safety or to provide additional service capacity, or both. Each service connection should have sufficient capacity to supply the building when the other service is shut off. The water pipes within the building should be connected to both service pipes, and each service pipe should be equipped with a check valve to prevent the drawing of water through the building into another main.

When a service connection of the type shown in Fig. 1-4 is made, a hole is tapped in the street main and the *corporation cock* is inserted. This can be done while water is under pressure in the main by means of a tool specially designed for this purpose. Such a connection is known as a *wet connection*. The size of the hole that can be made in the main is limited by its diameter. In some cities the service connection is made by the water department. In a few it is made by the plumber. The corporation cock should be threaded, not driven, into the hole in the water main. The connection usually consists of a brass valve of the ground-key type. A brass union soldered to the end of the gooseneck is threaded onto each corporation cock.

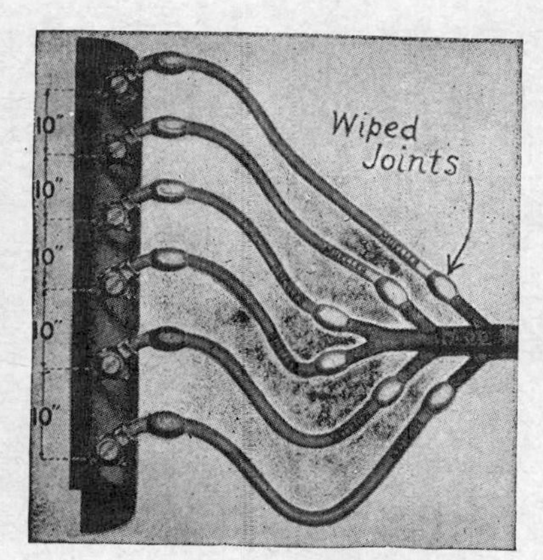

Fig. 3-3. Multiple gooseneck connection. (*Mueller Co.*)

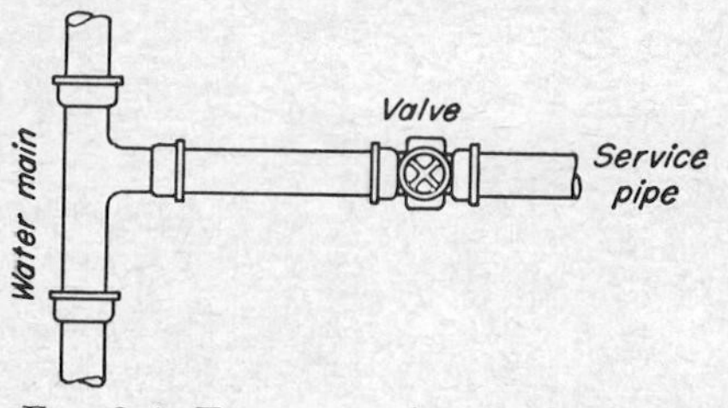

Fig. 3-4. Tee connection for large-size service pipe.

A lead gooseneck is often used to permit slight movements between water main and service pipe. The gooseneck may consist of about 18 in. or more of lead pipe to each end of which threaded brass unions have been soldered, or there may be a union at one end and a threaded nipple at the other. Such a gooseneck is illustrated in Fig. 3-2. The service pipe is connected to the end of the union opposite the corporation cock. Copper tubing is sometimes used for the gooseneck.

Where a service pipe larger than 2 in. in diameter is needed, a number of taps and goosenecks can be used, as illustrated in Fig. 3-3, or a tee connection can be made as shown in Fig. 3-4. The multiple connection shown in Fig. 3-3 is not recommended, although it is often used, because it increases the probability of troublesome breaks and leaks. It is to be noted that wiped joints are shown in Fig. 3-3 instead of stronger, more easily made, but possibly somewhat more expensive threaded brass joints as shown in Fig. 3-2. The capacities of all goosenecks in a multiple connection should equal or exceed the capacity of the service pipe. For example, a 2-in. service pipe with a net cross-section area of 3.36 sq in., as shown in Table II-46 in Appendix II, might be connected by five 1-in. goosenecks, each with a cross-section area of 0.864 sq in.

The valve shown in Fig. 1-4 on the service pipe under the parking, sometimes called the *curb cock* or *curb valve*, may be placed elsewhere along the pipe for accessibility and convenience in shutting off the water without entering the building. The curb cock should be placed in a cast-iron gate box, sometimes called a *Buffalo box*, permitting access to it by a key from the surface. Such valves are not universally used in addition to the corporation cock adjacent to the street main. In passing through the wall of the building the service pipe may be loosely encased in a larger pipe to allow movement between the service pipe and the building. If the wall is a footing or a foundation wall, the "National Plumbing Code"[1] requires the construction of a relieving arch over the pipe or an iron pipe sleeve two sizes larger to receive the service pipe or equivalent protection. The annular space between the pipe and sleeve should be filled or tightly calked with coal tar or asphaltum compound, lead, or other equally effective material.[2]

The service pipe should preferably be laid in a trench by itself at least 10 ft away from the building drain or building sewer, and if parallel to a bearing wall or footing, the pipe must be at least 3 ft away from it. The "National Plumbing Code" and other codes permit the laying of the water pipe in the same trench with the sewer when no other expedient can be followed, provided that the bottom of the water pipe is at least 12 in. higher than the bottom of the sewer and that the water pipe is laid on a solid shelf of ground to one side of the trench. It is further provided by the "National Plumbing Code" that the materials and joints of both water and sewer pipes shall have strength and durability under all known adverse conditions to prevent the escape of solids, liquids, and gases. It is not always necessary to excavate a trench or to open paved surfaces, as both water pipes and sewer pipes can be hand- or power-pushed under the ground for distances up to 100 ft or more.[3]

The diameter of a service pipe should depend on the rate of flow through it, its length, and the permissible loss of pressure due to flow through it. Some recommended sizes are shown in Table 3-7. Pressure-

[1] Sec. 2.10.1

[2] *Ibid.*, Sec. 2.13.1.

[3] See also *Air Conditioning, Heating, and Ventilating,* April, 1956, p. 114, and October, 1956, p. 126.

TABLE 3-7. RECOMMENDED SIZES OF SERVICE PIPES IN INCHES*
(Standard wrought pipe)

Class of building	Length of service-pipe main to meter, ft			
	100	50	25	10
A†	1¼	1	1	¾
B‡	1½	1¼	1¼	1
C§	2	1½	1½	1¼
D‖	2	2	1½	1¼

* Computed on basis of 20-ft loss of head from main to meter.

† An ordinary single-family dwelling—2 to 2½ stories and not more than 8 to 10 rooms, containing 1 bathroom, a kitchen sink, laundry trays, and garden hose.

‡ A 2-family house or larger dwelling, up to about 16 rooms, containing 2 bathrooms, 2 kitchen sinks laundry trays, and 1 garden hose.

§ A 4-apartment building, apartments not more than 6 rooms each. Building contains 4 bathrooms, 4 kitchen sinks, 4 sets of laundry trays, and 1 garden hose.

‖ A large apartment building containing not more than 25 apartments with a total of about 100 rooms; with full equipment of 1 bathroom and 1 kitchen, laundry trays for each apartment, and 2 hose connections for the building.

loss computations[1] are based on the principles outlined in Secs. 2-8 and 2-15. A helpful "rule of thumb" may be to provide a ⅝-in. service pipe for 7,500 sq ft of floor area and the equivalent of an additional ⅝-in. service for each additional 10,000 sq ft or fraction thereof. A larger size of service pipe may be needed where flushometer valves are used or where the water-supply pressure is low and roof or pneumatic storage tanks are not used. The trend toward the increased use of water and of appliances requiring higher water pressures and higher instantaneous rates of water supply[1,2] requires consideration of larger service pipes than are shown in Table 3-7.

3-9. Materials Used for Service Pipes. Among the materials used for service pipes may be found

1. Copper, either iron-pipe size or copper tubing
2. Lead, either unlined or lined with tin
3. Wrought iron or steel, either galvanized or plain, either unlined or lined with tin, lead, or cement
4. Brass
5. Cast iron, either unlined or lined with cement
6. Plastics

These materials are discussed in Chap. 4.

[1] See also Friction Loss in Service Pipe and Fittings, *J. Am. Water Works Assoc.*, June, 1956, pp. 744 and 752, and October, 1956, p. 1243.

[2] See also J. G. Carns, Jr., *J. Am. Water Works Assoc.*, October, 1956, p. 1243. *Ibid.*, April, 1955, p. 398.

3-10. The Private Water Supply. Where no public water supply is available or where its quality, quantity, or pressure is unsatisfactory, it will be necessary to install a private supply. Private water supplies may be obtained from wells or underground galleries, cisterns that collect rain-water, springs, or surface streams. These are named in the approximate order of their relative usefulness and suitability.

The average and maximum rates of use of water in private supplies are usually less than for public supplies because the private supply is used only for special purposes such as domestic, institutional, or industrial whereas the public water supply must deliver water for all purposes. Quantities of water used and rates of demand are discussed in Sec. 3-4, and data are given in Tables 3-2 to 3-6 and 3-8 to 3-11.

TABLE 3-8. RATE OF WATER USE FOR VARIOUS PURPOSES
(Expressed as percentages of rate of use for all purposes)

Purpose or use	Percentage of total daily consumption		
	Minimum	Maximum	Average
Domestic	20	50	35
Commercial and industrial	10	50	30
Public	5	15	10
Leakage and waste	15	40	25
Total	50	155	100

The pressures required in public water mains, stated in Sec. 3-6, are required also in a private supply. The pump must supply this pressure, or in the event the supply is to be by gravity, the height of the source above the highest fixture in the building should be equal to the sum of all friction losses in the pipe leading to the building and in the plumbing in the building during the maximum rate of demand.

3-11. Location of Wells. To assure that water will be found when attempting to locate a well, experience in the locality is the best guide, aided by observations of the topography and knowledge of the geology of the region. The divining rod may be as good as a guess. Wells should be constructed as far as possible away from sources of contamination. Minimum distances recommended in the "National Plumbing Code"[1] are shown in Table 3-12. The minimum depth below the ground surface at which water may be taken from a water-bearing stratum, under the restrictions of the "National Plumbing Code," is 20 ft, with 30 ft preferred.[2] The well should be uphill from any source of pollution,

[1] Sec. A.3.2.
[2] Sec. A.4.1.

TABLE 3-9. WATER USAGE RATES FOR SPECIAL PURPOSES*

City	Population in thousands	Gallons per day per meter			
		Domestic	Industrial	Commercial	Public
Altoona, Pa.	55.5	155	12,370	1,486	
Beloit, Wis.	18	116	44,000	311	2,436
Corning, N.Y.	14.9	354	10,420	430	1,712
Fort Smith, Ark.	28	114	4,855	2,233	2,673
Holyoke, Mass.	60	1,948	4,990	4,990	3,177
Jefferson City, Mo.	13.5	147	32,560	448	2,830
Jersey City, N.J.	300	1,500	7,570	500	14,700
Kokomo, Ind.	20	121	3,733	8,550	1,907
Madison, Wis.	27	187	1,988	298	2,076
Marinette, Wis.	14.6	66	2,430	568	785
New Orleans, La.	360	166	†	3,617	2,841
Portsmouth, Va.	75	143	11,127	187	2,150
Quincy, Ill.	38.6	143	4,583	329	1,295
Richmond, Va.	23	107	4,937	2,953	17,071
Rochester, N.Y.	250	183	11,425	6,777	8,040
Scottdale, Pa.	8.5	540	15,930	1,060	6,700
St. Louis, Mo.	730	1,127	17,679	1,959	4,608
Spokane, Wash.	104	2,186	31,400	3,600	59,840
Washington, D.C.	353	189	27,348	2,500	4,017
Wichita, Kan.	54.5	104	3,923	560	2,162

* From Report of Committee, American Water Works Association, 1915.
† Included in commercial.

TABLE 3-10. COLD-WATER AND HOT-WATER SUPPLY FOR RESIDENTIAL TYPES OF BUILDING
(Flush-tank supply)*

Building types as to number and kind of fixtures	Kind of demand			Total fixture units		Total demand, gpm	
	Bathrooms	Number kitchen sinks	Groups of 1 to 3 laundry trays	Main and cold-water branch	Hot-water branch	Main and cold-water branch	Hot-water branch
A	1	1	1	11	6	12	8
B	2	1	1	17	9	16	10
C	3	2	1	25	12	20	12
D	4	4	2	38	20	25	16
E	8	8	3	73	37	35	24
F	16	16	4	140	69	52	34

* From "Plumbing Manual," p. 44, National Bureau of Standards, 1940.

TABLE 3-11. RATES OF USE OF WATER BY PLUMBING FIXTURES*
(Rates in gallons per minute)

	Number of fixtures installed													
	1	2	4	8	12	16	24	40	50	80	100	150	200	300
Water closet:														
Flush tanks		8	12	16	20	30	38	50	58	80	95	140	215	290
Flushometers		36	46	61	72	83	100	123	148	180	200	250	295	375
1-in. flush valves†	*30*	*50*	*80*	*120*	*140*	*160*	*200*	*300*						
Urinal:														
Flush valve		18	23	30	36	42	50	64	72	90	100	125	148	188
¾-in. flush valve	*25*	*37*	*45*	*75*	*85*	*100*	*125*	*175*						
Lavatories and wash sinks		6	10	15	18	20	25	30	36	51	60	75	85	115
	4	*8*	*12*	*24*	*30*	*40*	*48*	*75*						
Shower baths and wash fountains	*5*	*10*	*20*	*40*	*60*	*80*	*120*	*200*						
Service sinks	*15*	*25*	*40*	*64*	*84*	*96*	*120*	*200*						

* Rates of usage by other fixtures in gallons per minute are wash sink, *4*; drinking fountain, *1*; water closet with low tank, *5*.

† Includes both water closets and urinals with 1-in. flushometer-type valves.

Figures in the table in roman type are from "Methods of Estimating Loads in Plumbing Systems," National Bureau of Standards, Dec. 16, 1940. Figures in italics are from Water Demand by Plumbing Systems, *Heating & Ventilating*, November, 1954, p. 138. Discrepancies between the recommendations indicate variations among localities and among standards. The designer must be guided by local ordinances and by personal judgment.

TABLE 3-12. PROPER DISTANCE FOR A WELL FROM A SOURCE OF CONTAMINATION*

Source	Distance, ft†	Source	Distance, ft†
Sewer	50	Subsurface disposal field	100
Septic tank	50	Seepage pit	100
Subsurface pit	50	Cesspool	150

* From "National Plumbing Code," Sec. A.3.2.

† These distances constitute minimum separation and should be increased in areas of creviced rock or limestone or where the direction of movement of the ground water is from sources of contamination toward the well.

but the fact that it is so located is not an assurance that underground strata sloping from the source of pollution toward the well may not carry pollution into the well. If the underground strata dip away from the well toward the source of pollution or the bottom of the well is higher than that source, danger therefrom is remote. In general, the top of the well should be at least 2 ft above the highest known flood record of nearby water. Adequate protection of the top of the well is essential to prevent contamination by droppings from the surface. Wood and other porous or leaky covers are unsatisfactory. One form of approved construction, with all surface drainage leading away from the well, is shown in Fig. 3-5.

3-12. Well Construction. Three types of well are used for private water supplies: the dug well; the driven, bored, or jetted well; and the

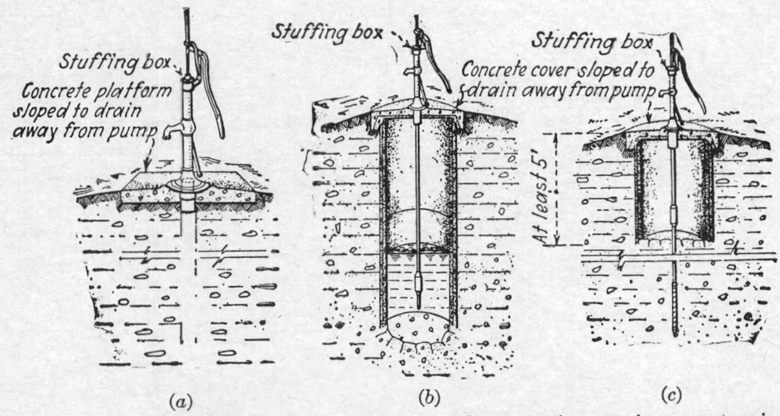

FIG. 3-5. Methods of dug-well construction and protection against contamination recommended by U.S. Public Health Service. (*Public Health Rept. Reprint* 952, 1924.)

drilled well. Finished wells of these types are shown in Fig. 3-5. Details of the connection between the base of the pump and the top of the well are shown in Fig. 3-6.[1]

3-13. Dug Wells. The dug well, as the name implies, is excavated by digging with a shovel. Such wells for small, private water supplies are usually circular, 3 or 4 ft in diameter, and are rarely over 50 to 60 ft deep. Dug wells are classified as shallow wells. They should be lined with masonry blocks or stones laid up without mortar in the joints except near the top of the well. The amount of water available from them is usually small, seldom exceeding 20 to 50 gpm.

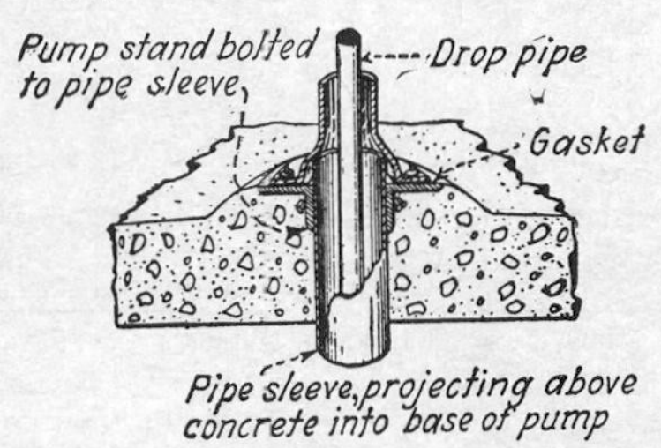

FIG. 3-6. Detail of construction at tops of wells shown in Fig. 3-5.

3-14. Driven Wells. A driven well is constructed by driving into the ground a pipe, the lower end of which has been drawn to a point and the sides of which have been perforated for a short distance above this point. A 1¼- or 2-in. galvanized-iron pipe, perforated with 1,500 to 2,000 $\frac{1}{16}$- to ⅛-in. holes near the lower pointed end, may be able to supply water at a rate greater than 300 gal per hr if pumped continuously.

[1] "Public Health Reports," U.S. Public Health Service, Reprint 952, 1924.

Larger pipes, up to about 6 in. in diameter, with more holes, should supply water at a rate increasing about as the square of the well diameter. The diameter of the holes in well piping should not be increased, but the number of holes may be increased indefinitely, provided that the strength of the perforated pipe is not reduced.

In driving a well into the ground the well point, as shown in Fig. 3-7, is attached to a short length of drive pipe and is driven into the ground by successive blows of a wooden maul or other instrument that will not flatten the upper end of the driven pipe. Additional lengths of pipe are coupled onto those already driven to reach the desired depth. Such wells must be driven in sand or other porous material free from stones or rock. The difficulty of driving increases more rapidly than the square of the diameter, and such wells are seldom driven to depths greater than 30 to 50 ft.

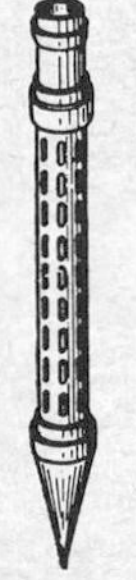

Fig. 3-7. Well point for driven well.

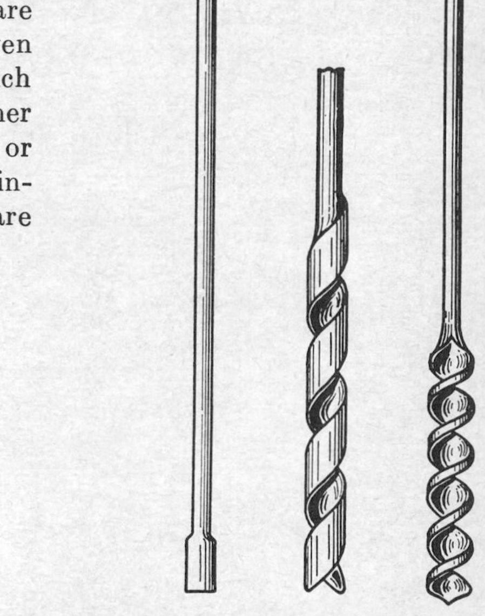

Fig. 3-8. Hand-boring tools for bored wells.

3-15. Bored Wells. Wells may be hand-bored or machine-bored in soft material with such boring tools as are illustrated in Fig. 3-8. If a bored well must be cased, the casing (Fig. 3-9) is driven down to follow closely after the auger. When the desired depth is reached, the auger is withdrawn permanently and a screen, such as is shown in Fig. 3-10, is lowered through the casing to the bottom of the well. A lead ring at the top of the screen is expanded against the casing to form a watertight seal. If no casing is used during the boring, a screen and pipe are lowered into the hole after the boring tool has been finally withdrawn. When the tools shown in Fig. 3-8 are used, bored wells may be constructed of large diameter and to depths greater than dug or driven wells, but only in relatively soft, granular materials. Wells may be bored in rock or other hard material to depths measured in miles by the use of diamond drills and other heavy equipment. Such wells are, however, usually bored for oil. Potable water is seldom obtained at depths greater than 1,000 or 2,000 ft and usually at shallower depths.

3-16. Jetted Wells. A well hole may be advanced into soft ground by jetting at a lower cost and higher speeds than by other methods of construction if the conditions for jetting are favorable. When a well is sunk in this manner, the casing is driven into the ground at the start slightly in advance of the jet. The jet nozzle on the end of a length of hose is lowered inside the casing. Water is pumped into the hose as the nozzle is pushed down in the casing. The displaced material rises in the casing and is discharged into a settling basin on the ground surface. The casing usually sinks of its own weight. It should follow closely behind the jet to prevent collapse of the sides of the hole and to provide for the escape of the water and dislodged solids leaving the well hole.

3-17. Drilled Wells. Wells in rock or coarse or hard materials may be advanced by percussion tools, the process being called *drilling*. The construction of drilled wells requires special equipment and ex-

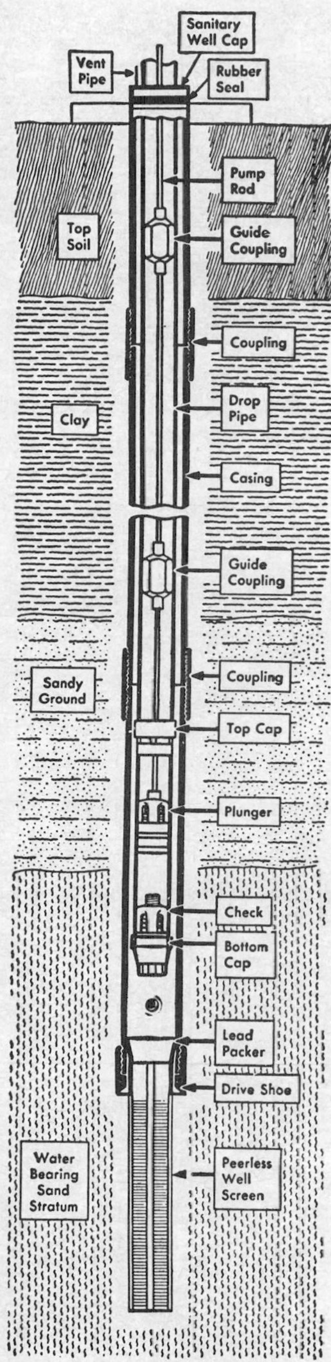

FIG. 3-9. Details of equipment of a drilled well.

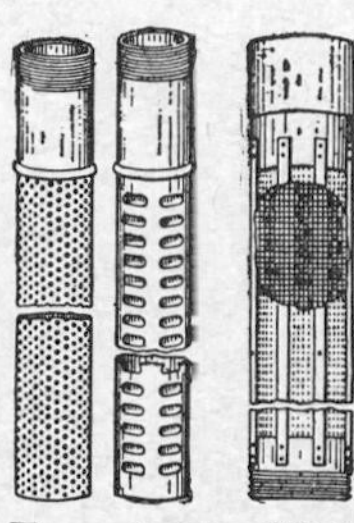
FIG. 3-10. Well screens.

perience not within the field of plumbing. Equipment of a completed well is shown in Figs. 3-9 and 3-12.

3-18. Well Screens. The screen, which goes in the bottom of a well, is usually made of brass or other strong, nonabsorbent, and corrosion-resistant material. Types of well screen are illustrated in Fig. 3-10. Well screens must be selected so that the opening in the screen is smaller than the particles, composing the aquifer, that it is desired to exclude from the well. The total area of the openings should provide an entrance velocity of water entering the well when being pumped at rated capacity less than the velocity necessary to carry the smallest excluded particle. Lifting velocities of water to move grains of sand are given in Table 3-13. Screens

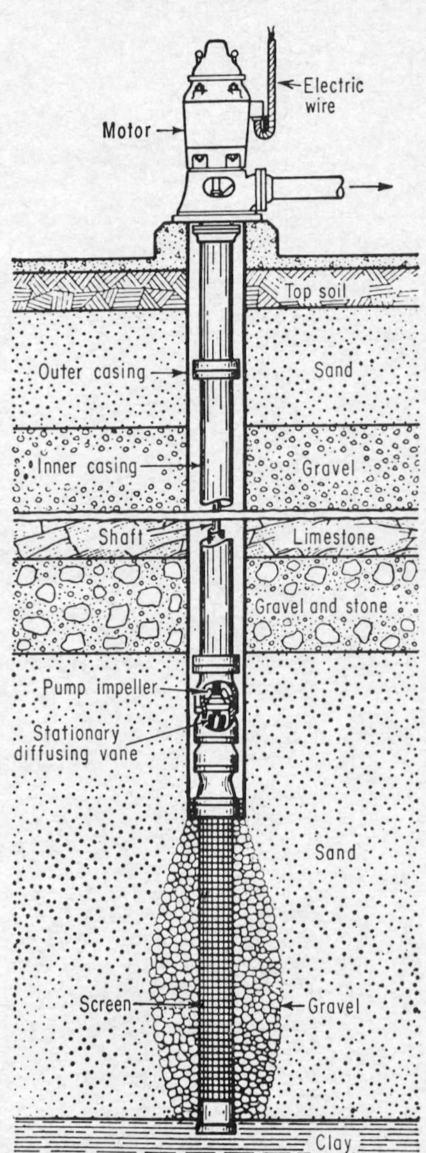

FIG. 3-11. Gravel-packed well.

TABLE 3-13. LIFTING VELOCITIES OF WATER

Diameter of sand grains, mm	*Lifting velocity of water, fps*
Up to 0.25	0.0 –0.10
0.25–0.50	0.12–0.22
0.50–1.00	0.25–0.33
1.00–2.00	0.37–0.56
2.00–4.00	0.60–2.60

are sometimes selected with openings slightly larger than the finest particles of the material in the aquifer. The finest particles are drawn into the well and are expelled from it with the discharged water. The larger particles remaining in the aquifer form a coarse screen encircling the well screen. The largest practicable openings in the screen should be selected, since small openings tend to corrode and clog more rapidly than large openings. Sometimes the effect of a large screen with large openings is obtained by packing gravel around the outside of a coarse metal, concrete, or other screen that is about the diameter of the well casing. A section through a gravel-packed well is shown in Fig. 3-11.

3-19. Well Equipment. If a pipe other than the casing is used for the removal of water from a well, as shown in Figs. 3-9 and 3-12, it may be called the *drop pipe* or the *eductor pipe*. The casing alone can be

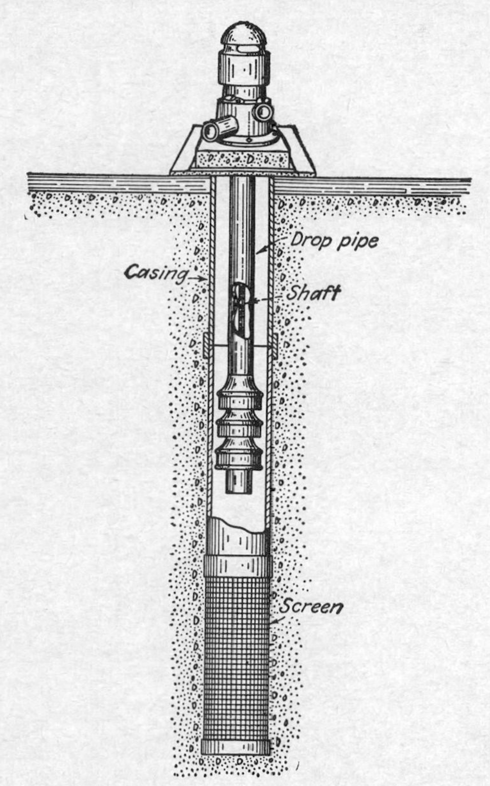

FIG. 3-12. Typical deep-well equipment.

used for removing water from a well when the casing is 2 in. or less in diameter. If the pump valves are submerged or placed below the ground surface for frost protection, as shown in Fig. 3-5, the casing can be used as an eductor pipe or, as is common in larger wells, the equipment can be installed as shown in Fig. 3-12.

The capacities of some small well pumps are shown in Table 3-14.

3-20. Protection of Wells from Frost. When wells are constructed in cold climates, care should be taken to protect the eductor pipe and the pump valves from freezing. This can be done by placing the pump valves below the frost line in a pit, as shown in (*b*) and (*c*) in Fig. 3-5, in such a way that when pumping has stopped water will drain back into the well—not into the pit. The bottom of the pit should be above the ground-water level, and a drain should be provided to prevent the accumulation of water in the pit.

3-21. Cisterns to Collect Rain Water. Rain water stored in a cistern and properly protected from contamination is used as a satisfactory source of water in regions where other sources are unavailable, where soft water is desired, or where other conditions are unsatisfactory. In some climates sufficient storage is necessary to supply water through a dry

TABLE 3-14. APPROXIMATE CAPACITIES OF DEEP-WELL RECIPROCATING PUMPS

Inside diameter of pump cylinder, in.	Length of stroke, in.	Gallons per stroke		Gallons per revolution, double-acting cylinder pump
		Single-acting	Two-stroke	
2	6	0.08		
	14	0.19		
2¼	6		0.103	0.19
			0.40	
3	6	0.18		
	14	0.43		
3¼	12		0.43	0.79
	30		1.075	
4	6	0.33		
	14	0.76		
4¾	12		0.92	1.69
	30		2.30	
5¾	12		1.35	2.47
	30		3.375	

season or through the season of subfreezing temperatures. Water in the cistern may become contaminated by ground water leaking into the cistern, through the overflow of surface water into the cistern, or through the deposition of the eggs of flying insects in the water. Cisterns offer an attractive environment for the breeding of mosquitoes. They must therefore be covered and made insectproof, but at the same time a vent must be provided to permit egress and ingress of air as the cistern is filled and emptied.

The rain-water catchment area, usually the house roof, should occasionally receive attention, particularly for the cleaning off of leaves, birds' nests, dead animals, etc. The first rush of rain water on any catchment area will probably carry with it into the cistern much undesirable material. This can be diverted by the use of a bypass valve in the rain-water leader, as shown in Fig. 3-13. Otherwise, a rain-water filter should be installed as described in Sec. 17-2.

The size of the cistern is fixed by the roof or catchment area available and by the number of persons using the water. When water for all purposes must be supplied from the roof and the annual rainfall is about 40 in., which is well distributed between seasons, the water supply must be conserved and approximately 500 sq ft of catchment area must be available per person. When the rain-water supply is depended on for soft water only and some source of water exists for other purposes, a

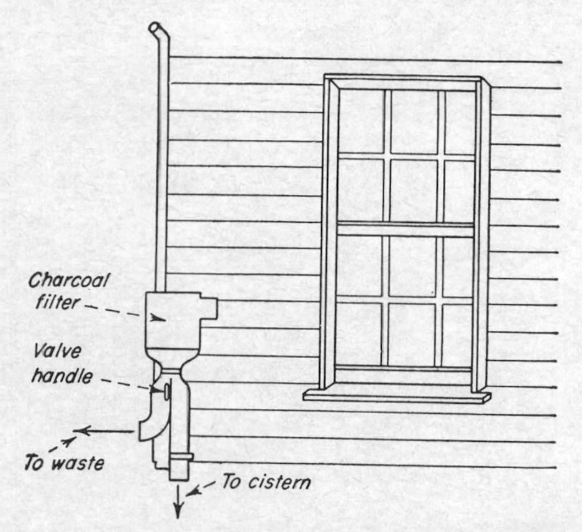

Fig. 3-13. Bypass valve and filter in rain-water leader.

cistern of 500- to 1,500-gal capacity should supply the ordinary needs of the average-size family. Rates of runoff from roofs are discussed in Sec. 13-10.

3-22. Springs and Infiltration Galleries. The source of all underground water is rain water that has soaked into the ground until it has reached the water already in the ground. The surface of the ground water is called the *ground-water table.* The ground-water table is not horizontal. In porous ground, if the ground water is in motion, the ground-water table follows approximately the contour of the ground surface, as shown in Fig. 3-14, sloping toward the nearest surface outlet, which may be the sea, a river, or a spring. Only in limestone regions and in fissured rock are there large underground streams that flow under the hydraulic conditions of open-channel flow, as in a surface stream.

Springs can be divided into four classes for convenience in considering methods for their development:

1. Overflow of the ground-water table into surface streams, as shown in Fig. 3-14

2. Underground flows encountering outcropping impervious strata, as shown in Fig. 3-15

3. Artesian springs, which are underground flows that are confined to an aquifer by a superimposed impervious stratum, the spring water breaking through the stratum near or at the spring (see Fig. 3-16)

4. Underground streams in fissured rock

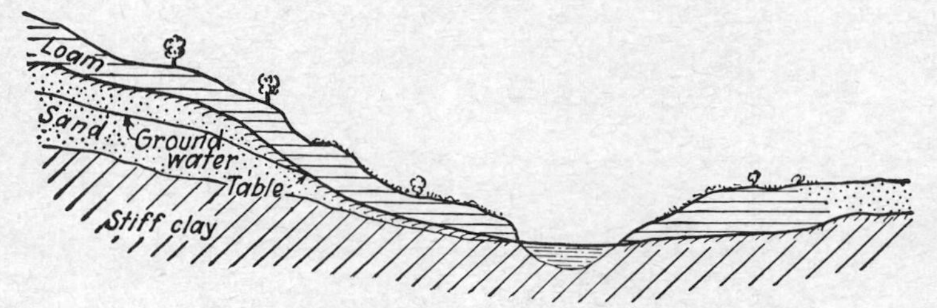

FIG. 3-14. Overflow of ground water into surface stream.

FIG. 3-15. A surface spring.

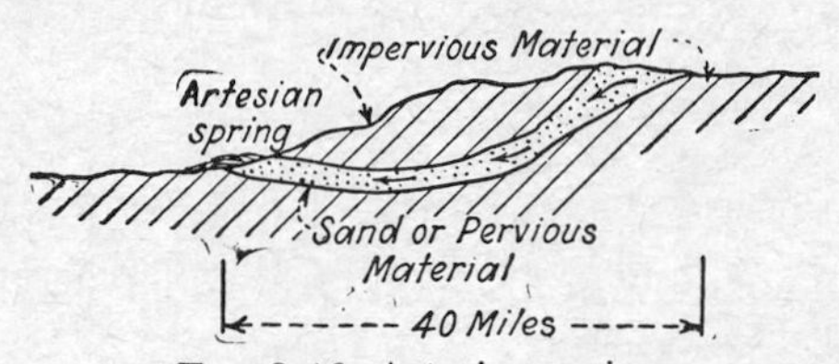

FIG. 3-16. Artesian spring.

The first two types of spring are best developed by constructing a trench or by laying a line of drain tile at right angles to the direction of flow of ground water and by discharging the water into a sunken barrel or other receptacle that may serve as a pump suction pit. This type of construction is known as an *infiltration gallery*. A satisfactory arrangement is shown in Fig. 3-17. Springs of the third type can best be developed by the construction of a deep well at the site. Little can be done to increase the flow from the fourth type of spring except to clear away the debris from about the mouth of the spring.

3-23. Surface Sources. Surface sources rarely supply a potable water without the construction and operation of treatment works. It is safe to assume that all surface water is polluted and unfit for human consumption without filtration and disinfection. Usually, the cost and the magnitude of the problems involved in the development of a satisfactory surface source of water supply restrict their use principally to public and to large industrial supplies. These are not within the scope of the activities of the plumber.

Where a surface supply must be used, as for a temporary camp or remote home, a satisfactory intake may be constructed by sinking a barrel into the stream or lake near shore or, if adequately protected

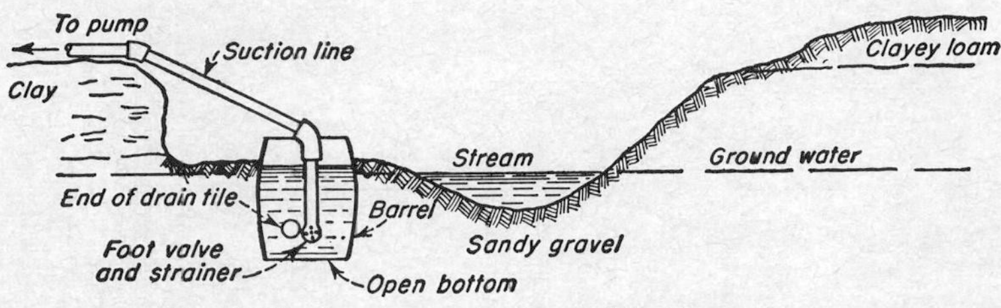

Fig. 3-17. Development of an infiltration gallery.

against boats and other floating objects, out in the stream. Care should be taken when sinking the barrel that water will flow into it at low stages of surface elevation, and the intake pipe should be placed but slightly above the bottom of the barrel. In a large body of water the barrel may be completely submerged and may be protected by piling rocks around and over it.

A barrel sunk inshore from the edge of a stream or lake that will force water to travel underground to it for a distance of 25 ft or more and that is otherwise protected against pollution may assure water of safe, sanitary quality. It is to be remembered that the natural direction of flow of underground water is toward a stream, and hence a barrel or well so located near the edge of a river or lake will supply water of better sanitary quality than that in the nearby river or lake.

CHAPTER 4

MATERIALS, DIMENSIONS, AND TYPES OF PIPES, FITTINGS, AND EQUIPMENT

4-1. Materials Commonly Used in Plumbing. The materials most commonly used in plumbing include cast iron, which is used in water pipes, drainage pipes, traps, vent pipes, and sewers, and wrought iron or steel, which is used mainly for water pipes and drainage pipes, particularly in tall buildings. Cast iron, wrought iron, steel, and malleable iron are used for fittings and appurtenances for water, drainage, and vent pipes. Although ferrous metals may be attacked by many corrosive substances, their relatively low cost, satisfactory structural characteristics, satisfactory protective measures against corrosion, and other desirable qualities make them the most widely used materials in plumbing. Alloys of iron, such as stainless steel and Durion, are highly resistant to corrosion but, unfortunately, are expensive.

Copper and brass are used for water pipes, flush pipes from tanks to fixtures, traps, faucets, and valves and in sheet form for safe wastes, flashings, drip pans, and strainers. Brass is made in various grades from red brass with high copper content to yellow brass containing more zinc. Red brass with 85 per cent copper is more highly resistant to corrosion, normal ranges in use lying between 67 and 85 per cent copper. Brass is used for valve parts and for those parts of mechanisms the corrosion of which might affect their functions. Bronze is highly resistant to corrosion and is used for the parts of mechanisms, as is brass. The resistance to corrosion of bronze and Tobin bronze (copper-tin-zinc) is greater with the higher proportion of tin. Tobin bronze is harder than brass and also resists corrosion better.

Copper may be affected by such dissolved gases as oxygen, carbon dioxide, hydrogen sulfide, ammonia, and chlorine. Although it is less affected by them than is steel or wrought iron, the use of copper does not assure freedom from corrosion under all conditions. The effect of the electrolysis of copper pipe in contact with water is to give the water a blue color. The use of copper to convey either hot water or cold water is hygienically safe, since it is insufficiently soluble in potable water to be poisonous to man. It may, however, affect the growth of some fish and plants.

Lead is used for cold-water pipes, traps, and waste pipes and in sheet form for flashings and safe wastes. It should not be used for hot-water pipes, as explained in Sec. 8-6. Lead is used also for making joints in some cast-iron pipes, for wiping joints between lead pipes, and in the making of solder. Its advantages and disadvantages are discussed in Sec. 4-14.

Zinc is used for galvanizing and in sheet form for flashing. Although zinc will corrode, it will protect ferrous metals from corrosion when both are in the same electrolytic solution. Zinc is attacked by some soft waters and by such gases as carbon dioxide when dissolved in water, but it is not attacked by chlorine. Its soluble salts, in the ordinary concentrations encountered, are harmless to man, so that zinc products are suitable for use in contact with potable water.

Tin is used for lining iron pipes that are to be exposed to particularly corrosive water, and it is used in sheet form for flashing. Tin, zinc, and lead are used for lining pipes to prevent corrosion. Tin mixed with lead is used to make solder. Tin is nontoxic when in contact with potable water.

Outstanding desirable characteristics of aluminum are light weight and resistance to corrosion. Its appearance is pleasing, and it can take and hold a high polish. In sheet form aluminum is used for flashing, safe drip pans, tank linings, roof drainage gutters and downspouts, and other conditions suitable for other sheet metals such as tin and zinc. It is used also for special applications for conveying water and corrosive and hot liquids, whether or not under light pressure, where its applications may be competitive with brass and copper. It is, however, relatively costly in some applications.

Rubber is used in tubing and pipes, either as hard or as soft rubber, to convey fluids to which it is corrosion-resistant, such as chlorine gas or aqueous chlorine solutions. Soft rubber is used in hose for conveying water and other fluids such as steam where flexibility is required. Rubber, asbestos, and other materials are used for gaskets in pumps and valves and in the packing of valves and faucets.

Vitreous ware, enameled iron, and slate are used for plumbing fixtures. Cement and sand are used for making joints in clay pipe and with stone aggregate for making concrete and concrete pipe. Concrete and asbestos-cement pipes are used for conveying water and sewage and as drainage pipes. Wood, hard rubber, and plastics are used for making toilet seats.

Asbestos, hair felt, wool felt, mineral wool, mica, magnesia, cork, and other materials are used for heat insulation and for other purposes. Lead, cast iron, brass, clay, and fiber are all used for pipes and traps conveying acid wastes. The life of some of these materials may be short, although vertical pipelines coated with coal tar have been known to last many years.

Materials used for plumbing fixtures are discussed in Sec. 15-15.

In addition to a choice of the quality of material to be used for any particular purpose, existing standards permit a choice in the strength or weight of various pipes and fittings. These choices are discussed in the section of this book devoted to the particular pipe or fitting in which a choice is available. Other materials are used in plumbing, and the materials described above are used for purposes other than those mentioned.

4-2. Standard Specifications. Standard specifications for materials and equipment, as listed in Table II-1 in Appendix II, have been prepared by various authorities such as the American Standards Association (ASA), the American Society for Testing Materials (ASTM), the Federal Specifications Board (FS), the Division of Simplified Practice, U.S. Department of Commerce (SPR), the American Society of Mechanical Engineers (ASME), the American Water Works Association (AWWA), and others. In the introduction to the index of Federal Specifications (FS), issued by the Procurement Division of the U.S. Treasury Department, revised to 1943, it is stated:

Commercial Standards: 1. Outstanding examples of commercial specifications voluntarily proposed and supported as a basis for marketing by the industries concerned are the Commercial Standards established through the procedure of the Division of Trade Standards of the National Bureau of Standards.

Simplified Practice Recommendations: 1. Closely allied with the commercial types, grades, sizes, etc. . . . are the Simplified Practice Recommendations proposed by industry and established under the auspices of the Division of Simplified Practice of the National Bureau of Standards.

2. While Simplified Practice Recommendations differ from Federal Specifications in that they originate with, and are voluntarily supported by, industry itself, nevertheless the programs of simplification should serve as helpful guides to procurement officers.

4-3. Cast-iron Water Pipe and Fittings. Cast-iron water pipe is manufactured in accordance with a number of standard specifications and in a number of different weights and lengths. The dimensions shown in Table 4-1 are for pipes used in waterworks distribution systems. Pit-cast pipe is less generally used because of its greater weight and cost compared with centrifugally cast pipe, but fittings that are pit-cast according to the AWWA specifications of 1908 are used almost exclusively with cast-iron water pipe. When cast-iron water pipe is selected for any particular service, the internal pressures and other stresses to which it may be exposed should be compared with the characteristics given to the different classes of pipes and fittings in standard specifications, some of which are given in Tables II-2 to II-17 and Figs. II-1 to II-12.

TABLE 4-1. SOME STANDARD THICKNESSES FOR CENTRIFUGALLY CAST* AND PIT-CAST† IRON PIPE‡

Inside diameter, in.	Centrifugally cast				Pit cast			
	Thickness, in.		Pounds per 12-ft length		Thickness, in.		Pounds per 12-ft length with bell	
	150 psi	250 psi	150 psi	250 psi	130 psi	260 psi	130 psi	260 psi
4	0.34	0.38	195	220	0.48	....	280	
6	0.37	0.43	315	350	0.51	0.61	430	531
8	0.42	0.50	475	545	0.56	0.71	625	802
10	0.47	0.57	640	760	0.62	0.80	850	1,114
12	0.50	0.62	810	990	0.68	0.89	1,100	1,474
14	0.55	0.69	1,060	1,320	0.74	0.99	1,400	1,905
16	0.60	0.75	1,320	1,635	0.80	1.08	1,725	2,358
18	0.65	0.83	1,595	2,015	0.87	1.17	2,100	2,872
20	0.68	0.88	1,860	2,365	0.92	1.27	2,500	3,448
24	0.76	1.00	2,480	3,200	1.04	1.45	3,350	4,707

* Federal Specifications WW-P-421.

† American Water Works Association, 1908.

‡ From H. E. Babbitt and J. J. Doland, "Water Supply Engineering," 5th ed., McGraw-Hill Book Company, Inc., New York, 1955.

Three types of end are standard on cast-iron water pipe and fittings: bell-and-spigot, flanged, and screwed. Bell-and-spigot ends are used principally for underground work and in buildings where there is ample space for the larger dimensions of the fittings and for calking operations. Flanged ends are used within buildings where space is limited, the pipe sections being held together by bolts, sometimes with gaskets between the flanges and sometimes with watertight, machine-faced flanges. American Standard flanged fittings for steam pipe, adopted by the ASME, Mar. 20, 1914, have small radii and permit work in confined space. Threaded or screwed ends are used in small pipes in general practice, although they are available in pipes up to 8 and 10 in. in diameter. Screwed cast-iron fittings for wrought-iron or steel pipe are widely used. The dimensions of such fittings are stated in Table II-14 and in Fig. II-11.

Flanged fittings and threaded fittings are more expensive than bell-and-spigot fittings, they require more accurate cutting and fitting, and they lack the adaptability of bell-and-spigot pipe to the slight changes in dimension or direction that are sometimes met in installation. When properly fitted, however, they are easier to handle, and they can be installed more quickly than bell-and-spigot ends.

The use of any fitting other than standard may be costly and time-consuming in design, manufacture, and installation. The use of non-standard fittings should, therefore, be avoided, and ingenuity exercised in the selection of standard fittings to fit the installation being designed.

4-4. Cast-iron Soil Pipe and Fittings. Cast-iron pipe that is manufactured primarily for use as drainage pipe in plumbing is lighter than cast-iron water pipe. It is generally known as *soil pipe*. This use of the term should not be confused with its use to designate pipe carrying human wastes.

Three weights of soil pipe are recognized as standard: standard, medium, and extra heavy. The weights and dimensions of each are given in Appendix II. The choice between them requires judgment; categorical recommendations of general application cannot be properly made. When a first-class, permanent job is desired, where the building may be subject to vibration or settling, or where corrosive attack may be expected, extra-heavy pipe should be used. Extra-heavy pipe is less likely to have flaws or sand holes, hubs are less likely to split in calking, or pipes to split when being cut. Extra-heavy pipe should be used for house drains and house sewers where cast-iron pipe is to be used. In temporary locations, in which a high standard is not required, standard pipe can be used. Cast-iron water pipe should be used for pipes to pass through foundation walls or where the pipes are subjected to internal bursting pressures or to high external loads.

Cast-iron soil pipe is used for drainage pipes 2 in. in diameter or larger for roughing-in of stacks and branches except in buildings of more than about eight stories or where temperature changes may cause appreciable movement of the building or pipe, as in greenhouses. Wrought-iron or steel pipe is used for these conditions.

Cast-iron fittings with either bell-and-spigot or threaded ends or a combination of such ends are manufactured for soil pipe. The threaded ends are tapped with a standard thread to fit wrought pipe. The use of bell-and-spigot ends on soil pipe is common in plumbing. It requires less precise cutting and fitting than for threaded pipe, and pipes and fittings are less expensive. More skill is required, however, in making the joints. Threaded joints are neat in appearance and can be installed more quickly. Successful cutting of cast-iron soil pipe with hammer and chisel requires training and skill. A special tool is available which simplifies the work.

Threaded cast-iron fittings, known as *drainage fittings* or *recessed fittings*, are made with a shoulder and recess and are so tapped that the pipe screws up tightly to this shoulder, making a continuous interior surface of the same diameter as the pipe. These fittings are illustrated in Fig. II-51. These fittings are known also as *Durham* fittings or pipe. Durham fittings are preferable to other fittings that do not provide the smooth, interior waterway. However, threaded fittings do not resist corrosion so well as bell-and-spigot ends, so that the life of the recessed fittings may be shortened.

Standards for cast-iron drainage fittings are given in Appendix II.

Styles of commonly used cast-iron soil fittings, as illustrated in Appendix II, together with other fittings manufactured but not illustrated, are so many as to make possible, by the exercise of ingenuity, almost any reasonable installation without the use of special fittings. The use of cast-iron fittings that are not generally manufactured is to be avoided in the interests of time and economy.

The tar coating of cast-iron soil pipe is customary before installation for the sake of convenience, appearance, and protection against corrosion. It is not always practiced, however. An objection to its application before the pipe has been installed is the possibility of concealing flaws.

4-5. Wrought Pipe and Fittings. Galvanized wrought-iron or steel pipe is properly used for drainage, soil, waste, and vent pipes and for the conveyance of water and other fluids that are not corrosive to it. It is widely used in plumbing. Wrought-iron and steel pipe, whether or not galvanized, corrode relatively rapidly underground, particularly when surrounded by ashes, especially hot, moist ashes.

Since the method of manufacture and the materials in wrought-iron and steel are similar, since good-quality puddled steel has better lasting qualities than inferior wrought iron, and since there is difficulty in distinguishing between wrought iron by appearance, there is confusion in the trade between wrought-iron and steel pipe and difference of opinion concerning the relative qualities of the materials. Standard dimensions of pipes made from each material are the same, so that in plumbing installations wrought-iron and steel pipes are used interchangeably and they are sometimes called *wrought pipe*. Standard specifications for each are given in Table II-1.

In the manufacture of wrought iron, pig iron is melted in contact with slag, which is mainly iron oxide, under the action of an oxidizing flame. The melted iron becomes more pasty as it is stirred or puddled in the furnace. When of proper consistency the puddled ball is removed from the furnace and rolled into flat sheets with straight edges and of the thickness of the finished pipe. The flat sheets are known as skelp. The skelp is cooled and is sheared into about 18-ft lengths. Wrought iron is made also from scrap iron consisting of materials of differing composition with a resulting product of decidedly inferior quality. The more uniform the quality of the metal, the greater its resistance to corrosion. Knowledge of the method of manufacture of wrought-iron pipe aids in an understanding of its characteristics as compared with steel pipe.

Standard dimensions of wrought pipe are shown in Table II-46. Pipes larger than 12 in. are known by their outside diameter, and all pipes of the same nominal size have the same outside diameter regardless of their thickness. In the trade, pipes that are within 5 per cent of the weights shown in Table II-46 are known as *full-weight* pipes; pipes that vary more than 5 per cent are known as *merchant* pipes. Merchant pipes are usually lighter than full-weight pipes but seldom fall more than 8 per cent below full weight. They are sufficiently strong for most plumbing purposes.

Couplings and nipples are the only pipe fittings made of wrought iron or steel. Other iron fittings are made of cast iron or malleable iron. Various sizes and styles of wrought-iron or steel couplings and nipples are illustrated in Fig. II-55. The threads on these fittings are commonly cut right handed, but fittings with either left-hand or right- and left-hand threads can be obtained. The use of right- and left-hand threads is explained in Sec. 7-26. The unthreaded portion of a sleeve coupling, as shown in Fig. II-55, protrudes over the threaded portion of the pipe, which does not come into contact with the threads of the coupling. In this manner the weakest portion of the pipe is protected. The dimensions of wrought couplings are shown in Table II-50. Extra-heavy or hydraulic couplings and double extra-heavy hydraulic couplings are available for use on extra-heavy or double extra-heavy pipe. Nipples are really short lengths of wrought pipe. They can be made any length from a close nipple, which has a length of twice the standard length of threads on a pipe, to the full length of a pipe. Manufacturers usually carry nipples in stock in lengths varying ¼ in. from 3½ to 6 in. and by ½ in. from 6 to 12 in.

Wrought pipe with threaded ends has advantages over cast-iron pipe. Among the advantages are the following: There are fewer joints per unit length. The joints are tighter and stronger, and the alignment of the pipe is maintained better. Flanged joints can be used to connect lead pipe by bossing out the lead as a gasket between two flanges. The joints are not affected by ordinary expansion and contraction as are leaded joints in cast-iron pipe. The pipe can be cut to exact length in the shop with full confidence that it will fit on the job. With good cutting tools the pipe can be cut and fitted more rapidly and with less labor than bell-and-spigot pipe. It can be installed further in advance in the construction of a building, and it does not occupy so much space in floors and partitions. It is more expensive in first cost than cast-iron pipe, and in drainage work it is customary to use recessed fittings, which are also more expensive.

4-6. Steel Pipe. Seamless steel tubing is made in sizes from ⅛ to 24 in., as shown in tables in Appendix II. It is manufactured in accordance with the standard specifications listed in Table II-1. Galvanized-steel pipe, although less resistant to corrosion than galvanized wrought pipe, is used in plumbing for the same purposes. Neither steel nor wrought pipe, whether or not galvanized, should be buried underground, as explained in Sec. 4-5. Pipes made of steel plates, with either welded or riveted joints, are made in sizes up to many feet in diameter. Spiral-welded steel pipe is available in sizes from 6 to 36 in. and larger. Corrugated-steel pipe, used mainly in sewers and culverts, is available in many different diameters and also in oval sections that are suited particularly to underground locations with limited headroom or cover. Some dimensions of corrugated-steel pipe are given in Table II-54.

4-7. Fittings. Fittings for various kinds of pipe are illustrated in Appendix II. Any of the fittings shown can be used for conveying potable water and other fluids free from suspended solids, but fittings in drainage pipes should have a smooth, continuous interior surface and no abrupt changes in direction. Hence, such special fittings as the "soil-pipe fittings" shown in Figs. II-15 to II-49 and the drainage fittings in Fig. II-51 are manufactured. Such fittings are more cumbersome and expensive than more simple fittings, and they are not ordinarily used on water-supply and gas pipes. The use of the double hub, shown in Fig. II-43, should be limited because of the roughness of the connection and its liability of clogging. The use of the double-hub fitting can be avoided by the use of double-hub pipe.

Plain, threaded fittings should be made of cast iron, malleable iron, brass, or copper. Fittings used on wrought pipe may be of the cast-iron, recessed variety, shown in Fig. II-51, and tapped with a standard pipe thread. Threaded, drainage fittings should be made of cast iron, malleable iron, brass, or copper with smooth interior waterway and the threads tapped in solid metal. All cast-iron fittings used for the distribution of potable water should be galvanized. All malleable-iron fittings should be galvanized where used for permanent installations.

Fittings of iron, mild steel, brass, or copper should be of equal quality to the pipe of the same material, and they should have a thickness and weight corresponding to pipe of the same diameter. The minimum thickness of metal on threaded parts must not be less than the minimum thickness permissible on pipe of the same size.

Sizes of fittings are expressed in terms of the diameter of the pipe to which they connect. Fittings without branches and with the same inlet and outlet diameter are expressed in terms of the diameter. If the inlet and outlet are not the same, the size of the larger opening is stated first, then of the smaller opening. If there are branches, the sizes on the run are given with the larger diameter first, then the smaller diameter on the run, then the larger diameter on the branch followed by the smaller diameter on the branch. For example, tees and crosses would be sized as indicated in Fig. 4-1. An outside thread may be specified as *male*, and an inside thread as *female*.

4-8. Malleable-iron Fittings. The confusion in the trade between the terms wrought iron and steel and the difficulties of distinguishing between the two materials exist also, but to a lesser degree, with the term *malleable iron*. Malleable iron is used for the manufacture of fittings. It is not used for the manufacture of pipe.

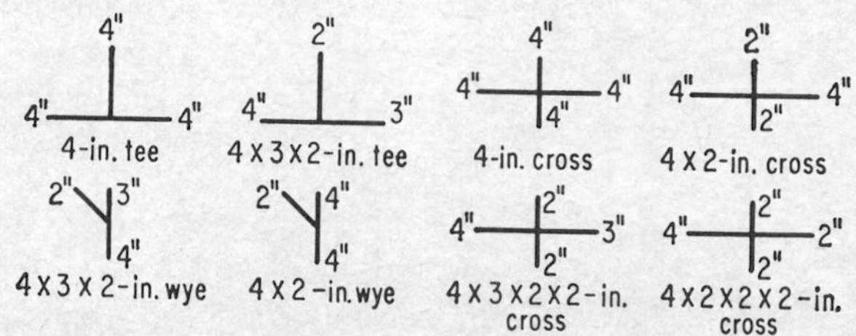

FIG. 4-1. Method of dimensioning tees and crosses.

Malleable iron is a cast iron of such quality that it has toughness permitting it to be bent or pounded to a slight extent. It cannot be forged, and it can be broken by the blow of a hammer. In the manufacture of malleable-iron castings the first casting is made up of white cast iron containing the ordinary impurities. The casting is then made malleable by prolonged annealing. Malleable cast iron is stronger than ordinary cast iron.

The standards under which malleable-iron fittings are manufactured are listed in Table II-1, and dimensions are given in Table II-52. Fittings are illustrated in Figs. II-56 and II-57.

4-9. Brass and Copper Fittings. Brass and copper pipes are used in conveying water and other fluids because of their resistance to corrosive attack and their durability, attractive appearance, and smoothness, which results in a low friction coefficient. The corrosion of brass and copper pipe, especially that due to electrolysis caused by stray electric currents, is known to have undesirably affected the quality of the water.[1] The loss of heat from copper pipe when conveying hot water may be only about 60 per cent of the loss from black steel pipe of the same size. Copper tubing is suitable for waste and vent pipes where cost is not the final consideration. The use of 3-in. tubing in prefabricated plumbing

[1] See also R. E. Ebert, Copper Dissolution . . . by Stray Electric Currents, *J. Am. Water Works Assoc.*, December, 1956, p. 1547.

assemblies permits their use with 2- by 4-stud-supported walls. A lightweight copper drainage tube known as DWV is used for this purpose, as shown in Table II-55. Information concerning brass and copper pipe is given in Tables II-55 to II-62. Brass or copper pipe should not be used for conveying alkaline liquids because of the low resistance of such pipes to corrosion thereby.

Standard specifications for brass and copper tubing are listed in Table II-1, dimensions are given in Tables II-55 to II-62, and illustrations are shown in Figs. II-10, II-58, and II-59. It may be said that, in general, the strength of standard and extra-heavy brass pipe corresponds to the strength of standard and extra-heavy wrought pipe. Brass pipe is made plain, polished, or nickel-plated and of hard, soft, or medium temper, the last being the most suitable for plumbing work.

Brass fittings are made with either threaded or flanged ends, in two weights, standard and extra heavy, and of cast brass with dimensions similar to fittings for steel pipe. Standard fittings are designed for working pressures up to 125 psi, and extra-heavy fittings for pressures up to 250 psi. Copper fittings are not manufactured. Fittings are threaded or flanges drilled in accordance with iron-pipe standards as given in Secs. 7-26 to 7-28. Practically all fittings available in malleable iron or flanged cast iron are available also in brass. Nickle-plated fittings are generally on a brass base. They are used because of their appearance, since they have no other advantage over brass. They have the disadvantage, however, that the plating sometimes wears unevenly, giving an untidy appearance. Cast-brass fittings may be finished with a polished surface. The polishing of the surface serves only to improve its appearance.

4-10. Seamless Copper Water Tube. Seamless copper water tube is specified, under ASA H 23.1–1956, as follows:

This specification covers seamless copper tube especially designed for plumbing purposes, underground water services, drainage, etc., but also suitable for copper coil water heaters, fuel oil lines, gas lines, etc. For refrigeration and air conditioning the following is recommended: (*a*) When it is desirable to use sweat fittings hard tube should be used, and (*b*) When it is necessary to use soft tube with sweat fittings, then rounding and sizing tools should be used.

There shall be 3 types of copper water tube with principal uses as follows:[1] *Type K* for underground service, general plumbing and heating purposes involving sewer service conditions and for gas, steam, and oil lines. *Type L* . . . for interior applications involving general plumbing and heating purposes. *Type M* . . . sizes 2½ to 12 in. . . . for use with soldered fittings only, for interior applications involving general plumbing and heating purposes. Sizes 1½ to

[1] See Table II-55.

12 in. inclusive, recommended for use with soldered fittings only, for water, vent, and soil lines and for other interior, non-pressure applications.

Dimensions, weights, and tolerances . . . this specification shall be as prescribed in . . . ASTM . . . B 251 . . . .

4-11. Calking Ferrules. Brass calking ferrules should be of the best quality red brass with weights and dimensions as given in Table II-60. Seamless copper ferrules may be used in lieu of cast brass provided that they correspond in size and weight.

4-12. Soldering Nipples and Bushings. Soldering nipples should be of brass, iron-pipe size, or of heavy cast red brass of not less than the weights given in Table II-62.

4-13. Pipe and Hose Threads. Specifications for the cutting of threads for pipe and for hose are listed in Table II-1, and dimensions are given in Tables II-42 to II-45. Threads are illustrated in Figs. II-53 and II-54. The American or Briggs standard was adopted by the American Society of Mechanical Engineers on Dec. 29, 1886, and later, with some modifications, by the American Standards Association. It has been extended and revised since its original adoption by trade usage to conform to the conditions shown in Fig. II-53 and listed in Tables II-43 and II-44. Details of the Federal Specifications for Standard Pipe Threads are given in Table II-42. The following conditions are standardized under the American or Briggs standard:

The depth of the thread $H = 8/10N$; where N is the number of threads per inch.

The pitch of the threads increases roughly with the diameter.

The conically threaded ends of pipe are cut at a taper of $\frac{3}{4}$ in. diameter per foot of length, that is, at a slope of 1 in 32 to the axis of the pipe.

The thread is perfect for a distance L from the end of the pipe. The value of L is

$$L = \frac{0.8D + 4.8}{N}$$

where D is the outside diameter in inches.

Then come two threads perfect at the root or bottom but imperfect at the top, and then come three or four threads imperfect at the top and bottom. These last do not enter into the joint at all but are incident to the process of cutting the threads.

The thickness of the pipe under the root of the thread at the end of the pipe equals

$$T = 0.0175D + 0.025 \text{ in.}$$

In the threading of pipes or fittings it is important, to secure tight joints, that a sufficient length of thread be cut to make good contact and to develop the full strength of the pipe or fitting. The length of threads that should be in contact is given in Table II-44.

Data on the dimensions of standard hose threads are given in Table II-45, and standard specifications for hose and hose couplings are listed in Table II-1.

4-14. Lead Pipe and Sheet Lead. The use of lead pipe and of sheet lead is declining in plumbing partly because of the availability of better materials and partly because of the skill required in the use of lead. Its use, however, is far from obsolete, so that the qualified plumber should be acquainted with its characteristics and applications.[1] Among the advantageous features of lead pipe are its flexibility and durability. It does not transmit sound readily. It will withstand almost an unlimited amount of vibration without injury, and it is practically noncorrodible by potable water supplies or by domestic sewage. Corrosion is resisted partly by the formation of an insulating coating of lead carbonate or lead oxide. Lead is resistant also to attack by both alkali and acid.

Objections to the use of lead include cost, the skill required in its manipulation, and its toxicity in contact with otherwise potable water. Lead is so flexible that if lead pipe is not well and continuously supported, especially in horizontal projection, the pipe will sag, producing pockets conducive to the deposition of sediment and to clogging. For this reason horizontal runs of lead drainage pipe greater than about 6 ft are prohibited by some codes. It is claimed that rats will gnaw holes through lead pipe, and it is certain that workmen sometimes drive nails into it. It will not always stay in position or remain the same length, as it "creeps" under alternating stresses. Lead pipe may soften sufficiently when conveying hot water to burst under pressure.

Leadwork in the making of joints[2] is described in Chap. 12.

Lead may be corroded by soft waters containing dissolved oxygen, carbon dioxide, or organic acids. Alkaline waters form insoluble lead carbonate which may protect the lead from further attack. Lead is not affected by dissolved chlorine. Because of its solubility and the solubility in water of some of the products of corrosion and because of their poisonous properties, the use of lead, lead products, or lead paints should be avoided in water pipes and water containers, particularly in soft-water regions and in hot water. Tests made by the Massachusetts State Board of Health have shown lead content as high as 3 to 5 ppm in natural waters and an increase of 50 to 100 per cent and even more after the water has been standing in contact with lead. Since 0.5 ppm is considered dangerous to health, the use of lead in contact with potable water should be prohibited. The most common uses of lead pipe are as a flexible connec-

[1] See also *Air Conditioning, Heating, and Ventilating*, March, 1958, p. 83.

[2] See also *Domestic Eng.*, July, 1956, p. 187.

tion, called a *gooseneck*, between a water-supply main in the street and the house service pipe; as a trap in the drainpipe from plumbing fixtures; as a flexible connection between a roof gutter and a rain-water leader; and as a flexible connection between a water closet and the drainage pipes of the plumbing system. Sheet lead is used for roof flashings and for drip pans or safe wastes. Specifications for the composition and dimensions of commercial lead products are given in Tables II-63, and II-64 and lead traps and a pipe bend are shown in Figs. II-62 and II-63.

Various weights of lead pipe have been given trade names, as listed in Table II-63. The size of lead pipes is designated by the inside diameter. The following recommendations are made for the class of lead pipe to use for particular conditions:

For soil, waste, vent, or flush pipes—*light*
Water-supply pipes for normal pressure above ground—*strong*
Water-supply pipes for normal pressure under ground—*extra strong*

Where unusually high internal pressures are to be carried, the thickness of the lead pipe should be computed using the working strength of lead as 300 psi in tension.

A typical lead gooseneck and the manner of its installation are illustrated in Figs. 3-2 and 3-3, a lead connection between a roof gutter and a rain leader is shown in Fig. 13-6, and lead water-closet bends are shown in Fig. 15-10.

Recommended weights of sheet lead for various purposes are shown in Table 13-6, and the weights of sheet lead for various thicknesses are shown in Table II-70.

4-15. Composition of Solder. The material of which solder is made is of such importance in securing success in joint wiping, leadwork, and sheet-metal work that the composition of solders should be understood by the plumber. Some standard specifications for lead products and solders are listed in Table II-1, and information on the composition of some lead products and of solders is given in Tables 12-2, 12-3, II-63, and II-64.

4-16. Vitrified-clay Pipe. Vitrified-clay pipe is used principally for house sewers and rarely for house drains. Its use aboveground in the interior of buildings is commonly prohibited by plumbing codes, and some codes prohibit its use within a building. Vitrified-clay pipe is used in English plumbing practice; their installations withstand the customary air and water tests possibly because of the satisfactory bituminous joints made.

The use of vitrified-clay pipe within a building is prohibited because of the uncertainty that the joint will be and will remain watertight and gastight. The brittleness and inflexibility of the pipe throw the strain of any movement into the joints. If the joints are made of cement, cracking results and the joint is no longer tight.

Standard dimensions for vitrified-clay pipe and other clay pipe are given in Tables II-65 and II-66. Fittings are illustrated in Fig. II-60. Double-strength pipe and pipes with extra-wide and -deep bells are also used. The dimensions and weights of these pipes are given in Appendix II. Extra-heavy pipes are used where unusually heavy stresses are anticipated but the stresses are insufficiently heavy to demand the use of cast-iron pipe. Deep and wide sockets are used for cement joints in wet trenches. Three-foot lengths of pipe are preferable to shorter lengths in straight runs of pipe, but standard 2- and 2½-ft lengths are more convenient for fitting to special dimensions and for laying around curves.

Advantages in the use of vitrified-clay pipe are its durability, its relatively low cost as compared with metal pipes, and the simplicity of its installation. Clay pipe is unaffected by ordinary corrosive substances and many highly corrosive compounds that may be found in sewage. Clay is not affected by most acids or alkalies. It is the cheapest material available, except possibly some cement products, for pipe of equal strength and carrying capacity, and it can be installed satisfactorily by unskilled workmen.

The procedure in making a cement joint in bell-and-spigot pipe is described in Sec. 12-24.

4-17. Cement Pipe and Concrete Pipe. Cement pipe and concrete pipe are suitable for and are used under conditions similar to those for vitrified-clay pipe, except that cement and concrete pipe may be more subject to corrosion by acids and they have been badly affected by the decomposition of stale sewage where the concrete has been subjected to gases and vapors therefrom. Because the use of concrete pipe for such small sizes as are needed for ordinary building sewers is no more economical than vitrified-clay under most conditions and because of its susceptibility to corrosion, the use of cement and of concrete is not recommended for such purposes. Concrete is used successfully, however, in the construction of septic tanks in which the concrete is kept well submerged and protected from corrosive gases.

Reinforced-concrete pipe to carry water under pressure is manufactured under standard specifications, as indicated in Table II-1. Such pipe is used almost exclusively in long pipelines and aqueducts and not in plumbing. Standard dimensions are given in Tables II-67 and II-68, and a section of a pipe is shown in Fig. II-61.

4-18. Asbestos-cement Pipe. Asbestos-cement pipe is composed of asbestos fiber and portland cement combined under pressure into a dense, homogeneous material. It is manufactured under such proprietary names as Century, Eternite, Roxite, and Transite. In waterworks practice and in sewerage its use has been confined mainly to pipes in aqueducts, distribution systems, and collection systems. Dimensions are given in Table II-69.

TABLE 4-2. PROPERTIES OF POLYVINYL CHLORIDE PIPE[a]

Physical properties of polyvinyl chloride			Temp-pressure relationship				Max operating pressure at 75°F, psi		
				Temp factor K[b]		Size, in.	Std wt, Schedule 40[d]	Extra heavy, Schedule 80[e]	
Property	Normal[b] impact	Normal[c] impact	F°	Normal impact	High impact			Plain end	Threaded end
Weight, lb per cu ft	86.0	84.2	10	1.43	1.43	½	410	575	330
Tensile strength, psi at:									
−40°F	12,200	.....	30	1.29	1.29	¾	335	470	285
+32°F	8,900	7,800	50	1.16	1.17	1	310	435	255
+78°F	6,400	6,000	70	1.03	1.04	1¼	255	260	220
+140°F	4,100	3,600	100	0.84	0.83	1½	230	325	205
Compression, psi, at +78°F	9,600	8,500	120	0.73	0.45	2	195	280	190
Dielectric strength, volts per mil	1,413	1,100	130	0.66	0.27	3	185	260	170
Coefficient of thermal conductivity	0.105	0.135	150	0.53	....	4	155	225	160
Coefficient of linear expansion	9.36	18.0	...	....	....	6	125	195	150

[a] See also W. P. Chapman, *Air Conditioning, Heating, and Ventilating*, July, 1956, p. 58; Anon., How to Join PVC Pipe, *Heating, Plumbing, Air Conditioning*, August, 1957, p. 115, and *ibid.*, December, 1955, p. 114.

[b] Normal impact has greater strength and chemical resistance but less impact resistance. High impact is suited for installations requiring excellent chemical resistance and a high degree of toughness even at low temperatures. Properties are best suited to handling corrosive materials at moderate pressures and temperatures.

[c] Temperature factor for use in equation $P = kp$, where P is the maximum operating pressure at operating temperature in pounds per square inch and p is the maximum operating pressure at 75°F.

[d] Threaded joints are not recommended for Schedule 40 pipe. The pipes in this column are plain end.

[e] Figures given here are for normal impact. Figures for high impact are 82 per cent of these figures.

4-19. Fiber Pipe. The use of bituminized-fiber pipe[1] is permitted by some plumbing codes,[2] and the material and pipes are manufactured under standard specifications, as listed in Table II-1. Properties of polyvinyl chloride (PVC) pipe are shown in Table 4-2. Some dimensions of polyethylene pipe and bituminous pipe are given in Table 4-3, and some details are shown in Fig. 4-2.

[1] *Domestic Eng.*, July, 1956, p. 187.

[2] "The Uniform Plumbing Code for Housing," Housing and Home Finance Agency, February, 1948.

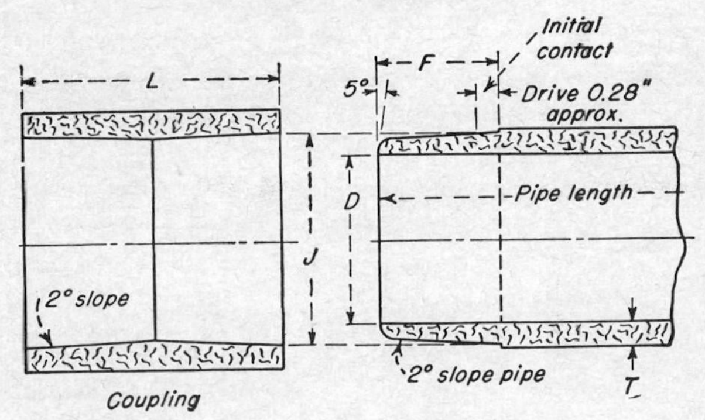

FIG. 4-2. Bituminized fiber pipe.

TABLE 4-3. DIMENSIONS OF BITUMENIZED FIBER PIPE AND OF POLYETHYLENE CHLORIDE PLASTIC PIPE*

Bituminized fiber pipe†

Characteristic‡	Nominal size, in.				
	2	3	4	5	6
Diameter, D, min inside	2.00	3.00	4.00	5.00	6.00
Wall thickness, T	0.23	0.28	0.32	0.41	0.46
Coupling length, L	2.90	3.42	3.92	3.92	3.92
Coupling diam, J	2.470	3.448	4.493	5.726	6.782
Pipe joint length, F	1.43	1.69	1.94	1.94	1.94
Laying length,§ ft ±1 in	5	8	8	5	5

Polyethylene chloride pipe¶

Characteristic	Nominal diameter, in.				
	½	¾	1	1¼	1½
Outside diameter	0.84	1.05	1.315	1.660	1.900
Inside diameter	0.822	0.924	1.049	1.380	1.610
Working pressure at 72.4°F, psi	104	86	81	68	61
	2	2½	3	4	6
Outside diameter	2.375	2.875	3.50	4.50	6.625
Inside diameter	2.067	2.489	3.068	4.026	6.065
Working pressure at 72.4°F, psi	52	51	50	42	31

* See also Commercial Specifications CS 197-57.

† See Fig. 4-2 and Commercial Standard Specifications 116 to 144.

‡ All dimensions in inches unless otherwise noted.

§ 2-, 5-, 6-, and 8-in. sizes may be supplied also in 8- or 10-ft lengths. See also *Air Conditioning, Heating, and Ventilating*, September, 1957, p. 114.

¶ Comes in coils from 100 to 500 ft.

4-20. Plastic Pipes.[1] Plastics used for pipe include Saran, Tenite, Plastitube, and polyvinyl chloride.[2] The materials are useful for water-supply pipes and for the conveyance of some corrosive liquids at a moderate temperature and pressure. Saran, among the more commonly used of the materials,[3] is a flexible, yellow, translucent plastic that melts at 310°F; will withstand water temperatures up to 170°F continuously; is hydraulically smooth; has a heat conductivity of 0.00022 times that of copper; has a coefficient of expansion of 0.000083; and weighs about 125 lb per cu ft. Black Saran pipe is available in sizes from ½ to 4 in. It will withstand bursting pressures up to 1,500 psi, depending on the diameter, thickness, and temperature. Joints may be threaded as for metal pipe, or they can be made by autogenous welding.

Among the advantages claimed for plastics, particularly for Saran, may be included freedom from corrosion, freedom from damage due to freezing of water in a closed pipe, acid resistance for solutions up to about 10 per cent concentration, light weight compared with cast iron, easy bending and joining, and adequate strength. It is unaffected by age, sunlight, or weather; it is resistant to shock; it is a good electric insulator; it is resilient and flexible; and in smaller sizes it can be transported in coils. Among the objections may be included lack of wide experience with it and the fact that it can be penetrated by nails.

Plastitube is rigid and comes in 12-ft lengths. Plastiflex is flexible and comes in 60-ft coils in sizes from ⅛ to 2 in. outside diameter. Tenite is a similar substance, but it is about as combustible as bulk newspaper.

4-21. Plastic Toilet Bowls and Other Fixtures.[4] Reinforced plastics are claimed to give satisfaction as toilet bowls, lavatories, and some other fixtures. Their odor retention has been measured and found low, and their strength and durability within the needs of plumbing. They should compete economically with ceramics even where the saving in weight is not important.

4-22. Sheet Metal. Sheet metal is used in plumbing for rain-water gutters and leaders, safes and safe wastes, flashings, roof extensions of vent pipes, and other purposes. It is used also in warm-air heating for

[1] See also "A Study of Plastic Pipe for Potable Water Supplies," from Society of Plastic Industry, New York, and Plastic Pipe Use to Increase, *Heating, Piping, Air Conditioning*, December, 1954, p. 89.

[2] See also W. P. Chapman, *Air Conditioning, Heating, and Ventilating*, July, 1956, p. 58, and How to Join PVC (Polyvinyl Chloride) Pipe, *Heating, Piping, Air Conditioning*, August, 1957, p. 115.

[3] See also F. M. Dawson and A. A. Kalinske, Studies Relating to the Use of Saran for Water Pipes in Buildings and for Service Pipes, *J. Am. Water Works Assoc.*, August, 1943, p. 1058.

[4] See also S. B. Swenson and H. Graves, Investigation of Plastics for Toilet Bowls *Modern Plastics*, December, 1956, p. 164.

furnace pipes and register bases, etc. The types of sheet metal available include black iron, steel, galvanized iron, tin, zinc, copper, brass, lead, and aluminum. Table II-70 and others give the standard thicknesses and weights of these metals.

Black iron or steel sheets consist of uncoated metal just as it comes from the rolls. It is seldom used in plumbing except for the most temporary installations. Brass and aluminum are seldom used in plumbing work on account of their cost and the suitability of less expensive metals. Galvanized iron, zinc, tin, and copper are the most commonly used sheet metals. Recommended weights and thicknesses of sheet metal to be used for different purposes are listed in Table 13-6. Methods of using sheet metal are discussed in the sections devoted to the particular use to which the metal is to be put.

Sheet lead, according to the "National Plumbing Code" is to be ". . . not less than the following: for safe pans, not less than 4 lb per sq ft. For flashings of vent terminals, not less than 3 lb per sq ft. Lead bends and lead traps shall not be less than 1/8-in. wall thickness."

4-23. Hangers and Escutcheons. Various types of pipe hanger and pipe support are discussed in Secs. 7-11 and 12-2. They are illustrated in Fig. 7-10.

4-24. Color Scheme for Piping. A standard scheme of color bands[1] to designate uses of piping for various purposes is outlined in Table 4-4.

TABLE 4-4. COLOR BANDS FOR PIPING
(ASA A 13.1–1956)

Purpose of pipe	Initial letters	Band color	Letter color	Outside diam. of pipe covering, in.	Width of color band, in.	Size of letters, in.
Fire protection	F	Red	White	3/4 to 1 1/4	8	1/2
Dangerous	D	Yellow	Black	1 1/2 to 2	8	3/4
Safe	S	Green	Black	2 1/2 to 6	12	1 1/4
Protective	P	Blue	White	8 to 10	24	2 1/2
				above 10	32	3 1/2

METALLIC CORROSION

4-25. Metallic Corrosion.[2] The effect of metallic corrosion is so important in the selection of plumbing materials and in the maintenance of plumbing that an understanding of its causes is desirable. Corrosion may be explained by such hypotheses as galvanic or bimetallic action,

[1] ASA A 13.1–1956.

[2] See also H. H. Uhlig, "Corrosion Handbook," John Wiley & Sons, Inc., New York, 1948.

hydrogenation, electrolysis, chemical reaction, rusting, direct oxidation, and biologic action.

4-26. Galvanic or Bimetallic Action. A metal when immersed in an electrolytic solution, i.e., a solution that has the property of conducting electricity, such as a weak solution of sodium chloride (common salt) in water, tends to go into solution. In some metals this tendency is greater than in others, and such metals corrode more easily. Where two dissimilar metals are immersed in the same electrolytic solution, there is an interchange of atoms carrying an electric charge, that is, of ions, between them. There is a flow of particles (ions) of the metal with the higher electrode potential (the anode) to the metal with the lower electrode potential (the cathode). The anode corrodes; the cathode is protected. Some metals and alloys are arranged in Table 4-5 in the descending order of their electrode potentials. Conditions accelerating electrolytic corrosion by speeding the migration of ions include higher temperatures, the release of gases from solution, diffusion and turbulence, and the lowering of viscosity.

TABLE 4-5. METALS AND ALLOYS ARRANGED IN THE ORDER OF THEIR RELATIVE ELECTROLYTIC SOLUTION PRESSURES*

Magnesium (Mg), −2.34†	Cast iron	Copper (Cu), +0.345 to +0.522
Magnesium alloys	Nickel (Ni), −0.250	Bronze
Aluminum (Al), −1.67	Lead-tin	Copper-nickel
Zinc (Zn), −0.762	Lead (Pb), −0.126	Silver (Ag), +0.800
Iron (Fe), −0.440	Tin (Sn), −0.136	Mercury (Hg), 0.854
Cadmium (Cd), −0.402	Hydrogen (H), 0	Platinum (Pt), +1.2
Steel and iron	Brass	Gold (Au), +1.42 to 1.68

* The flow of current due to dissociation is from the metal higher on the list, which acts as the anode and corrodes, to the metal lower on the list, which acts as the cathode and does not corrode.

† The figures represent the standard electrode potentials at 25°C.

Bimetallic or galvanic action is not the explanation of all corrosion, nor will one of two dissimilar metals placed in the solution of an electrolyte always corrode. The effect will depend on the nature and concentration of the electrolyte in the solution; on the nature of the metals involved, particularly if they are alloyed; on other conditions such as internal strain, turbulence in the solution, temperature of the metal and its environment, and aeration; and on protective coatings formed by the metals involved. The rate of corrosion is greatly affected by the temperature, under some conditions the rate at 180°F being 10 times and the rate at 210°F being 100 times the rate at 50°F.

4-27. Hydrogenation. Water is a weak electrolyte that is broken down or dissociated into positively charged particles of hydrogen, known as hydrogen ions (H^+), and negatively charged hydroxyl ions (OH^-). The hydrogen ions, acting similarly to positively charged metallic particles, will migrate to and plate out on a metallic surface immersed in water. To maintain electrical balance, negatively charged particles of metal are discharged into the water, causing metallic corrosion. In this case it is the cathode that corrodes. The film of hydrogen forms a protective coating on the metal that requires additional electrolytic solution pressure or voltage to discharge metallic particles into solution. The additional voltage is known as *overvoltage*.

4-28. Electrolysis. If two metallic electrodes are immersed in an electrolytic solution and a direct electric current is passed through the cell thus formed, positively charged particles of the anode will travel to the cathode, release their charge, and plate out on the cathode. The anode is corroded. This form of corrosion is not uncommon in plumbing as a result of stray currents from electrical equipment seeking a path along water pipes to the ground. At points where these currents leave the metal, corrosion occurs.

4-29. Chemical Reaction. If a metal is immersed in an acid, a chemical reaction occurs, releasing hydrogen and forming a salt of the metal. The metal is corroded.

4-30. Rusting. Rusting is a common form of corrosion in which the metal is exposed to an atmosphere containing moisture or to a liquid, especially water, containing oxygen in solution. A chemical reaction occurs in which the water acts as an acid, releasing hydrogen and oxidizing, or rusting, the metal. The action is accelerated in an electrolytic solution by the removal of hydrogen ions. It is limited to some extent by the coating of rust which acts as an insulator to prevent contact between the metal and oxygen.

Oxygen is important in electrolytic corrosion because of its activity in depolarizing the cathode by removing hydrogen as it is formed on the surface of the cathode. Depolarization by oxygen permits continued corrosion by bimetallic or galvanic action. The same result is effected by agitation of water in contact with metal, since depolarization is enhanced by removal of hydrogen from the cathode. This partly explains the more rapid corrosion of metals exposed to high velocity or turbulent flow of water.

4-31. Direct Oxidation. Metals in an atmosphere containing oxygen, particularly if they are heated, will combine directly with the oxygen in the atmosphere. A common example of this is the formation of mill scale, or magnetite (Fe_3O_4). During this type of corrosion, oxides of iron are formed that tend to spread on the metallic surface. The rust deposit on the metal tends to inhibit further corrosion from this cause.

4-32. Biological Action. Bacteria that cause metallic corrosion include the sulfate-reducing bacteria, the iron-consuming bacteria, and the sulfur-consuming bacteria. A form of blue-green alga has been suspected of causing corrosion by depolarizing the metal by removal of the natural protective coating of hydrogen in hydrogenation. No bacterium attacks the metal directly.

Sulfate-reducing bacteria, functioning in the complete absence of oxygen, react with the sulfates and organic compounds containing sulfur, in the soil surrounding buried metal, to produce hydrogen sulfide. This substance then combines with the metal, in the presence of moisture, to form compounds of sulfur, with the release of hydrogen, which is again used as bacterial food.

Iron-consuming bacteria remove compounds of iron that are in solution, releasing chemical substances that combine with more iron to put it into solution. The consumed compounds of iron are deposited by the bacteria as ferric hydroxide (iron rust) forming a sheath or tubercule around the central core. The effect is known as *tuberculation.*

4-33. Selective Corrosion of Alloys. In the selective corrosion of an alloy one metal dissolves out of the alloy more rapidly than the other or others. The so-called *dezincification* of brass is a common manifestation of this phenomenon. It is assumed that the brass dissolves as a whole but that the copper particle is redeposited, leaving the metal of low strength and composed of a spongy mass of copper. The phenomenon is limited to yellow, or high-zinc, brasses in contact with waters, particularly hot waters, carrying free mineral acids or acid-forming salts in the presence of oxygen.

4-34. Methods of Retarding Corrosion. Metallic corrosion may be retarded by (1) cathodic protection, (2) care in manufacture of the metal through control of its purity and composition, (3) application of coatings or linings, (4) control of the environment of the metal, and (5) treatment of the water. Methods for the treatment of water to retard corrosion are discussed in Sec. 4-52.

4-35. Cathodic Protection. Cathodic protection is provided by inserting in the corrosive water a metal whose electrode pressure is greater than that of the metal to be protected or by placing the metal to be protected in a direct electric circuit such that the current flows from the anode of a d-c generator through the surrounding ground or through the contained water into the protected metal which acts as a cathode. Such a circuit is shown in Fig. 4-3.

The insertion of a magnesium rod into the corrosive water in a pipe or storage tank creates an electrolytic cell in which the magnesium ions (Mg^+) of the anode go into solution and flow through the water to be deposited on the metal to be protected which acts as the cathode. Unfortunately a reaction between the magnesium and some waters, especially some hot, soft waters, has created foul odors and discoloration of the water. The difficulty has been overcome by removing the magnesium rod.[1]

In forced cathodic protection, as shown in Fig. 4-3, voltages used range between 1½ and 30 and the current is in the order of 1 to 5 ma per sq ft of surface area of metal protected.

4-36. Care in Manufacture. Since metallic impurities tend to create galvanic couples, resulting in corrosion, purity and homogeneity of metal will tend to minimize corrosion. Hence, the purer the metal, the less will be the tendency toward corrosion.

4-37. Alloys and Resistant Materials. Alloys of iron or steel with copper, nickel, or chromium are more resistant to corrosion than iron or

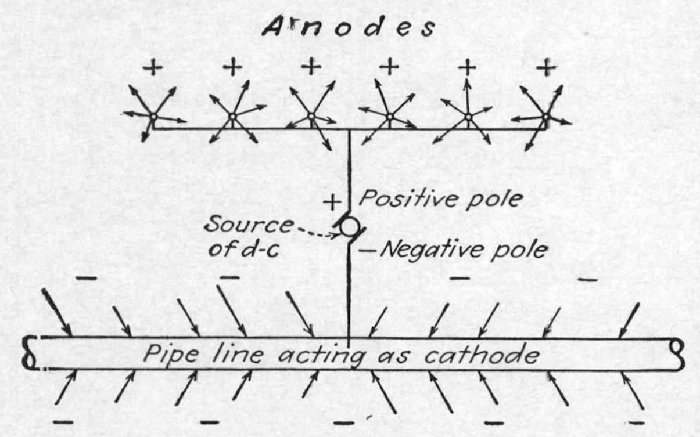

FIG. 4-3. Electrical circuit for cathodic protection.

steel alone. Everdur metal is a lightweight, corrosion-resistant alloy consisting of copper and silicon with other metals. Various compositions of the metal are used for tanks, gates, weirs, screens, bolts, tubing, and conduits. Special processes for the production of rustless metal are successful, and pipes and fittings of such material are available.

Highly corrosive substances, including acids, may be conveyed sometimes in cement-lined, glass-lined (vitreous), or rubber-lined pipe. Tarred, extra-heavy cast-iron pipe; cast iron enameled inside; vitrified tile; chemically pure lead; fiber; asbestos; and other materials are used with success. High-silica pipe is resistant to acids but is more susceptible to the action of alkalies, particularly soda wastes. Pipes and fittings of high-silica iron and of the same styles and dimensions as standard cast-iron pipes are manufactured. A fiber pipe is manufactured that is resistant to both acids and alkalies, but it is structurally weak, and its use is confined to drain pipes under chemical laboratory sinks and similar locations.

[1] See also discussion in *Plumbing and Heating J.*, October, 1952, Plumbing and Heating Section.

4-38. Coatings and Linings. Coatings and linings used for the protection of iron and steel surfaces include tar or asphaltic materials; enamels; lacquers; resins; zinc, or galvanizing; metalizing; plastics; and paints. Each has shown some success in retarding corrosion.

4-39. Bituminous Coatings and Linings. Coal-tar enamel coatings are used extensively for the protection of steel pipe, and the Angus Smith coating, introduced about 1866, is used for the protection of cast-iron pipe. The Angus Smith coating consists of a mixture of gas tar, Burgundy pitch, oil, and resin, into which the pipe is dipped while hot. About 1931, molten bituminous enamel was introduced. It is applied hot to the inside of the pipe while the pipe is spinning around its longitudinal axis. The application of enamel coatings to the exterior of pipe is accomplished by various devices that pour enamel on the pipe and spread it with a suitable drag, overlapping spiral paths being followed by the drag or drags along the revolving pipe. Most coated steel pipe is also wrapped, frequently with asbestos felt saturated with bitumen. It may be finished with a wrapping of heavy paper. The objection to some coatings, especially to heavy tar coatings placed on the metal before it has been inspected, is the possibility that flaws in the metal can be covered.

4-40. Resins and Lacquers. Resins, when applied to metallic surfaces, are sometimes called *lacquers.* They may be the natural exudations from plants, or they may be chemically synthetized. They form a highly satisfactory protective coat when applied to a metallic surface.

4-41. Paints.[1] Paints for the protection of metal from corrosion should be made of a metallic pigment, preferably zinc when the paint is to be in contact with water or lead if the paint is not to be submerged. The pigment is mixed with oil, usually linseed oil, that will dry to form a thin veneer filled with the metallic pigment.[2] Zinc is desirable both because of its nontoxic qualities and because of its higher electrode pressure in relation to most metals used in plumbing. A black paint with a pitch base and silicone additive is said to be applicable to hot surfaces up to about 750°F.[3]

Varnishes and lacquers for a japanned or enameled surface may be allowed to dry cold or may be baked on. Such coatings are pleasing in appearance, but they are usually easily broken.

Aluminum and gilt paints, particularly the former, may be successful as protective coatings, but they are relatively expensive. Their appear-

[1] See also J. O. Jackson, Experimental Studies of Tank Coating Materials, *J. Am. Water Works Assoc.*, September, 1941, p. 1,553.

[2] See also C. G. Sward, Paints for Waterworks, *J. Penn. Water Works Operators Assoc.*, vol. 8, p. 92, 1936.

[3] See also *Air Conditioning, Heating, and Ventilating,* March, 1956, p. 152, and August, 1956, p. 134.

ance is sufficiently attractive to result in their frequent use for the exterior of exposed structures, such as towers and tanks, the paint being sprayed rather than brushed on the metal.

4-42. Plastic Coatings.[1] Baked phenolic and vinyl-type plastics are used for pipe coating. The former, applied to pipes more than about 2 in. in diameter, have been found satisfactory at 212°F. Vinyl-type plastic is applicable to smaller pipes, but it cannot be used at temperatures above 150 to 180°F.

4-43. Vitreous Coatings. Vitreous coatings for metal are prepared by running molten silicate into cold water. The resulting material is ground in a pugmill with clay and water. When it has been worked to the required consistency, the metal is coated with it by dipping or brushing, and when dried, it is baked onto the metal.

4-44. Zinc Coating and Galvanizing. Galvanizing consists of applying a zinc coating to the metal. This can be done by cleaning (pickling) the metal in acid and then dipping it in molten zinc, by electroplating the metal with zinc, or by applying zinc by the sherardizing process, in which the hot iron is revolved in a drum containing zinc dust. This process produces the best results, since the threads on threaded pipe are still useful after sherardizing, which is not always the case after galvanizing by other processes.

As a result of tests by the National Bureau of Standards it has been concluded that over a 10-year period the rates of loss of weight by galvanized steel were from one-half to one-fifth of the rate of loss by bare steel.

4-45. Metallic Plating. Tin or lead plating or coating is applied in a manner similar to the hot-dip process of galvanizing, except that the iron is passed through the rolls at the same time that it is in the bath of molten tin or lead. The use of lead may be unsatisfactory in a paint or coating, for sufficient lead may be dissolved to be injurious to the health of the water consumers.

Nickel and copper plating are usually applied electrolytically to the cleaned metal. Nickel, copper, and tin coatings are sometimes used in the protection of iron, but because of their relatively low electrode pressure, when the coating is once broken, galvanic action is set up between them and the iron, resulting in more rapid corrosion than if it had not been coated.

4-46. Metalizing. Metalizing consists of spraying a coat of molten metal on any surface, metallic or otherwise, that will hold to it. With careful surface preparation, proper selection of metal, and care in application, good surface protection can be obtained. The process consists

[1] See G. D. Lain, What's New in Pipe Coatings, *Heating, Piping, Air Conditioning,* January, 1954, p. 164.

essentially of passing a wire of the desired metal through a flame and driving the molten metal in a fine spray onto the prepared surface by means of an air blast. Any metal that can be obtained in wire form can be used. The bond is formed by minute particles of the metal penetrating into and holding in irregularities in the prepared surface that must not be smoothly polished before metalizing.

4-47. Plastic Tape[1] and Plastic Films. Three types of plastic tape are used to wrap around pipe to prevent corrosion. They are pressure-sensitive polyethylene chloride with a 1-mil thickness of solvent-spread adhesive; pressure-sensitive, oriented polyethylene tapes with a 4-mil calendered, synthetic adhesive; and polyvinyl chloride backing laminated to a 15-mil butyl rubber film. They are known, respectively, as vinyl, polyethylene, and vinyl-butyl tape. A coating of paraffin plastic has been machine-applied hot to pipes wrapped in kraft paper. The coating has been compressed to a $\frac{1}{32}$-in. wax layer and a 2-mil plastic film.

4-48. Installations to Avoid Corrosion. Pipe should be installed to avoid corrosive substances or conditions; for example, wrought pipe, particularly when carrying steam or hot water, should not be placed in ashes that are occasionally wet, nor should it be placed below a cinder-concrete floor. Electrolysis can be avoided in aboveground water pipes by the strategic location of dielectric couplings in the pipe and by installing the pipe in locations free from stray electric currents.[2]

4-49. Corrosivity of Water. Pure water is a weak electrolyte. It permits the migration of electrically charged particles in all electrical phenomena of corrosion, it is necessary in the growth of bacteria, and it is involved in chemical reactions. In general, the greater the mineral impurities in water, the greater is its "strength," or chemical activity, as an electrolyte. However, soft water may be more corrosive than hard water, partly because of its greater solvent powers, low hydrogen-ion concentration, lack of electrolytic buffer action, and for other reasons. Hot water may be many times more actively corrosive than the same water when cold.

4-50. Deactivation of Water. The corrosive qualities of water to be carried in a pipe can be diminished or removed by "deactivation." The deactivation of water to remove active oxygen and other corrosive elements and compounds is accomplished by passing the water, usually when hot, through a tank, known as a *deactivator*, containing iron filings. A deactivator is illustrated in Fig. 4-4. It requires practically no attention in its operation except for the renewal of the iron filings once in 2 or

[1] See also F. Buck, Tape Coating for Distribution Piping, *Gas Age*, May 17, 1956, p. 18, and Anon., Wax-plastic Films Protect Pipes, *Chem. Eng. News*, Mar. 5, 1956, p. 1,132.

[2] See also *Domestic Eng.*, September, 1956, p. 26.

3 years, and its cost is so small that its installation in regions containing corrosive waters is economical. A somewhat similar effect to the deactivation of water can be obtained by boiling the water in a vacuum or an open tank, thus driving off the dissolved gases. This latter method is impracticable in most installations.

4-51. Solubility of Metals in Water. The solubility of metals and metallic compounds in water is a consideration of importance primarily from the viewpoint of its effect on health and to a lesser extent upon the durability of the pipe. The solubility of the metals is dependent, to a great extent, upon the quality of the water. Pure water containing

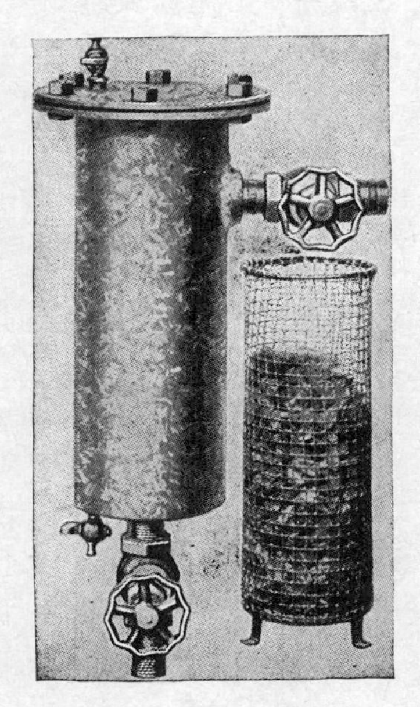

FIG. 4-4. A deactivator.

gases in solution, particularly oxygen and carbon dioxide, will dissolve metals more rapidly than hard waters already containing calcium and magnesium in solution. Rain water, soft and saturated with carbon dioxide, is particularly corrosive. Salty water and sea water are corrosive because of the activity of dissolved chlorides and their good electrolytic properties. Not only is the solvent property of hard waters reduced, but a scale is deposited on the walls of the pipe which retards solution of the metal. Surface waters containing organic acids also possess relatively high solvent properties. Because of the effect of materials already in solution in natural waters on the solution of metals, it is not feasible to express, with accuracy, the solvent properties of natural potable waters on metals.

The metals most commonly used for conveying water are iron, copper, zinc, tin, and lead. The solution of iron in water has no deleterious effect on health. It does, however, have an effect on the durability of the metal. The presence of iron in water is undesirable because of the taste and the staining of clothing and of plumbing fixtures. Iron is objectionable when present in a greater concentration than above 0.2 ppm.

The solubility of tin in water is so slight as to be negligible from the viewpoint of either health or durability. For this reason tin is used for conveying and storing water that is highly solvent or that may stand in contact with the metal for long periods.

Copper is sometimes slightly solvent in water. It has been found in waters flowing through brass pipe in a concentration of 0.2 to 0.3 ppm, although such concentrations are unusually high. In this concentration it is not dangerous to health.

Zinc is sufficiently soluble to have a slightly injurious effect on health under unusual circumstances. It can be dissolved from galvanized coatings, from brass, from paint, and from the pure metal. It has been found in concentrations as high as 8 or 9 ppm; the concentration after standing in galvanized or brass pipe may be high.

Lead is sufficiently soluble in water to offer a real menace to health, and for this reason its use in contact with potable water should be restricted if not prohibited. Since 0.5 ppm is considered dangerous to health, the use of lead in water pipe or in contact with potable water should be prohibited. Neither lead nor zinc paints should be used on surfaces in contact with potable water. Asphaltum, silicate, or enamel paints are preferable.

4-52. Treatment of Water. The corrosivity of water can be reduced chemically (1) by throwing calcium carbonate, known also as *carbonate hardness*, out of solution and causing it to precipitate on exposed metal and to form a protective coating thereon; (2) by diminishing the number of electrified particles of hydrogen, known as *hydrogen ions*, so that they will not be available to replace other electrically charged particles of the metal (a process known as increasing the pH); and (3) by deaerating the water so that oxygen is not available to react with the protective hydrogen film formed in corrosion by hydrogenation.

Water is generally not corrosive when there is sufficient calcium carbonate in solution to cause a slight precipitation and a deposition of calcium carbonate on metallic surfaces in contact with the water. If there is insufficient calcium carbonate in solution so that any protective carbonate film will be dissolved by the water, the corrosion of exposed metal will be increased. When water is in good "carbonate balance," it will tend to lay down a thin protective layer of calcium carbonate.

The precipitation of calcium carbonate, or the restoration of "carbonate balance," can be accomplished by "threshold treatment," which involves the use of sodium hexametaphosphate, known commercially as *Calgon.* A dose of 30 to 60 grains per 1,000 gal is usually satisfactory; too concentrated an application may be harmful. Lime may be added to increase the pH, but the exact amount required can be determined only by careful chemical control or by trial and error. Lime is used also in water softening, and soft water is normally more corrosive than hard water. Hence, the uncontrolled application of lime to prevent corrosion may be more harmful than helpful. Similarly, the deaeration of water to control corrosion requires knowledge of the chemistry involved and should be attempted only by one trained in the field or under regulated proprietary procedure.

The corrosivity of water may be affected by:

1. *Aeration.* Corrosion is increased by the increase of dissolved oxygen and possibly by the increase of dissolved carbon dioxide.

2. *Addition of lime* Carbonate balance is changed. Corrosivity may be decreased by the increase in pH, or the corrosivity may be increased by removing calcium carbonate from solution.

3. *Addition of alum.* This is added in routine filtration practice. It may tend to increase corrosivity.

4. *Applying heat.* Heating water increases corrosivity.

4-53. Harmful Wastes. Corrosive and harmful wastes and wastes at a temperature higher than about 140°F should not be discharged into drainage pipes not specially designed to receive them. Under no circumstances should such wastes be discharged into plumbing systems to which ordinary domestic plumbing fixtures are connected because of the greater dangers when the seal of traps thereon becomes broken.

CHAPTER 5

SMALL PUMPS FOR WATER AND FOR SEWAGE

5-1. Pumped and Gravity Supplies. Water must be pumped from wells, surface-water intakes, and other sources to be delivered into a storage reservoir, an equalizing reservoir, or pipes of a distribution system or a plumbing system. Where a collecting reservoir may be installed at a low elevation, a high-lift pump will be required to force the water to the plumbing fixtures at higher elevations. In a gravity supply the source is at an elevation higher than that at which the water is to be used.

5-2. Power and Pumps.[1] Among the types of pump in use for pumping water or sewage may be included reciprocating or displacement pumps, centrifugal pumps, rotary pumps, air-lift pumps, air-displacement pumps, ejectors, jet pumps, and hydraulic rams.

Power for small water supplies is developed by hand, windmill, falling water, gasoline or its equivalent gas, and sometimes hot air. Steam is seldom used. Electricity is in common use for the transmission of power to electric motors for driving pumps.

5-3. Power Required for Pumping. The horsepower of the engine or motor required to drive a pump can be computed from the expression

$$\text{hp} = \frac{LQ \times 8.3}{33{,}000E}$$

where hp = horsepower to be delivered by the motor to the pump

L = total lift for the pump, ft. The lift is to be computed as explained in Sec. 5-4

Q = rate of discharge from the pump, gpm

E = product of the mechanical efficiency of the pump and the belt, gearing, or shafting by which the power is transmitted from motor to pump

The efficiencies of various forms of power transmission are stated in Sec. 5-29.

5-4. Total Lift. The total lift of a pump is equal to the sum of the pressure head below atmospheric shown in the suction gage, the pressure

[1] See also T. G. Hicks, "Pump Selection and Application," McGraw-Hill Book Company, Inc., New York, 1957.

head shown on the discharge gage, the friction head losses through the piping and the pump, and the difference in elevation between the suction and discharge gages. The total lift is equal to the vertical height to which the water is lifted plus friction losses and velocity heads. The suction lift, discharge head, and total head are indicated in Fig. 5-1.

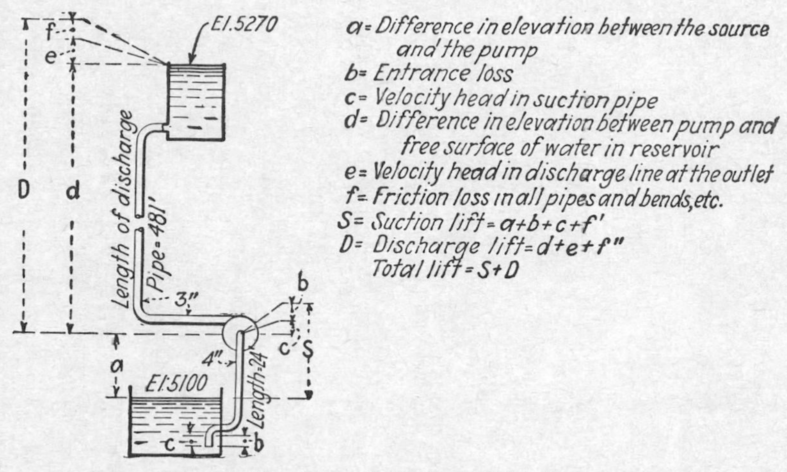

FIG. 5-1. Suction and discharge lifts of a pump.

Suction lift is discussed in Sec. 2-13. Head losses due to friction in pipes and fittings are discussed in Sec. 2-15. The following is an example of the determination of total lift.

Example. Determine the total lift for the pump shown in Fig. 5-1 if the pipe is of steel with a Hazen and Williams coefficient of 140 and the rate of pumping is 100 gpm.

Solution: The equivalent rate of flow, applicable in Fig. 2-6, must be $^{100}/_{140}$ or 78 gpm

The friction loss in feet per 1,000 ft is, from Fig. 2-6, 30 ft

Then head loss in discharge pipe is 30×0.481 or 14.4 ft

Head loss due to friction in suction pipe is 8×0.024 or 0.2 ft

Friction in bends in suction pipe, from Table 2-10, is $3 \times 0.043 \times 2.3$ or 0.3 ft

Friction in bends in discharge pipe, from Table 2-10, is $3 \times 0.128 \times 2.3$ or 0.9 ft

Friction in open gate valve, from Fig. 2-14, is $(30 \times 1.8)/1{,}000$ or negligible

Velocity head in suction pipe is $3.75^2/64$ or 0.1 ft

Velocity head in discharge pipe is $4.7^2/64$ or 0.3 ft

The total lift is, therefore,

$$(5{,}270 - 5{,}100) + 14.4 + 0.2 + 0.3 + 0.9 + 0.1 + 0.3 = 186.2 \text{ ft}$$

5-5. Reciprocating Pumps. The pitcher pumps and the cylinder pumps, shown in Figs. 5-2 and 5-3, illustrate two types of small, hand-driven, single-stroke suction pumps. They are suitable for use in cisterns and dug wells where there is but little or no suction lift and the elevation of the discharge is not higher than the point of application of power. A single-acting force pump that will lift water by suction and discharge it at an elevation above the pump is shown in Fig. 5-2. Such single-acting pumps must be primed at each use unless the moving plunger or diaphragm, shown at *A* in the figure, fits snugly or is submerged as shown in the nonfreeze cylinder pump. In this pump the cylinder is placed below the frost line, and water in the pump barrel drains back into the cistern or well.

Single-stroke or single-acting pumps will deliver water on one stroke only, usually the upstroke of the moving valve in the pump. This results in an intermittent flow of water from the pump and an uneven application

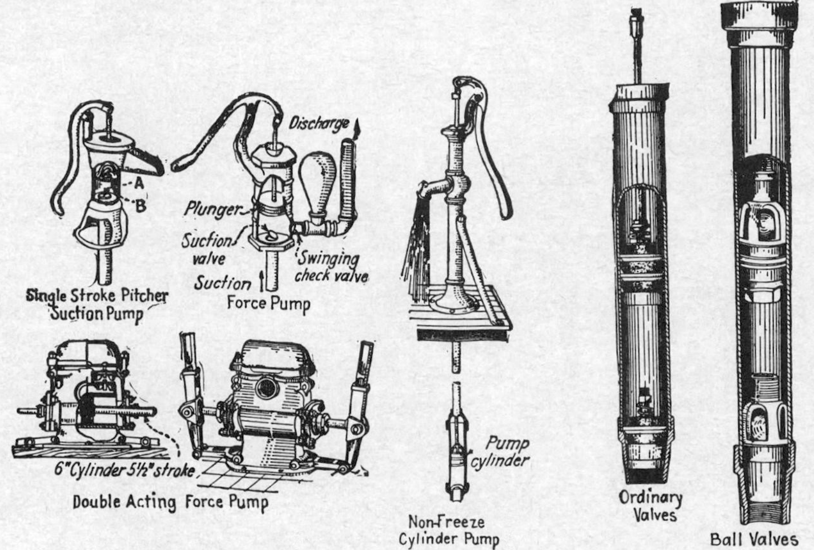

Fig. 5-2. Types of hand-operated pumps.

Fig. 5-3. Details of pump cylinders.

of power during one revolution. In a double-acting pump, water is drawn into and discharged from the pump on both the upstroke and the downstroke. A double-acting or two-stroke well pump is described in Sec. 5-7. A double-acting force pump is shown in Fig. 5-2.

The pitcher pump shown in Fig. 5-2 operates as follows: When the handle is pressed down, the plunger *A* rises; the plunger fits so closely to the walls of the cylinder that a vacuum is created in the chamber below, causing the lower valve *B* to open to admit water. The rise of plunger *A* also forces water out of the top of the pump. As plunger *A* descends, valve *B* closes, preventing the escape of water into the well and leaving the barrel of the pump filled with water, some of which will be discharged on the next upstroke of plunger *A*. The plunger *A* is shown descending in the figure.

In the maintenance of these pumps difficulty is sometimes caused by the wearing of the packing on the moving plunger, by leakage through the valves, or by the catching of some object in the pump so as to hold the valve open. When the packing becomes worn or the valves leak, the pumps will not prime themselves. They can be primed by filling the pump barrel with water so as to cover the moving plunger, or the packing and valves can be repaired to make the pumps self-priming.

A smoother rate of discharge than will be given by a single-acting pump can be obtained by the use of an air chamber on the discharge line, a differential plunger pump, a two-stroke pump, or a pump with a double-acting cylinder. Such types of pump are more commonly power-driven. In the design of power-driven double-acting pumps the best efficiency is obtained when the work on the upstroke and the work on the down-stroke are the same.

Capacities of hand-operated reciprocating pumps cannot be expected to exceed about 5 gpm, reducing as the lift increases to a practicable limit of about 100 ft. Power-driven reciprocating pumps are available in large capacities, with discharge pressures limited only by the strength of the parts of the pump and the power available. If the discharge valve of a reciprocating pump is closed during operation, pressure relief should be provided to avoid damage to the pump or motor.

5-6. Air Chambers. An air chamber can be used on the suction pipe or on the discharge pipe of a reciprocating pump to diminish fluctuations in the rate of flow and to minimize water hammer. An air chamber is shown on the discharge pipe of the small force pump in Fig. 5-2. When water is discharged from the pump, air is compressed in the chamber on the discharge pipe. When the discharge valve closes, as on the return stroke of a single-acting pump, or the rate of discharge slackens, the compressed air in the chamber expands and expels water that has previously entered the chamber, thus aiding in the maintenance of a smoother rate of discharge and cushioning water hammer. The capacity of the air chamber should be about three times the volume of water discharged on the upstroke of the pump. Air must occasionally be supplied to the chambers as it dissolves in or is carried away by the water. Air can be admitted to the chambers by draining them when the pump is not operating or by opening a small valve on the suction side of the pump when the pump is operating. The air admitted in this manner is caught in the air chambers.

5-7. Two-stroke Well Pumps. A two-stroke well pump is illustrated in Fig. 5-4. It consists of two single-stroke pumps, one within the other, with independent rods. It operates as follows: Pump rod D is solid and moves within the hollow rod C. The rods C and D move in opposite directions, one moving down as the other moves up. As D moves up, valve *B* is closed and the water in chamber *E* is discharged through valve *A*, one-half of it being discharged from the pump. As rod *C* moves up, the water that has previously passed through valve *A* is discharged from the pump and the chamber *E* is refilled through valve *B*.

5-8. Differential Plunger Pump. A differential plunger pump is shown in Fig. 5-5. In its operation, water is discharged on both the upstroke and

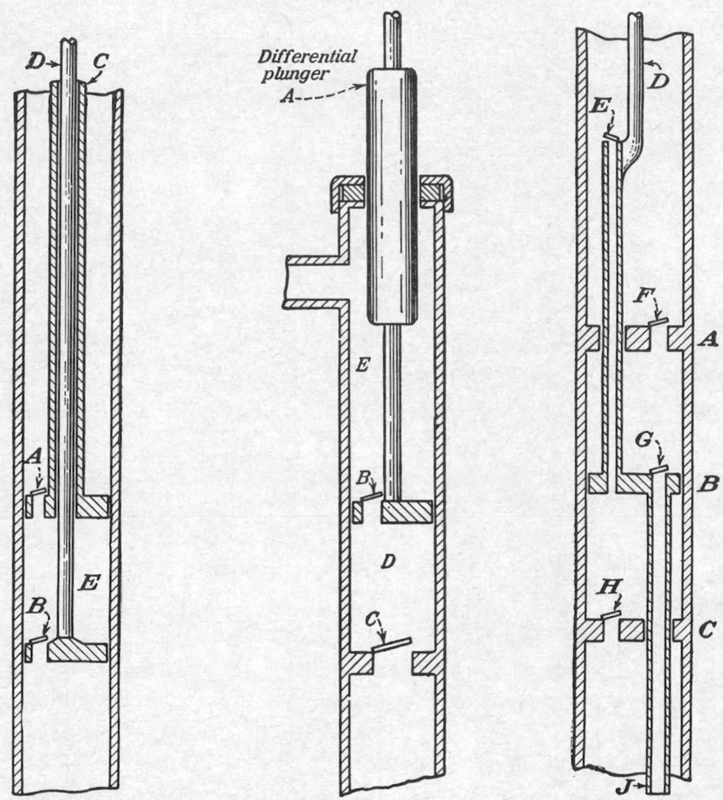

FIG. 5-4. Two-stroke deep-well reciprocating pump.

FIG. 5-5. Differential-plunger deep-well reciprocating pump.

FIG. 5-6. Double-acting-cylinder deep-well reciprocating pump.

the downstroke of the pump, but it is sucked in on the upstroke only. It operates as follows: As the differential plunger *A* and the valve *B* move upward, water is drawn into chamber *D* through valve *C*. Valve *B* is closed, and some of the water in chamber *E* is discharged from the pump. It is not all discharged, because some must remain to fill the space previously occupied by plunger *A*. On the downstroke, valve *C* is closed and there is no intake of water. The water in chamber *D* passes through valve *B* into chamber *E* from which a portion of it is forced by the entering plunger *A*. For smooth operation the size of the differential plunger should be such that the work done on the upstroke equals that done on the downstroke.

5-9. Double-acting Cylinder Pump. A section through a double-acting cylinder pump is shown in Fig. 5-6. *A* and *C* represent fixed partitions across the working barrel. *F* and *H* are check valves in these partitions. *B* is a moving partition, or plunger, attached to the pump rod *D*. *E* and *G* are check valves over the channels passing through the pump rod. On the downstroke *F* and *H* are closed; *E* and *G* are open. Water is being drawn into the pump through *J* and is being discharged through *E*. On the upstroke, *E* and *G* are closed; *F* and *H* are open. Water is being drawn into the pump through *H* and is being discharged at *F*. A disadvantage of the double-acting pump becomes evident when long rods are used. The thrust of the downstroke tends to bend the rods against the sides of the well. It has an advantage over the differential plunger pump, however, in that water is both taken in and discharged on both strokes of the pump.

5-10. Capacities of Reciprocating Well Pumps. Power-driven, reciprocating well pumps are not often used because of superior characteristics of centrifugal, jet, and air-lift pumps. Small, reciprocating well pumps are used because of availability, relatively low first cost, and suitability for hand operation. Some capacities of reciprocating well pumps are listed in Table 3-14.

5-11. Air-lift Deep-well Pumps. Air-lift pumps are used principally in wells with capacities greater than 25 gpm. In general, the greater the capacity of an air lift, the greater its efficiency. Air-lift pumps have the advantage over all other types of pump, except possibly the deep-well jet pump, in the simplicity of their parts and in their freedom from moving parts in the well. They are able to discharge more water from a well than any other type of pump that can be placed in the well, provided that there is water available in the ground. They give long service, low maintenance cost, and reliability. Their efficiency is low, however, and they usually necessitate digging the well deeper than would otherwise be needed because of the submergence required at the end of the air pipe. If the water must be raised to an appreciable height above the ground surface, additional pumping equipment is desirable at the surface, since the air lift is not suitable for the discharge of water under pressure. The aeration of the water may increase its corrosiveness, or it may be advantageous by precipitating dissolved minerals, such as iron, from the water. The efficiencies of air lifts are in the neighborhood of 25 to 33 per cent, the higher efficiencies being secured at the lower lifts.

The use of an air lift should be limited to conditions where efficiency can be sacrificed for the sake of low first cost and reduction in maintenance expense, where it is desired to increase the output of a well, where greater reliability is desired than could be obtained through some other available equipment, where the well is too crooked to permit the use of rods or shafting in the well, or where the water must be lifted more than about 200 ft in the well.

A section through an air lift is shown in Fig. 5-7. Compressed air enters the bottom of the eductor pipe through a diffuser apparatus called the *footpiece*. The water is lifted from the well because the weight of the column of the mixture of air and water in the well is balanced by the weight of the "solid" column of water outside the well. Piston or plunger action of the rising air bubbles in moving the water in the well is negligible. In order to lift water from a well there must be a minimum height of water column outside the well, or in other words, the footpiece must be deeply submerged. The ratio of the depth of the water D outside the well to the total lift, $D + h$, is called the *ratio of submergence* or, more commonly, the *submergence*. The submergency should be measured when the pump is in operation, since it will not be the same when the pump is idle. The level of the water in a well during operation cannot be predicted accurately. The lift and required submergence are assumed, and adjustments are made after the well has been tested. The ratio of $D/(D + h)$, as recommended by the Sullivan Machinery Company, is given in Table 5-1.

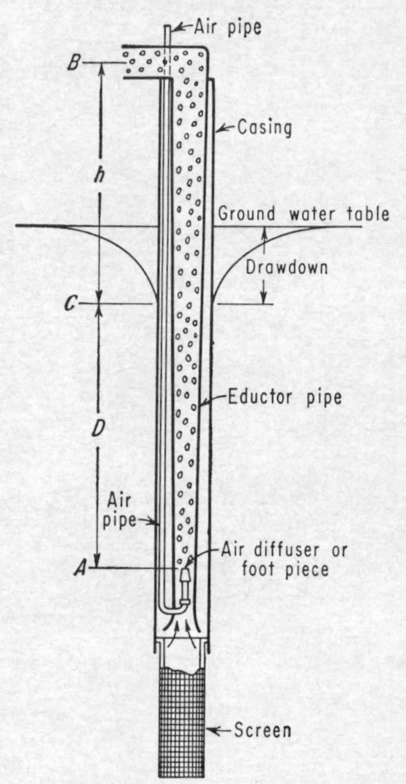

FIG. 5-7. Diagram of an air-lift pump.

The sizes of the parts of the Pohle air lift are given in Table 5-2. Data furnished by the American Steam Pump Company on air consumption by air lifts are given in Table 5-3. Other data concerning air-lift operation are given in Table 5-4.

The equipment of an air lift includes the air compressor, an air receiver for storing compressed air to make the operation of the compressor and air lift smoother, the necessary air piping, and a footpiece to distribute the air in small bubbles at the bottom of the eductor pipe. By the addi-

TABLE 5-1. SUBMERGENCE OF AIR-LIFT PUMPS
(*Sullivan Machinery Company*)

Lift, ft	Submergence, per cent	Lift, ft	Submergence, per cent	Lift, ft	Submergence, per cent
Up to 50	70–66	100–200	55–60	300–400	43–40
50–100	66–55	200–300	50–43	400–500	40–33

TABLE 5-2. POHLE SIDE-INLET AIR LIFT, CAPACITIES AND PIPE SIZES

Air pipe, in.	Water pipe, in.	Size of well, in.	Capacity, gpm*	Approximate free air, cfm			Air pipe, in.	Water pipe, in.	Size of well, in.	Capacity, gpm*	Approximate free air, cfm		
				Lift, ft							Lift, ft		
				50	100	200					50	100	200
½	1	3	7	2.3	4.5	7.7	1½	3½	7	120	39	78	132
¾	1½	4	20	6.5	13	22	1½	4	8	160	52	104	177
1	2	4½	35	11	23	29	1½	5	9	250	81	162	276
1	2½	5	60	20	39	66	2	6	10	350	115	226	386
1¼	3	6	90	29	58	99							

* Maximum for ordinary lifts.

TABLE 5-3. APPROXIMATE AMOUNT OF FREE AIR REQUIRED TO ELEVATE WATER BY AIR LIFT
(Based on submergence of 60 per cent. *American Steam Pump Company*)

Lift, ft	20	30	40	50	60	80	100	120	140	160	180	200	250
Air pressure, psi	13.5	20	27	34	40.5	54	67.5	81	94.5	108	121.5	135	168
Cubic feet free air per gpm	0.310	0.350	0.387	0.422	0.457	0.522	0.585	0.642	0.697	0.755	0.810	0.862	0.988

TABLE 5-4. AIR-LIFT EQUIPMENT

Air requirements					Sizes of pipe		
Lift, ft	Submergence, ft	Free air*	Compressed air*	Air pressure, psi	Flow, gpm	Eductor pipe, in.	Air pipe, in.
20	47	0.26	0.197	10.8	75–100	3½	1¼
50	100	0.338	0.244	23.1	100–150	4	1½
100	122	0.563	0.308	28.1	150–250	5	2
200	200	0.86	0.366	46.1	250–375	6	2
					375–650	8	2½
					775–1000	10	2½

* Cubic feet per minute of air per gallon per minute of water.

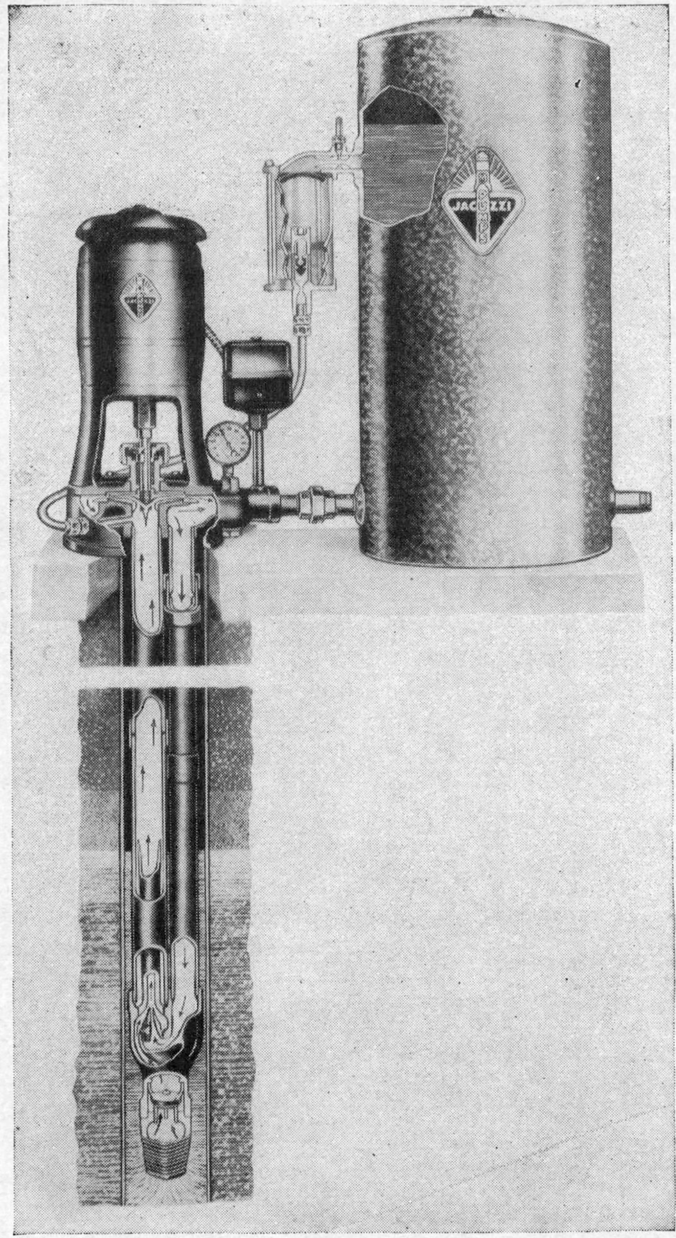

FIG. 5-8. Typical cross section of two-pipe jet pump with pump set directly over the well. (*Jacuzzi Bros., Inc.*)

tion of a *booster* on the upper end of the eductor pipe, the air is separated from the water, and the pressure of the air and the kinetic energy of the high velocity of the water issuing from the well are converted into pressure to raise the water to a higher elevation. The compressed air is released from the booster through a valve, which is throttled by trial, to obtain the best operating conditions.

Where a reciprocating air compressor is used, an air chamber with a capacity of about fifteen times the piston displacement should be placed between the compressor and the well, or two chambers may be used on a long air pipe, one relatively small at the compressor and the larger near the well. This serves to collect moisture and to aid in dissipating heat due to the compression of the air and in maintaining a smooth flow of air to the well. Water should be drained from the air chamber periodically. All compressed-air pipes and equipment should be protected against freezing temperatures because of the moisture that accumulates in them.

5-12. Jet Well Pumps. A jet well pump is illustrated in Fig. 5-8. It consists of a centrifugal pump and motor at the ground surface and a jet below the water in the well. The centrifugal pump has two discharge pipes. One discharge pipe sends water down the well to the jet, through which it is discharged at a high velocity. This serves to draw additional water from the ground surrounding the well and to mix it with the water leaving the jet. This mixture, of greater volume than that entering the jet, now rises from the well and enters the suction of the centrifugal pump. The other discharge pipe from the centrifugal pump sends water to the point of discharge from the well. The rate of discharge through the discharge pipe must, of course, equal the rate at which water is drawn into the well.

Jet well pumps are widely used in low-capacity installations. They are available up to about 70 gpm at a 30-ft lift to about 20 gpm at a lift of about 150 ft. Common horsepower ratings are 3/4 and 1 for two impellers, giving about 60 psi, and 1 1/3 hp or over for three-stage jet, giving 90 psi.

Jet pumps for draining sumps are discussed in Sec. 13-7.

5-13. Centrifugal Pumps.[1] Centrifugal pumps are suitable for the pumping of either water or sewage. They have many advantages over other types of pumps, among which may be included

1. Simplicity of parts, installation, and operation
2. Low first cost and low cost of operation
3. Small space requirements
4. Silent and pulseless discharge
5. Satisfactory efficiencies

[1] See also A. J. Stepanoff, "Centrifugal and Axial Flow Pumps," John Wiley & Sons, Inc., New York, 1948.

6. Adaptability to be driven by electric motor or internal-combustion engine

7. Availability in a wide range of characteristics

Centrifugal pumps are not self-priming without special design or equipment, they are not suited to operation under suction lift, and there are limitations to their performance that require the selection of a pump for the conditions under which it is to be operated. The highest efficiency of almost any centrifugal pump is limited to a rather narrow range of discharge rate and pressure, as explained in Sec. 5-15. The advantages of centrifugal pumps so greatly outweigh their disadvantages that they are used almost exclusively where their limitations will permit.

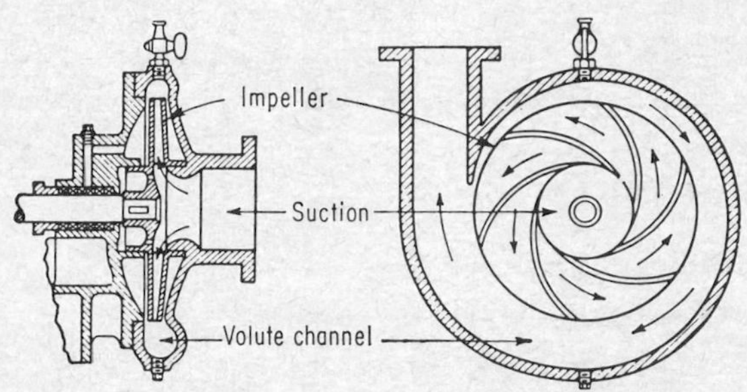

Fig. 5-9. Volute-type centrifugal pump.

Parts of a centrifugal pump are illustrated in Fig. 5-12. In operation, only the pulley (9), the shaft (8), and the runner (5), all rigidly attached to each other, are in motion. The water enters the pump casing at the center, point (2) in the figure, where it encounters the rapidly revolving vanes of the runner or impeller. The centrifugal force thus brought into action forces the water rapidly and with increasing velocity to the periphery of the revolving vanes, called *impeller blades.* Here the velocity of the water is reduced in the discharge channels and the velocity energy is changed to pressure. The blades of the impeller do not push the water from the pump as the blades of a paddle wheel push a boat.

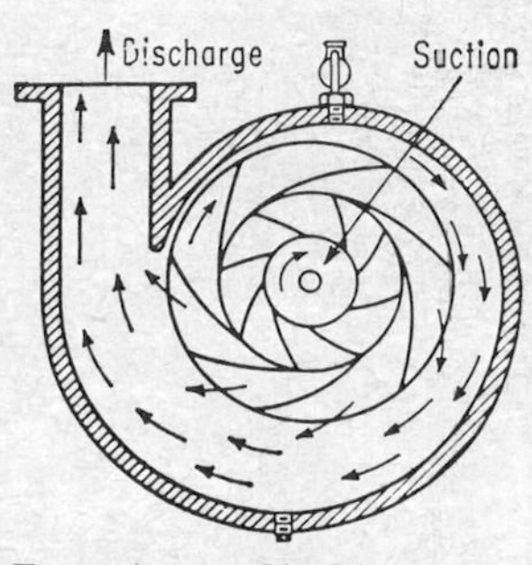

Fig. 5-10. Turbine-type diffuser-vane centrifugal pump.

The two types of centrifugal pump are the *volute pump* and the *turbine pump.* Each is shown in Figs. 5-9 and 5-10. In a volute pump the water leaving the impeller is discharged into a volute or snail-shell-shaped casing, the cross-sectional area of which increases directly as the rate of water flowing through it. A constant velocity of discharge is thus maintained. In a turbine pump the casing of the pump is circular, and the impeller is surrounded by diffuser vanes forming water passages that gradually enlarge, slowing down the velocity of the water and converting the velocity energy into pressure. The volute pump is simpler, lower in first cost, less subject to clogging, and better suited to small sizes. The turbine pump may be more efficient, particularly in the larger sizes. It is not suitable for pumping sewage.

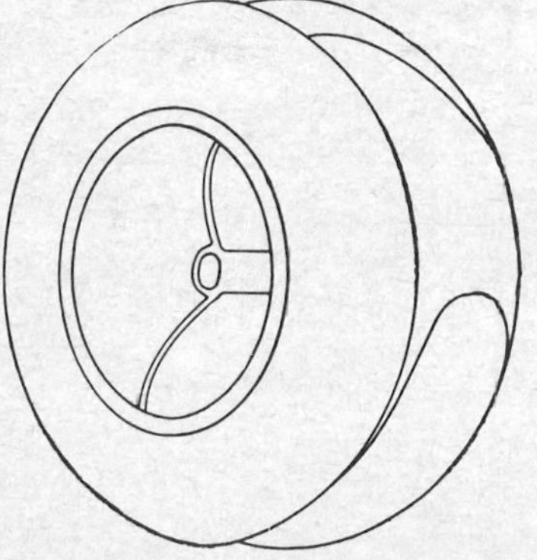

FIG. 5-11. Enclosed impeller for a centrifugal pump.

The return of water from discharge to suction parts of the pump by passing between an open impeller and the casing, sometimes called *slip* or *slippage*, can be diminished by enclosing the impeller in side plates, called *shrouds*, as shown in Fig. 5-11. An open impeller is shown at 5 in Fig. 5-12. Packing rings or labyrinth packing rings, as shown in the figure, may be used between the periphery of the shroud and the casing to diminish slippage. To prevent the escape of water or air from or into the pump, a packing gland is needed, as shown also in Fig. 5-12. In

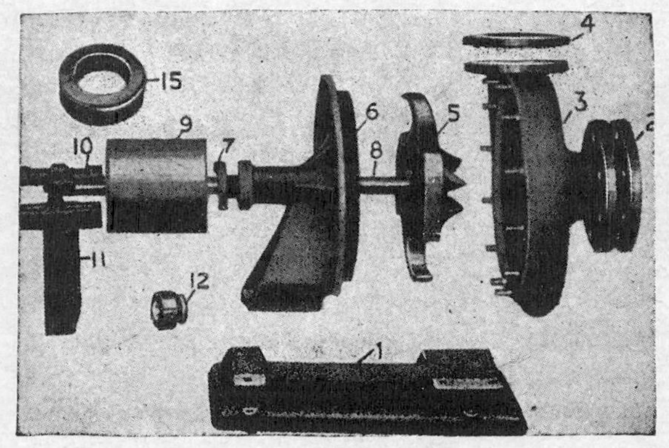

1. Base
2. Suction companion flange
3. Volute with studs
4. Discharge companion flange
5. Runner
6. Cover
7. Gland
8. Shaft
9. Pulley
10. Box cap
11. Bracket box
12. Set collar for type *A* or *B*
15. Horizontal ball bearing

FIG. 5-12. Parts of a centrifugal pump. (*American Well Works.*)

operation, the packing is tightened or loosened by means of the nuts shown. Since this packing should fit loosely, the passage of air past it may be prevented by directing a continuous small stream of water on it during operation. A pet cock or other small opening should be provided at the top of the pump casing to permit the draw-off of air that has accumulated in it.

Pumps with enclosed impellers, that is, with shrouds, may be more efficient than pumps with open impellers, but they are less suited to pumping liquids containing solids, such as sewage.

The capacities of centrifugal pumps, other things being the same, vary directly as the number of impellers operating on the same shaft in parallel. The discharge pressure varies as the number of impellers operated at the same speed in series, the discharge from one impeller entering the suction of the next higher stage. Pumps with two or more impellers on the same shaft and in series, that is, with one impeller discharging into the suction of the next stage, are known as *multistage pumps*. The dimensions, speed of revolution, and other conditions affecting a centrifugal pump are determined by the designer. No simple rule is available by which a purchaser can determine the probable performance of a pump from its over-all dimensions. The capacity of a pump can be estimated by multiplying the cross-sectional area of the discharge pipe by the assumed velocity of flow therein, usually about 10 fps.

The efficiencies of centrifugal pumps in domestic water-supply service are relatively low, as indicated in Table 5-5. Unfortunately, even such low efficiencies are not attained unless the pump is designed for the particular service to which it is put.

TABLE 5-5. SIZES, CAPACITIES, AND EFFICIENCIES OF CENTRIFUGAL PUMPS

Size of pump, in.	Capacity, gpm		Efficiency	Size of pump, in.	Capacity, gpm		Efficiency
	10-ft velocity	12-ft velocity			10-ft velocity	12-ft velocity	
1	25	...	27	5	612	734	59
1½	55	...	35	6	881	1,058	62
2	98	...	43	8	1,567	1,880	65
3	220	264	50	10	2,448	2,938	67
4	392	470	55	12	3,525	4,230	69

5-14. Information Required from Pump Purchasers. Information is usually required by manufacturers from pump purchasers, such as (1) the capacity of the pump; (2) characteristics of the suction head, the discharge head, and the head-discharge relations under which the pump is to be operated; (3) type or types of power to be used; (4) space limitations; (5) foundation or setting conditions; (6) efficiency limitations; (7) characteristics of the fluid to be pumped; (8) is the pump to be right-handed or left-handed; (9) double or single suction; and (10) other information affecting the operation or performance of the pump.

5-15. Characteristics and Capacities of Centrifugal Pumps.[1] The limitations or characteristics of a centrifugal pump should be understood before it is selected for any particular service, and these characteristics should be suited to the service for which the pump is selected. The hydraulic characteristics may be expressed as a series of curves or graphs whose abscissas are the rate of discharge Q from the pump and whose ordinates, corresponding to these abscissas, are, respectively, the discharge pressure or head H at a known speed of operation, the power P required to drive the pump, and the efficiency E. The graph showing the relation

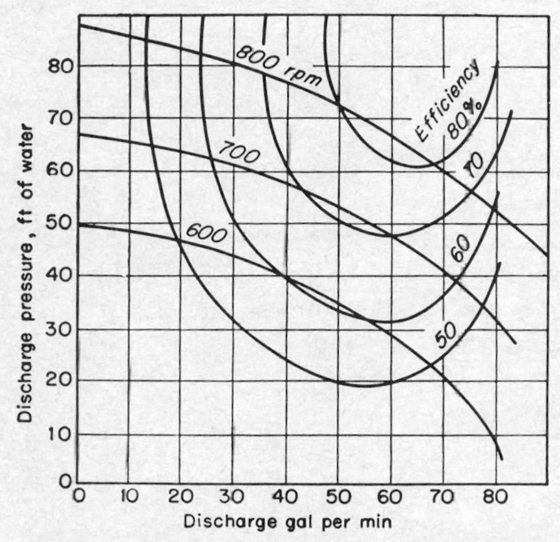

FIG. 5-13. Characteristics of a centrifugal pump.

between Q and H at a constant speed is sometimes referred to as *the characteristic.* Characteristics of centrifugal pumps are shown in Figs. 5-13 and 5-14. A study of Fig. 5-13 shows that for the highest efficiency the pump should be operated at about 800 rpm with a discharge rate of 60 gpm and a discharge pressure of 67 ft. A pump with a flat characteristic and a wide plateau at the highest efficiency is most desirable for varied conditions of operation.

The fact that there is a limit to the maximum discharge pressure that can be developed at any particular speed, even with the discharge valve shut, is an important safety protection in the operation of a centrifugal pump. So long as the pump is full of water, no damage can occur to the pump or motor through excess pressure when operating with the discharge

[1] See also I. J. Karissik and R. Carter, *Air Conditioning, Heating, and Ventilating,* May. 1956, p. 77.

valve closed. As the discharge valve is progressively opened beyond maximum discharge pressure, the rate of discharge increases, the discharge pressure falls, and the power required increases to a maximum, after which it falls off. The increase in the power required as the discharge valve is opened offers a hazard in the operation of a centrifugal pump caused by the overloading of the motor. An important limitation of a centrifugal pump is that at any particular speed there is only one rate of discharge that can be obtained for one discharge pressure. A change in the rate of discharge at a given discharge pressure or a different discharge pressure for a given rate of flow can be obtained by varying the

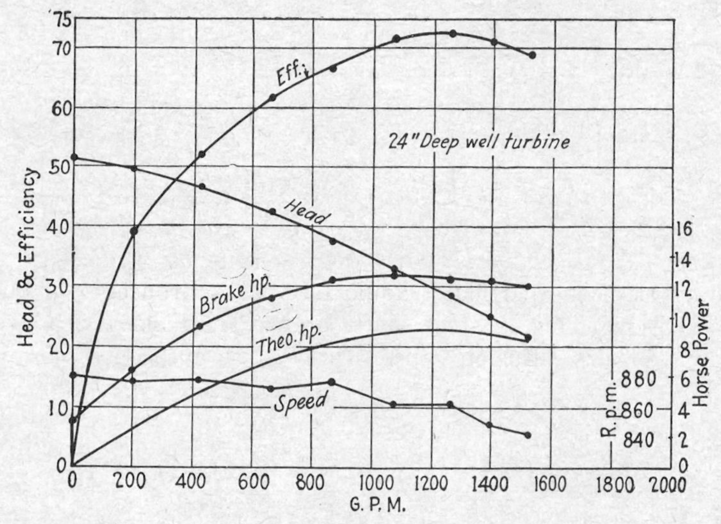

FIG. 5-14. Characteristics of a centrifugal pump.

speed. For example, in Fig. 5-13 the rate of discharge at a total discharge pressure of 60 ft will be increased from 34 to 70 gpm if the speed is increased from 700 to 800 rpm, or the total lift, at a rate of discharge of 50 gpm will be decreased from 72 to 35 ft by reducing the speed from 800 to 600 rpm.

Approximate or "rule-of-thumb" relations among speed, discharge pressure, rate of discharge, and power are, when speed N varies,

Q varies directly as N
H varies directly as N^2
P varies directly as N^3
Q varies directly as $H^{1/2}$
P varies directly as $H^{3/2}$

Knowledge of these approximate relations is useful for estimating the effect of changing the speed of operation in a pump. Small centrifugal pumps for water-supply purposes normally operate at speeds less than 1,800 rpm, with the highest efficiencies being found at the lower speeds.

Centrifugal pumps are manufactured in all capacities from less than 1 gpm up to thousands of gallons per minute and in discharge heads, up to thousands of feet, limited only by the strength of the materials involved. They are usually limited to a maximum lift of about 300 ft per stage and to not more than six stages in one pump.

5-16. Efficiencies of Centrifugal Pumps. The efficiency of a centrifugal pump under various conditions is important in its selection. If it is to be operated continuously, it may be economical to pay a somewhat higher first price for a pump of higher efficiency. Small-sized stock pumps may be expected to give efficiencies between 50 and 70 per cent; larger sized pumps are reported to have efficiencies up to 90 per cent.

5-17. Materials Used in Centrifugal Pumps. Standard-fitted centrifugal pumps are equipped with a cast-iron casing, steel shaft, bronze shaft sleeves (if any), a cast-iron impeller, and cast-iron diffusion vanes (if any). Such pumps are suitable for ordinary service when moving noncorrosive liquids. Pumps handling sewage or corrosive liquids should be bronze-fitted or all-bronze and should be made of special composition to resist the effect of the particular liquid pumped. Bronze-fitted pumps have a cast-iron casing, a steel shaft, bronze shaft sleeves, a bronze impeller, and bronze diffusion vanes (if any). In all-bronze pumps, all parts that are in contact with the liquid are made of bronze, except that a bronze-covered steel shaft or Monel-metal shaft may be used.

5-18. Power for Driving Centrifugal Pumps. Because of the relatively high speed of operation of most small centrifugal pumps, because of the relatively slow starting torque in relation to full-load torque required, and for other reasons, centrifugal pumps are frequently driven by electric motors belt-, gear-, or shaft-connected to the pump. Internal-combustion engines are less frequently used but are satisfactory for the purpose. Reciprocating steam engines are seldom used for driving centrifugal pumps, and steam turbines are used for only the largest pumps.

If the motor and the pump are on the same shaft, the speed of the motor and of the pump must be the same and a flexible connection must exist in the shaft. The purpose of this connection, or coupling, is to allow temperature changes and movements of the shafts without mutual interference. The coupling should not be depended on to correct for misalignment in the shafts. Where the pump and motor are connected by a belt or gear train, the two machines need not have the same speed, the gear ratio or the diameter of the pulleys being relied upon to adjust the speeds of the pump and motor. The relation between the speeds can be expressed as $N_m/N_p = D_p/D_m$, where N_m and N_p are the motor and pump speeds, respectively, and D_m and D_p are the diameters of the motor pulley and of the pump pulley, respectively. For example, if the speed of the pump is 690 rpm and that of the motor is 1,140 rpm, the diameters of the pulleys should be in the ratio of 1,140:690 = 1.65, the larger pulley being on the slower-moving machine; that is, a 3-in. pulley on the motor will call for a 5-in. pulley on the pump under the above conditions.

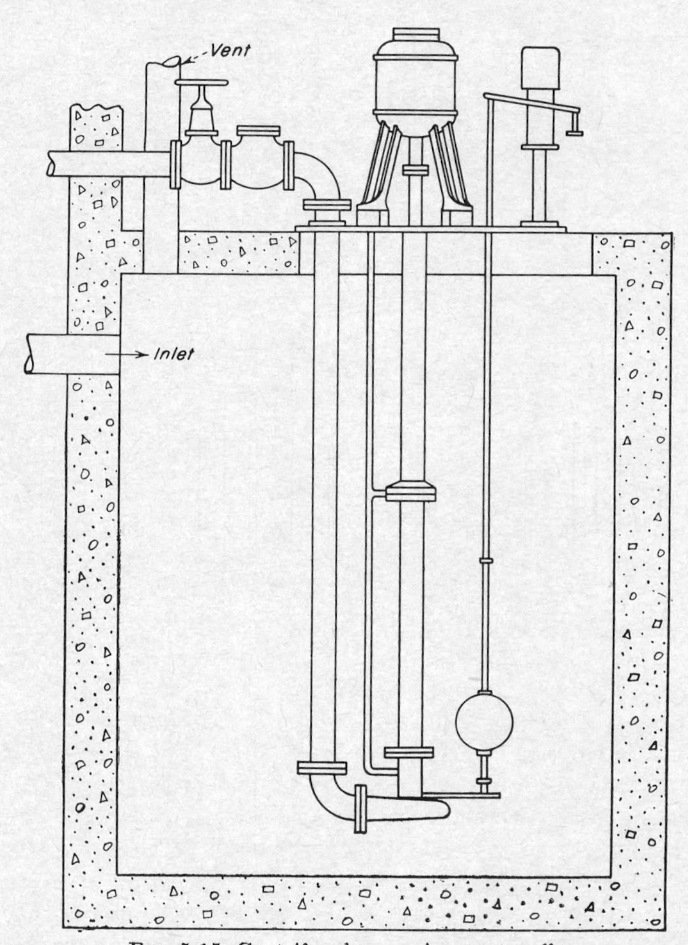

FIG. 5-15. Centrifugal pump in a wet well.

5-19. Setting of Centrifugal Pumps. In the installation of a centrifugal pump, conditions to be considered include accessibility without hazards from or inconvenience to traffic in the building and adequate space, illumination, and ventilation to permit inspection, maintenance, and repairs. Less obvious conditions include suction lift, which should be avoided or reduced to a minimum; alignment of pump and motor on the same shaft when direct drive is used; adjustability of pump and motor on foundation to allow adjustment of alignment, belt tension, or other power drive; rigid foundation to hold alignment of pump and motor or rigid support to maintain their relative positions; and separation of foundation of heavy equipment from the building structure to minimize or avoid vibration.

Suction lift may sometimes be avoided by submerging the pump in the wet well, somewhat as shown in Fig. 5-15. A pump should be so installed

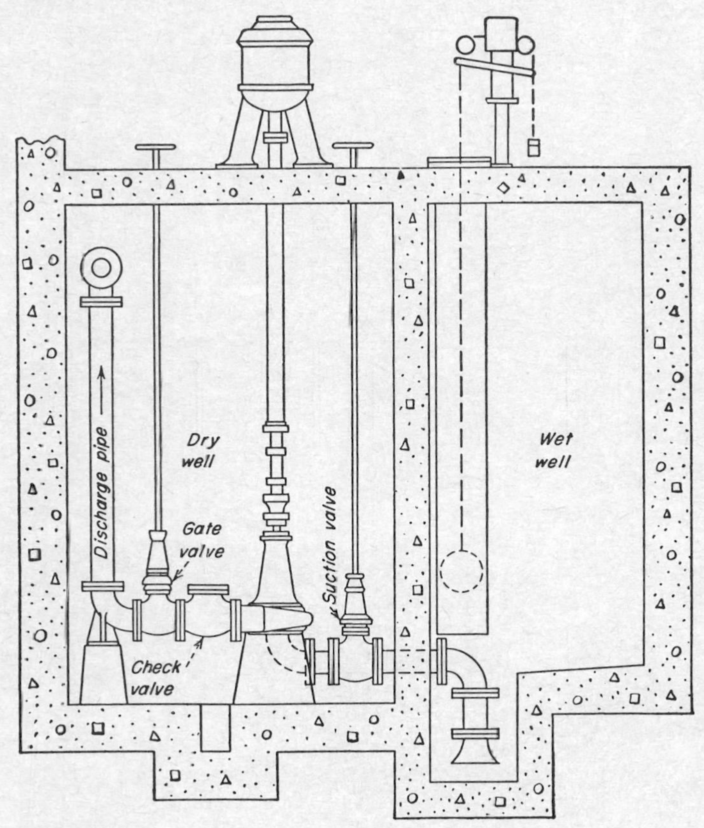

FIG. 5-16. Centrifugal pump in a dry well. (*American Well Works.*)

only when it can be easily removed for maintenance purposes. Such an installation is restricted to relatively light installations in which the pump, motor, and frame constitute a unit. Another method of avoiding suction lift is to place the pump in a dry well at an elevation below the water level in an adjoining wet well, the pump suction pipe being under pressure and leading through the wall separating the two wells, somewhat as shown in Fig. 5-16. The use of a vertical shaft makes it unnecessary to place a directly connected electric motor in the dry well where dampness, illumination, accessibility, and ventilation may be unsuited to good maintenance. A long, vertical shaft between pump and motor is undesirable because of the difficulties with bearings and in maintaining alignment and lubrication.

Heavy pumps and motors may be supported on a concrete foundation independently of the building structure. The equipment is fastened to the foundation by bolts set into the concrete. Lighter equipment may be fastened directly to the building structure, possibly with provisions for avoiding the transmittal of vibration. Small pumps and motors may be supported on a timber or metal frame, somewhat as shown in Fig. 5-17, holding the pump and motor as a single piece of equipment that can be moved with relative ease.

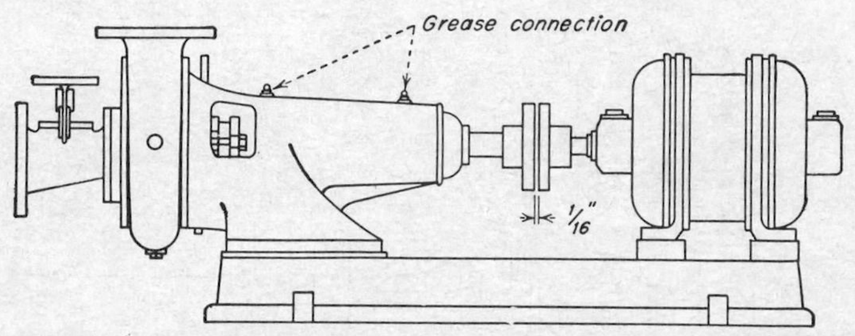

FIG. 5-17. Small centrifugal pump and motor on a common steel base.

5-20. Piping for Centrifugal Pumps. Where suction lift cannot be avoided, the suction pipe should be made as short and as straight as possible, and large enough to minimize friction losses. Velocities of less than about 2 fps are satisfactory. Since the slightest leak of air into the suction pipe may be cumulative, causing loss of prime or reduction of pump capacity, special care should be taken to assure that the suction pipe is airtight. A satisfactory installation is illustrated in Fig. 5-16. Note that in Fig. 5-17 the reducer connection on the suction is eccentric, so that the top of the suction piping is installed in such a way that no air pocket can accumulate in it. Long-sweep elbows should be used where possible. Suction pipes and discharge pipes should slope toward the suction pit or to other drainage outlets to permit the emptying of the pipes. It may be desirable to place a screen on the lower end of the inlet pipe to prevent the entrance of solid particles into the suction pipe. The area of the openings in the screen should be at least double the cross-sectional area of the suction pipe. It may be necessary to place a foot valve between the screen and the lower end of the suction pipe to permit priming of the pump.

The discharge pipe should be equipped with a gate valve and with a check valve between the gate valve and the pump, somewhat as shown in Fig. 5-16. The check valve prevents the backflow of water through the pump. Water hammer resulting from the sudden closing of the check valve may be minimized by the installation of an air chamber, by the use of a slow-closing check valve, or by other means. However, the valve should close with sufficient rapidity to prevent injurious reversal of motion in the pump and motor. The gate valve is required when it is necessary to shut off the discharge line or to throttle the pump discharge.

5-21. Pump Priming. The priming of a centrifugal pump that has suction lift can be accomplished in one of three ways: by a bypass around the discharge valve which, when open, will admit water from the discharge pipe into the pump; by admitting water into the pump from some

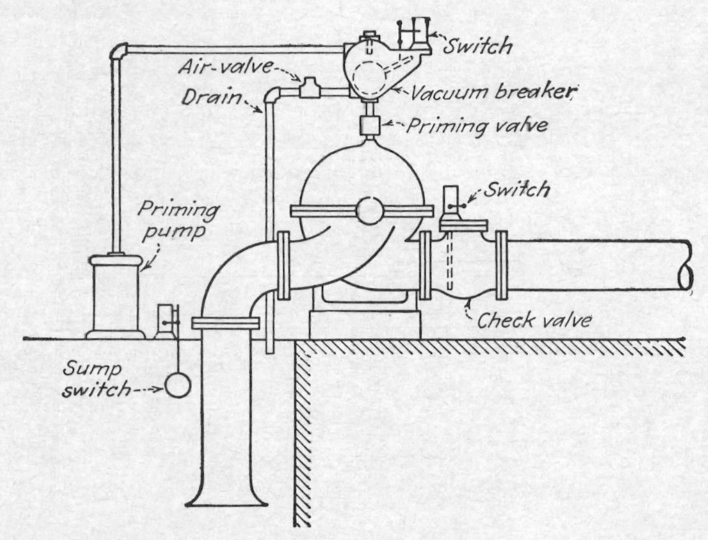

FIG. 5-18. Self-priming centrifugal pump and automatic control system.

other source than the discharge from the pump, such as the water-distribution system; and by drawing air from the suction line and the pump casing by means of a vacuum pump. It may be possible to install a successful automatic primer to function as needed.

An automatic device for priming a centrifugal pump is shown in Fig. 5-18. Note that no foot valve is required on this device. Air rising through the suction pipe may pass through the pump and vacuum breaker, ultimately escaping from the apparatus through the air-relief valve on the top.

5-22. Operation of Centrifugal Pumps. In starting a centrifugal pump the procedure depends, to some extent, on the capacity and characteristics of the pump and motor. Pumps requiring no more than 2 or 3 hp for full-load operation can usually be started under full load without special precautions. Under other circumstances it may be necessary to start the pump with the discharge valve closed, opening it slowly as the motor gains speed and picks up the load, or the motor may be brought to speed independently of the pump, which is then started, with discharge valve closed, by throwing in a clutch, after which the discharge valve is opened.

In the routine operation of centrifugal pumps, considerations include

1. *Lubrication and Cooling of the Bearings.* Lubrication may depend on grease cups, which should be kept filled and adequately turned down, or on oil cups that should also be kept filled and dropping. The bearings may bind or run hot; if so, the packing glands should be loosened sufficiently to allow water to drip from them. Cool water may be run on the packing glands continuously to cool the bearings and to aid in preventing air from being sucked into the pump.

2. *Maintenance of Prime and Avoidance of Entrained Air.* All possible points of leakage of air into the pump should be closed, and if air or gases accumulate in the pump during operation, the air-relief valve at the top of the casing should be opened. Bearings should be kept constantly wet by directing a stream of water on them.

3. *Noises.* A rattling noise that sounds like sand in the casing may indicate sand in the casing, entrapped air, or cavitation. If the admission of a small amount of air into the suction stops the noise, it was probably due to cavitation. If the release of air from the casing stops the noise, it was probably due to entrapped air. If neither expedient stops the noise, it is probably due to a solid object in the pump. Pounding or sharp knocks indicate fluttering or slamming of the check valve due to intermittent discharge resulting from periodic loss and restoration of prime.

4. *Maintenance of Delivery Pressure.* If the casing of the pump becomes warm while the pump is running and no water is discharged, it means that the discharge pressure is too high for the speed of the pump. This may be a result of stoppage in the discharge line or of operating with too high a discharge lift. If the pump remains cool while running without discharge and neither air nor water comes from the relief valve on top of the casing when the valve is opened, the pump has probably lost its prime. Corroborative evidence will be warm or hot bearings and the absence of water dripping from them (if they are normally operated loose enough to permit the escape of water).

5. *Freezing.* When the pump is shut down and is exposed to freezing conditions, it should be drained.

5-23. Turbine and Other Vertical-shaft Well Pumps. Turbine pumps of the type shown in Fig. 5-19 are widely used in wells 6 in. in diameter and larger, with the pump submerged in the well and the electric motor at the surface of the ground. In one type, less commonly used, the electric motor is submerged in the well below the pump runners. A number of runners, or stages, are usually placed in series, one above the other, to provide the required capacity within the limited space available. The selection of the speed and number of stages should be made with the aid of trained advice and, if possible, the manufacturer's guarantee of performance. Efficiencies of up to 90 per cent are reported, but less than

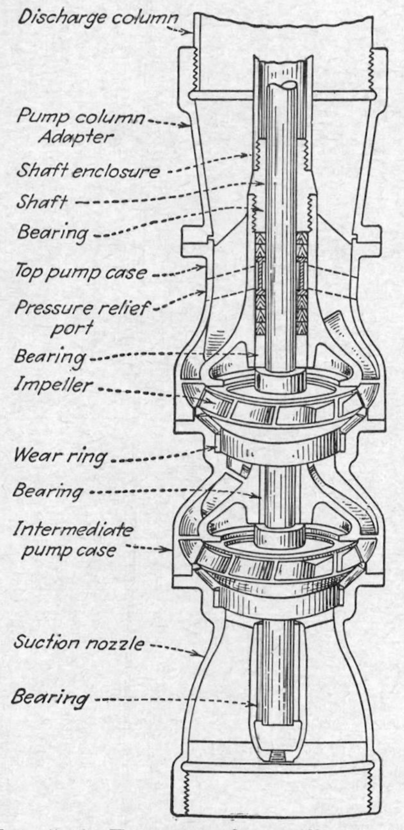

FIG. 5-19. Two-stage deep-well centrifugal pump. (*From Water Works & Sewerage, April*, 1946, *p*. R-88.)

FIG. 5-20. Multistage, deep-well turbine pump with submerged motor. (*Published through the courtesy of Aermotor Company.*)

75 to 80 per cent is to be expected in pumps with capacities of less than about 200 gpm.

Vertical-shaft pumps, which are similar in appearance to the turbine pump but which have the characteristics of a helical or axial-flow pump, are available in lifts of up to 600 ft with capacities of up to 50 gpm.

These and other types of vertical-shaft pumps can be used satisfactorily for lifting water from a well and pumping it under pressure into the distribution system or the supply pipes of a building.

5-24. Submersible Pumps.[1] A submersible pump is a pumping unit combining a centrifugal pump and an electric motor in a single container. The unit is lowered into the well with the motor below the pump, somewhat as shown in Fig. 5-20. The outstanding advantage of the submersible pump is the absence of the vertical shaft and the simplicity of the unit. However, when inspection or repairs are necessary, the well must be shut down and the pumping unit withdrawn from it. Submersible pumps are widely used in small or moderate capacities at voltages of 120 or higher.

5-25. Rotary Pumps. Rotary pumps, such as those shown in Fig. 5-21, are displacement pumps. The blades of the rotor fit tightly together and to the casing. As the blades revolve, water is displaced from the pump casing and is pushed through the discharge pipe. A high vacuum can be created in the suction pipe because of the tight-fitting parts.

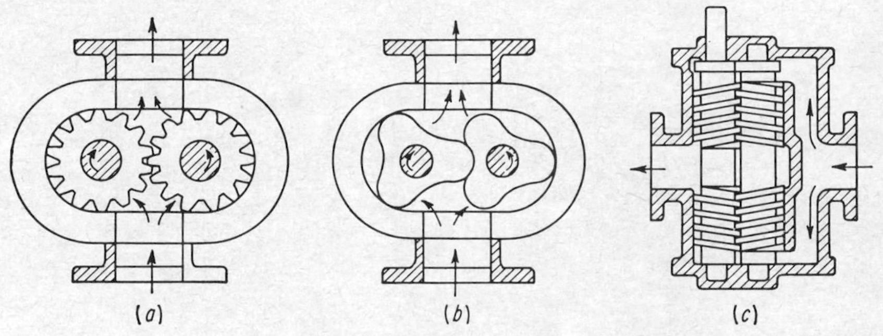

FIG. 5-21. Rotary pumps.

Although rotary pumps have no valves, the close fit of the rotor blades makes them suited only to the pumping of liquids that are free from suspended solids. They are less foolproof than centrifugal pumps because the closing of the discharge valve while the pump is in operation will result in damage to some part of the pumping equipment unless pressure relief is provided. Rotary pumps are self-priming, and no foot valve is required on the suction line. The same care in the setting of the pump is required as in the setting of a centrifugal pump, and a check valve and a gate valve are required in series on the discharge pipe.

Rotary pumps are used in plumbing mainly in small sizes on automatic and auxiliary apparatus. Their use as main water-supply pumps is not common because of their relatively low efficiency and hazardous operating characteristics. They are unsuited for pumping sewage or liquids containing abrasive sediments.

Data concerning rotary pumps are shown in Table 5-6.

[1] See also J. W. McConaghy, The Submersible Deep Well Pump, *Plumbing and Heating J.*, October, 1952, p. 90.

TABLE 5-6. DATA ON ROTARY PUMPS

Blackmer rotary pumps*												Roots' small-capacity rotary pumps†				
Pump number	Pipe size, in.	Rpm	Capacity, gpm	Suction lift, ft	Maximum pressure, psi	Pump number	Pipe size, in.	Rpm	Capacity, gpm	Suction lift, ft	Maximum pressure, psi	Displacement per revolution, gal	Rpm	Gross gpm	Horsepower‡ at maximum gpm	Suction and discharge pipes, in.
0	½	600	6	3	75	6	2½	250	100	15	100	0.06	750	45	0.3	1¼
1	¾	500	12	10	75	8	3½	200	200	18	100	0.10	600	60	0.4	1¼
2	1	500	20	10	75	10	5	175	350	20	75	0.175	475	83	0.6	1½
3	1½	400	35	12	75	12	6	150	500	20	75	0.25	450	112	0.8	2
4	2	400	50	12	100	6 spc.	2	250	100	15	40	0.50	400	200	1.3	3
												0.95	340	320	2.1	4

* From catalogue of Blackmer Rotary Pump Company, Petroskey, Mich.
† From catalogue of P. H. and F. M. Roots Company, Connersville, Ind.
‡ Based on 25-ft discharge head.

5-26. Hydraulically and Pneumatically Driven Pumps. Hydraulically driven pumps are used where two water supplies are available, one of relatively poor quality but in large quantity and the other of good quality but possible restricted quantity. This condition exists in localities where the public water supply is hard and it is desired to raise soft water from cisterns to roof or attic tanks. The pump is driven by the public hard-water supply. The hard water used for driving the pump may be run to waste, or it may be made available, at a reduced pressure, for flushing fixtures, cooling, watering the lawn, or other purposes.

The pressure and volume required to operate such pumps depend on the quantity of water and the height it is to be lifted. Under ordinary conditions the efficiencies of these pumps cannot be expected to exceed 50 per cent, and the areas of the pistons and the stroke of the pump should be designed to fit the conditions to be met in service. It is possible, with these pumps, to discharge water at a higher pressure or to a higher elevation than is available in the water supply being used for power purposes.

Pneumatically driven pumps operate on the same principle as the hydraulically driven pumps, compressed air being used instead of water to actuate the pump. Such pumps are seldom used for household supplies, since they are suitable only where compressed air is available at a low cost for the transmission of power.

5-27. Hydraulic Rams. A hydraulic ram is a pump that is actuated by water hammer created in the drive pipe. The pump can be set only in a flowing stream where a drop of 3 or 4 ft or more is available. It can be used to raise flowing water to any reasonable height, possibly as great as 600 ft above its own level, or flowing water of unsatisfactory quality can be used to lift water of satisfactory quality. Among the advantages of the hydraulic ram may be included simplicity, durability, no need for lubrication, no cost for power, and automatic and continuous operation, attention being required only occasionally to replace worn parts. Although rams are set to discharge directly into water-supply pipes, vibrations and pulsating flow can be avoided by placing a closed air chamber on the discharge pipe or by discharging into a storage reservoir, elevated or pneumatic, with provision made for overflow. The rate of discharge from a ram can be controlled between limits, after a ram has

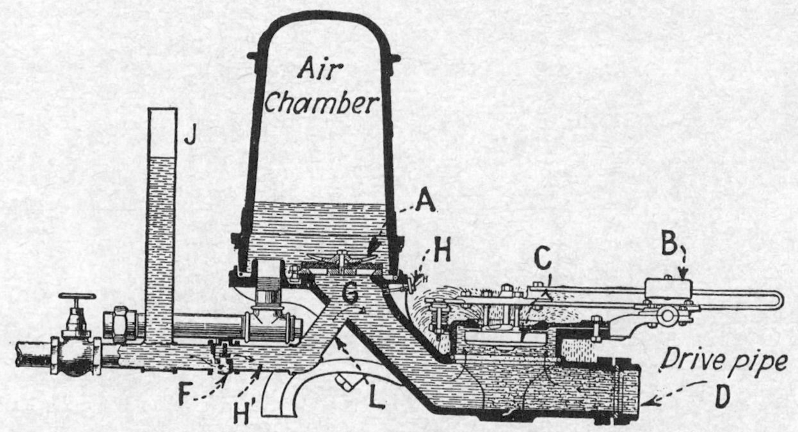

FIG. 5-22. Rife double-acting hydraulic ram.

been set, by adjusting the cycle rate. The cycle, or number of strokes per minute, can be varied between about 40 and 300; the more rapid the strokes, the lower is the rate of discharge and the less the amount of water wasted in the operation of the ram.

A hydraulic ram is illustrated in Fig. 5-22. When operating to lift the same quality of water that is used in the drive pipe, pipe *L* in Fig. 5-22 is not used. No water enters at *G*. Under these conditions it operates as follows: Water enters the ram through the drive pipe and wastes out of the ram through valve *C*. The water gains such a velocity that valve *C* suddenly slams shut. The sudden closing of the valve creates sufficient pressure through water hammer to force valve *A* open, thus discharging a small quantity of water through discharge valve *A*. The air chamber on the discharge pipe prevents the development of a high pressure in the discharge pipe and reduces fluctuations in the rate of discharge. A short chamber of large diameter is more effective than a long, narrow chamber.

After the pressure impulse resulting from the closing of valve *C* has been expended there is a tendency for the water to surge backward up the drive pipe. This surge opens a small valve *H* and admits a small quantity of air, which is required to replenish the air in the air chamber. The surge, with the aid of weight *B*, also serves to open valve *C*. Valve *A* is automatically closed by the pressure of the water in the discharge pipe. The cycle of action is repeated indefinitely.

When pipe *L* is connected as shown in Fig. 5-22, the ram will pump one quality of water by means of water of another quality. The pure water at *H′* comes from a source at least 18 in. higher than the less pure supply at *D*. When valve *C* is open, impure water is wasting from it and pure water is flowing through the check valve at *F* into the drive chamber at *G*

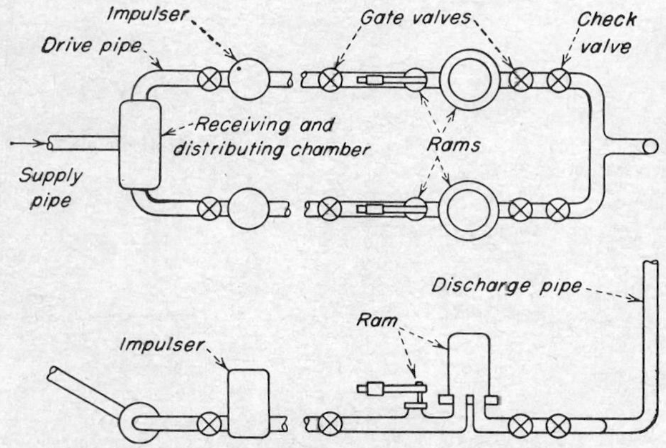

FIG. 5-23. Installation of multiple hydraulic ram.

and is also wasting through the valve at *C*. When *C* slams shut, valve *F* is also shut, and pure water passes through valve *A*. There is thus a slight waste of good-quality water through valve *C*, but no impure water passes through valve *A*.

An appreciable volume of water passing into the drive pipe is not lifted. The amount not lifted may be expected to exceed the amount lifted at the minimum drive lift of 3 ft at which a ram will work. The higher the lift, the greater is the ratio of water wasted to water lifted.

A typical installation of a hydraulic ram is shown in Fig. 5-23. In the selection of a hydraulic ram the information that must be obtained includes (1) the rate of flow of water during various seasons of the year, (2) the demand for water during the corresponding periods, (3) the available fall of water for driving the ram, (4) the height of desired lift, and (5) the greatest possible length of drive line. The length and diameter of the drive pipe are important, because upon them depends the successful operation of the ram. The lower the available driving head, the longer and larger must the drive pipe be to produce the required water hammer. In general, the length of the drive pipe is about 7 times its fall; it may vary between 5 and 10 times, depending on the conditions of delivery. Where the drive pipe is too long, pressure relief may be provided by the installation of an impulser, as shown in Fig. 5-23. The impulser consists of a standpipe or closed air chamber, which allows water that is above it in the drive pipe to run into the impulser when water hammer is created in the drive pipe between the impulser and the ram. During the next cycle the impulser will return water to the drive pipe.

The diameter of the drive pipe is usually about twice the diameter of the delivery pipe. A number of rams may be installed in parallel to deliver into the same header or manifold, but each ram will probably require its own drive pipe. Some data on hydraulic rams are given in Table 5-7.

TABLE 5-7. SIZES AND CAPACITIES OF HYDRAULIC RAMS*

Rife Engine Company					Markey Machinery Company				
Number	Drive pipe, in.	Delivery pipe, in.	Gpm to operate	Least permissible fall, ft	Number	Drive pipe, in.	Delivery pipe, in.	Minimum gpm to operate	Maximum discharge, gpm
10	1¼	¾	2–6	2	1	1	½	4	1/16–1¾
15	1½	¾	6–12	2	1½	1½	¾	8	⅛–2¾
20	2	1	8–18	2	2	2	1	15	¼–5½
25	2½	1	12–28	2	3	3	1½	25	⅝–16
30	3	1¼	20–40	2	4	4	2	45	1–35
40	4	2	30–75	2	6	6	3	90	3–65
80	8	4	150–300	...	9	9	4	200	7–140
120	12	5	375–700	...	12	12	5	300	13½–450
...	....	...	....	...	18	18	8	600	22½–900
...	....	...	....	...	24	24	10	1,000	450–1,600

EFFICIENCIES OF RIFE HYDRAULIC RAMS

Ratio of discharge lift to driving head.......	2½	3	18	23	30
Efficiency, per cent........................	75	70	67	60	50

* From information given in catalogues.

5-28. Windmills. Windmills are used for driving well and cistern pumps of relatively small capacities. The speed of revolution of the wind wheel, its diameter, the pitch of the vanes, the ratio of the gears, and the speed of the pump are fixed by the designer. The purchaser specifies the required performance which should be guaranteed by the seller. Some windmill-pump capacities are stated in Table 5-8, and wind pressures at various velocities are stated in Table 5-9. Gasoline motors or electric motors may be connected to windmill-driven pumps to supplement the wind on calm days.

TABLE 5-8. CAPACITIES OF WINDMILLS, GPM*

Diameter of mill wheel, ft	Velocity of wind, mph	Revolutions of wheel per min	Elevation, ft raised						Useful horsepower developed
			25	50	75	100	150	200	
8½	16	70–75	6.2	3.0	....	....	....	....	0.04
10	16	60–65	19.2	9.6	6.6	4.8	....	....	0.12
12	16	55–60	34.0	18.0	11.9	8.4	5.7	....	0.21
14	16	50–55	45.1	22.6	15.3	11.2	7.8	5.0	0.28
16	16	45–50	64.6	31.7	19.5	16.2	9.8	8.1	0.41
18	16	40–45	97.7	52.2	32.5	24.4	17.5	12.2	0.61
20	16	35–40	125.0	63.8	40.8	31.2	19.3	16.0	0.78
25	16	30–35	212.4	107.0	71.6	49.7	37.3	26.7	1.34

* U.S. Wind Engine and Pump Company.

TABLE 5-9. PRESSURE OF WIND AGAINST PLANE SURFACES. ACTION ON WINDMILLS

Velocity of wind, mph	Pressure, psi*	Description of wind	Effect on windmill
3	0.027	Calm	Will not move
8	0.19	Light air	Just starts
13	0.5	Light breeze	Pumps well
15	0.67	Rated breeze	
18	1.08	Gentle breeze	Excellent work
23	1.75	Moderate breeze	Excellent work
28	2.6	Fresh breeze	Maximum results
34	3.9	Strong breeze	Too fast, should be folded back out of service
40	5.3	Moderate gale	
48	7.7	Fresh gale	
56	10.4	Strong gale	
65	14.0	Whole gale	
75	18.8	Storm	
90	27	Hurricane (tornado)	Wrecked

* $P = 0.0032V^2$.

5-29. Power-driven Pumps. Power-driven pumps, or, as they are more commonly known, *power pumps*, are driven by gasoline engines, electric motors, or some less-frequently used motive power. They are connected to the motor by means of belt, gears, or shafting. Belt connection makes possible ease in the adaptation of motors and pumps of diverse speeds, but the space occupied is greater than for other forms of connection. Gears assure positive operation of the pump, but they are sometimes noisy, require constant lubrication, and are so inflexible that a sudden stoppage in the discharge pipe will result in rupture of some part of the pipe or mechanism. Direct-connected pumps and motors on the same shaft give the most efficient driving mechanism between the pump and motor, but they necessitate the use of a pump and motor of the same speed.

FIG. 5-24. Rumsey triplex power pump.

The most common type of power pump is the triplex pump illustrated in Fig. 5-24. It consists of three reciprocating single-acting water cylinders. Such pumps have a long life, will give excellent service, are suitable for varying loads without material variations in efficiency, and are self-priming. They are noisy and require attention and lubrication. They must be secured to a firm foundation, and care should be taken in the location of the suction pipe to avoid bends, sags, and air pockets. The discharge pipe should be equipped with a gate valve and a check valve, the latter being nearer to the pump. The mechanical efficiency of belt-connected triplex power pumps is about 65 to 75 per cent, the efficiency of the belt alone being about 95 to 97 per cent. The efficiency of geared connections is about 87 to 93 per cent.

CHAPTER 6

STORAGE TANKS

6-1. Storage Tanks for General Water-supply Purposes. When the pressures in a water supply are irregular, uncertain, or intermittent, some method is required to obtain water during periods of inadequate pressure. Inadequate pressures during the hours of greatest demand are not unusual in tall buildings. The difficulty can be overcome by the installation of either a gravity storage tank or a pneumatic storage tank. Tall buildings should be equipped with their own booster pumps, possibly supplemented by storage tanks on intermediate floors to avoid too high a pressure on floors between tanks. The even pressures provided by storage tanks give more desirable service to fixtures and to automatic equipment, particularly to water-closet flushometer valves.

Storage tanks are used for the storage of water for fire protection, as explained in Sec. 6-2, and for the storage of hot water, as described in Sec. 8-20.

6-2. Gravity Storage Tanks. A gravity storage tank may be located at any convenient point on the roof, within the building, on an independent tower, or on a hill. A tower or hill location avoids the need for strengthening the building to support the tank and removes the hazards to the building inherent in a tank on the roof or inside the building. It is desirable that the tank be placed as near as convenient to the building to be supplied and at a height not less than 15 to 20 ft above the highest fixture or equipment to be supplied. One type of gravity storage tank suitable for placing on a tower is shown in Fig. 6-1, and the piping connections for a house tank are shown in Fig. 6-2. An air gap may be required at the tank outlet and overflow to avoid backflow into the tank. The air gap required by the "National Plumbing Code" (Sec. C.2.5) is shown in Fig. 6-2*a*.

Tanks placed within buildings are usually made of steel and are rectangular or circular in shape. Cylindrical tanks may be less costly per unit volume of storage capacity, but rectangular tanks are more economical of space. Wood tanks are more subject to leakage, when compared with steel tanks, because the alternate filling and emptying may permit drying and shrinkage of the wood staves, but they are less subject to

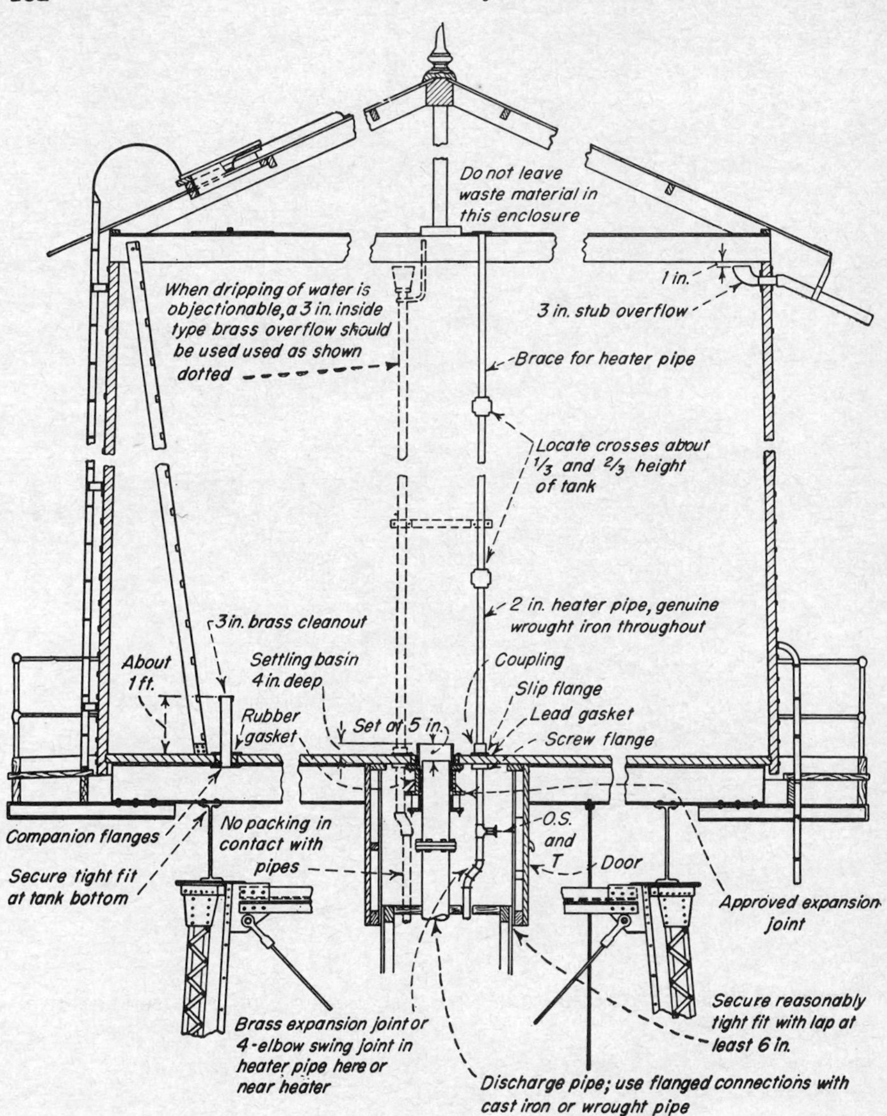

FIG. 6-1. Gravity storage tank for exterior tower. NOTE: Interior ladder is undesirable in cold climates. (*From National Board of Fire Underwriters.*)

difficulties arising from condensation of atmospheric moisture. Wood tanks should be constructed so that there is no leakage and no necessity for calking when the staves are soaked. Calking when dry is undesirable because it increases the stress on the hoops when the wood is wet. Such tanks may be painted on the outside but not on the inside. Exterior painting is done as a protective measure, but the work must be carefully done to cover the wood under the hoops and hoop saddles. Steel tanks are more durable than wood if well maintained. They can be painted both inside and out or otherwise protected against corrosion, as explained in Sec. 4-34. Steel tanks are less liable than wood tanks to sudden, complete failure as a result of corrosion. The corrosion of steel is progressive, so that small holes and leaks develop first. The failure of the hoops on a wood tank may be sudden, and the collapse complete.

Water is delivered to a storage tank during periods of adequate pressure in the water supply. When the pressure is reduced, water runs from the tank into the plumbing pipes. The water supply to the tank may come through the same pipe that is used to take water away from the tank, or

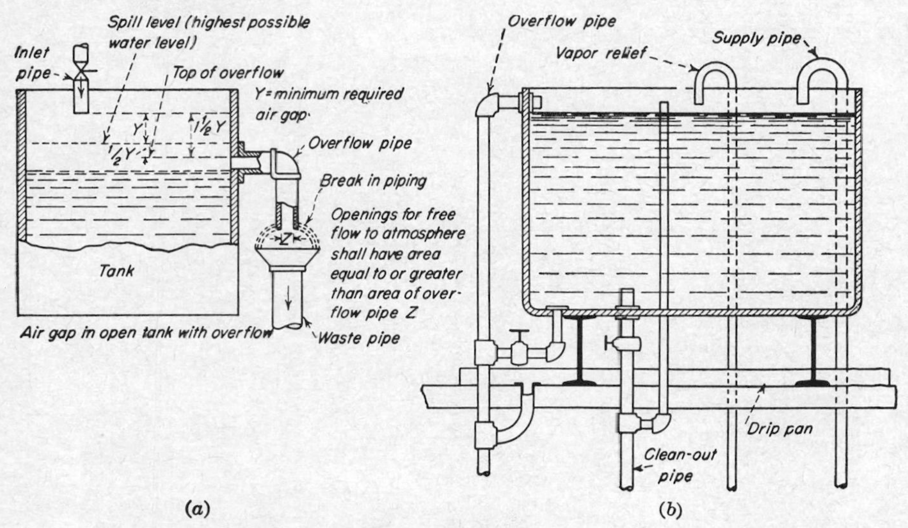

FIG. 6-2. Open storage tanks.

it may come through a different pipe. If the same pipe is used to supply water to the tank and take water away from the tank, it is said that the tank is "floating on the line." In this case the service pipe should be equipped with a check valve to prevent backflow into the water mains, and a riser pipe of generous size should lead to the tank. The riser pipe should terminate 6 to 12 in. above the bottom of the tank to allow the undisturbed accumulation of sediment in the tank between cleanings. The riser pipe may enter the tank at any convenient elevation, or it may pass over the top of the tank. If the riser pipe terminates at any appreciable elevation above the bottom of the tank, extra work is required to lift the water into the tank when the tank is not full. The branch pipes of the plumbing system should take off from the riser pipe. The system has the advantage of utilizing the highest pressures available and of utilizing water that may be of a more desirable temperature than it would have if it had been stored in the tank.

If separate supply and take-off pipes are used and the riser pipe delivers water above the top water level in the tank, no check valve is required on the supply pipe from the water main and a cross connection between the public water supply and the plumbing pipes cannot exist. The continuation of the supply pipe over the edge of the storage tank is not advantageous in the utilization of available pressures, however, because water will be delivered into the tank only at the pressure required when the tank is full, regardless of the elevation of the surface of water in the tank. In both the one-pipe and the two-pipe systems the pipe taking water from the tank should extend at least 6 to 12 in. above the bottom of the tank to permit quiescent settling and to avoid carrying sediment into the plumbing. Water in the tank below the top of the outlet pipe is not available for use in the plumbing. The pipe opening should be protected by a strainer to prevent the entrance into the pipe of objects that fall into the tank.

To avoid overflow, the tank should be equipped with an automatic valve, usually float controlled, to shut off the inlet when the tank is full. Regardless of the provision of such a valve, the tank should be equipped with an overflow pipe capable of taking water away " . . . fast enough to keep the water level from rising more than half the minimum required air gap, as shown in Table 1-4, said distance to be measured above the top of the overflow."[1] The Code states (p. 34):

8.3.3. Overflow pipes for gravity tanks shall discharge above and within 6 in. of a roof or catch basin, or they shall discharge over an open, water-supplied sink. Adequate overflow pipes properly screened against the entrance of insects and vermin shall be provided.

If the storage tank is placed on the roof or inside the building, provision must be made to conduct away water of condensation or leakage or, in case of failure of the tank, to conduct away water without damage to the building. Leakage may be cared for by setting the tank in a shallow, watertight container that drains into the overflow pipe. A pipe controlled by a valve should be provided to allow the drainage of all water, sludge, and sediment from the lowest point in the tank. This pipe may be connected also to the overflow pipe, provided that the discharge of sludge and sediment will not be objectionable. Such drain pipes should not be connected to a sanitary sewer without an adequate air gap in the line, even though the outlet of the pipe is trapped, because the water seal in the trap may evaporate and allow sewer air to enter the tank. Sizes of drain pipes recommended by the Code are given in

[1] From the "Uniform Plumbing Code for Housing," Housing and Home Finance Agency, Technical Paper 6, Par. C.2.7, p. 68, February, 1948.

TABLE 6-1. RECOMMENDED SIZES OF DRAIN PIPES FROM GRAVITY STORAGE TANKS*

Tank capacity, gal less than...	750	1,500	3,000	5,000	7,500	over 7,500
Drain-pipe size, in...........	1	1½	2	2½	3	4

NOTE: Each drain line shall be equipped with a valve of the same diameter as the drain pipe.
* From "The Uniform Plumbing Code for Housing," Sec. 8.3.5., Housing and Home Finance Agency, February, 1948.

Table 6-1. The National Plumbing Code states (Sec. 10.8.6) that the drain pipe is to be equipped with a quick-opening valve of the same diameter as the pipe. An indicator, sometimes called a *telltale*, should be provided to show from some convenient point of observation where the surface of the water is in the tank.

Tanks placed outside the building should have a watertight and fireproof roof designed to exclude birds and flying insects. The purpose of the roof is to avoid pollution of the water by birds, insects, and dust and dirt in the air and to aid in protection against freezing. A double roof provided with an insulating dead-air space, such as is shown in Fig. 6-1, is particularly effective for this purpose. The roof must be adequately vented to permit rapid filling and emptying of the tank without the creation of a pressure or a vacuum under the roof. The riser pipe to an outside tank must be equipped with an expansion or flexible joint to permit movement between the tank and the pipe. In a cold climate the outside riser pipe must be insulated and the tank must be protected against the effects of freezing. Methods of insulating riser pipes are illustrated in Fig. 6-3. Tanks have collapsed owing to the formation of a heavy ice cap that has thawed enough to fall to the bottom of the empty tank. This hazard can be minimized by constructing the tank with vertical walls or with walls tapering inward toward the top not more than about ½ in. per ft of water depth and by avoiding any obstruction, such as interior ladders and structural members, to prevent the free rise and fall of the ice cap as the tank fills and empties. Outside tanks can be protected against freezing by constructing them with double walls with an intermediate dead-air space or nonabsorbent insulating materials between the walls or by running hot water or steam through a coil of pipe placed in the tank or discharging the hot water or steam directly into the tank. Insulated walls may be highly unsatisfactory owing to the freezing of water that has leaked from the tank into the insulation and because of the difficulty of making repairs to the exterior of the tank.

6-3. Pneumatic Storage Tanks. A pneumatic storage tank is an airtight tank filled partly with water and partly with air. The tank may be connected to the pump and to the distribution system by a single pipe, as indicated by the diagram in Fig. 6-4. The direction of flow in this pipe depends on the relative pressures in the tank and in the plumbing pipes. It is possible by means of a combination of an electric-motor-driven pump and such a tank to maintain a limited variation of pressure in the plumbing of a building. Such a combination for a household water supply is shown in Fig. 6-5. Reciprocating or rotary displacement pumps are recommended for such a service. A centrifugal pump can be used only

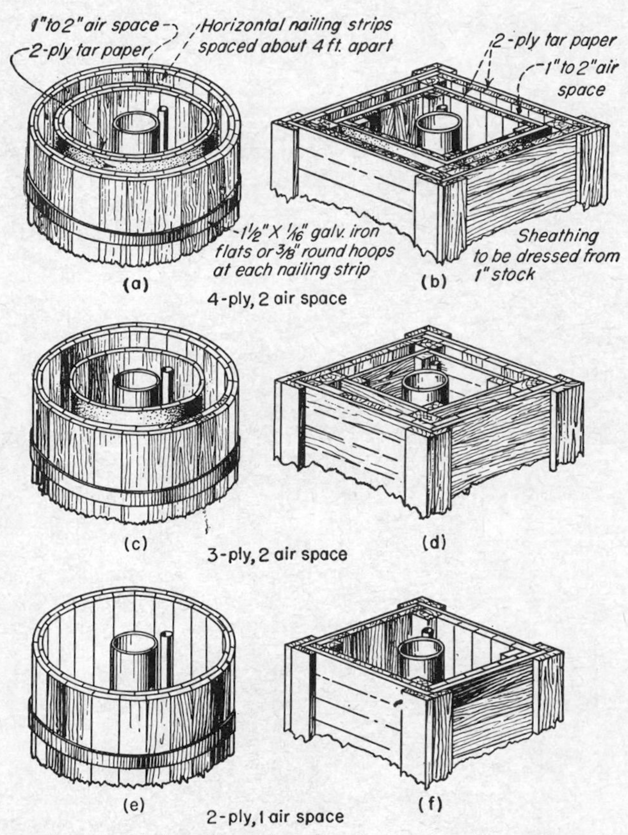

Fig. 6-3. Methods of insulating riser pipes for elevated tanks. (*From National Board of Fire Underwriters.*)

when there is positive pressure on the suction and an auxiliary air compressor is provided. When the pressure in the storage tank drops, the motor is automatically started to raise the pressure to the predetermined limit at which the motor is stopped. Such hydropneumatic equipment is widely used in small water supplies. Internal-combustion engines, automatically or manually controlled, can be used where electric power is not available.

This method of storing and equalizing flow overcomes some of the objections to storage in elevated tanks. The pneumatic tank can be placed on the ground within an outbuilding or in the basement of the

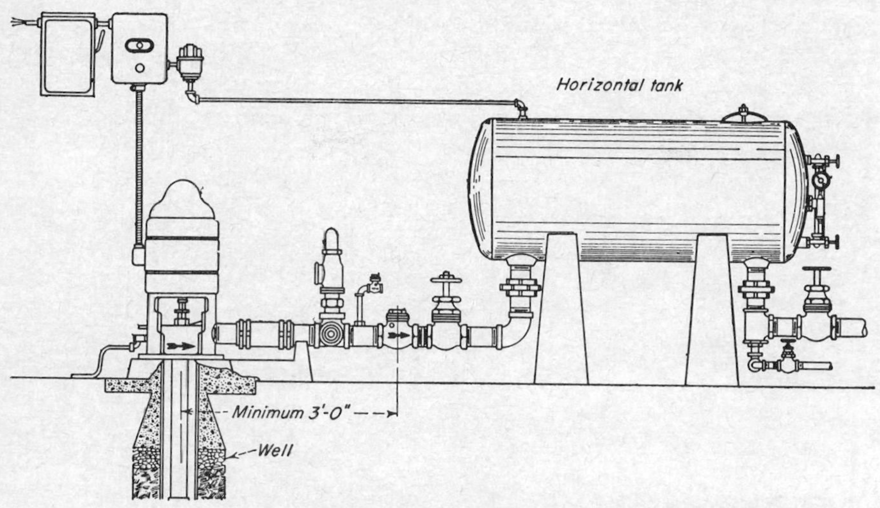

Fig. 6-4. Pneumatic tank.

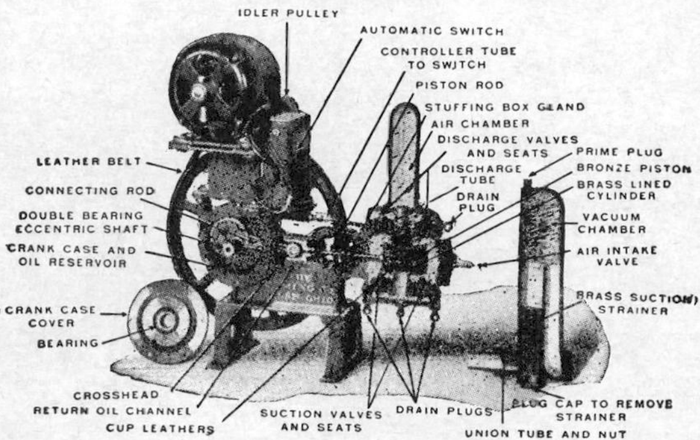

Fig. 6-5. Electric-pneumatic pumping system. (*The Deming Co.*)

building to be supplied, avoiding any special structural strengthening of the building and the unsightliness of the exposed tank and protecting the tank against high temperature, freezing, and contamination. Disadvantages of such tanks are that only about three-fourths to two-thirds of the tank capacity is available for storing water, the remainder of the capacity being occupied by compressed air, and that the maintenance or replenishment of air is necessary. This may be difficult. Air is lost by dissolving in the water, the rate of solution increasing with the pressure. Tall, narrow tanks minimize air loss by exposing a relatively small water

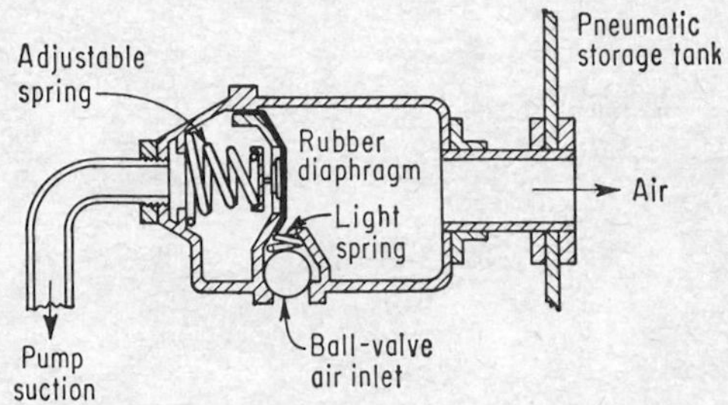

FIG. 6-6. Snifter valve on pneumatic storage tank.

surface to air. Air may be drawn into the system by such means as an air inlet or "snifter valve" on the pump suction. The valve admits a limited amount of air under a slight suction, by aspirating air through a high-velocity jet, and by other means. A sketch through a snifter valve is shown in Fig. 6-6.

The tank should be so designed that at the maximum pressure desired the air occupies about one-third of the volume of the tank. A pressure-relief or safety valve must be provided on the top of the tank to prevent the development of a dangerously high pressure, and a vacuum-relief valve is needed to avoid collapse of the tank or backflow into the water system.[1] Variations between maximum and minimum air pressures under various lifts are shown in Table 6-2.

TABLE 6-2. AIR AND WATER PRESSURES IN PNEUMATIC STORAGE TANKS

Minimum discharge pressure, psi*	0	10	20	30	40	50	60	70	80	90	100
Maximum discharge pressure, psi†	30	60	90	120	150	180	210	240	270	300	330
Ratio maximum to minimum pressure	∞	6	4.5	4	3.75	3.6	3.5	3.4	3.3	3.3	3.3

* Tank full of air.
† Tank two-thirds full of water and one-third full of air.

Wide variations in pressure are undesirable and uneconomical, but they are unavoidable in this type of storage. Partly for this reason,

[1] See also "National Plumbing Code," Sec. 10.8.8.

additional capacity should be provided where the system is to be used for fire protection or for a garden hose.

In determining the required water-storage capacity of the pneumatic tank the same principles apply as for the gravity tank. Some information on manufacturers' standards for pneumatic storage tanks is given in Table 6-3. The capacity of the tank should not be less than about 30 gal for a small household. It should be increased in proportion to additional fixtures beyond about five up to a capacity of about 70 gal, which should be sufficient for a large household. In any installation the pump should not be thrown on and off more often than once in about 15 to 30 min.

TABLE 6-3. MANUFACTURERS' STANDARD DIMENSIONS FOR PNEUMATIC TANKS, IN INCHES

Vertical tanks

Size, in.	Capacity, gal	Top to upper water-gage faucet	Top to center water-gage faucet	Bottom of tank to inlet	Opposite side to drain	Diameter of drain	Bottom to discharge	Diameter of outlet	Diameter of inlet
16 by 36	32	8½	13½	6	3	1	9	1	1
16 by 48	42	17	13½	6	3	1	9	1	1
20 by 60	82	23	13½	6	3	1	9	1	1
24 by 60	120	23	13½	6	6	1	9	1½	1½
30 by 72	220	29	13½	6	6	1	9	1½	1½
36 by 72	315	29	13½	6	6	1	9	2	2

Horizontal tanks

Size, in.	Capacity, gal	Water-gage opening	Center of water-gage opening	Inlet to outlet opening	Drain to inlet	Edge to drain	Diameter of drain opening	Diameter of outlet	Diameter of inlet
36 by 120	530	½	13½	8	8	8	1	2	2
42 by 120	720	½	21½	8	8	8	1	2	2
48 by 120	940	½	21½	10	8	8	1	2	2
48 by 288	2,260	½	21½	10	8	8	1	3	3
60 by 252	3,000	½	21½	10	8	8	1	3	3
72 by 288	5,000	½	21½	10	8	8	1	3	3

6-4. Storage-tank Capacity. The capacity of a storage tank to provide water when there is no supply should be at least equal to the volume of water that is to be consumed during the no-supply period. Conditions involved in determining this volume include the number of consumers, the duration of the time when there is no supply, and the rate of consumption during this time. The minimum capacity of the tank should be the product of these two items. Where neither can be determined precisely, it may be sufficient to provide about 300 gal of storage volume for each 1,000 sq ft of floor space in the building. This should be sufficient to provide for a 12-hr shutdown of the water supply. The tank capacity may be reduced proportionally for shorter periods of shutdown. Whenever in doubt, it is better to provide too much than too little storage.

CHAPTER 7

WATER-SUPPLY PIPES IN BUILDINGS

7-1. Water-supply Systems. The basic principles of hydraulics and of plumbing outlined in preceding chapters are to be applied in the design and installation of water-supply pipes in buildings. Plumbing pipe installations are more or less standardized for most buildings, but special piping layouts and equipment are required in such places as hospitals, hotels,[1] motels,[1] and some industrial plants. For most residences and commercial buildings water enters the plumbing through the service pipe, as shown in Fig. 1-4. A stop-and-waste valve, as described in Sec. 14-27, should be inserted on the service pipe on the building end to permit shutting off the water and draining it from the plumbing pipes in the building. The location of a meter near this spot on the service pipe is discussed in Sec. 7-18. The water-supply pipes should branch from the building end of the service pipe or the meter in the building into risers, intermediate pipes, and pipes to individual fixtures.

Two systems of water-supply piping for small rural residences are shown in Fig. 1-2*a* and *b*. A system for an urban residence is shown in Fig. 1-2*c*. Water-supply pipes for the lower floors of a large urban residence or an apartment building are shown in Fig. 7-1*a* and *b*. Systems of supply pipes suitable for tall buildings are shown in Fig. 7-2. The open storage tanks and the pressure-regulating valves shown on intermediate floors, in Fig. 7-2, prevent excessive pressures in the pipes supplying fixtures on lower floors. The lowest four or five floors may be supplied directly from the city mains.

Dual water-supply systems are sometimes installed in buildings where one of the supplies is hard, warm, or otherwise unsatisfactory for all purposes. The installation of two systems, one of which is or may become unsatisfactory for all purposes as unfit for human consumption, should be avoided because of the probability that a cross connection will be made.

7-2. Water-supply Systems for Tall Buildings. Pumps to increase the pressure from the public water supply so that the upper floors of tall

[1] See also L. Blendermann, Plumbing Systems for Hotels, *Air Conditioning, Heating, and Ventilating,* December, 1957, p. 90, and January, 1958, p. 77, and Plumbing Systems for Motels, *ibid.*, October, 1957, p. 109.

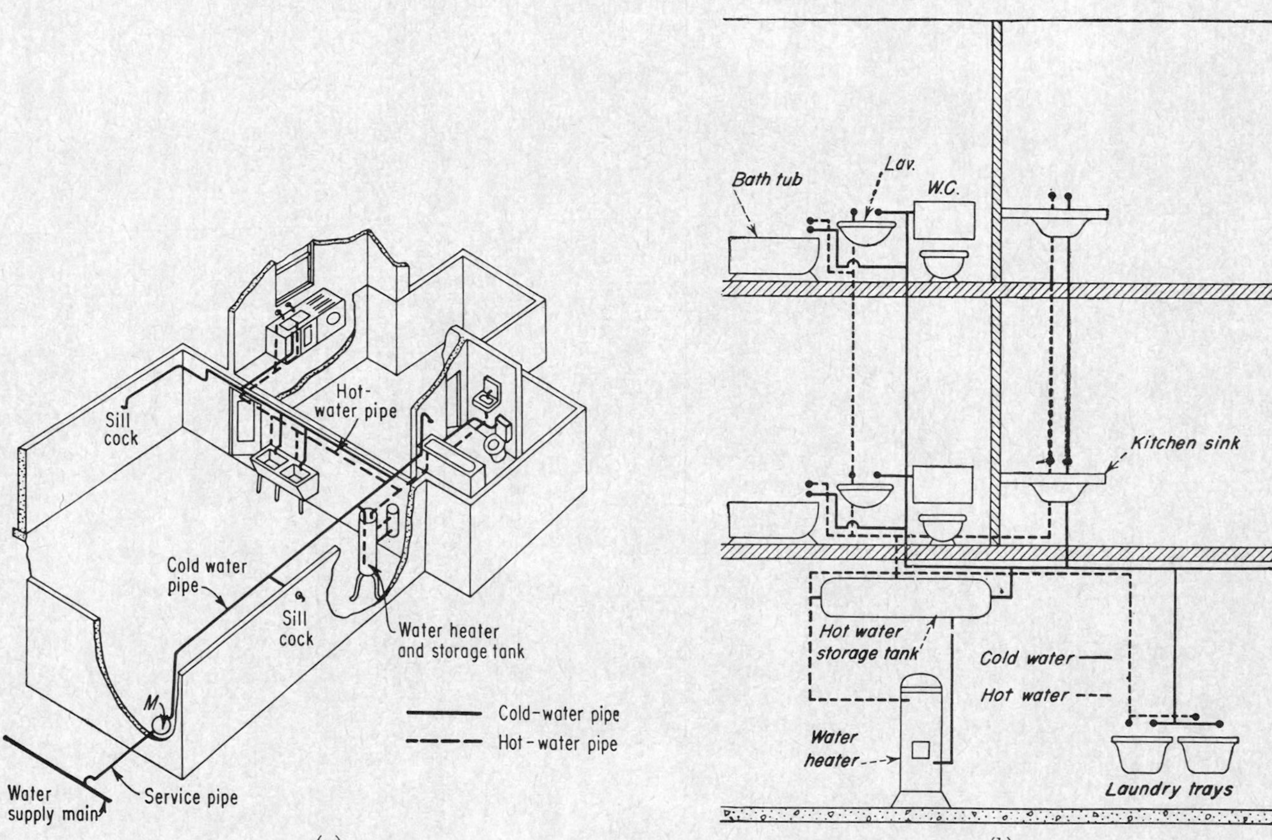

FIG. 7-1. (*a*) Water-supply piping for a one-family urban residence; (*b*) water-supply piping for a small apartment building.

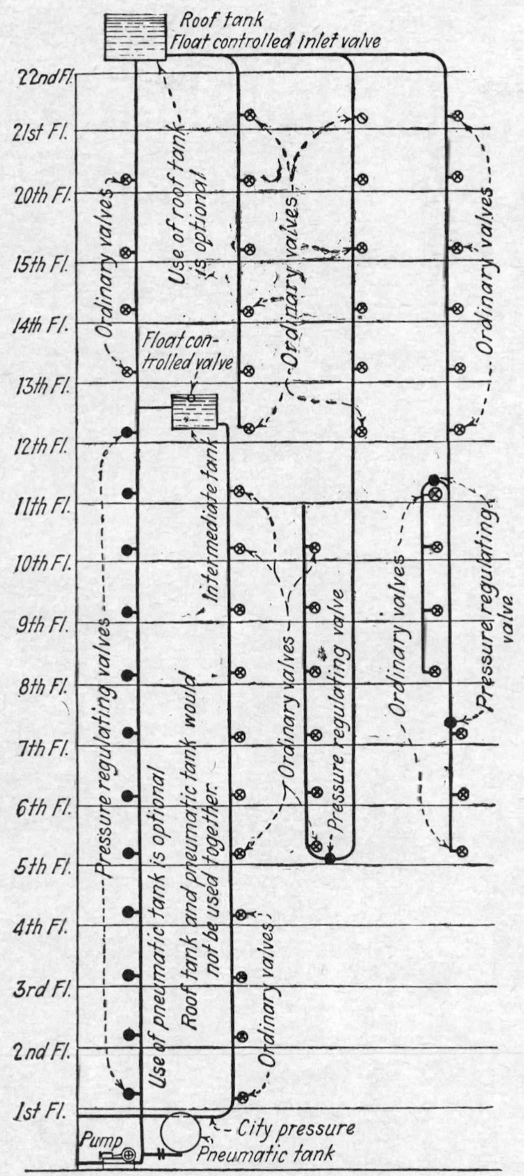

FIG. 7-2. Systems of supply pipes for tall buildings.

buildings can be served are required where the pressure in the public water supply is inadequate. The pumps may be located in the basement of the building or at such an elevation that their suctions will be under pressure from the public water supply, in order that their lift may be limited to the difference between the required discharge pressure and the supply pressure. If the public water supply is discharged into an open reservoir at a low elevation in the building and pumps draw their supply from this reservoir, advantage cannot be taken of the pressure of the public water supply. In extremely high buildings it may be desirable to have pumps on various floors to limit the pressures within the pipes of the building.

The building may be divided vertically into pressure zones, with a storage tank,[1] a pressure-regulating valve, or a pump to control each zone. An arrangement of this type[2] is shown in Fig. 7-2.

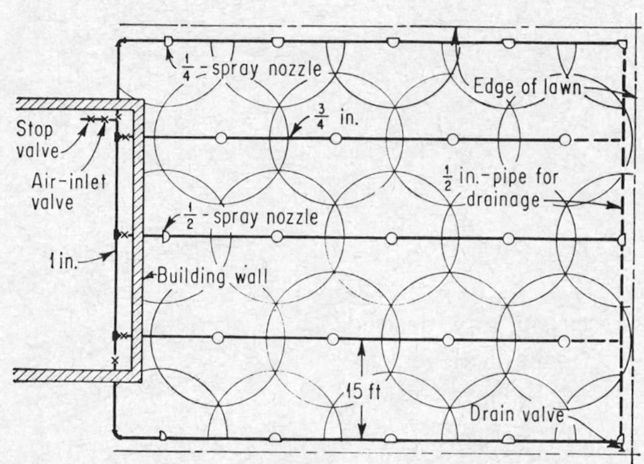

FIG. 7-3. Lawn-sprinkling layout.

7-3. Water Supplies for Special Purposes. *Lawn Sprinklers.* A suggested layout of water pipes for a lawn-sprinkler installation[3] is shown in Fig. 7-3. Principles to be observed in design and installation include sloping all pipes to a low point at which a drain valve is located and from which water can be drawn. A type of drainage valve is available which closes automatically when water pressure is turned on and opens when the water pressure is turned off in the pipes. Locate sprinkler nozzles where desirable, usually at the apexes of equilateral triangles whose sides are as long as the desired distance between nozzles. The proper distances depend on the available pressure and the type of nozzle. Information must be obtained from the manufacturer.[4] Types of nozzles used include

[1] See also G. R. Jerus, Roof and Zone Tanks for Multi-story Buildings, *Air Conditioning, Heating, and Ventilating*, March, 1958, p. 68.

[2] For additional layouts see *Air Conditioning, Heating, and Ventilating*, February, 1957, pp. 111 and 112.

[3] See also *Domestic Eng.*, December, 1956, p. 84.

[4] See also *Domestic Eng.*, January, 1956, p. 80, and December, 1956, p. 84.

fountain, bubble, mist, and fog. Special nozzles may be provided for special plants and trees or to throw any reasonable shape of spray desired. Fountain nozzles with circular sprays are most common. Semicircular sprays may be used at the edge of the lawn. Arrange nozzles in groups with a control valve for each group to provide flexibility in operation; place nozzles just far enough below the surface to avoid interference with lawn mowing.

Copper or plastic pipe has been found satisfactory, since the conditions just beneath sod are usually highly corrosive to iron pipe. Time-controlled devices[1] are available that will open and close the valves on any desired schedule. Mechanical equipment, either manually or power driven, is available for sod cutting and shallow trenching required.

Buildings under Construction.[2] The provision of water is necessary for buildings under construction for drinking, fire protection, and general services. Provision must be made also for disposal of the used water and of human wastes. The responsibility for these services is usually the plumber's. The principal water-supply pipes of the permanent installation and the principal drainage pipes can be installed during the construction period, with temporary service installed on the upper floors as the building rises. Pipes for the fire-protection supply should rise with the

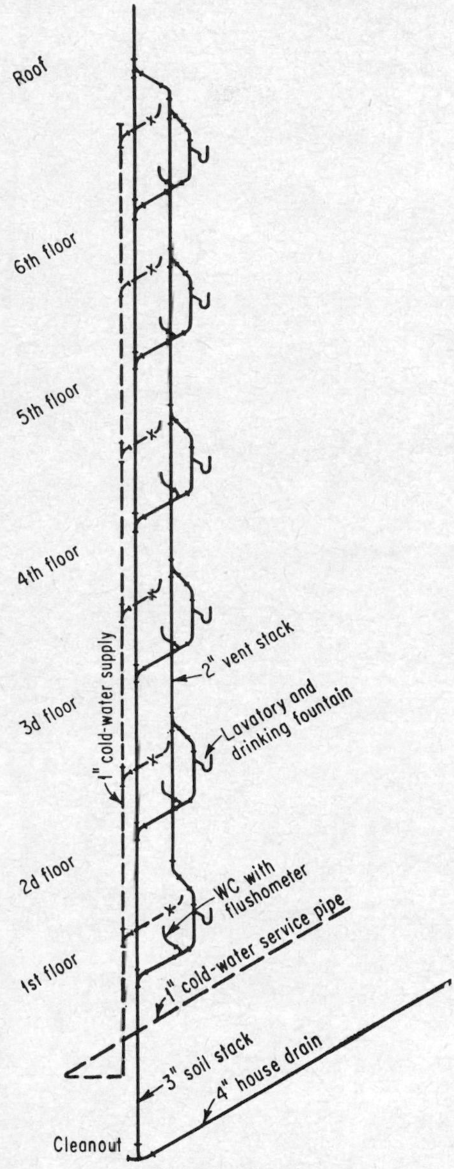

FIG. 7-4. Roughing-in for temporary plumbing during construction.

[1] *Domestic Eng.*, January, 1956, p. 80, and December, 1956, p. 84.

[2] See also L. Blendermann, *Air Conditioning, Heating, and Ventilating*, January, 1956, p. 84.

building, but they should be used for no other purpose than fire protection. Temporary water-supply piping has been installed on some building jobs, somewhat as indicated in Fig. 7-4.

7-4. Water-supply Pipe Materials. Characteristics of materials used in plumbing are discussed in Chap. 4. Materials used for water-supply pipes within a building include wrought iron and steel, usually galvanized; brass; and copper. Lead, tin, and zinc are sometimes used in pipes, in pipe linings, and in solder when wiped joints are made. Galvanized-iron and -steel pipes give satisfactory service, experience with them is wide, and their first cost is relatively low when compared with other acceptable materials. Copper and brass may be more resistant to corrosion than are galvanized pipes, and they are sufficiently smooth to permit some designers to feel that it is safe and economical to use brass and copper pipes one size smaller than would be required if galvanized pipes were used. Cast iron is used primarily for water-supply pipes laid underground in sizes 4 in. and larger in diameter. It is seldom used for water-supply pipes within a building.

Pipes are available for water supply in various strengths to resist internal pressures, as shown in Appendix II. Extra-heavy pipes and valves and the special equipment of fixture fittings are required when water-supply pressures above 100 psi are to be resisted. Such high pressures can be prevented by the use of gravity supply tanks or by the use of pressure-regulating valves. Both of these expedients for controlling pressures are useful also in controlling water-hammer pressures in pipes.

7-5. Sizes of Water-supply Pipes.[1] Recommended sizes of water-supply pipes to individual fixtures are shown in Tables 3-5 and 7-1. The sizes of water-supply pipes in buildings can be selected by the following procedure:

1. Select the material and a trial size for each pipe, valve, and fitting to be shown on the plans. When such a selection is made, the judgment and experience of the designer must be relied on. For example, a reasonable selection of pipe sizes would show pipes progressively smaller from the service pipe to the fixture, the sizes of pipes to fixtures being suggested in Tables 3-5 and 7-1.

2. Determine the demands, in fixture units, on each pipe, using Tables 3-4 to 3-6, 3-10, and 3-11, and the actual flow from intermittently used fixtures from Fig. 3-1. Then add the flow to special outlets such as hose bibs, air-conditioning equipment, or industrial equipment.

3. Determine the loss of pressure from various critical points in the piping system to one or more critically located fixtures. The friction loss in straight pipe can be found from Fig. 2-6 and Table 2-12, and head losses in fittings can be found in Tables 2-10 to 2-14 and Fig. 2-14. Judg-

[1] See also Figure Pipe Sizes the Simple Way, *Power Eng.*, October, 1955.

TABLE 7-1. SIZES[a] OF BRANCH WATER-SUPPLY PIPES TO PLUMBING FIXTURES

Fixture	U.S. Dept. of Commerce	Housing Code[b]		National Plumbing Code[c]	Sizes for different rates of head loss. Head loss per unit L[d]			W. S. L. Cleverdon[e]
		Cold	Hot		$0.5L$	L	$5L$	
Bathtub:								
4 ft long	½	½	½	½	1	¾	½	½–⅝
7 ft long	...	...	...	...	1¼	1	¾	
Bidet	...	½	½	...	⅜	⅜	⅜	½
Drinking fountain	...	⅜	...	⅜	⅜	⅜	⅜	
Foot bath	...	...	...	...	½	⅜	⅜	½
Garden hose	...	...	...	½	¾	¾	½	
Hot-water heater	½	...	...	...	¾	½	½	⅝–¾
Kitchen sink:								
Small	½	½	½	½	¾	½	⅜	½–⅝
Hotel	...	...	...	¾	¾	½	⅜	
Laundry tray	½	½	½	½[f]	½	½	⅜	½–¾
Pantry sink	...	...	...	...	⅜	⅜	⅜	⅜
Shower	...	½	½	½	¾	½	½	
Slop sink	...	...	...	½	¾	½	⅜	½–⅝
Urinal:								
Flushometer	...	1	...	¾	...	...	...	½
Flush tank	...	½–¾	...	½	⅜	⅜	⅜	⅜
Wash basin	⅜	⅜	⅜	...	⅜	⅜	⅜	⅜
Water closet:								
Flushometer	...	1	⅜	...	1½	1¼	1	1
Flush tank	⅜	...	...	⅜	½	⅜	⅜	⅜

The "Housing Code" recommends supply pipes for other fixtures, stating the size of the cold-water pipe first, then the size of the hot-water pipe, in inches, as follows (where only one size is given it refers to the cold-water pipe): bathroom group, consisting of one shower or tub, one lavatory, and one water-closet flush tank, ¾ and ¾; bedpan washer, 1; combination fixture, ½ and ½; dental cuspidor, ⅜; electric water cooler, ½; surgeon's sink, ½ and ½; soda fountain bar sink, ⅜ and ⅜; flush rim sink, 1 and ½; service sink, ½ and ½; pot or scullery sink, ¾ and ¾; pedestal urinal, 1; wall lip urinal, ½; trough urinal, every 2 ft, ½.

[a] In inches.

[b] "The Uniform Plumbing Code for Housing," Technical Paper 6, Housing and Home Finance Agency, February, 1948.

[c] "National Plumbing Code," Sec. 10.14.2. Additional fixtures specified by the "National Plumbing Code" are combination fixture, ½; dishwasher, domestic, ½; flush-rim sinks, ¾; and wall hydrant, ½. Only cold-water pipe sizes are given.

[d] L = length of pipe, ft.

[e] In *Plumbers Trade J.*, vol. 72, p. 867.

[f] One, two, or three compartments.

ment must be depended on for the selection of the critical points. If the pressure loss under the assumed conditions is too large or too small, the sizes of pipes and fittings should be changed and the computation repeated until satisfactory pressures are found at all fixtures.

The foregoing procedure may be difficult to follow. It is used principally in making approximate checks of a design. An illustrative example is solved in the next section.

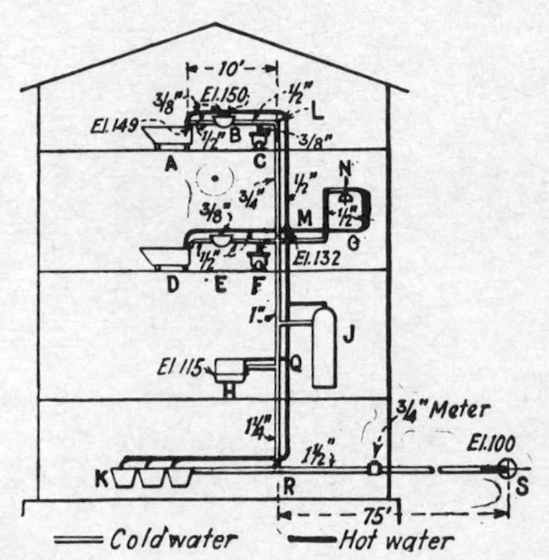

FIG. 7-5. Illustration for example in Sec. 7-6. Double line is cold-water pipe. Heavy line is hot-water pipe.

7-6. Pipe-size Determinations. A procedure for the selection of water-supply pipe sizes for a plumbing system is illustrated in the following problem:

Example. Determine the probable pressure, in pounds per square inch, in Fig. 7-5, at the base of the cold-water faucet in the bathtub on the top floor under the following conditions: Pressure in the street main, at S, is 50 psi; the following fixtures are in use simultaneously: kitchen sink at H, flush valves on water closets at C and F, and both bathtubs at A and D. Galvanized, wrought-iron pipe is used throughout the building.

Solution: The rate of flow, from Table 3-4, in gallons per minute is equal to one kitchen sink, 5; two bathtubs, 16; two water closets, 10. Total 31.

Head loss through 75 ft of 1½-in. service pipe, from Table 2-12, is $[0.038 + (0.069 - 0.038)6/10]75 = 4.3$ ft.

Head loss through the ¾-in. meter, from Fig. 14-44, is $21 \times 2.3 = 48.3$ ft.

Head loss through the tee at base of riser pipe, from Table 2-10, is $[0.369 + 1(0.502 - 0.369)/5]2.3 = 0.9$ ft.

Flow through tee on top floor to one water closet and one bathtub is 13 gpm. The tee is ½ in. in diameter. Head loss through it, from Table 2-10, is $[1.6 + 3(3.6 - 1.6)/5]2.3 = 6.4$ ft.

Head loss through 1¼-in. pipe from R to Q, with a length of 12 ft and a flow of 31 gpm, is, from Table 2-12, equal to $[0.09 + 6(0.16 - 0.09)/10]9 = 1.2$ ft.

Head loss through 1-in. pipe from Q to M, for an assumed length of 20 ft and a flow of 31 gpm minus the kitchen sink, equals 26 gpm is, from Table 2-12, equal to $[0.33 + 1(0.60 - 0.33)/10]20 = 7.2$ ft.

Head loss through ½-in. pipe from C to A for an assumed length of 8 ft and a flow of 8 gpm, from Table 2-12, equals 5.52 ft.

Head loss through ½-in. pipe from M through L to C for an assumed length of 22 ft and a flow of 22 gpm less one bathtub and one water closet equals $26 - 5 - 8 = 13$ gpm is, from Table 2-12, $[0.95 + 3(1.89 - 0.95)/5]22 = 33.2$ ft.

Head loss through ½-in. pipe from C to A for an assumed length of 8 ft is, from Table 2-12, $[0.29 + 3(0.95 - 0.29)/5]8 = 5.7$ ft.

Static difference in head is 49 ft. Total head loss is, therefore, $4.3 + 48.3 + 0.9 + 6.4 + 1.2 + 7.2 + 5.5 + 33.2 + 5.7 + 49.0 = 161.7$ ft or 70 psi.

This is greater than the pressure in the street main. It is evident that a larger meter and larger piping must be used to reduce the pressure losses. New pipe sizes must be selected and the analysis repeated until a reasonable head loss is found.

The following example is modified from an example presented in the "Housing Code":[1]

Example. Suppose that one of the branches from a service pipe is required to supply cold water to three water closets flushometer-valve-operated, two bathtubs, and three lavatories. Using the pressure available for friction loss in pipes as 8.1 psi per 100 ft of pipe, the computations would be as shown in Table 7-2.

TABLE 7-2. RESULTS OF APPLYING FIXTURE UNIT

Fixture type	Fixture units (Table 3-4 and Fig. 2-6)	Demand (Fig. 3-1)	Pipe size (Fig. 2-6)
3 water closets with flush valve....	3 × 6 = 18		
2 bathtubs......................	¾(2 × 2) = 3		
3 lavatories.....................	¾(3 × 1) = 2.25		
Total..........................	23.25	38	1½

Water demand for sill cocks or fixture branches taken off the service pipe is frequently the cause of inadequate water supply to the upper-floor fixtures. This should be prevented by sizing the water-distribution system so that the pressure drops from the street main to all fixtures are the same or by installing a throttling valve for the lower fixtures and reducing the flow.

The following is the solution of another example in the determination of pipe sizes:

Example. Determine the size of a riser pipe to serve eight water closets, four urinals, 12 lavatories, eight bathtubs, four shower baths, and eight kitchen sinks. Table 3-5 shows that these fixtures call for pipes of the following sizes, respectively: 1¼, 1, 1, 1½, 1¼, and 1½ in. It is found from Table 2-16 that these are equivalent, respectively, to the following number of ½-in. pipes: 10.9 + 6.2 + 6.2 + 17.4 + 10.9 + 17.4 = 69. A 2½-in. pipe is equivalent to 65.5 ½-in. pipes. It will be considered that one 2½-in. pipe is the nearest pipe with capacity equal to all these pipes; hence a 2½-in. riser pipe should be used if no allowance is to be made for the improbability of the simultaneous use of all fixtures, as discussed in Secs. 3-4 and 15-18.

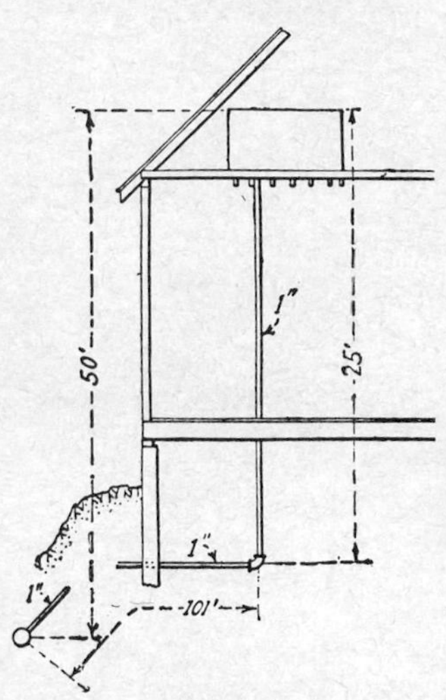

FIG. 7-6. Illustration for example.

If the conditions for which it is desired to select a pipe are not covered in Table 3-5, it will be necessary to compute the size of the riser or intermediate pipe. An illustrative solution will be given:

[1] "The Uniform Plumbing Code for Housing," Housing and Home Finance Agency, February, 1948.

Example. Compute the sizes of service and riser pipes for the roof storage tank shown in Fig. 7-6 so that the rate of discharge will be 50 gpm when the pressure in the main is 42 psi.

Solution: The height that the water must be lifted is 50 ft, equivalent to 21.7 psi. The permissible loss of pressure in the pipeline, valves, and fittings is, therefore, the difference between 42.0 and 21.7 psi, equal to 20.3 psi. The loss of head in the valves and fittings will be assumed as 2.0 psi, leaving a net loss of 18.3 psi in the pipe. In the 126-ft length of pipe this gives a loss of 12.9 lb per 100 ft. The nearest size of pipe, from Fig. 2-6, that will discharge 50 gpm with a head loss of 12.9 lb per 100 ft, or 300 ft per 1,000 ft, is 1½ in. in diameter.

Pressure losses to all parts of an extensive plumbing system should be computed to make certain that the design is good and that there are no portions in which the pipes are too small. If so extensive a study is undesirable, the approximate sizes suggested in Table 7-3 may be found suitable.

TABLE 7-3. FIXTURE UNIT TABLES FOR DETERMINING WATER-PIPE AND METER SIZES*

For installations using flush-tank closets

Meter and service pipe, in.	House supply and branches, in.	Maximum length, ft	Pressure in main, psi			
			30–45	45–60	60–75	75–90
¾	½	50	5	7	10	12
		100	4	6	8	9
		150	3	5	6	7
¾	¾	50	17	27	33	38
		100	12	17	22	27
		150	10	14	18	22
¾	1	50	28	42	55	60
		100	21	34	41	50
		150	17	29	35	41
1	1	50	35	51	74	98
		100	27	37	47	61
		150	19	32	38	46
1	1¼	50	51	94	135	140
		100	38	66	96	123
		150	32	53	75	96
1½	1¼	50	76	150	208	260
		100	48	90	128	163
		150	37	65	92	120
1½	1½	50	137	240	340	380
		100	92	168	240	300
		150	70	130	183	236

* Taken, by permission, from "Uniform Plumbing Code," adopted October, 1947, by Western Plumbing Officials Association.

7-7. Sill Cocks, Hose Connections, and Small Hydrants. The rate of demand at a sill cock or hose connection is frequently so great as to be the determining factor in the calculation of pipe size or the thief of water from the upper floors in an unsatisfactory design. The demand by these outlets is often so great that the importance of their consideration in design must be emphasized.

Illustrations of hose connections and a sill cock, or small wall hydrant, are shown in Fig. 7-7.

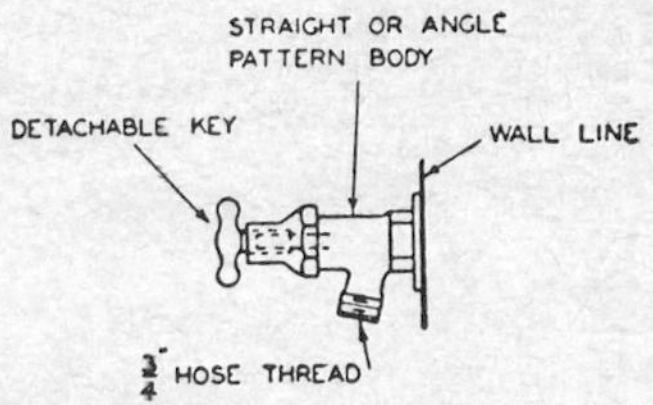

FIG. 7-7. Lawn faucet or sill cock.

7-8. Installation of Water-supply Pipes. Among the first steps in the installation of plumbing may be included laying the service pipe and connecting it to the water main or other source of supply. The supply and drainage pipes within the building are placed as the erection of the building framework proceeds, preceding the placing of the floors and walls.

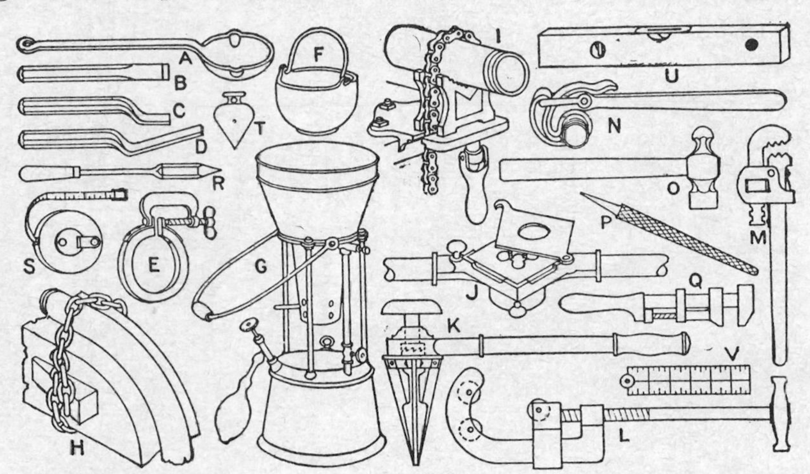

A, pouring ladle; *B*, cold chisel; *C*, calking iron; *D*, yarning iron; *E*, asbestos or rubber pipe jointer; *F*, melting pot; *G*, gasoline blast furnace; *H*, home-made pipe bender; *I*, pipe vise; *J*, stock and die for threading pipe; *K*, pipe reamer; *L*, three-wheel pipe cutter; *M*, 14-in. pipe wrench; *N*, brass pipe wrench; *O*, hammer; *P*, file; *Q*, monkey wrench; *R*, soldering copper; *S*, measuring tape; *T*, plump bob; *U*, spirit level; *V*, measuring rule.

FIG. 7-8. Essential tools needed in plumbing work. (*U.S. Dept. Agr. Farmer's Bull.* 1426.)

The work done at this stage of construction is known as *roughing-in*. Some plumber's tools are shown in Fig. 7-8 and are listed in Table 7-4.

The layout of the water-supply piping in small buildings is relatively simple, as it usually includes only one riser for cold water and one riser for hot water, with necessary fixture branches. In apartment buildings, hotels, large residences, and hospitals there may be a large number of

TABLE 7-4. PLUMBER'S TOOLS NOT SHOWN IN FIG. 7-8

Flaring tool, for flaring pipe
Hole saw, for cutting holes for pipe
Masonry drills
Plunger, or "plumber's friend," for clearing stopped pipes with air compression or suction. See Fig. 18-1
Sewer cleaning tools, shown in Fig. 18-3
Trouble lamps
Valve seating or reseating tools

riser pipes. Simplicity of layout for any number of pipes is essential to economical operation and maintenance. Riser pipes for different purposes may branch from a single distributing manifold, as is shown in Fig. 7-9, or fewer and larger pipes may be used with more frequent branches to mains and fixtures. The layout is dependent more on the method of operation and control than on cost. Easy control by the operating engineer or janitor should be the principal consideration. In some institutions, such as mental hospitals and prisons, control valves and delicate equipment should be placed out of reach of the inmates. Each riser pipe should be equipped with a stop-and-waste valve, and if possible, it should be labeled so that the supply to the riser can be cut off without affecting other portions of the building and the bad effects of the manipulation of the wrong valve can be minimized. Pipes branching from the riser and branches to fixtures should be equipped with valves that are labeled or whose purpose is evident by their location. In so far as it is possible, valves and pipes should be placed in a visible and accessible location.

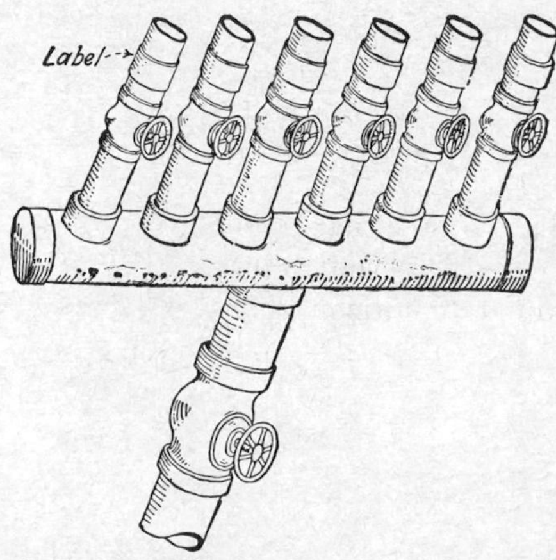

FIG. 7-9. Manifold and riser pipes.

7-9. Location of Pipes. In the location of pipes in the floors, walls, and partitions of a building, their accessibility should be considered. It is undesirable to install a pipe in such a place that an attempt to gain access to it will result in damage to the building. A good plan is to group pipes in *runs* within the walls. These are vertical shafts to which access is possible through doors or easily removable panels. Horizontal partitions at reasonable intervals will prevent the shaft from acting as a chimney in the event of fire. The pipe runs should be lined with waterproof material to prevent moisture from penetrating the walls, and they should be provided with "safe" drains at various levels in the building to catch leakage or water of condensation that may collect on the colder pipes in humid weather. Hot-water pipes, cold-water pipes, drainage pipes, gas pipes, electric conduits, and similar pipes can be grouped in a pipe run. It is undesirable to place hot and cold pipes too close together, as the temperature of one may affect the temperature of the other. The most objectionable effect may be heating of the cold-water supply. The outside of the pipe run should be paneled or otherwise covered so as to be easily and inexpensively uncovered and recovered.

In cold climates neither cold-water nor hot-water pipes should be located in outside walls unless adequately insulated. Hot-water pipes in which there is inadequate circulation freeze more quickly than cold-water pipes because of their higher freezing point due to the absence of dissolved gases from the water.

7-10. Pipe Alignment and Slopes. Pipes should be laid as straight as possible to minimize friction losses, and they should be free from upward bends and downward dips to avoid air pockets and sediment traps. Good alignment is necessary especially where threaded joints are to be made. Slight misalignment can be adjusted in bell-and-spigot joints. A flexible hose coupling[1] will make possible the adjustment of some poor alignment, but its use is usually not permissible under most codes.

Water-supply pipes should be sloped at not less than about $\frac{1}{4}$ in. per ft to a point of drainage where a valve is placed to permit emptying the pipe. Horizontal pipes should be only approximately horizontal, the slope being toward the stop-and-waste valve or intermediate drain valve. The slope of the pipe should be more than enough to place any support below the pipe sag between it and the next higher support, as explained in Sec. 7-11.

7-11. Pipe Supports.[2] The breakage of pipes caused by movements of the building can be avoided by placing the pipe in supports that will permit some motion between the pipe and the support. This is possible in all the supports shown in Fig. 7-10 except M, which is known as a *drop elbow*. The pipe is threaded into it, and it is screwed to its support. Pipes should be adequately supported to prevent sagging. The movement of structural members in wood buildings is conducive to sagging of pipe supports and to stress in pipes. Pipe should not be rigidly attached to the building, as movement between the pipe and its support should be possible. Vertical timber supports, such as shown at G in Fig. 7-10, are preferable to horizontal beams, as the movement of columns is usually less than that of beams in wood-framed buildings. The hook support shown at P is not altogether desirable because of the possibility that the hook will sag. Since pipe may vibrate in its supports, some form of soft

[1] See also *Domestic Eng.*, February, 1957, p. 148.

[2] See also Detail Sheet in *Air Conditioning, Heating, and Ventilating*, July, 1956, and March, 1956, p. 144.

material, such as felt, or steel springs[1] should be used to dampen the motion or otherwise deaden the sound. Vertical pipe supported from the ground should be supported on a concrete, brick, or stone pier or a cast-iron pedestal. Wood is not desirable because of its tendency to rot.

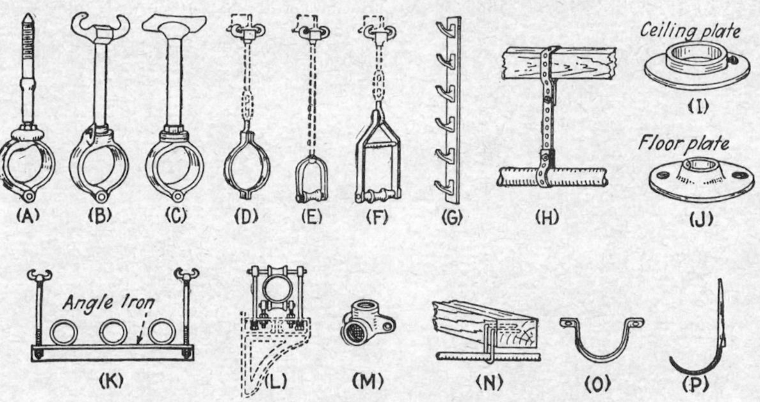

Fig. 7-10. Types of pipe support.

Lead pipe should be supported for its entire length wherever sagging is possible. Cast-iron and vitrified-clay pipe, when laid horizontally or sloping, should be supported at each joint; wrought, brass, and copper pipe, when horizontal or approximately so, should be supported at least every 10 ft.

The strength required of the support depends on the strength of the pipe and on the load carried. A rule of thumb, satisfactory for small plumbing installations, is to support with a ½-in. rod for 5-in. pipe and a ⅜-in. rod for 4-in. pipe and smaller.

7-12. Horizontal Distance between Pipe Supports.[2] Horizontal or sloping pipes must be supported to prevent sags that will produce traps to allow the accumulation of deposits in them or will allow the retention of air or vapor in high spots. Each length of pipe should be independently supported; it should not depend on the neighboring pipe for support. The proper slope on which supports are placed and the distance between supports can be computed so that each point of support is lower than the nearest upstream point of sag, as indicated in Fig. 7-11. Information for the computation is given

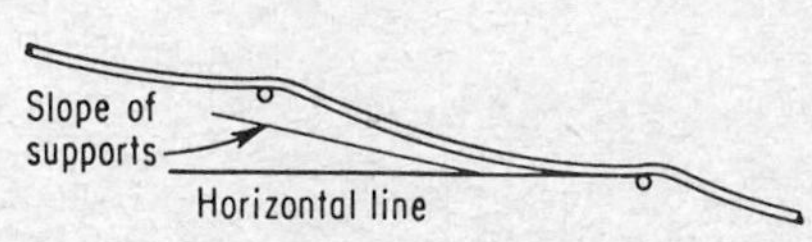

Fig. 7-11. Showing slope of pipe supports required to prevent pocket in sagging pipe.

[1] See also *Air Conditioning, Heating, and Ventilating*, February, 1956, p. 128.

[2] For Spitzglass formula and solutions see *Heating, Piping, Air Conditioning*, March, 1957, p. 135.

in Table 7-5. The use of the table requires knowledge of the permissible deflection between supports. Distance for deflections other than those shown in the table can be interpolated logarithmically.

TABLE 7-5. APPROXIMATE DISTANCES BETWEEN SUPPORTS FOR KNOWN DEFLECTIONS OF STEEL PIPE FILLED WITH WATER*
(Distances in feet)

Deflection, in.	Pipe sizes, in.												
	3/4	1	1 1/2	2	2 1/2	3	3 1/2	4	5	6	8	10	12
1/16 = 0.0625	...	5	9	13	14	15	16	17	18	20	23	25	27
1/8 = 0.125	6	7 1/2	12	14	16	17	18	20	22	23	26	29	32
3/16 = 0.1875	8	10	14	16	18	20	21	22	25	27	30	33	35
1/4 = 0.25	9 1/2	11	16	17	19	21	23	24	26	28	32	35	38

* Data in the table have been computed from information given by G. Metry in *Heating, Piping, Air Conditioning*, January, 1958, p. 198.

Vertical pipes are usually supported at each floor, the support being arranged so that the settlement of the building will allow the pipe to lift out of the support, as illustrated in Fig. 7-12. A joint, or coupling, is shown at the support, as is desirable. Pipes passing through floors, particularly through concrete floors, should do so through loose sleeves. The pipes should not be encased in concrete or plaster, yet the passageway should be too small to permit the passage of pests such as rats, cockroaches, or silverfish.

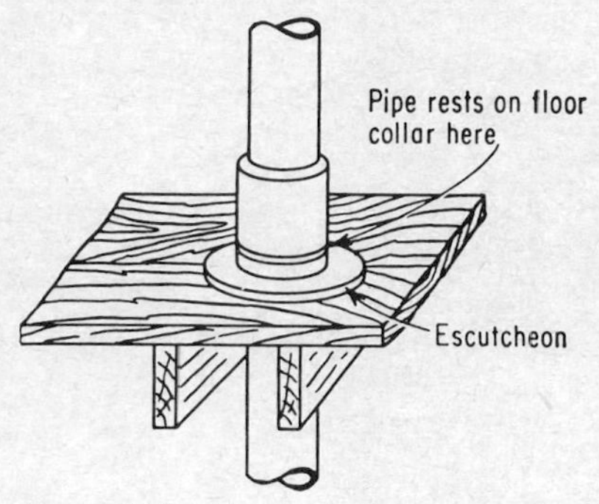

FIG. 7-12. Support for a vertical pipe.

7-13. Movement of Buildings. The movements of buildings are of such magnitude and have resulted so often in the breakage of plumbing pipes that provision should be made to avoid the damage even in the smallest buildings. The movements of buildings are due to the settling of the foundation, to wind pressures, to expansion and contraction resulting from temperature changes, and to the drying or moistening of lumber in the structure. No exact measure can be placed on any of these for any building, but that movements are appreciable is attested by everyday experience in frame buildings when the plaster cracks, the floor boards draw away from the baseboards, doors will not close, and windows jam. Measurements of tall buildings show them to be many inches out of plumb in some cases.

As it is not possible to prevent such movements, the plumber provides for them in the installation of the pipes by the use of expansion joints, the location of the pipes, and in other ways that are discussed in the sections pertaining to the installation of pipes.

7-14. Shrinkage of Lumber. In drying, lumber shrinks appreciably, the greater change occurring across rather than along the grain. The average values for cross-grain shrinkage are shown in Table 7-6. Lumber will swell and shrink repeatedly as it absorbs moisture and as moisture evaporates from it.

TABLE 7-6. APPROXIMATE SHRINKAGE OF TIMBER IN DRYING

Timber	Transverse shrinkage*	
	Per cent	Approximate size to nearest 1/16 in. of a dry timber that was 2 by 10 in. when green
Light conifers (soft pine, spruce, cedar, cypress)	3	2 by 9 11/16
Heavy conifers (hard pine, tamarack, yew), honey locust, box elder, old oaks	4	2 by 9 9/16
Ash, elm, walnut, poplar, maple, beech, cherry, sycamore, and black locust	5	1 15/16 by 9 1/2
Basswood, birch, chestnut, blue beach, young locust	6	1 15/16 by 9 3/8
Hickory, young oak, red oak	10	1 7/8 by 9

* The longitudinal shrinkage of all timber is usually less than 1/10 per cent.

7-15. Pipe Stresses. Among the causes of failure of pipes can be included defective material, weak pipes, excessive pressure, movements of the building, freezing, and water hammer. Poor material can be avoided by purchasing material known to fulfill standard specifications.

It is not usually necessary for the designer to compute bursting and other stresses in plumbing pipes. With knowledge of the internal pressure to be expected, the proper strength of pipe specified therefor under ASA or other standards, as given in Appendix II of this book or elsewhere, is selected, no computation being necessary. Where pipe supports are adequate and movements of the building due to pipe expansion and contraction are properly provided for through expansion joints and flexible or sliding supports, stresses due to applied loads need not be computed. However, in the design of piping for some industrial plants the computation of pipe stresses may be necessary.[1]

[1] See also S. W. Spielvogel, "Piping Stress Calculations Simplified," 5th ed., published by author, Lake Success, N.Y., 1955; S. Crocker, "Piping Handbook," McGraw-Hill Book Company, Inc., New York, 1945; and "Piping Design and Engineering," The Grinnell Co., 1951.

7-16. Leaks in Pipes. A leak may occur in a pipe in the walls of a building, in the ground, or elsewhere in such a manner that water escapes into ground or sewer without detection or into the building, with resulting damage. In either case financial loss may be large. The cost of the loss of water through leaky pipes has been variously estimated. The principal factors controlling this are the size of the hole through which the leak is occurring and the pressure of the water. Under normal conditions it can be estimated that the losses are as shown in Table 7-7.

TABLE 7-7. APPROXIMATE WASTE OF WATER THROUGH LEAKS

Size of opening, in.	Gallons per day	Gallons per month	Cost per month at 10 cents per 1,000 gal
1/4	12,000	360,000	$36.00
3/16	8,000	240,000	24.00
1/8	4,000	120,000	12.00
1/16	1,000	30,000	3.00
1/32	400	12,000	1.20

The location of a leaking pipe within the walls or floor of a building requires experience and ingenuity, which should be supplemented by knowledge of the piping installation. It may be difficult to locate the source of a leak because, unfortunately, water does not always appear near the point at which it is leaving the pipe. Tearing out walls and floors is expensive. It can sometimes be avoided by the use of instruments, such as Darley's leak locater, which depends on the principle of the electric coil and magnetic field for the location of pipe. In costly buildings X rays and radioactive isotopes have been used to locate pipes and leaks.

The location of leaks in pipes is discussed in Sec. 18-10.

7-17. Noises in Plumbing.[1] Roaring, rumbling, crackling, rattling, whistling, and other sounds occur in all plumbing pipes—supply, vent, and drainage. Noises are due to cavitation, temperature and other stress changes, vibrations, and other causes. Jet action of water, particularly if it contains undissolved air, as the water passes through a small orifice at a high velocity sets up turbulence and cavitation which create noise. Noise is not likely to occur where velocities are below about 10 fps, but higher velocities may be expected in small leaks, meters, flush tanks, flushometer valves, slow-closing valves, and loose washers. The fluttering of a loose washer can produce a roar worthy of a steamboat whistle.

[1] See also F. W. McGhan, Noise in Domestic Water Systems, *Plumbing and Heating J.*, January, 1952, p. 48, and W. L. Rogers, Experimental Approach to the Study of Noise and Noise Transmission in Piping Systems, *Trans. Am. Soc. Heating Ventilating Engrs.*, vol. 59, p. 347, 1953, and vol. 62, p. 39, 1956.

Loud noises or even slight whispers detectable only in the silence of the night may be caused by temperature changes, load changes, or movements of the building. Water hammer causes sudden changes of pressure within the pipe, making the pipe strike against its supports or adjacent surfaces and set them in vibration. The water hammer itself is silent. Sounds resulting from it are not difficult to detect and diagnose. Although the cause of the noise may be understood, its prevention may present difficulties.

The condensing of water vapor in heaters and pipes causes noises that are typical and are easily recognized.

Noises may be transmitted rapidly and for long distances through pipes[1] and through the building structure. The transmission of sound

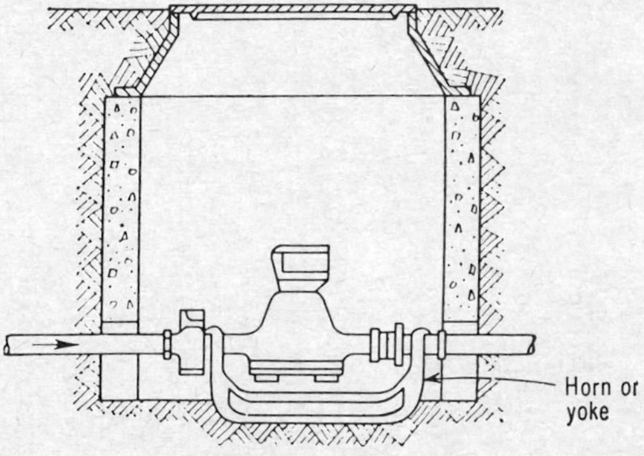

Fig. 7-13. Outside meter setting (Badger type of meter). The horn or yoke is used to aid in holding the meter in position and to relieve it from strain.

through solids is far more rapid than through air, so that the location of the cause of the noise may not be easy.

Noises can be avoided to some extent by precautions in design and installation. Some principles include the following: Avoid velocities above about 10 fps; select valves and equipment known to be silent; use flexible supports and soundproof material, such as felt or rubber, between contact surfaces. Remedies for noises after installation are discussed in Sec. 18-28.

7-18. Location and Setting of Meters.[2] Water meters may be located conveniently on the service pipe either inside or outside the building. Inside settings are more-or-less standard in climates where protection from freezing is necessary. An outside setting, such as is shown in Fig. 7-13, may be chosen for one or more of the following reasons:

[1] See also L. Blendermann, Noise Transmission in Piping Systems, *Air Conditioning, Heating, and Ventilating*, March, 1958, p. 98.

[2] See also Recent Meter Setting Developments, *J. Am. Water Works Assoc.*, November, 1957, p. 1459.

1. Convenience of location.
2. Fear of the possibility of illicit connections.
3. Recording all water entering premises, including leakage in service pipe.
4. Convenience in reading and servicing without need for entering building.
5. Water can be shut off without entering building. This is especially convenient in the event of fire. The meter is protected against destruction by fire.

However, the advantages of lower meter-reading costs with outside settings may be lost through greater installation costs and maintenance costs. Other disadvantages of an outside setting include possibilities of flooding, freezing, and inconvenience from covering of snow, ice, or excavated material.

Meters in an outside location in cold climates can be protected against frost by placing them in a vault, well below the frost line. The vault should be waterproof, adequately drained, with a double cover and dead air space between the covers. Additional insulation such as sawdust or mineral wool may be placed over the meter where conditions are most severe. The vault must be large enough to permit the inspection, reading, and minor repairs of the meter. Under some conditions meters have been equipped with an extension dial so that the meter can be placed deep in the ground while the dial, connected to the meter by an extension arm, can be placed near the ground surface for easy reading. The additional mechanism involved is an undesirable feature of such a setting.

A desirable setting is shown in Fig. 7-14. Inside locations are widely used because of lower first cost, convenience, protection from frost, and other reasons. When inside, the meter should be accessible, easily read and serviced, and accessible for repairs. A disk meter should be placed with the disk in a horizontal position to assure even wear of the disk. The meter should be located at a point remote from any dangers in the room. It should be placed not more than 2 or 3 ft above the general alignment of the water service pipe in order to avoid the accumulation of air or other gases. The meter should not be placed in a dip in the supply pipe because of the danger of the accumulation of grit or other detritus in the meter. Both the meter and the connecting pipe should be strongly supported to protect the meter and to avoid noise. If the water service pipe enters the building at an inconvenient elevation, a vertical turn can be made in the pipe to bring the meter about waist high above the floor without creating an air pocket or grit trap. As a space-saving expedient meters are sometimes placed in a vault beneath the floor. If such a location is chosen, there must be assurance that the vault will not be flooded and that it is adequately drained.

Meters of almost all types must be set so that they will always be full of water when operating. This condition is assured when the discharge is into a pipe or container constantly filled with water. Water should be admitted slowly to the meter to avoid damage from water hammer created if air is expelled too rapidly from the meter. Where the meter discharges into an open tank, the end of the discharge pipe from the meter must be submerged or the pipe must rise above the meter, allowing the water to

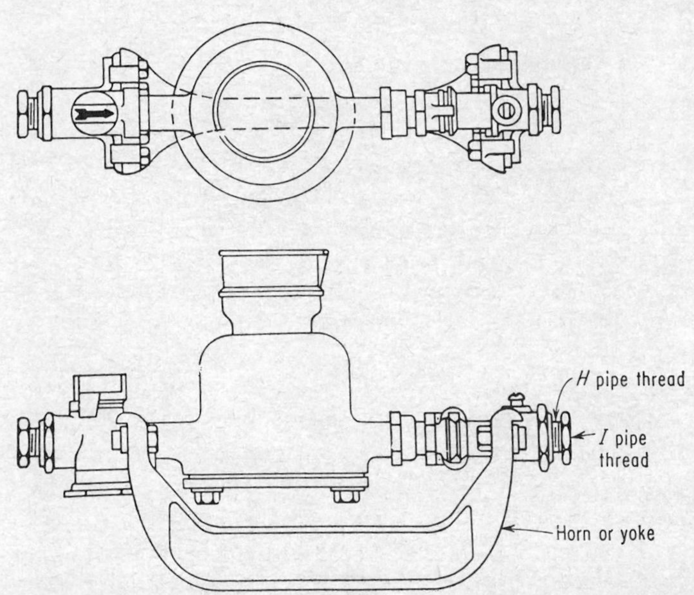

FIG. 7-14. Setting for household meter (Badger type of meter).

fall through the air into the tank. If a float-controlled valve is located in the tank, the valve should be placed above the meter to avoid water hammer or other shock to the meter.

The grounding of electrical equipment on water pipes is a widespread practice that is permissible with restrictions.[1] When the water meter is removed from the pipe, the path of the electric current to ground is broken unless an electric jumper is inserted across the gap. The possibility of uncomfortable or even fatal electric shock should lead the plumber to see that a copper wire, No. 8 or larger, spans the gap that will be left in the pipeline by the removal of the water meter.[2]

[1] See "Uniform Plumbing Code for Housing," Housing and Home Finance Agency, February, 1948, and F. M. Dawson and A. A. Kalinske, Studies Relating to the Use of Saran for Water Pipes in Buildings and for Service Pipes, *J. Am. Water Works Assoc.*, August, 1943, p. 1058.

[2] See also National Electric Code Standard of the National Board of Fire Underwriters, ASA approved in 1953.

Typical settings of displacement meters are shown in Figs. 7-13 and 7-14. Details to be noted in a meter setting include

1. The stop-and-waste valve on the water-main side of the meter and the gate valve on one side of the meter. Where only a stop valve, instead of a stop-and-waste valve, is used on the water-main side of the meter, a stop valve must be placed on the house side of the meter and provision must be made to permit drainage of the water pipes in the building when the water is shut off.
2. The provision for testing the meter in place. The removable connections shown are replaced by the test meter during the test.
3. The electric jumper.
4. The unions and flanges for connections. They are used for ease in connecting and disconnecting the meter.

Pipe dope or other material that will harden on the threads should not be used in setting a meter. The use of red or white lead is to be avoided, since it may interfere with the operation of the disk or clog the strainer. If used, it should be put on the outside rather than the inside threads of the connection. The use of fish traps and other flow-obstructing devices is to be avoided.

A check valve is sometimes placed on the house side of a meter to prevent hot water or other damaging liquids from backing through the meter. The use of a check valve is dangerous unless provision is made for pressure relief in the plumbing.

Before a water meter is put into service, the air in it should be slowly displaced by water to avoid damage to the meter by water hammer.

7-19. Expansion of Pipes. The expansion of water-supply and drainage pipes, particularly of hot-water pipes, is sufficient to require attention in design. The coefficients of expansion of the various materials used in pipes are given in Table 7-8.

TABLE 7-8. COEFFICIENT OF EXPANSION OF PIPES

Metal	Coefficient of expansion per °F	Change in length of 100 ft of pipe for 100°F change of temperature, in.
Wrought iron	0.00000686	1/16
Steel	0.0000061	1/16
Cast iron	0.0000059	1/16
Copper	0.0000095	3/32
Brass	0.0000104	7/64
Lead	0.0000159	5/32

The movement of a brass hot-water pipe 100 ft long when the temperature changes from 40 to 212°F is equal to

$$0.0000104 \times 100 \times (212 - 40) \times 12 = 2\tfrac{1}{8} \text{ in.}$$

approximately. It is evident that the movements are appreciable and must be cared for. This can be done by the use of swing joints in threaded pipe or with flexible bends, both of which are shown in Fig. 7-15. Supports in which the pipe can move should be used. Many such supports are shown in Fig. 7-10. Expansion joints should be used in hot-water pipelines at least every 50 ft, and the pipe should be well supported between expansion joints. The pipes should be free to change in length at the expansion joints.

Expansion in cast-iron pipe is usually not provided for because of the relatively low coefficient of expansion of cast iron and the conditions of its installation. When it is used as a cold-water supply pipe, the changes in temperature are slight, and when it is used as a drainage pipe, the

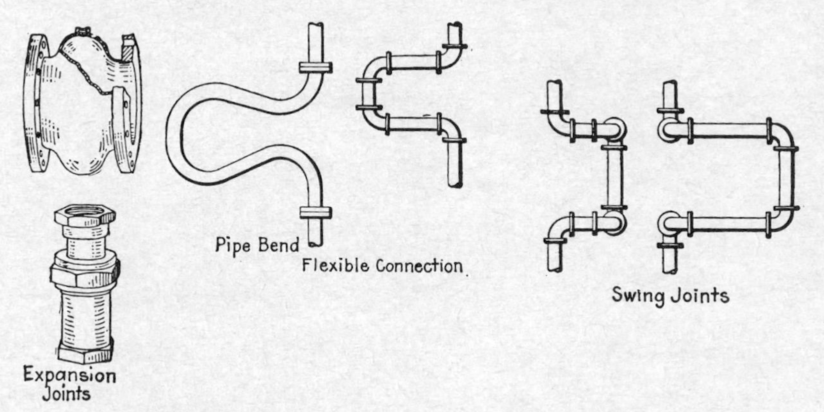

FIG. 7-15. Provisions for expansion in pipe.

normal changes of temperature in a building are within 50 to 60°F and the building and pipe are subjected to the same temperatures so that they tend to move together. The increase in length of 100 ft of cast-iron pipe with a 50° change of temperature will be about ⅜ in.

In order to minimize the movements of drainage pipes due to temperature changes, the discharge of hot wastes into them is usually prohibited.

7-20. Insulation of Pipes.[1] Insulation is used to prevent loss of heat from hot-water pipes, to prevent the freezing of hot water and cold water in pipes, and to prevent condensation of moisture in the air onto cold-water pipes. Insulation adds no heat to the water. If the temperature of the water drops below freezing and remains there long enough, ice will form regardless of the amount of insulation. A valuable effect of insulation, which diminishes changes of temperature in water pipes, is to minimize expansion and contraction.

[1] See also C. T. Littleton, "Industrial Piping," McGraw-Hill Book Company, Inc., New York, 1951.

Insulating materials include hair felt, wool felt, asbestos, asbestos air cell, magnesia, mineral wool, cork, and sawdust. Any material that may catch, absorb, or hold moisture is unsatisfactory for insulating purposes. Sawdust is, in general, unsatisfactory for this reason. A resilient "urethane" foam has been used with satisfaction for temperatures up to 200 to 250°F. The material can be cut with a knife and can be wrapped around the pipe to be insulated. One form of pipe-insulating material is illustrated in Fig. 7-16, and some dimensions are given in Table 7-9. Specifications are mentioned in Table II-1.

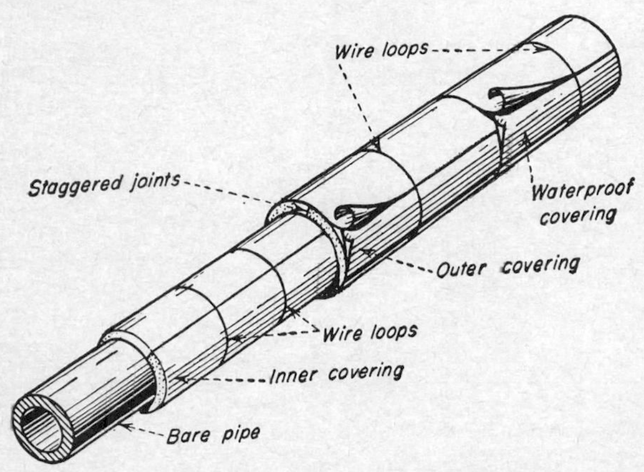

FIG. 7-16. Double thickness of pipe protection with weatherproof covering.

Satisfactory insulating will be attained by the use of 1-in. asbestos covering or of a standard thickness of 85 per cent magnesia pipe covering.

7-21. Pipe Connections. Connections between pipes that are to be used under pressure are made with screwed couplings, screwed unions, flange unions, and calked joints in bell-and-spigot pipes and in other ways. Mechanical joints and rubber-packed or filled joints have been used more commonly since the Second World War, partly because of the comparative scarcity of lead. Other types of joint, including those used in drainage pipes or pipes carrying fluids under low pressure, are discussed in Sec. 12-16.

Flexible and expansion joints of the type shown in Fig. 7-15 are used where movements of the pipe and of the building are to be cared for. Methods and materials used in making joints are discussed in Sec. 12-16.

In laying out a piping system the designer should consider the convenience of the plumber in making joints and installing supports.

7-22. Pipe-joining Materials and Their Uses. Pipe joints may be classified in accordance with the manner in which the pipes are connected, as, for example, bell-and-spigot, threaded, and flanged; techniques used in making the joints, as, for example, wiped or soldered, calked, and slip;

TABLE 7-9. THICKNESSES OF PIPE INSULATION IN INCHES

Pipe size, in.	Carey Corp. 85 per cent magnesia					Johns-Manville 85 per cent magnesia					Eagle Pitcher Co. ‡
	Cold water *	Temperature, °F									
		150 250	250 350	350 450	450 550	100 199	200 299	300 399	400 499	500 599 †	
1½ and less	1	1	1	1½	2	1	1	1	1½	1½	1
2	1	1	1	1½	2	1	1	1½	1½	2	1
2½	1	1	1	1½	2	1	1	1½	2	2	1
3	1	1	1	1½	2	1	1	1½	2	2	1
3½	1	1	1	1½	2	1	1½	1½	2	2	1
4	1	1	1½	2	2½	1	1½	1½	2	2½	1
4½	1	1	1½	2	2½	1	1½	2	2	2½	1
5	1	1	1½	2	2½	1	1½	2	2	2½	1
6	1	1	1½	2	2½	1	1½	2	2½	2½	1
8	1	1	1½	2	2½	1	1½	2	2½	3	1
10	1	1	1½	2	2½	1	2	2½	2½	3	1
12	1	1½	2	2½–3	3	1	2	2½	3	3	1

* To prevent freezing.
† 85 per cent magnesia not recommended above 600°F.
‡ Hot-water and low-pressure steam.

or the material used, as, for example, lead, cement, bituminous, etc. Some joints are best suited for pipes conveying fluids under pressure, as in water-supply pipes; others for conduits conveying fluids at approximately atmospheric pressure, as in drainage pipes; and some joints may be suitable at any pressure ordinarily encountered in plumbing. Types of joint are listed in Table 7-10. Joints most suitable for drainage pipes are discussed in Sec. 12-16. In addition to the joints listed in the table slip joints requiring no additional material to complete them are used and a smooth, threadless joint has been announced which grips adjacent pipe ends through a brass clutch ring.[1]

7-23. Bell-and-spigot Joints in Cast-iron Pipe. When a bell-and-spigot joint is made in a cast-iron pipe, either water-supply or drainage pipe, the pipes are first supported in position. An oakum gasket from 1½ to 1 in. thick is calkled tightly into the bell and around the spigot with a calking tool and hammer such as are shown at *C* and *D* in Fig. 7-8. The calking must be accomplished with skill, since there is danger of breaking the fitting if it is struck too hard. After the oakum is placed and the pipes are in a vertical position with the bell end up, the lead may be poured in either of two ways:

[1] See *Air Conditioning, Heating, and Ventilating,* May, 1957, p. 115.

TABLE 7-10. MATERIALS USED IN PIPE JOINTS

Pipe material	Type of joint	Materials used
Cast iron to cast iron	Bell-and-spigot	Oakum and lead wool, or poured lead
Cast iron to cast iron, wrought pipe, brass, or copper	Flanged	Rubber, copper, or asbestos gasket
Wrought pipe to wrought pipe	Threaded	Graphite, lead compounds, or lampwick
Brass or copper to brass, copper, cast iron, or wrought pipe	Flanged	Machined, rubber, asbestos, or copper gasket
Brass or copper to brass, copper, cast iron, or wrought pipe	Threaded	Graphite, lead compounds, or lampwick
Lead to lead, brass, or copper*	Wiped	Solder
Vitrified clay to vitrified clay or cast iron	Bell-and-spigot	Cement, bituminous materials, sulfur, and sand
Cast iron to cast iron	Bell-and-spigot (rust)	Iron filings, sal ammoniac, sulfur
Wrought pipe to wrought pipe	Threaded (rust)	Sal ammoniac, sulfur
Cast iron to wrought pipe	Threaded	Graphite, lead compounds, lampwick

* Lead cannot be soldered to cast iron or wrought pipe. A brass ferrule with a threaded end is soldered into the lead pipe and screwed or calked into the iron pipe or fitting.

1. The lead is heated until it is cherry red. A small amount of lead is then dipped out of the pot with a ladle, and a thin layer is poured into the joint and calked with quick, light taps. When the first layer has hardened, a second pour is made to fill the hub completely slightly above the rim. If too much lead is used, calking will not expand the lead at the bottom of the pour and lead will be wasted. The outer edge of the lead is then calked with light taps proceeding slowly around the rim of the hub. This last calking is done immediately after the lead has hardened in the joint. It usually hardens sufficiently as it is poured so that a joint can be calked within a few minutes after pouring. The inside edge is calked in a similar manner, and the surface of the lead is smoothed to finish the appearance of the joint.

2. The joint is first calked with oakum, after which the entire space is filled with molten lead and calked. Calking is necessary to compensate for shrinkage of the lead as it cools.

The amount of lead and oakum required for joints is shown in Table 7-11. The lead is melted over a gasoline furnace and is poured into the joint from an iron ladle or dipper such as is shown in Fig. 7-8. The ladle should be warmed before being dipped into the molten lead, and care should be taken to keep the ladle, the furnace pot, and the lead free from dirt.

TABLE 7-11. WEIGHT OF LEAD AND OAKUM IN CALKED JOINTS. APPROXIMATE QUANTITIES TO BE USED IN ESTIMATING

Diameter, in.	Pounds of lead per joint		Pounds of oakum per joint		Diameter, in.	Pounds of lead per joint		Pounds of oakum per joint	
	Soil pipe	Water pipe*	Soil pipe	Water pipe*		Soil pipe	Water pipe*	Soil pipe	Water pipe*
2	1½–2	....	0.21		7	5¼–7	....	0.73	
3	2¼–3	....	0.31		8	6–8	13¼	0.83	0.44
4	3–4	7½	0.42	0.21	10	7½–10	16	0.94	0.53
5	3¾–5	....	0.52		12	9–12	19	1.25	0.61
6	4½–6	10¼	0.63	0.31					

* Thickness of lead 2 in.

Where the pipes are in such a position that the bell is not facing vertically upward, an asbestos gasket is clamped tightly around the pipe to fit snugly against the bell, as shown in Fig. 7-17, so that its ends do not quite meet at the highest point. Molten lead is poured into the opening, and the gasket is removed as soon as the lead has cooled sufficiently. Care must be taken to have the joint dry before lead is poured into it, as otherwise steam might be generated and the lead blown from the joint with explosive violence.

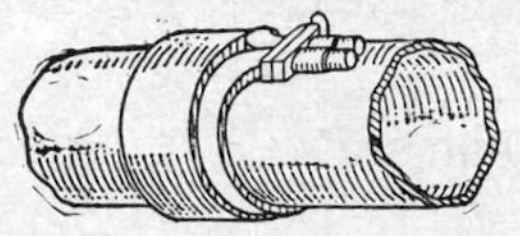

FIG. 7-17. Asbestos joint runner.

Oakum used in bell-and-spigot joints in cast-iron pipes and vitrified-clay pipes consists of loose hemp or material frayed from ropes. It is usually soaked in a tar preparation or creosote in order to preserve it and to render it less absorbent of moisture.

Lead for bell-and-spigot joints in cast-iron pipes is pure cast lead, lead wool, shredded lead, or leadite. Lead wool and shredded lead are calked cold into the joint. They have the advantage that they can be used under water or in wet places where molten lead cannot be used. Greater skill and more labor are required in making such joints. In making these joints the lead wool is calked into the joint in strands about ½ in. in diameter and 1 to 3 ft long, rolled compactly, and tapered at each end. Each strand is calked separately, and the joints between strands are staggered. Leadite is a composition of iron, sulfur, slag, and salt, finely ground and thoroughly mixed. The material is melted and poured like lead at a temperature of about 400°F. It weighs 118 lb per cu ft as compared with lead weighing 708 lb per cu ft. Lead requires a temperature of about 327°F to melt it. Leadite is said to be easier to manipulate than lead; when cast in a joint, it is hard and vitreous; it is more elastic than lead; it does not squeeze out in case of pipe movements as does lead; it requires no calking; and, finally, it is less expensive than lead.

In roughing-in a new piece of work the pipe fitter must be careful that he does not install the piping so that

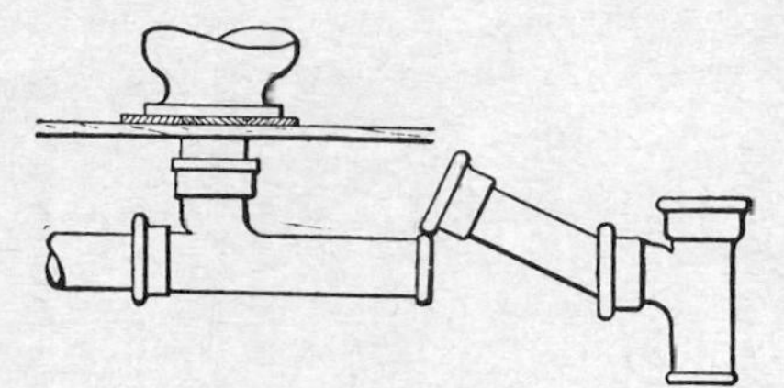

Fig. 7-18

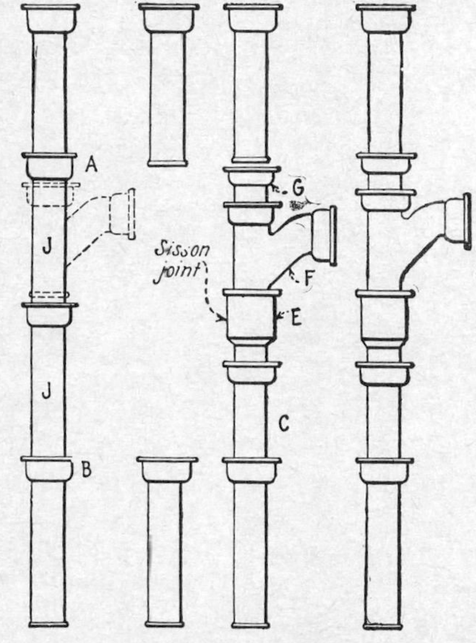

Fig. 7-19. Method of inserting a Sisson joint.

one piece of bell-and-spigot pipe or a fitting must be used to connect two pieces already in place. The remaining piece cannot be connected, as shown in Fig. 7-18. Bell ends should be pointed upward or horizontally, not vertically downward, and they should be kept away from obstructions so as to make calking easier.

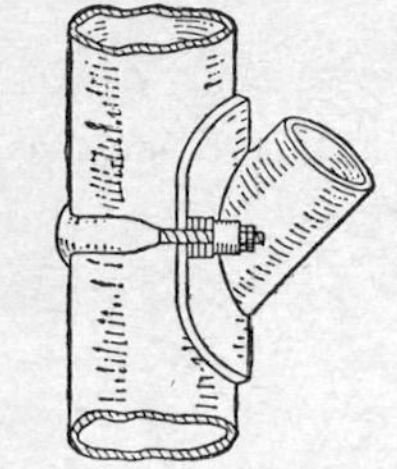

Fig. 7-20. Saddle fitting.

In some cases it is necessary to insert a fitting into an existing line of pipe, as shown in Fig. 7-19. The position of the new fitting is shown in dotted lines. The first step in inserting the fitting is to melt the lead from joints *A* and *B* and to smash pipes *J*. Then calk in the proper length of pipe *C*. Pieces *E*, *F*, and *G*, cut to proper length, are then slipped loosely into place. Piece *E* is a Sisson insertable joint that has a hub double the ordinary depth. Piece *E* is then calked to *C*. *G* and *F* are pulled up into place, and all joints are calked.

Another method of making the connection would be with a saddle fitting, as shown in Fig. 7-20. The use of saddle fittings is not, however, recommended because of dangers from leaking and clogging.

In planning piping with bell-and-spigot joints, a space must be left around the joints of at least 6 in., and a length of 2 ft or more is desirable for convenience in calking the joint.

7-24. Joints between Cast-iron and Other Piping. Joints between cast-iron, wrought-iron, steel, or brass piping may be either screwed or calked. The end of the threaded pipe should have a ring or half-coupling screwed onto it to form a spigot when a calked joint is to be made. Joints between iron and clay pipes are discussed in Sec. 12-26.

7-25. Joints between Lead and Other Piping. Joints between lead and cast-iron, steel or wrought-iron piping should be made with a calking ferrule, soldering nipple, or bushing. When joined to cast-iron pipe a brass ferrule may be wiped onto the lead pipe and calked into the cast-iron pipe; when joined to wrought-iron pipe a brass solder nipple wiped onto the lead pipe may be threaded and screwed into the wrought-iron pipe or fitting.

7-26. Threaded or Screwed Joints. Threaded joints require care in the cutting of the pipe and threads. The pipe should be cut square and reamed. To ream a pipe is to remove the burr often left by the cutting tool on the inside of the pipe. The threads are then cut according to the standards given in Sec. 7-27. Before the pieces to be connected are joined, an outside thread should be smeared with a lubricant but not so much that it will protrude into the pipe and, after hardening, obstruct it. Graphite compounds of pipe dope used on joints are generally proprietary articles consisting of finely ground graphite held in suspension or mixed with heavy oil or a light lubricating grease, together with other substances. Such material is applied to the outside threads only and lubricates and protects the threads so that a tighter joint can be secured than without the use of the lubricant. It protects the threads against corrosion, and as it hardens, it prevents leakage. Smooth-On products are iron compounds manufactured by the Smooth-On Company. They include joining and leak-stopping compounds for joining pipes, repairing breaks, and stopping leaks in pipes and castings. Red lead, white lead, or both, mixed with oil or grease, are also in common use as a lubricating, preserving, and tightening compound for threaded joints and is as satisfactory as the graphite mixtures. A thin sirup formed by mixing red lead or litharge with boiled oil may be applied to the threads with a paint brush. Lamp wicking wrapped about the outside threaded end of a pipe is used to some extent in securing tight joints. It is neither a lubricant nor a preservative. It makes a satisfactory joint.

Joints in threaded pipes are made with couplings such as shown in Fig. II-55, Appendix II. Where two pipes in position are to be joined, the connection is made in any one of three ways: by means of a box union, a flange union, or a right-and-left threaded pipe coupling. The method of joining a box or flange union is evident from the illustrations.

The positions of the couplings for making a right-and-left-hand coupling connection are shown in Fig. 7-21. The coupling at the left has a right-hand thread, and that at the right a left-hand thread. The number of threads cut on the pipes should be just sufficient to make the couplings tight when turned up. Couplings and unions should be used generously in pipelines to permit taking down portions of the pipe without dismantling large portions of the piping installation.

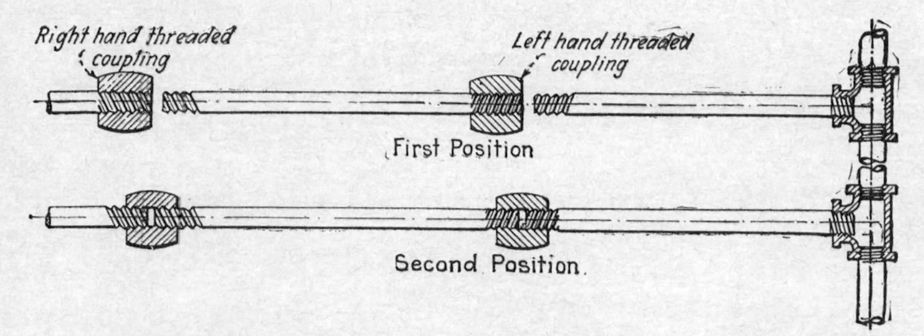

FIG. 7-21. Right-handed and left-handed threaded connection.

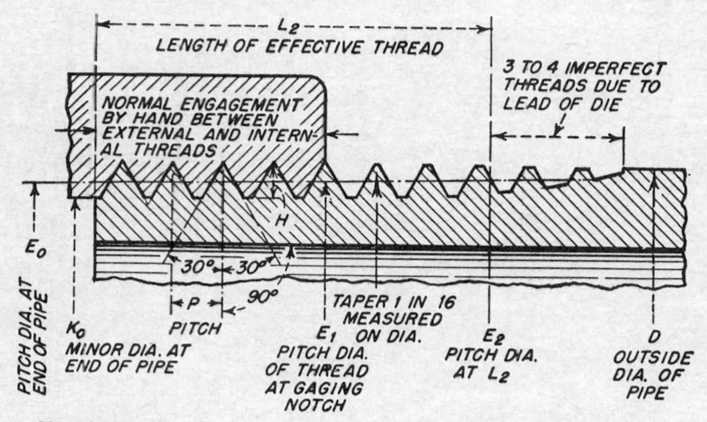

FIG. 7-22. Standard pipe threads (see Table II-43 and Fig. II-53).

In connecting pipes and couplings with screwed joints it is essential that they be accurately in line before the threads will catch. Since this difficulty increases rapidly with increase in diameter and length of pipe, adequate working space must be allowed in the layout of piping to facilitate the making of connections.

7-27. Standard Pipe Threads. The American, or Briggs, standard for pipe threads was adopted by the American Society of Mechanical Engineers on Dec. 29, 1886,[1] by the American Standards Association as B 33.1, and as Federal Specifications GGG–P 351. The more recent American National taper pipe thread, ASA B 2.1–1945 or FS–GGG–P 351a, is

[1] Standard Pipe Threads, *Trans. ASME*, 1886, p. 312.

commonly followed. Additional information on pipe threads is given in Appendix II. The Briggs standard has subsequently been revised by trade usage to conform to the conditions illustrated in Fig. 7-22.

7-28. Flanged Joints. Pipes may be joined by bolted flanges of the types shown in Figs. II-9 and II-12 and in Table II-13. The flanges may be machine-ground to a watertight fit, or gaskets may be used to form a watertight joint. Gasket materials that are used include rubber, rubberized cloth, asbestos composition, and copper. Cloth, rubberized compositions, and similar materials are made up about $\frac{1}{16}$ in. thick. The copper is corrugated and is made from No. 27 gage sheet metal. Gaskets

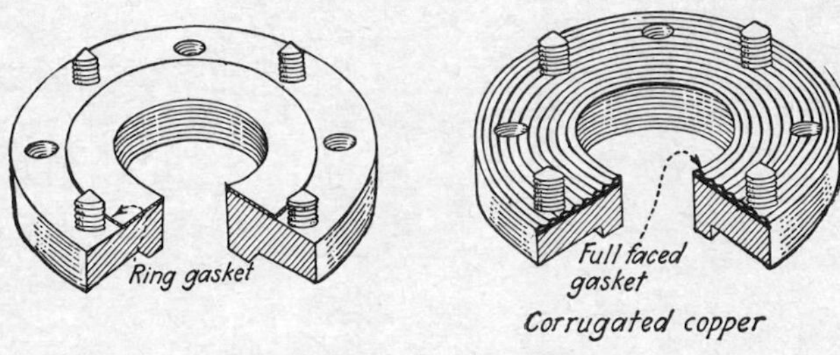

FIG. 7-23. Metal gaskets.

are called *ring gaskets* or *full-faced gaskets.* A ring gasket is shown in Fig. 7-23. It consists of an annular ring that fits inside the bolt circle and outside the outer diameter of the pipe. A full-faced gasket, shown in the same figure, is an annular ring covering the full face of the flange, having holes for the bolts. Gaskets are also used in those types of malleable-iron unions which do not have ground seats. Such gaskets are usually of rubber. All the gasket materials mentioned will make watertight joints. Considerations of economy and durability must be balanced in the selection between the various materials. Copper gaskets are high in first cost, but their life is practically unlimited. Rubber gaskets cannot be used on hot-water pipe.

Additional information concerning the dimensions of standard flanges is given in Table II-17 and Fig. II-12.

7-29. Rust Joints. Rust joints are made in bell-and-spigot cast-iron pipe or threaded iron pipe where it is desired to make a tight, permanent joint that will not be affected by the liquid passing through the pipe. In bell-and-spigot pipes the joint is first calked with oakum and is then filled with a mixture of powdered sulfur, 1 part; sal ammoniac, 1 part; and iron filings, 98 parts. To make a threaded joint rust, a solution of sal ammoniac alone is used.

7-30. Sweat Joints. A sweat joint is made by heating the pipes to be joined after they are in position with a blowtorch and allowing molten solder to flow into the small space provided between the pipes. This type of joint is used on copper tubing, pipes, and fittings that are slightly recessed or "rifled" to retain the solder in the joint.

7-31. Flared Joints. Flared joints are used in soft copper tubing and are made somewhat as indicated in Fig. 7-24.

7-32. Dimensioning and Measuring Pipe. In dimensioning pipelines on drawings the distance along the center line of the pipe is shown, the distance between pipes being shown to the center lines of the pipes. When pipes are cut to fit such dimensions, care and knowledge of standards are necessary.

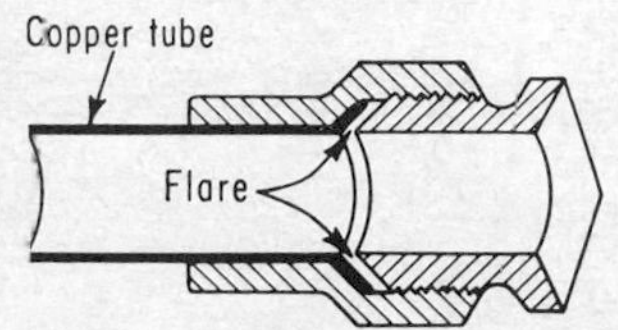

FIG. 7-24. Flared joint.

When pipe is cut where threaded joints are to be used, the length can be determined from the expression

$$L = D - (F_1 + F_2 - 2T)$$

The significance of the letters is shown in Fig. 7-25. Standards for sizes of fittings are given in Appendix II, Tables II-50 to II-52.

Example. As an illustrative problem, assume that in Fig. 7-25

$$D = 7 \text{ ft } 3\tfrac{1}{2} \text{ in.}$$

that the pipe is 2½-in. standard wrought pipe, and that it is connected into a malleable, 90° elbow and a wrought coupling.

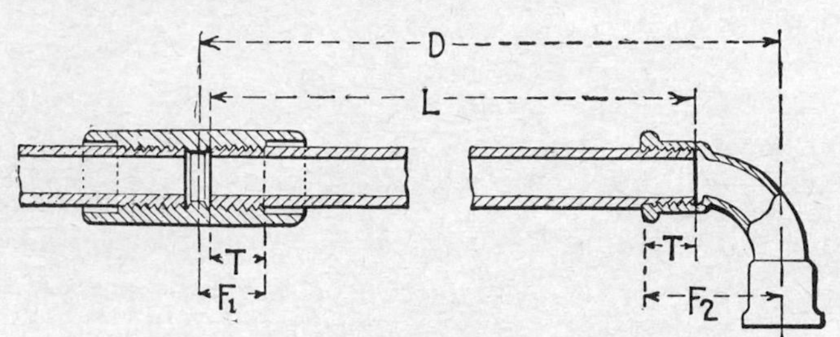

FIG. 7-25. Pipe measurement and threaded joints.

Then, from Table II-50, the value of F_1 is 1⁹⁄₁₆ in.; from Table II-51, F_2 is 2¹¹⁄₁₆ in., and from Table II-43, T = ⁹⁄₆₄ in. The value is

$$L = (7 \text{ ft } 3\tfrac{1}{2} \text{ in.}) - (1\tfrac{9}{16} \text{ in.} + {}^{2}\tfrac{9}{16} \text{ in.} - 2 \times 1\tfrac{9}{64} \text{ in.}) = 6 \text{ ft } 9 \text{ in.}$$

When threaded pipe is to be used with flanged fittings, a flange must be screwed onto the end of the pipe. The cut length of the pipe will then be ⅛ to ¼ in. less than $D - F_2$, as shown in Fig. 7-26.

If, as before, D = 7 ft 3½ in., then, from Table II-13, F_2 = 5 in. and, therefore, L = 6 ft 10¼ in.

In cutting and fitting bell-and-spigot pipe the problem is again slightly different, as shown in Fig. 7-27. The dimensions of the pipe and fittings, as shown in Fig. 7-27, are given in Table II-31. The difference between the sum of the known dimensions of uncut pipe and the fittings and dimensions 7 ft 3½ in. between center lines will give the length of the cut pipe to be 11½ in., as shown in the figure.

The determination of the length of pipe in angular directions sometimes presents problems that can be solved with precision only with a knowledge

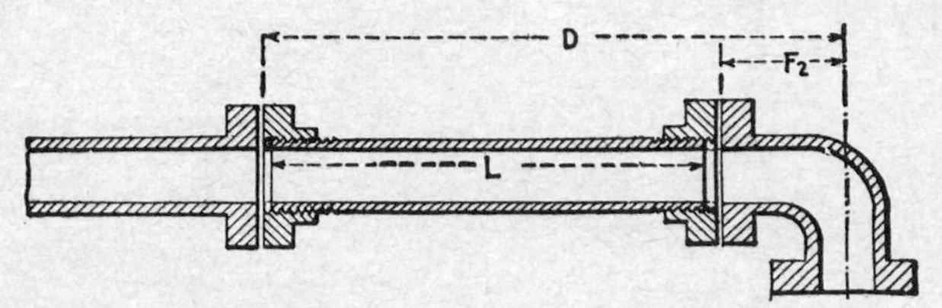

FIG. 7-26. Pipe measurement and flanged joints.

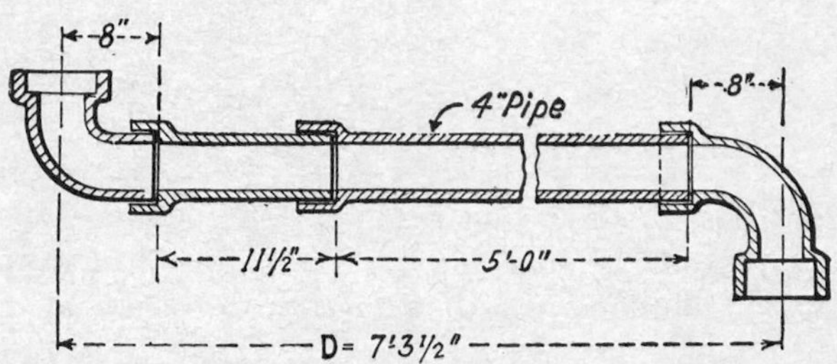

FIG. 7-27. Pipe measurements and bell-and-spigot joints.

of geometry and trigonometry but for which a satisfactory solution can usually be reached by "rule of thumb." Where a precise result is not desired, the use of Table 7-12 will generally be found satisfactory.

TABLE 7-12. LENGTH OF PIPE MEASURED ON AN ANGLE
(For sketch see Fig. 7-28)

Angle α degree of offset	Number of inches to add to each foot of length A to get length L	Number of inches to add to each foot of length B to get length L
5⅝	110⅜	0³⁄₃₂
11¼	37⅜	0¼
22½	19⁵⁄₁₆	1
30	12	1⅞
45	5	5
60	1⅞	12

For example, the following problem will be solved by the application of the principles of geometry and by the use of Table 7-12.

Example. It is required to join two vertical lines of pipe that are 18 in. apart by a pipe at an angle of 45° using galvanized malleable fittings as shown in Fig. 7-28. What length of pipe should be cut to fit between the two vertical pipes?

From the geometrical relations of the figure, the distance D is the hypotenuse of a 45° right triangle whose base is 18 in.; hence the length $D = \sqrt{(2)(18)^2}$ or $25\frac{1}{2}$ in. The length L is then determined as shown previously and in this case will be $20\frac{7}{8}$ in.

The same problem will now be solved by the use of Table 7-12. It is evident from the table that for every foot of distance between the parallel pipes, the length of D should be 17 in.; hence, in this case, the length D should be 1.5×17 or $25\frac{1}{2}$ in.

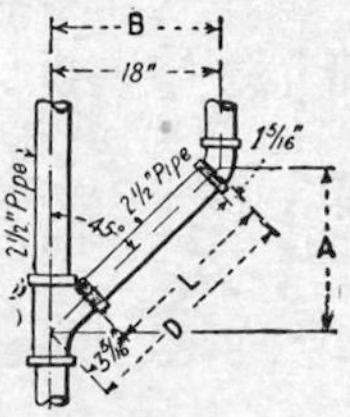

Fig. 7-28. Pipe measurements on an angle.

7-33. Cleanouts. Cleanouts are desirable on pipelines in order that clogging materials can be removed. They are used more commonly on drainage pipes than on water-supply pipes. A satisfactory type of cleanout consists of a plugged tee, a cross, or a wye, somewhat as shown in Fig. 7-29. Cleanouts for drainage lines are discussed in Sec. 11-29.

7-34. Prevention of Water Hammer.[1] The theory of water hammer is explained in Sec. 2-19. Its most common causes in plumbing are the sudden closing of valves or faucets, particularly of the automatic self-closing type and the quick-closing types, or through other methods of quickly stopping the flow of water in a pipe. Water hammer may be caused also by displacing air from a closed tank or pipe from the top, by the condensation of steam in water in a closed pipe, by reciprocating pumping machinery, by the sudden stoppage of a pump, and by other means. Water hammer can be prevented when a closed tank or pipe is being filled by filling it from the bottom, allowing the air to escape from the top. Steam and water should not be allowed to come into contact in a closed pipe. To this end, downward dips in steam pipes should be avoided or drained. The installation of an air chamber or air chambers may control water hammer, which would otherwise result from the sudden stoppage of the flow of water in the pipes.

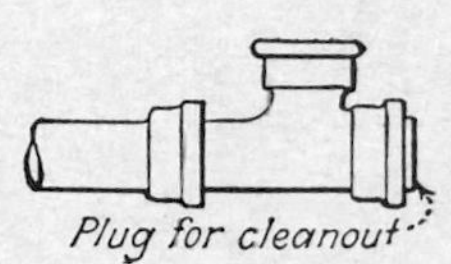

Fig. 7-29. Plugged tee for cleanout.

Other methods of avoiding water hammer include the use of slow-closing valves and faucets, of mechanical compensating devices, and of pressure-reducing valves. Other types of water-hammer arresters are shown in Figs. 7-30 and 7-31. Air chambers should be installed near the valve that is causing water hammer and, if possible, in a vertical position over the top of a riser pipe, as shown in Fig. 7-31.

The air chamber should have a capacity of at least 1 per cent of the total capacity of the pipeline in which the water hammer is occurring.

[1] See also L. Blendermann, Water Hammer Arresters, *Air Conditioning, Heating, and Ventilating*, November, 1957, p. 77.

The purpose of placing the chamber in the position shown in the figure is twofold: (1) Air in the riser pipe will be trapped in the chamber, aiding in keeping air in the chamber, and (2) it will receive the full thrust of the

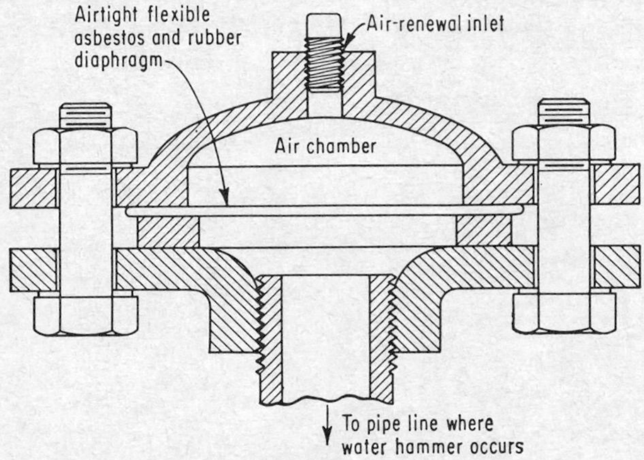

FIG. 7-30. Water-hammer arrester.

pressure from the vertical pipeline and will be more effective in its operation. Provision should be made for renewing the air in the chamber. This can be done by the use of a stop-and-waste valve and a pet cock or larger valve, as shown in Fig. 7-31.

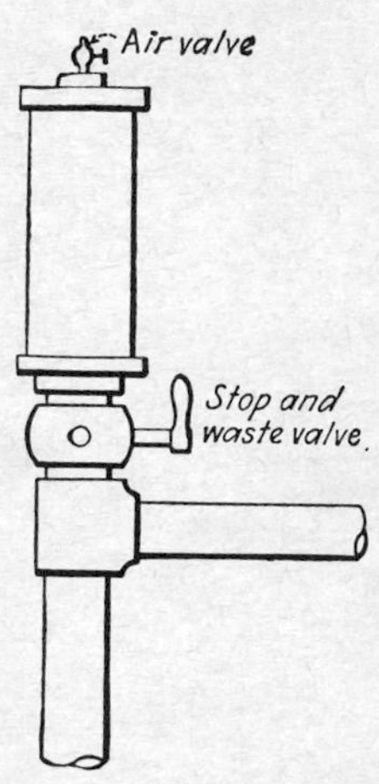

FIG. 7-31. Air chamber to control water hammer.

If the water hammer is created in the water main in the street, the house plumbing may be protected by locating an air chamber on the service pipe as it enters the building.

A mechanical compensating mechanism operates on a similar principle to the air chamber, except that the cushioning effect is supplied by a coiled spring within the chamber instead of by compressed air. Such a mechanism usually requires no attention for maintenance.

The installation of a pressure-reducing valve on the supply line to the source of the water hammer will result in a reduction of the velocity of flow in the pipe. It will be seen from the formulas for water-hammer pressure that the intensity of pressure varies directly with the velocity of flow in the pipe. The use of large-sized pipe will also cut down the velocity of flow, but to determine the exact size necessary to reduce the water hammer to any particular amount would require tests of an existing installation and a knowledge of mechanics and hydraulics.

7-35. Causes and Prevention of Freezing. The freezing of water in pipes is due to too long an exposure of the water in the pipe to a temperature below 32°F. It is to be noted that both time and temperature are factors in the freezing of pipes, because it takes time to remove sufficient heat from the water and its surroundings. Plumbing will sometimes pass through severe cold snaps without freezing, whereas a milder temperature of longer duration, possibly accompanied by high winds, will have a more serious effect. Unfortunately, when water freezes it expands with a force that cannot be resisted by any pipe manufactured. Water expands on freezing about one-twelfth of its volume, i.e., 12 cu ft of water will become 13 cu ft of ice. This increase in volume usually causes the bursting of the pipe.

The occasional failure of pipes to burst on freezing is explained by the conditions under which freezing has taken place. In order to burst a pipe the freezing water must be confined. If there is no confinement, the expanding ice pushes along the pipe without damage. This will explain why a pipe in a warm place may burst when freezing occurs in the basement. The expanding ice exerts a high pressure, which finds the weakest spot in the line. This may not be near the seat of the trouble. There is no value in trying to thaw a pipe by heating it near the point of bursting. The pipe must be thawed where the ice has formed.

Pipes can be protected against freezing by locating them in inside partitions and in other warm and protected places, by keeping the building warm when water is in the pipes, by covering the pipes with insulation, and by draining water from the pipes when the building is not heated. The thawing of frozen pipes is discussed in Sec. 18-18.

7-36. Condensation of Moisture on Pipes and Fixtures. Condensation of moisture on pipes and fixtures occurs when the temperature falls below the dew point of the surrounding air. This condition exists during periods of high atmospheric humidity and temperature. It results in the dripping of water from cold-water pipes, water-closet flush tanks, and other containers of cold water. The volume of water so condensed may be appreciable and may cause both nuisance and damage.

Condensation can be prevented by insulating the cold-water pipes and containers to prevent the contact of warm, humid air with them. It may be more important to insulate cold-water pipes and containers to avoid condensation than to insulate hot-water pipes to prevent loss of heat. Damage from condensation may be avoided by placing drip pans, with proper drainage outlets, beneath pipes and containers that are inadequately insulated.

7-37. Disinfection of Water-supply Equipment. Plumbing pipes and equipment that are to be in contact with potable water should be cleaned and disinfected before use. Surfaces should be scrubbed clean when accessible. After cleaning, the surfaces should be placed in contact with a water solution of chlorine containing not less than 50 ppm of available chlorine for a period of not less than 6 hr. If the concentration is 100 ppm, the time of contact can be reduced to 2 hr.[1]

7-38. Tests of Water-supply Pipes. Tests of the water-supply pipes should be made after roughing-in and before walls, ceilings, and floors are completed. The tests may be made with either water or air under pressure in the pipes. The tests are made by closing all openings and developing the desired pressure within the piping system. The pressures maintained during the test should be 50 to 100 per cent higher than the maximum pressure to which it is expected the system will be subjected under normal operation, and it should not be less than 100 psi of water pressure or 35 psi of air pressure. The increased pressure can be obtained by attaching a hand force pump to the plumbing system and raising the pressure the required amount. The amount of additional water required by the force pump after the pipes are filled is slight, usually less than 5 gal. The force pump should be equipped with a pressure gage. After the required pressure has been reached and the pump has stopped, a drop in the pressure will indicate a leak in the system. The pipes are then inspected by eye and ear for the leak. A similar procedure is followed after the fixtures have been installed and the plumbing system completed. The use of water rather than air in the test is usually more satisfactory, since the detection and location of leaks are easier. However, air can be used in freezing weather and the pressures throughout the system are more uniform than with water. The procedure to be followed in the air test when temperature changes affect the observations is explained in Sec. 12-30.

Tests on the water-supply pipes are seldom required in plumbing codes, but they should be made, nevertheless, in the interest of the owner and to assure good work on the part of the contractor.

[1] See also "National Plumbing Code," Sec. 10.9.

CHAPTER 8

HOT-WATER SUPPLIES[1]

8-1. Quantity and Temperature of Hot Water.[2] Information on rates of demand for hot water is given in Tables 3-10, 8-1, and 8-2. Rates of hot-water use in apartments vary with the grade of the apartment and other conditions. If the figures in Table 8-1 are taken to represent grade C apartments, then the ratios of use in grades A, B, and C apartments can be expressed as 1.67, 1.33, and 1.0.

A temperature at the faucet between 120 and 130°F is generally satisfactory for normal, residential needs. On washday, in dishwashing machines, and for other special purposes the required temperature may be higher. Temperatures can be maintained as high as 160°F in apartment buildings and in food-preparation and -serving establishments. In some industries the highest possible temperature at atmospheric pressure may be required. Approximately one-half as much water at a temperature of 180° is required to deliver as many heat units for domestic use as would be delivered by water at 125°F. However, much of the heat might be wasted, so that a temperature of 180° is generally considered uneconomical for domestic uses and may possibly scald the users.

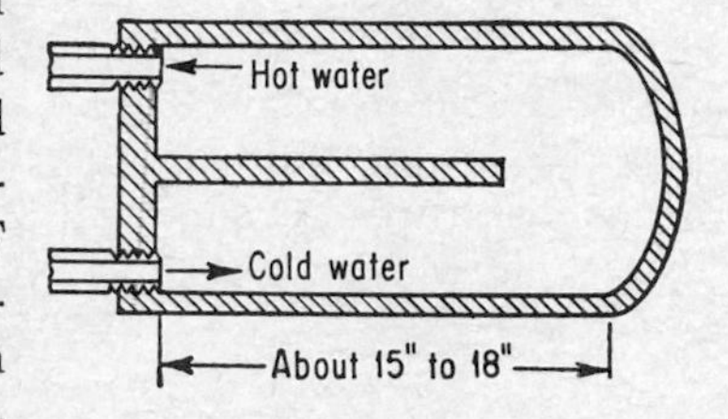

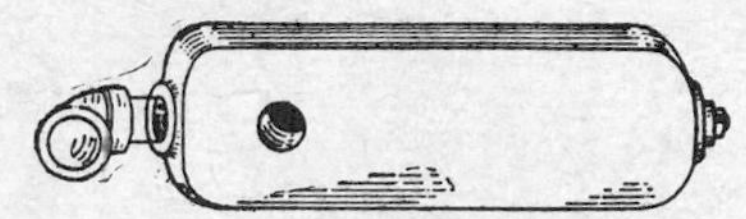

FIG. 8-1. Water backs.

8-2. Methods of Heating Water. Water for domestic use may be heated by water backs in kitchen ranges, as illustrated in Figs. 8-1 and 8-2; by water heaters containing a coil, as illustrated in Fig. 8-3; by a water coil placed in a furnace; by a gas or oil heater; by an electric coil or hot point; and by less common methods.

Water heaters can be classified as direct or indirect or as fuel fired,

[1] See also F. M. Reiter, Design Data for Service Hot Water, a series beginning in *Air Conditioning, Heating, and Ventilating*, February, 1956, p. 90, and W. Hutton and W. M. Dillon, "Hot Water Supply," published by *Plumbing and Heating J.*, 1951.

[2] See also M. B. Mackay, Large Volume Water Heating, *Am. Gas J.*, vols. 164 and 165, 1946.

TABLE 8-1. RECOMMENDED CAPACITIES OF HOT-WATER HEATING EQUIPMENT TO SUPPLY FIXTURES IN VARIOUS TYPES OF BUILDINGS*
(Gallons per hour per fixture)

Fixture	Apartment house	Club	Gymnasium	Hospital	Hotel	Industrial plant	Laundry	Office building	Public bath	Private residence	School	YMCA
Private lavatory	3	3	3	3	3	3	3	3	3	3	3	3
Public lavatory	5	8	10	8	10	15	10	8	15	...	18	10
Bathtub	15	15	30	15	15	30	...	...	45	15	...	30
Dishwasher	15	30	..	30	30	30	...	...	...	15	15	
Foot basin	3	3	12	3	3	12	...	...	...	3	3	12
Kitchen sink	10	20	...	20	20	20	...	...	...	10	10	20
Laundry stationary tubs	25	35	...	35	35	...	42	...	...	25	...	35
Laundry revolving tubs	75	75	...	100	150	...	75 to 100	...	100	75	...	100
Pantry sink	10	20	...	20	20	...	...	...	...	10	20	20
Shower	100	200	300	100	100	300	...	...	300	100	300	200
Slop sink	20	20	...	20	30	20	10	15	15	15	20	20
Dishwasher	200 gal per hr at 180° per 500 people											
Dishwasher†	300 gal per hr at 180° per 500 people											
Total water for all fixtures likely to be drawn at one time	35	60	80	75	60	90	100	20	100	50	25	
Per cent heater capacity‡	20	50	80	60	50	90	100	15	100	50	25	
Storage capacity in per cent of	100	75	50	50	25	50	25	100	50	100		
maximum heating capacity§	35	40	40	45	45	50	50	40	50	30	40	

Apartments. One kitchen and one bath

Number of families	up to 25	25–50	50–75	70–100	over 100
Gallons per hour per family	35	30	35	20	15

Swimming pools

Fill in 24 hr. Refilter every 24 hr. 50,000-gal pool good for 100 persons.

* A. Buenger, *J. Am. Soc. Heating Ventilating Engrs.*, vol. 26, p. 701, 1920, except as noted. See also F. M. Reiter in *Ind. Gas*, November, 1954, p. 8.

† Computed at final temperature of 150°F.

‡ The figures in this line are from discussion by Perry West in *J. Am. Soc. Heating Ventilating Engrs*, vol. 27, p. 253, 1921.

§ The figures in this line are from Reiter, *loc. cit.*

TABLE 8-2. ESTIMATED RATES OF USE OF HOT WATER*

Type of building	Gal per day	Type of building	Gal per hour
Dwelling, single-family:		Industrial plant:	6
No bath	30–60	Lavatory	12
One bath	60–100	Shower	225
Two baths	100–200	Slop sink	20
Dwelling, large family, two baths	200–300	Dishwasher	20–100
Apartment:		Demand factor	0.40§
One bath	60–125	Office:	
Two baths	100–200	Lavatory	6
Three baths	150–250	Slop sink	15
Hotel:		Demand factor	0.3§
Room and bath	20–70	Hospital:	
Room and lavatory	5–10	Lavatory	6
Restaurant:		Shower	75
Hand dishwashing, per meal	1–2	Slop sink	50–150
Machine dishwashing, per meal	1½–4½	Demand factor	0.20§
Laundry, per 100 lb dry clothes	80†	Hotel:	
Washing machines:‡		Lavatory	8
100-gal tank	150	Shower	75
150-gal tank	200	Slop sink	30
200-gal tank	300	Dishwasher	50–200
Industrial building, per occupant	5	Demand factor	0.25§
Office building, per occupant	5		
Commercial, per worker	2		
Manufacturing, per worker	5		

* Much of the information in this table is from F. M. Reiter in a series of articles in *Ind. Gas*, 1954.

† In addition, 100 gal of cold water, 30 lb of steam to heat the water, and 1 kwhr of electrical energy. These minimum figures are exceeded by 50 per cent by many average hand laundries and by 100 per cent by small mechanically equipped laundries.

‡ From F. M. Reiter, *Air Conditioning, Heating, and Ventilating*, February, 1954, p. 90.

§ Probable number of fixtures in use simultaneously.

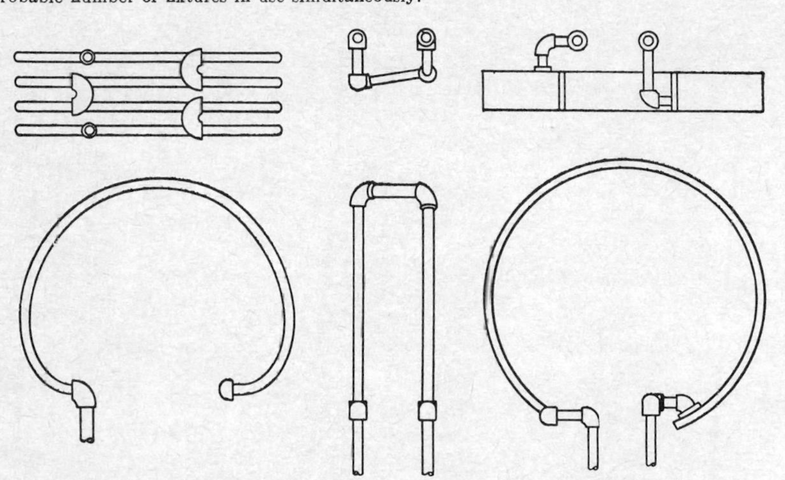

FIG. 8-2. Furnace coils and water back for coal-fired furnace.

electrically heated, or steam heated. In a direct heater, water comes in contact with a hot surface from which the heated water passes into a storage tank or directly to use. In the indirect heater a separate and enclosed water supply, from which no water is withdrawn, is heated and circulated as hot water or steam through a heating coil or other heat-transfer surface that is submerged in the water to be heated for the hot-water supply. Indirect heaters may be storage or tankless, or they may be external or internal with respect to the boiler.[1] Some indirect heaters are shown in Figs. 8-4 and 8-12. A study of comparative costs of heating water in a storage tank and in a tankless heater[1] shows the tankless heater to use about 30 per cent less gas under the conditions of the test.

FIG. 8-3. Water heater containing a coil. (*L. O. Koven Bros.*)

TABLE 8-3. CAPACITIES OF HOT-WATER HEATING APPARATUS RECOMMENDED FOR DWELLINGS AND APARTMENT HOUSES*

Number of families	Capacity, gal per hr	Coal heater: Grate area, sq ft	Coal heater: Heating surface, sq ft	Coal heater: Flue diameter, in., by height, ft	Coal heater: Coal used in 8-hr firing period, lb	Gas heater: Cubic feet of gas per hour	Steam heater: Pounds of steam per hour	Steam heater: Area of steam pipe required, sq ft	Steam heater: Btu added to water per hour	Steam heater: Required storage-tank capacity, gal	Steam heater: Circulating pipe size between heater and storage tank, in.
1	10	0.25	5.0	8 by 10	10	21	10.5	0.3	10,030	30	1
2	20	0.33	6.6	8 by 20	21	42	21.0	0.9	20,060	60	1¼
4	40	0.60	12	8 by 20	38	84	41.8	1.9	40,100	120	2
10	100	1.5	30	10 by 40	96	209	104	4.7	100,260	300	3
15	150	2.2	45	12 by 20	143	313	157	7.0	150,390	450	4
20	200	3.0	59	12 by 30	190	418	209	9.3	200,500	600	4
30	300	4.5	90	12 by 60	286	626	313	14.0	300,800	900	5
50	500	7.4	148	16 by 30	475	1,044	522	23.3	501,300	1,500	6
90	900	13.5	268	18 by 50	860	1,800	940	42.0	902,400	2,700	8

* W. S. L. Cleverdon, *Plumbers' Trade J.*, vol. 73, p. 164, 1922.

The direct heater, either fuel fired or electrically heated, is common in domestic installations. Indirect heaters[2] using steam coils are more common in large installations.

[1] See also L. L. Hill and W. S. Harris, Comparative Performance of Indirect Water Heaters, *Trans. Am. Soc. Heating Air Conditioning Engrs.*, vol. 62, p. 317, 1956, and L. Blendermann, External Indirect Water Heaters for Service Hot Water, *Heating and Ventilating*, November, 1954, p. 112.

[2] See also Know Your Basic Controls for Basin Hot-water Heating Boilers, *Power*,

Heaters are classified also as fast recovery or slow recovery. The recovery time of a heater is discussed in Sec. 8-9.

Gas, oil, and electric heaters are suited to automatic control and to heating water quickly. Solar energy is also used, in favorable climates, for heating water.[1] For quick heating, the water must be distributed among many small pipes or other heating units subject to a high temperature. Coils of copper tubing are commonly used with gas, and cast-iron boxes, called *water backs*, and iron-pipe coils are used with coal-fired heaters. Capacities of water heaters expressed in terms of hot water delivered are shown in Tables 8-3 and 8-6, and a comparison of types of hot-water heater is shown in Table 8-4.

TABLE 8-4. COMPARISON OF VARIOUS TYPES OF HOT-WATER HEATERS*

Type of heater	Heat input, Btu per hr	Storage capacity, gal	Gas consumption per month, cu ft	
			50-gal withdrawal daily	Standby, no withdrawal
Multiflue, slow recovery	5,500	45	3,460	760
Reversible flue, slow recovery	4,500	28	3,670	1,200
Instantaneous 2.5 gpm	120,000	28	3,670	400
Internal single flue, adjustable recovery	9,000	30	3,790	1,100
Internal single flue, adjustable to low heat input	2,500	30	3,800	1,000
Adjustable, quick recovery	25,000	22	4,190	1,600
Conversion burner, slow recovery	4,200	30	4,390	1,700
Side-arm coil, quick recovery	30,000	32	4,970	1,800
Side-arm coil, quick recovery	21,000	32	5,370	1,900
	4,200	30	5,960	2,300

* From F. M. Reiter, Design Data for Service Hot Water, *Air Conditioning, Heating, and Ventilating*, a series of articles beginning February, 1956, p. 90.

8-3. Overheating and Explosions. When water is heated in a closed container to a temperature above 212°F, the temperature and the pressure will rise and some water will vaporize if there is space for it to do so. Relations among temperature, pressure, and water vapor are shown in Table 8-5. If there is insufficient volume for vaporization to take place and the expansive force of water is exerted against the tank, pressures will be created that cannot be resisted by any plumbing material. Pressure relief is necessary to prevent an explosion. If the pressure in a tank containing water as a liquid at a temperature above 212°F is suddenly released, the water will vaporize with explosive violence. The energy

November, 1956, p. 124, and E. S. Ross and others, Water Heater Performance, *Plumbing and Heating J.*, March, 1953, p. 50.

[1] See also E. A. Farber and others, *Air Conditioning, Heating, and Ventilating*, October, 1957, p. 75.

TABLE 8-5. TEMPERATURE, VOLUME, AND BOILING POINT OF WATER

Temperature, °F	Relative volume	Relative density	Weight per cu ft, lb	Temperature, °F	Relative volume	Relative density	Weight per cu ft, lb	Approximate gage pressure, psi	Absolute pressure	Boiling point, °F	Approximate gage pressure, psi	Absolute pressure	Boiling point, °F
32	1.00000	1.00000	62.418	140	1.01690	0.98339	61.381	−14	1	102.018	17	32	254.002
35	0.99993	1.00007	62.422	145	1.01839	0.98194	61.291	−13	2	126.302	19	34	257.523
39.1	0.99989	1.00011	62.425	150	1.01889	0.98050	61.201	−12	3	141.654	21	36	260.883
40	0.99989	1.00011	62.425	155	1.02164	0.97882	61.096	−11	4	153.122	23	38	264.093
45	0.99993	1.00007	62.422	160	1.02340	0.97714	60.991	−10	5	162.370	25	40	267.168
46	1.0000	1.00000	62.418	165	1.02589	0.97477	60.843	− 9	6	170.173	27	42	270.122
50	1.00015	0.99985	62.409	170	1.02690	0.97380	60.783	− 8	7	176.945	29	44	272.965
52.3	1.00029	0.99971	62.400	175	1.02906	0.97193	60.665	− 7	8	182.952	31	46	275.704
55	1.00038	0.99961	62.394	180	1.03100	0.97006	60.548	− 6	9	188.357	33	48	278.348
60	1.00074	0.99926	62.372	185	1.03300	0.96828	60.430	− 5	10	193.284	35	50	280.904
62	1.00101	0.99899	62.355	190	1.03500	0.96632	60.314	− 4	11	197.814	37	52	283.381
65	1.00119	0.99881	62.344	195	1.03700	0.96440	60.198	− 3	12	202.012	39	54	285.781
70	1.00160	0.99832	62.313	200	1.03889	0.96256	60.081	− 2	13	205.929	41	56	288.111
75	1.00239	0.99771	62.275	205	1.04140	0.9602	59.93	− 1	14	205.604	43	58	290.374
80	1.00299	0.99702	62.240	210	1.0434	0.9584	59.82	0	‖	212.00	45	60	292.575
85	1.00379	0.99622	62.182	212	1.0444	0.9575	59.76	0	15	213.067	47	62	294.717
90	1.00454	0.99543	62.133	230	1.0529	0.9499	59.26	1	16	216.347	49	64	296.805
95	1.00554	0.99449	62.074	250	1.0628	0.9411	58.75	2	17	219.425	51	66	298.842
100	1.00639	0.99365	62.022	270	1.0727	0.9323	58.18	3	18	222.424	53	68	300.831
105	1.00739	0.99260	61.960	290	1.0838	0.9227	57.59	4	19	225.255	55	70	302.774
110	1.00889	0.99119	61.868	298*	1.0899	0.9175	57.27	5	20	227.964	57	72	304.669
115	1.00989	0.99021	61.807	338†	1.1118	0.8994	56.14	7	22	233.069	59	74	306.526
120	1.01139	0.98874	61.715	366‡	1.1301	0.8850	55.29	9	24	237.803	61	76	308.344
125	1.01239	0.98808	61.654	390§	1.1444	0.8738	54.54	11	26	242.225	63	78	310.123
130	1.01390	0.98630	61.563	...	......	......	.....	13	28	246.376	65	80	311.866
135	1.01539	0.98484	61.472	...	......	......	.....	15	30	250.293	67	82	313.576
											69	84	315.250
											71	86	316.893
											73	88	318.510
											75	90	320.094
											77	92	321.653
											79	94	323.183
											81	96	324.688
											83	98	326.169
											85	100	327.625
											90	105	331.169
											95	110	334.582
											100	115	337.874
											105	120	341.058

* Steam at 50 psi.
† Steam at 100 psi
‡ Steam at 150 psi
§ Steam at 205 psi
‖ Exact = 14.69 psi absolute

released when water turns to steam at atmospheric pressure is far greater than that released when steam suddenly expands from the same pressure to atmospheric pressure. Although steam issuing from a hot-water faucet is dangerous, superheated water issuing from a faucet will expand into steam with explosive violence. If temperatures are allowed to rise

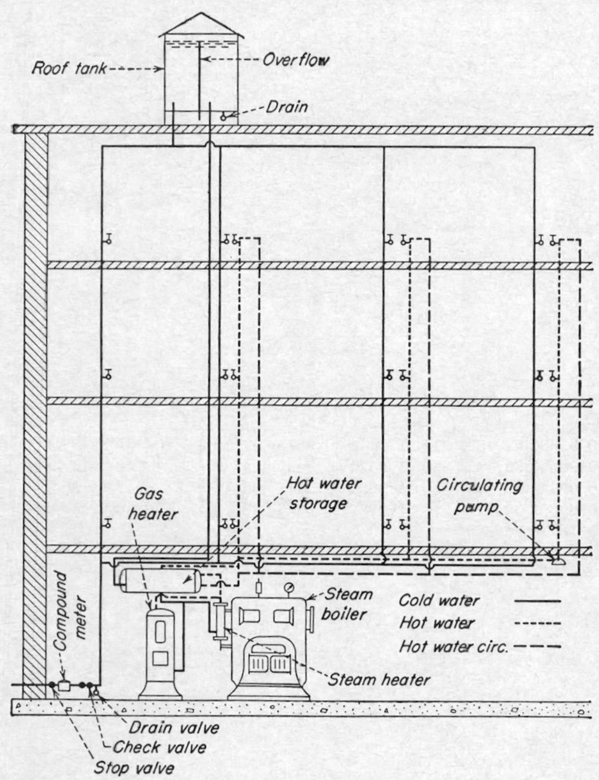

FIG. 8-4. Hot-water and cold-water distribution with both forced and gravity hot-water circulation and an expansion tank.

above 324°F, lead or solder may soften sufficiently to release pressure with a resulting explosion.

The overheating of water heaters can be controlled by thermostatically controlled valves that shut off the heat source at a predetermined level. The overloading of a tankless heater can be controlled by a water-flow regulator[1] without moving parts that limits the flow of cold water into a heater to its rated capacity. The overheating of water by coal heaters can be prevented by a device that opens the damper on the heater when the temperature of the water is not correct. The operation of this device

[1] See *Domestic Eng.*, April, 1957, p. 72.

depends on the generation of steam in it when the temperature reaches 212°F. Steam will not be generated in the heater at this temperature because of the greater pressure in the heater. The generation of steam in the device causes the closing of a damper and cutting down the inten-

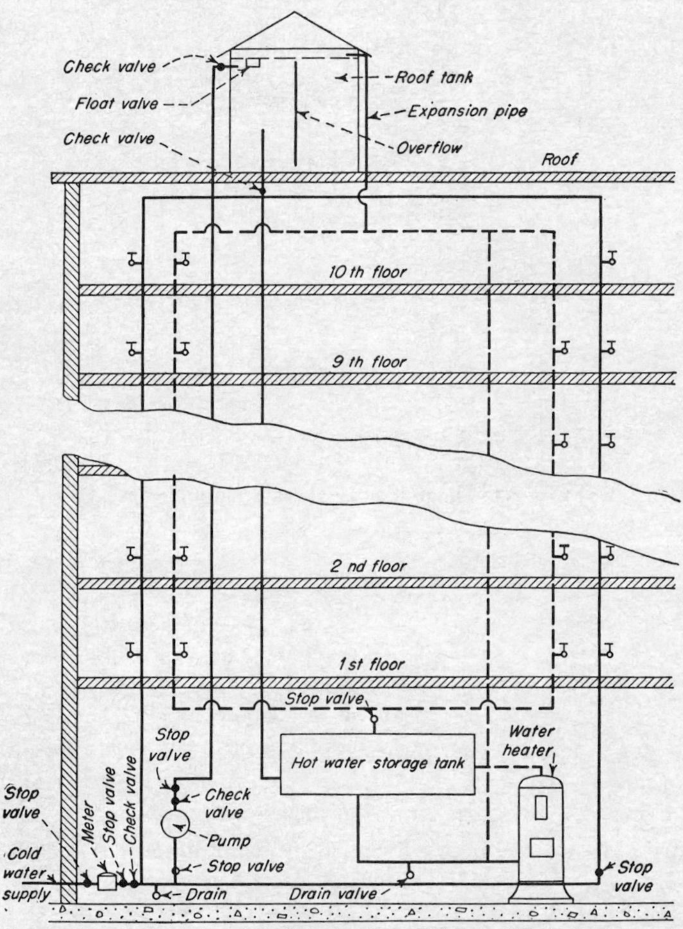

FIG. 8-5. Hot-water and cold-water distribution with gravity hot-water circulation and city or pumped cold-water pressure.

sity of the fire. Although the temperature of the water in the heater is higher than desirable, the generation of dangerous temperatures is avoided.

It is sufficient to heat water to 120 to 150°F for most domestic purposes, although 180° may be called for in dishwashing, washing clothes, and other special purposes. At these temperatures there is no danger of explosion.

8-4. Pressure and Temperature Relief. Pressure relief can be provided by permitting the expansion of hot water into the cold-water lines, by the use of pressure-relief and temperature-relief valves, and by such

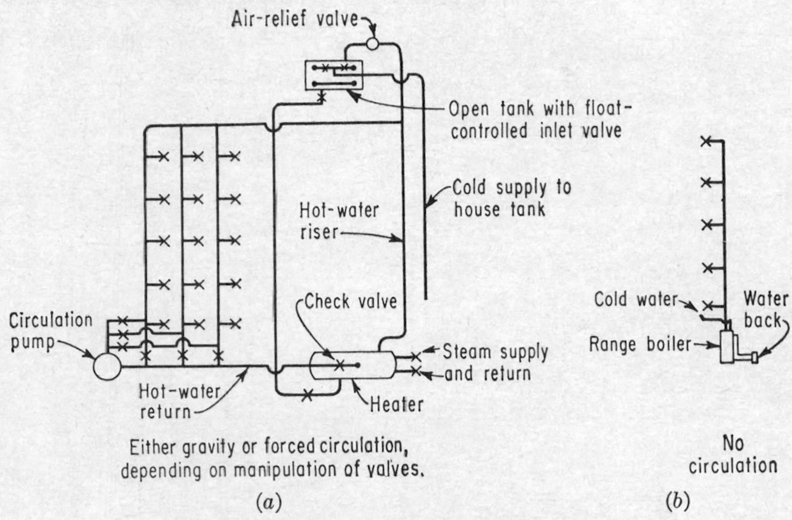

FIG. 8-6. Diagrams of hot-water-supply pipes.

installations as are shown in Figs. 8-4 to 8-6, involving the use of expansion tanks. Expansion into the cold-water line may not give relief if the cold-water line is clogged, as, for example, if it is frozen, the cold-water supply valve has been closed, or a check valve has been installed on the line. The use of a check valve without a relief valve is universally prohibited. The most certain relief, however, is given through a hot-water expansion tank, one form of which is shown in Fig. 8-7. Some dimensions are given in Table II-80 in Appendix II.

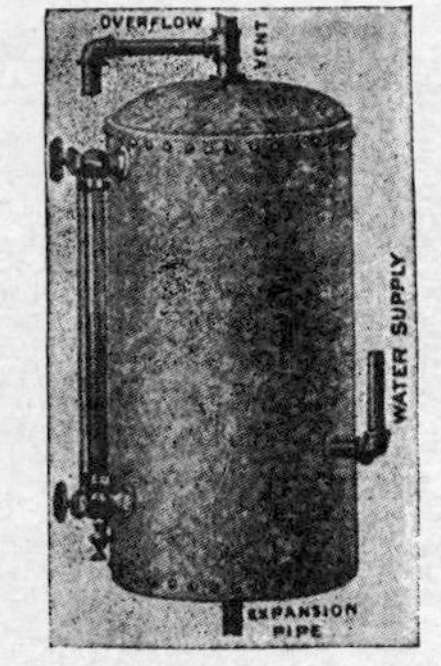

FIG. 8-7. Hot-water expansion tank. (*Wm. B. Scaife and Sons.*)

The principles of the pressure-relief and the temperature-relief valves have been combined with those of a vacuum valve in the valve shown in Fig. 8-8. The discharges from temperature-relief valves, pressure-relief valves, and expansion tanks should be cared for as indirect wastes and should be so protected that scalding cannot occur when there is a discharge.

Temperature control assures greater safety than pressure relief because no steam can be created at or above atmospheric pressure if the temperature of the water is kept below boiling. A temperature-control valve opens to waste water when the temperature becomes excessive. The valve should, therefore, be placed at the highest point in the storage tank where the hottest water will accumulate.

8-5. Hot-water Heating Controls. The automatic control of the temperature in a hot-water system is desirable for reasons of safety, convenience, and economy. In one- or two-family installations it may be unnecessary, but in large buildings it is essential. The method of temperature control depends somewhat on the method of heating. For example, in a coal-fired heater the draft is controlled by a thermostat which shuts down the draft as the water approaches the maximum allowable temperature and increases the draft as the water temperature drops. Such a device is shown in Fig. 8-9. Other controls are shown in Figs. 8-10, 8-15, and 8-16.

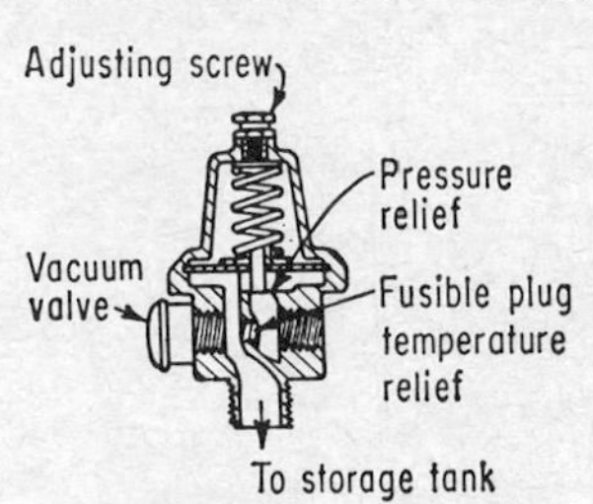

FIG. 8-8. Pressure-temperature-vacuum relief valve.

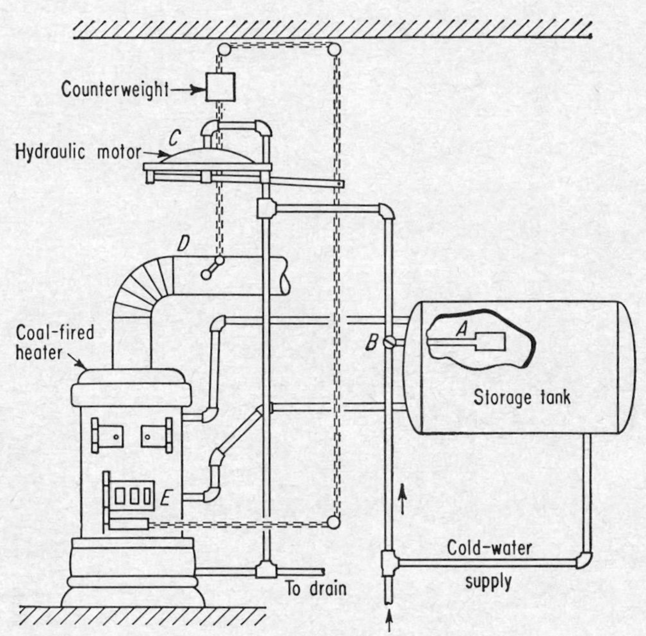

FIG. 8-9. Hot-water heater control. Control operates as follows: When the temperature of water at A in storage tank rises above desired figure, thermostat opens valve B transmitting water pressure to hydraulic motor at C. Chain is dropped, and dampers at D and E are closed. When temperature at A falls, valve B is closed, water is drained from motor, and counterweight causes dampers to open.

8-6. Hot-water Tank and Pipe Materials. Hot-water storage tanks used for domestic supplies are made of either galvanized iron or copper. The former is usually the stronger, particularly against collapse resulting from a partial vacuum in the tank, but the copper tank is more durable, is usually free from corrosion, and will not cause the staining of clothing or of plumbing fixtures as iron storage tanks sometimes do.

Brass or copper pipes are best suited to convey hot water because of their resistance to corrosion, which is more active in hot than in cold water. Steel or iron pipes are satisfactory, though they may be badly corroded unless the water is inactive. Lead should not be used for hot water because it dissolves too rapidly and it softens so much that at higher temperatures it is not safe, particularly at the joints, from internal pressure.

The installation near the heater of a chamber containing silicate of soda through which the hot water flows is an aid to preventing the corrosive action of hot water on iron. The dissolved silicate forms a protective coating over the corroded places in the iron.

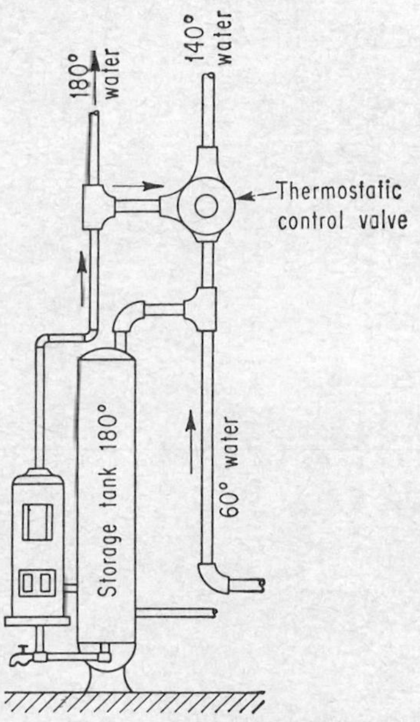

Fig. 8-10. Water heater delivering waters at different temperatures by means of a thermostatic control valve.

8-7. Insulation of Hot-water Tanks and Pipes. Hot-water storage tanks are sometimes covered with an asbestos jacket to prevent loss of heat and to avoid heating the room in which the storage tank stands. The saving effected by insulation is appreciable. Conclusions reached from tests made at the University of Pittsburgh are:

1. A saving of about 30 per cent of the total amount of gas usually burned can be effected by insulating a 30-gal hot-water boiler.

2. A considerable saving in gas can be effected by insulating all exposed hot-water piping.

3. The tank jacket will hold the heat in the water for a considerable time after the fire has been shut off.

Hot-water storage tanks in residences are seldom covered except when the storage tank is placed in the basement, where the unsightly covering is not so prominent as it would be in the kitchen. Storage tanks used with automatic heaters are usually covered to conserve heat.

Loss of heat can be reduced by covering the pipes and the storage tank with insulation, sometimes called *lagging*. Insulating materials are discussed in Sec. 7-20. Pipe should be covered with a thickness of about 1½ in. Some thicknesses of insulation are shown in Table II-77. Beyond these thicknesses the insulating effect is not materially increased.

The loss of heat from bare pipes may be reduced 75 to 80 per cent by covering them, as indicated in Fig. 8-11, but in a residence the actual loss of heat is so small that hot-water pipes are not always insulated.

8-8. Capacities of Domestic Water Heaters.[1] A domestic water heater should be capable of delivering the peak load during a 2-hr period

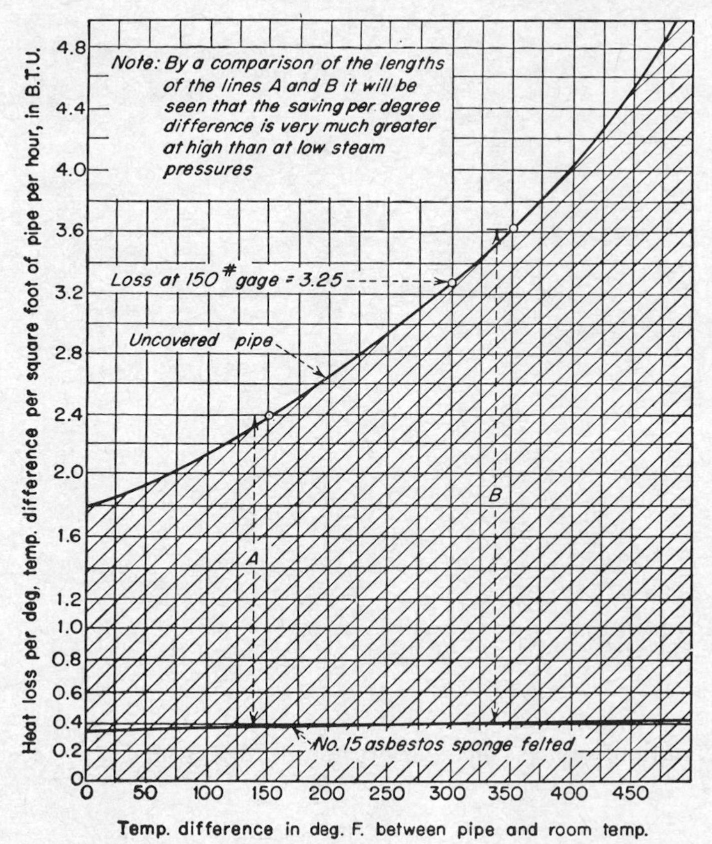

FIG. 8-11. Chart showing heat loss from bare steam pipe and the saving that can be secured by using a good covering. (*From Fuel Economy in Operation of Hand Fired Power Plants, Univ. Illinois Eng. Expt. Sta. Circ.* 7, 1918.)

at a temperature not less than 150°F and should have a storage capacity of not less than 30 gal. Capacities of hot-water heaters for other purposes are based on their estimated loads. Information concerning them is given in Tables 8-3 and 8-6.

8-9. Recovery Capacity. Water heaters are rated in accordance with their recovery capacity, i.e., the number of gallons of water raised 60°F

[1] See discussion of report on Building Research Advisory Board Report in *Domestic Eng.*, July, 1956, p. 99. See also E. A. Farber and others, *Air Conditioning, Heating, and Ventilating*, October, 1957, p. 75.

TABLE 8-6. CAPACITIES OF WATER HEATERS*

Tank size, nominal, gal	Input, Btu per hr	Recovery rate, gal per hr†	Gallons of water available at end of			
			1 hr	2 hr	3 hr	4 hr
20	24,000	22.3	36	59	81	103
30	27,000	25.1	46	71	96	121
40	34,000	31.6	60	91	124	155
50	40,000	37.2	72	109	147	184
60	47,000	43.7	86	129	173	217
75	60,000	55.8	99	139	199	250
100	78,000	72.3	103	150	225	283

* Based on data from E. A. Clifford, *L-P Gas*, November, 1950.

† Temperature rise of 90°F, with thermal efficiency of 70 per cent.

per hour in storage heaters and per minute in instantaneous heaters. The capacity of a heater can be expressed as

$$\frac{R \times 60}{T} \tag{8-1}$$

where R = recovery capacity

T = rise in temperature, °F

The recovery capacity of a heater is expressed as

$$R = \frac{BE}{8.33 \times T} \tag{8-2}$$

where B = heat input, Btu per hr

E = efficiency

Recommended capacities of water-heating equipment are shown in Table 8-1, data on rates of hot-water use are given in Table 8-2, and recommendations concerning hot-water storage tanks are given in Table 8-7.

TABLE 8-7. SOME STORAGE-TANK CAPACITIES FOR HOT-WATER HEATERS

Storage tank capacity, gal	15	20	30	45	60	75
Gallons hot water per day	30–60	60–100	100–200	200–300	300–450	450–600
Applications	Small family, small home	Average family, one bathroom	Average family, one and two bathrooms	Large family and small commercial	Large homes and small apartment buildings	Large homes and small apartment buildings

8-10. Selection of Water Heaters. The selection of the type of heater to be installed must be based on a study of the requirements, the heater performance, its first cost, and the cost of maintenance and of fuel. Coal-fired heaters with coils in the furnace or a water back in the kitchen range will probably be most economical in the cost of fuel. They may not, however, be the most desirable, because of the nonuse of coal-fired equipment in present-day residences, especially during warm weather. Where hard water, or water high in dissolved solids, is heated, the deposition of scale will lower the effectiveness of the heater and may clog the pipe coil or the water back, giving unsatisfactory service or causing an explosion. Heating coils and water backs should be installed, therefore, so that scale can be removed from them or they can otherwise be cleaned or removed and replaced with new equipment. All water-heating equipment must be constructed to avoid the accumulation of a pocket of air or of steam within it. An outlet pipe which is too low or a baffle improperly placed may create a vapor trap, resulting in noise, a cracked water back, a split coil, or an explosion.

In an "Investigation of the Performance of Automatic Storage-type Gas and Electric Water Heaters"[1] it was concluded that:

> The one-hour delivery studies showed an average recovery efficiency, i.e. ratio of heat utilized in heating water to heat input, for the gas water heaters to be 76 per cent compared with 93 per cent for the electric water heaters. However, on a comparable figure, namely gallons delivered per gallon of actual tank capacity, the gas heaters showed a value of 1.88 as compared to 0.94 for the electric, or twice as great an output for the gas heaters. The required input per gallon of hot water was found to be 1000 Btu for the gas and 800 Btu for the electric.
>
> The 2-hr delivery studies showed the same recovery efficiencies as the 1-hr studies, whereas the delivery per hour per gallon of actual tank capacity was found to show 1.50 to 0.62, or about 2½ times as much output from gas as from electric heaters. The required heat input was the same as that found in the 1-hr tests.

8-11. Coal-fired Heaters. Coal-fired water heaters are shown in Figs. 8-4, 8-5, and 8-12. Their capacities as water heaters can be expressed in terms of the grate area. Recommended sizes and other data are given in Table 8-3. Ordinarily a heater can be expected to burn about 3 to 6 lb per hr of coal per square foot of grate surface, depending on the construction of the heater, on the fuel, and on the draft. From 6000 to 8000 Btu will be transmitted to the water per pound of coal, with a heat efficiency of about 70 per cent. Since 1 Btu is the amount of heat necessary to raise the temperature of 1 lb of water 1°F, the approximate capacity of the heater can be calculated from the above information.

[1] E. F. Hebrank, *Univ. Illinois Eng. Expt. Sta. Bull.* 436, 1956.

Such calculations can be expected to be within 50 to 150 per cent of the correct amount, and heaters should be designed for overcapacity, since they frequently fail to give the expected results.

Coal-fired water heaters are not suited for use in small houses during warm weather because of the undesirable heating of the residence. However, in large residences, apartment buildings, and other large establishments with a coal-fired, central, space-heating system equipped also to

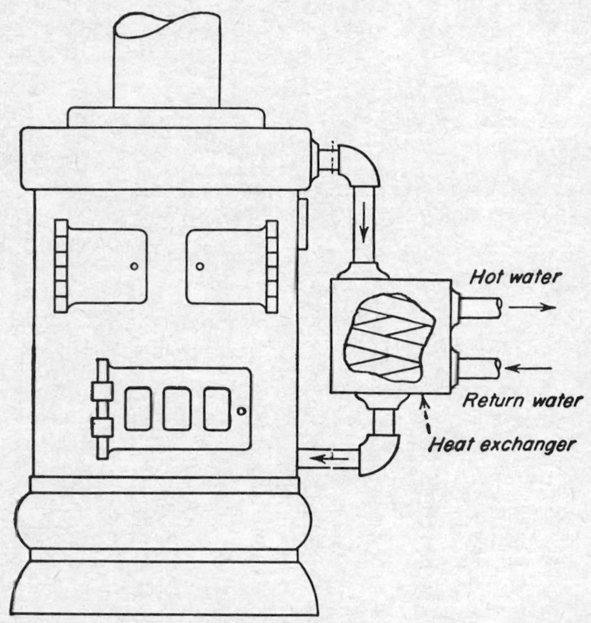

FIG. 8-12. Indirect water heater using hot water from a boiler in heat exchanger.

heat water, it has been found economical and satisfactory to operate the boiler during warm weather just sufficiently to heat water only. A side-arm heater for this purpose is illustrated in Fig. 8-12.

The following is the solution of a problem in the determination of the capacity of a coal-fired water heater:

Example. If water is delivered to a heater at a temperature of 50°F, how many gallons per hour at a temperature of 150°F will be delivered by the heater? The heater has a grate area of 2 sq ft, and the areas and materials of the heating surfaces are assumed to be adequate for the absorption of the heat from the fuel. The 2 sq ft of grate surface will permit the combustion of 8 lb of coal per hour.

Solution: Assuming the fuel to contain 12,000 Btu per lb and the heater to be 50 per cent efficient, the Btu delivered to the water will be $12{,}000 \times 0.5 \times 8 = 48{,}000$ Btu per hr. 48,000 Btu will raise 480 lb of water 100°F. 480 lb of water is approximately 58 gal. Therefore, the heater will have a capacity of 58 gal per hr of water from 50 to 150°F.

The design of the combustion chamber of the heater should be such as to provide a capacity of at least ½ to 1 cu ft for each square foot of grate area, dependent on the kind of fuel used and the frequency of firing. Anthracite coal requires less combustion space than bituminous coal. About one-half of the volume of the combustion chamber should be occupied by the fuel. For small heaters the area of the flue is taken at about 12 to 15 per cent of the grate area. For large heaters it may be economical to apply the principles of chimney design to the design of the flue. An adequately designed hand-fired heater may require firing three or four times a day. Stoker-fired heaters operate satisfactorily with attention only once a day.

8-12. Sizes of Water Backs and Pipe Coils. The area of the surface of water back or heating coil to be exposed to the fire depends on the

TABLE 8-8. CONSTANTS FOR COMPUTATION OF STEAM-COIL SIZES IN WATER HEATERS

Condition of steam	Kind of coil	Constant*	Condition of steam, psi	Kind of coil	Constant*
Exhaust	Iron pipe	10	25	Copper pipe	30
Exhaust	Copper pipe	15	50	Iron pipe	25
5 psi	Iron pipe	15	50	Copper pipe	40
5 psi	Copper pipe	22	75	Iron pipe	33
25 psi	Iron pipe	20	75	Copper pipe	50

* The constant is the figure by which the number of gallons of water heated per hour is to be divided to give the square feet of coil.

capacity of the water-storage tank, the temperature of the fire, and the material of the water back or coil. Water backs are usually made of cast iron and have a heating area exposed to the fire of approximately 2 to 2½ sq in. or more per gallon of tank storage capacity or 5 to 8 sq in. per gal per hr raised to 150°F. An oversize heater is undesirable because steam will probably form in it. There is a greater probability that the area will be too small.

Cast-iron water backs are generally unsafe at pressures higher than 75 psi. The approximate sizes of heating coils to be used are given in Table 8-8. The size of the heating coil can be computed from the following expression:

$$A = \frac{8.3Q}{f} \tag{8-3}$$

where A = area of inside surface of the pipe coil exposed to the fire, sq ft
Q = quantity of water flowing through the coil, gal per hr
f = coefficient of heat exchange. For iron, f = 145 to 200; for brass or copper, f = 220 to 300

The relative effectiveness of brass and iron for heating coils, as allowed by this formula, should be noted.

If the water back or heating coil is made too large for the capacity of the storage tank, the water will be overheated. It will turn to steam when a faucet is opened, and a crackling noise will probably be heard in the pipes. The proportioning of the storage tank and the heater requires a knowledge of the conditions of operation of the system, but fortunately a wide range of difference in capacity is possible with successful results.

Coils may be more efficient than water backs, but the latter may be preferable under relatively low-pressure conditions because of their comparative freedom from clogging due to scale deposits. Where pressures may exceed 75 psi, water backs are not safe and pipe coils should be used.

When a heating coil is used in a furnace, the coil should be buried in the hot coals of the fire. Such a coil usually consists of two lengths of straight pipe and a return bend. The coil or water back should not be placed in the gas chamber above the fire, since the heat from the fire will be less effective in this location.

8-13. Sizes of Steam Coils and Pipes. The sizes of coils in heaters using steam coils immersed in water or coils of water surrounded by steam are determined by the amount of water to be heated, the temperature of the steam, and the material of which the coils are made. The area of the inside surface of a heating coil can be found from the expression

$$A = \frac{Q(T_1 - T_2)}{f\left(T_s \dfrac{T_1 - T_2}{2}\right)} 8.3 \qquad (8\text{-}4)$$

where A = area of inside surface of heating coil, sq ft

f = coefficient of heat transmission. For iron, $f = 200$; for brass and copper, $f = 300$. It is expressed as the number of Btu transmitted per hour per square foot of surface

T_1 = temperature of hot water, °F

T_2 = temperature of cold water, °F

T_s = temperature of steam, °F

Q = amount of water to be heated per hour, gal

The temperature of steam at various significant pressures is shown in Table 8-5. The steam coil should be so arranged in the tank that water of condensation will drain out and heated water will rise to and leave at the highest point in the coil. The number of pounds of steam used can be computed from the expression

$$P = \frac{8.3Q(T_1 - T_2)}{C_1 - C_2} \qquad (8\text{-}5)$$

where P = steam delivered to the heater, lb per hr
C_1 = original heat content of the steam, Btu per lb
C_2 = heat content of the steam or the condensed steam in leaving the heater, Btu
T_1 and T_2 are as in Eq. (8-7)

The heat content of steam at various temperatures will be found in tables showing the properties of saturated steam. Ordinarily, it is most economical to condense all the steam delivered to the heater and reduce it to approximately the same temperature as that of the hot water leaving the heater. An illustrative example will be given.

Example. How many pounds of steam at 15 psi gage pressure will be required to heat 30 gal of water from 50 to 150°F?

Solution: In the preceding formula, $Q = 30$, $T_1 = 150$, $T_2 = 50$, $C_1 = 1{,}158.3$, $C_2 = 150.3$; therefore,

$$P = \frac{(8.3)(30)(150 - 50)}{32 + 1{,}158.3 - 150.3} = 23.9$$

The approximate horsepower of a boiler to supply steam is equal to $P/30$.

The size of the steam-supply line can be estimated from the expression

$$d = 13.5\sqrt{\frac{Q}{V}} \tag{8-6}$$

where d = diameter of steam supply pipe, in.
Q = rate of flow of dry steam, cfm
V = velocity of flow of steam, fpm. For low-pressure steam use V about 500, and for high-pressure steam not more than 6,000

8-14. Gas Heaters.[1] Gas heaters, such as are shown in Fig. 8-13, include manually operated, automatic, and instantaneous heaters. Some heaters are equipped with a thermostat which controls the fuel supply so that when the water falls below a predetermined temperature, the fuel is automatically turned on. In some types the hot-water storage tank is well insulated to economize in the use of fuel. Instantaneous heaters are arranged so that the opening of a faucet on the hot-water pipe will increase the flow of fuel, which is ignited by a continuously burning pilot light to heat the water to from 120 to 130°F. The possibility that the pilot light will die out offers a source of danger in the use of automatic appliances which depend on a pilot light. Gas and oil heaters are dangerous, and

[1] See also *Air Conditioning, Heating, and Ventilating*, August, 1956, p. 113. American standards for gas water heaters are ASA Z 21.10.1–1956, and sidearm-type heaters are ASA Z 21.10.2–1956. Tubing and fittings to convey gas must withstand 1000°F without deformation. For a study of gas water-heating costs see *Gas Age*, Dec. 15, 1955, p. 26.

they should be designed to prevent the accumulation, in a confined space within the heater, of a large volume of an explosive mixture. Openings for air not less than 36 sq in. in area should be provided near the floor and near the ceiling of any compartment in which a gas heater is located. No open-flame heater should be installed in a sleeping room or in any other small room.

The provision of a ventilating flue is desirable to carry off the products of combustion and as a safety factor to carry off unburned gas. The fuel

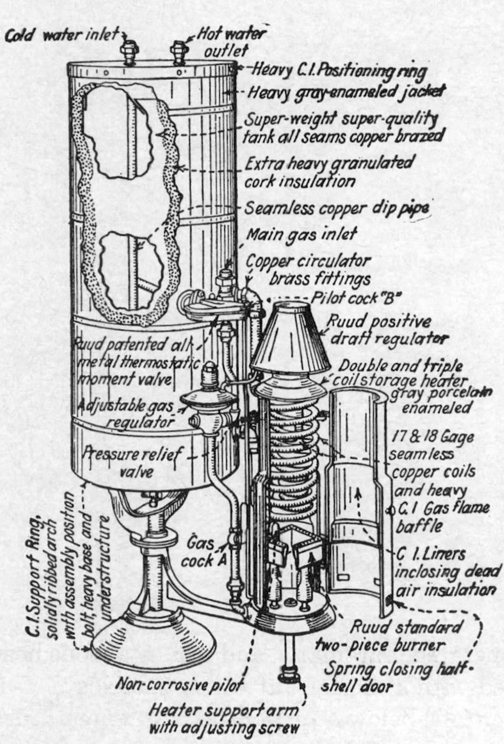

FIG. 8-13. Rudd gas water heater.

and water valves should be interlocked so that the fuel valve cannot be turned on unless the water is also turned on and is in the heater coils. The burning of gas produces water of condensation that is immediately evaporated in continuous heaters and passes up the flue. In instantaneous heaters, however, some of the water is precipitated, and drip pans with proper wastes must be provided for its disposal.

A water-pressure-relief valve set slightly above the maximum pressure anticipated in the water-supply pipes should be placed on the water piping of the heater whenever it is possible to manipulate valves on the wate' pipes in such a manner as to provide no other pressure relief.

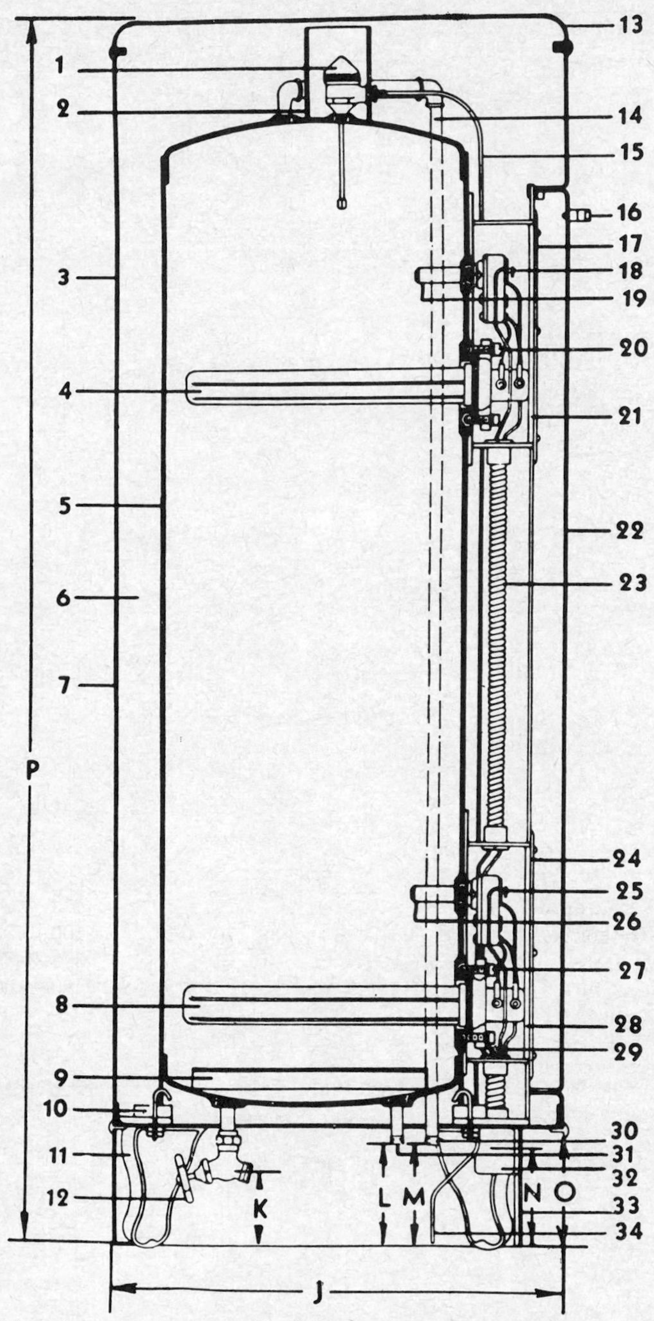

FIG. 8-14. Electric water heater. (1) Safety relief valve, (2) relief-valve casing, (3) heater cabinet, (4) upper heating element, (5) storage tank, (6) Fiberglas insulation, (7) exterior finish, (8) lower heating element, (9) antimixing baffle—shields cold-

The amount of gas required for heating water can be computed from the expression

$$G = \frac{8.3Q(T_1 - T_2)}{H_g E} \tag{8-7}$$

where G = gas required, cu ft
Q = water to be heated, gal
T_1 = final temperature of water, °F
T_2 = original temperature of water, °F
H_g = heat delivered by combustion of 1 cu ft of gas, Btu
E = efficiency of heating coils in absorbing heat

In 1 cu ft of illuminating gas there are 500 to 600 Btu; in 1 cu ft of natural gas, 1000 Btu; in 1 cu ft of butane at 60°F, 3275 Btu; and in fuel

TABLE 8-9. APPROXIMATE INPUT RATINGS OF COMMON GAS APPLIANCES*

Appliance	*Input rating, cu ft per hr*
Domestic gas ranges	
4-burner top	125
6-burner top	215
Domestic hot plates or laundry stoves, per burner	25
Automatic storage water heaters	
Slow recovery	5–20
Quick recovery	30–140
Instantaneous water heaters, per each 2 gpm capacity	150
Domestic circulating water heaters	50–75
Gas boilers	130–10,000
Gas steam radiator, per section	4
Domestic room heaters, radiant heater	
Per single radiant	4
Per double radiant	8
Conversion burners	160–800
Unit heaters	100–1,800
Refrigerators	3.8–7.8
Warm-air furnaces	80–1,000
Floor furnaces	30–160

* From "Standards for Piping, Appliances, and Fittings for City Gas," Pamphlet 54, National Board of Fire Underwriters, Sept. 1, 1943.

water inlet and minimizes mixing of incoming cold water with hot water in the tank, (10) wood block heat stops—prevent heat conduction from tank to legs of heater, (11) pressed-steel legs, (12) tank drain valve, (13) cabinet top, (14) heat trap—consists of ¾-in. fittings and pipe which runs along outside of tank and through base of heater and cabinet and prevents heat loss from stored hot water, (15) relief-valve drain tube—connects relief valve to base of heater, (16) door panel handle, (17) access panel, (18) upper thermostat—adjustable from 120 to 170°F, (19) thermostat well, (20) mounting bolts, (21) insulating pad, (22) cabinet door panel, (23) flexible metal conduit, (24) lower access panel, (25) thermostat, (26) thermostat well, (27) mounting bolts, (28) insulating pad, (29) power leads, (30) hot-water outlet (base connection to service line)—¾-in. iron pipe, (31) cold-water inlet—¾-in. iron pipe, (32) outlet box for electrical connections, (33) base panel, (34) relief-valve drain tube.

oil and in gasoline there are 19,500 and 20,750 Btu per lb of fuel, respectively. It is assumed in the above expression that the pipe coils and the combustion chamber are properly designed to permit the absorption of heat by the water. With good design the heat efficiency of the heater should be between 80 and 95 per cent. Some gas consumption rates with such equipment are stated in Tables 8-3 and 8-9.

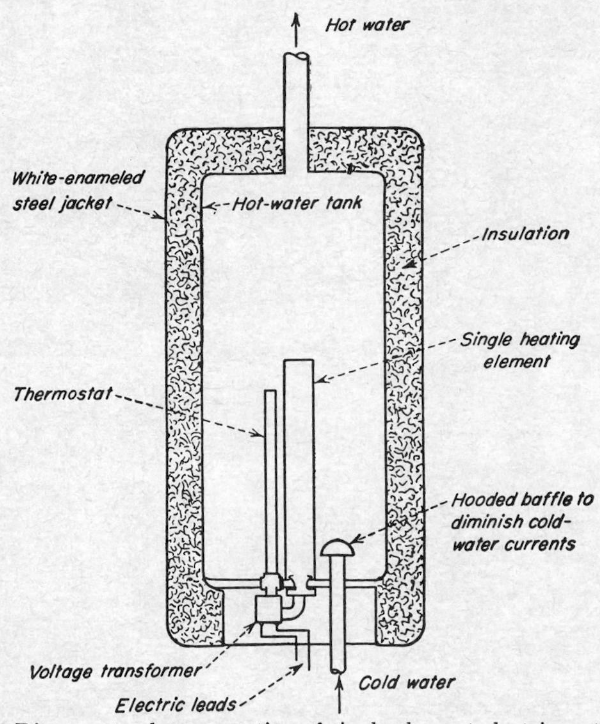

FIG. 8-15. Diagram to show operation of single-element electric water heater.

8-15. Instantaneous Heaters. Instantaneous or tankless heaters[1] heat water as fast as it is required and at the same rate as cold water is supplied to the heater. The efficiency and the heat-unit delivery of such heaters must be sufficient to deliver the water at the desired rate and temperature, raising the water to this temperature in the time required to pass through the heating coils. The heater unit, without the storage tank, shown in Fig. 8-13 typifies a gas-fired instantaneous heater. The combination of the heater with a storage tank reduces the required capacity of the heater.

8-16. Electric Heaters.[2] An automatically controlled electric water heater is shown in Fig. 8-14, a diagram showing the principles of operation is given in Fig. 8-15, and a bimetallic thermostat for its control is shown

[1] See also *Domestic Eng.*, February, 1956, p. 69.

[2] See also ASA C 72.1–1949.

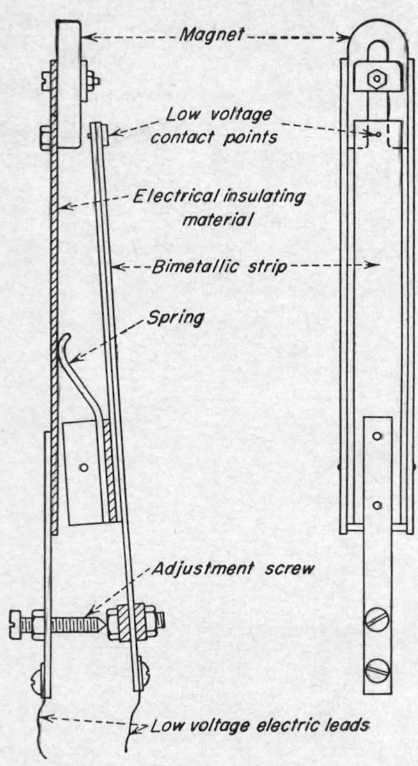

FIG. 8-16. Bimetallic thermostat for water-heater control.

TABLE 8-10. NATIONAL ELECTRICAL MANUFACTURERS ASSOCIATION STANDARDS FOR ELECTRIC WATER HEATERS

Single-element heaters			Twin-unit heaters		
Tank size, nominal gal	Wattage rating		Tank size, nominal gal	Wattage rating	
	50 series	20 series		Upper unit	Lower unit
30	1,500	.....	30	1,000	600
40	2,000	.....	40	1,250	750
52	2,500	1,000	52	1,500	1,000
66	3,000	1,250	66	2,000	1,250
80–90	3,000	1,500	80–90	2,500	1,500
			110	3,000	2,000
			120	4,000	2,500
			140	4,000	3,000

in Fig. 8-16. When the temperature falls below a predetermined limit in the heater, the current is automatically turned on. The predetermined limits for the highest and lowest temperatures of the water are adjustable by the householder. To obtain favorable electric rates for water heating, electric heaters can be controlled to receive current only during off-peak periods, a large, well-insulated storage tank being provided to carry over between periods when power is not available. Electric water heaters have thermal efficiencies as high as 83.5 per cent.

The "Standards of the Electrical Manufacturers Association for Electric Water Heaters"[1] contains the information shown in Table 8-10. The amount of current consumed by an electric heater can be estimated from the following expression

$$C = K\,\frac{Q(T_2 - T_1)}{E} \tag{8-8}$$

where C = current or power required, kwhr
Q = volume of water heated, gal
T_1 = initial temperature of water, °F
T_2 = final temperature of water, °F
E = heat efficiency of the heater
K = coefficient depending on units used. When units suggested herein are used, $K = 0.00275$

8-17. Immersion Heaters. An electrical immersion heater consists of an electrical heating element immersed in the hot-water storage tank or other water container.[2] Immersion heaters are efficient, flexible, and convenient. A heating element can be inserted relatively easily into the storage tank of an inadequate hot-water heating system to increase its capacity.

Another type of immersion heater passes hot gases through tubes submerged in tanks, usually horizontal cylindrical tanks, containing the water to be heated. Such heaters are applicable principally to loads above about 100 gal per hr.

8-18. Oil Heaters. Water heaters fired with oil or gas may be classified as:

1. The external tank heater, in which the flue products pass around the whole outside area of the tank
2. The vertical flue heater, which has a straight tube built into the storage tank
3. The reversible flue heater, which is equipped with an immersed tube or coil
4. The instantaneous water heater

[1] "Standards of the Electrical Manufacturers Association for Electric Water Heaters," Publication 66, 1930–1931.

[2] See also Immersion Heaters, *Elec. Rev.*, Nov. 25, 1955, p. 1,044.

The first three are illustrated in Fig. 8-17, and the instantaneous heater is shown in Fig. 8-13. The possible combinations of methods of ignition, feed, and control of oil heaters are indicated by the information in Table 8-11.

The use of gasoline in water heating is universally prohibited because of the dangers connected with it.

Oil heaters resemble and are operated similarly to gas heaters, except in the details of flame control.[1] The amount of oil required can be computed from the expression similar to that shown for gas heaters, as follows:

$$O = \frac{8.3Q(T_1 - T_2)}{H_0 E} \qquad (8\text{-}9)$$

where O = amount of fuel oil required, gal

H_0 = heat delivered by combustion of 1 gal of oil, Btu

and other nomenclature is as given in Eq. (8-7) for the expression for gas required. Oil heaters may be less efficient than gas heaters, with efficiencies varying between 10 per cent in summer and 55 per cent in winter.

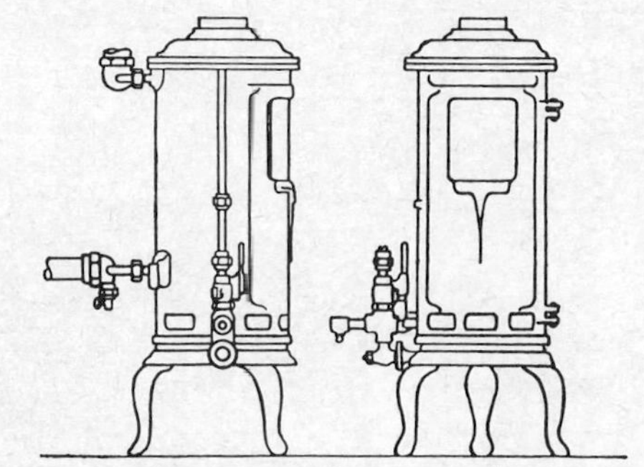

Fig. 8-17. Automatic-storage large-volume multicoil heater.

The installation of an oil heater usually requires the installation of an oil storage tank and oil piping. This must be done under special specifications[2] commonly covered in the plumbing or building code.

Table 8-11. Control, Ignition, Feed, and Ventilation of Gas and Oil Water Heaters

Control: Either manual or automatic
Ignition: Electric spark; pilot light; match or taper
Oil preparation: Air atomizing; emulsifying; pressure nozzle; rotary; vaporizing
Draft: Forced; natural
Operation: Continuous; intermittent

8-19. Installation of Water Heaters. Where large quantities of hot water are required[3] water-heating equipment should be installed in duplicate to permit repairs, and where water heating and space heating are done for a building by the same fire[4] in winter, the water can be heated

[1] "Standards for the Installation of Oil-burning Equipment," National Board of Fire Underwriters, March, 1947.

[2] See also C. H. Burkhart, Oil Tank Installation and Piping, *Plumbing and Heating J.*, June, 1954, p. 96.

[3] See also F. M. Reiter, Hot Water for Large Volume Needs, a series beginning in *Ind. Gas*, June, 1954.

[4] See also V. R. Berry, Team-up Control of Water and Space Heating, *Elec. World*, June 24, 1957, p. 128.

during the summer by auxiliary heating equipment. Pipes and valves must be installed to permit flexibility in the operation of the heating system and repairs to units without shutdown and to avoid the dangers associated with the heating of water.[1] Relief must be provided for pressures that may build up in the hot-water heating system, as pointed out in Sec. 8-4.

8-20. Hot-water Storage Tanks. Storage tanks for hot water are required except where "instantaneous" or tankless heaters are used. The capacity of the hot-water storage tank should be fixed by the number

TABLE 8-12. HOT-WATER STORAGE-TANK CAPACITIES

Number of persons served	5	25	50	75	100	150	200	250	350	500
Capacity of tank, gal	30	125	200	250	330	450	550	625	800	1,000

of persons using it and the purpose to which it is to be put. For ordinary domestic supplies there should be about 30 gal per family of four or five persons, provided that the heating plant is sufficient to raise the temperature of all water used in 1 hr in a dwelling at least 40°F or 50°F in an apartment building. Storage allowances for various numbers of persons are suggested in Tables 8-7 and 8-12. Reiter[2] suggests the following formula:

$$Q = K(M - CH) \tag{8-10}$$

where Q = storage capacity, gal
K and C = constants
M = maximum 24-hr demand, gal
H = heater capacity, gal per hr

Information on values of K and of C is given in Table 8-13. The storage capacity required is affected both by rate of use of water and by heater capacity; the greater the rate of use and the lower the heater capacity, the larger the required storage capacity. Capacities of hot-water storage tanks are stated in Tables II-78 and II-80. Tables 8-1 and 8-2 show the approximate proper allowance for hot-water heating in residence and apartment buildings. Dimensions of expansion tanks and of storage tanks are shown in Tables 8-7, II-78, and II-79.

It is to be noted that hot water, unlike cold water, is stored in pressure tanks. The storage of hot water in open tanks would entail loss of too much heat and of water by evaporation.

Storage tanks and all equipment for the heating or storage of hot water[3] must be equipped with a pressure-relief valve of such size that its dis-

[1] See also Know Your Basic Controls for Hot-water Heating Boilers, *Power*, November, 1956, p. 124.

[2] F. M. Reiter, *loc. cit.*

[3] "National Plumbing Code," Sec. 10.16.1.

charge rate will limit the pressure rise for any given heat input to 10 per cent of the pressure rise at which the valve is set to open. A temperature-relief valve relieves pressure when the water temperature reaches a fixed limit. It is rated on its Btu capacity, so that at 210°F it is capable of discharging sufficient hot water to prevent further temperature rise.[1] The functions of pressure-relief and of temperature-relief valves are sometimes combined in one valve known as a T-P valve. The "National Plumbing Code"[2] states that temperature-relief valves shall be so located in the tank as to be actuated by the water in the top one-eighth of the tank served and in no case more than 3 in. away from such tank. Pressure-relief valves can be located adjacent to the equipment they serve. There shall be no check valve or shutoff valve between a relief valve and the heater or tank for which it is installed. The outlet from such valves is to be connected to the drainage system as a direct waste. The storage tank should be equipped, in addition, with a blowoff valve at the bottom to permit the sediment that may have accumulated there to be blown off and also to permit water to be drawn off for any other purpose. The discharge from the blowoff valve should lead into a sink or other fixture. It should not be connected directly to the house drainage pipes below a trap.

TABLE 8-13. CONSTANTS IN FORMULA (8-10)

Usage	*K*	*C*	Usage	*K*	*C*
Hotel	0.4	15	1-meal restaurant	0.6	5
Apartment house	0.4	10	2-meal restaurant	0.4	10
Office building	0.26	7.5	Luncheonette	0.42	4.5
Average residence	0.4	10			

A vacuum-relief valve is desirable on any hot-water storage tank, and one must be placed on a copper tank to prevent collapse due to a vacuum as shown in Fig. 2-10. The relief valve should be placed at the top of the tank and in no case more than 3 in. away from the storage tank.[3]

8-21. Collapse of Storage Tanks. The collapse of a hot-water storage tank or the siphonage of its contents as a result of the creation of a vacuum in the tank or pipes must be guarded against. A collapsed tank is illustrated in Fig. 2-10. Siphonage may result when the main water-supply pipe is shut off and a faucet on the supply line is opened in the basement or there is a shutoff valve of the stop-and-waste pattern below

[1] See "Uniform Plumbing Code for Housing," Housing and Home Finance Agency, February, 1948, and "National Plumbing Code," Sec. 10.16.2.

[2] "National Plumbing Code," Secs. 10.16.4 and 10.16.5.

[3] See also Basic Controls for Hot-water Heating Boilers, *Power*, November, 1956, p. 124.

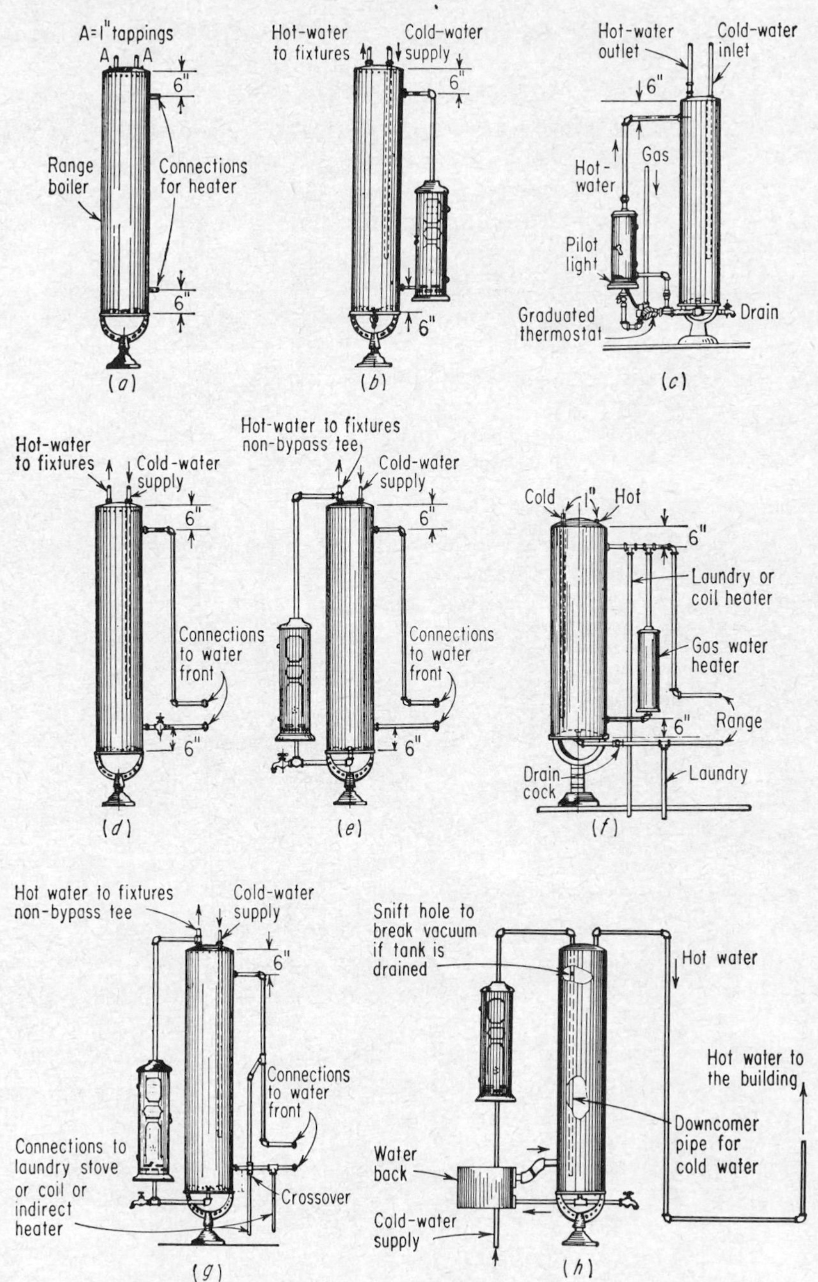

FIG. 8-18. Hot-water heaters and storage-tank connections. (*a*) Showing location of tappings; (*b*) showing method of connecting range boiler to gas water heater; (*c*) showing method of connecting range boiler to gas water heater with graduated thermostat; (*d*) showing method of connecting range boiler to water front; (*e*) showing method of

the tank. The tank may be collapsed by the vacuum created, or the heating coil or the water back may be cracked because of being emptied of water before the fire is extinguished. The danger of the collapse of a storage tank can be reduced by boring a small hole in the supply pipe near the top of the tank. This hole should have an area equal to one-fourth of the area of the cold-water pipe. The cold-water pipe should not extend down into the tank below the level of the top of the water back or heater coil in order to avoid drawing the water therefrom in the event of the siphoning of the water from the tank. The siphonage of the tank can be prevented, also, by placing a vacuum valve on the cold-water supply pipe near the tank or by placing a check valve on the cold-water supply pipe. Either of these devices should be installed in a building where there are faucets or other outlets on the cold-water line below the hot-water storage tank. A check valve should not be used unless some provision is made to relieve excess pressure in the storage tank.

8-22. Cathodic Protection of Storage Tanks. The cathodic protection of some hot-water storage tanks is sometimes recommended[1] because of the high corrosivity of some waters when heated. The insertion of a magnesium anode extending from the top of the tank to within about 3 in. of the bottom may be effective. The subject is discussed also in Sec. 4-35.

8-23. Connecting Heaters and Storage Tanks. The Committee on Simplified Practice of the U.S. Department of Commerce has proposed a standard method for connecting up hot-water storage tanks, which is illustrated in Fig. 8-18. It is to be noted in the figure that "water front" is used in the same sense as "water back" is used elsewhere in this text. Other methods of installing storage tanks are also shown in Fig. 8-18.

An explosion may result from the improper connection of a heater and a storage tank. An explosion may result if (1) cold water can enter the heater directly and, by suddenly cooling the heater, cause it to crack and explode and (2) water may be siphoned or drained from the heating system backflowing into the water-supply pipes. If the cold water is admitted by means of a pipe passing through and terminating near the bottom of the storage tank, the tank will be emptied when there is a failure of the cold-water supply. It would not be possible for this to happen under the conditions shown in Fig. 8-18*h* unless the vacuum opening, shown near the top of the tank, becomes clogged. A vacuum valve attached to the

[1] See *Domestic Eng.*, July, 1956, p. 99.

connecting range boiler to water front and gas water heater; (*f*) showing arrangement of tappings and method of connecting to gas water heater, water front in kitchen range, and laundry heater; (*g*) showing method of connecting to water front, gas water heater, laundry stove, or indirect heater; (*h*) connections to water back and gas heater with descending hot-water pipe.

outside of the storage tank, as shown in Fig. 8-8, will serve as a supplementary safeguard. It is a requirement, therefore, that the downcoming cold-water supply tube in the storage tank should have a 1/8- to 1/4-in. hole in it within 6 in. of the top of the tank to break any vacuum that might otherwise be created when the tank is emptied.

Cold water should not enter at the bottom of the tank, nor should the downcomer tube in the hot-water storage tank terminate too close to the bottom, to avoid stirring up sediment. The entrance of sediment into the heater can be minimized by an inlet connection above the tank bottom, where feasible. The available storage capacity of the tank is reduced by such a connection in proportion to its height above the tank bottom. This height should, therefore, not be greater than about 6 in. The hot-water pipe from the heater to the storage tank can be connected at any convenient distance less than 6 in. below the top of the tank. A connection above the top of the tank as in Fig. 8-18*e* will supply hot water promptly from the heater, but if the capacity of the heater is limited, cold water may be drawn from it rather than hot water from the storage tank. The connection at a distance of 6 in. below the top of the storage tank is preferable except for a flash heater.

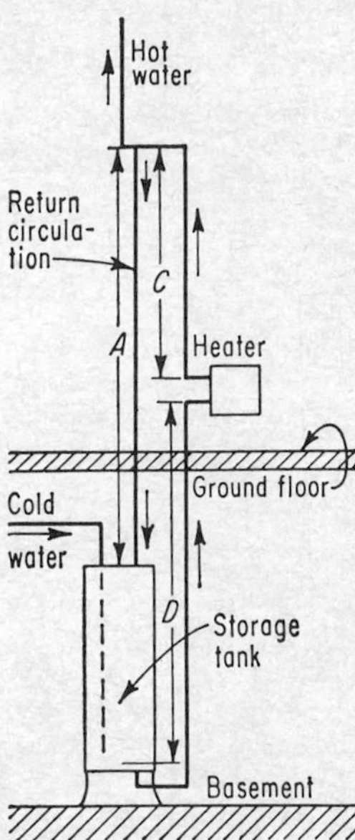

FIG. 8-19. Arrangements of pipes when water heater is above water-storage tank.

The installation of a vacuum valve on the cold-water supply to the storage tank prevents such a condition as is illustrated in Fig. 2-10.

The hot-water storage tank will usually not drain when the supply pipes are closed and the drain valve is opened unless air is admitted to the tank or to the hot-water pipes by opening a hot-water faucet somewhere in the building above the elevation of the hot-water storage tank. The drain pipe in the storage tank should terminate as in an indirect waste, as discussed in Sec. 11-30.

Since water is a poor conductor of heat, it cannot be heated successfully by applying heat at the top of a storage tank. The heat must be applied at the bottom with hot water taken off at the top. Water heated at the bottom rises to the top, where it is stored ready for immediate use. If the water is heated outside the storage tank, as is sometimes done, the heated water is conveyed to the top of the tank in a pipe outside it, as is shown at *b* and *h* in Fig. 8-18. Hot water to fixtures can be drawn from the top of the storage tank, as shown in Fig. 8-18*c*. Other methods of connecting water heaters and storage tanks are shown also in Fig. 8-18.

It is undesirable to place a storage tank below the heater because hot water tends to rise and it is difficult to force water down into the storage tank. Circulation will take place, however, if the hot-water pipe from the heater is connected to the downcomer pipe into the tank at a height as far as or farther above the heater than the heater is above the bottom of the storage tank, as shown in Fig. 8-19, with the length *C* as high as practicable and at least five to ten times length *D* where possible. Pipe *A*

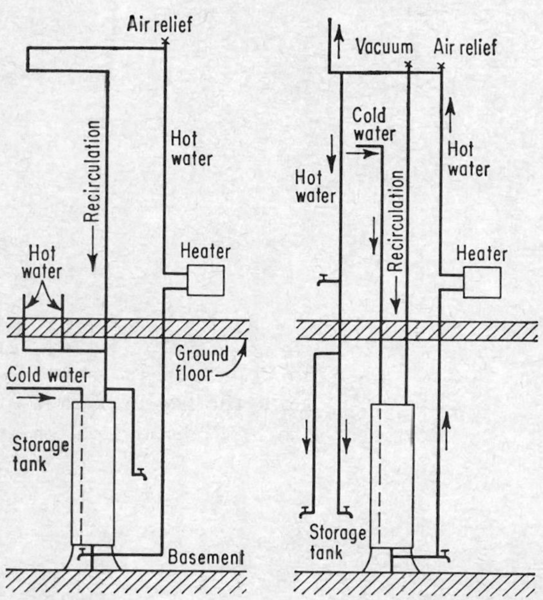

FIG. 8-20. Piping connections for water-storage tank at lower elevation than hot-water heater.

should be smaller than pipe *C*. Other pipe connections under similar conditions are shown in Fig. 8-20.

A high-pressure steam boiler should not be connected directly into the water-supply pipes of a plumbing system. It is desirable also to provide a tank for at least a 6-hr supply so that in the event of a shutdown of the water supply there will be sufficient water to avoid danger of an explosion of the boiler.

8-24. Multiple Heaters and Storage Tanks. The use of two water heaters is not uncommon in large buildings. In winter water can be heated in the coal-fired furnace and in summer by a gas heater. Where there are only two heaters and one storage tank, the heaters should be connected independently to the storage tank, with, as nearly as possible, the same friction loss in the circulating pipes from the storage tank to each heater. The connection of the pipe from the gas heater to the hot-water riser tank above the storage tank, as shown in Fig. 8-18*e*, has the advantage of drawing the hottest water from the heater or from the storage tank because the hottest water will rise first. The connection is known as a quick-heating connection. Where heaters are connected in series, as shown in Fig. 8-21, the arrangement is inefficient because when both fires *A* and *B* are going, the water entering *B* is already heated and does not absorb heat from the fire, and when the fire at *B* is not going, the water is chilled in passing through the cold heater.

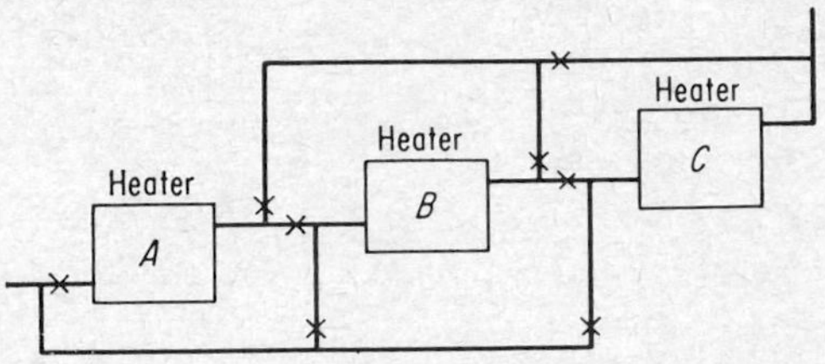

FIG. 8-21. Piping and valves for one, two, or three water heaters connected in series.

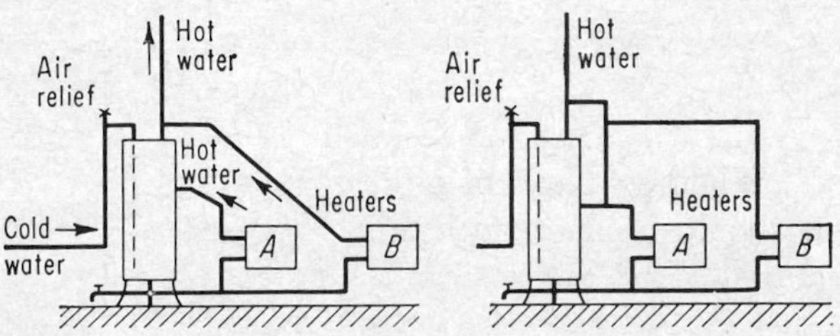

FIG. 8-22. Two heaters in parallel feeding one storage tank.

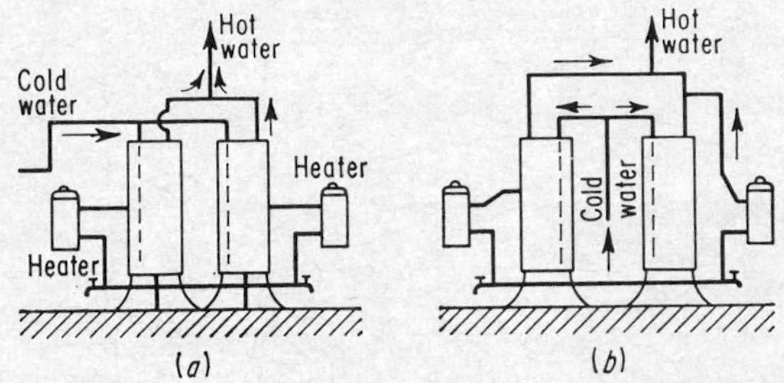

FIG. 8-23. Two heaters and two storage tanks in parallel.

The connection of heaters in parallel, as shown in Fig. 8-22, offers no serious difficulties. If there are two heaters and two storage tanks on the same level, they can be connected independently to the heaters, as shown in Fig. 8-23, and if on different levels, as shown in Fig. 8-24. Principles applicable include the following: (1) Do not circulate hot water through a cold heater. (2) Arrange piping and valves so that any combination of heater and tank can function together but independently of others. (3) Use as few valves as possible so arranged that no combination of closed and open valves can confine water to result in an explosion. (4)

Hot-water pipes should rise and return pipes should descend continuously. (5) Friction losses due to circulation should be equalized among tanks and heaters. (6) Where conditions will not permit the use of open expansion tanks, pressure-relief valves must be used and air relief must be provided through a faucet or an air-relief valve at the highest point. (7) To avoid an air-lock in a horizontal pipe connecting hot-water risers or circulating pipes, either the pipe should be below the highest frequently used fixture or an air-relief valve must be placed at the high point in the circulating pipe. (8) Where the combination is such that the storage tank is below the heater, the principles given in Sec. 8-23 should control. (9) Where pipe runs from heaters to storage tanks are long, use rising slopes instead of horizontal runs. (10) Where pipes conveying hot water from different heaters join, make the connection with long-radius ells and 45° wyes in the direction of flow. Do not use 90° tees. When a layout of piping has been completed, it should be checked against the principles enumerated above and in Sec. 8-30.

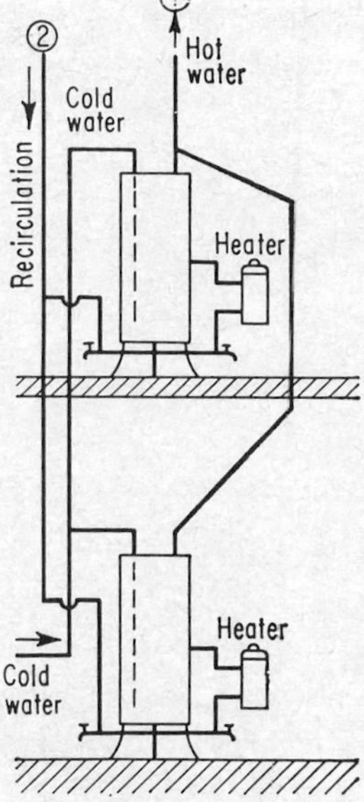

Two or more storage tanks can be connected to one or more heaters, as shown in Fig. 8-25.

FIG. 8-24. Two heaters and storage tanks in parallel and on different floors. Numerals 1 and 2 refer to piping shown in Figs. 8-29 and 8-30.

8-25. Multiple-temperature Hot-water Systems. Water can be supplied at different temperatures and in different piping systems from the same heater. Such a system may be useful in restaurants, laundries, industrial plants, and other establishments. The water is raised to the

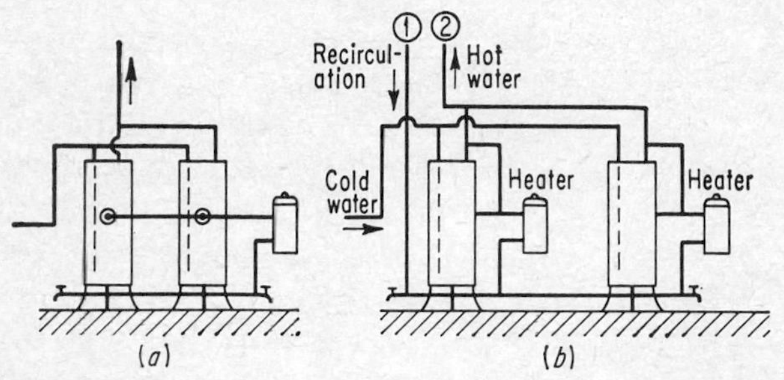

FIG. 8-25. (*a*) Two storage tanks, one heater, no circulation, no "quick" hot water. (*b*) Two storage tanks and heaters, recirculation and "quick" hot water. Numerals 1 and 2 refer to piping shown in Figs. 8-29 and 8-30.

highest temperature desired by the water heater. The water then flows to the high-temperature storage tank and distribution system. Water to be used at a lower temperature flows from the same heater to a mixing tank, where it is mixed with cold water to attain the desired temperature. The cooler water then flows to its storage tank and distribution pipes. Control is usually automatic and involves the use of thermostatically controlled valves,[1] as shown in Fig. 8-10.

8-26. Heating with Steam. It may be economical in large buildings, factories, and institutions with widely scattered buildings requiring hot

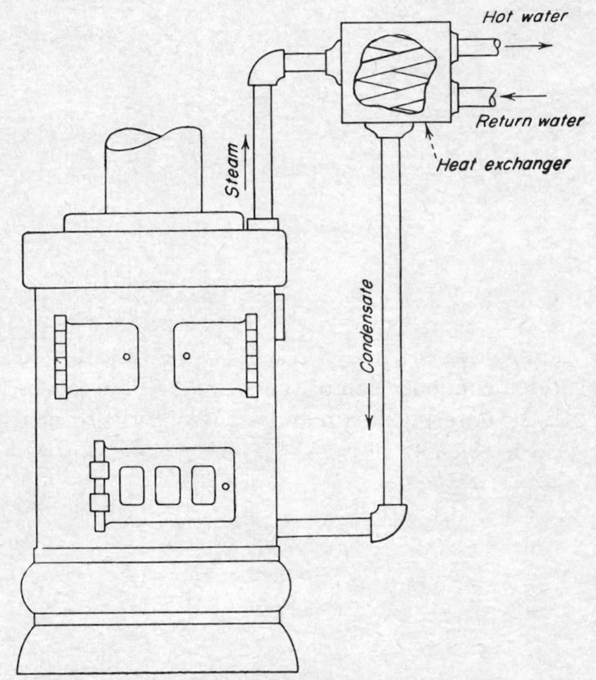

FIG. 8-26. Indirect water heater using steam from boiler in heat exchanger.

water to heat the water by steam near the points at which the water is to be used. Under such conditions an indirect steam heater may be installed, somewhat as shown in Fig. 8-26. A central water-heating system with a hot-water-distribution system, compared with a system that heats water locally by steam, involves the additional costs and heat losses in operation that might make it impossible to keep the water sufficiently hot at the point of use. Exhaust steam in power plants can be used sometimes for heating water with little or no additional cost for fuel.

8-27. Mixing Steam with Water. The heating of water by mixing steam with it is simple, economical, and easily controlled automatically.

[1] See also L. Blendermann, *Air Conditioning, Heating, and Ventilating,* February, 1956, p. 102, and January, 1958, p. 99.

An illustration of such a heater is shown in Fig. 8-27. The steam is discharged into the water through a perforated pipe, preferably of brass. A check valve and a stop valve should be placed on the steam pipe, to prevent water from entering the steam pipe when the water pressure exceeds the steam pressure and to permit the steam to be shut off.

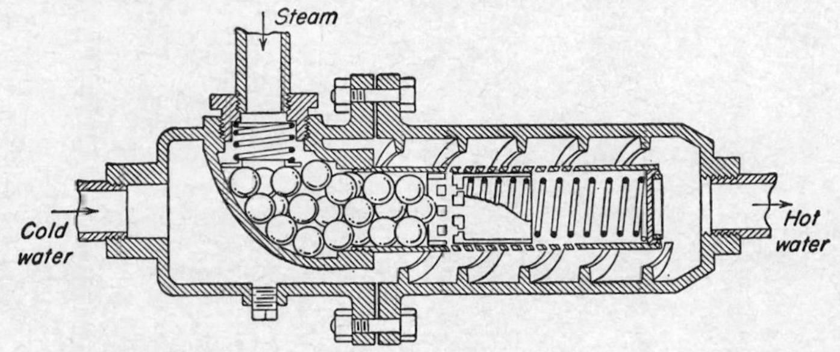

FIG. 8-27. Water heater mixing steam directly with cold water.

The openings through which the steam passes into the water should have an area of about eight times the cross-sectional area of the steam pipe. An objection to this method of heating water arises from the noise created by the sudden condensation of the steam. This can be overcome by mixing air with the entering steam and arranging the discharge so that it spreads out into a cone as it rises. If the velocity of the entering

TABLE 8-14. CAPACITY OF DIRECT-STEAM WATER HEATERS
(Steam and water mixed)

Gallons of water per hour* raised 100°F	Pounds of steam required per hour at gage pressure, psi					Diameter of steam pipe, in.	Diameter of air pipe, in.
	10	20	40	60	80		
5	0.70	0.69	0.69	0.68	0.68	1/4	1/8
100	1.40	1.38	1.38	1.36	1.36	1/4	1/8
200	2.8	2.76	2.76	2.72	2.72	1/4	1/8
300	4.2	4.1	4.1	4.1	4.1	1/4	1/8
400	5.6	5.5	5.5	5.0	5.0	1/4	1/8
500	7.0	6.9	6.9	6.8	6.8	1/2	1/4
1,000	14.0	13.8	13.8	13.6	13.6	1/2	1/4

* 1 gal raised 100°F = 830 Btu.

steam is sufficient and the air pipe is placed in the center of the stream of steam, sufficient air will be entrained by injector action to quiet the condensation of the steam.

Devices for mixing steam, water, and air for water heating are manufactured[1] in various capacities, as shown in Table 8-14. Expedients

[1] See also *Air Conditioning, Heating, and Ventilating,* August, 1956, p. 112.

for mixing steam and cold water silently include breaking the steam up into small particles by specially designed orifices with many minute openings, by increasing the surfaces of contact as shown in Fig. 8-27, by causing a spiral or helical motion of the mingling streams, by mixing air with the steam, or by a combination of these expedients. The minimum pressure at which steam can be discharged into the tank in order to draw in air by aspiration is

$$S = \frac{H^2}{2} \tag{8-11}$$

where S = steam pressure, psi
H = depth of water, ft

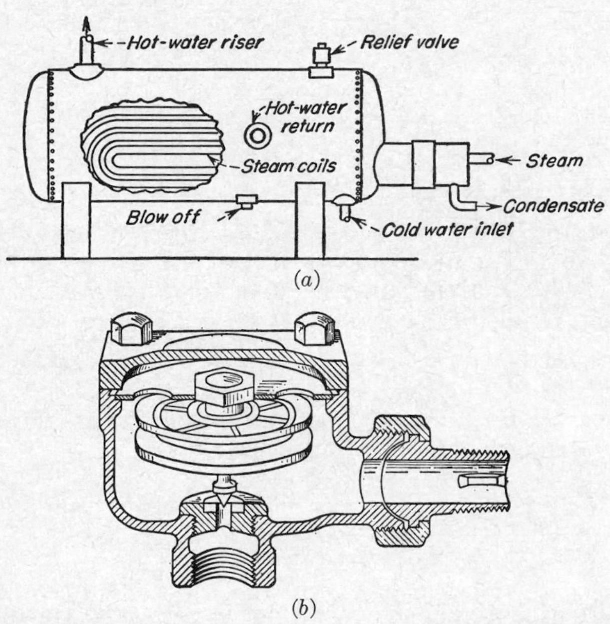

FIG. 8-28. (a) Steam-coil water heater; (b) condensate trap.

Where air is mixed with the steam in a closed tank, an air-relief valve or other device must be provided to permit the escape of the air. Devices are manufactured for mixing steam and water under pressure in a small mixing chamber in silence without the addition of air. In these devices the jet of steam is broken up into fine bubbles by a steam-spray nozzle. This mixer or heater is placed on the feedwater line to the hot-water storage tank, the arrangement being simple and compact.

8-28. Heating Water through Steam Coils. Two types of equipment are used to transfer heat from steam to water through pipe coils. In one type the coils containing the water are surrounded by steam held in a closed container, and in the other type, steam within the coils is surrounded by water held in a closed container. Both are illustrated in Fig. 8-28.

The amount of heat required can be computed from the volume of water to be heated and the degree of temperature it is to be raised. A rule of thumb allows 1 sq ft of coil surface for every 15 gal of storage-tank capacity. The area of surface[1] of coil required can be computed from the information given in Table 8-15 and Fig. 8-11. Coils in tanks less than about 50 gal capacity may be ¾ in. or larger; between 50 and 200 gal, use 1-in. pipe coils; between 200 and 300 gal, use 1¼-in. pipe; and above 300 gal, use 1½-in. pipe coils.

TABLE 8-15. APPROXIMATE HEAT LOSS FROM BARE PIPE AT PIPE SURFACE TEMPERATURES SHOWN IN THE TABLE WITH SURROUNDING AIR AT 70°F
(Btu per hour per linear foot of bare steel pipe surface)*

Pipe diameter, in.	Pipe surface temperature, °F					
	90	100	120	150	180	210
½	0.37	0.40	0.41	0.44	0.48	0.52
¾	0.54	0.55	0.58	0.62	0.66	0.70
1	0.64	0.68	0.72	0.78	0.83	0.89
1¼	0.74	0.77	0.81	0.88	0.95	1.01
1½	0.86	0.88	0.94	1.03	1.12	1.22
2	1.06	1.11	1.19	1.32	1.43	1.56
2½	1.29	1.33	1.43	1.57	1.72	1.81
3	1.53	1.59	1.68	1.86	2.02	2.19
3½	1.73	1.79	1.92	2.11	2.31	2.50
4	2.02	2.08	2.22	2.40	2.61	2.79
5	2.34	2.44	2.60	2.86	3.11	3.38
6	2.56	2.70	2.99	3.41	3.86	3.96

* Heat losses from horizontal, tarnished type L copper tube are about 60 per cent of the figures in this table.

Heaters containing water in pipe coils inside a steam boiler are used where the same fire is used for space heating and for water heating. Heaters with steam coils inside water tanks can be connected with an auxiliary source of heat for use during warm weather, the equipment being arranged somewhat as shown in Fig. 8-4. Under any conditions the steam coil should be placed near the bottom of the tank to avoid thermal stratification of the water in the heater. For the same reason cold water enters near the bottom of the tank to contact the heater coils, rises, and is drawn off at the top as hot water. Condensed steam from the steam coil may discharge to waste through a steam trap, as shown in Fig. 8-28; it may be pumped back into the boiler; or it may be mixed

[1] See also J. F. Davis, Rate of Heat Transfer through Metal Walls of Pipes, *Chem. Eng.*, August, 1951, p. 122.

with the hot water. The steam trap should be set below the heater to permit the thorough draining of water of condensation from the steam tubes. When water heaters are connected with steam boilers, adequate stop valves should be placed on the connecting pipes to permit repairs to either the heater or the boiler.

Copper is preferred for steam coils because of its relatively high rate of heat transmission and because of the low pressures ordinarily involved. It is not suitable for use with high-pressure steam. The cylindrical water container may be of steel, wrought iron, or cast iron, the latter being preferred because of its greater durability.

8-29. Auxiliary Steam Heaters. In one satisfactory form of steam water heater that combines space heating with water heating from the same fire, brass pipe coils are placed below the water line in the steam boiler. These coils are connected to a set of coils in an auxiliary water heater placed outside the steam boiler and lower than the hot-water storage tank. Water heated in the coils in the steam boiler circulates through the coils in the auxiliary, or sidearm, heater, warming the surrounding water, which then rises to the hot-water storage tank. An installation of this type is shown in Fig. 8-12.

Such an installation is advantageous in heating water regardless of the requirements for steam for space heating. In hot weather the steam-heating supply valve can be closed, and the boiler used for water heating only.

8-30. Hot-water Piping Systems.[1] Hot-water piping can be laid out as a pressure system in which the pressure is provided from a source outside the building, or the pressure may come from a tank open to the atmosphere and located at a high point in the building, as shown in Figs. 8-4 and 8-5. In each of these systems the piping may depend on non-continuous circulation by gravity as shown in Fig. 8-6*b*, continuous gravity circulation as shown in Fig. 8-6*a* without the circulating pump, or forced circulation as shown in Fig. 8-6*a* with the pump. Among the features of each system are the following: (1) The pumping system is simple, requiring less pipe than the tank system. (2) In the tank system no pressure can be built up to cause an explosion. (3) Noncontinuous circulation requires the least amount of piping, but cold water may be drawn for some time before hot water reaches the faucet, and heat is lost from hot water standing in the pipe. A noncontinuous layout should not be used where the hot-water pipes are long and there are numerous fixtures to be served. It is satisfactory only for short pipes between the heater and a hot-water faucet. (4) Continuous gravity circulation should be used where noncontinuous circulation is unsuitable. However,

[1] See also Balancing Service Hot Water Systems, *Air Conditioning, Heating, and Ventilating*, March, 1957, p. 90.

it is not suitable for more than two- or three-story buildings or for more than two or three apartments. (5) Continuous forced circulation can be used satisfactorily for the largest installations.

8-31. Hot-water-pipe Layouts. In laying out hot-water piping systems the pressures of the hot water and the cold water should be made equal at each fixture, especially where mixing faucets are to be used. Otherwise there is the possibility that the higher-pressure water will force itself into the lower-pressure supply when the mixing faucet is opened to both supplies. Commonly the hot-water pressure is lower than that of the cold water owing to the more circuitous route followed by the hot water. This difficulty can be partly overcome by the use of larger and smoother pipes and long-radius fittings on the hot-water lines. On the other hand, sudden demands for cold water, as by flushometer valves, may so reduce the cold-water pressure as to draw hot water into the cold-water pipes. Such conditions can be avoided by an analysis of the system to balance the head losses in each supply.

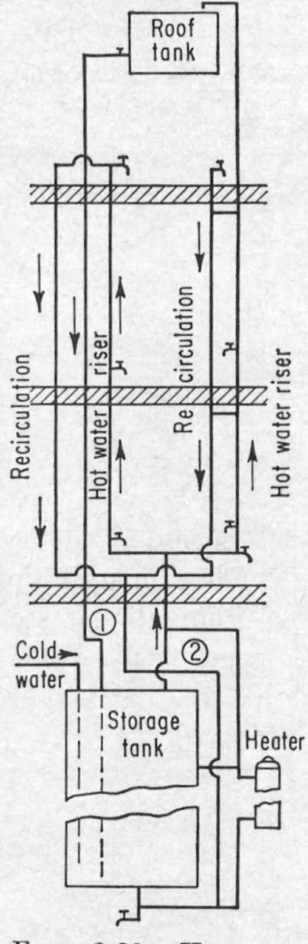

FIG. 8-29. Hot-water distribution and circulation. One heater and storage tank, two risers and a roof tank. Numerals 1 and 2 refer to corresponding numbers in Figs. 8-24 and 8-25.

When circulation is maintained by gravity, the following principles should be considered in design: (1) Hot-water pipes should rise continuously from heater to faucets. Only return, circulating water should descend. Exceptions to this rule permit a short, descending spur from a riser pipe to a single fixture, and in a multistory building a few fixtures on each floor may be supplied from the descending return pipe. (2) The riser pipe should have two or three times the cross-sectional area of the return pipe, and no riser should be less than ¾ in. in diameter if of galvanized iron. (3) Riser pipes should rise continuously to the highest point in the system, and return pipes should descend continuously to the bottom of the heater, as shown in Fig. 8-29, or into the storage tank, as shown in Fig. 8-30. (4) Multiple circulating loops connected to the same heater must be designed with approximately equal head differential from all causes. Otherwise the flow of hot water will be uneven among the loops. The condition can be remedied by the adjustment of valves, by trial, at the base of each riser, or it can be avoided by proper design before the pipes are installed. (5) Air relief should be provided at the high point or points in the form of an air-relief valve, a faucet, or, in a tank system, a pipe connected at the high point or points and terminating with an open end above the highest water level in the tank. (6) Provision must be made for wasting water which may escape from the air-relief valve in the form of an indirect waste. Since the water may be scalding, there should be protection against possible injury.

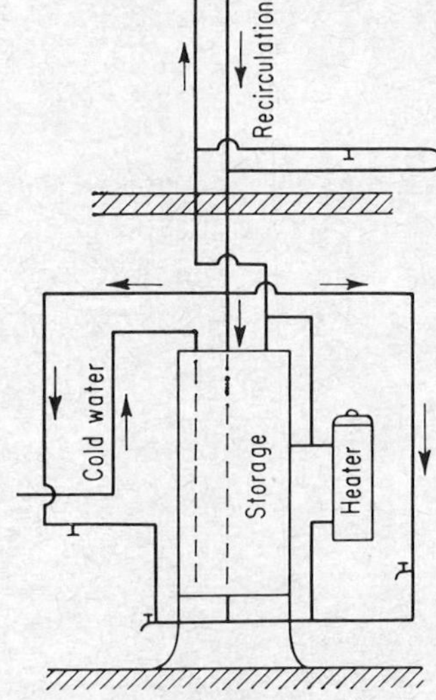

Fig. 8-30. Hot-water distribution and circulation. One heater and storage tank. Closed circuit.

The velocity of flow of water in the circulating pipes depends on the difference in weight of the water in the riser and the return pipes. The weight of a column of water can be formulated as

$$W = HS \tag{8-12}$$

where W = weight of column of water, 1 sq ft in cross section, lb
H = height of column, ft
S = specific gravity, as given in Table 2-1

The head of water causing flow is

$$H_1 = \frac{H_a S_a - H_b S_b}{W_T} \tag{8-13}$$

where H_1 = the head of water causing flow, ft
H_a and H_b = heights, respectively, ft
S_a and S_b = specific gravities, respectively, of water columns in riser and return pipes
W_T = weight of 1 cu ft of water at average temperature in the two pipes, lb

The velocity of flow will be $\sqrt{2gH_1}$, where g is the acceleration due to gravity, usually taken as 32.2 fps^2. Information on limiting lengths of gravity hot-water circulating systems is given in Table 8-16.

The tank in a tank-distribution system should be placed as high as possible, within reason, above the water heater to assure good water pressure and to avoid a vapor lock due to the combination of steam and low pressure in the supply pipe.

When circulation is forced, as by a circulating pump such as is shown in Fig. 8-6a,[1] the capacity of the pump can be determined by dividing the

[1] See also *Domestic Eng.*, February, 1956, p. 71.

total estimated heat losses or requirements, expressed in Btu, by 10, on the assumption that the return water is 10° cooler than in the heater. The quotient will give the pounds of water to be circulated. The head pumped against should be 5 psi greater than the cold-water pressure, and to this should be added the friction loss in the pipes.

TABLE 8-16. MAXIMUM LENGTHS OF VARIOUS HOT-WATER LINES THAT MAY BE INSTALLED WITHOUT CIRCULATING LINES*

Pipe diameter, in.	Maximum length, ft	Feet of pipe containing 1 gal hot water	Gallons in pipe length given	Heat loss from 100 ft of pipe, 1000 Btu
½	150	63.2	2.4	2,296
¾	90	36.1	2.5	2,851
1	50	22.3	2.3	2,585
1¼	30	12.8	2.4	4,514
1½	23	9.4	2.4	5,191

* From F. M. Reiter, Hot Water for Large Volume Needs, *Ind. Gas*, November, 1954, p. 9.

8-32. Hot-water Supplies in Tall Buildings. Layouts of hot-water supplies in tall buildings follow the principles outlined in the previous section. A noncirculating supply is unsatisfactory in a tall building because of the length of piping involved. Well-balanced gravity or forced-circulating systems must be provided. Systems are shown in Figs. 8-4 to 8-6.

In tall buildings the water pressure may be divided into zones with separate heaters for each zone, or one heater may serve two or more zones with a pressure tank for each zone, as shown in Fig. 8-31. One heater may serve two zones by means of a double storage tank. This device consists of one tank within the other. The outer tank only is connected to the heater and usually to the pressure zone on the lower floors. The water in the inner tank is connected to the pressure zone on the upper floors and receives its heat from the water in the outer tank. The inner tank must be constructed to resist full external water pressure from the outer tank when the inner tank is empty, as such a condition may occur during operation. Water in the outer tank must not have access to the top or bottom of the outside of the inner tank to avoid unbalanced vertical forces when either tank is empty and the other is full. Double storage tanks are infrequently used.

In systems with a number of risers the even distribution of the hot water among the risers may be difficult. There should be a valve at the bottom of each riser the manipulation of which by trial may give a satisfactory distribution of water, or a circulating pump may be put on the system or on different risers.

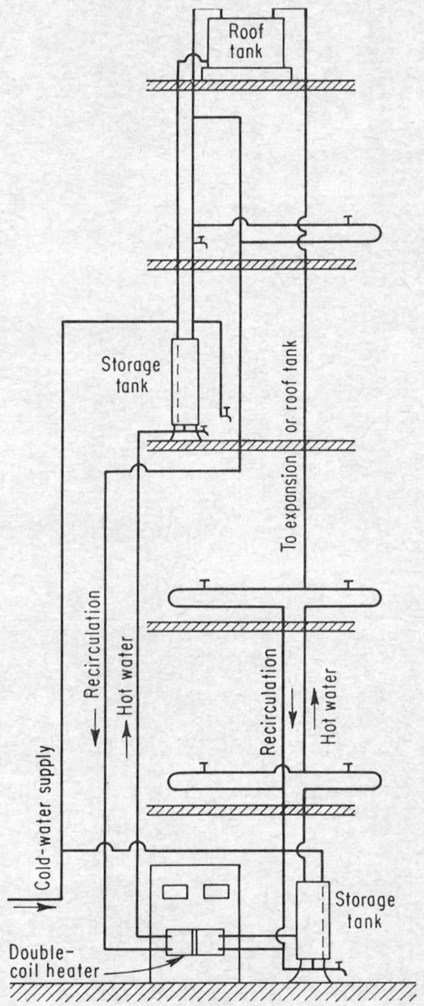

FIG. 8-31. Two-zone heating in tall building with double-coil heater, two storage tanks, and roof tank, with circulation.

The backflow of cold water into a return circulating pipe can be prevented by a check valve, as shown in Fig. 8-32. Such check valves may impede flow in the desired direction, and unless sensitive, they may not function to stop low rates of backflow. The use of a flow-control valve may be preferable.

Allowance should be made for longitudinal expansion[1] in hot-water pipes in lengths greater than 60 to 100 ft. The longitudinal expansions of pipes for various temperatures and materials are listed in Table 8-17.

8-33. Closed-circuit Soft-water Heating. It is possible to heat hard water with soft water through a heat exchanger. The same soft water is circulated continuously between the heater and the heat exchanger, thus avoiding troubles from lime deposits in the heater.

8-34. Snow-melting Systems.[2] Snow can be removed from the surfaces on which it has fallen by circulating warm water or warm antifreeze solution through pipes embedded about one diameter beneath the surface. Steps in the design may involve (1) layout of piping, (2) selection of pipe materials, (3) selection of circulating liquid, (4) calculation of heat losses, (5) selection of heater, (6) selection of circulating pump, and (7) calculation of pipe sizes.

[1] See also L. Blendermann, Expansion in Hot-water Piping Systems, *Air Conditioning, Heating, and Ventilating*, July, 1957, p. 71.

[2] See also W. P. Chapman and S. Katunich, Heat Requirements of Snow Melting Systems, *Trans. Am. Soc. Heating Air Conditioning Engrs.*, vol. 62, p. 359, 1956; W. P. Chapman, Calculating Heat Requirements of a Snow Melting System, a series starting in *Air Conditioning, Heating, and Ventilating*, September, 1956; H. R. Henke, Snow Melting System, *Domestic Eng.*, April, 1956, p. 96; and Anon., A Snow Melting System Hookup, *Air Conditioning, Heating, and Ventilating*, February, 1958, p. 410.

In the layout of piping it is desirable (1) to place the pipes as near to the surface as possible, usually within about one pipe diameter, and (2) to place the pipes about 12 in. apart and slope them to an accessible drainage

TABLE 8-17. LINEAR EXPANSION OF WROUGHT METALS

Temperature, °F	Steel	Iron	Brass	Copper	Temperature, °F	Steel	Iron	Brass	Copper
0	0	0	0	0	100	0.77	0.80	1.17	1.10
20	0.15	0.155	0.25	0.24	120	0.915	0.96	1.41	1.36
40	0.30	0.31	0.45	0.45	140	1.015	1.11	1.69	1.57
60	0.455	0.475	0.67	0.66	160	1.235	1.29	1.91	1.77
80	0.63	0.63	0.92	0.87	180	1.40	1.46	2.15	2.00

point, preferably within the building. If the pipe is embedded in concrete or otherwise protected against external corrosive attack, galvanized wrought iron or steel can be used; otherwise copper is suitable. Because of the possibly high temperatures involved plastic pipe should not be used unless it is adequately heat-stable.

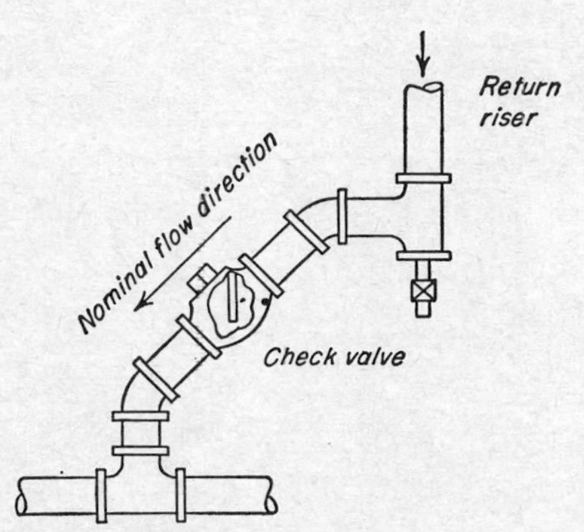

FIG. 8-32. Check valve used on hot-water return lines.

In the operation of the system an antifreeze solution should be used whose characteristics are known at the time of the design. For example, a 50 per cent aqueous solution of ethylene glycol has a friction coefficient 50 per cent greater than water at 100°F and 100 per cent greater at 0°F. Its specific heat, being lower than water, requires the circulation of 30 per cent more liquid for the same heating effect. The addition of a rust inhibitor and its periodic renewal are necessary where glycol antifreeze solutions are used with ferrous-metal pipes.

CHAPTER 9

CHILLED WATER, GAS, AND OTHER FLUIDS

9-1. Cooling of Drinking Water. Drinking water can be cooled by flowing through pipe coils placed in a compartment chilled by a refrigerant or by ice, or the coils can be submerged in chilled brine. The cooling coils

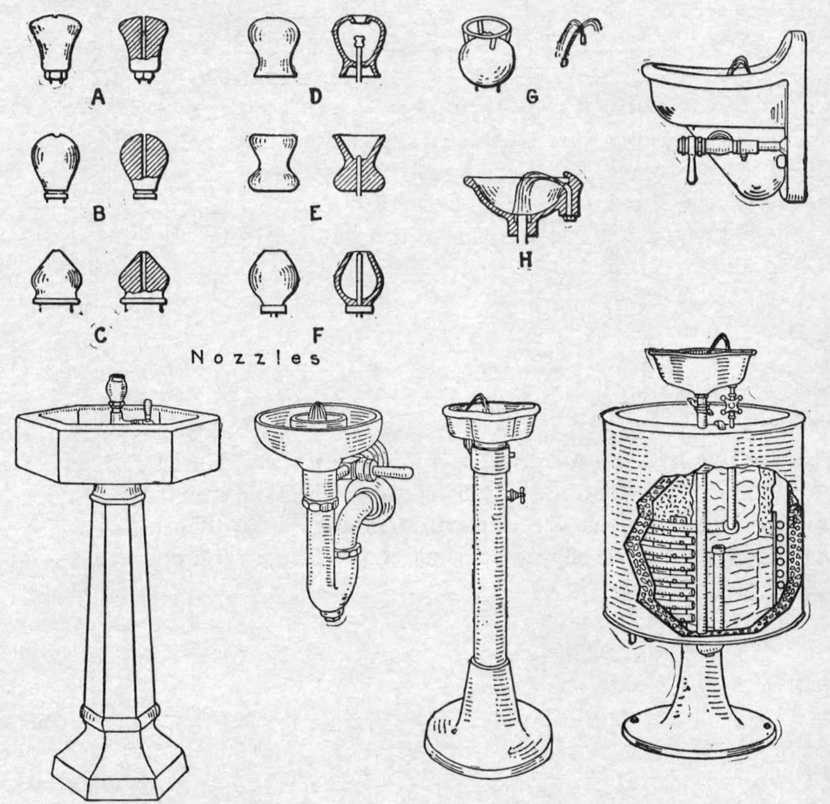

FIG. 9-1. Types of bubbling drinking fountain and nozzle.

should be submerged at the bottom of the container, somewhat as shown in Fig. 9-1, and both the coils and the container should be arranged so that water can be drained from them. Block tin is the best material to be used for the pipe coils. Neither copper nor lead should be used because water may stand a long time in contact with the metal. Iron is not dangerous, but it may impart rust to the water.

The size of the compartment, the length of the pipe coils, and the amount of ice or refrigeration to be used depend on the amount of chilled water to be supplied and on the temperatures of the incoming and outgoing waters. Hence, only the most general recommendations for minimum requirements can be given. An increased demand for water on an excessively hot day may increase the ice consumption or its equivalent three- or fourfold, requiring high overload capacity. In an ice-chilled compartment there should be at least 3 cu ft of space for ice and 3 sq ft of area in the coils for each drinking fountain supplied. For a small fountain, infrequently used, allow 2½ sq ft of coil surface in the cooling chamber; for a more frequently used fountain, allow 9 sq ft; and for a fountain subject to heavy demand, allow 15 sq ft. Eight feet of 1-in. pipe will expose about 2½ sq ft of surface. From 50 to 100 lb of ice or its equivalent will be used daily per fountain supplied. One hundred pounds of ice or its equivalent will cool about 50 gal of water from 75 to 45°. It is generally unnecessary to cool water below about 45°F and impracticable to chill it below about 40°F with ice. Electric refrigerators are rated on the basis of the equivalent weight of ice melted.

In parks and other outdoor places the water can be chilled in an underground pit lined with concrete or brick and provided with drainage. Such pits are more economical than iceboxes or refrigerators aboveground and less ice or its equivalent in energy is consumed. Pits are satisfactory and economical except in temporary locations.

9-2. Distribution of Chilled Water.[1] In order that chilled water can be distributed successfully, provision must be made for continuous circulation of the water; otherwise, water incompletely chilled will flow from the faucet. Two installations of running chilled water that will assure continuous circulation are illustrated in Fig. 9-2. The pipes shown in these systems are so arranged that convection currents force the continuous circulation of chilled water. Such a method of circulation is not always satisfactory, however, because of the necessity for carrying ice to or placing the refrigeration on the top floor and because the difference in temperature between the waters in the two circulating pipes is so slight as to give little circulation. Satisfactory circulation is obtained by means of circulating pumps, as shown in the forced-circulation system. In some buildings it has been found more economical to chill the water in separate refrigerators on different floors rather than to circulate chilled water from one point to all fountains throughout the building. Most

[1] See also L. Blendermann, Drinking Water Systems, *Air Conditioning, Heating, and Ventilating*, May, 1957, p. 10, and Anon., Piping Connections for Chilled Water Systems, *ibid.*, August, 1957, p. 79.

recent installations chill the water separately at each fountain as fast as it is used.

Cleanouts and drains should be plentiful in the supply pipes so that the quality of the water can be maintained at a high standard. Pipes should be insulated, as explained in Sec. 7-20, to diminish the rate of heat absorption and to prevent sweating. This may become a serious problem in humid weather. In order to assure rapid circulation, the chilled-water pipes should be one or two sizes smaller than those recommended in Table 7-1.

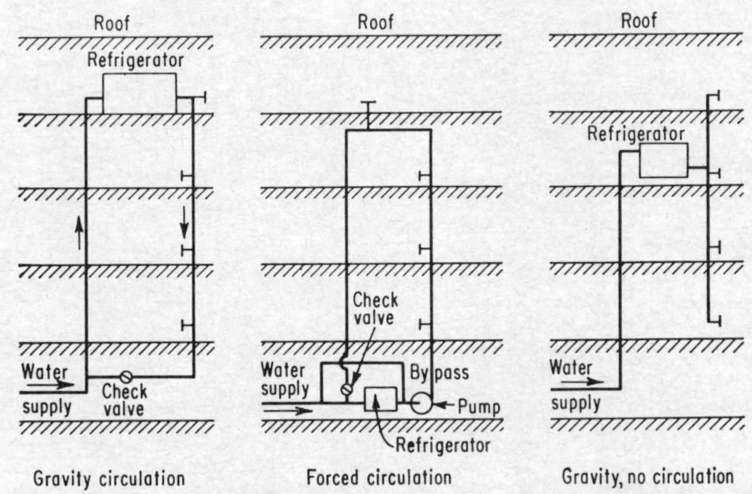

FIG. 9-2. Chilled-water distribution. See also Piping for Chilled Water Supplies, *Air Conditioning, Heating, and Ventilating*, August, 1957, p. 79.

9-3. Nature of Illuminating Gas. Illuminating gas is colorless and odorless, with a specific gravity about 0.6 that of air. It consists of natural gas, manufactured gas, or a mixture of them. The composition of various gases and other information concerning illuminating gases are given in Table 9-1. Its heat content lies between 500 and 1000 Btu per cu ft, depending on the mixture of the gases. Illuminating gas is dangerous to life and to property because of its poisonous, asphyxiating, and explosive properties. It is explosive in proportions of about 15 per cent gas to 85 per cent air by volume. For these and other reasons special precautions are required when gas is piped into a building. The regulations[1] of the National Board of Fire Underwriters provide a guide to good practice.

In order that gas leaks can be detected, some highly offensive, strongly odoriferous substance, such as ethyl mercaptan, is mixed with illuminat-

[1] "Regulations for the Installation, Maintenance, and Use of Piping and Fittings for City Gas," National Board of Fire Underwriters, 1943.

ing gas before it is sent into the distribution pipes. The detection of the odor serves as a warning of leaking gas. Other warning substances can be used. For example, croton aldehyde is useful as a nose and throat irritant so that when mixed with illuminating gas it will awaken a sleeper. Ethyl mercaptan will not do this.

TABLE 9-1. COMPOSITION OF GASES*

Name of gas	Composition per cent by volume							Heating value, Btu per cu ft		Ratio of net to total
	H_2	CO	CH_4	C_2H_6	C_2H_4	Other illuminants	Inert	Total	Net	
Hydrogen	100	....	....	....	...	...	....	318.5	269.2	84.5
Carbon Monoxide	....	100	....	....	...	...	....	316.1	316.1	100
Methane	....	....	100	....	...	...	....	994.1	895.2	90.1
Ethane	....	....	....	100	...	...	....	1757.0	1607	91.4
Ethylene	....	....	....	....	100	...	....	1572.0	1472	93.7
"Theoretical" water gas	50	50	....	....	...	...	....	317.8	293.0	92.3
Carburetted water gas	34.8	30.6	15.0	....	9.5	2.0	8.1	585.4	540.6	92.4
Coal gas	51.0	11.0	23.7	....	2.0	1.1	11.2	508.7	456.2	89.7
Natural gas	....	....	84.2	14.8	...	...	1.0	1098.3	992.6	90.4

* From Standards for Gas Service, *Natl. Bur. Standards Circ.* 405, 1934.

9-4. Ventilation Required. Adequate ventilation of gas appliances in enclosed spaces is necessary. It may be provided through connections to chimney flues, somewhat as shown for the water heater in Fig. 8-13, or through special ventilating flues. The latter are required for house-heating boilers, warm-air furnaces, and most domestic appliances with an input rating in excess of 5000 Btu per hr if the appliance is automatic and 50,000 Btu per hr if manually controlled. Flues are required if the input exceeds 30 Btu per hr per cu ft of space in the room containing the appliance. Such ventilation is required also to remove the odors, water vapor, and gases of combustion.

Vent sizes can be approximated from Table 9-2. Vents over domestic cooking equipment should have an area of not less than 36 sq in. and should lead directly to the outer air with approved natural or forced draft.

TABLE 9-2. SIZES OF VENTS FOR GAS APPLIANCES

Gas input, Btu per hr	75,000	100,000	200,000	300,000
Diameter of vent, in.	5	6	7	8

The products of combustion of illuminating gas may be highly corrosive and laden with water vapor. The vent pipes must be properly designed and installed, and the pipe materials should be corrosion resistant. Cement pipe, terra cotta, or vitrified-clay flues may be satisfactory. Copper, Monel metal, and other corrosion-resistant metals and alloys are also used.[1] A cleanout and a condensation drain should be provided

Table 9-3. Flue Capacities for Gas Appliances

Heat input, 1,000 Btu per hr	28	53	94	160	210	290	500	1,400	2,700	4,000
Diameter of flue, in	3	4	5	6	7	8	10	15	20	24

at the base of the flue. The flue connections to appliances should be above the appliance, they should be a foot or more above the bottom of the flue, and they should not protrude into the flue to form an obstruction. Some data on flue capacities for gas appliances are given in Table 9-3.

Table 9-4 can be used in computing capacities of rectangular flues equivalent to circular flues. It is inadvisable to construct flues with a ratio of long to short dimension greater than 2:1.

The amount of oxygen consumed by gas heaters is appreciable. For example, 9.4 cu ft of air are required in the combustion of 1 cu ft of natural

Table 9-4. Equivalent Capacities of Rectangular, Elliptical, and Circular Flues

(The smaller dimension of the rectangular flue and of the elliptical flue is equal to the diameter of the "equal diameter" circular flue)

Ratio of long to short dimension of flue	1.0	1.25	1.4	1.5	1.75	2.0
Ratio of capacity of rectangular to equal diameter circular flue	1.28	1.67	1.93	2.10	2.52	2.94
Ratio of capacity of elliptical flue to equal diameter circular flue	1.0	1.42	1.68	1.82	2.25	2.68

Table 9-5. Approximate Air Requirements in the Combustion of Hydrocarbon Gases

Btu per cu ft of gas	500	750	1,000	1,500	2,000
Cu ft of air per cu ft of gas burned	4.5	7.0	9.4	14.4	19.3

gas. The heat released will be about 1000 Btu. Other approximate requirements for air for the combustion of hydrocarbon gases are shown in Table 9-5. If the gas appliance is properly equipped with vent and flue, the ventilation of the room will be improved as a result of the draft induced.

[1] See also R. L. Stone, Using Insulated, Double-wall, All-metal, Type B Gas Vent Pipe, *Air Conditioning, Heating, and Ventilating*, July, 1957, p. 76.

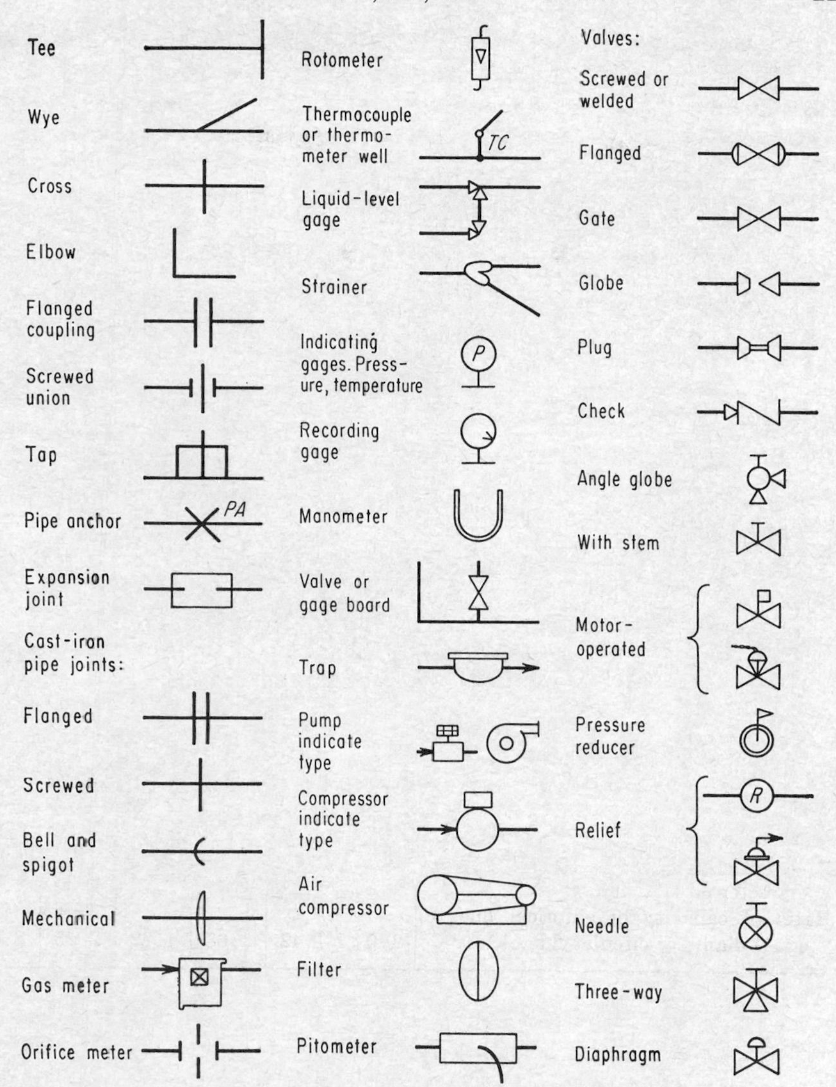

FIG. 9-3. Graphic symbols for gas piping, fittings, controls, and accessories. *(Quoted, by permission of the publishers, from "American Gas Journal Handbook," Sept.* 15, 1957.)

9-5. Gas-distribution Pipes and Service Pipes.[1] In the design and layout of gas-pipe installations standard symbols and nomenclature as shown in Fig. 9-3 should be used. Municipal gas supply is most commonly conveyed into the building through a service pipe with a connection

[1] See also W. Hole, "The Distribution of Gas," 4th ed., Ernest Benn, Ltd., London, 1921; Standards for Gas Service, *Natl. Bur. Standards Circ.* 405, 1934; and *Am. Gas J.*, Gas Handbook Issue, 1957.

similar to the connections used for water service pipes, as shown in Fig. 1-4. Pressures in gas-distribution pipes may be as high as 60 psi, but within a building, pressures are limited to less than 25 psi and are usually in the neighborhood of 4 to 6 in. of water column. A loss of pressure of

TABLE 9-6. SUGGESTED SIZES FOR GAS PIPING FOR DOMESTIC AND SMALL INDUSTRIAL SERVICE

Number of connected burners*	Greatest length of pipe, ft	Diameter of pipe, in.	Number of connected burners*	Greatest length of pipe, ft	Diameter of pipe, in.
3	26	3/8	200	200	2
6	36	1/2	300	300	2 1/2
20	60	3/4	450	450	3
35	80	1	600	500	3 1/2
60	110	1 1/4	750	600	4
100	150	1 1/2			

* Based on 6 cu ft per hr per burner.

TABLE 9-7. GAS PIPES AND GAS CONSUMPTION*

Diameter, in.	Building services		Domestic and small industrial		Automatic storage water heaters		Automatic water heaters		Gas engines		Commercial and industrial†	
	Length, ft	Gas flow, cu ft per hr	Length, ft	Gas flow, cu ft per hr	Length, ft	Water heated, gph	Length, ft	Water heated, gpm	Length, ft	Horse-power	Length, ft	No. of 1/2-in. outlets
3/8	...	....	30	20								
1/2	...	....	40	40	...	...	...	...	...	...	30	1
3/4	...	....	60‡	120	90	30	90	2	...	...	60	8
1	70	160	80	200	100	50	100	3	100	7	70	12
1 1/4	100	400	100	350	150	200	...	...	100	12	100	20
1 1/2	150	800	150	600	...	...	200	6	100	22	150	35
2	200	1,200	200	1,200	250	500	250	8	100	35	200	50
2 1/2	250	2,000	300	2,000	...	...	...	...	100	50		
3	300	3,000	450	3,000	...	...	...	...	100	100		
4	450	5,000	600	5,000	...	...	...	...	200	150		

* Rates of gas consumption, in cubic feet per hour: cooking range in a residence, 50; water heater, 40; grate or log, 30; laundry appliance, 30.

† Running line not less than 3/4 in.; drop outlet not less than 1/2-in.

‡ One 3/4-in. outlet may supply both a kitchen stove and a water heater.

about 4 in. of water is to be expected in flowing through the gas meter. The higher pressure is required for methane because of its greater heat content and the smaller gas orifice required in the adjustment of gas-air mixtures for gas appliances.

Recommended sizes for gas pipes are stated in Tables 9-6 and 9-7, and some rates of gas usage for different conditions are given in Table 9-7. Service pipes for domestic users should not be less than 1¼ in. in diameter unless a gas furnace is used, for which the minimum size of pipe should be 2½ in. Gas service pipes should be laid on a continuous upward slope from the gas main to a point near the foundation wall to avoid the accumulation of water of condensation in the service pipe. Before passing through the foundation wall the service pipe should loop upward and then downward with the top of the loop at or near the ground surface to prevent leaking gas from following along the outside of the pipe and entering the building. Where it is not possible to do this, readily accessible drip pipes, not less than 6 in. long, should be installed in such a manner that water can be drained from the service pipe but illicit connections cannot be made. Materials used for water service pipes, as discussed in Sec. 3-9, are suitable for use in gas service pipes.

9-6. Installation and Sizes of Gas Pipes.[1] Black iron, steel, or wrought pipe with screwed joints or welded joints is suitable for gas piping within a building, since illuminating gas is noncorrosive to such metals. The use of lead pipe or rubber tubing for the conveyance of gas is commonly prohibited. Pipes exposed to external corrosive attack can be protected as discussed in Sec. 4-34. The principles, methods, and materials applicable to installing, supporting, and connecting cold-water pipes are applicable to gas piping, with the following additional recommendations:

1. Slope all pipes toward accessible drip pipes, and place drip pipes accessibly at the base of every vertical pipe and every low point in horizontal lines.
2. Take branches off the top or sides of connecting pipes in order to prevent the accumulation of water in the branch when this becomes necessary.
3. Locate all pipe unions and valves accessibly. The use of unions and bushings on concealed work is undesirable. Cast-iron fittings may not be used on gas piping.
4. Locate gas pipes as accessibly as possible.
5. Support gas piping adequately. When pipes are run on an outside wall, place a furring strip between pipe and wall. When they are run on masonry walls, support may be with straps fastened to wood plugs embedded in the masonry. When gas pipe is embedded in concrete, the pipe should be covered with tar paper or other suitable covering, or better it should be laid inside a larger conduit that is in contact with the concrete.
6. It is good practice, but not compulsory, to provide outlets for both fuel and illumination, whether or not both are to be used, at the time of installation of the gas supply.

[1] See also L. Blendermann, Gas Services for Multi-story Buildings, *Air Conditioning, Heating, and Ventilating*, December, 1954, p. 106.

7. Extend outlets at least 1 in. through finished walls and ceilings. Outlets should be capped with threaded pipe caps until fixtures are installed.

8. No piping should be less than ¾ in. in diameter and, if embedded in concrete, not less than 1 in.

9. The gas meter and the lower end of gas riser pipes should be located in an accessible, ventilated, warm, dry space preferably with a ceiling height of 6 ft or more. Protection against freezing of water of condensation must be provided in unheated locations in cold climates.

10. Meters and riser pipes should be located some distance from electric meters, switches, fuses, and other equipment.

11. Water heaters, space heaters, etc., utilizing more than 50,000 Btu per hr input should have an independent pipe from the meter.

12. Outlets for cooking appliances should be ¾ in. in diameter or larger; for water heaters, not less than 1 in.; and for room space heaters, not less than ½ in.

The fact that gas is lighter than air results in an increase in gas pressure as it rises in a vertical pipe, as shown in Table 9-8.

TABLE 9-8. CHANGES IN PRESSURE IN GAS PIPES

Gain in pressure per 100 ft of rise in vertical pipe, in. of water	0.96	0.89	0.81	0.74	0.66	0.59	0.52	0.44
Sp gr of gas compared to air	0.35	0.40	0.45	0.50	0.55	0.60	0.65	0.70

Many formulas[1] are available for expressing flow in gas pipes. Pole's formula is

$$Q = 2{,}340 \sqrt{\frac{d^5 h}{SL}} \tag{9-1}$$

where d = diameter of pipe, in.
h = head loss of water, in.
L = length of pipe, ft
Q = flow of gas, cu ft per hr at atmospheric pressure
S = specific gravity of gas in relation to air

Some solutions of Pole's formula are shown in Table 9-9.

In determining the volumetric rate of flow of gas to an appliance, the input rating of the appliance, expressed in Btu per unit of time such as an hour, should be divided by the heat content of the gas. The quotient will be the required flow to the appliance in cubic feet per hour. Relations among head losses, pipe diameters, and rates of flow in the pipe are shown in Tables 9-10 and 9-11, using the specific gravity of gas with respect

[1] For Spitzglass formula and solutions see *Heating, Piping, Air Conditioning*, March, 1957, p. 135.

TABLE 9-9. FLOW OF GAS IN PIPES, ACCORDING TO POLE'S FORMULA, USING $S = 0.6$ AND $L = 100$
(Flow of gas is in cubic feet per hour)

Head loss in. of water (h)	Diameter of pipe, in.			
	¾	1	1½	2
0.1	47	100	263	563
0.2	66	141	373	796
0.3	81	173	456	974
0.4	92	200	526	1,128
0.5	104	223	589	1,260
0.6	114	244	645	1,380
0.7	124	264	697	1,490
0.8	132	282	745	1,590
0.9	140	300	790	1,690
1.0	148	315	832	1,780

TABLE 9-10. CAPACITIES OF PIPES TO CONVEY ILLUMINATING GAS*
(Flows are expressed in cubic feet per hour and are based on gas with a specific gravity of 0.6 in relation to air and to a pressure drop of 0.3 in. of water)

Length of pipe, ft	Diameter of pipe, in.									
	½	¾	1	1¼	1½	2	3	4	6	8
15	76	172	345	750	1,220	2,480	6,500	13,880	38,700	79,000
30	55	120	241	535	850	1,780	4,700	9,700	27,370	55,850
45	44	99	199	435	700	1,475	3,900	7,900	23,350	45,600
60	38	86	173	380	610	1,290	3,450	6,800	19,330	39,500
75	..	77	155	345	545	1,120	3,000	6,000	17,310	35,300
90	..	70	141	310	490	1,000	2,700	5,500	15,800	32,350
105	..	65	131	285	450	920	2,450	5,100	14,620	29,850
120	..	...	120	270	420	860	2,300	4,800	13,680	27,920
150	..	...	109	242	380	780	2,090	4,350	12,240	25,000
180	..	...	100	225	350	720	1,950	4,000	11,160	22,800
210	..	...	...	205	320	660	1,780	2,700	10,330	21,100
240	..	...	...	190	300	620	1,680	3,490	9,600	19,740
270	...	...	...	178	285	580	1,580	3,250	9,000	18,610
300	..	...	...	170	270	545	1,490	3,000	8,500	17,660
450	...	...	...	140	226	450	1,230	2,500	7,000	14,420
600	..	...	...	119	192	390	1,030	2,130	6,000	12,480

* From "Standards for Piping, Appliances, and Fittings for City Gas," Pamphlet 54, National Board of Fire Underwriters, Sept. 1, 1943.

to air as 0.6. The factors by which the rates of flow in Tables 9-10 and 9-11 must be multiplied for gases of other specific gravity are shown in Table 9-12.

TABLE 9-11. CAPACITIES OF GAS PIPES
(Losses of pressure are shown in inches of water per 100 ft of pipe, due to the flow of gas with a specific gravity of 0.6 with respect to air)*

Rate of flow of gas, cu ft per hr	Size of pipe, in.											
	3/8	1/2	3/4	1	1¼	1½	2	2½	3	3½	4	5
10	0.14	0.09										
20	0.59	0.19										
50	....	1.0	0.14	0.08								
100	....	....	0.59	0.16	0.04							
150	....	....	1.02	0.36	0.08							
200	....	....	2.20	0.65	0.15	0.06						
250	....	....	3.5	1.00	0.23	0.10						
500	....	....	....	4.1	0.90	0.40	0.09					
750	....	....	....	9.9	2.1	0.89	0.21	0.09				
1,000	....	....	....	....	4.0	1.6	0.38	0.15	0.05			
1,500	....	....	....	....	8.8	3.6	0.86	0.33	0.11	0.05		
2,000	....	....	....	....	....	6.3	1.50	0.60	0.18	0.08	0.05	
3,000	....	....	....	....	....	....	3.5	0.94	0.44	0.19	0.10	
5,000	....	....	....	....	....	....	8.8	3.6	1.2	0.54	0.34	0.08
6,000	....	....	....	....	....	....	....	5.4	1.7	0.76	0.40	0.13
8,000	....	....	....	....	....	....	....	....	3.0	1.3	0.72	0.23
10,000	....	....	....	....	....	....	....	....	5.2	2.1	1.2	0.36
20,000	....	....	....	....	....	....	....	....	....	8.2	4.8	1.4
30,000	....	....	....	....	....	....	....	....	....	....	10.0	3.2
40,000	....	....	....	....	....	....	....	....	....	....	....	5.7
50,000	....	....	....	....	....	....	....	....	....	....	....	9.0

* To determine head losses for other lengths of pipe, multiply the head losses in this table by the length of the pipe and divide by 100. For head losses due to flow of gases with specific gravity other than 0.6, use the figures given in Table 9-12.

TABLE 9-12. FACTORS BY WHICH FLOWS IN TABLE 9-11 MUST BE MULTIPLIED FOR GASES OF OTHER SPECIFIC GRAVITY*

Sp gr of gas	0.35	0.40	0.45	0.50	0.55	0.60	0.65	0.70
Factor	0.77	0.82	0.87	0.91	0.96	1.00	1.04	1.08

* From "Standards for Piping, Appliances, and Fittings for City Gas," Pamphlet 54, National Board of Fire Underwriters, Sept. 1, 1943.

After installation, gas pipes should be subjected to an air-pressure test, similar to that described in Sec. 12-30, with a pressure of about 3 psi. After fixtures are installed, the pipes should hold a pressure of at least 12 in. of water, or about 0.34 psi, for not less than 10 min. Gas pipe other than welded pipe should withstand an air pressure of 10 psi for 10 min or more. Welded pipe should hold a pressure of 50 psi for at least 10 min.

9-7. Setting Gas Meters. The gas meter should be installed inside the building, and it should be supported on or close to the wall through which the service pipe enters the building. The bottom must be free from extremes of temperature within a range of 50 to 90°F, and the room should be adequately ventilated. The meter should be rigidly supported at an elevation higher than the entering service pipe, which should turn up to enter the meter. A satisfactory setting of a gas meter is illustrated in Fig. 9-4. Gas meters are discussed in Sec. 14-39, and their capacities are given in Table 14-6.

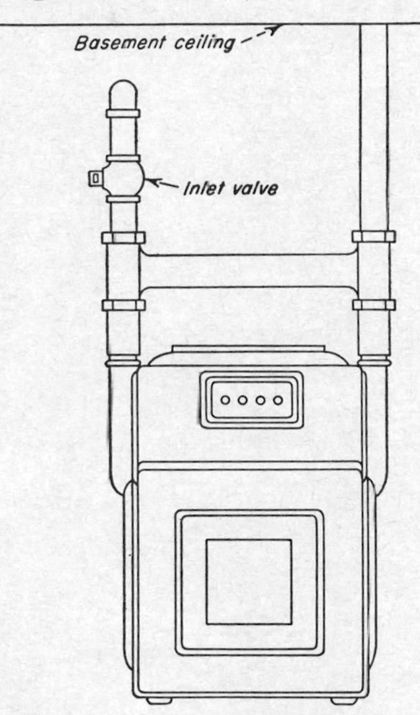

FIG. 9-4. Gas-meter installation.

9-8. Compressed-gas Installations.[1] Where public gas mains are not available, liquefied gas, such as butane, can be supplied in high-pressure cylinders. These cylinders should meet the requirements of the Interstate Commerce Commission for interstate shipment.[2] Connections for the installation of such gas, which is frequently called *bottled gas*, are shown in Fig. 9-5. The cylinders are painted in aluminum or some other heat-reflecting color, and where it is possible, they should be installed in the open air in a shady location protected from traffic, sources of heat, flame, and inclement weather, especially from snow and ice.

Piping should comply with the standards for the distribution of illuminating gas. The piping leading from the gas cylinders to the gas pipes in the building is commonly made of ¼-in. flexible copper or brass tubing with upset screwed joints with left-hand threads. A pressure-reducing valve should be installed on the gas piping between the gas cylinder and the shutoff valve outside the building. The pressure can be reduced to

[1] See also "Regulations for the Installation and Operation of Compressed Gas Systems Other than Acetylene for Lighting and Heating," National Board of Fire Underwriters, 1937.

[2] See also H. A. Campbell, "Freight Tariff No. 4," Bureau of Explosives, New York, 1939.

about 7 in. of water pressure in the house piping. A relief valve in the equipment should discharge into the open air at least 5 ft away from any opening into the building.

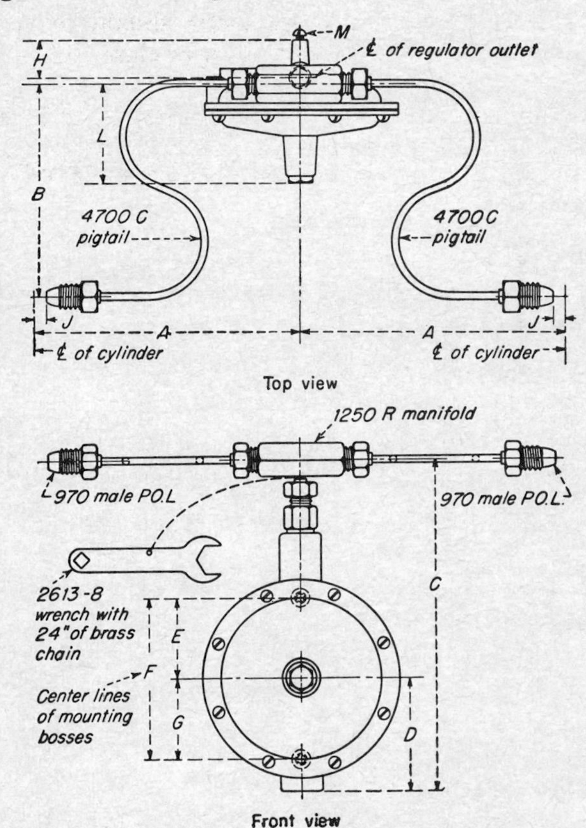

FIG. 9-5. Connections for bottle-gas installation. (*The Bastian-Blessing Co.*)

9-9. Compressed-air and Vacuum Pipes. The sizes of compressed-air[1] and vacuum pipes can be computed from the following formula:

$$d = 13.5 \sqrt{\frac{Q}{V}} \tag{9-2}$$

where d = internal diameter of pipe, in.

Q = rate of flow in pipe, cfm

V = velocity of flow in the pipe, fpm

The volume of compressed or of rarefied air Q can be expressed in terms of the equivalent volume of air at atmospheric pressure (free air). However

[1] See also R. J. Nemmers, How to Design and Maintain Compressed Air Piping Systems, *Heating, Piping, Air Conditioning*, August, 1937, p. 92.

expressed, the value of Q to be used in Eq. (9-2) is the actual rate of flow of gas through the pipe. This can be computed also by use of the expression

$$Q_p = \frac{Q_A P_A}{P_p} \tag{9-3}$$

where Q_p = rate of flow of compressed air or rarefied air in pipe. This is the value that is substituted in Eq. (9-2)

Q_A = rate of flow of free air, at atmospheric pressure, into compressor

P_A = atmospheric pressure, equal to 14.7 psi at sea level. Values at other elevations are shown in Table 2-7

P_p = absolute pressure (gage pressure + atmospheric pressure) in pipe, psi

Example. Determine the diameter of a pipe to carry compressed air under a pressure of 60 psi if the flow is equivalent to 15 cfm of free air.

Solution:

$$Q_p = \frac{Q_A P_A}{P_p} = \frac{15 \times 14.7}{60 + 14.7} = 2.95 \text{ cfm}$$

$$d = 13.5 \sqrt{2.95 \div 2{,}000} = 13.5 \sqrt{0.0015} = 0.525 \text{ in.}$$

Use a ½-in. pipe.

Friction losses in air pipes can be computed[1] from the expression

$$P_1 - P_2 = \frac{2fV^2L}{gdV} \tag{9-4}$$

where d = diameter of pipe, in.

$f = 0.0054 + 0.375 \dfrac{\mu\nu}{dV}$ for air or gas

g = acceleration due to gravity, 32.2 fps²

L = length of pipe, ft

P_1 = pressure at upper end of pipe, psi

P_2 = pressure at lower end of pipe, psi

V = velocity of flow gas, fps

μ = viscosity [see Eq. (9-5)]

ν = density

when the drop in pressure is small so that the change in density of the air is negligible. The value of μ is expressed as

$$\mu = 0.0165 + 2.5 \times 10^{-5}t \tag{9-5}$$

where t is the temperature in degrees Fahrenheit.

[1] See also J. F. Waters, Pressure Drop in Compressed Air Piping, *Heating, Piping, Air Conditioning,* February, 1958, p. 105.

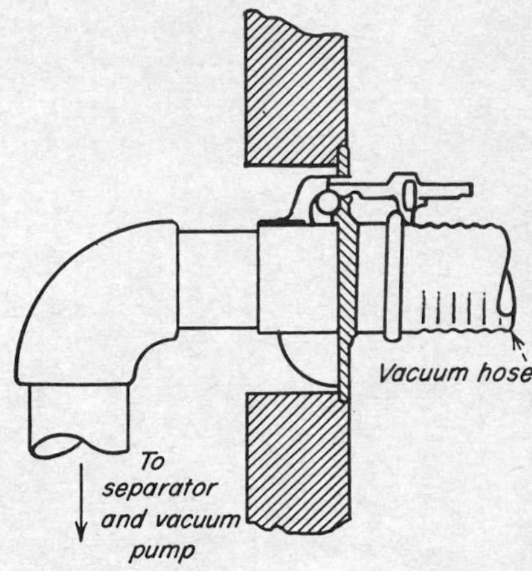

FIG. 9-6. Automatic self-closing vacuum connection that can be held open only when vacuum hose is attached.

If the drop in pressure is considerable, the following expression must be used:

$$P_1^2 - P_2^2 = \frac{0.1085 f V_a^2 L}{d^5} \tag{9-6}$$

where V_a is the velocity of flow in feet per second

Vacuum systems are maintained in large buildings mainly for cleaning purposes. Connections of the type shown in Fig. 9-6 are located at convenient points with an allowance of about 50 ft of suction hose. The vacuum to be maintained depends partly on the service required and on the size of hose used. For services other than carpet cleaning, a vacuum of 1 to 2 psi may be provided. Where 50 ft of 1¼-in. hose is to be used,

TABLE 9-13. SIZES OF PIPES FOR VACUUM-CLEANER SYSTEMS
(Figures in the table show the size of pipe, in inches, that should be used for the corresponding number of sweepers and length of pipe)

Length of pipe,* ft	Number of cleaners											
	1	2	3	5	8	10	15	20	25	30	40	50
100	2.2	2.4	2.7	3.25	3.9	4.2	5.0	5.5	6.0	6.5	7.5	8.6
250	2.5	2.8	3.3	4.0	4.75	5.2	6.2	6.7	7.4	8.0	9.1	10.2
500	2.8	3.3	3.75	4.5	5.5	6.0	7.1	7.8	8.6	9.75	10.5	11.7
750	3.0	3.5	4.0	4.75	5.8	6.3	7.5	8.3	9.2	10.0	11.2	12.4
1,000	3.1	3.6	4.1	4.9	5.9	6.5	7.7	8.7	9.6	10.4	11.6	12.75
1,500	3.2	3.7	4.25	5.2	6.1	6.7	8.0	9.1	10.1	10.7	11.8	13.2

* In computing head losses due to bends, allow a length of 10 ft of pipe as equivalent to one-quarter bend.

mainly for carpet cleaning, there should be a vacuum of 7 to 10 ft of water or 3 to 4½ psi. A 1-in. hose for carpet cleaning will require a vacuum between 9 and 9½ psi. The amount of free air moved per sweeper is about 30 cfm. Capacities for large buildings seldom exceed 90 cfm.

In the design of air pipes an allowance for velocity of about 3,000 fpm should be made. Depressions and sags in the pipe must be avoided, and a generous allowance of cleanouts should be provided. Some recommendations of pipe sizes are given in Table 9-13.

9-10. Distilled-water Systems.[1] Distilled water may be highly corrosive to ferrous-metal and some other metallic pipes. It is necessary, therefore, to use corrosion-resistant metal, such as stainless steel, or glass pipes to prevent corrosion and injury to the quality of the distilled water. The layout of distilled-water piping follows the rules for the layout of cold-water pipes.

[1] See also L. Blendermann, Distilled Water Systems, *Air Conditioning, Heating, and Ventilating,* August, 1957, p. 100.

CHAPTER 10

INTERIOR FIRE PROTECTION

10-1. Methods of Fire Protection.[1] Any method of fire protection involving the conveyance of water in pipes within a building falls into the field of plumbing. Water to extinguish fires within buildings may be supplied through (1) riser pipes or standpipes with hose connections, (2) automatic sprinklers, (3) storage tanks, and (4) pumps. For the best protection these may supplement one another. An additional source of water may be required. Water may be drawn from the public supply or from other sources, some of which may be polluted or otherwise nonpotable, such as sea water or polluted river water. Cross connections are a potential danger where a nonpotable water supply is available for fire protection regardless of whether or not it is used. Choice must be made between illness and death due to polluted water and property loss, injury, and death due to uncontrollable fire.

A standpipe with hose connection in a tall building, as shown in Fig. 10-1, may be fed from a storage tank, from pumps, or from a mobile pumping engine in the street connected as illustrated in Fig. 10-2. Automatic sprinklers are devices that discharge water automatically when the temperature of the air surrounding them reaches a predetermined figure. They are discussed in Sec. 10-11. Water spray nozzles[2] apply water in small particles forming a fog or mist, with fire-extinguishing characteristics superior to liquid water under some conditions.

In the design and in the inspection of fire-protection systems, the publications of the National Bureau of Standards, the recommendations of the National Board of Fire Underwriters, and the codes of the National Fire Protection Association should be used as guides.

10-2. Quantity and Rate of Demand for Water. Conditions to be considered in the procurement of water for fire protection include quality, quantity or rate of demand, pressures, and means of distribution. Ordinarily the public water supply meets all requirements, but in the event

[1] See also National Fire Protection Association, "Handbook of Fire Protection," 11th ed., 1954; L. Blendermann, *Air Conditioning, Heating, and Ventilating*, July, 1956, p. 92, and June, 1957, p. 112.

[2] See also "Water Spray Systems for Fire Protection," National Board of Fire Underwriters, 1954.

that it does not, additional water must be obtained. Under such circumstances quantity and rate requirements supersede quality restrictions, since fire may be as great a threat as disease to the safety of life and health. Even sea water or polluted water may be made available as a

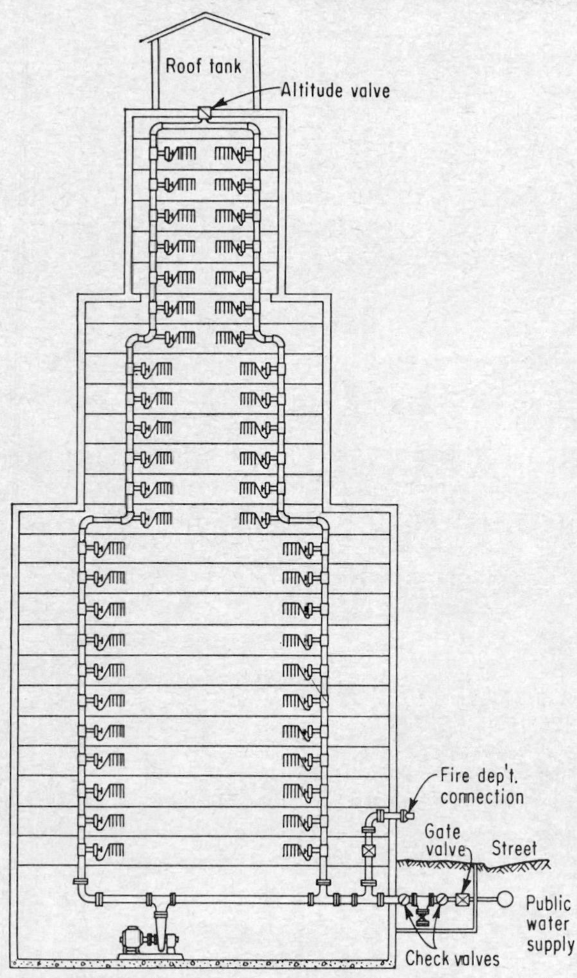

FIG. 10-1. Arrangement of a fire-protection standpipe system in a tall building. This is a modification of the installation shown in *Air Conditioning, Heating, and Ventilating,* June, 1957, p. 112.

last resort. However, the use of nonpotable water should be avoided, wherever possible, because of the dangers accompanying its misuse.

The rate at which water must be supplied rather than the volume of water to be used controls the design of pumps and pipe sizes. The volume of water available at the source should exceed the product of the rate of demand and the duration of the longest fire. Since estimates of these factors cannot be expected to give exact results, the regulations of the National Board of Fire Underwriters may serve as a guide. Rates of demand applicable in the design of interior fire-protection systems differ from those applicable in the design of a public water supply. Some rates

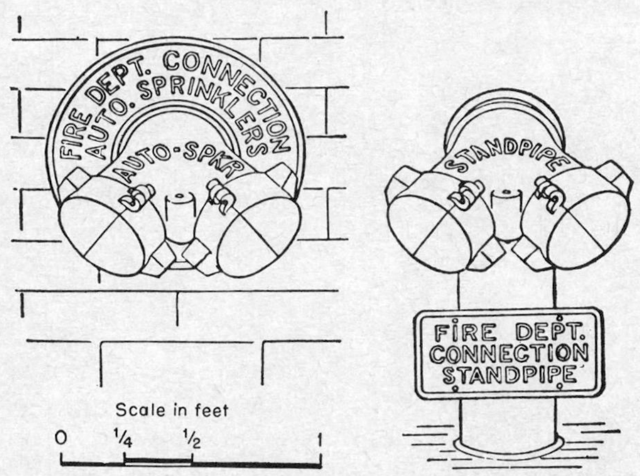

FIG. 10-2. Typical connections to automatic sprinkler systems and standpipes for use of outside fire departments. (*From National Fire Protection Association*, 1943.)

of demand for municipal fire protection are given in Table 10-1. The National Board of Fire Underwriters states, concerning interior fire protection:

Minimum supplies for use by fire department or specially trained men (2½-in. hose and 1⅛-in. nozzle) shall be calculated upon the basis of not less than 250 gpm for each standpipe riser. Capacity of supplies should be such that for a period of 1 hr there will be available a pressure of at least 50 lb at topmost 2½-in. outlet (not including roof outlet) while water is being discharged through 50 ft of 2½-in. cotton rubber-lined hose and a 1⅛-in. nozzle from the topmost outlet. When the supply is from fire pump or tank, the minimum sizes which should be recog-

TABLE 10-1. EMPIRICAL FORMULAS FOR RATES OF FIRE DEMAND BY CITIES

Name of originator or authority	Formula: Q = demand, gpm; P = population in thousands	Gpm per 100,000 population
Kuichling (on basis of fire streams of 250 gpm)	$Q = 700\sqrt{P}$	7,000
John R. Freeman	$Q = 250\left(\frac{P}{5} + 10\right)$	7,500
National Board of Fire Underwriters	$Q = 1{,}020\sqrt{P}\,(1 - 0.01\sqrt{P})$	9,180

nized are: approved fire pump, 500 gpm; pressure tank, 4,500 gal; gravity tank, 5,000 gal with bottom elevated 40 ft above highest hose outlet.

Minimum supplies for standpipes for use by occupants of buildings as first-aid fire protection shall be calculated on the basis of 100 gpm flowing with 25-lb pressure at highest hose outlet. This will afford two good first aid streams simultaneously. Supply may be from adequate city waterworks system, fire pump, from supplies for large standpipe systems, pressure tanks, and/or gravity tanks elevated 25 ft above highest hose outlet. Minimum supplies shall conform to those required for large (2½-in.) hose, except that when supply is furnished by a domestic gravity tank a minimum of 3,000 gal of water shall be reserved exclusively for fire protection.

The quality, quantity, and pressure of water for fire protection to be delivered by fixed or mobile pumps fed from hydrants at the ground level are based on the same principles as for standpipe systems. Interior sprinkler systems, described in Sec. 10-11, require less water, since they function for only a short time at the start of the fire. The sprinkler becomes useless when the fire spreads beyond the area protected by it.

10-3. Standpipes with Hose Connections.[1] Standpipes with hose connections furnish satisfactory means of providing water for fire protection on the upper floors of tall buildings. If the standpipes are kept filled with water, the method gives a means of providing water in adequate quality and rate at a moment's notice. The system should be designed for effective use by either amateur or trained fire-fighting personnel, or both.

A standpipe and hose system, such as is illustrated in Fig. 10-1, involves the installation of vertical riser pipes with hose connections at strategic points throughout the building. The use of hose makes it possible to locate the riser pipes conveniently without being so conspicuous as to detract from the appearance of the building and to keep them out of traffic lanes. The standpipes can be kept filled with water provided that there is no danger from freezing, bursting, or leaks or other objection. Under some conditions the dry-pipe system is used, with the expectation that the pipes will be filled with water immediately when needed. In some buildings the dry-pipe system is arranged to be filled through a hose connected to a mobile engine of the public fire department. If there are two or more standpipes in the same building, they should be interconnected to provide the greatest flexibility and efficiency. Both the standpipes and the interconnections should be provided with adequate stop valves to permit desired flexibility in control. Such valves should be prominently located and conspicuously labeled. Outside screw and yoke valves of the type shown in Fig. 14-10 are preferred because the position

[1] See also "Standpipe and Hose Systems," National Board of Fire Underwriters, 1952.

of the gate can be seen at a glance. The valves are normally left open. They are closed to cut off any portion of the system that has failed. The usefulness of the fire-fighting installation might be nullified if, as a result of a structural failure, water could escape from the water-distribution system and could not be shut off.

Sizes of standpipes should be sufficient to supply the number of fire streams that can be connected to them simultaneously. Computations

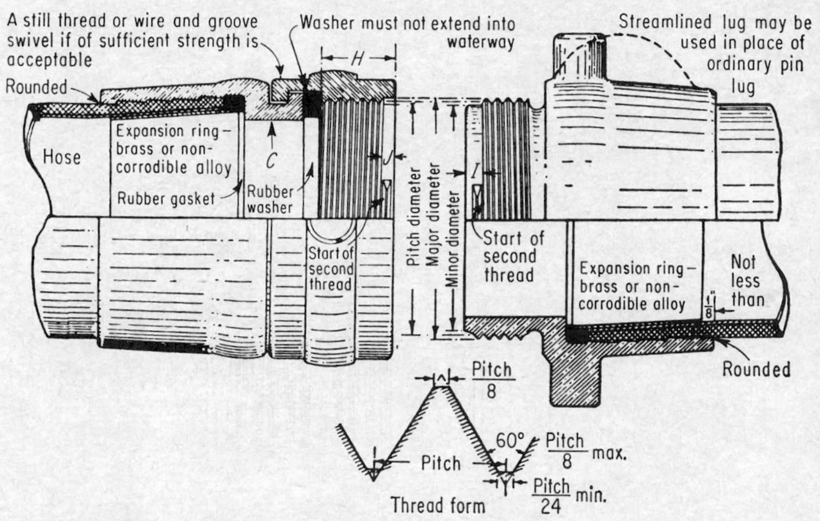

FIG. 10-3. Standard fire-hose coupling. (*From "Handbook of Fire Protection," reproduced by permission of National Fire Protection Association.*)

of head losses, based on the number and sizes of streams in simultaneous use and on the pressures desired at the nozzle, may follow the procedure given in Chap. 2.

Some recommendations of the National Board of Fire Underwriters are

For standpipes supplying two small fire streams, with ½-in. nozzles or smaller, with a combined discharge of not more than 100 gpm; and for buildings not more than four stories or 50 ft high, use 2-in. standpipes; and for buildings higher than four stories, use 2½-in. pipes. For standpipes supplying 2½-in. hose with 1- to 1⅛-in. nozzles; and for buildings not more than six stories or 75 ft high, use 4-in. pipes; and for higher buildings use 6-in. pipe.

Standpipes, especially those supplying large hose, should be located where they are least exposed to damage by fire and where they will be accessible and usable during a fire. Locations such as fire-resistant stairways or corridors and the outside of an exterior wall near a fire escape are suggested. In cold climates, only dry standpipes should be exposed to freezing locations. Standpipes that may be exposed to heat from fires should be insulated with fireproof insulation.

10-4. Fire Hose. Standard fire hose[1] is made of rubber-lined cotton, 2½ in. in diameter, capable of standing routine test pressures of 200 psi. Other acceptable hose may be unlined linen or rubber-lined or rubber-covered cotton. Unlined linen is used for hand hose left permanently

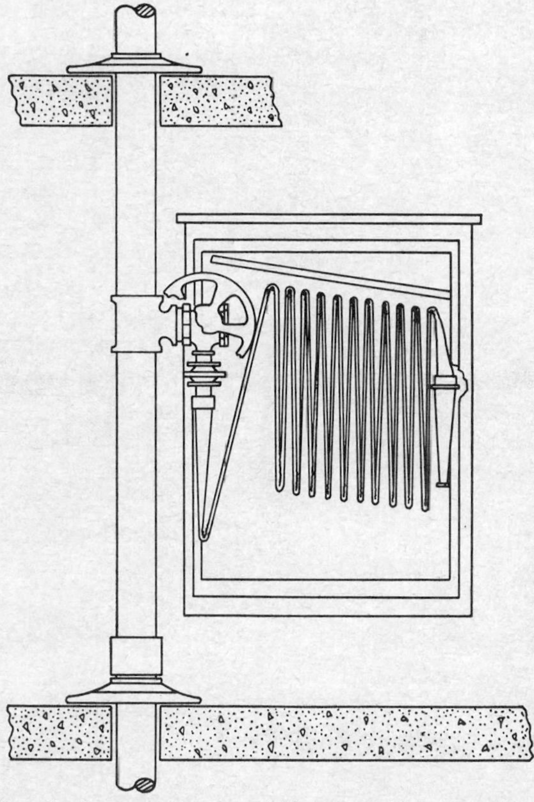

FIG. 10-4. Detail of fire-hose station for 1½-in. first-aid hose. (*Based on Air Conditioning, Heating, and Ventilating, June,* 1957, *p.* 114.)

connected to standpipes in buildings, sometimes called first-aid hose. These may be 1½-in. hose.

Hose couplings must meet specifications given in Table II-1 and shown in Fig. 10-3. Standard hose threads are discussed in Sec. 4-13.

Inside fire hose should, in general, be folded and supported somewhat as indicated in Fig. 10-4 so that it can be run out to full length without kinking before the water is turned on. A closet in which hose is stored should be ventilated, and the hose should be treated against mildew.

[1] See Table II-1 for specifications.

The length of 2½-in. hose provided at one outlet should not exceed 100 ft. This length, measured around obstructions and partitions, should make possible the use of a nozzle within 30 ft of every part of every floor in a properly protected building. The length of smaller, or first-aid, hose should not exceed 75 ft, with a nozzle within 20 ft of every part of the building. More than one standpipe may be necessary to fulfill these requirements.

10-5. Fire-hose Stations. A fire-hose station for 2½-in. hose, not more than 100 ft long, should be located conspicuously, close to a standpipe and not more than 6 ft above the floor. First-aid hose must be attached to a standpipe constantly filled with water. A first-aid hose station may be located in a room or in a corridor and connected by a branch pipe through the wall to a standpipe in a stair well or corridor, serving 2½-in. hose.

Fire hose may be supported on a "semiautomatic" or a "one-man" hose rack that permits one man to pull the nozzle from the rack toward the fire *after* he has turned on the hose valve. The hose must be folded so that it can be withdrawn from the rack or reel by pulling on the nozzle. This requires a special form of folding or winding. Some racks may require two men, one to handle the hose and the other to manipulate the hose valve. The hose valve is located close to the water-supply pipe and between it and the hose. A gate valve should be equipped with threads that are interchangeable with those used by the local fire department. American Standard threads, as described in Sec. 4-13, should be used where possible. If pressures in the standpipe may exceed 50 psi, a pressure-regulating valve should be placed between the hose and the standpipe, preferably on the hose side of the gate valve, to protect the hose against excessive pressures.

10-6. Standpipe Connections. Recommendations by the National Board of Fire Underwriters require that

Connections from gravity tanks (on buildings) and pressure tanks should be made to the top of the standpipe system except where the tanks are used as a supply to standpipes in several buildings or sections of a building, in which cases they should be made at the base of the standpipes. Connections to standpipes for the larger streams shall be at least 4 in. Those to standpipes for the smaller streams shall be at least 2½ in.

Connections from fire pumps and sources outside the building should be made at the base of the standpipes. The connections from each supply shall be large enough to deliver its full rated capacity without excessive friction losses.

Where two or more standpipes are installed in the same building or section of a building they should be (interconnected) at the bottom. Where standpipes in a single building are supplied by tanks they should be (interconnected) at the top.

10-7. Valves and Fittings for Standpipes. Recommendations by the National Board of Fire Underwriters require that

Connections to each water supply, except fire department hose connections, should be provided with an approved gate and check valve located close to the supply, as at tank, pump, and the connection from the waterworks system. Where the water supply feeds the standpipes in more than one building or section of a building, the check valves should be placed in a safe position in the underground connections or where not exposed to danger from fire or falling buildings.

Where there is more than one standpipe in a single building, each standpipe riser should be controlled by an approved gate valve located at the base of the riser.

Connections to public waterworks should, where feasible, be controlled by post indicator valves of an approved type and located not less than 40 ft from the building protected; or if this cannot be done, placed where they will be readily accessible in case of fire and not subject to injury. Where post indicator valves cannot readily be used, as in a city block, underground gates should conform to the above as far as possible and their locations and directions to open be plainly marked on the buildings. All post indicator valves should be plainly marked with the service they control.

Where the standpipes are supplied from a yard main or header in another building, the connection should be provided with an approved outside post indicator valve at a safe distance from the building or an approved indicator valve at the header.

Fire department hose connections shall be provided with an approved straightway check valve located in the building or valve pit, but not with a gate valve. Piping between the check valve and outside hose connections shall be arranged to drain automatically.

Gate and check valves shall be of the approved extra-heavy flanged pattern where the pressures are in excess of 150 psi or where the pressures are likely to be in excess of this amount, if required by the inspection department having jurisdiction.

The fittings in the standpipe connections should be of the extra-heavy pattern where the pressures are in excess of 150 psi or where the pressures are likely to be in excess of this amount, if required by the inspection department having jurisdiction.

Fittings should be of the flanged pattern for sizes in excess of 6 in. Screwed fittings may be used for the smaller sizes but are not recommended on account of the difficulty in making tight joints.

Fittings in the main water connections to standpipes shall be of the long-turn pattern.

Approved expansion joints should be provided where necessary.

10-8. Pipe Drains. Recommendations by the National Board of Fire Underwriters require that

The system should be provided with a system of drain pipes large enough to carry off the water from the open drains while they are discharging under pressure.

The drains should be so arranged as to be free from the possibility of causing water damage and not exposed to freezing. If practicable, the drain should be so arranged that the discharge will be visible from the point of operation on the drain valve.

10-9. Tests and Maintenance. Recommendations by the National Board of Fire Underwriters require that

At the time of installation standpipe systems shall be tested and proved tight at a hydrostatic pressure at least 25 per cent in excess of the highest normal working pressure to which they will be subjected. Tests should be for 2-hr duration and pressure should be taken at the base of the system.

NOTE: Where standpipes or connections are built in the walls or partitions the above tests should be made before they are covered in or permanently concealed.

Annual tests of standpipe systems shall be made by delivering the required quantity of water at the required pressure through hose lines from the topmost outlet of the standpipe.

NOTE: In old systems a tolerance of 10 per cent in quantity delivered is satisfactory; greater deficiencies demand attention.

Old dry systems should be tested with air at a pressure not exceeding 25 lb to determine their tightness before water is turned into them.

NOTE: This test is suggested to avoid water damage to buildings in the event that pipes have broken or become disconnected.

10-10. Bases of Sprinkler-system Design. The installation of a sprinkler system requires special planning in new-building design and usually involves extensive renovation in an existing building. Aside from the provision of pipe runs and supports, insulation against freezing and protection against other damage, and the arrangement of partitions and closets to permit action by the sprinkler heads, it is necessary, for example, to avoid drafts that might expose the sprinkler system to hot gases rising from a fire. The skill and the equipment necessary for the installation and maintenance of a sprinkler system are in the hands of the plumber. He is guided by architectural plans, the following features of which are required by the National Board of Fire Underwriters:

These plans . . . should . . . also indicate the location and size of the water supplies, connecting pipes, feed mains and risers, gate, check, alarm and dry-pipe valves, as well as the location, spacing, number and type of sprinklers.

Two grades of building risks are recognized by the National Board of Fire Underwriters. Class A standards are applicable to all buildings, unless otherwise stated. Class B standard systems are permitted in such "light-hazard" buildings as

Apartment houses	Dormitories	Museums
Asylums	Dwellings	Office buildings
Club houses	Hospitals	Schools
Colleges	Hotels	Tenements
Churches	Libraries	

Small stores in first floors and/or basements of the above listed occupancies, when not over 3,000 sq ft is occupied by any one store.

10-11. Automatic Sprinkler Systems.[1] Automatic sprinklers connected to a water-distribution system are devices on which a sprinkling nozzle is closed by a fusible plug that melts at a predetermined temperature above normal room temperature, releasing water to fall on the source of the heat. Fusible plugs are made to melt at a standard temperature of 160°F. If the metals in the plug are varied, temperatures up to 360°F can be reached before the plug melts. A sprinkler installation is shown in Fig. 10-5, standard symbols used are shown in Fig. 10-6, and the details of some automatic sprinkler heads are shown in Fig. 10-7.

An open-head system for the protection of the interior of a building is operated by an automatic valve controlled by thermostats distributed throughout the building. In this system not more than 5 open sprinkler heads are supplied by a 1½-in. valve, up to a maximum of 75 open, sprinkler heads supplied by a 4-in. valve. An open-head sprinkler system is used to supply a certain amount of water to protect the outside of a building from fires in neighboring buildings or from other exposure fires.

Sprinklers have the advantage of quickly supplying water to a fire before it has gained dangerous headway and of preventing the access of air to the fire by smothering it with water before it is well under way. Sprinklers can be equipped also to implement an alarm when one or more sprinklers discharge. The installation of such sprinklers and alarm may reduce fire insurance rates and may make the services of a night watchman unnecessary. The successful operation of sprinklers has resulted in their wide use. Their installation is required by law for specific hazards in some cities.

Undesirable features of sprinklers include unsightly appearance, possibilities of damage due to leakage, dangers from freezing and bursting, and unnecessary melting of plugs and the release of water on hot equipment that may, unwittingly, be placed beneath them. They provide no better defense against fire than the water supply to which they are connected. Without water they are valueless. Systems that are constantly filled with water will give no trouble from condensation in a humid

[1] See also "Sprinkler Systems," National Board of Fire Underwriters, 1955, 1956.

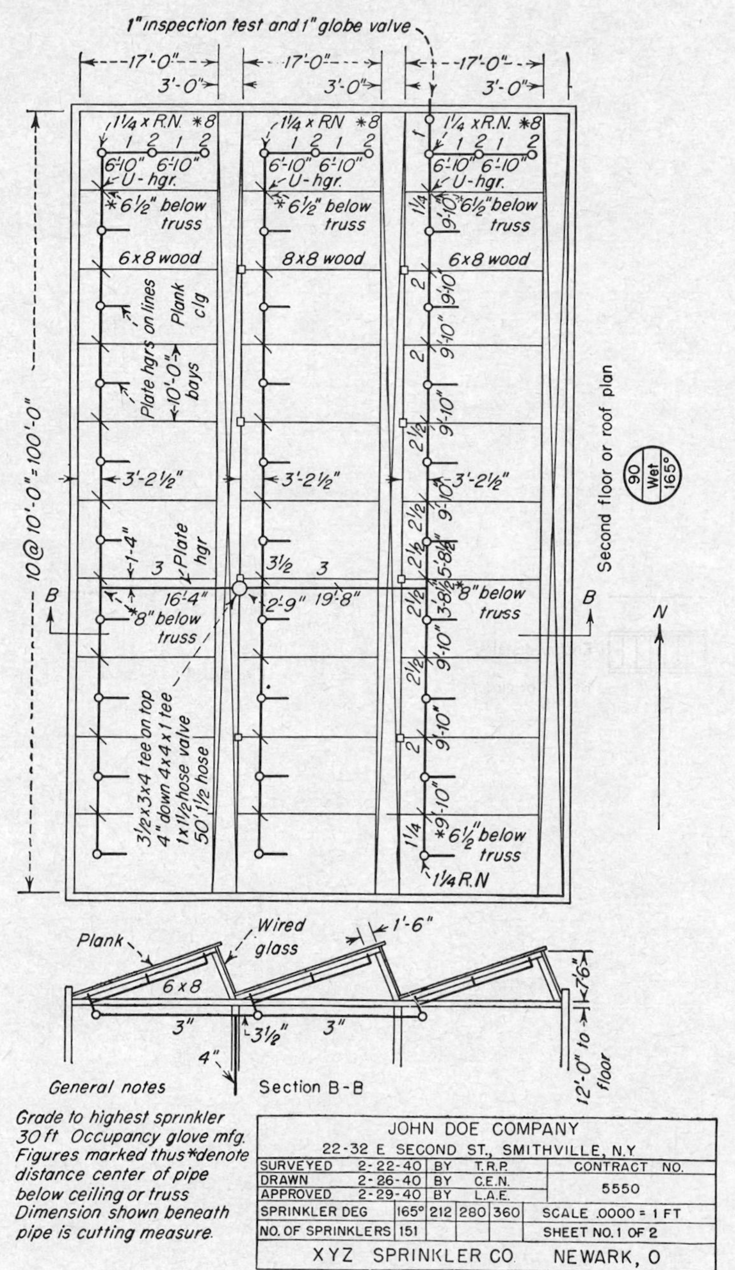

FIG. 10-5. Typical working plan of a sprinkler installation. (*From National Fire Codes of National Fire Protection Association*, 1943.)

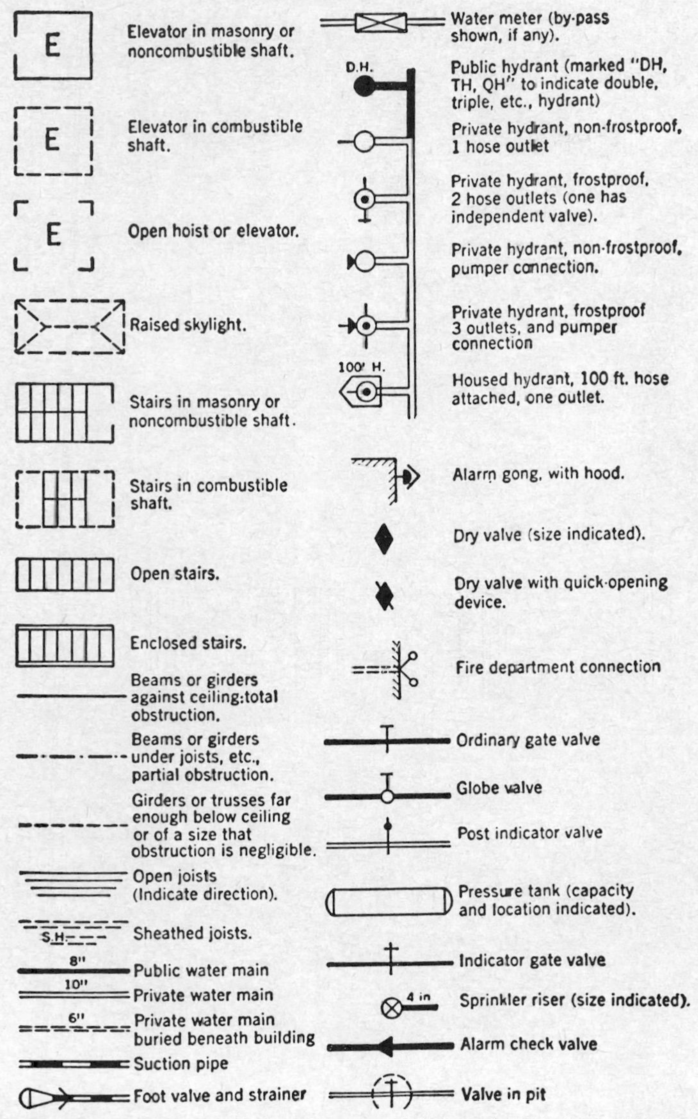

FIG. 10-6. Standard plan symbols recommended by the National Fire Protection Association, 1943.

environment because the water in the pipes is static and takes on the temperature of its surroundings. Almost all the foregoing undesirable features can be mitigated or avoided in design. For example, unsightliness can be avoided by concealing the pipes and supports behind a drop ceiling, allowing only the sprinkler heads to protrude; the building can be kept warm at all times to prevent freezing; and adequate warning signs prevent the improper installation of hot equipment beneath a sprinkler head.

Sprinkler systems should be used for no other purpose than fire protection because of the objections to the circulation of water through them. Hand hose, 1½ in. or smaller, with nozzles not larger than ½ in. may be properly connected to sprinkler systems provided that these hoses are used for fire protection only.

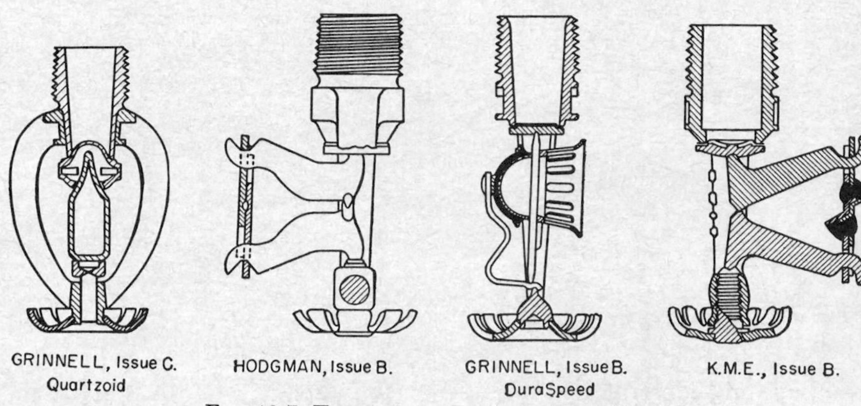

FIG. 10-7. Types of automatic sprinkler head.

10-12. Location of Sprinklers. The National Board of Fire Underwriters recommends that sprinklers shall be located throughout the premises:

> . . . including basement and lofts; under stairs; inside elevator wells; in belt, cable, pipe, gear and pulley boxes; inside small enclosures such as drying and heating boxes, tenter and dry-room enclosures, chutes, conveyor trunks; and all cupboards and closets, unless they have tops entirely open and are so located that sprinklers can properly spray therein. Sprinklers are not to be omitted in any room merely because it is damp, wet, or of fire-resistive construction.

Among possibly unexpected or overlooked places in which sprinklers should be located may be included telephone booths, beneath wide tables, under galleries, on stairways, under peaked roofs and gables, and in narrow corners.

Sprinklers should be placed:

> . . . parallel to ceiling, roofs, or the incline of stairs; but when installed in the peak of a pitched roof they shall be horizontal. Distance of deflectors from ceilings of mill or other smooth construction, or bottom of joists of open joist construction, should preferably be from 5 to 8 in., and should not exceed a maximum of 10 in. Minimum distance should not be less than 4 in. Note particularly that the rule for distance refers to the deflector of the sprinkler.
>
> In the case of fire-resistive buildings, the distance between the deflectors and ceilings may be increased where conditions warrant, *i.e.*, under panel ceilings.

Sprinkler heads should be placed in the position for which they are designed, either upright or pendant, and so that there is a clear space of at least 12 in. below the deflectors to permit the formation of a curtain of water.

The distance between sprinkler heads depends somewhat on the purposes and type of construction of the building. For example, in Class A protection for mill construction of smooth, solid plank and timber, involving 5- to 12-ft bays, one line of sprinklers should be placed in the center of each bay not more than the following distances apart:

8 ft in 12-ft bays	11 ft in 9-ft bays
9 ft in 11-ft bays	12 ft in 5- to 8-ft bays
10 ft in 10-ft bays	

Measurements should be taken from center to center of the timbers. The spacing in semimill construction, as the term is applied to plank and beam construction with narrow bays, is

> Under ceilings, the area per sprinkler shall in no case exceed 90 sq ft and the distance between sprinklers on a line and between lines should not exceed 10 ft.

In Class B mill construction, sprinklers may be spaced not exceeding 14 ft apart in any bay not over 14 ft wide, and in semimill construction, not exceeding 12 by 14 ft apart.

All supply pipes should be as centrally located as possible among sprinkler heads to obtain the minimum hydraulic head losses. Safety from freezing and from other causes of damage is more important, however, than the minimizing of head losses. No sprinklers should be used where the addition of water to materials in the building will increase conflagration or the danger to life and property. Among such materials may be included aluminum powder, calcium carbide, and quicklime.

10-13. Sizes of Sprinkler Supply Pipes. Sizes of pipes to supply sprinklers, as recommended by the National Board of Fire Underwriters, are shown in Table 10-2. Under Grade A conditions:

> Where practicable, it is desirable to arrange the piping so that the number of sprinklers on a branch will not exceed eight. All heads in excess of eight should be provided with a 2½-in. pipe where, in a few cases, it may be necessary to have more than eight heads on a branch line run.
>
> Where feed mains supply branch lines of only two sprinklers each, the conditions approach those of long single lines. Such feeds should usually be centrally supplied where there are over 8 or 10 branch lines. Lines up to 14 in number may be fed from the end, provided that 2½-in. pipe does not supply more than 16 sprinklers.

Under Grade B conditions no branch line with more than 16 sprinklers is allowed. The main supply pipes may not be less than 4 in. in diameter, and a pipe of this size is to be connected to the base of each 2½-in. or larger riser on a system having more than one riser.

The National Board of Fire Underwriters states:

There should be one or more separate risers in each building and in each section of the building divided by fire walls. Each riser should be of sufficient size to supply all the sprinklers on said risers or any one floor, as determined by the standard schedule of pipe sizes.

Note that, except in special cases, the main feeders are large enough to supply the sprinklers on one floor only. Where there are vertical openings permitting fire to spread between floors, the sizes of the riser pipes should be adequately increased over those shown in Table 10-2. Such increases may be based, by the designer, on studies of the hydraulic head losses permissible to supply the rates of flow and the pressures required throughout the system. A graphic schedule of sprinkler pipe sizes recommended by the National Fire Protection Association is shown in Fig. 10-8.

TABLE 10-2. WATER-SUPPLY PIPES TO SPRINKLERS

Diameter of pipe, in.	National Board of Fire Underwriters number of sprinklers				National Fire Protection Association number of sprinklers
	Grade A conditions	Grade B conditions	Open-head systems	Dry pipes with thermo-statically operated valves	
¾	1	1	...	...	1
1	2	2	2	...	2
1½	5	5	5	5	5
2	10	10	8	10	10
2½	20	...	16	...	20
3	36	...	36	36	36
3½	55	...	55	...	55
4	80	...	80	75	80
5	140	...	140	...	140
6	200	...	200	...	200
8	400				

10-14. Valves, Fittings, and Connections. Some special requirements, not applicable to all plumbing installations, concern the valves, fittings, and connections used on sprinkler systems. Valves should be located on sprinkler systems as follows:

1. A check valve on all sources of water supply, except where cushion tanks are used with automatic fire pumps. Gate valves on supply pipes

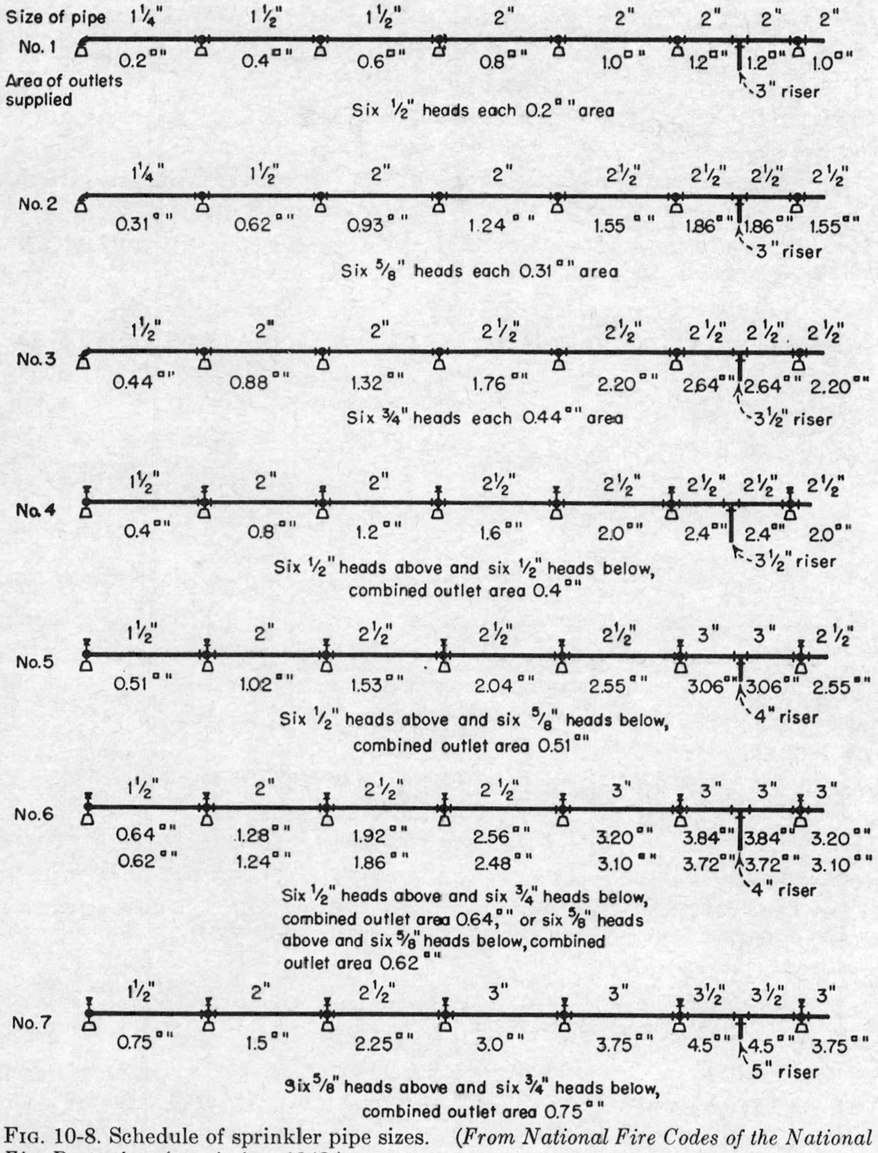

Size of pipe 1¼" 1½" 1½" 2" 2" 2" 2" 2"
No. 1
0.2□" 0.4□" 0.6□" 0.8□" 1.0□" 1.2□" 1.2□" 1.0□"
Area of outlets supplied
3" riser
Six ½" heads each 0.2□" area

1¼" 1½" 2" 2" 2½" 2½" 2½" 2½"
No. 2
0.31□" 0.62□" 0.93□" 1.24□" 1.55□" 1.86□" 1.86□" 1.55□"
3" riser
Six ⅝" heads each 0.31□" area

1½" 2" 2" 2½" 2½" 2½" 2½" 2½"
No. 3
0.44□" 0.88□" 1.32□" 1.76□" 2.20□" 2.64□" 2.64□" 2.20□"
Six ¾" heads each 0.44□" area
3½" riser

1½" 2" 2" 2½" 2½" 2½" 2½" 2½"
No. 4
0.4□" 0.8□" 1.2□" 1.6□" 2.0□" 2.4□" 2.4□" 2.0□"
Six ½" heads above and six ½" heads below, combined outlet area 0.4□"
3½" riser

1½" 2" 2½" 2½" 2½" 3" 3" 2½"
No. 5
0.51□" 1.02□" 1.53□" 2.04□" 2.55□" 3.06□" 3.06□" 2.55□"
Six ½" heads above and six ⅝" heads below, combined outlet area 0.51□"
4" riser

1½" 2" 2½" 2½" 3" 3" 3" 3"
No. 6
0.64□" 1.28□" 1.92□" 2.56□" 3.20□" 3.84□" 3.84□" 3.20□"
0.62□" 1.24□" 1.86□" 2.48□" 3.10□" 3.72□" 3.72□" 3.10□"
4" riser
Six ½" heads above and six ¾" heads below, combined outlet area 0.64□" or six ⅝" heads above and six ⅝" heads below, combined outlet area 0.62□"

1½" 2" 2½" 3" 3" 3½" 3½" 3"
No. 7
0.75□" 1.5□" 2.25□" 3.0□" 3.75□" 4.5□" 4.5□" 3.75□"
5" riser
Six ⅝" heads above and six ¾" heads below, combined outlet area 0.75□"

FIG. 10-8. Schedule of sprinkler pipe sizes. (*From National Fire Codes of the National Fire Protection Association*, 1943.)

outside the building should be located away from heat of the burning building as, for example, opposite a blank wall or, preferably, more than 40 ft from the building. The valve should be accessible for operation, and it should be equipped with an indicator to show the position of the valve.

2. Gate valves on each side of each check valve, except on fire department connections, and where gravity tanks and pressure tanks are the only source of supply. A gate valve is required on the pressure-tank connection only between the tank and the check valve, and not on the other side of this check valve.

3. A gate valve on the water-main side of fire department connections.

4. A check valve, well removed from fire hazard or other danger, on the discharge pipe from a pump subjected to such hazard.

5. A gate valve to control all fittings inside a building, except the drain tee, where there is a gravity tank supply above the roof.

Long-radius fittings should be used on supply mains 2½ in. and larger, except that reducing fittings of more than 1 in. reduction may be short radius, and long-radius fittings on risers should be flanged on one end.

All inside piping shall be installed by means of screwed or flanged fittings, except that welded joints may be allowed in special cases for risers and large feed mains.

Couplings are to be used only where it is unavoidable. Reductions in pipe sizes are to be made with one-piece reducing fittings, not with bushings.

10-15. Fire Department Connections.[1] A Siamese or wye-branch hose inlet pipe to the sprinkler system to be connected to by the fire department should be located accessibly outside of the building, somewhat as shown in Fig. 10-2. For Class A sprinkler systems, the pipe should be not less than 4 in. in diameter for pumpers and 6 in. for fire boats. Each fitting should be provided with a straightway check valve and with a drain for the piping between the hose connection and the check valve. Each hose connection should be identified by raised cast letters at least 1 in. in height, designating the appropriate service: **AUTO. SPKR.** or **OPEN SPKR.**

10-16. Test Pipes and Drain Pipes. Periodic tests of the sprinkler system are necessary to assure its functioning when needed. Tests are made by draining water from the sprinkler system through a test or drain connection not less than ¾ in. in diameter, and larger if the hydraulic head is appreciable. The discharge from each such test pipe should be visible. Test pipe for the main water supply should be not less than 2 in. in diameter. Drainage of any portion of or the entire system should be provided to allow the removal of test water when the pipes are drained. This can be done by the installation of air-inlet valves or plugged fittings at high points and of valve-controlled drainage outlets at low points.

[1] See also "Fire Department Hose Connections," National Board of Fire Underwriters, 1939.

The National Board of Fire Underwriters recommends a drain and valve connection for each group of 20 or fewer sprinklers.

No pipe in the sprinkler system should have a horizontal slope of less than ¼ in. in 10 ft. Drip or drain pipes for 4-in. and larger supply-pipe risers should be 2 in. in diameter or larger; they should be not smaller than 1½ in. for risers between 1½ and 4 in. and the same size as the riser for risers smaller than 1½ in. The drain pipes may discharge at any convenient point that is free from ice or other objectionable features, and

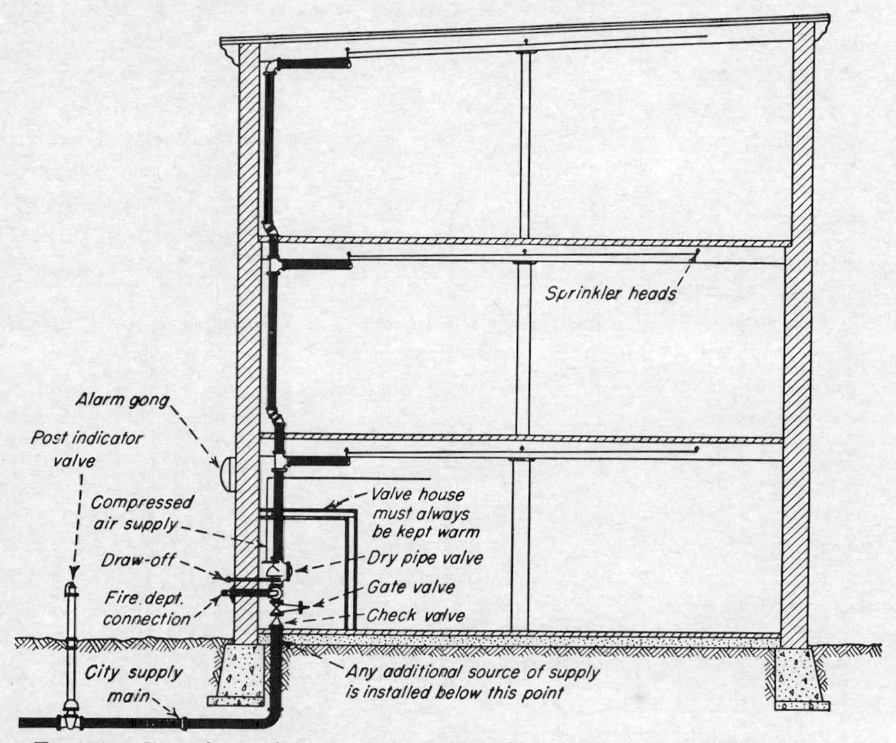

FIG. 10-9. Dry-pipe sprinkler system. (*Eng. News-Record, Nov.* 25, 1948, *p.* 57.)

they should not conduct cold air into the building or cause freezing of the sprinkler pipes. The drain pipes should not discharge into a sanitary sewer. Such sewer connections are undesirable because of the possibility of admitting sewer air into the building and the difficulty of maintaining a water seal in traps in the infrequently used drain pipes.

10-17. Dry-pipe System. A dry-pipe system is made up of a complete water-distribution system with sprinkler heads in which there is no water. One form of the system is illustrated in Fig. 10-9. The system is used partly to protect the interior of the building against hazards of burst and leaky pipes, to avoid freezing of water in the pipes, and for other reasons.

Water to smother or to extinguish a fire is turned into the water-distribution system either automatically or manually on the outbreak of fire or the sounding of an alarm.

Water must be provided and air exhausted through the sprinkler head quickly when the fuse melts in the plug. One form of quick-opening device is shown in Fig. 10-10. Most of the devices depend on an air chamber, a diaphragm normally under balanced pressure, and a restricted opening that prevents the quick release of the air pressure in the air chamber. When the sprinkler head opens and the air pressure in the distribution pipes drops slightly, there is an unbalanced pressure on the diaphragm which permits the air pressure in the distribution pipes to enter the intermediate chamber and open the discharge valve.

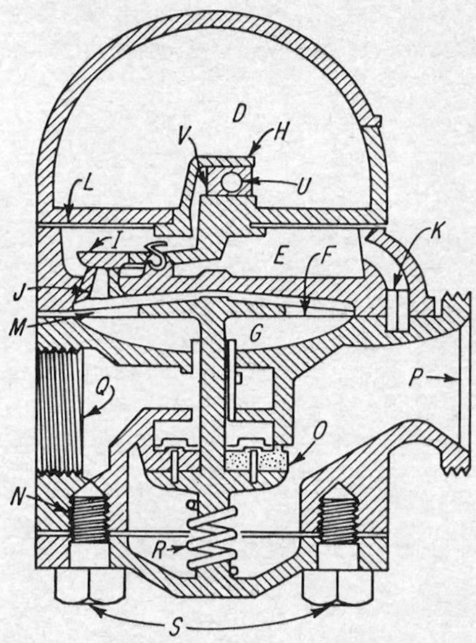

Fig. 10-10. Quick-opening sprinkler valve. *D*, upper chamber; *E* and *F*, air chambers above diaphragm *M*; *G*, exhaust chamber connected to intermediate chamber dry valve; *I*, auxiliary valve operated by diaphragm *L*; *O*, air-release valve; *P*, air inlet; *Q*, air outlet; *R*, slow-leak element. (*Rockwood Sprinkler Co. Exhauster, Model C.*)

One type of dry-pipe system involves the provision of risers and branches serving hydrants throughout the building. Water is supplied to the system through a connection provided for the fire department to pump water *into* the distribution pipes and risers in the building. Such a connection is shown in Fig. 10-2.

In another dry-pipe system the distribution pipes on which the sprinklers are placed are filled with air under pressure of about 12 to 20 psi. When a sprinkler head opens, the release of the air pressure opens the main water-supply valve to admit water to the system.

In the "preaction" system water is admitted to the system by a valve actuated by a thermostatically controlled device that functions in advance of the sprinkler head. In the pneumatic pressure-tank system air is held in the sprinkler piping under high pressure. Upon the fusing of a sprinkler head, the air pressure in the piping is released, permitting the air under pressure in the pneumatic storage tank to discharge water from the tank into the dry-pipe sprinkler system.

The pipes of a dry-pipe system should be so tight that not more than 10 psi pressure is lost between weekly renewals of the pressure. The

capacity of the air-compressing system should be sufficient to raise the air pressure to 40 psi in not more than 30 min. An air-pressure-relief valve should be provided to release at a pressure of more than 5 psi above the maximum pressure to be carried in the system. The air-supply pipe should connect to the dry pipe beyond the dry-pipe valve, and it should be equipped with a check valve between two shutoff valves.

The dry-pipe valve must be located accessibly, as near as practicable to the dry-pipe sprinklers that it serves, and it should be protected against freezing and mechanical injury. The valve may be placed in a frostproof pit or vault or in a room that is kept constantly above freezing. The National Board of Fire Underwriters recommends that

> . . . the point of connection to the riser shall be such that the capacity measured between the normal priming level of the air chamber and the connection will be at least 1½, 2, and 3 gal for 4-, 5-, and 6-in. risers, respectively. In all cases the connection shall be at the top or side of the pipe, and shall be located within the dry-pipe valve enclosure when such enclosure is provided.
>
> Dry-pipe valves controlling 300 sprinklers or less will not ordinarily require the attachment of a quick-opening valve. Not more than 600 sprinklers should be controlled by one dry-pipe valve.

Drainage should be provided for all parts of dry-pipe sprinklers, preferably back to the main drip, the pipes being laid on slopes of ½ in. or more in 10 ft. Where sprinklers that are too low to drain back to the main drip are supplied from a large pipe, such as a trunk main, the connection should be made near the top of the large pipe, looped back, or installed in a similar manner to prevent condensation from entering the dry pipe.

10-18. Pressure Gages. Pressure gages that are easily visible are used to show at a glance the pressure of water in the sprinkler system. These gages should be Bourdon gages with a dial of not less than 4½-in. and with a minimum limit of twice the normal pressure for which they are installed. A pressure gage should be located at each of the following points:

1. Main water-supply pipes or pipes
2. Above the alarm check valve, dry-pipe valve, or main drip
3. Below the alarm check valve, dry-pipe valve, or main drip
4. On the air pump supplying the pressure tank or on the tank
5. On each pipe from air-supply to dry-pipe system
6. At test pipes

Each pressure gage should be equipped with a control valve and a drain. A plugged tee should be provided on all gages to permit the connection of a test gage.

10-19. Sprinkler Alarms. Sprinkler-actuated alarms are useful in announcing a fire. Two types are approved; one is an electric gong and the other a water-motor gong. Both are set to sound when the flow of water through the alarm exceeds 10 gpm. The actuating device is placed on the main supply pipe of the sprinkler system so that it can be either disconnected or not actuated during a test of the sprinkler system. An alarm attachment should be installed on a dry-pipe system close to and on either the water side or the dry side of the valve. The electric gong may be placed in a prominent position inside or outside the building or in the fire department station. A water-motor gong must be placed near the sprinkler supply pipe to minimize the length of pipe leading to it.

10-20. Pumps and Tanks.[1] Where the water supply is inadequate in pressure and in quantity, adequate elevated storage or pressure storage must be installed. For Class A sprinkler systems either a gravity tank of 30,000 gal capacity with its bottom 20 ft above the highest sprinkler or a pressure tank of not less than 4,500 gal capacity located above the upper floor of the building is considered to be satisfactory. Where the source of supply is ample in quantity but not in pressure, a fire pump drawing from this supply will be satisfactory for Class A sprinkler systems provided that the capacity of the pump is not less than 500 gpm for sprinklers only and 750 gpm when it supplies hydrants also.

For Class B sprinklers a gravity tank is not recommended owing to the need for good pressure on widely spaced sprinklers with small mains and branches. However, if it is used, its capacity should exceed 5,000 gal and its bottom should be 40 ft or more above the highest fixture. The requirements for Class B and Class A pressure tanks are the same.

With regard to pumps the National Board of Fire Underwriters recommends

> Where public water is deficient in pressure but adequate in volume, an electrically controlled automatic centrifugal pump of not less than 250 gpm capacity, taking water from the city main, may be accepted as the only supply to sprinklers, only, however, if there is also provided an approved supervisory service or an approved proprietary system or their substantial equivalent.

10-21. Tests after Installation. The National Board of Fire Underwriters recommends, with respect to testing sprinkler systems after installation, that

> All systems should be tested at not less than 150 psi for 2 hr or at 50 psi in excess of the normal pressure when the normal pressure is in excess of 100 psi. Emergency tests of dry-pipe systems, under at least 60 psi air pressure, should be made at seasons of the year which will not permit testing out under water pressure.

[1] See also "Centrifugal Fire Pumps," 1957, and "Water Tanks for Private Fire Protection Service," 1939, National Board of Fire Underwriters.

Branches from underground mains to inside sprinklers should be flushed out before connecting the sprinkler riser.

Piping between the check valve in the fire department inlet pipe and the outside connection should be tested the same as the balance of the system.

To prevent the possibility of serious water damage in case of a break, pressure should be maintained by a small pump, the main controlling gate being meanwhile kept shut.

In the case of dry systems with a differential type of dry-pipe valve, the valve should be held off its seat during the test to prevent injuring the valve.

In dry systems an air pressure of 40 psi should be pumped up, allowed to stand 24 hr, and all leaks stopped which allow a loss of pressure of over 1½ psi for the 24 hr.

A working test of the dry-pipe valve should be made, if possible, before acceptance.

CHAPTER 11

DESIGN AND LAYOUT OF DRAINAGE PIPES, VENT PIPES, AND TRAPS

11-1. The Problem. In the design of the drainage pipes of a plumbing system it is desired to use the smallest pipes that will conduct waste water away rapidly without clogging, to place them in the most convenient location, and to avoid the production of siphonage and of back pressures. Such a desire requires the solution of a number of problems involving the principles of hydraulics and pneumatics. The solution of such problems is made difficult by the lack of a formula to fit any assigned conditions. Consequently drainage-pipe design is based on plumbing codes and successful experience, and it is sometimes affected by the results of laboratory tests and field investigations.

11-2. Parts of a Drainage System. The principal parts of a drainage system, which can be called its elements, are the traps, the vents, the drainage pipes, and the building drain and sewer; the rain-water gutters and leaders are sometimes not included as a part of the plumbing.

A *soil pipe* in a plumbing system is a drainage pipe that carries or is designed to carry human excrement. If a pipe is vertical, it is called a *stack.* If it is horizontal, it is called a *branch.* A *waste pipe* carries liquid wastes that do not include human excrement.

Vent pipes, with few exceptions, are attached to the drainage pipes near the traps and between the trap and the sewer for the purpose of admitting air to or taking air away from the drainage pipes. The vent pipes should lead to the outside air at some distance from any other opening into the building. Ventilating and drainage pipes at points near the traps assist in preventing the trap seal from being broken by air pressures in the drainage pipes.

11-3. Materials. Materials used for soil, waste, and vent pipes are cast iron, wrought iron, steel, copper, brass, and lead. Black iron or steel should not be used without protection against corrosion, as they will corrode rapidly unless protected. The "National Plumbing Code," Sec. 11.1.3, permits the use of galvanized steel or galvanized ferrous alloy, lead, or copper pipes or copper tubing, in addition to cast-iron pipe when underground or within a building. Threaded joints are to be coated and wrapped after installation. The building sewer may be of vitrified clay, concrete, or cast iron. The materials that are used for rain-water gutters and leaders are discussed in Sec. 4-22.

11-4. The Drainage System. The flow of water in the drainage pipes of a plumbing system is normally at atmospheric pressure, the direction of flow being from a higher to a lower elevation. The pipes are laid out so that the stacks are vertical, offsets seldom being used, and horizontal pipes are laid on slopes just steep enough to include self-cleansing velocities. The principles of design of drainage pipes are different from those of water-supply pipes that run under pressure greater than atmospheric and that may flow from a lower to a higher elevation in a building.

The drainage pipes of a plumbing system take waste water from the plumbing fixtures and deliver it to the sewer or some other outlet. On leaving the fixture, the waste water passes through a trap and then the branch, the stack, the building drain, and the building sewer and finally into the public sewer or other outlet. Since it is undesirable to permit air, odors, or vermin from a sewer to enter a building, some device for preventing this must be installed and the drainage pipes must be made tight.

The requirements for a drainage system can be summarized as follows:

1. It must carry the waste water rapidly away from the fixtures.
2. The passage of air, odors, or vermin from the sewer into the building must be prevented.
3. The drainage pipes must be gastight, airtight, and watertight.
4. The pipes must be durable and so well installed that slight movements of the building or of the pipe will not cause leakage.

Materials for the drainage system should be selected for strength and durability and to resist the corrosive action of wastes discharged into them. Ordinary drainage pipes of cast iron or wrought iron should not be exposed to acid wastes or to steam or hot water. Heat increases the rapidity of the evaporation of water from traps and causes the emission of foul odors, and the alternate expansion and contraction of pipes tend to loosen calked joints.

11-5. The Layout of the Drainage System. A piping layout shown on a plan should include a complete presentation of the location of both the supply and the drainage pipes. In order to plan a proper layout for a plumbing system, a designer must know how to make the user comfortable, must understand the phenomena occurring in the pipes, and must be acquainted with the practices of plumbing. Comprehensive general rules of value would be difficult to state because of the multitude of conditions that may arise. The "Plumbing Manual,"[1] the "Housing

[1] "Plumbing Manual," Report BMS66, National Bureau of Standards, 1940.

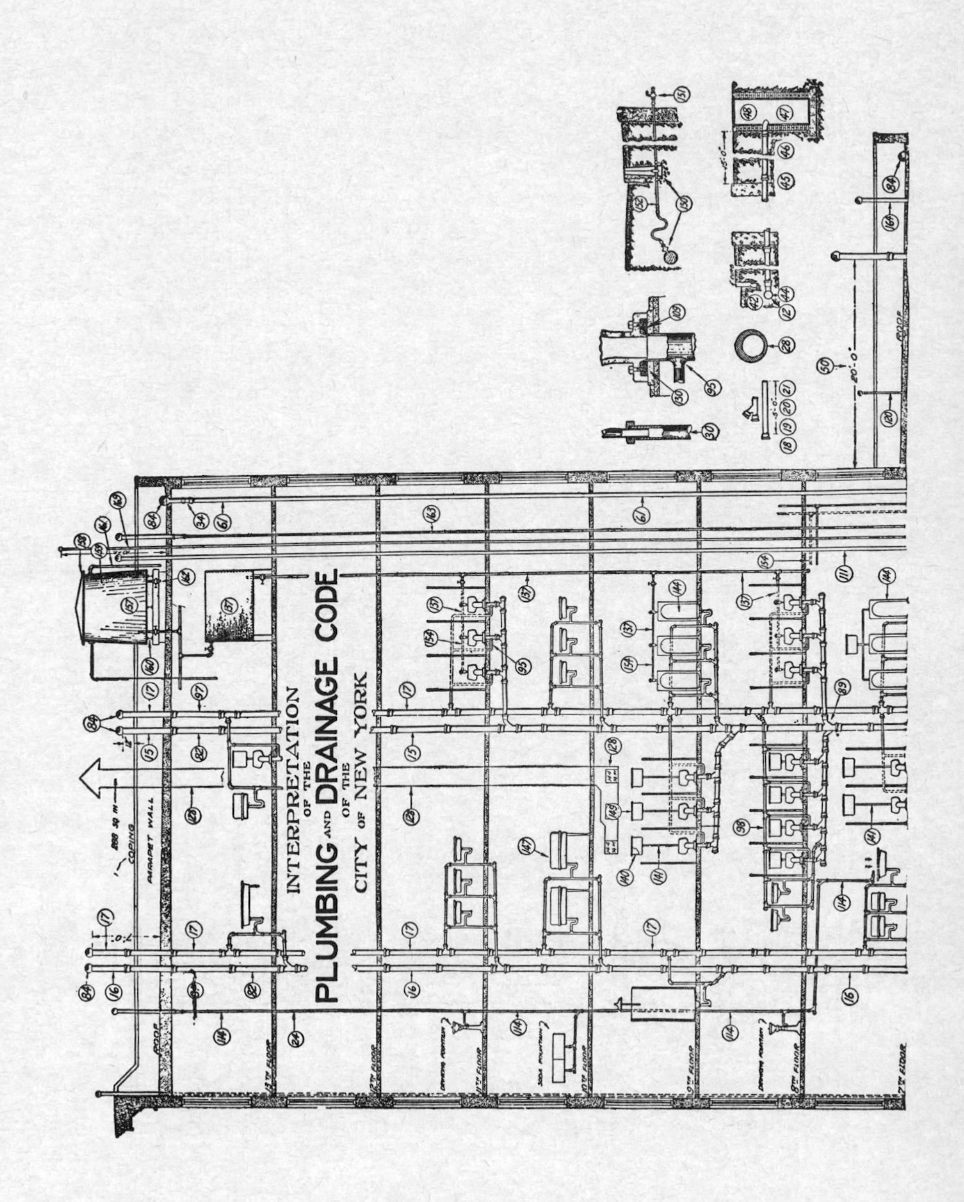

INTERPRETATION
OF THE
PLUMBING AND DRAINAGE CODE
OF THE
CITY OF NEW YORK
COPING
PARAPET WALL
ROOF

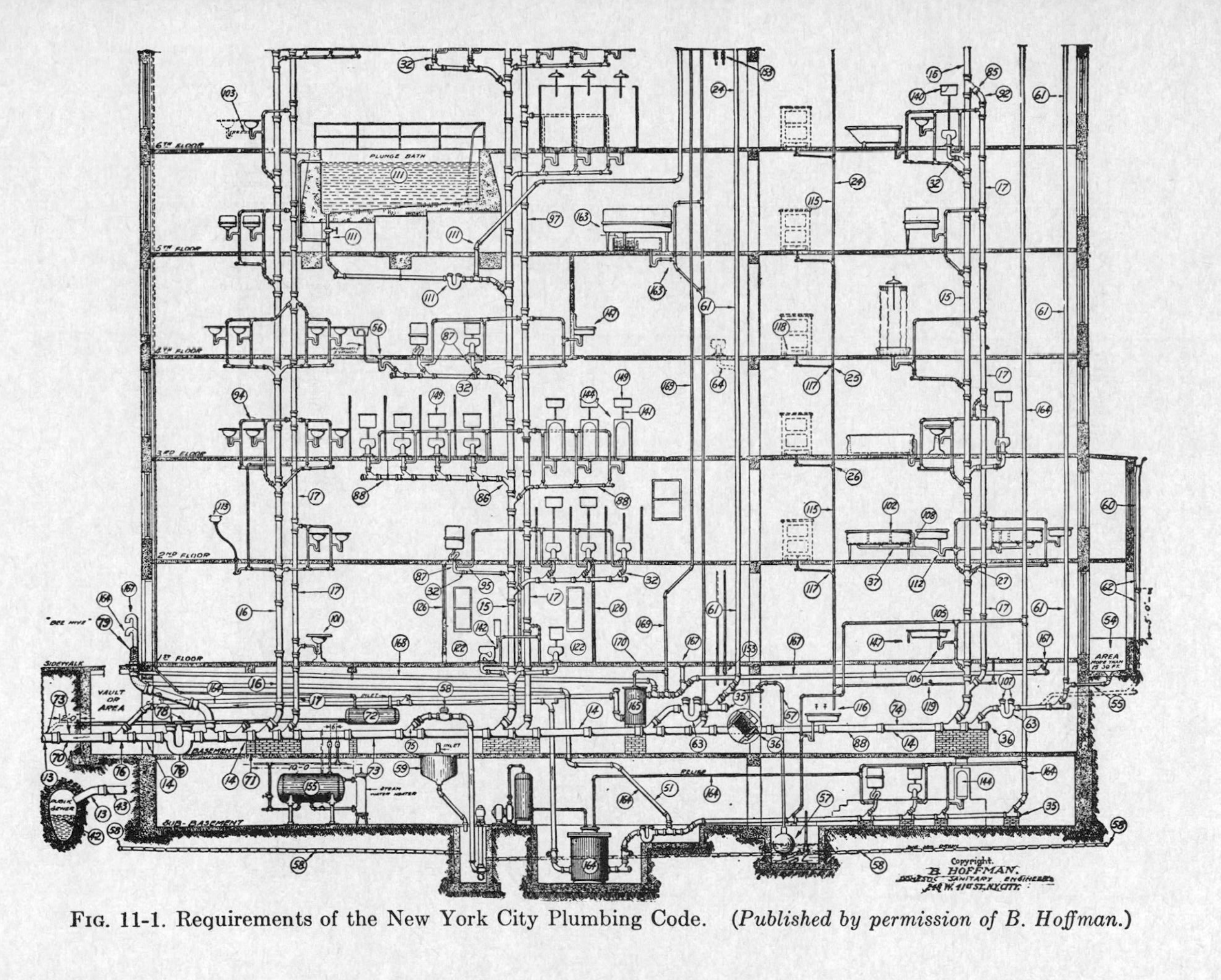

FIG. 11-1. Requirements of the New York City Plumbing Code. (*Published by permission of B. Hoffman.*)

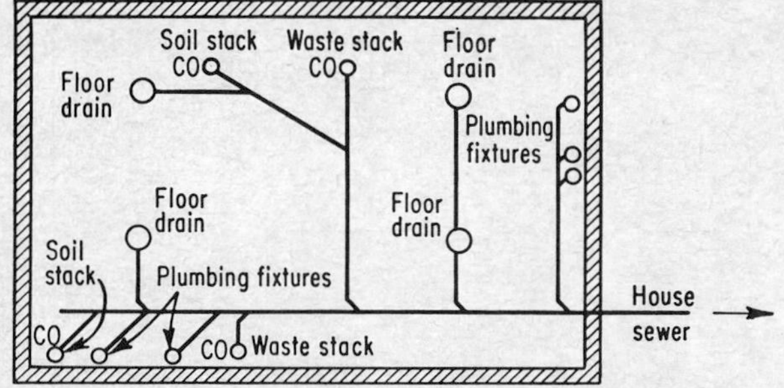

FIG. 11-2. House drain and connections in the basement of a residence.

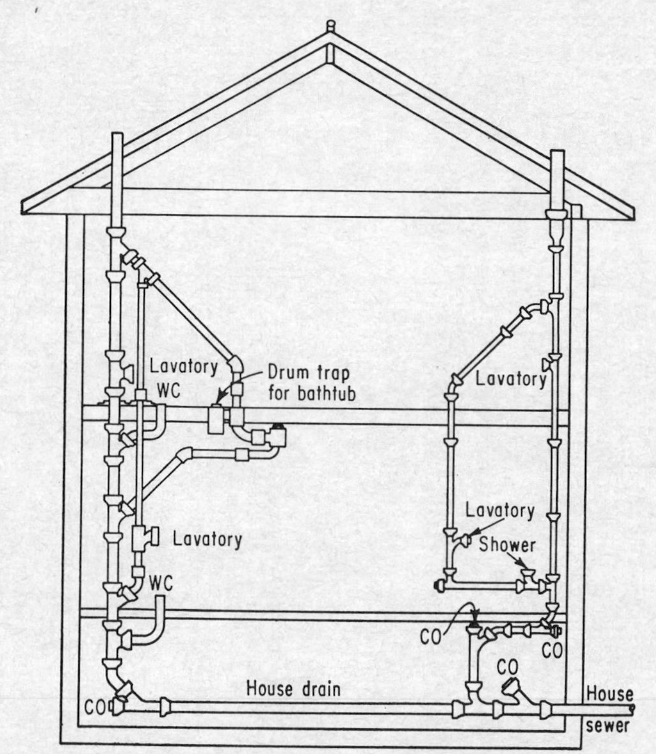

FIG. 11-3. Roughing-in of residential drainage pipes in plumbing.

Code,"[1] and the "National Plumbing Code" give helpful information for design. Principles and practices in drainage- and vent-pipe installation are illustrated in Figs. 11-1 to 11-8. A scheme of drainage pipes under the cellar floor of a residence is shown in Fig. 11-2, and a scheme of vertical pipes in a residence after roughing-in is shown in Fig. 11-3.

[1] "The Uniform Plumbing Code for Housing," Housing and Home Finance Agency, February, 1948.

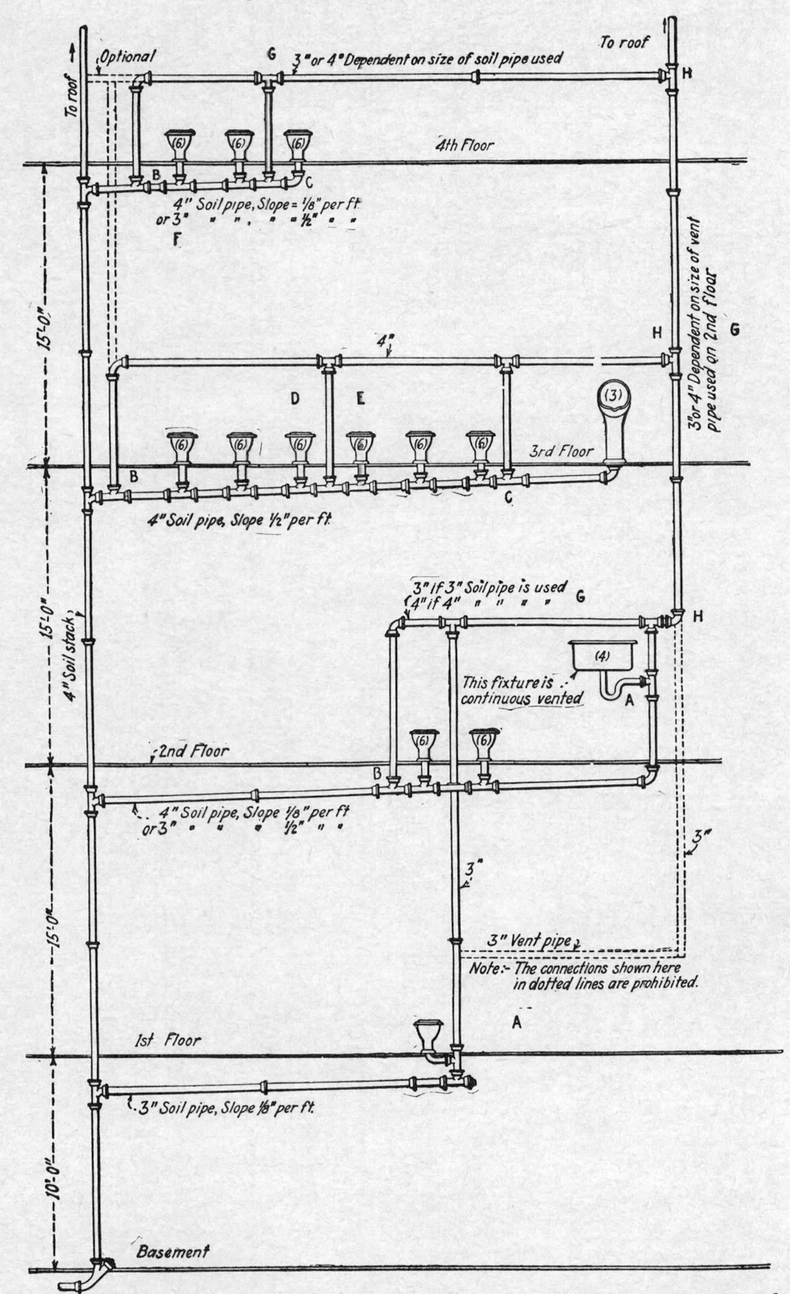

FIG. 11-4. Illustrating drainage- and vent-pipe installations. (The numbers in parentheses on the fixtures indicate fixture units.)

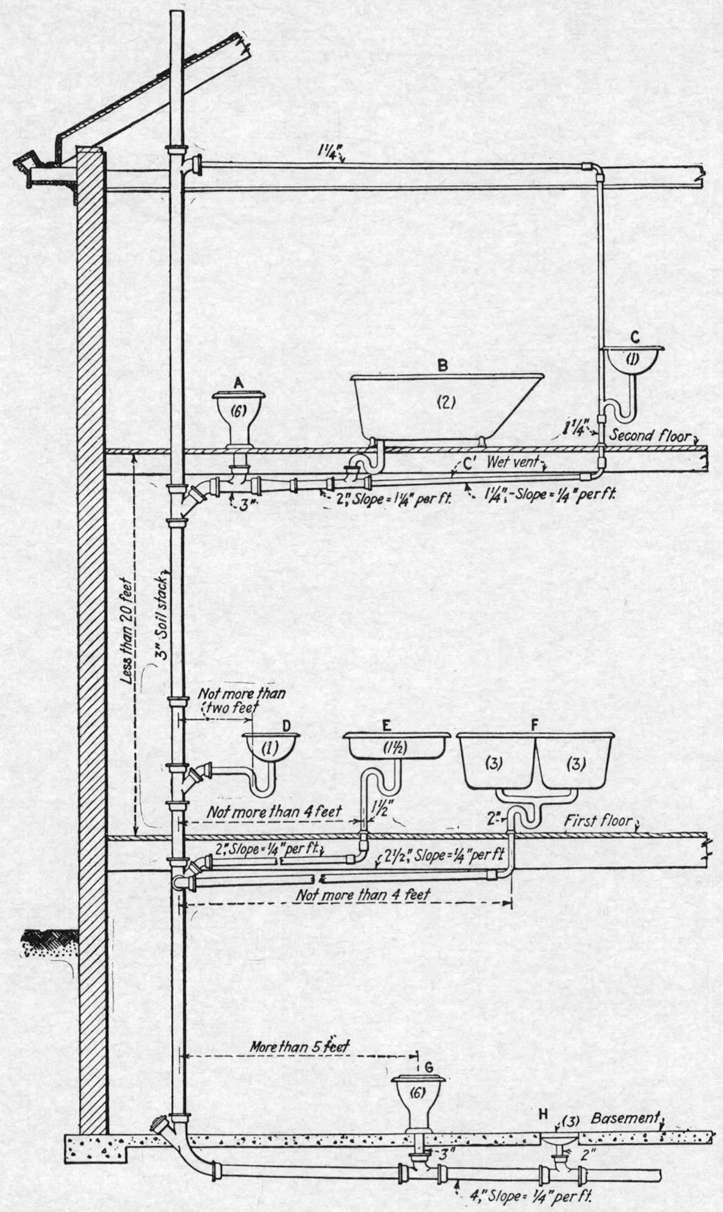

FIG. 11-5. Illustrating a wet vent and installations without vents. (The numbers in parentheses on the fixtures indicate fixture units.)

11-6. The Fixture Unit. In the design of drainage and vent pipes it is convenient to express the rate at which water or drainage is to be carried by the pipes in terms of *fixture units.* The fixture unit is based on the rate of discharge from an intermittently used fixture. It has been so

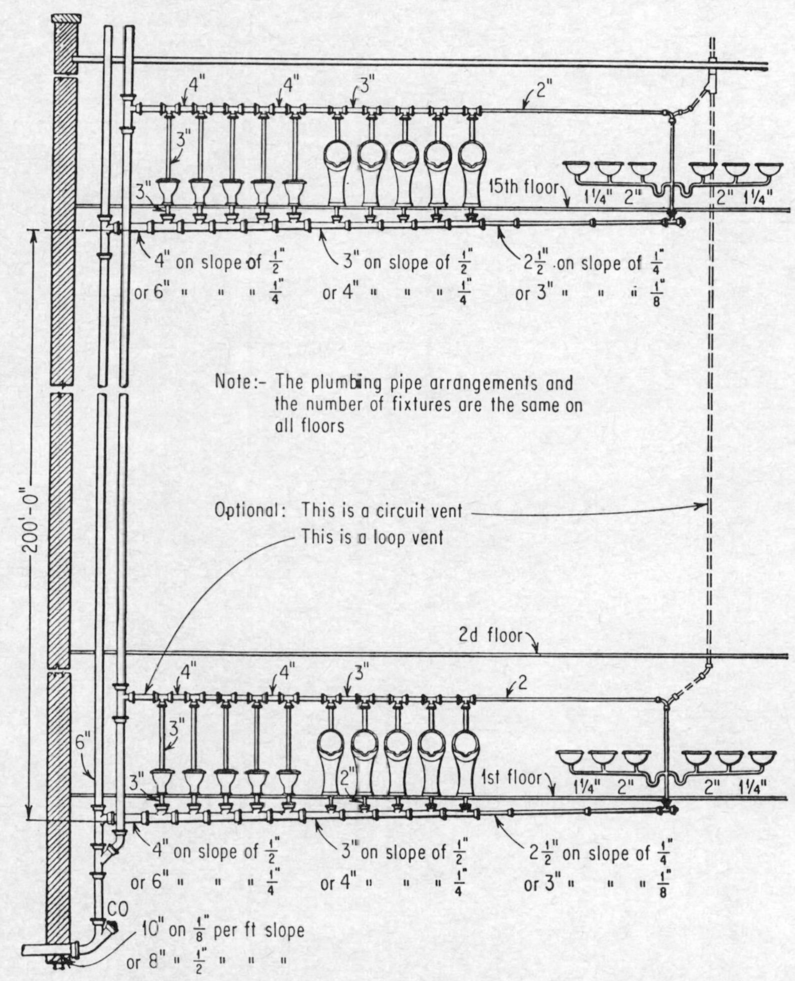

FIG. 11-6. Illustrating drainage- and vent-pipe installations in a tall building.

chosen that the load-producing values of the different plumbing fixtures can be expressed approximately as multiples of that factor. The lavatory is usually taken as the unit fixture, the rate of discharge from it being about 7.5 gpm or 1 cfm.

The number of fixture units equivalent to the rate of discharge from various fixtures are listed in Tables 11-1 and 11-2. When the rate of

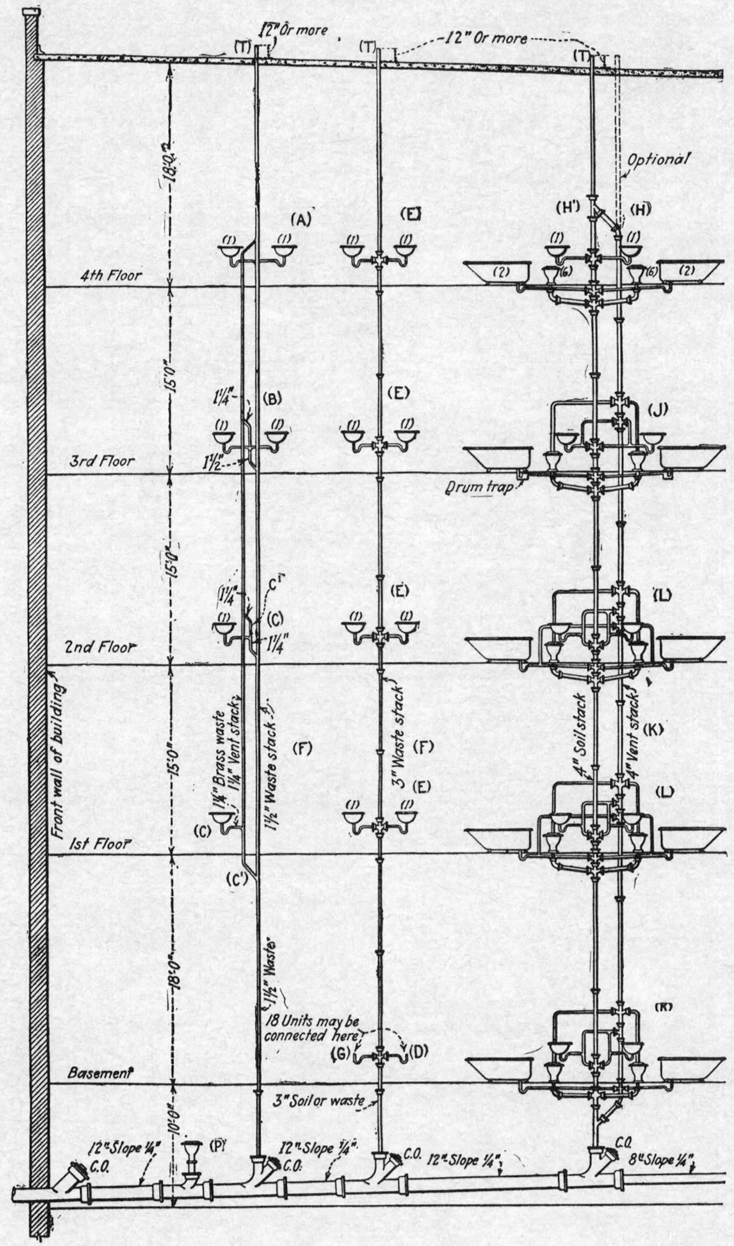

FIG. 11-7. Requirements of "Model Plumbing Code" recommended by Illinois State

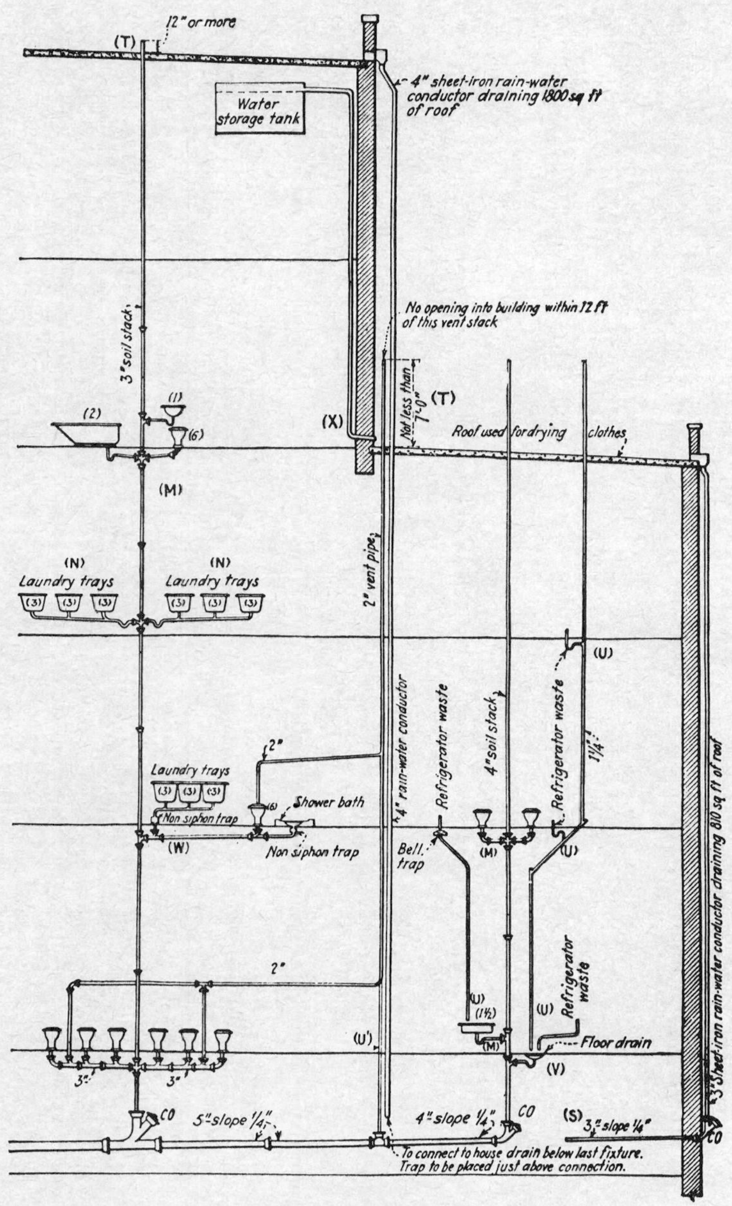

Department of Public Health. (The numbers in parentheses indicate fixture units.)

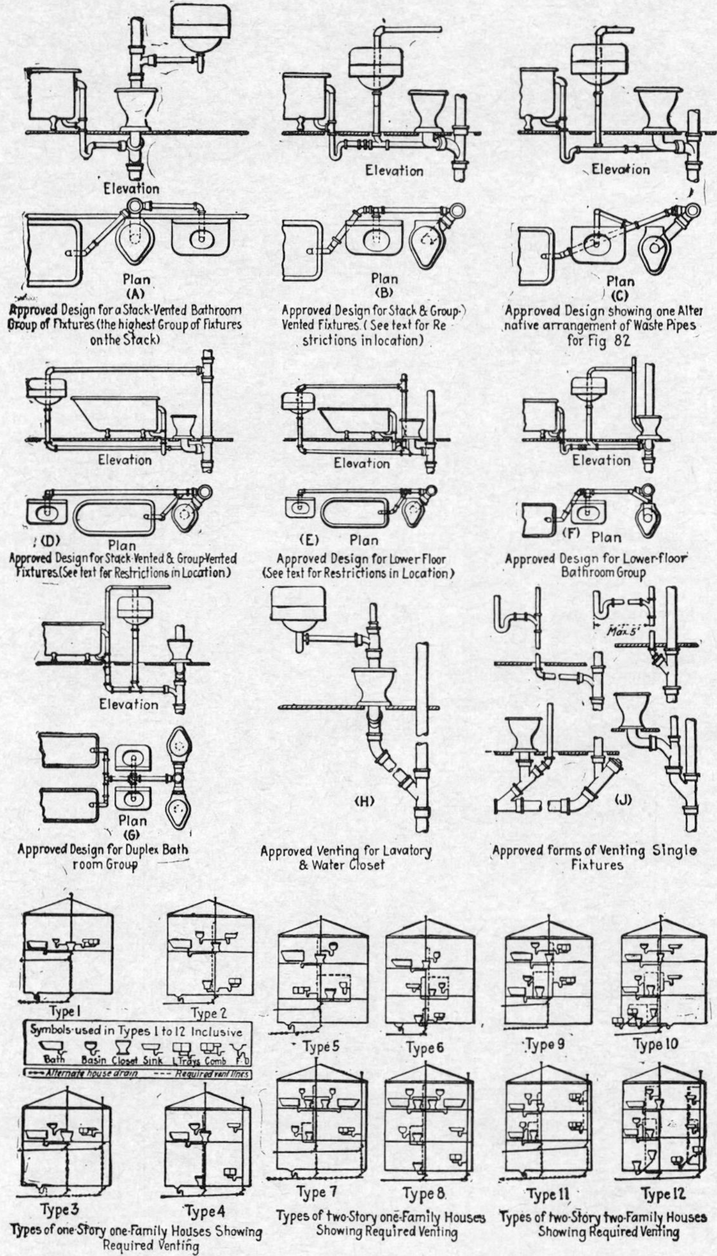

FIG. 11-8. Types of plumbing installation approved by Hoover Report.

TABLE 11-1. FIXTURE UNITS FOR VARIOUS FIXTURES[a]
(As stated by various authorities)

Fixture	Number of units: Housing Code[b]		Number of units: Plumbing Manual[c]		Number of units: National Plumbing Code[d]	Min. size of trap and fixture drain, in.: Housing Code[b]	Min. size of trap and fixture drain, in.: Plumbing Manual[c]	Min. size of trap and fixture drain, in.: National Plumbing Code
	Private	Public[e]	Private	Public[e]				
1 bathroom group: 1 lavatory, 1 W.C., 1 tub or shower	8	....	8[g]					
Bathtub[f]	2	4	2	4	2, 3	1½	1½	1½, 2
	3	6	3	6	...	2		
Bedpan washer	...	10	...	...	...	3		
Bidet	2	4	...	...	3	1½	...	2
Combination fixture	3	....	3	...	3[h]	1½	1½	1½
Dental unit or cuspidor	...	1	...	...	1	1¼	...	1¼
Dental lavatory	1	2	...	...	1	1¼	...	1¼
Drinking fountain	...	1½	...	...	½	1¼	1¼	1
Dishwasher	2	4	...	...	2	1½	1½	1½
Floor drain[j]	1	2	...	...	1[j]	....	2	2
Kitchen sink								
½-in. outlet[k]	2	4	2	...	2[l]	1½	1½	1½[m]
2-in. outlet[k]	...	5	...	...	3[k]	2	2	1½
Lavatory	1	2	1	2	1	1¼ or 1½	1¼	1¼
Barber; beautician	...	3	...	...	2	1½	...	1½
Surgeon	...	3	...	...	2	1½	...	1½
Laundry tray, 1 or 2 compartments	2	4	3[n]	...	2	1½	1½	1½[o]
Shower,[i] each head	2	4	2	4	2[p]	2	2	1½
Sink:								
Surgeon	...	3	...	...	3	1½	...	1½
Soda fountain; bar	...	2	...	...	...	1¼	1¼	
Flushing rim, flush valve	...	10	...	...	8	3	...	3
Service	...	3	...	...	2[q]	1½	2	2[q]
with standard trap	...	3	...	...	3	3	...	3
Pot or scullery	...	4	...	...	4	1½	...	1½
		5	...	...	...	2		
Slop	...	....	...	...	8[r]	....	...	3[s]
Urinal								
Pedestal	...	10	...	10	...	3	...	
Wall lip	...	5	...	5	4	1½	...	1½
Stall	...	5	...	5	4	2	2	2
Trough, for each 2 ft	...	....	...	2	2	1½	2	1½
Wash sink, circular or multiple, for each set of faucets	...	2	...	...	2	1½	...	1½[s]
Water closet	6	12	6	10	4[t]	3	...	3[s]
Sewage ejector or sump, for each 25 gpm	...	....	...	50				

[a] See also Table 3-4.
[b] The Uniform Plumbing Code for Housing, Housing and Home Finance Agency, February, 1948.
[c] "Plumbing Manual," Report BMS66, Subcommittee on Plumbing, Central Housing Committee, National Bureau of Standards, 1940.
[d] Fixtures not listed herein are to have unit values determined by the size of the drain, as follows: drain 1¼ in. and smaller, one unit; 1½ in., two units; 2 in., three units; 2½ in., four units; 3 in., five units; and 4 in., six units.
[e] In public buildings.
[f] Shower head over tub adds no fixture units.
[g] With separate shower stall, 10 fixture units.
[h] Combination sink and tray with food-disposal unit: four fixture units, with separate traps 1½ in.
[i] Stall or gang shower.
[j] Size to be determined by area of surface to be drained.
[k] With or without garbage disposal unit.
[l] Domestic.
[m] With waste food grinder.
[n] Two or three compartments.
[o] For each compartment.
[p] Group, per head, three units.
[q] With P trap.
[r] Pedestal, siphon, jet, blowout.
[s] Nominal.
[t] With flush tank. With a flushometer, eight units.

TABLE 11-2. FIXTURE UNITS FOR DRAINS, TRAPS, AND VENTS NOT LISTED IN TABLE 11-1

Pipe diameter, in.	Drains and traps		Vents
	Private building	Public building	
1¼	1	2	1
1½	2	4	8
2	3	6	18
3	5	10	72
4	6	12	384

TABLE 11-3. CAPACITIES OF BRANCHES, STACKS, AND BUILDING DRAINS*

Diameter of pipe, in.	Maximum number of fixture units that may be connected to:							
	Building drains or sewers				One horizontal branch	Stacks with not over three branch intervals	Stacks with three or more branch intervals	
	Fall per foot							
	1/16	⅛	¼	½			Per branch interval	Total in stack
1¼					1	2	1	2
1½					3	4	2	8
2			21	26	6	10	6	24
2½			24	31	12	20	9	42
3			27†	36†	20†	30†	11‡	60†
4		180	216	250	160	240	90	500
5		390	480	575	360	540	200	1,100
6		700	840	1,000	620	960	350	1,900
8	1,400	1,600	1,920	2,300	1,400	....	660	3,600
10	2,500	2,900	3,500	4,200	2,500	....	1,000	5,600
12	3,900	4,600	5,600	6,700	3,900	....	1,500	8,400
15	7,000	8,300	10,000	12,000	7,000			

* From "The Uniform Plumbing Code for Housing," Housing and Home Finance Agency, February, 1948. Minimum Size of Soil and Waste Stacks. No soil or waste stack shall be smaller than the largest horizontal branch connected thereto.

† Not over two water closets.

‡ Not over one water closet.

rainfall is taken as 4 in. per hr, 180 sq ft of roof or other surface-drainage area is equivalent to one fixture unit.

11-7. Capacity and Size of Soil and Waste Pipes. Water will flow through or down a stack faster than it will flow into it. The useful capacity of a stack is expressed, therefore, as the rate at which water will flow *into* it under standardized conditions. A common, but not standardized, definition is that the capacity of a stack is the rate of flow into it from a horizontal branch entering the stack when all the fixtures connected to the stack above this branch are discharging into the stack, a positive pressure is created in the horizontal branch immediately above its junction with the stack, and this positive pressure cannot be relieved by normal venting or by increasing the size of the building drain.[1] Partly because of the absence of a standard definition, stack capacities

TABLE 11-4. SIZES OF BUILDING BRANCHES, STACKS, DRAINS, AND SEWERS FROM "NATIONAL PLUMBING CODE"
(Showing maximum number of fixture units that may be connected)

Diameter of pipe, in.	Building drains[a] and sewers, NPC (Sec. 11.5.2)				Horizontal fixture branches and stacks, NPC (Sec. 11.5.3)			
	Slope, in. per ft				Any horizontal fixture branch[b]	One stack of 3 stories in height or 3 intervals	More than 3 stories in height	
	1/16	1/8	1/4	1/2			Total for stack	Total at 1 story or branch interval
1¼	.....	.....	....	....	1	2	2	1
1½	.....	.....	....	....	3	4	8	2
2	.....	.....	21	26	6	10	24	6
2½	.....	.....	24	31	12	20	42	9
3	.....	20[c]	27[c]	36[c]	20[d]	30[e]	60[e]	16[b]
4	.....	180	216	250	160	240	500	90
5	.....	390	480	575	360	540	1,100	200
6	.....	700	840	1,000	620	960	1,900	350
8	1,400	1,600	1,920	2,300	1,400	2,200	3,600	600
10	2,500	2,900	3,500	4,200	2,500	3,800	5,600	1,000
12	3,900	4,600	5,600	6,700	3,900	6,000	8,400	1,500
15	7,000	8,300	10,000	12,000	7,000			

[a] Includes branches of the building drain.
[b] Does not include branches connected directly to the building drain.
[c] Not over two water closets.
[d] Not over two water closets.
[e] Not over six water closets.

are not expressed uniformly. They are expressed in fixture units in Table 11-3.

The correct size of stack should be used for any given installation. Too large a pipe will add to the cost and may cause trouble because of the tendency of low flows to adhere to the sides of the pipe with resultant less-than-scouring velocities. Too small a pipe will not admit water

[1] See also R. S. Wyly and H. N. Eaton, *Plumbing and Heating J.*, February, 1953, p. 62.

fast enough, and the stack may clog due to the inability of large particles to pass through it or through its connections.

A 2-in. stack is the smallest that should be used for a waste stack, and 3 in. is the smallest that should be used for a soil stack. Tests and experience have shown these stacks to give satisfactory service in individual family residences. These sizes are advantageous also, as they are more easily concealed within buildings walls than the 4-in. size specified as a minimum in some plumbing codes. The size of stacks to be selected

TABLE 11-5. LIMITS IN CAPACITIES OF BUILDING DRAINS*
(Nonpressure drainage†)

Diameter, in.	Limits in fixture units							
	Primary branch				Secondary branch or main			
	Fall per ft, in.				Fall per ft, in.			
	1/16	1/8	1/4	1/2	1/16	1/8	1/4	1/2
2	.....	18	21	26				
3	.....	24	27	36	.....	90	125	180
4	.....	180	216	250	.....	450	630	900
5	360	400	480	560	600	850	1,200	1,700
6	600	660	790	940	950	1,300	1,900	2,700
8	1,400	1,600	1,920	2,240	1,950	2,800	3,900	5,600
10	2,400	2,700	3,240	3,780	3,400	4,900	6,800	9,800
12	3,600	4,200	5,240	6,080	5,600	8,000	11,200	16,000

* From "Plumbing Manual," Report BMS66, p. 49, National Bureau of Standards, 1940.
† Limits in capacities of building drains for *pressure drainage* are double the figures shown in this table.

for large buildings is based on the type, location, and numbers of fixtures connected to it. The effect of offsets on the size of stacks is discussed in Sec. 12-5.

The minimum size of a house drain or house sewer carrying human excrement is 4 in., and the smallest drainage pipe which does not carry human excrement and is installed underground is 2 in.[1] The sizes of larger pipes depend on the load to be carried.

The capacities of branches, stacks, vents, building drains, and building sewers are given in Tables 11-3 to 11-7, inclusive. The "Plumbing Manual"[2] states (p. 16):

b. In case the sanitary system consists of one soil stack only or one soil stack and one or more waste stacks of less than 3-in. diameter, the building drain and building sewer shall be of the same nominal size as the primary branch from the

[1] "National Plumbing Code," Sec. 11.5.7.
[2] *Op. cit.*

TABLE 11-6. CAPACITIES OF HORIZONTAL BRANCHES AND PRIMARY BRANCHES OF THE BUILDING DRAIN*

Diameter of pipe, in.	Permissible number of fixture units				
	Horizontal branch at minimum permissible slope or greater	Primary branch			
		Fall per ft, in.			
		1/16	1/8	1/4	1/2
1¼	1	.....	.....	2	2
1½	3	.....	.....	5	7
2	6	.....	.....	21	26
3†	32	.....	36	42	50
3‡	20	.....	24	27	36
4	160	.....	180	216	250
5	360	360	400	480	560
6	600	600	660	790	940
8	1,200	1,400	1,600	1,920	2,240
10	1,800	2,400	2,700	3,240	3,780
12	2,800	3,600	4,200	5,000	6,000

* From "Plumbing Manual," Report BMS66, p. 16, National Bureau of Standards, 1940.
† Waste only.
‡ Soil.

TABLE 11-7. SIZES OF HORIZONTAL STORM DRAINS FOR VARIOUS DRAINAGE AREAS
(Areas shown are in square feet)

Pipe diameter, in.	Fall of pipe, in. per ft						Pipe diameter, in.	Fall of pipe, in. per ft					
	1/8		1/4		1/2			1/8		1/4		1/2	
	Data from		Data from		Data from			Data from		Data from		Data from	
	I	II	I	II	I	II		I	II	I	II	I	II
2	250	.....	350	.....	500	.....	6	4,300	5,350	6,100	7,550	9,000	10,700
2½	450	.....	600	.....	900	.....	8	9,600	11,500	13,000	16,300	19,000	23,000
3	700	822	1,000	1,160	1,500	1,644	10	16,500	20,700	24,000	29,200	35,000	41,400
4	1,500	1,880	2,100	2,650	3,000	3,760	12	27,000	33,300	40,000	47,000	56,000	66,600
5	2,700	3,340	3,800	4,720	5,500	6,680	15	......	59,500	......	84,000	......	119,000

I. From "New York City Building Code," Sec. C26-1298.0. The capacities here refer to storm water drainage only. No sanitary sewage is included.

II. From "National Plumbing Code," Table 13.6.2. Based on a maximum rate of rainfall of 4 in. per hr "If . . . the maximum rate of rainfall is more or less than 4 in. per hr, then the figures for area must be adjusted by multiplying the figure in the table by 4 and dividing by the rate of rainfall in inches per hour."

soil stack as given in Table 11-6, except that (*d*) of this Section and the applicable rules [given here as (*f*) and (*g*)] relating to pressure drainage may apply when the prescribed conditions are complied with.

c. In case the plumbing system has two or more soil stacks, each having its separate primary branch, the number of fixture units for a secondary branch, the main building drain, or the building sewer of a given diameter and slope may be increased from the value given in Table 11-6 for a primary branch of the same diameter and slope to the value given in Table 11-5 provided that the increase is made strictly within the principles and rules of . . . [the following paragraphs].

d. In case there is no fixture drain or horizontal branch connecting directly with the building drain or a branch thereof, and the lowest fixture branch or horizontal branch connected to any soil or waste stack of the system is 3 ft or more above the grade line of the building drain, main building drain and building sewer may be increased [up to the limits shown for pressure drainage in the footnote in Table 11-5] provided the increases are made in accordance with the principles and rules given in . . . [the following paragraphs].

f. Section *c* provides for permissible limits in fixture units for secondary branches of the building drain, the main building drain, the building sewer of given diameter and slope, applicable when the building drain has two or more primary branches of 3-in. or greater diameter and when the lowest horizontal branch or fixture drain is less than 3 ft above the grade line of the building drain.

g. Section *d* and [the conditions for pressure drainage shown in the footnote in Table 11-5] provide for limits applicable when all fixture branches and horizontal branches connected directly to the stacks are 3 ft or more above the grade line of the building drain and no fixture branch or horizontal branch is connected to the building drain.

Table 11-6 is particularly applicable to comparatively simple systems with few stacks and branches but the limits may be applied to systems of any size without the restrictions governing the application of Secs. *c* and *d* above.

Table 11-5 is particularly applicable to the plumbing systems of buildings covering a relatively large area with several widely separated soil stacks. It will be noted that in accordance with the rules governing the application of the table, the limits for primary branches will determine the size of the pipe required unless the total length of all branches (primary and secondary) of the building drain is more than 40 ft.

Rules for Applying Section *c* and Table 11-5

1. Lay out the foundation plan of the building to scale as in Fig. 11-9, and mark thereon the position of all soil and waste stacks in vertical projection.

2. Lay out on the plan the most direct practical lines for the building drain and its branches that can be readily obtained by use of standard bend and wye fittings, observing requirements regarding bearing walls. Make, if possible, no turn in direction greater than 45° and not greater than 90° in any case, with no angle between branches greater than 45°.

3. Scale off the developed length of each primary and secondary branch to the nearest foot mark and mark its length on the plan.

4. Determine the total number of fixture units on each soil and waste stack, from Table 11-1, and note the numbers on the plan.

5. Determine the required size of each soil and waste stack in accordance with Secs. *b* and *d* above taking into account the restrictions as to limits in one branch interval in relation to the height of the stack system and the fact that each stack shall be at least equal in diameter to the largest horizontal branch connected thereto.

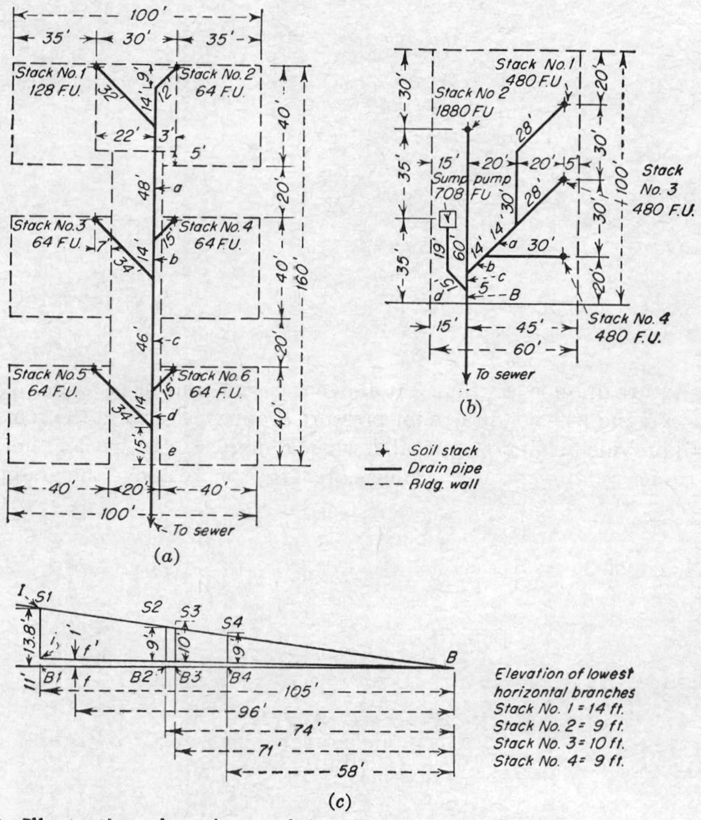

FIG. 11-9. Illustrative plans for applying Secs. *c* and *d* from the "Plumbing Manual" quoted in Sec. 11-7.

6. Determine the most convenient continuous slope between the minimum permissible and the ½-in. per ft fall that can be installed in relation to the elevation of the street sewer.

7. Determine the required size of each primary branch for the selected slope by Table 11-6, and note on the plan.

8. Starting with the two primary branches farthest upstream, compute the permissible number of fixture units for each successive secondary branch and the main building drain by means of the following equation:

$$FU = P\{1 + [a(\Sigma L - 40)]\} \qquad (11\text{-}3)$$

where FU = maximum permissible fixture units on the secondary branch in question

P = permissible fixture units on one primary branch of the selected slope, by Table 11-6

a = a numerical factor depending on the slope as shown in the following tabulation

ΣL = sum of the lengths of all primary and secondary branches greater than 10 ft in length upstream from the particular secondary branch for which a computation is being made

Values of (a)	*Fall, in. per ft*
0.010	1/16
0.007	1/8
0.005	1/4
0.004	1/2

9. The diameter of the building sewer shall be the same as for the main building drain if it is laid at the same slope as or at a greater slope than the building drain. If the building sewer is laid at a smaller slope than the building drain, its capacity in fixture units by Table 11-5 shall be equal to or greater than the total fixture units on the system.

10. The preceding rules assume that all primary branches enter the building drain from the side at the selected slope. If any primary branch, horizontal branch, or stack enters the building drain from the top or at a slope greater than that of the building drain, rule 8 for increase in permissible number of fixtures shall not be applied to the primary or secondary branch immediately upstream from such entry or to the secondary branch or main immediately downstream from each entry connection, but the lengths of an upstream secondary branch and of a downstream from the entry connection may be employed in evaluating ΣL for succeeding downstream branches and the main building drain.

11. Rule 8 also assumed a clear passage through the building drain at the selected slope. If any obstruction is built into the building drain as, for example, a running trap or a backwater trap, rule 8 shall not be applied to that part of the building drain in which such obstruction is set unless the last branch thereto enters the building drain at least 2 ft developed length upstream from such obstruction, and the limit in permissible fixture units on the main building drain shall not exceed the value given by the last computed limit by this rule except as provided in Sec. *d* [p. 274] for pressure drainage. This rule is not to be construed as prohibiting the use of a house trap or backwater trap when drainage conditions require it, or as prohibiting the connection of a primary branch or secondary branch within 20 ft from a house trap or backwater trap, but merely as a statement of the necessary modifications of the general rule if these devices are installed.

12. In no case shall the number of fixtures installed on a building drain of given diameter and slope exceed the limits given in Table 11-5 unless all fixture branches or horizontal branches are more than 3 ft above the grade line of the building drain, thus permitting the application of the principles and rules for pressure drainage as provided for in Sec. *d* [p. 274].

13. When the limit in permissible number of fixture units for a given diameter and slope of a building drain has been exceeded by the application of rule 8, or where a junction occurs, the value of P in the equation $FU = P\{1 + [a(\Sigma L - 40)]\}$ will become the value for the larger diameter and succeeding computations will proceed as by rule 8, except that if the larger branch is a primary branch there shall be no increase in the permissible number of fixture units for the secondary branch or main immediately downstream from the junction over the permissible number of fixture units for a primary branch of that diameter and slope unless the primary branch exceeds 20 ft in length.

Illustrative Computation. Referring to Fig. 11-9(a), assume that the building drain and building sewer are to be laid with a uniform fall of ⅛ in. per ft. By Table 11-5 the permissible load for a 4-in. primary branch with that slope is 180 fixture units, and by rule 8 the value of a for that slope is 0.007. Hence, the equation (rule 8) becomes: Permissible fixture units

$$FU = 180\{1 + [0.007(\Sigma L - 40)]\} \tag{11-2}$$

If the actual number of fixture units on each stack is as indicated in Fig. 11-9(a), all primary branches of the building drain may be 4 in. since the maximum is 128 on any branch and the limit by Table 11-5 is 180 fixture units. Starting with primary branches 1 and 2, the necessary computations for determining permissible limits for the various secondary branches a to d and on the main building drain e may be summarized as follows:

Branch	ΣL, ft	$P\{l + [0.007(\Sigma L - 40)]\}$	Actual fixture units
a	58	$180\{1 + [0.007(58 - 40)]\} = 202$	$128 + 64 = 192$
b	121	$180\{1 + [0.007(121 - 40)]\} = 282$	$192 + 64 = 256$
c	169	$180\{1 + [0.007(169 - 40)]\} = 342$	$256 + 64 = 320$
d	230	$180\{1 + [0.007(230 - 40)]\} = 419$	$320 + 64 = 384$
e (main)	278	$180\{1 + [0.007(278 - 40)]\} = 480$	$384 + 64 = 448$

It will be noted that for the main building drain e the computed limit, 480 fixture units, is greater than the limit for nonpressure drainage set in Table 11-5 for a 4-in. building drain with a uniform fall of ⅛-in. per ft, 450 fixture units. However, since the actual number of fixture units to be carried (448) is less than the limit (450), the rules would permit the use of a 4-in. drain at the assumed slope for the entire building, which corresponds to an eight-story apartment house with seven apartments on each floor.

Assume again that the same floor plan is employed for a nine-story building, increasing the total number of fixture units on the different stacks to 144 for stack 1 and to 72 for each of stacks 2 to 6, inclusive, and that the building drain is to be laid at a uniform fall of ¼ in. per ft. In the same manner as for the first illustration, the equation for computing permissible limits for a 4-in. branch at ¼-in. fall per ft becomes $FU = 216\{1 + [0.005(\Sigma L - 40)]\}$ and, for a 5-in. branch at the same slope, becomes $FU = 480\{1 + [0.005(\Sigma L - 40)]\}$.

Branch	ΣL, ft	$P\{1 + [0.005(\Sigma L - 40)]\}$	Actual fixture units
a	58	$216\{1 + [0.005(58 - 40)]\} = 235$	$144 + 72 = 216$
b	121	$216\{1 + [0.005(121 - 40)]\} = 303$	$216 + 72 = 288$
c	169	$216\{1 + [0.006(169 - 40)]\} = 355$	$288 + 72 = 360$
c	169	$480\{1 + [0.005(169 - 40)]\} = 790$	$288 + 72 = 360$
d	230	$480\{1 + [0.005(230 - 40)]\} = 936$	$360 + 72 = 432$
e (main)	278	$480\{1 + [0.005(278 - 40)]\} = 1{,}051$	$432 + 72 = 504$

Since for the secondary branches *a* and *b* the actual number of fixture units is less than either the computed permissible number or the set limit of 630 fixture units by Table 11-5, a 4-in. building drain with ¼-in. fall per ft may be installed to the beginning of branch *c*. For branch *c* the computed limit for a 4-in. drain is less than the actual number of fixture units; hence, the size of branch *c* must be greater than 4 in. in diameter and it becomes necessary to try 5 in. Obviously, as shown by the second computation for branch *c*, it would not be necessary for the engineer to make any further computation in the case illustrated to determine the sizes required for branch *d* and the main building drain *e*, since the actual number of fixture units (504) on the system does not exceed either the computed limit (790) for branch *c* or the fixed limit (1,200) for a 5-in. drain as given by Table 11-5. However, the complete summarized computation is given by way of illustration.

Application of these rules to drains of ½-in. fall per ft and for other sizes of drain will be made in the same manner for all sizes of primary branches and for all lengths of primary and secondary branches.

Table 11-5, second footnote, is most applicable to large buildings in which basement fixtures and possibly first-floor fixtures are to be drained into a sump in such a manner that all direct connections of fixture branches and horizontal branches will be materially greater than 3 ft above the grade line of the building drain. In computing the permissible limits for particular systems under rules (which see) [*sic*] applying to pressure drainage, both Sec. *c* and Sec. *d* apply.

Rules for Applying Sec. *d*[1]

1. Lay out the building plan to scale as for the application of Sec. *c* and scale off the developed lengths of each primary and secondary branch [Fig. 11-9(*b*)].

2. For each stack in the system, determine the elevation of the lowest horizontal branch above the grade line of the building drain.

3. Scale off the developed length of pipe from the intersection *B* of the branch with the building drain to the lowest horizontal branch connection to the stack farthest upstream. Also, determine the developed length from *B* to the lowest horizontal branch on each of the other soil and waste stacks.

4. Lay off to a convenient scale a horizontal line [B_1B in Fig. 11-9(*c*)] equal to or greater than the greatest developed length obtained by applying rule 3. Also lay

[1] In which the limits for pressure drainage stated in Table 11-5 are allowed.

off in order on line B_1B from point B all other developed lengths as obtained by the application of rule 3 and designate the points on line B_1B as B_2, B_3, etc. At each of the points B_1, B_2, B_3, etc., erect lines perpendicular to line B_1B. Also on line B_1B, or on B_1B extended if necessary, lay off a length Bf equal to that in which a total fall of 1 ft would be given for the slope involved and erect the perpendicular line ff' equal to 1 ft and draw the straight line Bf' to intersect the perpendicular line from B_1. Designate the intersection as i. Then on each of the perpendicular lines from B_1, B_2, B_3, etc., lay off lengths from the line Bf' equal to the elevation of the lowest branches on each of the corresponding stacks above the grade line and designate the resultant points as S_1, S_2, S_3, etc. Then draw a straight line from the point B such that at least one of the points S_1, S_2, S_3, etc., lies on the line and none of them lies below it. Extend the line to intersect the vertical line from point B_1, which intersection (designated as I) may be at or below point S_1, depending on the relative elevations of the lowest branch of the different stacks involved. Scale off the lengths B_1I and B_1i.

5. Select a value P from Table 11-5 for primary branches corresponding to the slope of the drain such that the product $P\sqrt{(B_1I)/4(B_1i)}$ is equal to or greater than the number of fixture units to be carried by a primary branch and the corresponding product $P_1\sqrt{(B_1I)/4(B_1i)}$ for the next smaller size is less than the number of fixture units to be carried by the primary branch in question. The larger of the two sizes will be the minimum permissible size for the primary branch in the particular case, provided that both the product of $P\sqrt{(B_1I)/4(B_1i)}$ and the limit for a primary branch of that diameter in Table 11-5, second footnote, are each equal to or greater than the number of fixture units actually carried.

6. To obtain the permissible load limits for secondary branches and the main building drain of the particular system, proceed as by rule 8, applying to Sec. *c*, starting in each case with the computed limits for primary branches as given by rule 5 above instead of the limits in Table 11-5.

7. If the slope of the building sewer is equal to or greater than the slope of the main building drain, it may be of the same nominal diameter as the main building drain arrived at by the application of the preceding rule 6. If at any point the slope of either the main building drain or the building sewer is decreased, the drain laid at the lower slope shall be increased in diameter, if necessary, so that its limit in capacity by Table 11-5, second footnote, at the slope laid, is equal to or greater than the load to be carried.

Illustration of Application of Rules for Pressure Drainage. For this illustration assume that the building has a base area 60 by 100 ft, as illustrated in Fig. 11-9(*b*); is 20 stories high; and will contain fixtures that total 3,528 units, distributed among four stacks, and a sump located as indicated in the figure. In this illustration the drain from the sump pump, entering the main building drain at point B, is the last branch. Now assume that the elevation above grade line of the lowest branch connected to each of the stacks is as follows: stack 1, 14 ft; stack 2, 9 ft; stack 3, 10 ft; and stack 4, 9 ft; and that the slope of the building drain is ⅛-in. fall per ft. The developed length of drain B to the lowest branch on stack 1 is $5 + 14 + 14 + 30 + 28 + 14 = 105$ ft. We now have the data for the scale, laying out Fig. 11-9(*c*) as prescribed by rules 3 and 4. From the figure, $B_1I =$

13.8 ft and $B_1 i = 1.1$ ft; from which

$$\sqrt{\frac{B_1 I}{4(B_1 i)}} = \sqrt{\frac{13.8}{4.4}} = 1.77$$

the factor by which the limits for primary branches as given in Table 11-5 are to be multiplied to determine the allowable loads within the limits for primary branches by Table 11-5, second footnote, for the particular building and conditions assumed. Ordinarily, the required minimum size for a primary branch can be predicted by comparing the permissible loads for pressure drainage, Table 11-5, second footnote, and the number of fixture units to be carried in the case at hand without going through the details of computations prescribed by rule 4. For example, in the present illustration the basic size for computing secondary branches will obviously be determined by the primary branch from stack 2, and the load (in this case, 1,880 fixture units) lies between the permissible load of 1,320 fixture units for a 6-in. primary branch with ⅛-in. fall per ft by Table 11-5, second footnote, and the limit in load, 3,200 fixture units, for an 8-in. primary branch at that slope. Hence, the secondary branch must be at least an 8-in. pipe and the trial computation becomes $1.77 \times 1{,}600 = 2{,}832$ fixture units.

Since 2,832 fixture units is less than the limit (Table 11-5, second footnote) for an 8-in. primary branch of ⅛-in. fall per ft under pressure, and greater than the actual number (1,880), the size required for the primary branch from stack 2 is 8 in. in diameter.

Similarly, for stacks 1, 3, and 4 the limits for a 5-in. primary branch at ⅛-in. fall per ft are 400 fixture units for nonpressure drainage (Table 11-5) and 800 fixture units for pressure drainage (Table 11-5, second footnote). Again, $1.77 \times 400 = 708$ fixture units is greater than the load to be carried (480 fixture units) and less than the limit (800 fixture units) for pressure drainage, and hence a 5-in. primary branch will be required from stacks 1, 3, and 4. Starting with stacks 1 and 3 and applying rule 5 for pressure drainage and then rule 8 for nonpressure drainage, the computations may be summarized as shown in the following table:

Branch	ΣL, ft	$P\{1 + [0.007(\Sigma L - 40)]\}$	Fixture units	Diameter, in.
a	86	$708\{1 + [0.007(86 - 40)]\} = 935$	960	6
b	130	$1{,}168\{1 + [0.007(130 - 40)]\} = 1{,}940$	1,440	6
c	190	$2{,}832\{1 + [0.007(190 - 40)]\} = 5{,}805$	3,320	8
d (main)	219	$2{,}832\{1 + [0.007(219 - 40)]\} = 6{,}380$	3,528	8

In this case there will be three different values of P (the permissible number of fixture units on the primary branch) to be used in the equation

$$FU = P\{1 + [a(\Sigma L - 40)]\} \tag{11-3}$$

$P_1 = 708$, applying to the secondary branch a for stacks 1 and 3;

$$P_2 = 1.77 \times 600 = 1{,}168$$

applying to the secondary branch *b* for stacks 1, 3, and 4; and

$$P_3 = 1.77 \times 1,600 = 2,832$$

applying to the primary branch from stack 2, to the secondary branch *c*, and to the main building drain *d*.

Summarizing, the requirements in sizes are

1. Five inches for primary branches from stacks 1, 3, and 4
2. Eight inches for primary branches from stack 2
3. Six inches for secondary branches *a* and *b*
4. Eight inches for secondary branch *c*
5. Eight inches for main building drain *d*

Since the number of fixture units carried by the main building drain and building sewer (3,320 + 208 = 3,528) does not exceed either the ultimate limit (5,600) for an 8-in. drain or the computed limit (5,805) for the secondary branch *c* immediately upstream from the junction point *B*, the main building drain *d* and the building sewer may also be 8-in. pipe if they are laid with not less than ⅛-in. fall per ft. It should be noted that the computed limit may sometimes exceed the ultimate permissible limit given by Table 11-5, second footnote, as in the case of the computation for branch *c*. In these cases the permissible limit in the table shall apply.[1]

11-8. The House or Building Drain. The required size of a building drain can be determined from the principles set forth in Sec. 11-7 and in Tables 11-3 to 11-7. Additional information is given in Figs. 11-10 and 11-11. Rain-water leaders may be at least as large as the smallest drains called for in Table 11-7. Where the flow in the house drain includes rain water together with the discharge from plumbing fixtures, Fig. 11-10 can be used in determining pipe sizes, or as stated in the "National Plumbing Code,"[2]

> When the total fixture-unit load on the combined drain is less than 256 fixture units, the equivalent drainage area in horizontal projection shall be taken as 1,000 sq ft. When it exceeds 256 fixture units, each fixture unit is to be considered the equivalent of 3.9 sq ft of drainage area.[3]

It is permissible to allow a building drain to run full provided there are no fixtures connected to it less than about 2 ft above the drain. An exception to this limitation is a sump pump, which may discharge under a slight pressure into the building drain.

The drain should be laid on a straight line in so far as is possible. Where a change of direction is necessary, it should be made with a long-

[1] "Plumbing Manual," Report BMS66, National Bureau of Standards, 1940.

[2] Secs. 13.7.2 and 13.7.3.

[3] This is based on a maximum rainfall rate of 4 in. per hr. If maximum rainfall rate differs, then figures from roof area must be adjusted by dividing the figures in this table by the assumed rainfall rate and multiplying by 4.

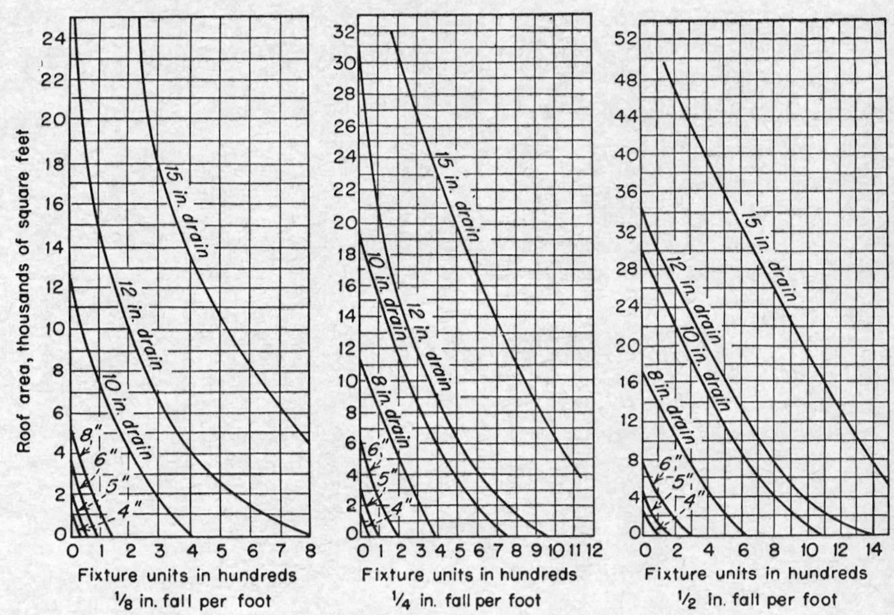

FIG. 11-10. Diagram for the determination of the size of the building drain.

radius fitting, and where possible, a wye or similar fitting should be used that will supply cleanout facilities. It is usually required that a cleanout be installed at the upper end of a house drain and also inside the house at the point where the drain passes through the foundation wall. Cleanouts are shown in Figs. 11-4 to 11-7 and others. A cleanout should consist of a fitting of the same size as the house drain or house sewer, with a 2-in. branch or larger extending above the basement floor or into other accessible area. There is no advantage in using a drain larger than is required, as the larger size is conducive to lower velocities and the stranding of solids due to the shallow depth of flow in it. The slope of the house drain should not be less than 1/4 in. per ft, and its diameter should not be less than the diameter of the largest stack or other pipe discharging into it.

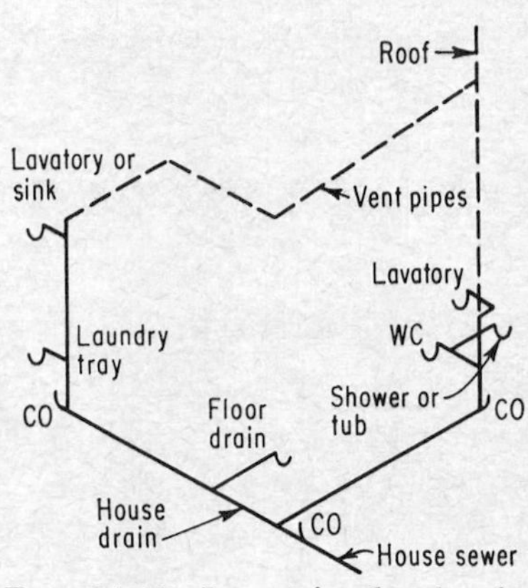

FIG. 11-11. Isometric sketch of drainage and vent pipes for a residence.

House drains are commonly placed 12 in. or more beneath the basement floor and may be laid out for a residence somewhat as shown in Fig. 11-2 to receive the discharges from floor drains and fixtures in the basement. Such drains should be extra-heavy, cast-iron soil pipe with calked joints, although some codes permit the use, under special conditions, of vitrified-clay pipe. The use of vitrified-clay pipe within a building is undesirable because of the difficulty of maintaining joints that are gastight. The use of wrought metal pipes, whether or not galvanized, is undesirable because of its inability to resist the highly corrosive conditions encountered. If the building drain is placed aboveground in the basement, cast-iron or wrought-metal pipe should be used with cast-iron drainage fittings or standard threaded fittings.

The building drain passes through the foundation wall of the building and terminates 4 to 5 ft beyond it in the building sewer discussed in Sec. 11-9. The opening in the building wall may be slightly larger than the drain pipe passing through it.

11-9. The House or Building Sewer. The building sewer extends from the terminal of the house drain to the common sewer or other point of discharge for the drainage from the plumbing system. The building sewer should be the same size or larger than the house drain, and preferably not less than 6 in. in diameter. The method for determining the required diameter of the house drain is explained in the preceding section. If the building sewer receives the discharge both from the house drain and from roof, yard, and other drains, the capacity of the sewer must be sufficient to carry the combined discharges. Required capacities of roof and surface-water drains are given in Tables 13-4 and 13-7 and Sec. 13-14.

The slope of the building sewer should be straight and unbroken and between ¼ and ½ in. per ft. A flatter slope may give too slow a velocity to convey suspended solids, and water may run so shallow on a steeper slope as to result in the stranding of solids in the sewer. Capacities of cast-iron and vitrified-clay pipes on various slopes are shown in Table 11-8. If the slope is greater than about 45°, however, solids will not strand. Hence, the building sewer may run on a straight slope between ¼ and ½ in. per ft until it reaches a point where it turns on a 100 per cent or greater slope to connect to the public sewer. It may be desirable to place a cleanout at this change in slope. The sewer should be placed in a straight, horizontal line from the end of the house drain to the connection with the public sewer or other outlet to facilitate cleaning. The depth of cover over the sewer and beneath a paved surface should not be less than 12 in. and beneath an unpaved surface not less than 18 in.

Materials in common use for building sewers are vitrified clay, cast iron, concrete, bituminized fiber, and asbestos cement.[1] Plastic pipe is

[1] See also "National Plumbing Code," Sec. 11.2.1.

TABLE 11-8. RATE OF FLOW THROUGH VITRIFIED-CLAY AND CAST-IRON PIPE FLOWING FULL
(Gallons per minute)

Diameter of pipe, in.	Cast-iron pipe							Diameter of pipe, in.	Vitrified-clay pipe						
	Slope								Slope						
	1/16 in. 0.5 per cent	1/8 in. 1.0 per cent	3/16 in. 1.5 per cent	1/4 in. 2.0 per cent	5/16 in. 2.5 per cent	3/8 in. 3.0 per cent	1/2 in. 4.0 per cent		1/16 in. 0.5 per cent	1/8 in. 1.0 per cent	3/16 in. 1.5 per cent	1/4 in. 2.0 per cent	5/16 in. 2.5 per cent	3/8 in. 3.0 per cent	1/2 in. 4.0 per cent
2	8.5	12	15	17	19	21	24	4	60	85	104	120	134	147	170
3	25	35	43	50	55	61	70	6	181	257	315	364	406	445	514
4	52	74	91	105	117	128	148	8	380	540	660	760	850	940	1,080
6	160	225	276	318	356	390	450	10	690	980	1,200	1,390	1,550	1,700	1,960
8	330	470	575	665	740	815	940	12	1,150	1,620	1,980	2,300	2,560	2,810	3,240
10	608	860	1,050	1,220	1,360	1,490	1,720								
12	1,000	1,420	1,740	2,010	2,250	2,460	2,840								

permitted by some authorities. Watertightness, prevention of root penetration, resistance to corrosion, and structural strength to resist external loads are all desirable characteristics.

11-10. Installation of the Building Sewer. An installation of a building sewer is shown in Fig. 11-12. When a public sewer system is available, every occupied building near by should be independently connected to it. There should be no intermediate connection to a building sewer downstream from the one house drain discharging into it. Where two or more house drains or house sewers discharge into a single house sewer, a stoppage of the house sewer may result in sewage flowing from one building into another. Where two or more buildings intended for occupancy are constructed in close proximity and the construction of a house sewer from

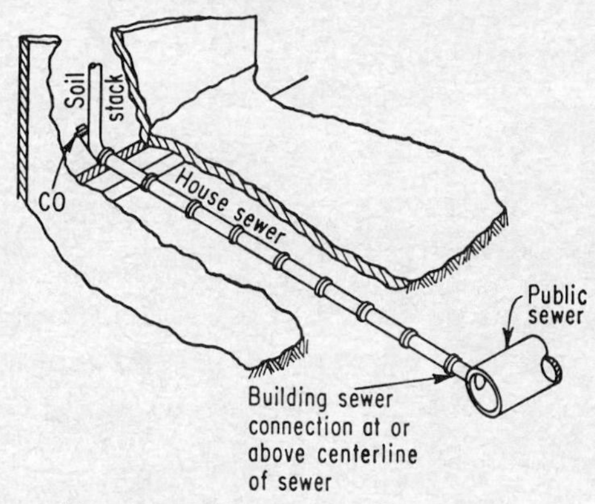

FIG. 11-12. House-sewer installation.

each building to the common sewer is undesirable, the design of the sewers for the adjacent buildings should be based on the principles of public sewer design, with manholes and other maintenance facilities properly provided. Unless this is done, backflows and overflows may result in damages creating hygienic, physical, legal, and other complications.

11-11. Laying the House Sewer.[1] Trenching, pipe laying, backfilling, and other steps involved in the laying of a house sewer should follow good sewerage practice. The trench should be as narrow as possible to minimize the magnitude of backfill and superimposed loads on the pipe. Where the depth of trench is less than 8 ft, a width of 18 in. has been found enough for laying 8-in. bell-and-spigot pipe.

When the excavation of the trench has come within one-half of the outside diameter of the pipe to the final elevation at which the sewer will be laid and if the material to be excavated is sufficiently firm, the remainder of the trench should be cut to conform to the shape of the outside of the lower half, with proper enlargements for each bell, to distribute the backfill and superimposed loads over the barrel of the pipe. If rock is encountered, it is removed to a depth of 3 to 6 in. below the bottom of the sewer, and the trench is refilled to grade with sand or well-tamped earth shaped to fit the pipe.

The trench should be excavated upgrade to facilitate drainage, but no water should be allowed to run through the sewer until the pipe has been firmly bedded and the joints have set. The trench should be opened for the shortest convenient length in advance of the pipe laying in order to avoid cave-ins and other hazards. Backfilling should be prompt after the pipe has been laid and tested or otherwise inspected.

If bell-and-spigot joints are to be used in vitrified-clay pipe, the sections of pipe that are to be contiguous in the trench should be fitted together on the surface and the relative positions marked with chalk so that the same position can be obtained in the trench. Pipes are usually laid with the bell end upgrade, since it is easier to make the joint in this position. Specifications for pipe laying state:

The ends of the pipe shall abut against each other in such a manner that there shall be no shoulder nor unevenness of any kind along the inside of the bottom half of the drain. They shall be laid so that the deviation of the center line of the pipe from a straight, horizontal line, between planned changes of direction, will not exceed ½ in., and so that the direction of the lowest point on the inside circumference of the pipe does not deviate more than 1/16 in. from the designed elevation. Vitrified-clay pipe should not be laid in wet trenches and cement joints should not be

[1] See also H. E. Babbitt and E. R. Baumann, "Sewerage and Sewage Treatment," 8th ed., John Wiley & Sons, Inc., New York, 1958.

made under water. Obstructions on the inside surface of the pipe that might interfere with flow should be removed before the pipe is laid.

All pipes laid in the trench should rest directly on firm, undisturbed ground. They should not be laid on backfill material. When it is desirable to place two pipes at different elevations in the same trench, the side of the trench should be benched back at least 18 in. to receive the higher pipe. Where a sewer and a water pipe are to be laid in the same trench, the procedure described in Sec. 3-8 should be followed. All pipes should be laid carefully in the trench in a straight line and on a smooth grade to facilitate cleaning and to avoid clogging. Where a deviation from a straight line is necessary, it should be made with curves, wyes, or other suitable fittings to avoid an abrupt change of direction. Long-radius curves can be made with straight pipe by placing each length slightly out of line at each connection.

A groove should be cut in the bottom of the trench for each hub in order to give the pipe a solid bearing for its entire length, and the soil should be well rammed on either side of the pipe. Joints in cast-iron pipes should be leaded and calked. The joints in vitrified-clay pipe or in concrete or in cement pipe should be cement, asphalt, or some form of poured joint, as described in Secs. 7-21, 12-24, and 12-25.

Where the drains may be subjected to a backflow of sewage, a backwater valve or other device should be installed to prevent the backing-up of sewage or water into the building. Whenever possible, building drains should be brought into the building below the cellar floor to ensure drainage of the cellar floor and of plumbing fixtures placed in the cellar.

11-12. Location and Slope of Drains. It is undesirable to run drains parallel to and less than about 3 ft from the bearing wall of a building owing to the danger of cracking the drain by movements of the building. All drains and sewers should be placed below the frost line. Where the elevation of the pipe at the building wall, at a public sewer connection, or at some other point in the line is known, the bottom of the ditch should be smoothed off parallel to a chalk line stretched tightly over it at any convenient distance above the two known elevations. Where a short line of sewer is to be laid on a fixed slope, the pipe can be laid to grade by constructing a straightedge from a plank about 12 ft long, one side diverging from the other at the desired slope of the sewer. When this plank is rested on edge on the pipes in the trench, the upper edge should be level if the sewer is on the correct slope. It is not satisfactory to test the slope of individual lengths of pipe because errors in reading slopes on the short lengths are magnified in the full length of the sewer.

11-13. Connection between Building Sewer and Public Sewer. Connections between house sewers and common sewers are made by means of wyes or slants previously installed in the common sewer, as shown in

Fig. 11-12, or by breaking through the existing common sewer and inserting the desired fitting. Connections with sewers 15 in. or smaller should be made through wye or tee branches provided in the common sewer for the purpose. Where such fittings have not been provided, a wye fitting can be placed in the common sewer in the following manner:

A section of the common sewer pipe should be removed by breaking it to pieces, care being taken not to disturb adjacent lengths of pipe. The wye fitting should then be inserted without chipping the pipe or branch where possible; where not possible, the upper half of the bell on the run of the wye branch to be inserted and the bell remaining on the main pipe and facing the opening should then be carefully removed, and the wye branch inserted wrong side up and revolved to bring the branch to the side for which it is intended, with the broken part of the bell up. The joint should then be cemented, the broken parts of the bells being well rounded over with a liberal amount of cement mortar. Methods of making cement joints are described in Sec. 12-24.

The building sewer must be connected to the proper public sewer. The separate or sanitary public sewer may receive only separate or sanitary sewage. No roof, yard, foundation, or storm drainage should be discharged in the sanitary sewer. Storm-water, surface, and subsurface drainage must be discharged into a storm sewer or a combined sewer. Hence, two building sewers may sometimes be required, one to convey the sanitary sewage and the other for storm water and similar clean-water discharges.

11-14. Materials for Building Drains and Sewers. The "National Plumbing Code" states:[1]

13.2.3. Underground Storm Drains. Building storm drains underground, inside the building shall be of cast-iron soil pipe.

13.2.4. Building Storm Drains. Building storm drains underground, inside the building, when not connected with a sanitary or combined sewer shall be of cast-iron soil pipe or ferrous-alloy piping except when approved by the Administrative Authority, vitrified-clay pipe, concrete pipe, bituminized-fiber pipe and asbestor-cement pipe, may be used.

13.2.5. Building Storm Sewers. The building storm sewer shall be of cast-iron soil pipe, vitrified-clay pipe, concrete pipe, bituminized-fiber pipe, or asbestos-cement pipe.

Similar provisions and practices are applicable to materials used for house drains and house sewers.

Where a sewer is to be laid at less distance than 10 ft from the exterior wall of a building, in bad ground, on a poor foundation, or where it will be subjected to vibration or to settlement, cast-iron pipe should be used.

[1] Sec. 13.2.

Where the ground is of sufficient solidity for a proper foundation or where special supports or a secure foundation are provided, vitrified-clay or concrete pipe or other substantial material may be used.

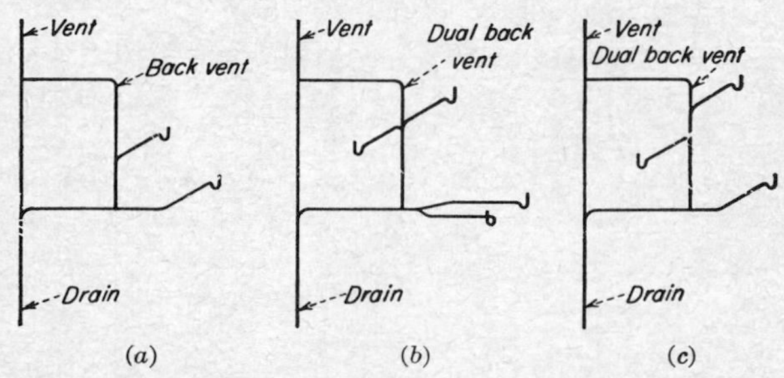

FIG. 11-13. Group vents for lavatories and bathtubs.

11-15. Types of Vent. Types of vent are defined in Appendix I and are shown in Figs. 11-13 to 11-17. The *continuous* vent, the *circuit* vent, and the *loop* vent are widely used types of vent. A circuit vent is shown in Fig. 11-18, and a loop vent is shown in Fig. 11-19. A continuous vent to a single fixture is shown in Fig. 11-20, and loop or circuit venting to a number of fixtures is shown in Figs. 11-6, 11-13 to 11-18, and 11-21. Continuous venting is probably the most satisfactory method of installing vents, since it is equally effective against back pressure, siphonage, and self-siphonage. It has the advantage of greater effectiveness in venting, minimum evaporative effect, and relative simplicity of installation. It is to be noted that the piping arrangement shown in Fig. 11-16 can be used on two floors only, since the addition of another floor would discharge waste into one of the vent pipes leading from a lower floor.

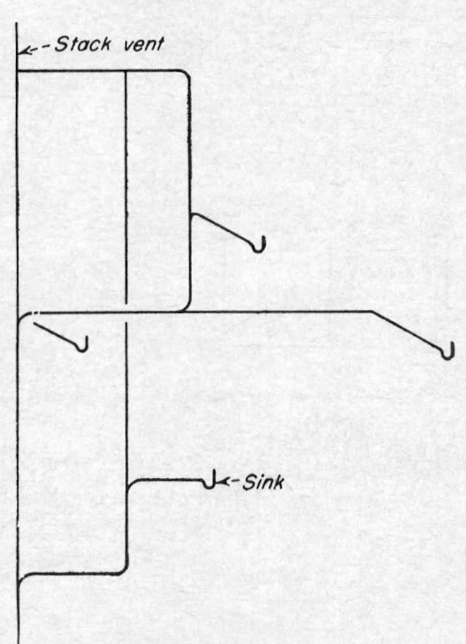

FIG. 11-14. Vent for sink.

Loop venting is illustrated for a battery of fixtures in Figs. 11-6, 11-13, 11-19, and others. Since waste water flows through the pipe that acts also as a vent pipe, with resulting possibilities of clogging, restrictions are placed on loop-vent designs. Some authorities disapprove of loop venting under all but the most exceptional circumstances. The use of loop venting is usually restricted to a battery of fixtures discharging into a branch drainage pipe with a slope of

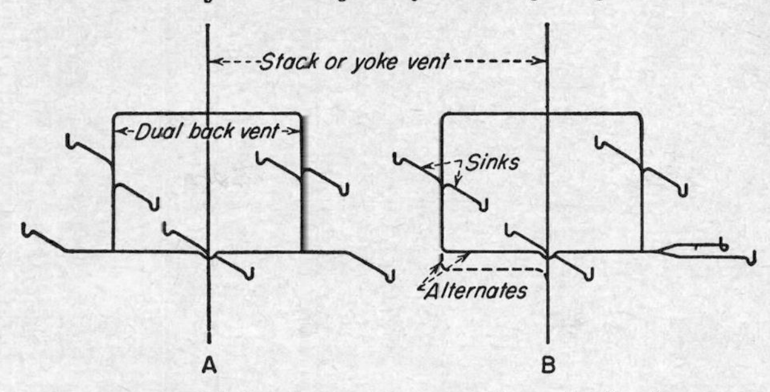

FIG. 11-15. Piping layout for one-story duplex house.

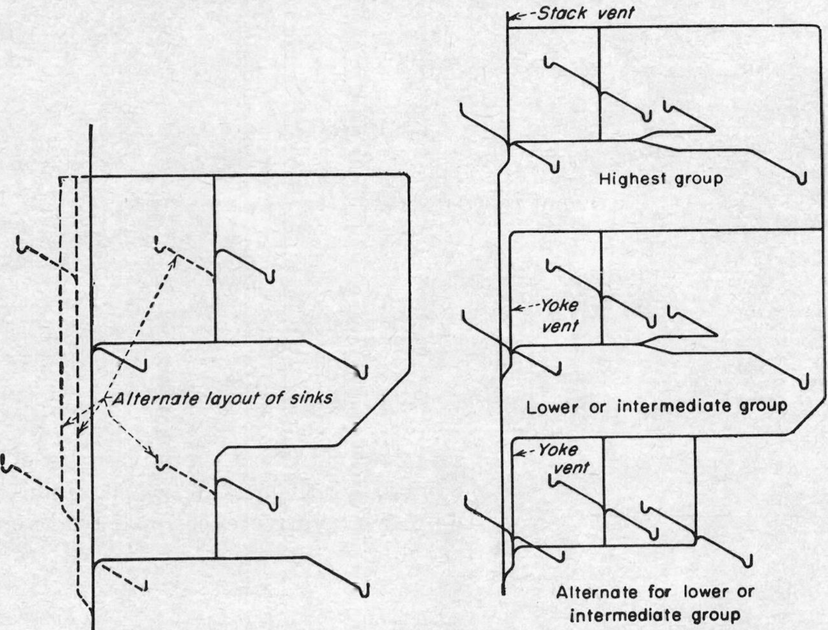

FIG. 11-16. Piping layout for bathrooms in each of two stories.

FIG. 11-17. Piping layouts for duplex apartments.

TABLE 11-9. CONDITIONS UNDER WHICH LOOP OR WET VENTING IS PERMISSIBLE

Diameter of horizontal waste or soil pipe, in.	Maximum number of fixture units when loop venting is used	Diameter of horizontal waste or soil pipe, in.	Maximum number of fixture units when loop venting is used
$1\frac{1}{2}$	2	5	30
2	3	6	72
3	9	8	210
4	18	10	500

not less than 1/4 in. per ft toward a stack; the number of fixtures discharging into the branch is limited, as shown in Table 11-9.

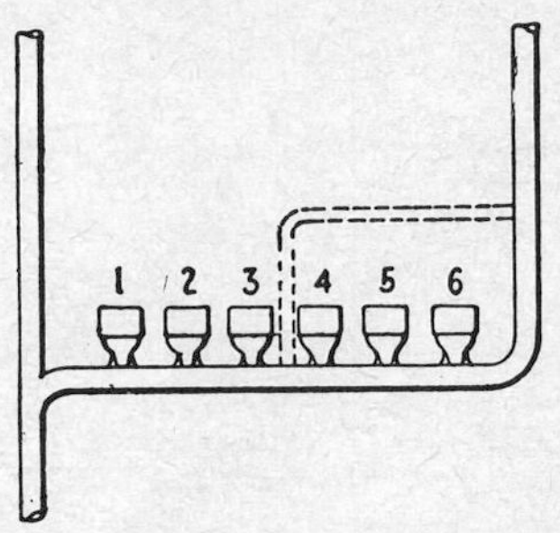

FIG. 11-18. Circuit venting.

The "National Plumbing Code" states:[1]

A branch soil or waste pipe to which two but not more than eight water closets (except blowout type), pedestal urinals, trap standard to floor, shower stalls, or floor drains are connected in battery, shall be vented by a circuit or loop vent which shall take off in front of the last fixture connection. In addition, lower-floor branches serving more than three water closets shall be provided with a relief vent taken off in front of the first fixture connection. When lavatories or similar fixtures discharge above such branches, each vertical branch shall be provided with a continuous vent.

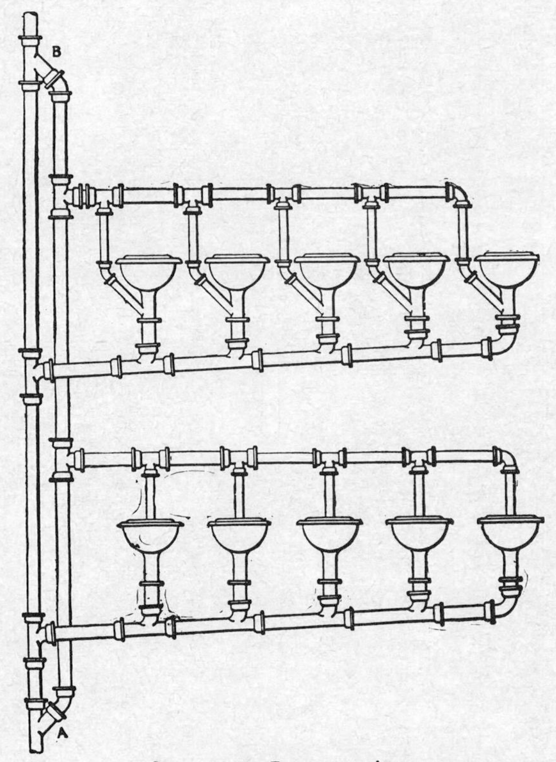

FIG. 11-19. Loop venting.

12.15.2. When parallel horizontal branches serve a total of eight water closets (four on each branch), each branch shall be provided with a relief vent at a

[1] Sec. 12.15.1.

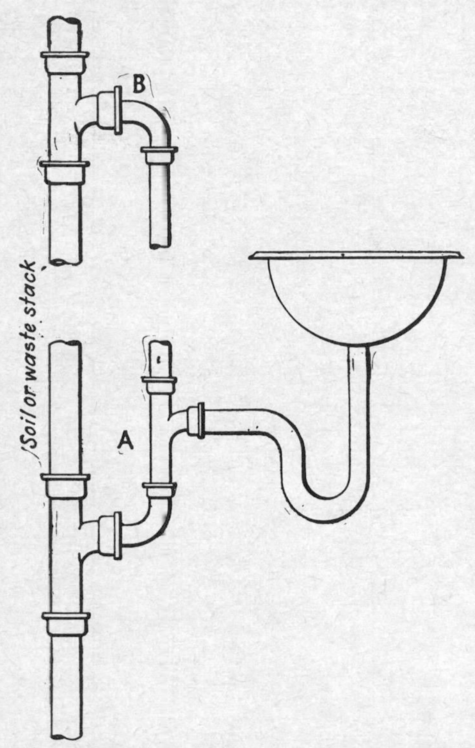

FIG. 11-20. Back vent and continuous vent.

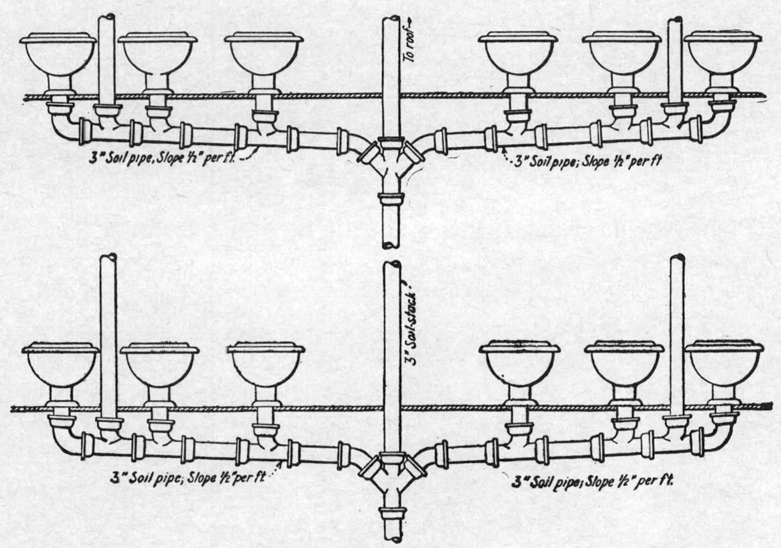

FIG. 11-21. Loop or circuit vent.

point between the two most distant water closets. Where other fixtures (than water closets) discharge above the horizontal branch, each fixture shall be vented.

12.15.3. When the circuit, loop, or relief vent connections are taken off the horizontal branch, the vent branch connection shall be taken off at a vertical angle from the top of the horizontal branch.

12.15.4. When fixtures are connected to one horizontal branch through a double wye or a sanitary tee in a vertical position, a common vent for each two fixtures back to back or double connection shall be provided. The common vent shall be installed in a vertical position as a continuation of the double connection.

In general, the fixture connection in loop venting, particularly for water closets, should be made at the side rather than at the top of the branch drainage pipe in order to avoid impeding the flow of air in the branch pipe. A vent should be taken off from the branch pipe between the two fixtures farthest away from the stack, as shown in Figs. 11-21 and 12-2 and on the two top floors in Fig. 11-4. It is desirable to have one fixture connected to the branch drainage pipe above the take-off on the circuit vent pipe, as shown in Fig. 11-22, in order that any obstruction falling into the branch pipe can be washed away by the discharge from the last fixture. Circuit venting under the conditions shown by the full lines in Fig. 11-18 is undesirable because if fixtures 4, 5, and 6 are discharged simultaneously, there is no vent for fixtures 1, 2, and 3, and their seals may be broken by siphonage. If fixtures 3, 4, and 6 are discharged simultaneously, the seal in fixture 5 may be blown by back pressure. The difficulties can be overcome by installing the vents as shown by the dotted lines and the full lines in the figure.

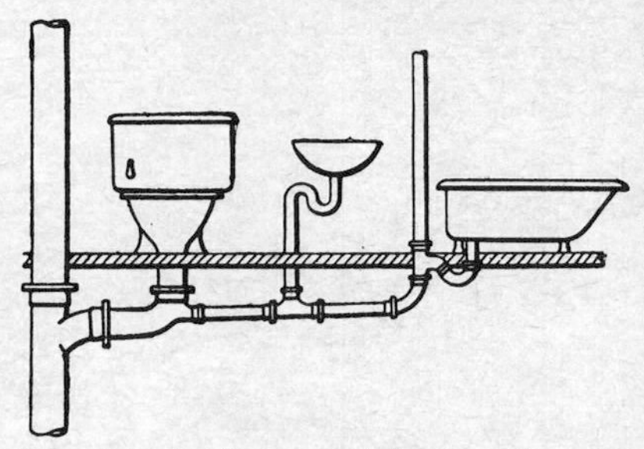

Fig. 11-22. An unsatisfactory installation of a wet vent. The lavatory is not a "resealing" fixture.

Where fixtures discharge into a soil or waste stack above its connection with a horizontal soil or waste pipe to which more than one fixture is connected and which is used as a circuit or loop vent, the horizontal soil or waste pipe that is used as a circuit or loop vent should be provided with a relief vent taken off between the soil or waste stack and the nearest fixture connection, as shown at *B* on the third floor in Fig. 11-4.

Where loop or circuit vents are used on different floors and such vents connected into the same vent stack, the connection between the branch vent, on all but the lowest floor, and the vent stack shall be made at a point higher than the fixtures on each floor, as shown at *H* in Fig. 11-4.

Loop and circuit venting is applicable principally to a battery of fixtures, especially fixtures that are free from self-siphonage, such as those with resealing traps, as shown in Fig. 11-23. Loop venting protects against siphonage and back pressure, and because of the relatively short length of piping usually involved, it may be simpler and less costly. In the event that a fixture without a resealing trap, such as a lavatory, is placed in a battery of loop-vented fixtures, the lavatory or other such fixture should be served by a continuous vent, as shown in Fig. 11-24.

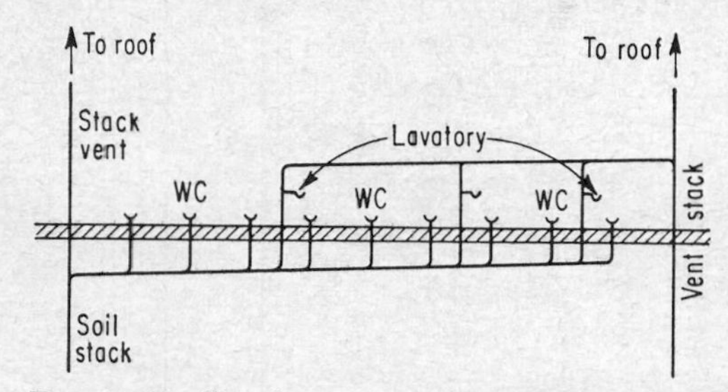

Fig. 11-23. Circuit venting with resealing traps.

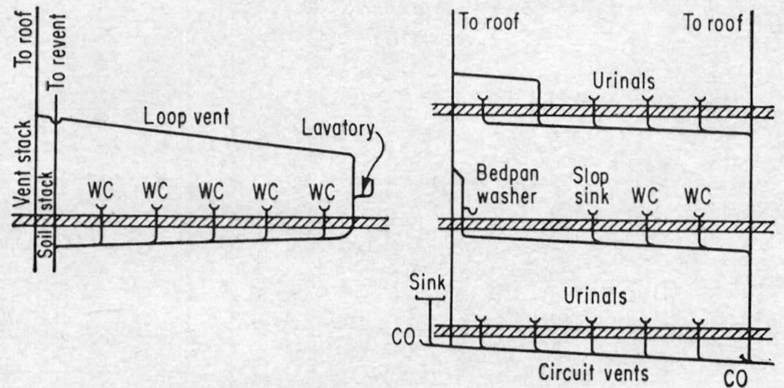

Fig. 11-24. Loop and continuous vents.

Circuit, loop, and continuous venting have their proper uses, but continuous venting is most effective where it can be used.

A *relief vent* is used primarily as a supplementary vent in loop or circuit venting. The "Plumbing Manual"[1] and some codes require the installation of a relief vent between the first fixture nearest the soil stack and the soil stack.

A *wet vent* is illustrated at C' on the top floor in Fig. 11-5 and in Fig. 11-22. A wet vent may properly be used where only relatively clean water flows through the vent pipe, where the simultaneous use of fixtures on the same drainage pipe is improbable, and where the trap so vented is a resealing trap. Water-closet discharge may not properly flow through a wet vent. A satisfactory wet-vent installation is shown in Fig. 11-5.

[1] Report BMS66, Subcommittee on Plumbing, National Bureau of Standards, 1940.

The installation shown in Fig. 11-22 is unsatisfactory because the lavatory is not a resealing fixture. A wet vent is not permissible on the first floor in Fig. 11-5 because of the possibility of the siphonage of the fixture at F by the discharge of the fixture at E. The use of a wet vent can, under some conditions, save expense in plumbing.

The "National Plumbing Code" states:

12.12.1. Single Bathroom Groups. A single bathroom group of fixtures may be installed with the drain from a back-vented lavatory, kitchen sink, or combination fixture serving as a wet vent for a bathtub or shower stall and for the water closet, provided that:

a. Not more than one fixture unit is drained into a ½-in. diameter wet vent or not more than four fixture units drain into a 2-in. diameter wet vent.

b. The horizontal branch connects to the stack at the same level as the water-closet drain or below the water-closet drain when installed on the top floor. It may also connect to the water-closet bend.

12.12.2. Double Bath. Bathroom groups, back-to-back on top floor, consisting of two lavatories and two bathtubs or shower stalls may be installed on the same horizontal branch with a common vent for the lavatories and with no back vent for the bathtubs or shower stalls and for the water closets, provided the wet vent is 2 in. in diameter, and the length of the fixture drain conforms to Table 11-10.

TABLE 11-10. DISTANCE OF FIXTURE TRAP FROM VENT
(From "National Plumbing Code," Sec. 12.9.3)

Size of drain, in.	1¼	1½	2	3	4
Distance from trap to vent, in.	30	42	60	72	120

12.12.3. Multistory Bathroom Groups. On the lower floors of a multistory building, the waste pipe from one or two lavatories may be used as a wet vent for one or two bathtubs or showers provided that:

a. The wet vent and its extension to the vent stack is 2 in. in diameter.

b. Each water closet below the top floor is individually back-vented.

c. The vent stack is sized as given in Table 11-11.

12.12.4. Exception. In multistory bathroom groups, wet vented in accordance with Par. 12.12.3, the water closets below the top floor need not be individually vented if the 2-in. waste connects directly into the water-closet bend at a 45° angle to the horizontal portion of the bend in the direction of flow.

It must be emphasized that wastes containing soil, grease, and large suspended solids should not flow through a wet vent.

A *crown vent* is a vent connected to a fixture waste pipe on top of the crown weir of the trap, as shown in Fig. 11-25. The crown vent differs from the continuous vent shown in Fig. 11-20 in the position of the connection between the vent pipe and the fixture branch. Because of its close proximity to the seal of the trap, it might be expected that such a vent would be more effective, but tests of the relative efficiencies of

TABLE 11-11. SIZES AND LENGTHS OF VENTS
("National Plumbing Code," Sec. 12.21.5)

Size of soil or waste stack, in.	Fixture units connected	Diameter of vent required, in.								
		1¼	1½	2	2½	3	4	5	6	8
		Maximum length of vent, ft								
1¼	2	30								
1½	8	50	150							
1½	10	30	100							
2	12	30	75	200						
2	20	26	50	150						
2½	42	...	30	100	300					
3	10	...	30	100	200	600				
3	30	...	...	60	200	500				
3	60	...	...	50	80	400				
4	100	...	...	35	100	260	1,000			
4	200	...	...	30	90	250	900			
4	500	...	...	20	70	180	700			
5	200	...	...	...	35	80	350	1,000		
5	500	...	...	...	30	70	300	900		
5	1,100	...	...	...	20	50	200	700		
6	350	...	...	...	25	50	200	400	1,300	
6	620	...	...	...	15	30	125	300	1,100	
6	960	...	...	...	...	24	100	250	1,000	
6	1,900	...	...	...	...	20	70	200	700	
8	600	...	...	...	...	...	50	150	500	1,300
8	1,400	...	...	...	...	...	40	100	400	1,200
8	2,200	...	...	...	...	...	30	80	350	1,100
8	3,600	...	...	...	...	...	25	60	250	800
10	1,000	...	...	...	...	...	.....	75	125	1,000
10	2,500	...	...	...	...	...	.....	50	100	500
10	3,800	...	...	...	...	...	.....	30	80	350
10	5,600	...	...	...	...	...	.....	25	60	250

continuous and loop vents show no pneumatic advantage of one over the other. Crown vents, however, are almost universally prohibited by plumbing codes because experience has shown that grease, hair, and other clogging materials splash into the vent pipe to accumulate there and finally close the vent and because of the proximity of the vent to the water surface there is some evidence that evaporation is increased.

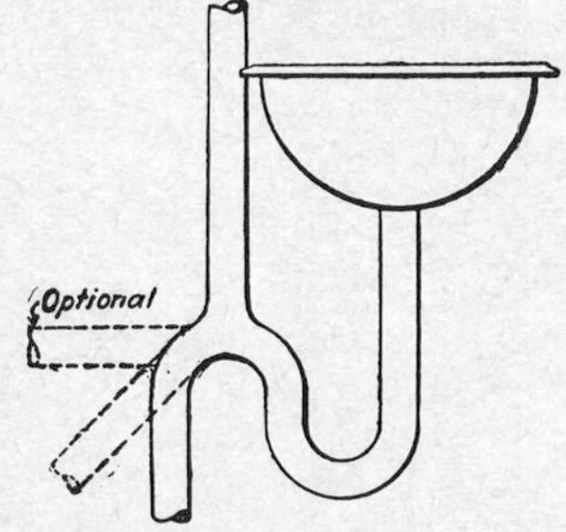

FIG. 11-25. Crown vent.

A *unit vent* or a *dual vent* is a single-vent pipe which serves two traps, as shown in Fig. 11-26*a*.

It is sometimes called a double waste-and-vent. An erroneous installation of a unit vent, in which self-siphonage may occur, is shown in Fig. 11-26*b*. The "National Plumbing Code" (Sec. 12.9.1) restricts the installation of a unit vent as follows:

A common vent may be used for two fixtures set on the same floor level but connecting at different levels in the stack, provided the vertical drain is one pipe diameter larger than the upper fixture drain but in no case smaller than the lower fixture drain, whichever is the larger and that both drains conform to Table 11-10.

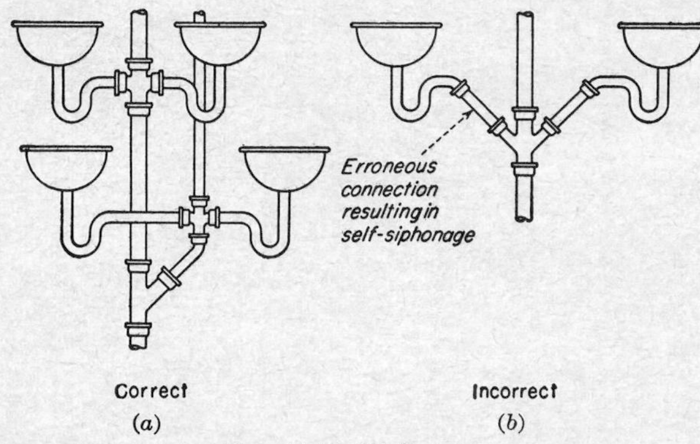

Fig. 11-26. Unit vents.

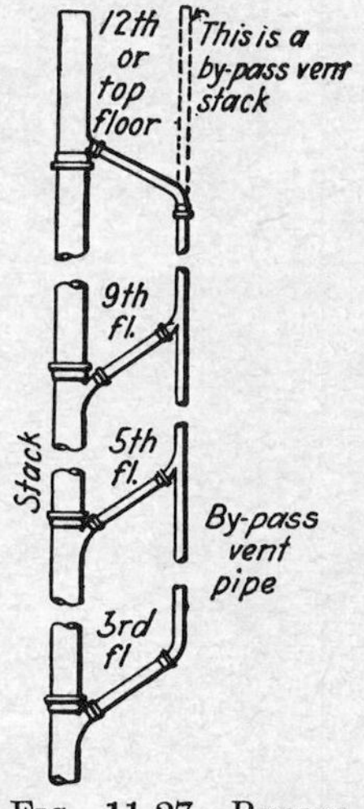

Fig. 11-27. Bypass vent.

A *bypass vent* is illustrated in Fig. 11-27. It is a vent stack parallel to a soil or waste stack with frequent connections at branch intervals between the two stacks. The helpful effect of such a vent in reducing pressures and siphonage in a tall stack is remarkable. Because of its connection between various parts of a stack, higher pressures are quickly transferred to regions of lower pressure, or vice versa, with consequent neutralization of both.

Utility vents are illustrated in Fig. 11-28. A utility-vent pipe loops above the overflow level of the vented fixture and is connected to a waste or soil stack at a lower elevation. The utility vent is used where the trap to be vented is located so that continuous or loop venting cannot be used. The use of a utility vent is undesirable where it can be avoided because such vents have the undesirable features of a wet vent.

11-16. Vents for Special Conditions. Most plumbing fixtures are set against a wall in which drainage and vent pipes can be concealed. However, fixtures such as laboratory sinks, operating tables, and dental lavatories can be set away from nearby walls, requiring special con-

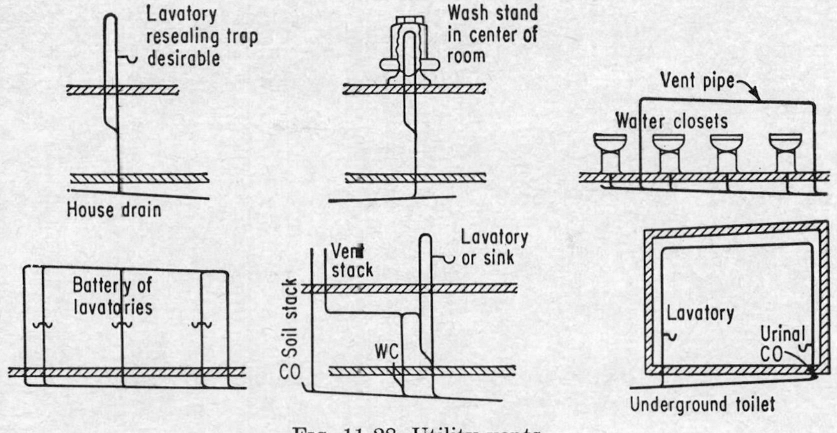

FIG. 11-28. Utility vents.

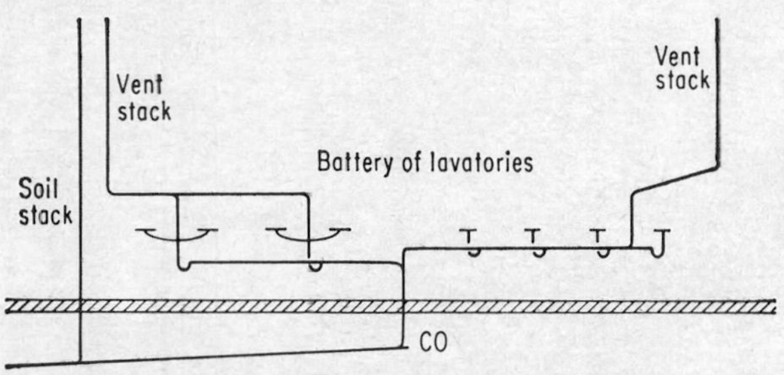

FIG. 11-29. Special installations of piping.

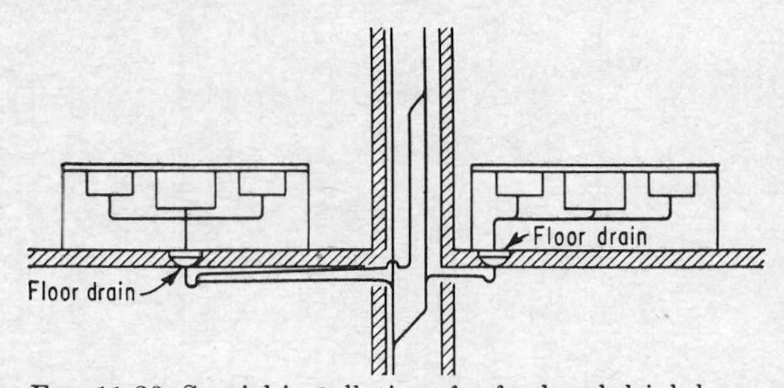

FIG. 11-30. Special installations for food and drink bars.

nections for vent and drain pipes. Some expedients are illustrated in Fig. 11-29. Under other conditions pipes may be badly located and other expedients called for. Some are shown in Figs. 11-29, 11-30, and 11-31.

11-17. Materials for Vent Pipes. Materials used for vent pipes are limited to cast iron, galvanized wrought iron or steel, ferrous alloys, lead, brass, and copper. Underground vents should be of cast iron, but if pipes of other material are used with threaded joints, the joints should be coated and wrapped after installation and test.[1] The "National Plumbing Code" (Sec. 12.1.7) states that other materials may be used for vent piping "when approved as such by the Administrative Authority."

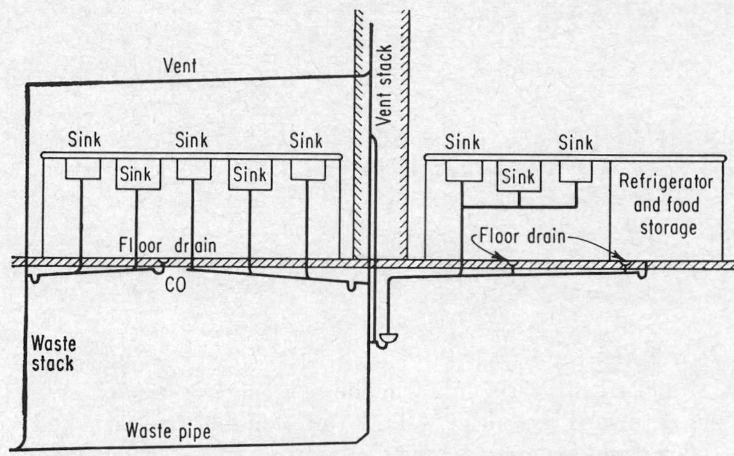

FIG. 11-31. Special food bar, refrigerator, and storage installations.

11-18. Distance of Vent Connection from Trap Seal. The best location of a vent is at the continuation of the vertical portion of the fixture branch, provided that it is within 48 pipe diameters, horizontally projected length, from the crown weir of the trap. The vertical distance should be such that the vent connection is[2] "... within the hydraulic gradient between the trap outlet and vent connection ...." The vent connection should not be placed closer than 2 pipe diameters horizontally from the crown weir nor more than 1 pipe diameter below it. The limitations are illustrated in Fig. 11-32. An illustration of a violation of the rule is shown in Fig. 11-26*b*. The "National Plumbing Code" (Sec. 12.9.1) states: "... the slope and developed length in the fixture drain from the trap weir to the vent fitting are within the requirements set forth in Table 11-10."

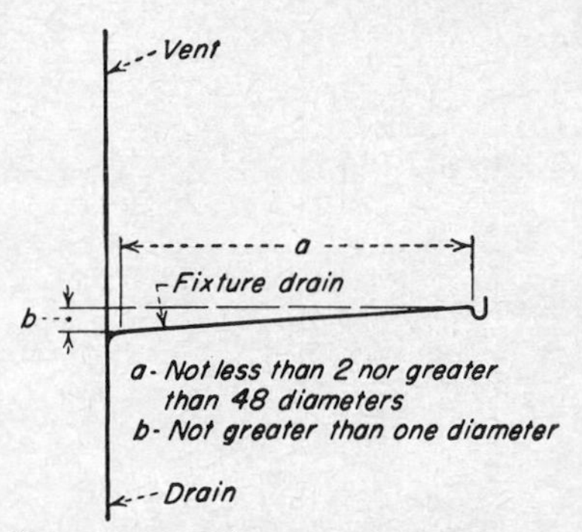

FIG. 11-32. Distance of trap from vent.

An exception to the foregoing rule applies to flat-bottom fixtures and to fixtures with resealing traps, such as water closets, whose functions

[1] "National Plumbing Code," Sec. 12.1.4.

[2] *Ibid.*, Sec. 12.11.1.

depend on the self-siphonage of the trap. Vents for such traps are discussed in Sec. 11-19.

Distances greater than those permitted are conducive to siphonage, self-siphonage, and reduction of the effectiveness of the vent. The smallest distance, within the limits stated, is desirable. A distance smaller than permissible tends toward the creation of the objections to a crown vent.

11-19. Unvented and Resealing Traps. Some fixtures are so designed that after the greater part of the body of waste water is discharged from the fixture, additional waste water is discharged at a rate too slow to cause self-siphonage. A water closet is an example of such a fixture, as explained in Sec. 15-29. Flat-bottom fixtures, such as bathtubs, kitchen sinks, and laundry trays, also discharge so that the last portion of the waste water running from the fixture will restore the seal of a self-siphoned trap. It is required that to restore the seal satisfactorily the fixture shall have not less than 200 sq in. of flat area in the bottom, sloping at not more than $\frac{1}{8}$ in. per ft toward the outlet. Traps on such fixtures are called *resealing traps*. Tests and experience have shown that rules for venting, as stated in Sec. 11-18, can be modified for resealing traps. The following has been taken mainly from the "Plumbing Manual":[1]

> The trap and fixture drain, not exceeding 2 in. in diameter, from a single flat-bottom fixture or from a group of not more than three such fixtures may be installed with a vertical section of the fixture drain not exceeding 24 pipe diameters in length at a distance not exceeding 10 pipe diameters from the crown weir of the trap, with a total length of sloping drain not exceeding 72 pipe diameters. It may be installed without vent, provided that the fixture is the highest on the stack or on a yoke-vented vertical extension of a horizontal branch. Examples of such installations are shown in Fig. 11-5.

Although resealing traps are immune from loss of seal by self-siphonage and are resistant to loss of seal by back pressure, their resealing property gives no resistance to loss of seal by siphonage. Only venting can give such protection. Hence, venting requirements should be applied to any fixture exposed to seal loss by siphonage. For example, the water closets in Fig. 11-5 are not vented, whereas those shown in Fig. 11-6 are vented.

Vents are not required for traps on rain-water leaders, back-water traps, or subsoil catch-basin traps, nor are they required on cellar-floor or area drains, provided that the drain connects with the building drain or a branch thereof at least 5 ft downstream from any soil or waste stack, that the length and fall of floor or area drain are within the limits of Sec. 11-18, that the load on the building drain or any of its branches does

[1] "Plumbing Manual," Report BMS66, National Bureau of Standards, 1940.

not exceed the limits given in Tables 11-5 and 11-6, and that the building drain is not subject to backwater effects.

The "National Plumbing Code" states conditions, in detail, under which it is permissible to install traps without individual vents. The Code states (Sec. 12.14.1) as follows:

One sink and one lavatory, or three lavatories within 8 ft developed length of a main-vented line may be installed on a 2-in. horizontal waste branch without reventing, provided the branch is not less than 2 in. in diameter throughout its length, and provided the wastes are connected into the side of the branch and the branch leads to its stack connection with a pitch of not more than ¾ in. per ft.

12.14.2. When fixtures other than water closets discharge downstream from a water closet, each fixture connection downstream shall be individually vented.

12.14.3. A fixture or a combination of fixtures whose total discharge rating is not more than three fixture units may discharge into a stack not less than 3 in. in diameter without reventing, provided such fixture connections are made above the connection on to the next highest water closet, or bathtub tee-wye, and the fixture unit rating of the stack is not otherwise exceeded, and their waste piping is installed as otherwise required in Par. 11.14.1.

12.13.1. . . . a group of fixtures, consisting of one bathroom group and a kitchen sink or combination fixture, may be installed without individual fixture vents, in a one-story building or on the top floor of a building, provided each fixture drain connects independently to the stack and the water closet and bathtub or shower-stall drain enters the stack at the same level and in accordance with the requirements in Table 11-12, except that

12.13.2. When a sink or a combination fixture connects to the stack-vented bathroom group, and when the street sewer is sufficiently overloaded to cause frequent submersion of the building sewer, a relief vent or back-vented fixture shall connect to the stack below the stack-vented water closet or bathtub.

The Code (Sec. 12.7.1) permits the installation of unvented traps serving sinks which are part of the equipment of bars, soda fountains, and counters where it is impossible to install a vent. It states: "When such conditions exist such sinks shall discharge into a floor sink or hopper which is properly trapped and vented."

11-20. Sizes of Vent Stacks. Rules for the determination of sizes of vent pipes are difficult to formulate because of the variety and number of conditions involved. In most of those plumbing codes in which sizes of vent pipes are specified, this difficulty is overcome by fixing arbitrary or empirical sizes for specific conditions. Although not necessarily the smallest sizes that will serve for the conditions, they give convenient sizes that are supposedly sufficiently large to control any pressure which is likely to develop. For example, the "National Plumbing Code" states that where a building drain is installed, there shall be at least one stack vent or vent stack carried full size through the roof not less than 3 in. in diameter or the size of the building drain, whichever is the smaller.

Tests to determine adequate vent sizes have been conducted at the National Bureau of Standards and at the University of Illinois. The results of these tests at the Bureau of Standards are to be found in the Hoover Report,[1] and the results of the tests at the University of Illinois

TABLE 11-12. NUMBERS AND LIMITS OF FIXTURE UNITS ON BRANCHES AND STACKS*

Diameter of pipe, in.	Fixture units on one branch	Fixture units on one stack	Limits in fixture units		
			In any branch interval for		Maximum on one stack
			One branch interval only, N	Two or more branch intervals, $N/2n + N/4$†	
(1)	(2)	(3)	(4)	(5)	(6)
1¼	1	2	1	1	2
1½	3	4	3	2	8
2	6	10	6	6	24
3	.....	.....	32	$16/n + 8$	80
3‡	32	48			
3§	20	30			
4	160	240	240	$120/n + 60$	600
5	360	540	540	$270/n + 135$	1,500
6	640	960	960	$480/n + 240$	2,800
8	1,200	2,240	1,800	$900/n + 450$	5,400
10	1,800	3,780	2,700	$1{,}350/n + 675$	8,000
12	2,800	6,000	4,200	$2{,}100/n + 1{,}050$	14,000

* From "Plumbing Manual," Report BMS66, pp. 15 and 47, National Bureau of Standards, 1940.

† N is the permissible number of fixture units for a stack having one branch interval only, and n is the number of branch intervals.

‡ Waste only.

§ Soil.

NOTES:

If the total fixture units are distributed on horizontal branches in three or more branch intervals (see definition in Appendix I) of the stack, the total number of fixture units on a straight soil or waste stack of a given diameter may be increased within the maximum limits given in the table.

Columns (2) and (3) are particularly applicable for buildings of one or two stories. The table may be applied safely, but not economically, to buildings of any height. Columns (4), (5), and (6) are especially applicable to taller buildings and to plumbing systems with relatively small horizontal branches.

It is essential in applying columns (2) and (3) that the number of fixture units in any one branch interval shall be in accordance with the quantity ($N/2n + N/4$), where n is the number of branch intervals and N is the permissible number of fixture units for a stack having one branch interval only, and that the total number of fixture units on the entire stack shall be within the limits of column (6).

are to be found in *Engineering Experiment Station Bulletins* 143 and 178. Subsequent to the publication of these reports, the "Plumbing Manual,"[2]

[1] "Recommended Minimum Requirements for Plumbing in Dwellings and Similar Buildings," National Bureau of Standards, July 3, 1923.

[2] "Plumbing Manual," Report BMS66, National Bureau of Standards, 1940.

the "Housing Code,"[1] and the "National Plumbing Code"[2] have been published. Some recommendations concerning the lengths of vents by the "Plumbing Manual" and the Housing Authority are given in Tables 11-11 and 11-12.

11-21. Sizes and Lengths of Vents. The "National Plumbing Code" states:

12.21.1. Length of Vent Stacks. The length of the vent stack or main vent shall be its developed length from the lowest connection with the soil stack, waste stack, or building drain to the vent stack terminal, if it terminates separately in the open air, or to the connection of the vent stack with the stack vent, plus the developed length of the stack vent from the connection to the terminal in the open air, if the two vents are connected with a single extension to the open air.

12.21.2. Size of Individual Vents. The diameter of an individual vent shall be not less than 1¼ in. nor less than one-half the diameter of the drain to which it is connected.

12.21.3. Size of Relief Vent. The diameter of a relief vent shall be not less than one-half the diameter of the soil or waste branch to which it is connected.

12.21.4. Size of Circuit or Loop Vent. The diameter of a circuit or loop vent shall be not less than one-half the size of the diameter of the horizontal soil or waste branch or the diameter of the vent stack, whichever is smaller.

12.21.5. The nominal size of vent piping shall be determined from its length and the total of fixture units connected thereto, as provided in Table 11-13. Twenty per cent of the total length may be installed in a horizontal position.

The "Plumbing Manual"[3] states:

Sec. 1014. Size and Length of Stack Vents. Stack vents shall be of the same diameter as the soil or waste stack, if the soil or waste stack carries one-half or more of its permissible load, or has horizontal branches in more than two branch intervals. If the soil or waste stack carries less than one-half its permissible load and has horizontal branches in not more than two branch intervals, the stack vent may be of a diameter not less and a length not greater than required by Table 11-13.

11-22. Relief Vents for Stacks. The "National Plumbing Code" states:

12.17.1. Stacks of More than 10 Branch Intervals. Soil and waste stacks in buildings having more than 10 branch intervals shall be provided with a relief vent at each tenth interval installed, beginning with the top floor. The size of the relief vent shall be equal to the size of the vent stack to which it connects. The lower end of each relief vent shall connect to the soil or waste stack through

[1] "The Uniform Plumbing Code for Housing," Housing and Home Finance Agency, February, 1948.

[2] "National Plumbing Code," ASA A 40.8–1955, American Society of Mechanical Engineers, 1955.

[3] *Op. cit.*

a wye below the horizontal branch serving the floor and the upper end shall connect to the vent stack through a wye not less than 3 ft above the floor level.

11-23. Sizes of Branch Vents. The "Plumbing Manual"[1] states (p. 21)

Sec. 1015. Size of Back Vents and Relief Vents. The nominal diameter of a (branch or) back vent, when required, shall be not less than 1¼ in. nor less than one-half the diameter of the drain to which it is connected; and under the conditions that require a relief vent for approved forms of group venting, the sum of the cross sections of all vents installed on the horizontal branches in one branch

TABLE 11-13. LIMITS FOR CIRCUIT AND LOOP VENTING*

Diameter of horizontal branch, in. (1)	Water closets, pedestal urinals, or trap-standard fixtures (2)	Fixture units for fixtures other than designated in column (2). (3)
2	none	6
3	2	20
4	8	60
5	16	120
6	24	180

* From "Plumbing Manual," Report BMS66, p. 20, National Bureau of Standards. 1940.

interval shall be at least equal to that of either the main vent or the largest horizontal branch in the branch interval.

Recommended sizes of individual vents are given in Table 11-11.

The "Housing Code"[1] states (p. 46) with respect to the length of branch vents:

10.13.2. Length of Branch Vent. The length of a branch vent shall be the developed length from its connection with the vent stack or stack vent to the fixture drain or horizontal soil or waste branch served by the branch vent.

11-24. Sizes of Circuit and Loop Vents. Limits for the sizes of circuit and loop vents are given in Table 11-13; limitations concerning their installation are given in Sec. 12-7; and their sizes are specified on page 21 of the "Plumbing Manual" as follows:

Sec. 1016. Size of Circuit and Loop Vents. *a.* The nominal diameter of a circuit or loop vent and the first relief vent as required by Sec. 233 shall be not less than one-half the diameter of the horizontal branch thus vented. Under the conditions that require a relief vent the sum of the cross sections of the circuit or loop and relief vents shall be at least equal to that of either the main vent required

[1] *Op. cit.*

or the horizontal branch. In determining the sum of cross sections for this requirement all relief vents connected to the horizontal branch may be included.

b. Additional relief vents, installed in compliance with Sec. 233, shall be not less in diameter than one-half that of the largest fixture branch connected to the horizontal branch.

A rule of thumb that meets most plumbing codes allows a 3-in. loop vent when the horizontal drainage pipe is 20 ft or less in length and draws four or fewer water closets or equivalent fixture units. Where more than four water closets or equivalent fixture units are served, the loop vent should be of the same diameter as the drainage pipe. A relief vent may be provided between every group of five or six fixtures, the diameter of the relief vent being equal to that of the drainage pipe.

11-25. Vent-pipe Grades and Connections. Drainage and vent pipes shall be installed in as direct and straight an alignment as possible. All vent and branch vent pipes should be free from drops or sags and should be graded and connected so as to drop back to the soil or waste pipe by gravity. All vent lines should be connected at the bottom with a soil or waste pipe in such a manner as to prevent the accumulation of rust, scale, or condensation in the bottom of the vent stack. Where vent pipes connect to a horizontal soil or waste pipe, the vent branch should be taken off above the center line of the pipe and, where possible, the vent pipe should rise vertically or at an angle of 45° or more from the horizontal to a point at least 6 in. above the flood level of the fixture it is venting before offsetting horizontally or connecting to the branch, main waste, or soil vent.

It is important that the upper end of a vent pipe shall be 6 in. or more above the flood level of the highest fixture vented. Otherwise, in the event of stoppage of the drain pipes, waste water would drain through the vent pipes, the stoppage of the drainage system would remain undetected, and the function of the vent pipes would be impeded.

The steep slope of the vent pipe is desirable to assure the dropping of any obstruction from the vent pipe into the drain pipe. Dips, humps, sharp changes in direction, negative slopes, and flat runs are conducive to clogging.

The connection of a revent with a stack can be made above the highest fixture and can be made by means of an inverted wye, as shown in Fig. 11-19, or a sanitary tee, as shown in Fig. 11-20. Where a bypass vent is constructed, the vent connection should be made as shown in Fig. 11-27 to prevent water from the soil or waste stack from entering the bypass-vent stack. No vent pipe should connect into a drain pipe on the building side of a trap, since this will provide an untrapped connection between the building and the sewer.

Where it is not possible to extend the vent above the fixture and the vent must be laid at an angle of less than 45° from the horizontal, a wet vent through which no human excrement will be discharged should be used or some provision should be made for flushing the vent frequently.

Where the bottom of a vent stack connects with any other pipe and no provision is made for automatically flushing the bottom portion of the vent stack, the unflushed portion of the vent stack should slope 45° or more from the horizontal. The purpose of this restriction is to prevent clogging at the base of the vent stack or pipe. To provide for flushing the base of the vent stack, fixtures can be connected at its base as described in Sec. 12-9. All main vents or vent stacks should connect, full size, at their base to the main soil or waste pipe at or below the lowest fixture branch and should extend undiminished in size above the roof or be reconnected with the main soil or waste stack above the highest fixture, as shown in Figs. 11-19 and 11-20. Connection above the flood level of the highest fixture is required to avoid the discharge of waste water into the vent pipes. It is required that the vent stack or the soil stack to which it is connected shall be carried full size through and terminate above the roof (1) to provide fresh air to the vent and drainage pipes, (2) to avoid dissemination of odors by terminating pipes within the building or below the roof, and (3) to minimize clogging which might be induced by diminishing the size of the vent pipe or roof stack.

When offsets or changes in direction are necessary, they should be made, where possible, at an angle of not less than 45°. Quarter bends and sanitary types of fitting should be used only to change from the horizontal to the vertical direction. All other changes in direction should be made by the use of appropriate "eighth" or "sixteenth" bends, wyes, half-wyes, wye-and-one-eighth bends, or similar types of fitting or connection in which the change of direction is gradual. Where changes in the size of pipe are necessary, they should be made with reducing or increasing fittings. These fittings should be pitched at an angle of not more than 45°, with the center line of the pipe between the two sizes.

Waste and soil pipes should not decrease in diameter in the direction of the flow of water, and vent pipes should not decrease in diameter in the direction from a waste or soil pipe to their connection or opening into the free air.

Piping should never be laid horizontal. So-called "horizontal" pipe should be laid at a uniform grade, preferably about 1/4 in. per ft and under no conditions at a grade of less than 1/8 in. per ft. Other requirements with regard to the grades and connections of vent pipes are stated in Chap. 12.

11-26. Fresh-air Inlets. Fresh-air inlets are used primarily in connection with a main trap or house trap or where the outlet of the house sewer may become submerged. A fresh-air inlet ventilates the plumbing system and prevents the development of excessive back pressure. The fresh-air inlet should not be used as a soil or waste stack, as a rain-water leader, or otherwise to convey liquids, in order that there may be no interference with its primary function. Fresh-air inlets are shown at points 79, 164, and 167 at the front of the first floor in Fig. 11-1. Where a main trap is used, the fresh-air inlet must be connected to the building drain above the main trap so as not to interfere with the cleanouts therein. The connection of the fresh-air inlet to the house sewer below the main trap would serve only to ventilate the common sewer and would serve no useful purpose for the plumbing system. Its connection to both sides of the main trap would render the main trap useless. In cold climates the fresh-air inlet should join the building drain at least 15 ft upstream from the main trap to avoid freezing of the water in the trap. Freezing in cold climates and evaporation in any climate are induced by fresh-air inlets. If the fresh-air inlet is used where the house-sewer outlet may become submerged and there is no main trap, the lower end of the inlet should be connected to the building drain or building sewer above the highest expected point of backup.

The upper end of the fresh-air inlet may terminate at the ground surface, or it may be carried to the roof. In closely built-up communities it is necessary to carry the inlet to the roof. If extended to the roof, it should be installed and protected as provided for vent stack terminals, as stated in Sec. 12-11. It is less desirable but much less expensive to terminate the upper end of the fresh-air inlet at or near the ground surface. If it is so terminated, the opening should face down, should be within about 4 in. of the ground surface, should be 15 ft or more from any opening, such as a door or window, in an occupied building, and should, if possible, be hidden in shrubbery or otherwise concealed and protected. In some cases the fresh-air inlet is terminated in a concrete, masonry, brick, or metal box about 18 in. square and deep enough so that the bottom of the box is at least 18 in. below the terminal of the inlet. The top of the box should be covered with a cast-iron or similar grating, level with the ground surface. The box must be drained to an outlet other than to the sanitary sewer. Wherever located the box containing the upper end of the fresh-air inlet must be protected with a substantial cover that cannot be easily removed but will permit the passage of air into or out of the vent pipe.

The size of the fresh-air inlet is based on the size of the building drain, as shown in Table 11-14.

11-27. Drainage Fittings. Information concerning the material, dimensions, and types of drainage fittings is given in Chap. 12 and Appendix II. The use of short-radius or right-angle-bend fittings is undesirable, since they are conducive to clogging and to the development of higher pressures than when long-sweep fittings are used. Long, sweeping curves and smooth, easy channels free from excessive roughness, sudden enlargements, or contractions are to be desired in any connection between the pipes in a drainage system. Because of the thin partitions, walls, or floors sometimes used in building construction, short-turn fittings must occasionally be used, but their use is to be avoided.

11-28. Prohibited Fittings and Connections. The use of certain types of fitting is sometimes prohibited on drainage and vent pipes because of the difficulties that experience has shown result therefrom. Among the

TABLE 11-14. SIZES OF FRESH-AIR INLETS

Size of house drain, in..........	4 or smaller	5 or 6	7 or 8	10 or larger
Size of fresh-air inlet, in.......	Same as house drain	4	6	8

excluded fittings are bands and saddles; traps with hidden partitions or tortuous passages; tees and crosses; double-hub fittings; single-tee or double-tee branch, tapped-tee branch, heel or side-outlet quarter bend when the inlet is placed in a horizontal position as a vent; fittings that provide a lodging place for solids; or any fitting or connection that has an enlargement, chamber, or recess with a ledge, shoulder, or reduction of the pipe area in the direction of the flow on the outlet or drain side of a trap.

The "Plumbing Manual" states (p. 32):

Par. 413. Increasers and Reducers. The prohibition of an enlargement chamber or recess with ledge, shoulder, or reduction in size of pipe . . . is not to be construed as prohibiting enlargements such as formed by standard wye or other standard branch fittings, as prohibiting the recess formed in the vertical run of fittings in soil or waste stacks in order to accommodate several branch connections unless a definite ledge is formed thereby, or as prohibiting the enlargement of a 3-in. closet bend to 4 in. at its inlet in order that a standard 4-in. floor flange may be used in making floor connections for water closets. It is to be construed as prohibiting any enlargement, recess, or ledge through faulty design of fittings or poor workmanship that would form a dam or other obstruction in the direction of flow.

The "National Plumbing Code" states (Sec. 2.4.1):

No fitting having a hub in the direction opposite to flow of the branch shall be used as a drainage fitting. No running threads, bands, or saddles shall be used in the drainage system. No drainage or vent piping shall be drilled or tapped.

It is stated in Sec. 2.4.2.:

A heel or side-inlet quarter bend shall not be used as a vent when the inlet is placed in a horizontal position.

The use of putty or rubber in making a connection is usually prohibited. The use of drive ferrules is not good practice, and combination lead ferrules should be used only when the calked joint can be made in the upright position. A fitting over which there is some controversy is the main or house trap. Some codes require it; others prohibit it; few are silent on the subject.

11-29. Cleanouts in Pipes. An easily accessible cleanout should be provided at or near the foot of each stack, at least at each alternate change in direction in a main soil or waste pipe, and within the building close to the outside wall through which the building drain passes. This latter cleanout should be made with a full-size wye or tee, somewhat as shown in Figs. 11-3 to 11-7, 11-33, and others. Other cleanouts should be of the same nominal size as the pipe in which they are inserted, up to 4 in., and for larger pipes the cleanout must be at least 4 in. in diameter, and it should be installed so that on pipes 3 in. or larger a clearance of at least 18 in. shall be allowed for rodding. A minimum clearance of 12 in. is required for smaller pipes.

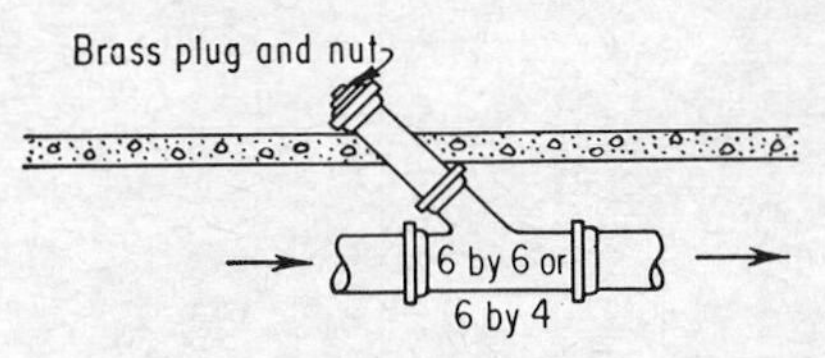

FIG. 11-33. Drainage cleanout.

The "National Plumbing Code" (Sec. 5.4.3) requires a cleanout at each change of direction of the building drain greater than 45°. In lieu of a cleanout at the base of a stack and where crawl space under the floor is less than 18 in., the Code permits the extension of the building drain to the outside of the building and terminated in an accessible cleanout or an accessible cleanout in the building drain not more than 5 ft outside the building wall. The Code (Sec. 5.4.1) limits the maximum distance between cleanouts on 4-in. or smaller drainage pipes to 50 ft and to 100 ft for larger pipes. The longer distance minimizes installation costs; shorter distances facilitate maintenance. The Code states (Sec. 5.7.1) that "A fixture trap or a fixture with an integral trap, readily removable without disturbing concealed roughing work, may be accepted as a cleanout equivalent, if there is no more than one 90° bend on the line to be rodded."

The bodies of cleanout ferrules should be made of standard pipe sizes, conforming in thickness to that required for pipe and fittings of the same metal and extending not less than $\frac{1}{4}$ in. above the hub. The cleanout cap or plug[1] should be of heavy red brass, not less than $\frac{3}{16}$ in. thick, and should be provided with a raised nut or recessed socket for removal. Both ferrule and plug should conform to the American National taper pipe thread standard, as stated in Secs. 4-13 and 7-27. Heavy lead plugs may be used for repairing a cleanout where necessary.

[1] See also Federal Specifications WW-P-401 and "National Plumbing Code," Sec 3.2.6.

All underground traps and cleanouts, except where cleanouts are flush with the floor, and all exterior underground traps should be made accessible by manholes with adequate metal covers giving access to the trap or cleanout. Any floor or wall connection of fixture traps where screwed or bolted to the floor or wall can be used as a cleanout.

11-30. Safe Wastes and Indirect Wastes. A safe waste is defined in Appendix I. Among the purposes of safe wastes is the protection of fixtures with relatively clean-water discharges from possible contamina-

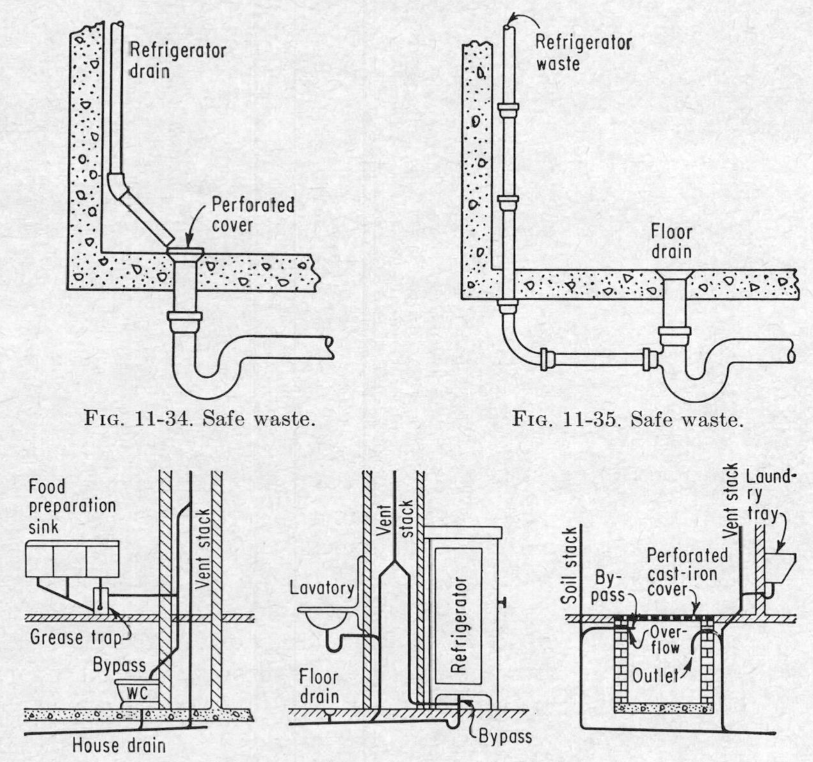

FIG. 11-34. Safe waste.

FIG. 11-35. Safe waste.

FIG. 11-36. Vent bypasses in plumbing.

tion by sewer air. Drainage of waste water carrying no organic wastes, as from water stills, sterilizers, tank overflows, expansion tanks, water-treatment devices, and plumbing safes, should discharge into a rain-water leader, into an indirect waste as shown for the refrigerator in Fig. 11-1, or into a drain pipe above a trap, as shown in Figs. 11-34 and 11-35.

Indirect waste pipes of appreciable length may be trapped and vented to avoid odors from long, exposed, drain pipes and the loss of trap seals. These vent pipes must not be connected to the vent pipes of the plumbing system, as a bypass would be created, as shown in Fig. 11-36. The "National Plumbing Code" (Secs. 9.3.1 and 9.3.2) requires a trap on an indirect waste exceeding 2 ft in length and limits the greatest length to 15 ft. The Code states that sheet copper used for safe pans shall not be less than 12 oz per sq ft. Indirect wastes can be used to safeguard the seal in an infrequently used trap.

11-31. Drainage from Food Containers. Drainage from food containers, such as refrigerators, should not be connected to a soil or waste pipe because of the possible exposure of the food to contamination by loss of trap seal or backflow of drainage. Drains from food containers are best discharged into a trapped sink or other trapped and vented fixture that is in normal use, as shown in Fig. 11-1, but not into a fixture used for domestic or culinary purposes. The "National Plumbing Code" (Sec. 9.4.1) requires that the air gap shown in the figure shall be at least twice the effective diameter of the drain served. Receptors for indirect wastes may not be located in a toilet room or in any inaccessible space such as a closet or storeroom.[1] The receptor is to be equipped with a readily removable metal basket over the discharge or the receptor outlet is to be equipped with a beehive strainer not less than 4 in. in height.[2] Such drains may profitably discharge in an accessible and observable place near the floor drain in a well-ventilated room. The discharge may contribute to the maintenance of the seal in an otherwise infrequently used trap.

Bars, soda fountains, dishwashers, and other fixtures in which food is handled should not be connected to soil pipes or discharge into a sink or another fixture. Their wastes should be handled as indirect wastes and connected to a drainage pipe above a trap, as shown in Fig. 11-31. Some codes, on the other hand, permit trapped connections of such wastes to plumbing waste pipes or soil pipes.

The waste pipe from a refrigerator, bar, soda fountain, or similar fixture should not be less than 1¼ in. in diameter for 1 opening, 1½ in. for 2 or 3 openings, and for 4 to 12 openings it should not be less than 2 in. It should be trapped at each opening, and cleanouts should be provided so that the pipe can be flushed. It may be desirable to continue such waste pipes through the roof except where the length of the waste pipe is small, as at *V* in Fig. 11-7.

Floor-drain waste pipes should be trapped and, if possible, not directly connected to a drainage pipe containing sewage unless provision is made for the renewal of the seal in the trap of the floor drain.

11-32. Blowoff Pipes from Steam Boilers. The exhaust, blowoff, sediment, or drip pipe from a steam boiler or other machine from which water at a high temperature or containing much sediment is drained

[1] "National Plumbing Code," Sec. 9.5.1.
[2] *Ibid.*, Sec. 9.5.3.

should not be directly connected to the ordinary pipes of the plumbing system to which other common fixtures are connected. Such pipes should discharge into the top of and above the level of discharge of a closed tank or condenser made of wrought or cast iron or steel. The tank should be provided with a relief or vent pipe not less than 3 in. in diameter extending to the outer air and not connected to any other pipe, as shown in Fig. 11-37.

The depth from the discharge level to the bottom of the tank should not be less than 24 in. The waste pipe from the tank should not be smaller in diameter than the inlet pipe or less than 3 in. in diameter. It should, when possible, connect to the house sewer or main sewer and not to the house drain. Such waste pipes need not be trapped, but they should be properly vented.

11-33. Fixture Drains and Traps. Types of trap in use on plumbing fixtures are shown in Fig. 11-38 and sizes required on fixtures are shown in Tables 11-1 and 11-2. The traps shown in Fig. 11-38 are made of lead; brass, either plain or nickel-plated; wrought metal, either plain or galvanized; and cast iron. Some standards for fixture traps are covered by Federal Specifications for Plumbing Fixtures, WW-P-541*a*, extract s, which are given in Appendix II and in this chapter.

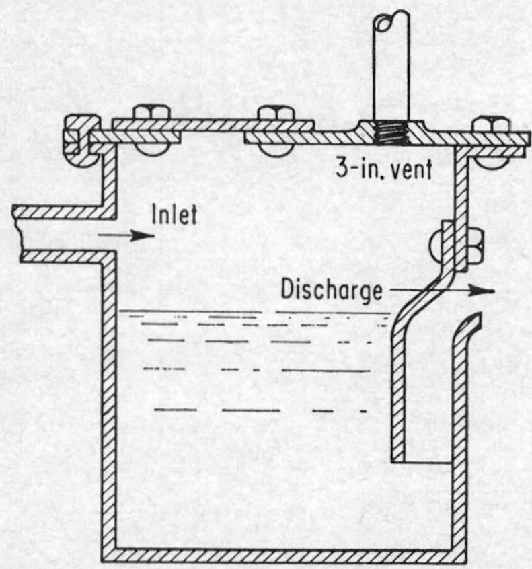

FIG. 11-37. Steam-boiler blowoff.

Standard dimensions of lead traps and sink traps are given in Table II-71 and of brass traps in Table II-72, Appendix II.

11-34. Performance of Traps. The primary purpose of a trap is to prevent the passage of air, odors, or vermin through it from the sewer into the building. Traps for special purposes, such as grease traps, sand traps, etc., are described in Secs. 11-44 to 11-47.

A trap must be self-cleansing; that is, it must permit the passage of liquid conveying solids in suspension without retaining the solids and becoming clogged. The two requirements work against each other in design, since the more complicated the passages, the better will the trap maintain its seal but the more likely is it to become clogged. The requirements of a "perfect" trap are listed as follows:

1. It must be able to pass used water freely without mechanical aid.
2. It must be able to prevent passage of sewer air in either direction whether sewage is flowing or not.
3. It must be self-cleansing.

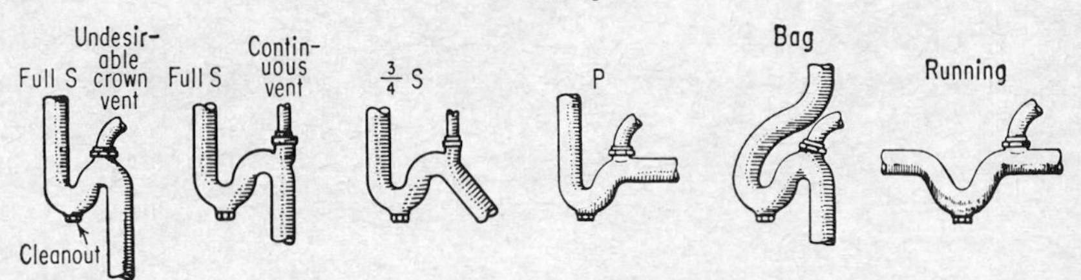

Lead traps. All styles are available, with and without cleanouts and vents

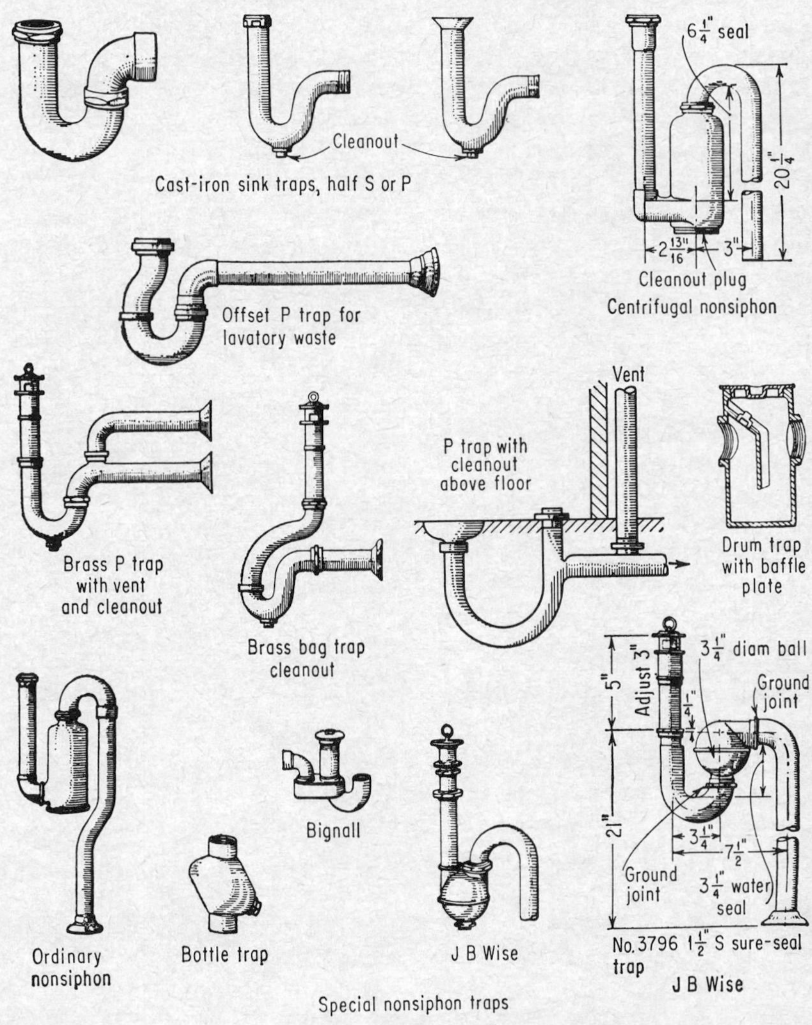

FIG. 11-38. Traps.

4. It must have a seal, preferably not less than 2 in. deep.
5. It must be strong and proof against leakage from it of gas or liquid.
6. It must have no mechanical or moving parts.
7. It should be provided with a cleanout or other means of access to interior in the event of chokage. The cover must be gastight and watertight.
8. It must have no recesses, cavities, or pockets that cannot be scoured by the flow of sewage through the trap.
9. It must have no internal projections to catch and hold hair, lint, bits of matches, etc., but must have a smooth inner surface every part of which is automatically scoured by the flow of sewage through the trap.
10. It must have no washers, gaskets, or packing on the sewer side of the seal that may decay and cause sewage or sewer air leakage.
11. Where necessary it must be back-vented.
12. It must not have concealed partitions, tubes, or other invisible parts if defects in construction might permit sewer air to enter the house.

Simple traps, such as S, ½S, and running traps, provide smooth, continuous passages for the water so that they are not easily clogged.

In most plumbing codes traps are required on all fixtures. In this connection the "National Plumbing Code" states (Sec. 5.1.1):

Plumbing fixtures, except those having integral traps, shall be separately trapped by a water-seal trap, placed as close to the fixture outlet as is possible. (*a*) Provided that a combination plumbing fixture may be installed on one trap, if one compartment is not more than 6 in. deeper than the other and the waste outlets are not more than 30 in. apart. (*b*) Provided that one trap may be installed for a set of not more than three single-compartment sinks or laundry trays or three lavatories immediately adjacent to each other in the same room, if the waste outlets are not more than 30 in. apart and the trap is centrally located when three compartments are installed.

11-35. Nonsiphon Traps. A nonsiphon trap offers much greater resistance to the breaking of its seal than is offered by a simple trap but, at the same time, it is not subject to clogging and is easily cleaned. Any trap in which the diameter is not greater than 4 in., the depth of seal is between 3 and 4 in., and the volume of water held in the trap is not less than 1 qt can be classified as a nonsiphon trap.

So-called nonsiphon traps have been designed with the purpose of creating traps with stronger seals than those found in simple traps. Some types of nonsiphon trap are illustrated in Fig. 11-38. Note that five different principles are involved in increasing the strength of seal in such traps: (1) Depth of seal is increased, (2) volume of water retained is increased, (3) passages in the trap are made more tortuous, (4) water is given a whirling motion in passing through the trap, and (5) moving parts are depended on to prevent the backflow of water. Each of these expedients will increase the strength of the trap seal, but all except the fourth are objectionable because of the increased tendency for the trap to clog.

When a nonsiphon trap is to be used, a drum trap with a whirling motion of the water is recommended as best suited under most conditions to resist breaking of the seal. A drum trap is commonly used under a bathroom floor on the drain pipe from the bathtub.

11-36. Trap Seal. The strength of the seal of a trap is closely proportional to the depth of the seal. Unfortunately, an increase in the depth of the seal increases the probability of solids being retained in the trap, and a limit of about a 4-in. depth of seal for traps that must pass solids has been imposed by some plumbing codes. The depth of seal most commonly found in simple traps is between 1½ and 2 in. The Hoover Report[1] recommends a minimum depth of 2 in. as a safeguard against seal rupture and a maximum depth of 4 in. to avoid clogging, fungus growths, and similar difficulties. Traps in rain-water leaders and other pipes carrying clean-water wastes only and which may be infrequently used should have seal depths equal to or greater than 4 in. The greater depth of seal is permissible in clean-water traps, as they are less likely to clog than other traps conveying suspended solids and putrefaction will not occur in clean water. These traps should not be accessible, however, to mosquitoes and other insects that may breed in them.

The increase in the volume of water retained in the trap helps but little to increase the strength of the seal, but it may reduce the velocity of flow through the trap and increase the probability of sedimentation therein.

11-37. Evaporation from Traps. The evaporation of water from traps is a cause of loss of seal, particularly where the fixtures are infrequently used. The principal factors affecting evaporation are velocity of air movement, percentage of moisture in the air, area of water surface exposed, and temperature. The conditions[2] are so variable among installations and at any one installation, and so little information is available that, in practice, it is sometimes considered that 1 in. of water will evaporate in 3 weeks from an unvented trap and twice as much in the same time from a vented trap. Nonsiphon traps, such as drum traps, with only a small surface exposure compared with the volume of water in the trap will lose their seal much less rapidly, retaining it for 6 months to a year in favorable conditions. The increase in the volume of water in a

[1] "Recommended Minimum Requirements for Plumbing in Dwellings and Similar Buildings," National Bureau of Standards, July 3, 1923.

[2] See also "New Facts on Plumbing Fixture Performance," *Domestic Eng.*, August, 1954, p. 127.

trap will help to minimize loss of seal by evaporation. Traps with large volumes of retained water can be used with good results on such locations as rain-water leaders, floor drains carrying no solid matter, and safe wastes. The evaporation of a trap seal on an infrequently used fixture is prevented by replacing the water in the trap with a nonvolatile liquid such as kerosene oil, as explained in Sec. 18-34.

11-38. Clogging Tendencies of Traps. From a practical standpoint the plumber is interested in the ability of a trap to keep itself clean. Although the so-called *hairpin trap* (now out of date) is very successful in maintaining its seal, no reliable plumber with experience would care to use it because of its propensity for clogging. The most successful traps in keeping themselves clean are those with the straightest, smoothest passages and with the highest velocities through them. The simplest P, S, and running traps fulfill these requirements. Since they are also satisfactory in resisting siphonage and self-siphonage, their use is generally to be recommended.

11-39. Undesirable Forms of Trap. Traps that depend on movable parts or concealed interior partitions are generally considered undesirable because of the danger of the movable part blocking open or of holes appearing in the partitions. Most of the nonsiphon traps, except the drum trap shown in Figs. 11-38 and 11-39, are undesirable forms of trap. The prohibition against the use of interior partitions does not apply to integral traps that are built in as an integral part of plumbing fixtures, as in a water closet. Traps that have covers over handholes on the sewer side of the trap which are held in place by lugs or bolts are not desirable because of the danger of leakage through the cover.

The bell trap, shown in Figs. 11-39 and 13-10, is used most commonly on floor drains. Its use is prohibited by the "National Plumbing Code"[1] because it is not a simple form of trap and its seal depends on the perfection of a single casting. Its use in floor drains is desirable only when the thickness of the floor is insufficient to permit the installation of a more desirable form of trap.

The creation of tortuous passages by baffling is probably the most effective method of increasing the strength of a trap seal, but it is objectionable because of the greater difficulty for solids to pass through the trap and because in the construction of such traps it is difficult to avoid leaving holes in the baffles, thus permitting the passage of gas through the trap. Such traps are universally prohibited by plumbing codes.

Traps with moving parts are manufactured. A ball trap is shown in Fig. 11-39. The ball trap has the advantage of a certain and permanent seal when it works properly, but the disadvantages of not being self-

[1] In Sec. 5.3.6.

cleansing and of not always closing properly render the trap unreliable. Ball traps are prohibited by most plumbing codes.

The "National Plumbing Code" prohibits the use of full S, bell, and crown-vented traps.

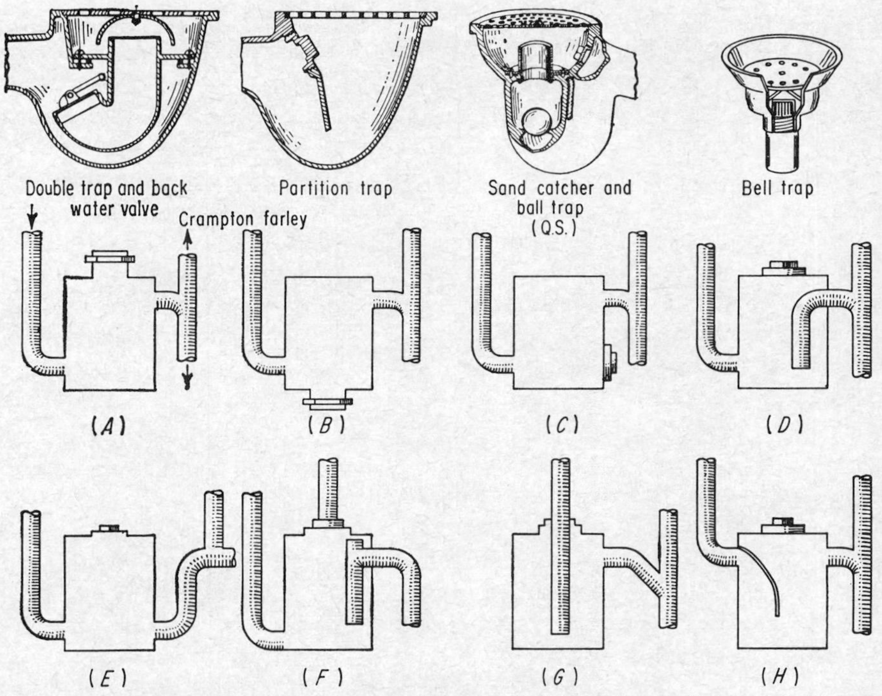

FIG. 11-39. Bell trap, partition trap, and mechanical trap for yard drains; bell trap for refrigerator drain; and various forms of drum traps.

Desirable features of drum traps include: Cleanout is accessible when properly placed, seal of some drum traps depends on water only, and the traps are not easily siphoned or blown. Some undesirable features include: The seal may depend on the tightness of a gasket, as in *A*, *D*, and *E* to *H*; it may be necessary to disconnect the vent pipe in order to clean the trap, as in *F*; deposits over the outlet pipe may cause clogging, as in *D* and *G*; if the vent clogs, *D*, *F*, and *G* will self-siphon; and in *H* the seal depends on a baffle and gasket.

The most desirable forms of drum traps are *B* and *C*, provided the cleanout is accessible and the seal depends on water depth only.

11-40. The Main Trap, or House Trap.[1] The main trap, or house trap, is a running trap placed on the building drain, usually near the building foundation, and below all other connections to the plumbing pipes and upstream from the building sewer. A main trap is shown at

[1] See also L. Blendermann, Providing Access for Building Traps, *Air Conditioning, Heating, and Ventilating*, September, 1957, p. 100.

point 76 in Fig. 11-1. The purpose of the house, or main, trap is to prevent the passage of odors, gases, and pests from the common sewer into the plumbing of a building.

Among the advantages claimed for a house trap are: (1) The prevention of the passage of air, odors, gases, and pests into the house plumbing; (2) a defense, supplementary to fixture traps, to prevent the passage of these things into the building. (3) The plumbing pipes are not used as ventilating ducts for the public sewer. That sewer air and gases do pass from the sewer into the house plumbing not protected by a main trap is incontrovertible. Combustible gases from a sewer have been seen to burst into flame at the top of a vent stack. Disadvantages claimed about house traps include: (1) The probability of the loss of seal in plumbing fixture traps because of increased back pressures due to the house trap. This danger may not be overcome by the fresh-air inlet discussed in Sec. 11-26. (2) A fresh-air inlet is required. It will add to cost, to possibilities of freezing and evaporation of water in the trap seals, and to the possibility of the escape of odors into occupied places, and it may not serve its purpose effectively. (3) Although the passage of pests from the sewer into the plumbing system is discouraged, it is not prevented.

Controversy over the main, or house, trap has been vigorous among sanitarians, sometimes generating more heat than light. Tests, experience, and the trend of practice indicate that the disadvantages outweigh the advantages and that a house trap or main trap should not be used. The "National Plumbing Code" prohibits its use except where required by the Local Authority.

11-41. Materials for Traps. Traps for bathtubs, lavatories, sinks, and similar fixtures should be lead, brass, cast iron, or malleable iron that is either galvanized or lined on the inside with porcelain or a similar substance. Concealed fixture traps should be of extra-heavy weight and of the recessed drainage type. Traps should have a full bore and smooth interior waterway, with threads tapped out of solid metal.

11-42. Cleanouts and Drains in Traps. Traps, with few exceptions, should be equipped with a submerged brass cleanout plug in the bottom of the trap, as shown in Fig. 11-38. The brass plug should be threaded, and the connection made watertight with a rubber gasket or a machine-ground joint. Whenever the plug or cover is removed, it is usually necessary to renew the gasket before replacing the plug or cover. Iron cleanout plugs should not be used, as they rust and cannot be removed. The main, or house, trap should have a heavy, cast, full-size cleanout on top on each side of the trap and, where possible, a brass cleanout plug in the bottom. Cleanout plugs in the bottom are unnecessary in traps buried underground.

Means of access must be provided to every cleanout.

11-43. Location and Setting of Traps. Traps should be set so that they are upright, level, well supported, and accessible. They should be set close to the fixture in order to avoid the exposure of foul interior surface of waste or soil pipe to the atmosphere in the room and to minimize the momentum of the mass of water moving from the fixture toward and through the trap.

Requirements that each fixture shall be trapped independently and exceptions thereto are discussed in Sec. 11-34 and are illustrated in Fig. 11-5. Double trapping is to be avoided because of the danger of air binding, as explained in Sec. 2-14. A double-trapped fixture is shown in Fig. 2-12. This limitation does not apply, however, to the installation of a grease trap or other interceptor in addition to a fixture trap, provided there is a vent on the drain pipe between the two traps. If the waste pipe from a double-trapped fixture becomes so filled with air and water that the illustrated conditions arise, the depth of water H in the fixture must be greater than h_1 in order that water can run through the discharge pipe. This may result in a overflow from the fixture or in the failure of the fixture to flush properly.

11-44. Grease Traps. A grease trap is a device installed in the waste pipe from one or more fixtures for the purpose of separating grease from the liquid and retaining the grease. The trap is so constructed as to cool the retained liquid sufficiently to precipitate some grease from solution and to permit suspended grease to rise to the top of the trap where the grease is held.

A grease trap is required in the waste pipes from sinks and other fixtures in which greasy foods are prepared and dishes are washed or from which other greasy wastes are to be expected. Sinks in restaurants, hotels, and institutional kitchens require grease traps. Household sinks usually do not because of the relatively small amount of grease wasted from them. A grease trap should not be placed on a pipe carrying human excrement because the solids retained in the trap will be increased, putrefaction will occur, and conditions when the trap is cleaned will be extremely obnoxious.

A grease trap may be located under the fixture served, in the basement or in a vault outside the building. No location is ideal. A location under the fixture is convenient, accessible, and free from frost and provides a short length of drain pipe from the fixture to the trap. The odor and unsightliness involved in cleaning the trap are objectionable, and the capacity of the trap may be restricted by the limited space available under the sink. A location in the basement removes from the kitchen the objectionable conditions arising when the trap is cleaned, and the trap will probably be protected from frost, but the drain pipe to the grease trap may be too long, resulting in the deposition of grease in it. A subsurface trap outside the building may be located near the sink. The location avoids odors and mess in the kitchen when the trap is cleaned, the size of the trap is not restricted, and freezing can be prevented by proper construction. The cost may be greater than an inside location, however.

To be effective a grease trap should be cooled. Grease traps may be water-cooled, air-cooled, or earth-cooled. Water-cooled and air-cooled

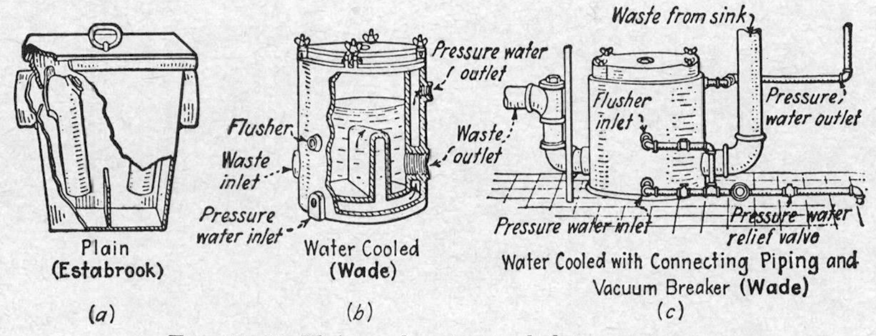

FIG. 11-40. Plain and water-cooled grease traps.

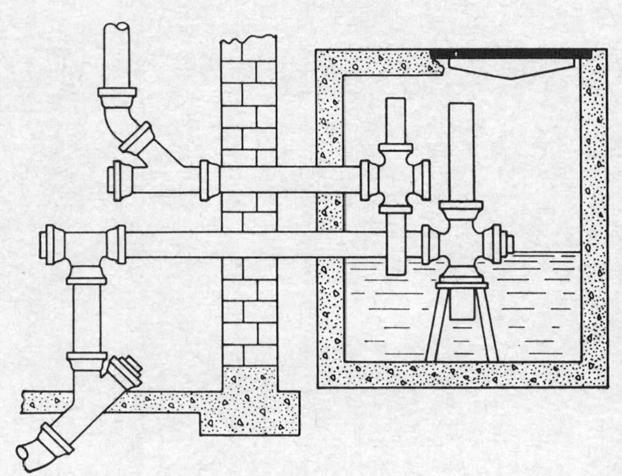

FIG. 11-41. Grease catch basin.

traps are shown in Fig. 11-40; an earth-cooled trap is shown in Fig. 11-41. If the trap-cooling system is connected to the potable cold-water supply, a vacuum breaker should be installed on the water-supply pipe near the trap, as shown in Fig. 11-40*c*, to avoid the undesirable effects of a cross connection. The water-supply pipe may lead from the trap to the hot-water heater and serve as its supply pipe.

The volume of water retained in an air-cooled grease trap should be at least twice the volume of the fixtures drained by the trap. A water-cooled trap may be smaller than an air-cooled trap for equal performance in the same location because of the greater rapidity of cooling. If the volume in the trap is too small, inadequate cooling will be provided and there may be inadequate volume for the retention of grease. Too large a volume may add unduly to the cost and inconvenience of the installation, but frequency of cleaning will be minimized.

Grease traps located inside a building should be constructed of iron, black, painted, enameled, or otherwise protected against corrosion. Traps or catch basins outside the building can be constructed of any suitable material, including brick or concrete, but the trap should be watertight below the water line and airtight above it, except through the required vent.

Both the inlet and the outlet pipes of the trap should enter and leave below the water line, respectively. A submerged inlet is required to avoid disturbance and comminution of the grease already in the trap, and a submerged outlet is necessary to prevent the escape of floating grease from the trap and, in addition, to serve the normal purposes of a plumbing trap.

Grease traps should be vented in accordance with the principles established for trap venting. It is to be noted that a grease trap violates the principles that a trap must be self-cleansing. For this reason a convenient and adequate cleanout must be provided and the trap must be cleaned periodically.

11-45. Interceptors. Interceptors are used to prevent the passage of such substances as grease, oil, and sand into the sewer and of large objects into the waste pipes.[1] Details of interceptors are discussed in Secs. 11-44, 11-46, and 11-47. Since they constitute a form of trap involving a water seal, interceptors must be designed to avoid air binding and self-siphonage. However, the use of an interceptor and a trap on the same drainage pipe is not to be construed as double-trapping of a fixture if a vent pipe is provided between them, as explained in Sec. 11-43.

11-46. Gasoline and Oil Traps. Gasoline and oil traps are constructed on principles similar to those of grease traps described in Sec. 11-44, but they are not water-cooled. The capacity of the trap should be at least double the hourly flow through it. The "Housing Code"[2] states (Sec. 5.7.3):

> *b.* Oil interceptors shall be constructed so as to be oiltight and of substantial construction, provided with necessary inlet and outlet connections and a separate 2-in. vent from top of the tank to the open air. Cover is to be easily removable; water- and gastight.

[1] See also *Architectural Record*, June, 1955, p. 215.

[2] "The Uniform Plumbing Code for Housing," Housing and Home Finance Agency, February, 1948.

c. Minimum capacity for an oil interceptor shall be 6 cu ft plus 1 cu ft for each vehicle serviced or 1 cu ft for each 100 sq ft of surface drained into the interceptor.

d. Floor drains discharging individually into an oil interceptor need not be trapped.

The details of the separator are shown in Fig. 11-42. A local vent should be provided as shown in the figure, leading from the highest point in the trap and terminating in the open air outside the building to remove and to discharge inflammable gases safely outside the building. This local vent should be at least 4 in. in diameter for a gasoline-and-oil separator in a large garage.

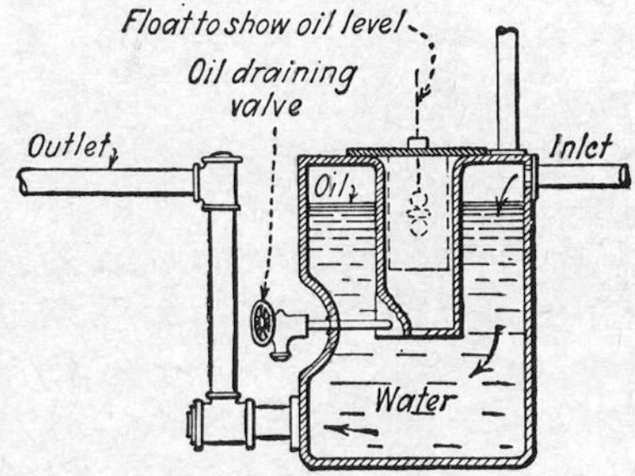

FIG. 11-42. Oil separator. (*Plumbers' Trade J., vol.* 76, *p.* 567.)

A separator should be located on the drain from public garages, on the drains from cleaning establishments, and in other places where explosive and inflammable materials may enter a sewer, but they are not usually required on the drain from a single-car garage in a residential area. No putrescible organic matter should be permitted to flow into an oil or gasoline separator.

11-47. Catch Basins or Sand Traps. Catch basins are used for removing sand, cinders, leaves, and other materials that would otherwise

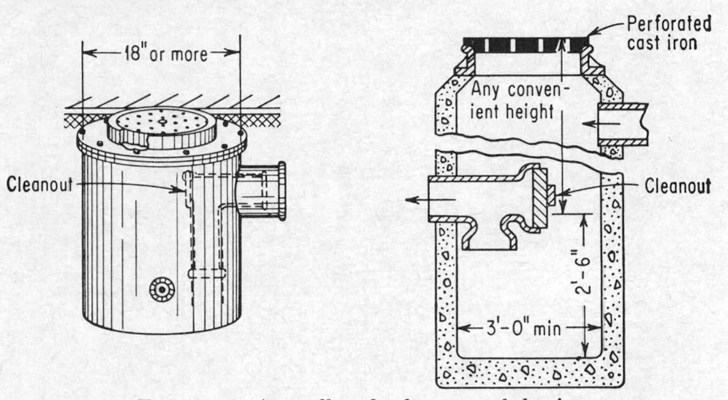

FIG. 11-43. A small and a large catch basin.

enter and might clog drain pipes. Catch basins are used principally on yard drains and roof drains and do not intercept sanitary sewage. Types of catch basins are shown in Fig. 11-43. When sanitary sewage is not intercepted, the catch basin may discharge untrapped into a storm-water outlet. If the basin discharges into a combined sewer or intercepts foul-smelling substances, the basin should be trapped with a depth of seal of at least 8 in. and vented as required for similar fixtures, with the vent terminating as required in Sec. 12-11. A removable gastight cover should be provided, and provision should be made for cleaning the basin.

Catch basins are constructed of masonry, concrete, vitrified clay, or cast iron. They may be of any shape conducive to ease in cleaning, such as rectangular, oval, or circular. Their size or capacity depends on the amount of detritus to be stored between cleanings and the rate at which water is passed through them. A rule of good practice for catch basins on yard drains requires that the smallest horizontal dimension shall be at least 20 in. and the depth of water retained below the outlet shall be at least 15 in.

Catch basins require periodic cleaning. A catch basin that does not require occasional cleaning serves no useful purpose. Catch basins may require attention also to prevent odors and oiling to prevent insects from breeding in them, especially mosquitoes. Insect breeding can be minimized by a long inlet pipe and the exclusion of light.

CHAPTER 12

INSTALLATION OF DRAINAGE AND VENT PIPES

12-1. Installation of Drainage Pipes. The drainage pipes of a plumbing system are installed in the walls and between the floors, in pipe runs, or in other concealed places within a building. Exterior installations are permitted in climates where there is no frost[1] or the pipes are adequately protected therefrom. The pipes should be made as accessible as possible to permit observation and repairs. Vertical pipes should be supported at each floor or branch interval by collars or iron hangers attached to the walls or floors of the building and caught under a pipe hub or coupling. The openings should be adequately ratproofed, the collars being of metal or other ratproof material. The base of each stack should rest on a masonry or cast-iron pier. Sloping pipes are supported within the wall or floor on joists or by such means as are described in Secs. 7-11 and 12-2. The house drain, when only slightly above the cellar floor, is supported on a masonry or metal pedestal at each joint but not more than 10 ft apart.

The "Plumbing Manual"[2] states (p. 9):

> Stacks should be supported at their bases and rigidly secured. Piping shall be installed without undue stresses or strains, and provision made for expansion, contraction, and structural settlement. No structural member shall be weakened or impaired beyond a safe limit by cutting, notching, or otherwise, unless provision is made for carrying the structural load.

Movements resulting from the expansion and contraction of cast-iron pipe due to temperature changes differ so little from movements of other parts of the building from the same cause that no special provision need be made for them. Allowance must be made, however, for movements of the building from other causes, such as settling of the foundation and the wind, in so far as changes in length are concerned, but the tapping and banging sounds caused by the movement of pipes through their supports should be suppressed by the methods explained in Secs. 7-17 and 18-28.

Horizontal drainage pipes should be run in as straight an alignment as possible and should be supported at intervals not exceeding 10 ft and

[1] "National Plumbing Code," Sec. 12.4.7.

[2] "Plumbing Manual," Report BMS66, National Bureau of Standards, 1940.

preferably at pipe joints. Minimum slopes for horizontal drainage piping are shown in Table 12-1.

12-2. Pipe Supports. All pipes should be supported so that their weight does not bear on a calked joint, except when the spigot end of one vertical pipe rests in the hub of the next lower vertical pipe. All soil and waste stacks should be thoroughly supported on concrete or masonry piers or should have substantial footrests at their bases; soil or waste stacks 10 ft or more in height should also be provided with floor rests or other substantial support at 10-ft or floor intervals. Brick supporting piers should be at least 18 in. square.

TABLE 12-1. MINIMUM SLOPES FOR HORIZONTAL DRAINAGE PIPES

Diameter of pipe, in.	1¼	1½	2	2½	3	4	5	6	8	10	12
Minimum slope recommended by "Plumbing Manual,"* in. per ft	¼	¼	¼	⅛	⅛	⅛	1/16	1/16	1/16	1/16	1/16
Minimum slope recommended by "Housing Code,"† in. per ft	¼	¼	¼	¼	¼	⅛	⅛	⅛	⅛	1/16	1/16

* "Plumbing Manual," Report BMS66, National Bureau of Standards, 1940.
† "The Uniform Plumbing Code for Housing," Housing and Home Finance Agency, February, 1948.

Connections of pipe supports or fixture settings with masonry, stone, or concrete backing should be made by expansion bolts without the use of wooden plugs.

Horizontal runs of pipe above the floor should be supported or anchored by wall brackets, iron hangers, or masonry piers at intervals not to exceed 10 ft for all materials except lead. Lead pipe should be supported for its entire length or at very close intervals.

Various methods for the support of pipes are discussed in Sec. 7-11. Many of these methods are applicable to the support of drainage and vent piping.

12-3. Grades, Alignment, and Junction of Pipes. Drainage and vent pipes should be installed in as direct and straight an alignment as possible. There should be no drops or sags, and the pipes should be graded and joined to drain to the outlet by gravity. The "National Plumbing Code" states (Sec. 12.6.2):

Vertical Rise. Where vent pipes connect to a horizontal soil or waste pipe, the vent pipe shall be taken off above the center line of the soil pipe, and the vent pipe shall rise vertically, or at an angle not more than 45° from the vertical, to a point at least 6 in. above the flood-level rim of the fixture it is venting before setting off horizontally or before connecting to the branch vent.

It is stated in Sec. 12.6.3:

Height above Fixtures. A connection between a vent pipe and a vent stack or stack vent shall be at least 6 in. above the flood-level rim of the highest fixture served by the vent. Horizontal vent pipes forming branch vents, or loop vents shall be at least 6 in. above the flood-level rim of the highest fixture served.

The "Plumbing Manual"[1] states (Sec. 202) and the "National Plumbing Code" states (Sec. 2.3.1), concerning changes in direction:

Changes in Direction. Changes in direction in drainage pipes shall be made by the appropriate use of cast-iron, 45° wyes, half wyes, long-sweep quarter bends, sixth, eighth, or sixteenth bends, by combinations of these fittings, or by the use of equivalent fittings or combinations.

The "National Plumbing Code" states further:

. . . single and double sanitary tees and quarter bends may be used in drainage lines only where the direction of flow is from the horizontal to the vertical.

The "Plumbing Manual"[1] states:

No change in direction greater than 90° shall be made in drainage pipes.

The "National Plumbing Code" states (Sec. 2.3.2):

Short sweeps not less than 3 in. in diameter may be used in soil and waste lines where the change in direction of flow is either from the horizontal to the vertical or from the vertical to the horizontal, and may be used for making necessary offsets between the ceiling and the next floor above.

Standard tees, quarter bends, and crosses (Fig. II-10, Appendix II) should not be used on drainage pipes, because the sudden turn of direction is conducive to clogging. Such fittings may, however, be properly used on vent pipes and on other piping conveying fluids that do not carry clogging substances.

12-4. Connections to Stacks. Connections to soil and waste stacks should be made with sanitary tee and wye fittings, wye branches, one-eighth bends, 45° ells, or as explained in Sec. 12-3. No double-hub, double-tee, or double-sanitary-tee branch should be used on horizontal soil or waste lines. All soil and waste stacks should be provided with correctly faced inlets for fixture connections.

12-5. Offsets in Stacks. Offsets in stacks cause pressures above and below atmospheric that are, for practical purposes, unpredictable. The pressures are discussed in the Hoover Report[2] and in the "Plumbing

[1] *Ibid.*

[2] "Recommended Minimum Requirements for Plumbing in Dwellings and Similar Buildings," National Bureau of Standards, July 3, 1923.

Manual."[1] The "National Plumbing Code" sums up the situation as follows:

11.6.1. Offsets of 45° or Less. An offset in a vertical stack, with a change of direction of 45° or less from the vertical, may be sized as a straight, vertical stack. In case a horizontal branch connects to the stack within 2 ft above or below the offset, a relief vent shall be installed in accordance with Par. 12.18.2.

11.6.2. Waste Stacks Serving Kitchen Sinks. In a one- or two-family dwelling only in which the waste stack or vent receives the discharge of a kitchen-type sink and also serves as a vent for fixtures connected to the horizontal portion of the branch served by the waste stack, the minimum size of the waste stack up to the highest sink branch connection shall be 2 in. in diameter. Above that point the size of the stack shall be governed by the total number of fixture units vented by the stack.

11.6.3. Above Highest Branch. An offset above the highest horizontal branch is an offset in the stack-vent and shall be considered only as it affects the developed length of the vent.

11.6.4. Below Lowest Branch. In the case of an offset in a soil or waste stack below the lowest horizontal branch, no change in diameter of the stack because of the offset shall be required if it is made at an angle of not greater than 45°. If such an offset is made at an angle greater than 45°, the required diameter of the offset and the stack below it shall be determined as for a building drain [Table 11-4].

11.6.5. Offsets of More than 45°. A stack with an offset of more than 45° from the vertical shall be sized as follows:

That portion of the stack above the offset shall be sized as for a regular stack based on the total number of fixture units above the offset.

The offset shall be sized as for a building drain on a ½-in. slope, as shown in Table 11-4.

The portion of the stack below the offset shall be sized as for the offset or based on the total number of fixture units on the entire stack, whichever is the larger [see Table 11-4].

A relief vent for the offsets shall be installed as provided in Chap. 12[2] and in no case shall the horizontal branch connect to the stack [within] 2 ft above or below the offset.

12-6. Connections at Base of Soil and Waste Stacks. The "National Plumbing Code" states (Sec. 12.19.1):

All main vents or vent stacks shall connect full size at their base to the building drain or to the main soil or waste pipe, at or below the lowest fixture branch. All vent pipes shall extend undiminished in size above the roof, or shall be reconnected with the main soil or waste vent.

[1] *Op. cit.*

[2] Of the Code. See Sec. 12-10 of this book.

The pressures in a plumbing system are affected by the type of connection or footpiece between the stack and the building drain or other horizontal pipe into which the stack may discharge. Tests have shown that back pressures may vary between zero when the stack is allowed to discharge wide open without obstruction up to a maximum when a short-turn 90° ell is used at the base of the stack. A long-sweep fitting will provide lower back pressures than any other fitting. The provision of a vent at the base of a soil or waste stack, in addition to the use of a long-sweep fitting, will greatly minimize back pressures produced.

12-7. Installation of Vents. Some examples of satisfactory practice in the installation of vents are shown in Figs. 11-1 to 11-7 and others. Some examples of unsatisfactory practice are shown in Appendix III.

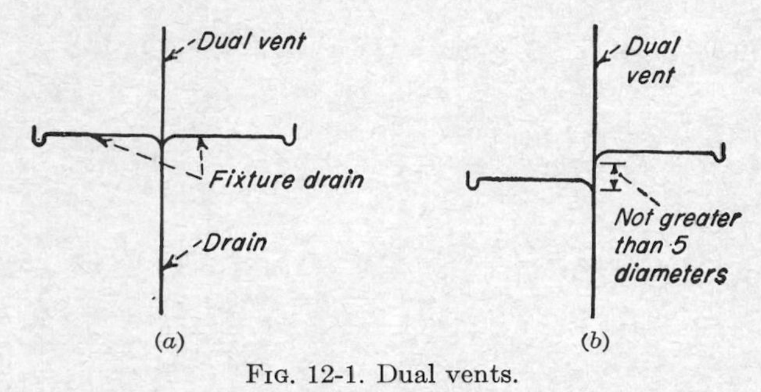

Fig. 12-1. Dual vents.

Figure 11-13*a* shows a permissible arrangement for a wet vent in which a lavatory and bathtub or shower are installed on the same branch without vent on the bathtub trap, provided that the vertical portion of the lavatory drain is not less than 1½ in. in diameter, that it connects with the bathtub drain in a vertical plane, and that the developed lengths of the lavatory and tub drains are within the limitations of Sec. 11-18.

Figure 11-13*b* shows a permissible arrangement for two lavatories and two bathtubs or shower stalls on the same branch with a unit vent for the lavatory traps and no vent for the bathtub or shower stalls, provided that the horizontal branch, except the separate fixture drains, are at least 2 in. in diameter and that the fixture drains for the bathtubs or shower stalls connect as closely as practicable upstream from the vent by means of a drainage wye.

Figure 11-13*c* shows a permissible installation of a lavatory, kitchen sink, and a bathtub or shower stall on the same branch, provided that the unit vent for the lavatory and sinks accords with the limitations of Fig. 12-1.

Figure 11-14 shows a permissible installation of two bathroom groups or one bathroom group and a kitchen sink or combination fixture *provided* they are installed on the highest branch or on a vertical yoke-vented branch not less than 3 in. in diameter, with no branch vents other than the yoke vent. A further provision is that each fixture drain connects independently to the soil stack or water-closet drain (closet bend) in the highest branch interval and that each fixture drain in all except the highest branch interval connects independently with the yoke-vented branch or with water-closet drains within the limits given in Sec. 11-18.

Figure 11-16 shows (1) a permissible installation of one bathroom group with group venting in accordance with the conditions in Fig. 11-13, where the horizontal branch is connected either to the soil stack at the same level as the water-closet drain or to the water-closet drain, or (2) a bathroom group and kitchen sink connected to the stack in the same

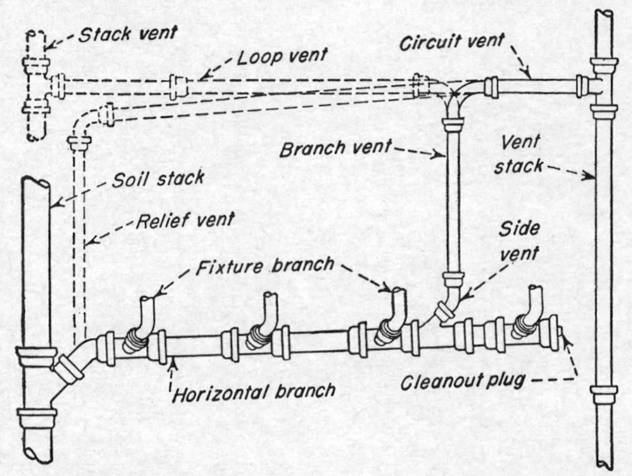

FIG. 12-2. Circuit and loop vents. (*The "Plumbing Manual."*)

manner as and with group venting in accordance with Fig. 11-13*c*, provided that either is installed in the same branch interval of a soil stack within the limits of the permissible fixture units allowed by Table 11-12 and provided further that a vent is installed from the water-closet branch drain in the third branch interval from the top and in each lower branch interval.

Figures 11-15 and 11-17 show permissible installations of two bathroom groups in accordance with Fig. 11-13*a* or *b* or two bathroom groups and two kitchen sinks with group venting in accordance with Fig. 11-13*c*, provided that a vent is installed for the second and lower branch intervals from the top.

Figures 12-2 and 12-3 show permissible installations of loop and circuit vents. Note that two lines of fixtures, back to back, are not to be circuit- or loop-vented on one branch but each line should be installed on a separate branch and circuit or loop vent. Good and bad practices in installing vents are shown in Fig. 12-4.

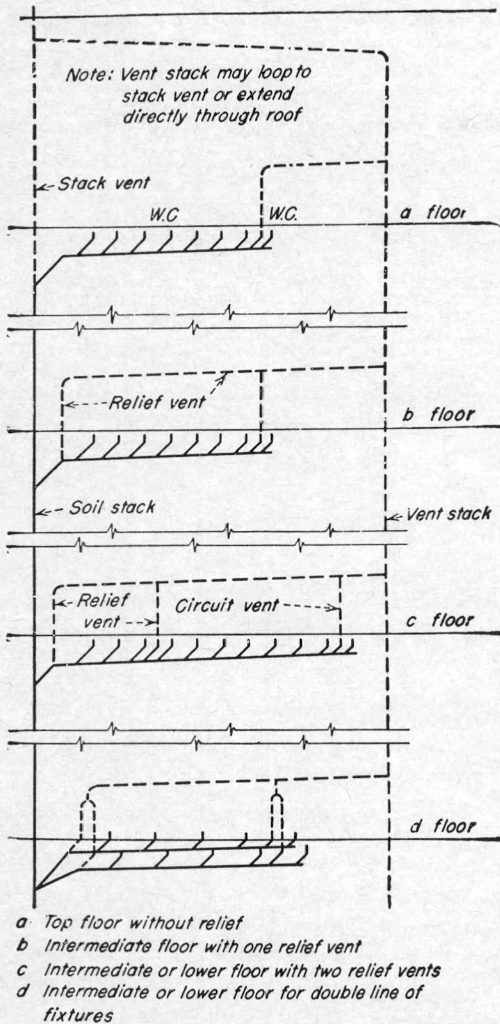

FIG. 12-3. Limits for circuit and loop vents. (*The "Plumbing Manual."*)

12-8. Circuit and Loop Vents. A battery of fixtures on the same floor, with circuit or loop venting, may be installed, to quote from p. 20 of the "Plumbing Manual"[1] as follows:

a. . . . on one horizontal branch with a circuit or loop vent connected to the horizontal branch in front of the last fixture drain, within the limits given in Table 11-13, provided that relief vents connected to the horizontal branch in front of the first fixture drain are installed as follows:

In each branch interval, if the total fixture units installed in the branch exceeds

[1] "Plumbing Manual," Report BMS66, National Bureau of Standards, 1940.

one-half the number given in Table 11-13, except that no relief vent shall be required in the highest branch interval of the system or in any branch interval if the total number of fixture units on the stack above the horizontal branch does not exceed the limits for one stack given in Table 11-13 and the number of fixtures on the circuit- or loop-vented horizontal branch does not exceed two for a 2- or 3-in. horizontal branch or does not exceed one-half the permissible number in column

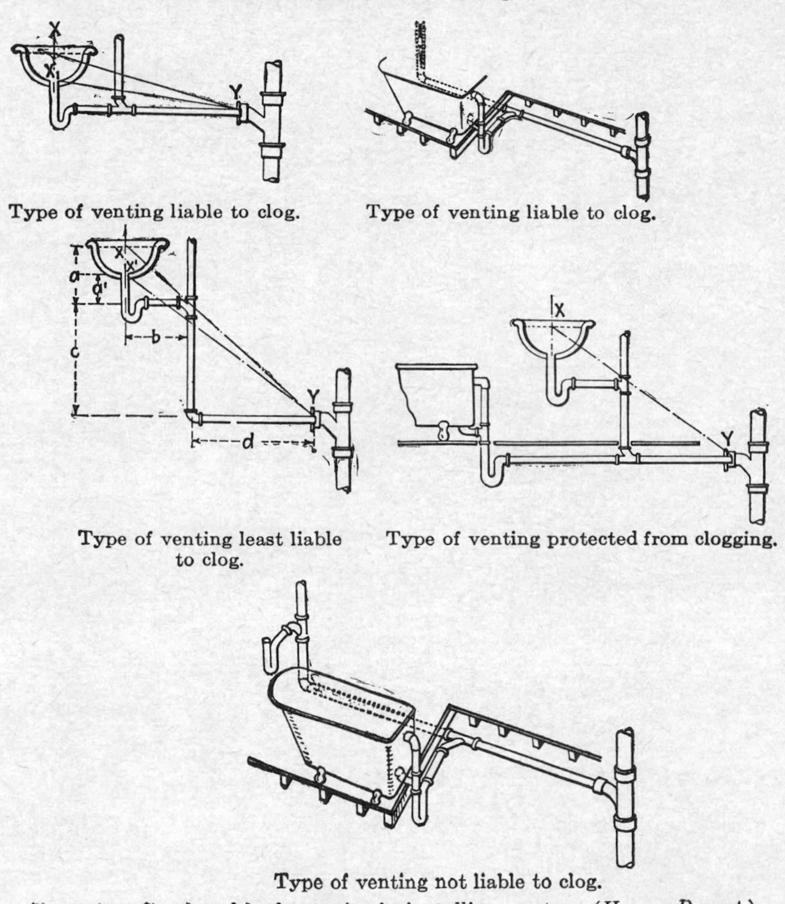

Type of venting liable to clog. **Type of venting liable to clog.**

Type of venting least liable to clog. **Type of venting protected from clogging.**

Type of venting not liable to clog.

Fig. 12-4. Good and bad practice in installing vents. (*Hoover Report.*)

2 of Table 11-13 for 4-in. and larger horizontal branches. A duel relief vent for two circuit- or loop-vented horizontal branches in the same branch interval may be installed.

b. The limits for circuit- or loop-vented horizontal branches may be increased to one and one-half times the values in Table 11-13 for 3-in. and larger branches when relief vents are installed so that there is a relief vent inside the first fixture drain, the number of fixture units or fixture units outside the last relief vent does not exceed the limits given in columns 2 and 3 of Table 11-13 and the number of fixture drains between two successive relief vents does not exceed two for each

3-in., three for a 4-in., five for a 5-in., or eight for a 6-in. or larger horizontal branch.

12-9. Fixture at Base of Vent Stack. The following is quoted from the "Plumbing Manual" (pp. 20 and 21):[1]

Sec. 1012. Fixtures at Base of Main Vent. A group of not more than three fixtures, none of which discharges greasy wastes, may be installed on a main vent or vent stack, below the lowest branch vent, provided that the load does not exceed one-half the allowable load by Table 11-6 on a horizontal branch of the same diameter as the main vent.

12-10. Vents for Stack Offsets. Venting is required to relieve the pressure created by stack offsets. The "Plumbing Manual" states (p. 21):

Sec. 1017. Relief Vents for Offsets. The relief vent required for an offset, as prescribed in Sec. 12-5,[2] shall be installed either as a vertical continuation of the lower section of the soil or waste stack or as a side vent connected to the lower section of the soil or waste stack between the offset and the next fixture or horizontal branch below the offset. The size of the required relief vent shall be determined as follows:

a. If the stack vent from the upper section of the soil or waste stack is equal to that of the upper section, the relief vent shall not be smaller than the main vent on the stack system.

b. If the stack vent from the upper section of the soil or waste stack is smaller in diameter than that section, it may be of the same diameter as the main vent required, in which case the diameter of the relief vent for the offset shall be equal to that of the lower section and shall be extended to the open air without reduction in size or may be connected to the main vent stack, provided that the one to which it is connected is of equal or greater diameter.

If the horizontal branches connect to any soil or waste stack between two offsets, each offset shall be vented as required in this section.

The "National Plumbing Code" states (Sec. 12.18):

12.18.1. Offset Vents. Offsets less than 45° from the horizontal, in a soil or waste stack, except as permitted in Sec. 12-6[2] shall comply with Pars. 12.18.2 and 12.18.3.

12.18.2. Separate Venting. Such offsets may be vented as two separate soil or waste stacks, namely, the stack section below the offset and the stack section above the offset.

12.18.3. Offset Reliefs. Such offsets may be vented by installing a relief vent as a vertical continuation of the lower section of the stack or as a side vent connected to the lower section between the offset and the next lower fixture or horizontal branch. The upper section of the offset shall be provided with a yoke vent. The diameter of the offset shall be provided with a yoke vent. The

[1] *Ibid.*

[2] Of this book.

diameter of the vents shall not be less than the diameter of the main vent, or of the soil and waste stack, whichever is the smaller.

Offset vents are discussed in Sec. 12-5 of this book.

12-11. Roof Extensions and Vent Terminals.[1] In the installation of a soil, waste, or vent stack the stack is carried through the roof as shown in Figs. 11-1, 11-3, and others, or the vent stack may be revented to the soil or waste stack as shown at H and H' in Fig. 11-7 and carried through the roof as at T in Fig. 11-7. The condition shown above the lavatories on the top floor in Fig. 11-7 is known as a *stack vent*. The vent must be carried full size or larger through the roof and should terminate within 6 in. above the roof in a cold climate unless the roof is accessible to the residents of the building. Where the roof is accessible, the vent stack should extend at least 5 ft above the roof; its top should be covered with a durable and strong metal screen to exclude birds or the dropping of objects down the pipe, and in cold climates the stack must be protected against frost. The protection of stacks against frost is discussed in Sec. 12-13. Where there is danger of pipe closure by frost[2] no pipe less than 4 in. in diameter should extend above the roof. Smaller vent stacks within the building are increased to 4 in. in passing through the roof.

The opening through the roof is made raintight, somewhat as indicated in Fig. 12-5, by the use of flashing, which is described in Sec. 12-13 and Table II-70. The "National Plumbing Code" states (Sec. 3.2.2) that sheet copper for vent terminal flashing should not weigh less than 8 oz per sq ft. On shingle roofs, the flashing is extended under at least two courses of shingles above the vent pipe. On sloping roofs that are covered with tar paper, similar construction is used. On flat roofs of tar, asphalt, or cement, the flashing is placed between the courses of the roofing material and the finishing course is placed over it. On metal roofs, the flashing is soldered to the metal.

The termination of a vent pipe in a chimney or similar flue is undesirable because such ducts are not always gastight, chimney gases and soot may enter the vent pipes, and unusual drafts and pressures may endanger trap seals.

The roof terminal of a stack should be more than 10 ft away, horizontally, from any building line, door, window, air shaft, or other opening through which air may enter a building, or if the pipe must terminate less than 10 ft from such an opening, the pipe shall be extended at least 2 ft above the opening. If extended through a wall the vent pipe should be

[1] See also "National Plumbing Code," Sec. 12.4.

[2] See also "Frost Closure of Roof Vents in Plumbing Systems," National Bureau of Standards, Building Materials and Structures Report 142, 1955, and *Air Conditioning, Heating, and Ventilating*, February, 1955, p. 144, and April, 1955, p. 98.

turned to provide a horizontal opening facing downward. The opening should be screened, and the joint should be calked or otherwise sealed.

12-12. Vent Headers. The "National Plumbing Code" states (Sec. 12.20.1):

Connections of Vents. Stack vents and vent stacks may be connected into a common vent header at the top of the stacks and then extended to the open air at one point. This header shall be sized in accordance with the requirements of Table 11-11, the number of units being the sum of all units on all stacks connected

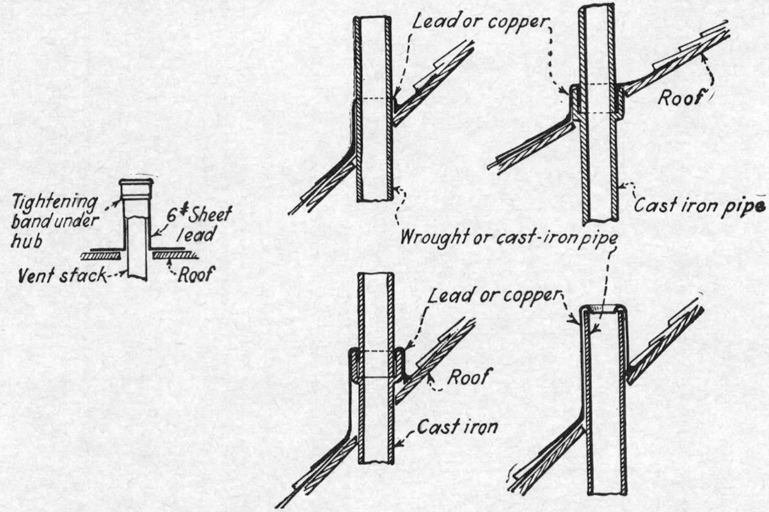

Fig. 12-5. Illustrating the use of flashing around a soil or vent stack passing through a roof. (*From Plumbers' Trade J., Apr.* 15, 1925.)

thereto and the developed length being the longest vent length from the intersection at the base of the most distant stack to the vent terminal in the open air as a direct extension of one stack.

12-13. Frost Protection of Stack Terminals. Protection against frost can be obtained by increasing the size of the pipe at least 12 in. below the roof and continuing the increased size through the roof. The increase in size should be made by the use of a long increaser, as shown in Fig. II-49. In very cold climates the flashing can be extended up outside the stack, leaving an annular space of 1 in. or more between the pipe and the flashing, as shown in Fig. 12-5. This air space is open at the bottom to the heat within the building. The upper end of the flashing can be bent over and calked into the bell at the top of the cast-iron stack, or it can be fastened to the pipe by a tightening band under the hub, as shown in the figure. Another method, less wasteful of heat from the building, is to encase the stack in a brick, tile, or other pipe protection and to pack the space between with a good insulating material. The stack opening should not be placed near a chimney on account of the danger of downdrafts carrying odors into the building.

12-14. Cowls. Cowls, as illustrated in Fig. 12-6, are sometimes placed on the top of stacks to prevent downdrafts or the dropping of objects down the stacks, as well as to increase the draft up the stack. Their general use is not recommended, however, since they stick and refuse to turn with the wind, thus increasing rather than preventing downdrafts. They are also more conducive to clogging by frost and to birds building

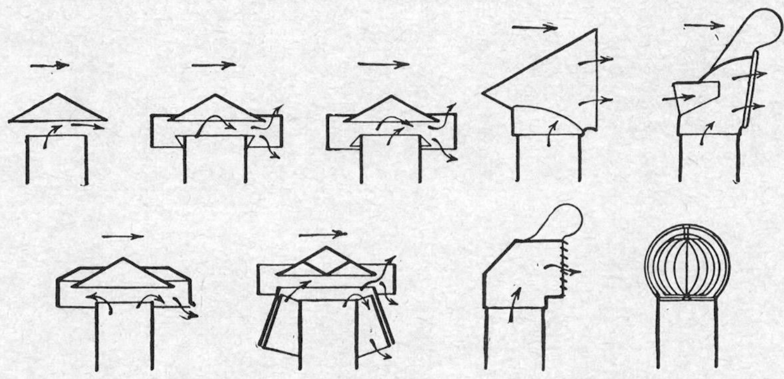

FIG. 12-6. Types of ventilating cowl used on plumbing stacks. (*From Plumbers' Trade J., Feb.* 15, 1925.)

nests in the cowl or the end of the stack. Power-driven roof ventilators are available in sizes capable of moving 200 to 2,000 cfm of air.

12-15. Effect of Wind. Wind blowing across the top of a stack seldom results in the breaking of a trap seal when it is properly vented. The Hoover Report[1] states (p. 128):

> In view of the small effects encountered it seems reasonable to assume that in small dwelling houses they will be taken care of by factors of safety referred to elsewhere.
>
> A considerable degree of protection would be afforded in tall buildings or other buildings exposed to very high winds by running the vent stacks and soil stacks separately through the roof and connecting them together above the highest fixture, thus giving an opportunity for relief of pressure differences above the fixture connections. Then only when the effect was of the same kind and intensity over both stacks would the full intensity reach the fixture traps.

12-16. Joints and Joint Materials. Materials and joints suitable for supply pipes, drainage pipes, and vent pipes are discussed in Sec. 7-21. Joints in drainage and vent pipes, not being subject to high pressures, can

[1] "Recommended Minimum Requirements for Plumbing in Dwellings and Similar Buildings," National Bureau of Standards, July 3, 1923.

be made differently from those in supply pipes. Joints are made tight in sewers to prevent liquid from leaking in or out and to prevent roots from entering. Watertight and roottight joints can be made economically.[1]

Types of joints in pipes under low pressure include solder or wiped joints, slip joints, and cement and various types of poured joints. Joints are made between metallic pipes also by brazing and welding and in copper or copper alloys, by hard facing.[2] Aluminum pieces are easily and economically joined.[3]

12-17. Prohibited Joints and Connections. The "National Plumbing Code" states (Secs. 4.7.1 and 4.7.2):

> Any fitting or connection which has an enlargement, chamber, or recess with a ledge, shoulder, or reduction of pipe area that offers an obstruction to flow through the drain is prohibited . . . . The enlargement of a 3-in. closet bend or stub to 4 in. shall not be considered an obstruction.

12-18. Solder or Wiped Joints.[4] Despite their long use in the development of the art of plumbing, wiped lead joints have fallen into disuse because of the greater ease, less skill, and greater economy possible in making many other kinds of joint. Solder or wiped joints are restricted to use on pipes that are not under high pressure. The "Plumbing Manual"[5] states:

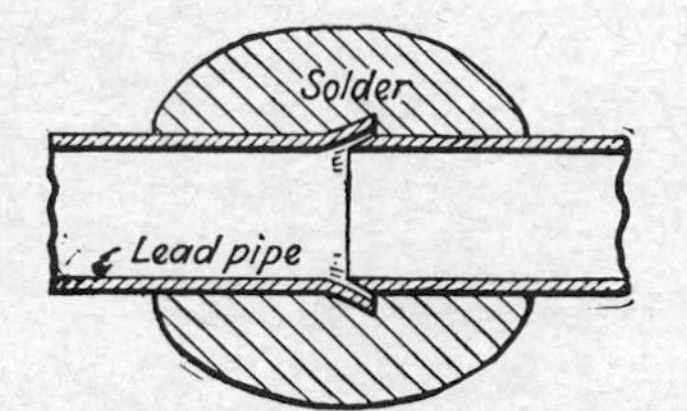

Fig. 12-7. A wiped joint.

> **Sec. 407. Wiped Joints.** Wiped joints in lead pipe, or between lead pipe and brass or copper pipes, ferrules, soldering nipples, bushings, or traps, in all cases on the sewer side of the trap and in concealed joints on the inlet side of the trap, shall be full-wiped joints, with an exposed surface of the solder on each side of the joint not less than ¾ in., and a minimum thickness of the thickest part of the joint of not less than ⅜ in. Where a round joint is made, a thickness of not less than ⅜ in. for bushings and flanges shall be provided.

A section through a wiped joint is shown in Fig. 12-7. To wipe a joint successfully requires skill and practice that no amount of reading will perfect.

[1] See also *Brick & Clay Record*, March, 1955, p. 57.

[2] See also *Am. Machinist*, vol. 99, p. 54, mid-November, 1956, and Jan. 2, 1956, p. 81; *Materials & Methods*, December, 1955, p. 152, and September, 1956, p. 158.

[3] See also *Iron Age*, Mar. 1, 1956, p. 71, and *Chem. Eng. News*, Jan. 30, 1956, p. 498.

[4] See also J. B. Mohler, Designing for Soldered Joints, *Machine Design*, Jan. 2, 1956, p. 81, and June 14, 1956, p. 123; *Iron Age*, May 10, 1956, p. 102; and A. I. Heim, Proper Procedure for Making Soldered Joints in Copper Water Tubes, *Plumbing and Heating J.*, February, 1952, p. 60.

[5] Report BMS66, National Bureau of Standards, 1940.

12-19. Solder.[1] Solder for wiping joints is an alloy of lead and tin, approximately in proportion of 3 parts of lead to 2 parts of tin for ordinary joints combining lead to lead or other metal. Other compositions of solder are shown in Table 12-2. The exact ratio of lead to tin depends to

TABLE 12-2. COMPOSITIONS OF SOLDERS
(Federal specifications*†‡)

Type of solder	Percentage of constituent, by weight					
Brazing*	Copper	Tin	Lead, max.	Iron, max.	Aluminum, max.	Zinc
A	49–52	None	0.50	0.10	0.10	Remainder
B	49–52	3–4	0.50	0.10	....	Remainder
C	68–72	None	0.30	0.10	....	Remainder
D	78–82	None	0.20	0.10	....	Remainder
Silver†	**Silver**	**Copper**	**Zinc**	**Phosphorus**	**Cadmium**	**Impurities**
0	19–21	44–46	33–37	....	....	0.15
1	44–46	29–31	23–27	....	....	0.15
2	64–66	19–21	31–17	....	....	0.15

Tin-lead‡	Tin plus lead	Tin	Antimony, max.	Copper, max.	Iron, max.	Bismuth, max.	Zinc, max.	Aluminum, max.	Others	Approx melting range, °F	
										Solids	Liquids
A	99.2	49–51	0.40	0.08	0.02	0.25	0.005	0.005	0.08	360	420
B	99.2	38–42	0.40	0.08	0.02	0.25	0.005	0.005	0.08	360	460
D	97.6	34–36	§	0.08	0.02	0.25	0.005	0.005	0.50	360–365	490–500
E	98.3	28–32	0.75	0.15	0.02	0.25	0.005	0.005	0.50	360	500–510
F	99.2	69–71	0.40	0.08	0.02	0.25	0.005	0.005	0.08	360	378
G	98.3	18–22	0.75	0.15	0.02	0.25	0.005	0.005	0.50	360	525–545
H	99.2	59–61	0.40	0.08	0.02	0.25	0.005	0.005	0.08	360	372

* QQ-S-551
† QQ-S-561b
‡ QQ-S-571a
§ 0.75–1.50

some extent on the preference of the workman. Too much tin makes a "fine" joint with smooth appearance and requires skill to make because of the tendency of the solder to stick to the soldering cloth. Too much lead makes a "coarse" joint with a grainy, porous appearance which may leak.

Fluxes used in soldering are shown in Table 12-3.

[1] See also L. S. Taylor, "Successful Soldering," McGraw-Hill Book Company, Inc., New York, 1943.

Plumber's soil, used in making wiped joints, is a composition of glue, lampblack, and water with a consistency of a thin paste. It is applied with a paint brush to the ends of the pipes to be joined. If black rubs off, there is too much lampblack. If the soil cracks and falls off when the pipe is bent, there is too much glue. Proprietary plumber's soil is available on the market.

TABLE 12-3. FLUXES

Flux	Used with	Metals to be joined
Resin	Copper bit or blow pipe	Lead, tin, copper, brass, and tinned metals
Tallow, unsalted	Blow pipe or wiping process	Lead, tin, or tinned metals
Sal ammoniac	Copper bit or blow pipe	Copper, brass, iron
Muriatic acid	Copper bit	Dirty zinc
Chloride of zinc	Copper bit or blow pipe	Clean zinc, copper, brass, tinned metals
Resin and sweet oil	Copper bit or blow pipe	Lead and tin metals
Borax	Blow pipe	Iron, steel, copper, and brass

12-20. Sweat Joints.[1] Sweat joints, also called brazed joints, are used on copper pipe. The ends of the pipes or fittings to be joined are first sandpapered to remove grease and dirt, and a small amount of nonacid flux is applied to the surfaces to be joined. The pipe is inserted into the fitting, and both are heated sufficiently to melt the solder. Solder is then applied through a small hole in the side of the fitting, or it can be applied to the face of the heated fitting before the pipe is slipped into it.

12-21. Slip Joints. A slip joint is made by slipping one pipe snugly into another. A lubricant or joint filler is sometimes spread on the outer surface of the inside piece before the other pipe is inserted into it. Since it is not possible to make such a joint airtight and watertight under all conditions, the joint should be used, according to the "National Plumbing Code" (Sec. 4.4.3), "only on the inlet side of the trap or in the trap seal and on exposed fixture supply." The Code goes on to state that "Ground-joint brass connections which allow adjustment of tubing provide a rigid joint when made up shall not be considered as slip joints."

Slip joints are easily made, they will take up vibrations or movements between fixture and the piping without injury to either, and as the water pressure, if any, is low, leakage does not occur.

12-22. Cup Joints. A cup joint is formed by inserting the square-cut or slanted-cut end of one pipe into the flared end of another pipe and filling

[1] See also J. I. Heim, How to Braze Copper Tube, *Air Conditioning, Heating, and Ventilating*, November, 1957, p. 72.

the space between them with solder. Such joints are commonly made only in lead pipe and for conditions of low or no pressure.

12-23. Joints in Vitrified-clay or Cement Pipes. Joints used in vitrified-clay and small concrete pipes include bell-and-spigot, push-together, rubber-ring, open, and proprietary joints. The bell-and-spigot is probably the most common. Materials used for filling the space in the bell-and-spigot joint shown in Fig. 12-8 include cement, sand, tar, rubber, asphalt, sulfur, and plastic gaskets.

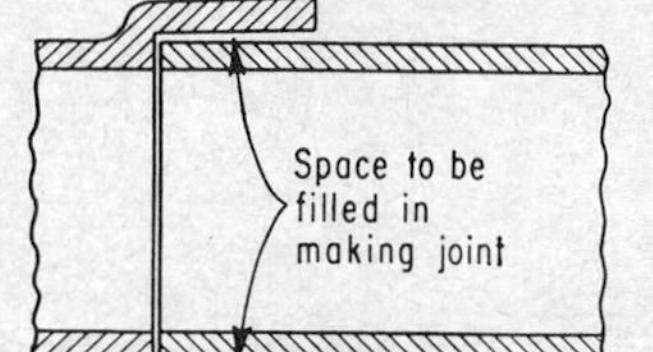

Fig. 12-8. Bell-and-spigot joint in clay pipe.

12-24. Cement Joints in Bell-and-spigot Pipes. Cement joints for bell-and-spigot-end pipes are made with portland cement either *neat*, i.e., cement and water only, or as a grout or mortar, i.e., a mixture of cement and sand with water mixed to the consistency of a stiff paste. Cement joints in bell-and-spigot pipes of any material are made as follows:

A length of closely twisted hemp or oakum gasket is selected to pass around the pipe and overlap about 2 in. The thickness of the gasket should be sufficient to support the spigot concentrically in the bell. The spigot and bell to be joined are cleaned and moistened, and the gasket, which has been soaking in cement grout, is laid in the bell for the lower third of the circumference of the joint and is covered with cement mortar. If it is desired to use less gasket material, the joint can be made by spreading over the lower third of the bell a layer of mortar stiff enough to support the inserted spigot concentrically. The spigot is inserted, carefully pushed into place until it touches the base of the bell, and adjusted by calking the gasket or mortar to assure that the spigot is concentric in the bell. A small amount of mortar is inserted in the annular space around the joint and firmly calked. The remainder of the joint is filled with mortar and beveled off smoothly at an angle of about 45° with the outside of the pipe.

After the joint is completed, the interior of the pipe should be scraped clean of mortar, and no foreign matter should be left in the pipe. A loosely fitting plug, called a "go devil," can be pulled forward to close the mouth of the pipe against the entrance of any object until the next section of pipe has been set.

The cement mortar used is a mixture of equal parts of portland cement and building sand mixed with just sufficient water to make the mortar workable. Within this limitation the less water used, the better the joint. The cement worker should wear rubber gloves while making the joint, as cement causes sores on bare hands.

12-25. Poured Joints in Vitrified-clay or Concrete Pipe. Poured joints in bell-and-spigot pipe are made with cement grout mixed to a thin consistency, sulfur and sand, and asphalt or bituminous compound made of vulcanized linseed oil, clay, and other substances, the resulting compound having the appearance of vulcanized rubber or coal tar. It is generally specified that the materials in poured joints in drainage pipes shall not soften at a temperature below 160°F and shall be insoluble in the wastes carried in the pipes.

When a poured joint is made with cement grout, the pipes are set in place and oakum or some other gasket material is calked into the bell of the joint. Grout is poured into the remaining space with the aid of a joint runner to hold it in place, somewhat as described for joints in cast-iron pipe in Sec. 7-23. The cement is allowed to set before the runner is removed. The joint is not particularly satisfactory, since it is likely to crack and is easily broken by movements of the pipe. It is more likely to be watertight than an ordinary cement joint because there is greater probability that the bottom of the joint is completed. Since the joints are usually made with the pipes in a horizontal position, a runner must be used around the bell of the pipe to hold the cement in place until it has set. The joint may be slightly more expensive than an ordinary cement joint made by hand.

Sulfur and sand are relatively inexpensive, are comparatively easy to combine, and will form a watertight and rigid joint stronger than the pipe itself. The rigidity and strength of the joint sometimes result in cracking the pipe, and the joint materials are objected to on that account. In making the mixture powdered sulfur and fine sand are mixed in equal proportions. It is essential that the sand be fine so that it will mix well with the sulfur and not precipitate out when the sulfur is melted. Ninety per cent of the sand should pass a No. 100 sieve, and 50 per cent should pass a No. 200 sieve. The mixture of sulfur and sand liquefies at about 260°F and does not soften at lower temperatures. It is poured into the joints while hot and liquid. This requires calking of the joint, the use of an asbestos gasket, and care in handling the hot compound.

Bituminous compounds fill nearly all the ideal conditions for a joint compound in bell-and-spigot drainage pipes outside a building except cost and ease in handling. If overheated or heated too long, they will carbonize and become brittle. When poured in cold weather, they do not stick to the pipe well unless it is heated before the joint is poured. On some work, joints have been poured under water with these compounds, but success is doubtful without skillful and experienced handling. An overheated compound will make steam in the joint, causing explosions that will blow the joint clean, and an underheated compound will harden before the joint is completed. The materials should be heated in an iron kettle over a gasoline furnace or other controllable fire until they just commence to bubble and are of a consistency of fine sirup. Only a sufficient quantity of material for immediate use should be prepared, and it should be used within 10 to 15 min after it has become heated. The ladle used should be large enough to pour the entire joint without refilling.

When a poured joint is made with a heated material, the pipes are first lined up in position. A hemp or oakum gasket is forced into the joint to fill the space of about ¾ in. An asbestos or other heat-resisting gasket such as a rubber hose smeared with clay, as shown in Fig. 7-17, is forced about ½ in. into the opening between the bell and the spigot, and the melted compound is poured down one side of the pipe, through a hole broken in the bell, until it appears on the other side and the hole is filled. In pouring cement-grout joints a paper gasket is used instead of the heat-resisting material. It is held to the bell and spigot by drawstrings. Greater speed and economy can be attained by joining two or three lengths of pipe on the bank and lowering them as a unit into the trench. The pipes are set up in a vertical position, and the joints poured without the use of an outer gasket. When a gasket is used, it should be removed as soon as possible to prevent sticking.

12-26. Joints between Cast-iron and Vitrified-clay Pipe. Joints between bell-and-spigot cast-iron and vitrified-clay pipe can be made where the cast-iron stack, house drain, or house sewer joins the vitrified-clay house sewer or common sewer. In making the joint the spigot end of the cast-iron pipe or fitting is thrust into the bell end of the clay pipe to the shoulder. Where the spigot end of a clay pipe is to join the bell end of a cast-iron pipe, a cast-iron increaser of the type shown in Fig. II-49 must be used, since the clay spigot is too large to be inserted into the cast-iron hub of the same nominal diameter. The iron pipe is held in position with well-calked oakum, and the joint is completed as described in Sec. 12-25.

12-27. Plastic Gaskets.[1] Polyvinyl chloride molded gaskets form a watertight and root-resistant joint for use in bell-and-spigot joints on specially prepared clay pipes. The plastic joint is made in the field by painting the inside of the bell with a plastic solvent and shoving the spigot with the gasket home in the bell. The plastic material is relatively inert to acid, alkali, oil, and grease, and it is flexible enough to allow some movement in the joint. A great advantage of the process is the lack of loss of time. The trench can be backfilled immediately after the pipe is in place, and water can be passed through the pipe. The making of the joint requires no special skill, and if necessary, the joint can be made under water. In some joints the gaskets are lubricated with soap solution and the pipes are pushed together with a pry bar. In other joints a threaded vinyl plastic ring is hot-molded into place. Joints are made also with corresponding hard-molded, phenolic, plastic rings.

[1] See also *Air Conditioning, Heating, and Ventilating*, March, 1956, p. 139.

12-28. Inspections and Tests. After completion of the roughing-in of a plumbing system, it is customary to test the system to locate leaky connections or faulty installations, and on the completion of the setting of the fixtures a further test is made to locate leaky traps or other unsatisfactory conditions. Where the best quality of work is desired, tests should be made during the progress of construction. This assures more careful workmanship, and where repairs are necessary, they are more easily made than after the completion of the job. A final test can be made after the fixtures have been installed but before the job is accepted.

Tests used are the water test, the air test, the odor test, and the smoke test. The first two can be used on the completion of the roughing-in, and the last two are used for finishing tests under low pressure.

12-29. The Water Test. For the water test all openings in the drainage pipes are closed except at the tops of the stacks. The openings can be closed by test plugs, one type of which is shown in Fig. 12-9, or by other means. The plug is inserted in the open end of a pipe or fitting so that the heavy rubber gasket fits snugly all around. Handle *A* is then screwed down while handle *D* is held motionless. Washer *B* is thus forced down onto the rubber gasket *C*, causing it to expand and press against the side of the pipe. A somewhat similar device, suitable for tests where the pressures exceed 50 ft of water, is held on the pipe by means of a collar clamped around the pipe.

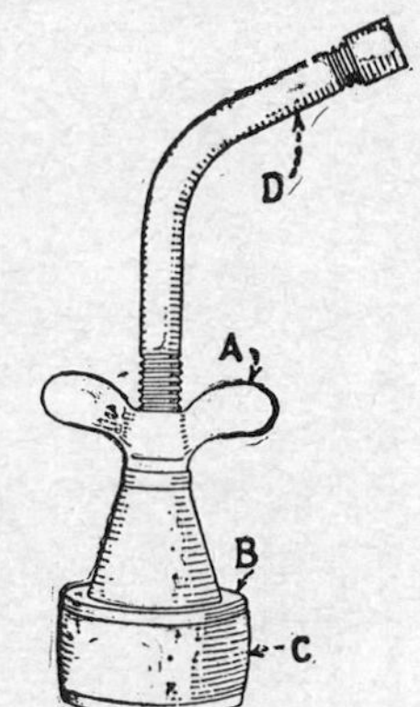

Fig. 12-9. Test plug.

Where test pressures are less than about 5 psi and the pipes are large, as in vitrified-clay and concrete pipes, other types of plug used include a bag of sand or clay around which oakum or similar material is lightly calked, a wooden diaphragm surrounded by an inflated inner tube for a tire and supported by a sand bag, a proprietary rubber diaphragm with flaring edge, and a quick-setting plug of plaster of paris and portland cement formed in the pipe.

When a test is to be made on a plumbing system already connected to the sewer and a main or running trap is in place on the house drain, the trap can be plugged by a special trap-testing plug that is usually supplied with a valve through which water can be discharged into the plumbing pipes or the trap can be filled with well-tamped clay and water put into the pipes through some other opening. Where there is no running trap, the house drain can be plugged by inserting a test plug through a cleanout opening or, as a final resort, a section of the pipe can be broken out.

When all openings, except at the tops of the stacks, are closed, water is run into the pipes until it overflows from the top of a stack. The water can be discharged into the top of a stack by means of a hose, through a connection temporarily soldered in a cleanout or other convenient opening, or through the hollow handle of specially made test plugs. The dropping of the level of the water after the pipes are full and the supply has been cut off will indicate a leak that must be found by inspection. During a satisfactory test the water level in the pipe being tested should remain stationary for not less than 15 min and preferably longer. All parts of the system should be subjected to a pressure of at least 10 ft of water. It is not desirable to use pressures over 30 to 40 ft of water. Higher stacks should be tested in sections of 10 to 40 ft. In very tall buildings this necessitates beginning at the top section. The top section is then connected to the next lower section, and the pipes are filled to a height 10 ft above the connection between the two sections. In this manner, no portion of the pipe will be subjected to a pressure of less than 10 ft of water.

12-30. The Air Test. For the air test, all openings are closed and an air pressure of at least 5 psi is exerted in the pipes for at least 15 min. The falling of the pressure, as shown by a sensitive gage attached to the pipes, will indicate a leak. An air leak can usually be detected by sound. Where but little or no sound is made and a leak is suspected, its location can be found by applying saliva or soap solution to the suspected spot. Air tests are useful in cold weather when there is danger of water freezing in making the water test. Another advantage of the air test as compared with the water test is the uniformity of pressure throughout the section under test. The air test is not so simple to apply as the water test, however, and the discovery of leaks is sometimes more difficult.

In making final tests after the fixtures are in place and the traps are filled with water, the highest permissible pressure to be used in a test is about 1 in. of water. If, after all openings have been closed, this pressure can be maintained for 15 min without additional pumping, the system can be considered to be airtight. There may be some difficulty in interpreting the pressure readings in a test because the pressure may fall slowly owing to cooling of the air and not because of a leak. If there is doubt, the air pressure should be restored to the original value without releasing air already in the pipes. If the rate of decrease of pressure is slower, the loss of pressure is probably due to temperature change and not to a leak. If the pipes are found not to be airtight by this test it is necessary to locate the leak.

12-31. The Smoke Test. For the smoke test, a thick smoke is made by burning oily waste, tar paper, or similar material in the combustion chamber of a smoke-test machine. The use of smoke made by mixing chemicals, such as ammonia and muriatic acid, is not recommended mainly because of the precipitation of such smoke particles in the pipes and the loss of their effectiveness. The smoke machine is connected to the lower end of the drainage system, and the bellows are then operated to fill the pipes with smoke. When the smoke issues from the top of the stack, the stack is closed with a test plug and the pressure run up to 1 in. of water and held until end of the test. A metal drum or tub with an open top is inverted in a larger tub of water and is connected with the piping to be tested. The air pressure in the piping and under the inverted tub is raised to about 1 in. of water. If the tub sinks, there is a leak to be found. To find it an inspection is made to locate smoke or odor issuing from the leak. Since the leaks are sometimes too small for the escaping smoke to be seen, windows and doors should be closed to retain the odor of the escaping smoke. The leak can be found by applying soap solution or saliva to suspected spots.

12-32. The Odor Test. The odor test can be made by the use of oil of peppermint, ether, or similar volatile and odoriferous substance. When this test is applied, the outlet end of the drainage system and all vent openings except the top of one stack should be closed. About 1 oz of oil of peppermint for every 25 ft of stack, but not less than 2 oz, is emptied down the stack, followed by a gallon or more of boiling water. The top of the stack is closed immediately, and the leak searched for by the sense of smell. Anyone who has recently handled peppermint should not enter the building until the search has been completed. The odor test is simpler than the smoke test, but results are not always satisfactory because there may have been insufficient pressure in the pipes to force the odor through the leak or leaks and it may be difficult to locate the leak after the odor is detected.

CHAPTER 13

SEWAGE AND DRAINAGE PUMPS AND DRAINS

13-1. Needs for Sewage and Drainage Pumps. It is not an unusual condition to find either the sanitary sewer or the storm sewer, or both, at a higher elevation than the lowest point within the building or on the property to be drained. In large cities the construction of subbasements below the elevation of city sewers is customary. Typical plumbing for such conditions is shown in Fig. 11-1. A pump is required to lift into the building drain or into the building sewer only that portion of the sewage or drainage originating below the elevation necessary for its flow into the sewer by gravity.

13-2. Types of Sewage and Drainage Pump. Three types of pump are commonly used for ejecting sewage or drainage from a premises: centrifugal, as shown in Fig. 5-15; air displacement, as shown in Fig. 13-1; and water or steam ejector. Piston, plunger, or other displacement pumps are not suitable for such service because of the solid particles or grit in sewage and in much of the surface drainage to be removed.

Sewage and drainage pumps in buildings are usually controlled automatically, the control being by means of a float that sets the pump into motion when the sump is full and stops the pump when the sump is empty. Apparatus for ejecting sewage should be installed in duplicate so that when one pump requires maintenance attention, the other can be in operation.

13-3. Capacities of Sumps and Pumps. The required capacity of the pump is affected by the rates of flow of drainage and by the size of the sump. Where there is no sump, the capacity of the pump or pumps must equal or exceed the maximum rate of drainage flow. Where the sump is large enough to permit the pump to operate continuously, the capacity of the pump must equal or exceed only the *average* rate of drainage flow. Since information on fluctuations in drainage flow rates is usually inadequate and the capacity of a sump is limited to a maximum retention period of 12 hr,[1] it is not practicable to select a pump that will run continuously. The rate of flow on which the capacity of the pump is based should equal the sum of all the fixture units draining into it with allowance

[1] "National Plumbing Code," Sec. 11.7.2.

for the nonsimultaneous use of fixtures as discussed in Sec. 15-18 and as shown in Fig. 3-1. For example, if all fixtures discharging simultaneously delivered 1,000 gpm, the actual flow on which the capacity of the pump should be based, according to Fig. 3-1, is 210 gpm. The head against which the pump discharges is the static lift plus friction and velocity heads.

In the ordinary installation in the basement of a building where an overflow of sewage or drainage might be disastrous, the pump should have

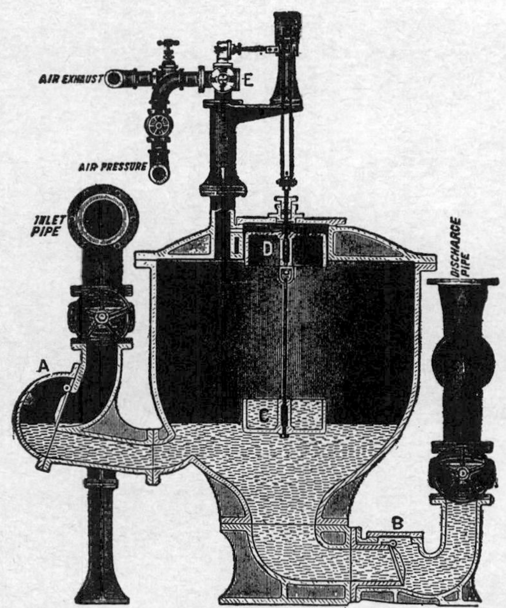

FIG. 13-1. Compressed-air sewage ejector. (*Yeomans Brothers Co.*)

a rated capacity equal to or greater than the estimated maximum rate of flow and the sump should be of such a size that it will require about 15 to 60 min to be filled under the maximum rate of flow, the longer period of storage being used for smaller installations. Where automatic control is not used, the sump should fill in not more than 12 hr and the sump should be empty when the pump stops. Pumps should be installed in duplicate in all but the smallest establishments, each pump having sufficient capacity to carry the maximum load. The "National Plumbing Code" states (Sec. 11.7.5) that duplex pumping equipment must be provided when the discharge is more than six water closets.

13-4. Automatic Sump Pumps. The operation of sump pumps for the removal of drainage from buildings should be automatic. A common type of automatic float-controlled centrifugal pump for a residential installation is shown in Fig. 5-15. The pump goes into operation when the rising float in the basin actuates the starting switch on the motor. The power circuit is broken when the float descends to the bottom of the basin. A check valve is necessary, and a gate valve should be provided

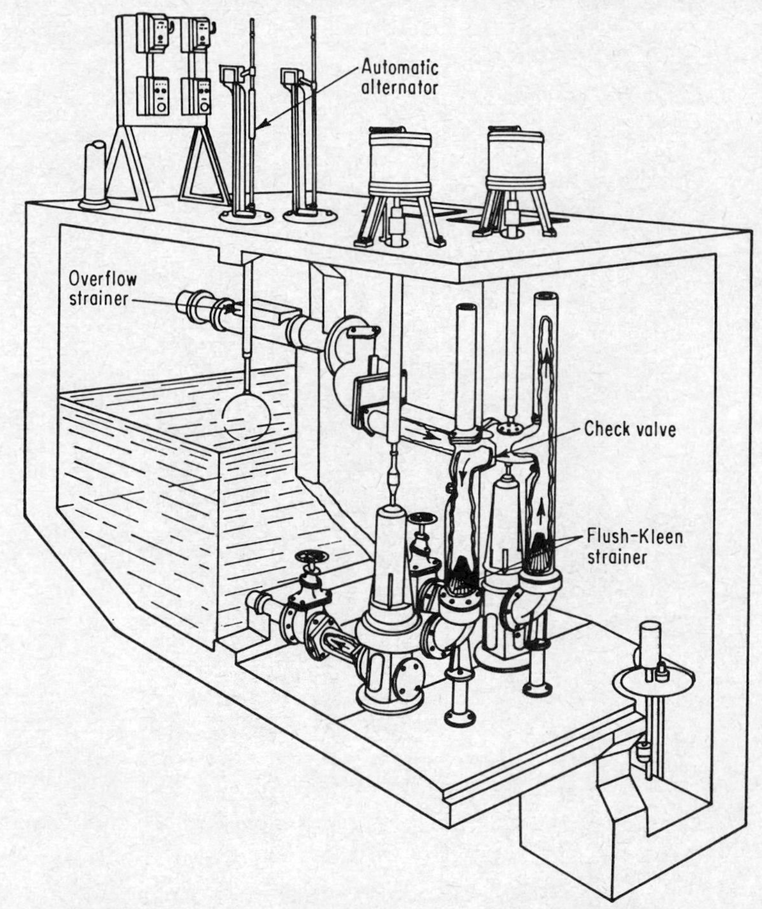

FIG. 13-2. Flush-Kleen pumps. When the pump on the right is discharging, sewage flows from the sewer through the discharge pipe of the pump on the left and into the wet well. When the wet well is empty, the pump on the right stops and the wet well refills. Then the pump on the left starts, closing the check valve on its discharge pipe, diverting sewage to the pump on the right, and throwing sewage from the wet well through the screen in its discharge pipe, which is thus cleaned. The discharge enters the building sewer or other conduit or discharge point. (*Chicago Pump Co.*)

on the discharge line. The check valve prevents backflow of sewage, and the gate valve, located between the check valve and the sewer, can be closed when repairs are necessary.

Sump pumps can be so arranged that when the pump is not operating, drainage flows through the screen on the *discharge* line of the pump, through the pump, and from the suction line into the sump. When the sump fills, the pump goes into operation automatically, closing a check valve on the drainage line. The direction of flow through the pump is reversed, and the screen on the discharge line of the pump is automatically flushed by the discharge from the pump. An arrangement of two such

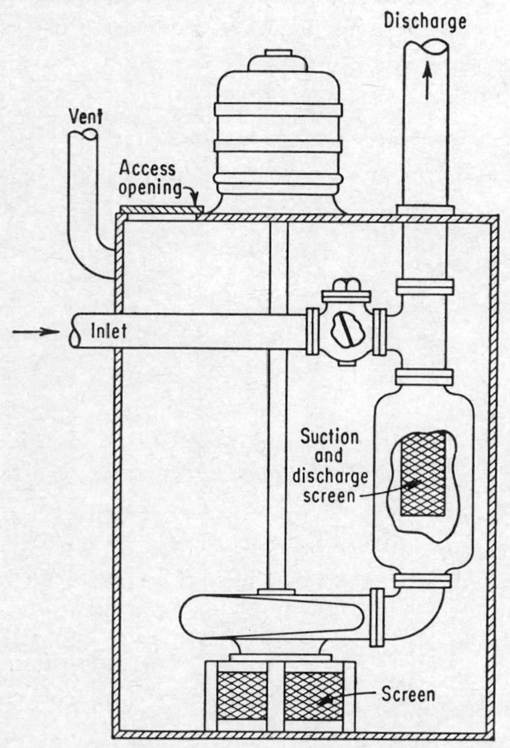

FIG. 13-3. Single, self-cleansing sump pump.

pumps which operate alternately is shown in Fig. 13-2, and a single pump of the type is shown in Fig. 13-3.

The sump pit can be constructed of concrete, tile, or metal, somewhat as shown in Fig. 13-4. The sump pit should be vented to the air outside the building to avoid the creation of nonatmospheric pressures in the sump and to remove odors from the building.

13-5. Air-displacement Pumps. The air-displacement sewage pump or compressed-air sewage ejector, such as is shown in Fig. 13-1, is often used for pumping sewage from buildings. Among its advantages for this service may be included the following: It is enclosed, thus preventing the escape of sewer air into the building. Its operation can be made completely automatic so that it goes into service only when there is need. The few moving parts in contact with sewage require little or no attention and no lubrication, and they are not easily clogged. Under normal conditions it will function for long periods of time without attention. A disadvantage of this type of pump is the need for an air compressor.

The pump shown in Fig. 13-1 operates as follows: Sewage enters the reservoir through the inlet pipe at the left, the air displaced being slowly expelled through the air exhaust, which is open to the air through a vent pipe. The rising sewage lifts the float *D*, which opens the valve in the

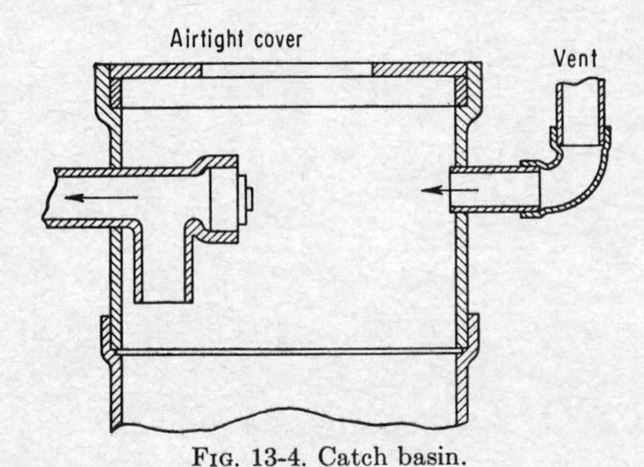

FIG. 13-4. Catch basin.

pipe above the reservoir when the reservoir fills. The opening of the valve admits compressed air to the reservoir. The air pressure closes the air exhaust and the inlet valve at the left and ejects the sewage through the discharge valve *B* and pipe at the right. As the chamber *C* drops with the descending sewage, it shuts off the air supply and opens the air exhaust through the small pipe at the top center. Sewage is prevented from flowing back into the reservoir by the check valve in the discharge pipe. The air vent from the compressor should be carried to the open air as a vent terminal[1] as provided in Sec. 12-11 and should be sized in accordance with Tables 11-11 to 11-13.

The capacity of air tanks and the intensity of air pressure should be so proportioned and the air compressors so designed that there will always be a sufficient volume of air at a pressure sufficient to raise all the sewage or water in the receiving sump at least 50 per cent higher than the maximum lift required.

The capacities of air compressors are stated in Table 13-1. These pumps are available in capacities from 50 to 1,000 gpm. The capacity of the air compressor required is dependent on its efficiency, the height of lift, and the quantity of sewage to be lifted. The following formula

[1] See *ibid.*, Sec. 11.7.7.

will give the approximate capacity of the air compressor required:

$$A = \frac{S(34 + h)}{255}$$

where A = free air per minute drawn in by air compressor, i.e., capacity of compressor, cu ft

h = lift against which sewage is to be discharged, ft

S = rate of flow of sewage during discharge, gpm

It is customary to place an air receiver between the compressor and the pump to permit the use of a smaller compressor which will operate for a longer period of time. The compressor may be driven by an electric motor or hydraulically by the building water supply. The water used

TABLE 13-1. APPROXIMATE CAPACITIES OF AIR COMPRESSORS REQUIRED TO OPERATE AIR-DISPLACEMENT PUMPS

Water pressure, psi	Cu ft free air per gal water	Water pressure, psi	Cu ft free air per gal water	Water pressure, psi	Cu ft free air per gal water	Water pressure, psi	Cu ft free air per gal water	Water pressure, psi	Cu ft free air per gal water
5	0.179	40	0.497	70	0.815	100	1.043	130	1.316
10	0.224	50	0.588	80	0.861	110	1.134	140	1.407
20	0.315	60	0.679	90	0.952	120	1.225	150	1.498
30	0.406								

by the hydraulically driven air compressor can be made available for other uses.

13-6. Centrifugal Pumps. Centrifugal pumps are frequently used for pumping drainage water, other than sanitary sewage, from an "open" sump. A typical pump installation is shown in Fig. 5-15. Because the sump is "open," that is, the air in the sump is open to the atmosphere in the building, such an installation is not suitable for pumping sanitary sewage from a building unless the pump is placed outside. To remove sewage from a building, the centrifugal pump and sump must be totally enclosed or the sump must not be in the building. If sanitary sewage is to be pumped, a special type of nonclog impeller is required unless a self-cleaning type of screen is used on the combined inlet-discharge pipe, as explained in Sec. 13-4 and as shown in Figs. 13-2 and 13-3.

13-7. Water-ejector Sump Pump. A drainage pump operated by a water ejector is illustrated in Fig. 13-5. The ejector must not be connected to a water supply used for any domestic purpose. The ejector operates because the velocity of water passing through the throat of the jet is retarded by the expansion of the stream beyond the jet. The stream expands because of the increase in size of the channel beyond the throat. The expansion of the jet creates a vacuum which causes suction on the inlet pipe. The velocity energy of the water is converted into pressure energy which is available for lifting water from the sump. Such a pump is simple and reliable and requires little or no attention in operation and no lubrication, and the power for its operation can be obtained from the public water supply. For this reason it is particularly adapted to use in small buildings where the mechanical equipment must be of the simplest kind and where the cost of water is low. The capacities of Penberthy hydraulic ejectors are given in Table 13-2. The discharge lift of the pump is limited to about 12 to 18 ft, and the discharge line should be arranged to prevent the backflow of air. This can be done by submerging the end of the discharge pipe or by other means. The operation of the pump requires about 4 to 5 psi pressure on the power water-supply line for each foot of lift and about 1 gal of supply water for each gallon of water discharged for lifts up to about 8 ft. In order to obtain the best efficiency it is recommended that the ejector be set so that all the lift is in the suction pipe and that the total lift is less than 8 ft.

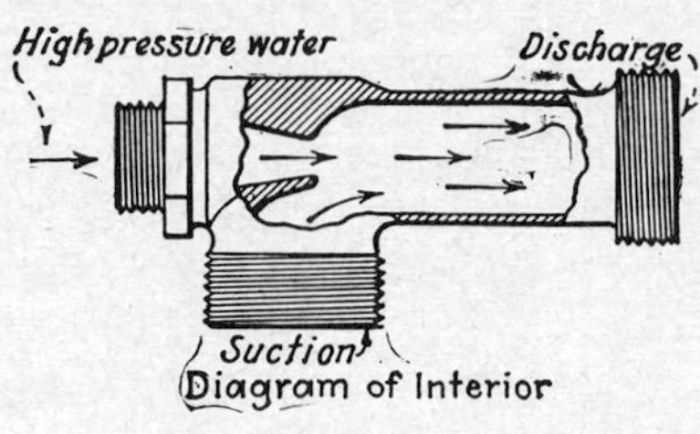

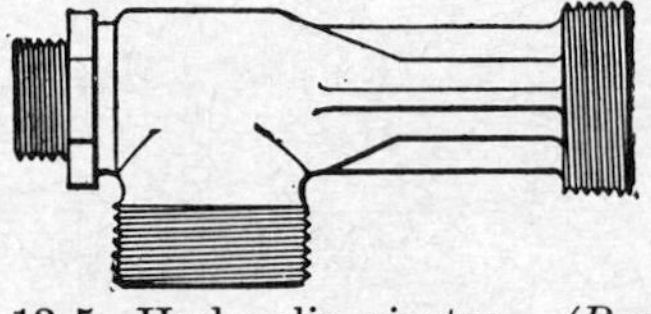

FIG. 13-5. Hydraulic ejector. (*Penberthy Injector Co.*)

TABLE 13-2. CAPACITIES OF PENBERTHY HYDRAULIC EJECTORS

Size number		62	63	64	65	66	67	68
Water-supply inlet, in.		3/8	1/2	3/4	1	1 1/4	1 1/2	2
Suction pipe, in.		3/4	1	1 1/4	1 1/2	2	2 1/2	3
Discharge pipe, in.		3/4	1	1 1/4	1 1/2	2	2 1/2	3
Length, in.		5	6 3/4	8	9	10 1/2	11 1/2	13 1/4
Capacity, gal per hr	40 psi	500	900	1,500	2,100	3,000	4,200	6,000
	60 to 80 psi	750	1,000	2,000	3,400	4,500	6,000	9,000

Capacities are given with 10-ft head and include the operating water. The ejector suction should not exceed 22 ft. The total lift, suction plus discharge, can be about 1 ft for every 4 or 5 lb of water pressure.

13-8. Ventilation of Closed Sumps. In the construction of "closed" sumps for the reception of sewage or drainage, ventilation must be provided to avoid the development of back pressures. This is particularly essential when air-displacement pumps are used, since there may be a discharge of sewage or drainage in the house-drainage pipes when the inlet valve to the sump is closed or submerged.

Each sump should be provided with a vent pipe connected to the highest practicable point in the tank unless the tank is of the air-displacement type that requires a completely closed chamber. When the design of the receiving tank is such that the house drain entering the tank may be closed at any time, a vent pipe should be provided on the house drain as near as possible to but below the lowest fixture, soil, or waste-pipe connection. The size and terminal arrangements of the vent pipe should be in accord with the requirements for vents as stated in Sec. 12-11.

13-9. Rain-water Gutters and Leaders.[1] Rain-water gutters and leaders are provided to catch the rain water falling on roofs or other catchment areas above the ground and to conduct it to some point of discharge, generally the ground surface or an underground drain. It is usually forbidden to discharge such drainage into the sanitary sewer. Materials for rain-water gutters and leaders are discussed in Sec. 13-11.

Rain-water leaders can be placed either inside or outside a building. When inside they are sometimes called *conductors*. When outside they may be called *leaders*. Among conditions favoring an inside location are convenience, safety, appearance, and freedom from freezing. Most rain-water leaders are placed on the outside of buildings, however, because of the lower cost.

Rain water can be collected in a cistern in rural or remote districts where no public water supply is available or soft water is desired. The rain-water leader can empty directly into the cistern, which will be provided with an overflow, or a valve, as shown in Fig. 3-13, can be used on the leader to divert any portion of the roof water that is not wanted, such as from the first rush of a storm which is laden with dirt from the roof or when the cistern is already full.

13-10. Sizes of Gutters and Leaders. Sizes of gutters and leaders are listed in Tables 13-3 and 13-4 and of gutters alone are shown in Table 13-5.

All roof or other watershed areas are to be computed as the horizontal projection of the area. Recommendations for gutter, leader, and drain sizes are based on a rainfall of 4 in. per hr unless otherwise stated. If another rate of rainfall is assumed in design, the roof areas are to be divided by 4 and multiplied by the assumed rainfall rate expressed in inches per hour.

The sizes of rain-water leaders, as given in the Hoover Report[2] and shown in Table 13-4, are based on the diameter of circular leaders, and the sizes of gutters shown in Table 13-5 are based on semicircular sheet-

[1] See also L. Blendermann, Storm Drainage System for Flat Roofs, *Air Conditioning, Heating, and Ventilating*, August, 1956, p. 78.

[2] "Recommended Minimum Requirements for Plumbing in Dwellings and Similar Buildings," National Bureau of Standards, July 3, 1923.

metal gutters with the top dimension given. Other shapes should have the same cross-sectional area. Outside leaders above the frost line should be one size larger than inside conductors, as given in Table 13-3. Some codes require that rain-water leaders shall be at least as large as the drains if placed on the steepest slope shown in Table 11-7.

TABLE 13-3. SIZES OF CIRCULAR AND SEMICIRCULAR ROOF GUTTERS AND INTERIOR LEADERS[a]

Maximum area of roof, sq ft		Gutter diameter, in.		Vertical leaders NPC, Sec. 13.6.1[e]	
Hoover Report[b]	"Plumbing Manual"[c]	Hoover Report and "Plumbing Manual"[d]	Of interior leader Hoover Report[b] only, in.	Equivalent diameter, in.[e]	Maximum projected roof area, sq ft
Up to 90	170	3	1½	2	720
91–270	360	4	2	2½	1,300
271–810		4	2½	3	2,200
811–1,800	625	5	3	4	4,600
1,801–3,600	960	6	4	5	8,640
	1,380	7[f]	5	6	13,500
3,601–5,500	1,990	8[g]	6	8	29,000
5,501–9,800	3,600	10			
	6,800	12			

[a] Gutters and leaders of other shapes should have the same cross-sectional areas as are required for the gutters and leaders shown in this table unless otherwise noted. Outside conductors to be one size larger to the frost line.

[b] "Recommended Minimum Requirements for Plumbing in Dwellings and Similar Buildings," National Bureau of Standards, July 3, 1923.

[c] Fall of 1/16 in. per ft or less. "Plumbing Manual," Report BMS66, National Bureau of Standards, 1940.

[d] *Ibid.*

[e] Based on a maximum rainfall rate of 4 in. per hr. If the maximum rainfall rate differs, then the figures for roof area must be adjusted by dividing the figures in this table by the assumed rainfall rate and multiplying by 4.

[f] This size is not shown in the Hoover Report.

[g] This size and larger to be hung with approved wrought-iron hangers.

Recommendations of the Copper and Brass Research Association[1] concerning rain-water gutters and leaders include the following principles:

1. Make leader same size as its outlet, for its full length, to avoid stoppage by leaves, ice, etc.
2. Base runoff computations at 8 in. per hr. [Note that the "National Plumbing Code" and most other standards are based on 4 in. per hr.]
3. Maximum spacing of leaders is 75 ft. A safe rule is 150 sq ft of roof area to 1 sq in of leader area. The figures can be changed to meet local conditions.

[1] "Copper Flashings," 2d ed., The Copper and Brass Research Association, 1925.

4. No leader should be less than 3 in. (circular), where there is possibility of leaves, etc., passing into it; 2-in. leaders are permissible for porches and decks.

5. Size of gutter depends on: (*a*) number and spacing of outlets, (*b*) slope of roof, and (*c*) style of gutter.

6. The best type of gutter has minimum depth equal to half and the maximum depth not exceeding three-quarters of the width. Thus the width becomes the

TABLE 13-4. SIZES OF RAIN-WATER LEADERS*

Type of leader	Area, sq in.	Leader size, in.	Area of roof drained, sq ft	Type of leader	Area, sq in.	Leader size, in.	Area of roof drained, sq ft
Plain round	7.07	3	1,060	Polygon octagonal	6.36	3	955
	12.57	4	1,885		11.30	4	1,695
	19.63	5	2,945		17.65	5	2,650
	28.27	6	4,240		25.40	6	3,810
Corrugated round	5.94	3	890	Plain rectangular	3.94	1¾ × 2¼	590
	11.04	4	1,660		6.00	2 × 3	900
	17.72	5	2,660		8.00	2 × 4	1,200
	25.97	6	3,895		12.00	3 × 4	1,800
					20.00	4 × 5	3,000
Square corrugated	3.80	1¾ × 2¼	570		24.00	4 × 6	3,600
	7.73	2⅜ × 3¼	1,160				
	11.70	2¼ × 4¼	1,755				
	18.75	3¾ × 5	2,820				

* Based on 150 sq ft of roof area for each square inch of leader area.

TABLE 13-5. SIZE OF SEMICIRCULAR RAIN-WATER GUTTERS
("National Plumbing Code," Sec. 13.6.3)

Diameter of gutter, in.*	Maximum projected roof area for gutters of various slopes, sq ft — Slope, in. per ft: 1/16	1/8	1/4	1/2
3	170	240	340	480
4	360	510	720	1,020
5	625	880	1,250	1,770
6	960	1,360	1,920	2,770
7	1,380	1,950	2,760	3,900
8	1,990	2,800	3,980	5,600
10	3,600	5,100	7,200	10,000

* Gutters other than semicircular may be used provided they have an equivalent cross-sectional area.

deciding factor in proportioning its size. There is no reason for a gutter deeper than three-quarters of the width except for ornamental purposes.

Assuming that this proportion is observed the gutter may be referred to by its width only.

A gutter smaller than 4 in. wide is to be avoided. In common practice 4-in. gutters are seldom used for they are difficult to solder and increase the labor cost. The gutter may be the same size as the leader it serves, but, of course, cannot be smaller.

Half-round gutters are most economical in material and ensure a proper proportioning of width and depth.

Safe rules for determining the size of gutters are

1. If spacing of leaders is 50 ft or less, use a gutter the same size as the leader, but not less than 4-in.

2. If spacing of leaders is more than 50 ft, add 1 in. to the leader diameter for every 20 ft (or fraction) additional spacing on peaked roofs.

3. For flat roofs add 1 in. to the leader size for every 30 ft of additional gutter length.

For ordinary residence construction 3- or 4-in. round and 2- by 3-in. or 2- by 4-in. rectangular leaders will generally suffice; 5-in. half-round gutters meet practically every requirement.

A safe system to follow in mill building design is that of the American Bridge Company. Their specifications provide as follows:

Span of roof	*Gutters*	*Leaders*
Up to 50 ft	6 in.	4 in. every 40 ft
50–70 ft	7 in.	5 in. every 40 ft
70–100 ft	8 in.	5 in. every 40 ft

NOTES. 1. Round leaders should not be less than 3 in. in diameter.

2. Rectangular leaders should not be smaller than 1¾ by 2¼ in. (This is commonly called "2" sq in.)

3. Gutters should not be less than 4 in. wide.

4. Gutters should have a fall of not less than 1 in. in 16 ft.

5. Scuppers should be provided for all roofs with a parapet wall built around them. This precaution prevents an overloading of the roof due to stoppage of the outlet.

6. All outlets should be provided with screens or strainers.

13-11. Materials for Roof Gutters and Leaders. The "National Plumbing Code" states (Secs. 13.2.1 and 13.2.2):

Inside Conductors. Conductors placed within a building or run in a vent or pipe shaft shall be of cast iron, galvanized steel, galvanized wrought iron, galvanized ferrous alloys, brass, copper, or lead.

Outside Leaders. When outside leaders are of sheet metal and connected with a building storm drain or storm sewer, they shall be connected to a cast-iron drain extending above the finished grade, or the sheet-metal leader shall be protected against injury.

The Code proceeds to state (Sec. 13.5.1):

Material. Roof drains shall be of cast iron, copper, lead, or other acceptable corrosion-resisting material.

Gutters are usually constructed of sheet metal. Cast iron and galvanized wrought iron or steel, brass, copper, and lead are used for leaders within the building or in an inner or interior court or ventilating pipe shaft when these leaders are connected with the house drain or sewer. Where inside conductors are connected to the house drain, sewer material, installation, and other features must meet the limitations applicable to soil and waste pipes. Outside leaders should be constructed of sheet metal with longitudinal corrugations rather than of cast iron or wrought pipe because the sheet metal is not so seriously affected by freezing and by temperature changes. Cast iron is sometimes properly used for a distance

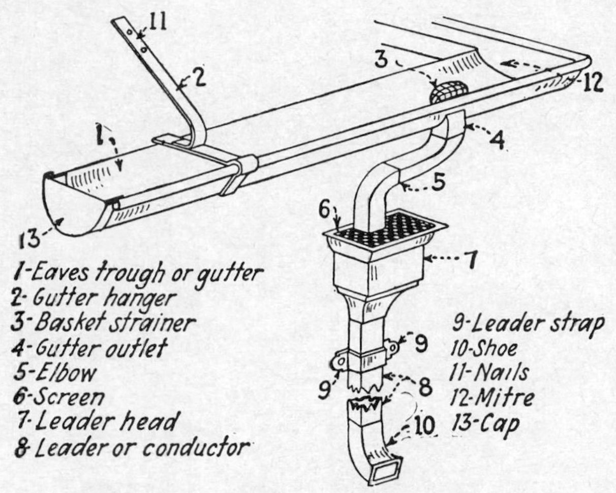

FIG. 13-6. Roof gutter and rain-water leader.

of 1 to 5 ft above the ground surface on account of the hard knocks to which this portion of the leader may be subjected. The entire length of an outside leader should not be made of cast iron, because of damage not only to the pipe itself by freezing but also to the calked joints. The lead will be forced from the joints by the alternate freezing and thawing of the wet oakum beneath it. Wrought iron with threaded joints is better than cast iron with calked joints if a sturdy pipe must be used. Sheet-metal leaders are most commonly used.

The support of the gutter and leader and the connection between the gutter and the leader are important in their erection and maintenance. Installations illustrating good practice and the names used for the parts of sheet-metal gutters and leaders are shown in Fig. 13-6. The thicknesses of sheet metal recommended by different authorities are given in Table 13-6. Recommendations of the Division of Simplified Practice of the U.S. Department of Commerce[1] are:

[1] Simplified Practice Recommendation 29, approved Jan. 1, 1925.

All eaves trough, conductor pipe, shoes, miters, and accessories, including gutters, valleys, ridge rolls, and so on, when made of copper, to be not lighter than 16 oz per sq ft (0.0216 in.).

TABLE 13-6. THICKNESS OF SHEET METAL FOR VARIOUS PURPOSES
(Thickness in inches)

Purpose	Galvanized iron	Copper or brass	Lead
Roof gutters	0.016	0.0215§	
		0.0156¶	
		0.040†	
Outside leaders	0.016	0.040†	
		0.0215§	
		0.0156¶	
		0.064	
Flashing	0.016	0.040†	0.135*
		0.021‡§	0.101‡
Local vents	.....	0.016	
Safes	.....	......	0.101†
Tank lining	.....	0.016	

Authority; * Dibble. † Hoover. ‡ Cosgrove. § Copper and Brass Research Assoc. ¶ Div. of Simplified Practice, U.S. Department of Commerce.

In discussing roof drains, the Bureau of Standards states:

Roof Drains: Body shall be galvanized cast iron with female threaded outlet and have clamping ring for roofing felt; antitilting, removable dome strainer and large sediment chamber; gravel stop; bronze post; brass screws.

Roof Drains for General Use on All Types of Flat Roof: Drains shall be cast iron, japanned, with slag stop, double drainage feature, flashing rings, etc. Domes and flashing rings shall be secured with brass or bronze bolts.

Roof Drains for Use in Cornice Gutters: Drains shall be cast iron, japanned, with outlet for calking.

The connection between the leader and the gutter should be made of brass, copper, or lead pipe and should be corrugated transversely so as to be flexible to allow for inevitable movements between the leader and the gutter. No. 19 B & S gage copper or brass pipe or extra-light lead pipe should be used for the flexible connection. Similar corrugated flexible joints should be installed at intervals along the pipe so that there is at least one fold for each $\frac{1}{4}$ in. of movement of the pipe. In general, one fold every 50 ft, with never less than two folds, is satisfactory. Outside leaders are subject to wide and rapid changes of temperature, the extreme range being sometimes greater than 125°F.

The opening into the leader from the gutter should flare out at a small angle to twice the diameter of the leader and it should be covered by a strainer to prevent the entrance of leaves, birds, and small animals. The "National Plumbing Code" states (Secs. 13.5.2 and 13.5.3):

Strainers. All roof areas, except those draining to hanging gutters, shall be equipped with roof drains having strainers extending not less than 4 in. above the surface of the roof immediately adjacent to the roof drain. Strainers shall have an available inlet area, above roof level, of not less than 1½ times the area of the conductor or leader to which the drain is connected.

Flat Decks. Roof drain strainers for use on sun decks, parking decks, and similar areas, normally serviced and maintained, may be of the flat surface type, level with the deck and shall have an available inlet area not less than 2 times the area of the conductor or leader to which the drain is connected.

Provision should be made to permit water to overflow from the gutter and to prevent its accumulation on the roof in case the leader becomes blocked.

The choice between copper and galvanized- or painted-iron gutters is a matter for the judgment of the designer, who should consider the relative cost and durability of the two materials.

13-12. Installation of Rain-water Leaders. Recommended connections between a roof gutter and a rain-water leader are shown in Fig. 13-6. The right-angle turn immediately below the roof provides flexibility to allow for differential expansion and contraction between the leader and the building. This is important, since the variations of temperature to which leaders may be subjected are sometimes as great as 125°F.

Trapping of rain-water leaders connected to a house drain, house sewer, or other sanitary sewer is essential except where the leader is made of cast iron with calked joints and meets the specifications for stack terminals stated in Sec. 12-11. Traps on inside conductors should be accessible inside the building. Sheet-metal leaders connected to a sanitary sewer should be trapped because of the possibility of gas leakage through the poorer joints and seams that exist in such pipes. The trap should have a large water capacity and a seal at least 4 in. deep. It may be desirable to discharge the safe waste from a frequently used fixture through the leader to assist in maintaining the trap seal during drought. A sand trap can be used on a leader if no sanitary sewage flows through it. Care should be taken to protect the trap on the leader from freezing by placing it within the building or by burying it below the frost line. Where the trap is buried, access to it should be provided to permit inspection and cleaning.

In order to prevent overflow into a fixture, no rain-water leader should connect to a house drain above any fixture that is less than 5 ft higher than the house drain. The connection between the house drain and the leader should be made by means of not less than one length of cast-iron pipe extending vertically at least 1 ft above grade line. Along driveways without sidewalks the leaders should be placed in niches in the walls, protected by wheel guards, or enter the building through the wall at 45° slope at least 12 ft above grade. The lower end of a rain-water leader is illustrated in Fig. 13-7.

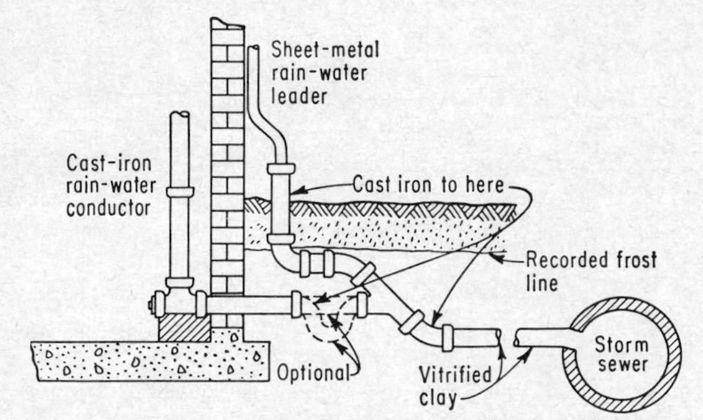

FIG. 13-7. Discharge ends of rain-water leaders and conductors.

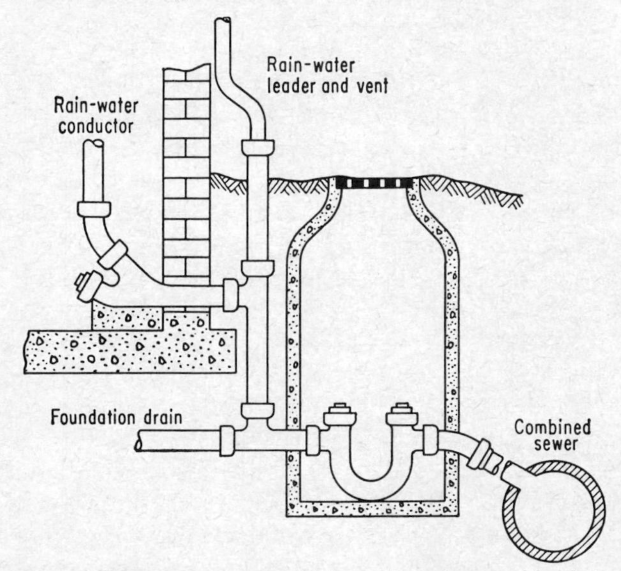

FIG. 13-8. Clear-water wastes discharged onto a combined sewer.

13-13. Clean-water Drainage. Rain-water leaders, yard drains, foundation drains, and other clean-water discharges can be connected to other clean-water drains from the same building to discharge finally into a storm sewer or to a sanitary sewer which is intended to be a combined sewer. It is advantageous to the householder to connect to a combined sewer, since the construction of but one house sewer is thus required, the total cost of plumbing can be reduced, and the heavy discharge of rain water will serve to flush the house sewer. The use of a combined sewer may be prohibited, however, necessitating the connection of rain-water leaders and other clean-water wastes to a storm sewer. Where legally permissible and otherwise feasible, clean-water drains should be trapped and discharged into a combined sewer, as shown in Fig. 13-8. The traps should be protected from frost, as mentioned in the preceding section. As many drains as is feasible should lead into one trap to give greater assurance of maintenance of seal during drought. Trap seals should meet the requirements given in Sec. 11-36.

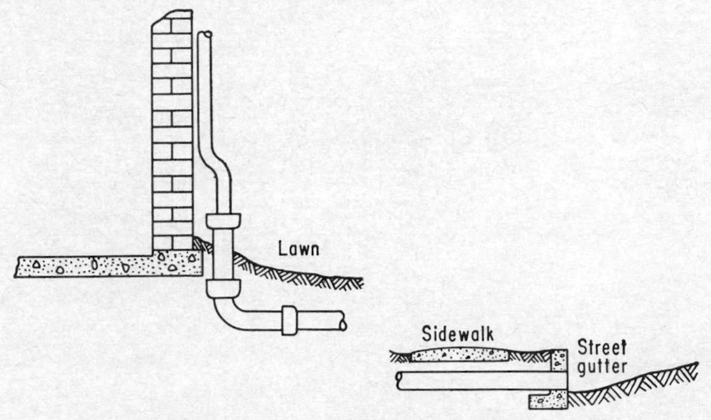

FIG. 13-9. Rain-water discharge into street gutter in a mild climate.

When there is no sewer, rain-water leaders can discharge onto the ground surface provided the slope of the surface will carry the water away from the building or, where freezing does not prevent, the rain-water drain can pass beneath the sidewalk and parking to discharge into the street gutter, as shown in Fig. 13-9.

13-14. Yard and Area Drains. The principal purpose of yard and area drains is to carry off rain water as rapidly as it falls. All roofs and paved areas, yards, courts, and courtyards should be drained into the storm-water sewer system or the combined sewer system, but not into sewers intended for sanitary sewage only. When drains used to carry off storm water are connected to the combined sewer system, they should be effectually trapped unless they terminate as provided for stacks in Sec. 12-11. A single trap can serve a number of drains, and all traps should be protected from frost by setting within a building or otherwise. Where no sewer is accessible, the discharge can pass, untrapped, under the sidewalk into the street gutter unless prohibited by local ordinance.

Yard and area drains are similar to floor drains except that they should be more substantial in construction and should have a greater capacity for draining off water. The construction must be substantial, and the foundation laid deep, not only to resist the shock of passing vehicles but to resist the heaving action of frost. Types of yard drain are shown in Figs. 11-39 and 13-10. The use of the bell trap shown in the figure is prohibited by the "National Plumbing Code" in Sec. 5.3.6. The traps shown in the figure are made of cast iron, the cover being perforated for the admission of water and the exclusion of debris. If possible, the traps should not discharge into drains carrying sewage, and under such conditions no trap is necessary. Where they must discharge into a sewer, however, they should be trapped with a deep-seal trap with generous water capacity and some clean-water waste should be led into the trap.

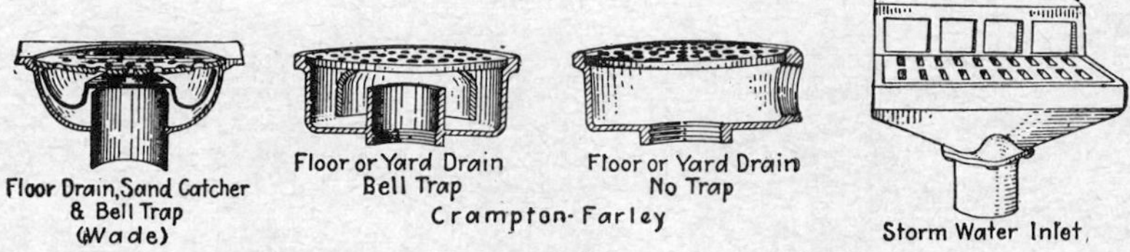

FIG. 13-10. Types of floor and yard drain.

The trap should be located within a building or beneath the frost line to protect it from frost. It is desirable also to include a sand trap or catch basin, somewhat as shown in Fig. 11-43.

Federal specifications WW–P–541a (1940) state, concerning area drains:

E-39f. Area drains shall be of heavy cast iron and shall conform to the following:

1. The double drainage pattern, Fig. II-104, shall have an integral seepage pan for embedding in floor construction and seep holes providing adequate drainage from pan to drain pipe, and shall have a cast-iron, nontilting grate. Minimum thickness of body $5/16$ in. Size 3- and 4-in. drain pipe as called for with bottom outlet for inside calked connection. Strainer shall be perforated or slotted.

2. The plain pattern with bottom outlet, Fig. II-105, shall have a hinged cast-iron cover, either perforated or slotted. Outlet shall be for inside calked connection to 3-in. drain pipe.

3. The plain pattern with side outlet, Fig. II-107, shall have a hinged cast-iron cover, either perforated or slotted, and spigot outlet for 3-in. drain-pipe hub.

The capacity of the yard or area drain should depend on the rate of rainfall. An empirical rule suggested in the Hoover Report[1] for estimating the rate of runoff is to consider the rate of rainfall as 4 in. per hr. This will give a rate of flow from each 100 sq ft of impervious area of,

[1] "Recommended Minimum Requirements for Plumbing in Dwellings and Similar Buildings," National Bureau of Standards, July 3, 1923.

closely, 4 gpm. When this is converted into equivalent fixture units, 180 sq ft of drainage area is to be taken as equivalent to one fixture unit. Having determined the rate of flow to be carried by the drain, its size can be found in Table 11-8. Recommendations for drain sizes as made in the "Plumbing Manual"[1] are shown in Table 13-7.

TABLE 13-7. SIZES OF HORIZONTAL STORM DRAINS
(Maximum roof area,* in square feet, that can be drained by horizontal storm drains on various slopes)

Diameter of drain, in.	From "Plumbing Manual," Report BMS66, National Bureau of Standards, 1940				From "National Plumbing Code,"† Table 13.6.2		
	Slope, in. per ft						
	1/16	1/8	1/4	1/2	1/8	1/4	1/2
2	......	......	350	500			
2½‡	......	480	670	960			
3	......	750	1,050	1,500	822	1,160	1,644
3½‡	......	1,100	1,550	2,200			
4	......	1,550	2,150	3,100	1,800	2,650	3,760
5	1,800	2,700	3,600	5,400	3,340	4,720	6,680
6	3,000	4,200	6,000	8,400	5,350	7,550	10,700
8	5,900	8,700	11,900	17,400	11,500	16,300	23,000
10	9,800	15,200	19,600	30,400	20,700	29,200	41,400
12	15,900	24,700	31,800	49,400	33,300	47,000	66,600
15	......	......	......	......	59,500	84,000	119,000

* Based on a rate of rainfall of 4 in. per hr. If the maximum rate of rainfall is more or less than 4 in. per hr, then the figures for roof area, given in this table, must be multiplied by 4 and divided by the maximum rate of rainfall assumed.

† No yard or building drain should be less than 3 in. in diameter.

‡ 2½- and 3½-in. cast-iron soil pipe and fittings and 3½-in. drainage fittings are not generally available.

A catch basin should be placed at the inlet to the drain, and the inlet to the catch basin should be protected by a heavy strainer. The capacity of the inlet is as important as the capacity of the drain pipe. If the area of the openings in the inlet strainer is double the cross-sectional area of the inlet drain, no trouble should result from the size of the strainer. Where no catch basin is used or where the catch-basin outlet is not trapped and the drain pipe is connected to the house sewer, drain, or other sanitary sewer, the yard drain should be trapped and access to the trap and the cleaning of the drain pipe should be provided for. The same care should be taken to prevent the freezing of the seal of this trap and for maintaining it as is described for the seal of a rain-water leader in Sec.

[1] "Plumbing Manual," Report BMS66, National Bureau of Standards, 1940.

13-12. A helpful arrangement in maintaining the seal is to connect the yard drain to the rain-water leader above the trap.

13-15. Cellar Drainage. There may be two types of drain beneath a cellar floor: one to remove ground water that might otherwise press against or seep upward through the floor and one to remove water falling on the surface of the floor. The first may consist of open-joint 4- or 6-in. drain tile laid from a few inches to 18 in. below the cellar floor. Such pipes should be placed not more than about 10 to 15 ft apart on centers, the more porous the soil, the greater the distance apart and the deeper the cover. The pipes should connect and terminate in a down-turned cast-iron quarter bend with its end submerged about 3 in. or less in a watertight collecting well from which the outflow is a few inches above the face

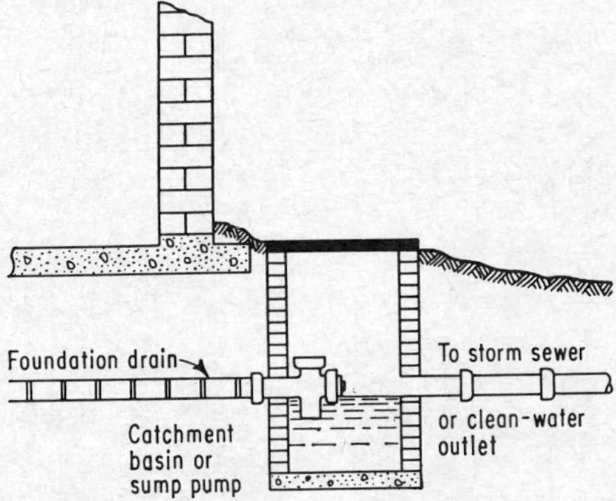

Fig. 13-11. Foundation and subsoil drainage outlet, suitable for a climate where water is below the frost line.

of the inlet elbow, somewhat as shown with a tee in Fig. 13-11. The 4- or 6-in. outflow pipe from this well may join the foundation drain or may discharge independently into the storm- or surface-water sewer. It should not discharge into a sanitary or a combined sewer. The collecting well provides a convenient location for a pump, as described in Sec. 13-1, in the event the outlet sewer is too high to receive the flow from the well by gravity.

The surface of the cellar or basement floor should slope at a grade of about ½ per cent, i.e., ¹⁄₁₆ in. per ft, toward a floor-drain inlet of the type shown in Fig. II-104. The distance of flow of water across the floor should be less than about 16 ft. Each floor-drain inlet should be protected by a bell trap or better. The outlet from the trap should discharge into the sanitary sewer. The floor drains may receive any drainage falling on the floor, including the discharge from washing equipment not otherwise connected to the plumbing.

13-16. Subsoil or Foundation Drains. Subsoil or foundation drains are laid close to and outside the foundations of a building to prevent water from entering the building. They are laid somewhat as shown at 58 in Fig. 11-1. The arrangement shown in Fig. 11-1 is objectionable when there is danger of sewage backing through the building sewer. Even a backwater valve may not prevent such difficulties because of clogging of the valve so that it is held open.

The "National Plumbing Code" states (Sec. 13.1.5):

> **Subsoil Drains.** Where subsoil drains are placed under the cellar or basement floor or are used to surround the outer walls of a building they shall be made of open-jointed or horizontally split or perforated clay tile, or perforated bituminized fiber pipe or asbestos cement pipe, not less than 4 in. in diameter. When the building is subject to backwater, the subsoil drain shall be protected by an accessibly located backwater valve. Subsoil drains may discharge into a properly trapped area drain or sump. Such sumps do not require vents.

The foundation drain pipe should be laid at about the elevation of the basement floor and not closer than 3 ft to the foundation wall.[1] If more than 3 ft below the nearest edge of the footing, the pipe should be as far away horizontally as it is vertically below the bottom of the footing. The pipe should be surrounded with gravel or other loose, nonabsorbent material, and the trench should be backfilled with this or similar material to a depth of at least 1 ft and preferably more. The remainder of the trench to within a short distance of the ground surface should be filled with a porous material that will intercept any ground water and carry it to the drain tile. The drain may be laid completely around the building or only on those sides from which ground water is expected. The drain is connected, where possible, to the storm sewer or other surface-water drain. Otherwise it is connected to the building sewer through a trap and a vitrified wye connection where no other point of discharge is possible. If there is a possibility of sewage backing up into the foundation drain, it should be protected by an accessible and otherwise properly located backwater valve. The valve should be higher than the sewage in the building sewer to minimize the possibility of clogging materials in the sewer becoming lodged on the valve seat. The connection of a subsoil drain to a sanitary or a combined sewer should be prohibited.

Materials for drains and sewers are discussed in Sec. 11-14, and the installation of a building sewer is discussed in Sec. 11-10.

[1] "National Plumbing Code," Sec. 2.8.1.

CHAPTER 14

VALVES, FAUCETS, METERS, AND OTHER APPURTENANCES

14-1. Definitions and Descriptions. Valves, faucets, and stopcocks are used for controlling the flow into, through, and from pipes. The use of the terms *valve, faucet, cock, bibb, stopper, tap*, etc., has created such confusion that there is an attempt to confine them all to one or two terms. The desired terms are *faucet* for plumbing fixtures and *valve* for other

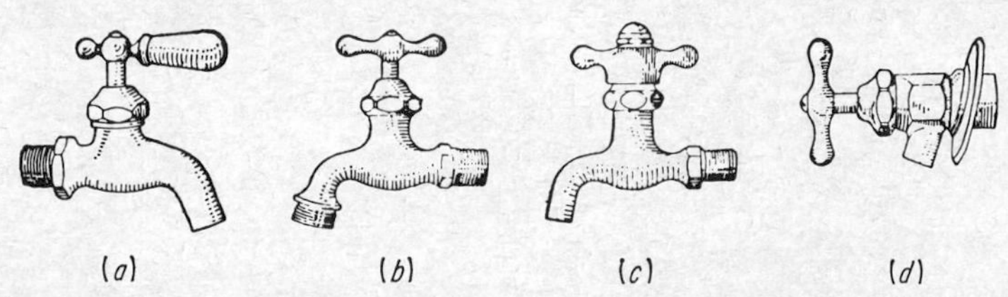

FIG. 14-1. Sink and laundry-tray faucets. (*a*) Wolff sink faucet, smooth nozzle, china lever handle; (*b*) Mueller sink faucet, hose thread, tee handle; (*c*) Mueller sink faucet, smooth nozzle, self-closing; (*d*) laundry tray faucet, smooth nozzle, inside thread.

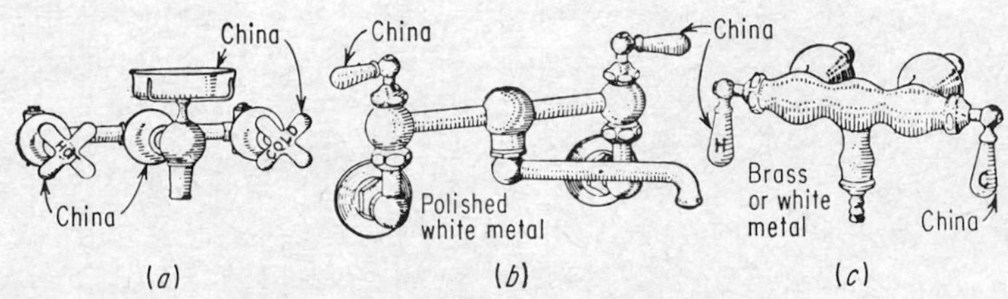

FIG. 14-2. Bathtub faucets. (*a*) Mueller bathtub combination; (*b*) Mueller sink combination, swinging nozzle; (*c*) Mueller bathtub combination.

locations. This does not seem to cover all possibilities, however, so that the terms *valve, faucet,* and *cock* have been used in this book with the special meanings given in the glossary in Appendix I.

14-2. Materials for Valves and Faucets. Valves are made of malleable iron, plain or galvanized; of brass, rough, polished, plain, or nickel-plated; of bronze; of cast iron with plain, brass, or bronze parts; and of less corrodible metals for special conditions. Standards for valves prepared by various authorities are listed in Table II-1. Faucets for household

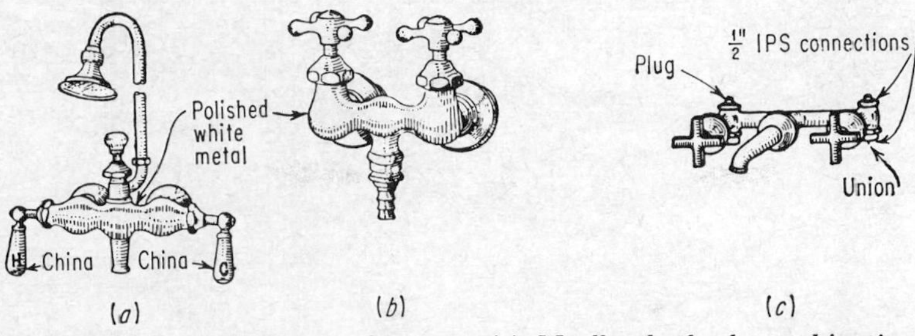

FIG. 14-3. Bathtub and shower faucets. (*a*) Mueller bathtub combination with shower attachment; (*b*) Mueller bathtub combination; (*c*) bath supply fitting for recessed and corner tubs.

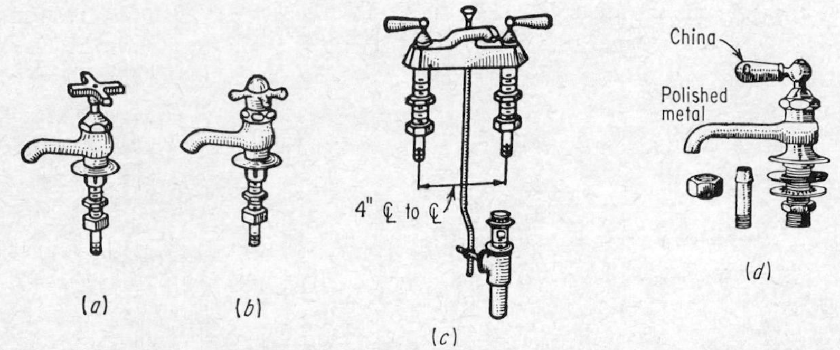

FIG. 14-4. Lavatory faucets. (*a*) Single lavatory faucet, regular; (*b*) single lavatory faucet, self-closing; (*c*) double lavatory faucet and drain fitting; (*d*) Mueller lavatory faucet.

FIG. 14-5. Special-type kitchen-sink faucet. (*a*) Double kitchen-sink faucet; (*b*) single faucet, shank and flange type. Note: flange may have setscrew or may be threaded to shank with concealed threads.

installations are usually made of brass, plain or nickel-plated, although there is a tendency to use faucets made of a white metallic alloy, since they stand polishing better than nickel-plated brass. Types of faucet are shown in Figs. 14-1 to 14-9. The faucets are generally available with either inside or outside threads and in sizes up to 1 in. Pressure losses and discharge rates are given in Table 14-1.

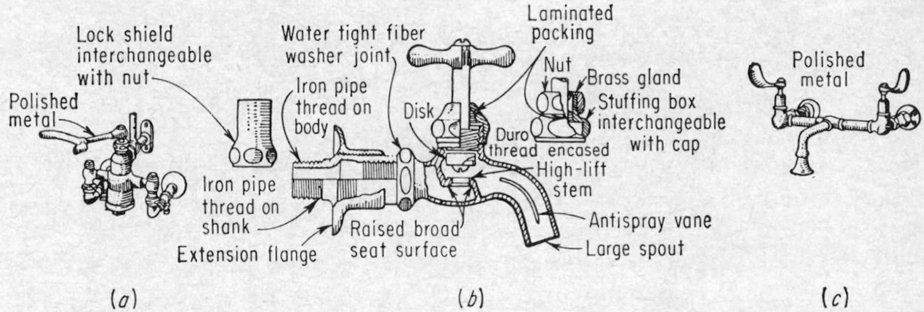

FIG. 14-6. Special faucets. (*a*) Mott wall-type knee-action supply valve with stirrup handle, check valves, loose-key graduated stop valves, connections to wall and wall brace; (*b*) details of compression sink faucet; (*c*) Mott quick-opening elbow-action compression hot and cold supply faucet has lever handles, supply nozzle with removable rose spray, and connections to wall.

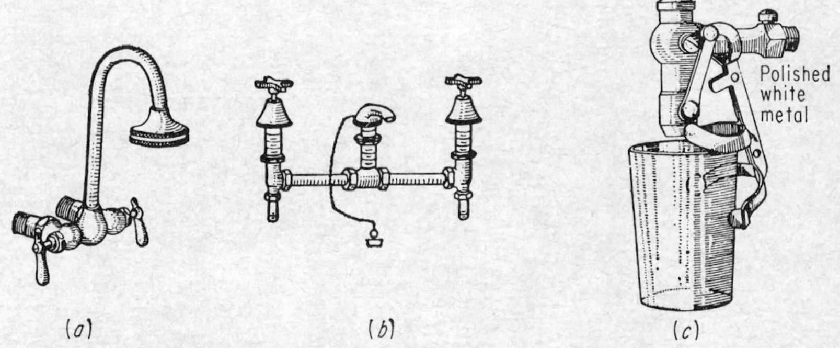

FIG. 14-7. Special faucets. (*a*) Wolff wash sink faucet; (*b*) combination lavatory supply fitting; (*c*) Mueller self-closing faucet for filling glasses.

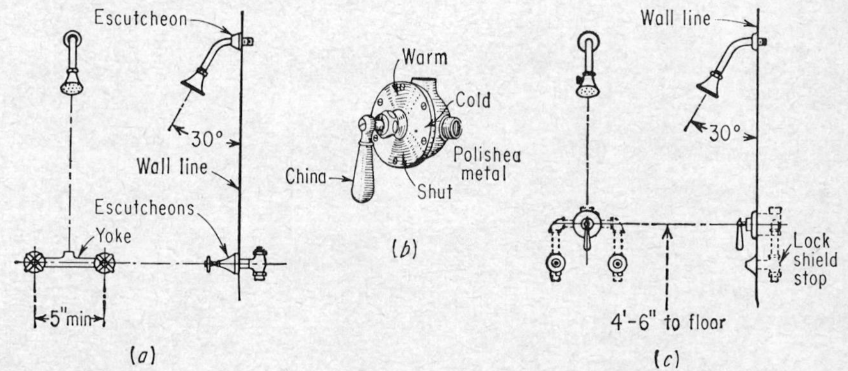

FIG. 14-8. Shower valves. (*a*) Concealed shower with hand valves; (*b*) Mott shower mixing valve; (*c*) concealed shower with mixing valve.

Most valves, excepting some special types, are available in all sizes up to and including 12 in. Gate valves are made with bell-and-spigot, flanged, and threaded ends in all sizes up to 12 in. and in larger sizes with bell-and-spigot or flanged ends. Other types of valve are made with threaded ends, and some valves are made with either threaded or flanged ends. Valve dimensions are not the same for all manufacturers, and no generally recognized standard has been adopted. Practically all threaded valves are equipped with inside threads, and some valves are equipped with one inside and one outside thread. Valves can be obtained with both ends

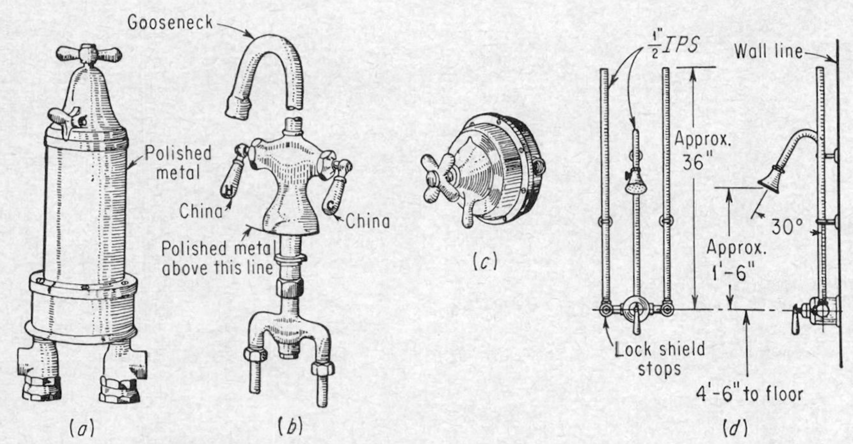

FIG. 14-9. Shower valves and thermostatic mixing valves. (*a*) Leonard shower mixing valve, thermostatic control; (*b*) pantry sink faucet, high gooseneck; (*c*) Leonard shower mixing valve, thermostatic control; (*d*) exposed shower with mixing valve.

threaded with outside threads. Valve handles may be made of the same materials as the valve, or they may be made of wood or lined with wood, porcelain, or other material. Handles are made in the shape of wheels or as straight shafts.

Valves are made tight against the flow of liquids or gases through them, either by a gasket of softer material than the valve or by a ground-faced metal-to-metal seat. The gasket is pressed tightly against the seat by the screwing down of the valve stem. Where the valves are controlled by handles that must be turned or lifted, soft packing is usually necessary around the valve stem to prevent leakage. This packing is held in place by a compression cup or gland that is screwed down to it. The packing glands are shown in various illustrations of valve sections. Packing can be renewed in some valves without shutting off the liquid or gas controlled by the valve, but it is usually necessary or safer to shut off the pressure on the valve before the packing is removed and renewed.

Three types of valve commonly used on plumbing are gate valves,

TABLE 14-1. RATES OF DISCHARGE FROM FAUCETS

Manufacturer	Type and condition of faucet	Test number	Discharge in gpm for pressure on faucet of			
			Inches of water	Psi		
			6	5	30	90
Wolverine	¾-in. compression sink faucet, wide open	1	1.8	8.1	20.0	33.4
	¾-in. compression sink faucet, three-fourths open	1	1.8	....	19.5	32.8
	¾-in. compression sink faucet, one-half open	1	1.8	7.6	19.0	32.9
	¾-in. compression sink faucet, one-fourth open	1	1.5	7.0	17.4	29.9
Wolverine	½-in. compression sink faucet, wide open	2	1.4	6.0	14.8	24.5
Mueller	½-in. ground-key sink faucet, wide open	5	2.2	9.5	23.4	36.4
Mueller	½-in. self-closing compression faucet, wide open	8	0.6	2.6	6.8	11.7
Mueller	½-in. compression laundry-tray faucet, wide open	12	1.4	6.3	17.3	25.3
Mueller	Compression wash-basin faucet, wide open	13	1.7	5.0	11.9	21.3
Mueller	1-in. ground-key sink faucet, wide open	15	7.2	30.7	78.9	118.8
Mueller	1-in. compression sink faucet, wide open	16	3.5	12.7	39.9	64.8
Wolverine	Combination laundry-tray faucet Catalogue No. 640. Both outlets open	17	2.3	9.6	22.4	38.6
Wolverine	Combination laundry-tray faucet Catalogue No. 640. Either hot or cold, wide open	17	1.4	6.1	14.4	24.8
Wolverine	Combination compression bathtub faucet. Both sides open. Nozzle attached. Catalogue No. 635	18	....	8.0	20.4	34.4
Mueller	Combination compression sink faucet with swinging nozzle. Both sides open	21	0.97	4.6	12.2	21.4
	Combination compression sink faucet with swinging nozzle. Hot open, cold closed	21	0.72	3.3	8.6	15.1
	Combination compression sink faucet with swinging nozzle. Cold open, hot closed	21	0.69	3.1	8.2	14.5
Mueller	Combination compression high gooseneck pantry-sink faucet, No. 3822. Hot and cold open	22	0.88	4.6	11.6	20.9

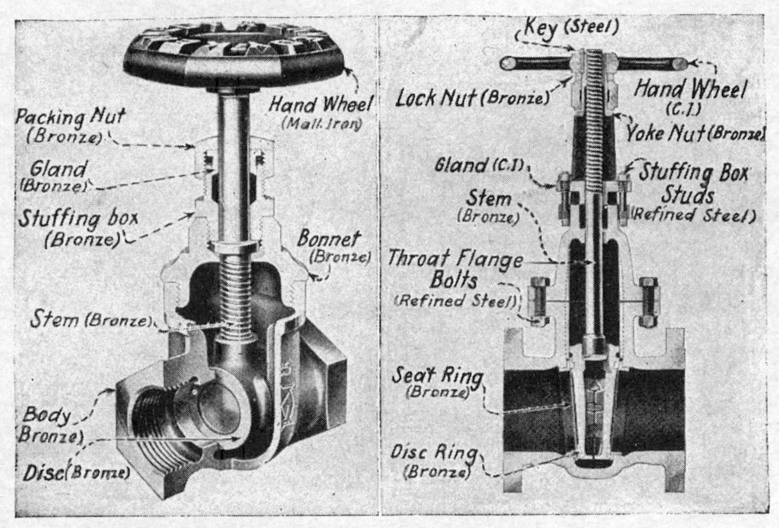

Non-rising stem — Outside screw and yoke

FIG. 14-10. Sections through two types of gate valve.

globe valves, and ground-key valves. No standard dimensions have been adopted by any generally accepted authority. Where roughing-in dimensions are desired, they must be obtained from the manufacturer. Losses of pressure through valves are shown in Tables 2-13 and 2-14 and in Fig. 2-14.

FIG. 14-11. Quick-closing gate valve.

14-3. Gate Valves.[1] Two types of gate valve are shown in Fig. 14-10. The mechanism consists of two ground-faced metal disks that fit against a double ground-faced metal seat. The valve is closed by turning the handle attached to the stem, which forces the disk down onto the seat. The nonrising stem is the type commonly used on small pipes. A gate valve is a satisfactory type because of the full waterway opening provided and the absence of packing around the valve seat. Although the valve can be placed on the pipe in any position and with either face against pressure, it is usually desirable, particularly in large valves, to place the stem vertical, either up or down, to provide even wear on the disk edges. Packing is required about the valve stem. The packing is kept in place by means of the stuffing box, or gland, shown in Fig. 14-10. A quick-

[1] See also B. Krueger and G. B. Pamero, Gate, Globe, and Check Valves, *Air Conditioning, Heating, and Ventilating*, December, 1957, p. 63.

closing gate valve, shown in Fig. 14-11, can be completely closed or opened by a short movement of the valve handle. Dimensions of gate valves are given in Tables II-74 to II-76.

14-4. Globe Valves. Two globe valves are shown in Fig. 14-12. The gasket or disk is forced down upon the seat of the valve by turning the handle, thus shutting off the flow of fluid. If the valve leaks at this point, the gasket must be replaced or the packing around the stem of the valve must be renewed. This is done after the pressure has been shut off in the pipe by removing the bonnet and slipping a new gasket into place or by replacing the packing. Globe valves are made with a ground-faced metal disk fitting against a ground metal seat as shown in Fig. 14-12. An objection to the use of these valves is the difficulty of stopping leaks when once started. The valves are more suitable on steam and hot-water lines than valves depending on gaskets for their tightness.

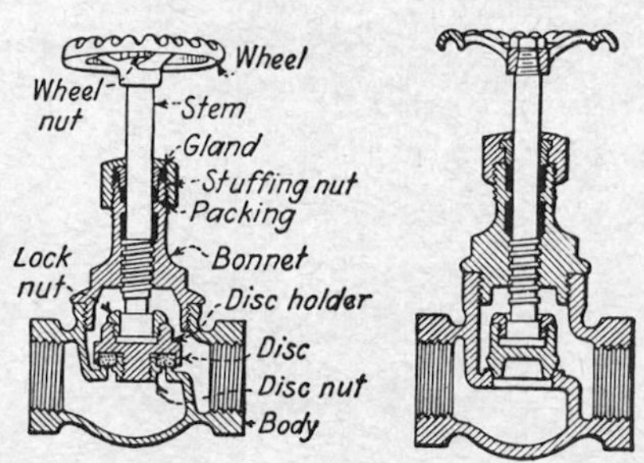

Fig. 14-12. Sections through globe valves.

In placing a globe valve on a pipe it is desirable, but not essential, that the valve be placed so that the flow is upward through the orifice and that the disk is screwed down against the pressure in shutting the valve. Globe valves are used to a great extent on plumbing because of their relatively low cost and despite the high loss of pressure through them. Ordinary-packed globe valves are not suitable on hot-water and steam pipes. A special form of packing must be used. An added objection to globe valves is the impossibility of draining pipes on which they are located because the high valve seat creates a trap almost as high as the center of the pipe, unless the valve is placed on its side.

An angle valve, shown in Fig. 14-13, is a special type of globe valve that does away with some of the objections to the straight globe valve and is suitable for use on a 90° change of direction in the pipe. The opening through an angle valve is usually larger than through the straight globe valve, the passage through the valve is straighter, and it offers no obstacle to the drainage of the pipe. Angle valves are found useful in close work where but little space is available. Their use replaces one fitting on the pipeline.

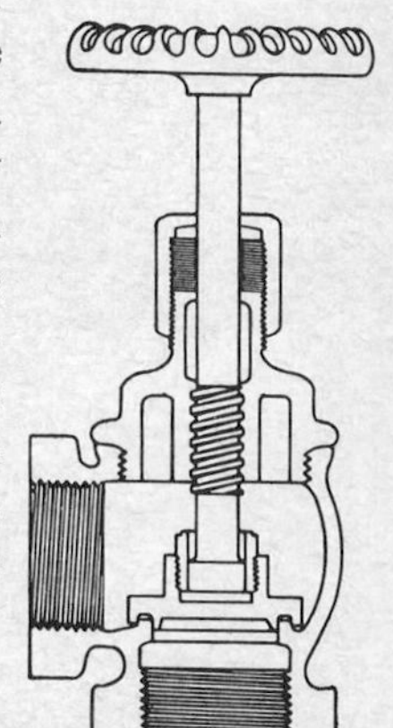

Fig. 14-13. Section through an angle valve.

14-5. Ground-key or Plug Valves. A ground-key valve is shown in Fig. 14-14. A corporation valve, a curb valve, and a meter valve or stop-and-waste valve, shown in Figs. 14-15 to 14-17, are types of ground-key valve commonly used. The principal advantages of this type of valve include the clear and unobstructed waterway when the valve is open, the absence of soft packing to wear out, its suitability for hot water, and its ability to open wide or close tightly by a one-quarter turn of the handle. There are two principal objections to the plug valve: One arises from the possible water hammer that may result from too quick manipulation of the valve. The other is the difficulty resulting when the key becomes worn; the valve may leak or jam and it is difficult to make repairs.

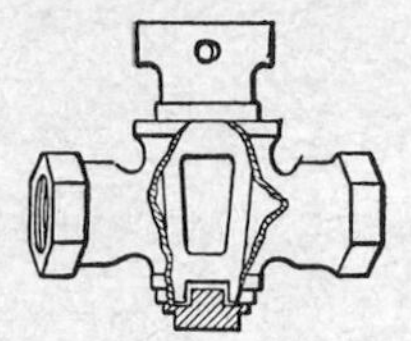

Fig. 14-14. Section through a ground-key valve.

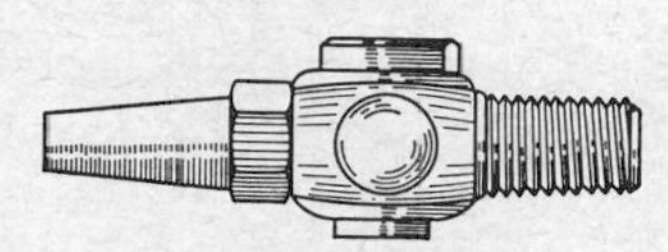

Fig. 14-15. Corporation valve or stop.

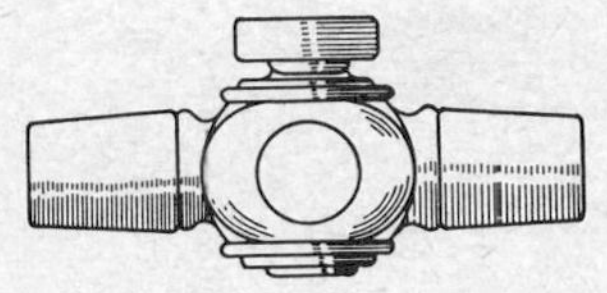

Fig 14-16. Curb stop.

Such valves are used as corporation valves and curb cocks and are popular as faucets for kitchen sinks.

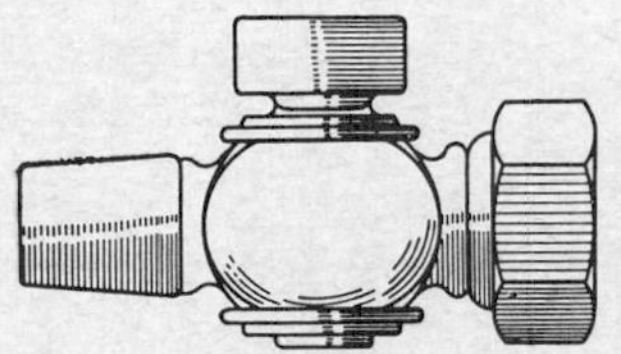

Fig. 14-17. Meter stop.

14-6. Check Valves. Check valves are used to prevent backflow or reverse flow in a pipeline. In the installation of a check valve care should be taken that it is placed in the correct position and that a valve designed for a horizontal pipe is not placed on a vertical pipe or that a vertical-pipe valve is not placed on a horizontal pipe, etc. Four types of check valve are shown in Fig. 14-18, and a swing check valve on a hot-water line is shown in Fig. 8-32.

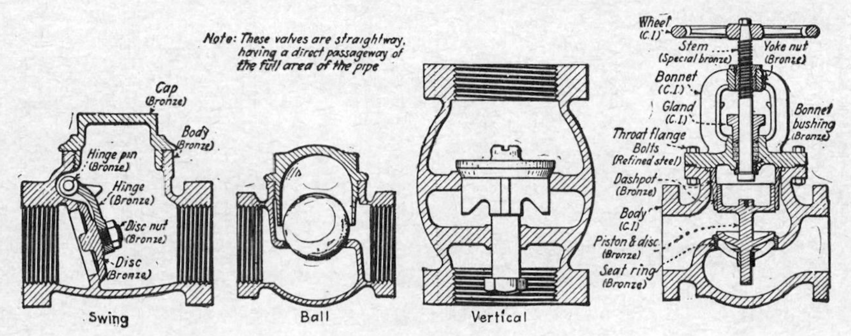

Fig. 14-18. Types of check valve.

14-7. Backwater Valves. A backwater valve for a drainage pipe, illustrated in Fig. 14-19, is one form of check valve acting in open-channel flow. In operation the reversal of flow will force the disk to close tightly, cutting off flow. The valves are used on drain pipes and are reliable as long as no objects become jammed under the seat. Under such conditions the valves are useless, and they should not, therefore, be used on pipelines with liquids carrying large suspended objects. A characteristic of this type of valve is the fact that the disk is closed except when liquid flow is passing through it. The valve stops the passage of air or gas and may serve as a deterrent to rats and other living pests.

FIG. 14-19. Backwater valve for drain.

14-8. Balanced Valves. A balanced valve, illustrated in Fig. 14-20, is used on pipelines operating under such high pressures that the opening of a gate valve or the closing of a globe valve would be difficult. Balanced valves are used also on automatic-control devices. The valve has an inlet and an outlet end, and it must be placed in the pipeline with its inlet upstream. Pressure is then exerted equally against both disks, and to move the disks it is necessary to overcome only the friction of the valve parts and the weight of the mechanism.

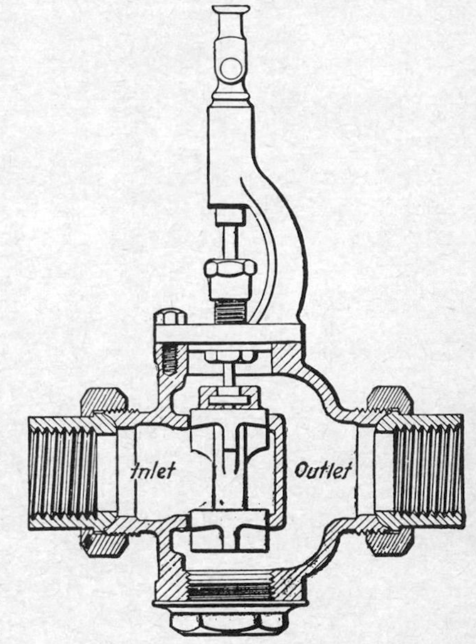

FIG. 14-20. Section through a balanced valve. (*Mason Regulator Co.*)

14-9. Pressure Regulators. Pressure regulators can be placed on water-supply lines where the supply pressure is greater than desired in the plumbing system. Pressure regulators should be used where the pressure may exceed 80 to 90 psi for an appreciable period of time. Such a condition often arises in tall buildings and occasionally in hilly cities. Pressure regulators can be made to give any difference of pressure desired, but the regulators available on the market are commonly limited to a minimum of 14 psi for a supply pressure of 40 psi and a minimum of 40 psi for a supply pressure of 200 psi. Between these limits almost any desired pressure can be obtained.

The principle of operation of one type of pressure regulator is illustrated

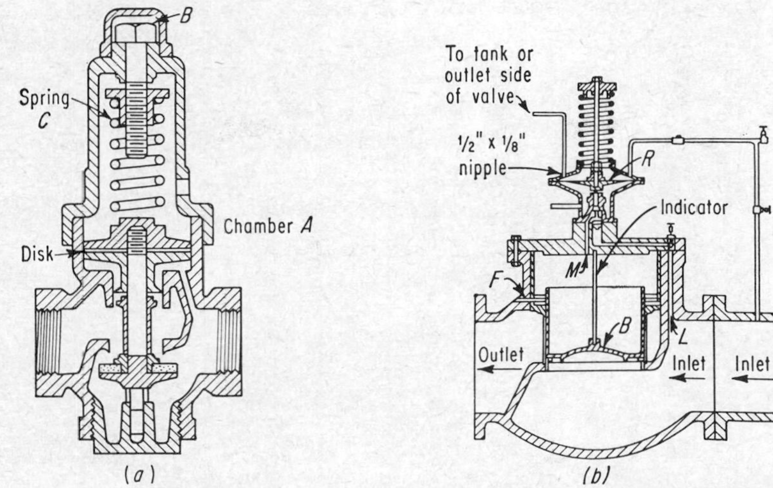

Fig. 14-21. Pressure-regulator valve and altitude valve. (*a*) Regulator; (*b*) altitude valve. (*Courtesy of Golden Anderson Valve Specialty Co.*)

in Fig. 14-21*a*. The desired low pressure is fixed by turning the nut at *B*, thus opening or closing the balanced valve that is supported on the flexible disk and spring *C*. In operation, if the pressure below the valve tends to increase beyond the fixed amount, the pressure in the chamber *A* is increased, forcing the disk up and partly closing the valve. If the pressure below the regulator becomes too low, the pressure on the flexible disk is relieved and the spring *C* opens the balanced valve in the regulator.

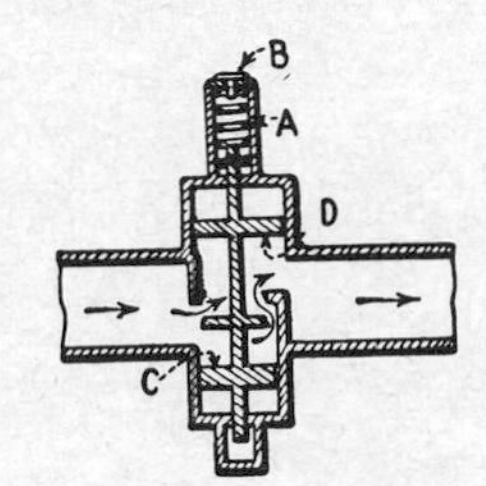

Fig. 14-22. A pressure regulator.

Another type of regulator is shown in Fig. 14-22. This is a less expensive type, but it is suitable only for small pipes. It operates as follows: The high pressure from the inlet side operates against disk *C* (the lower disk) to hold the valve open. The low pressure on the outlet side of the valve operates against the disk *D* (the upper disk), which is larger in area than disk *C*. The two pressures, with the aid of spring *A*, balance each other to hold the valve open in the correct position. Desired differences in pressure are obtained by adjusting spring *A* by turning screw *B*.

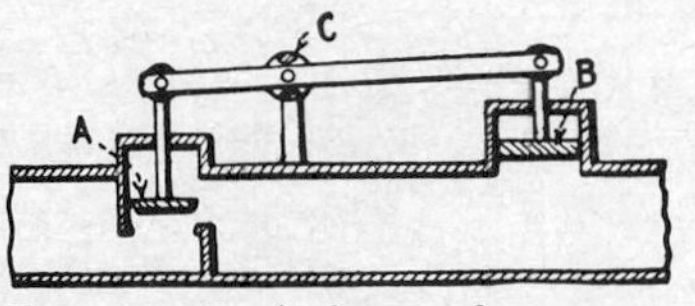

Fig. 14-23. A form of pressure regulator.

A third type of regulator is shown in Fig. 14-23. It operates as follows: The high pressure operates against the small disk *A* to open it, and the low pressure operates against large disk *B* to close the valve at *A*. The location of the fulcrum *C* and the areas of disks *A* and *B* can be arranged to give the desired reduction in pressure.

Wherever pressure-reducing valves are installed on the service pipe of a building, a bypass should be provided to use when the pressure regulator is out of order.

14-10. Pressure-relief or Safety Valves. Pressure-relief or safety valves are used to prevent or to relieve dangerously high pressures in plumbing. They are essential on hot-water supplies, as discussed in Sec. 8-4. Safety or pressure-relief valves are illustrated in Figs. 8-9,

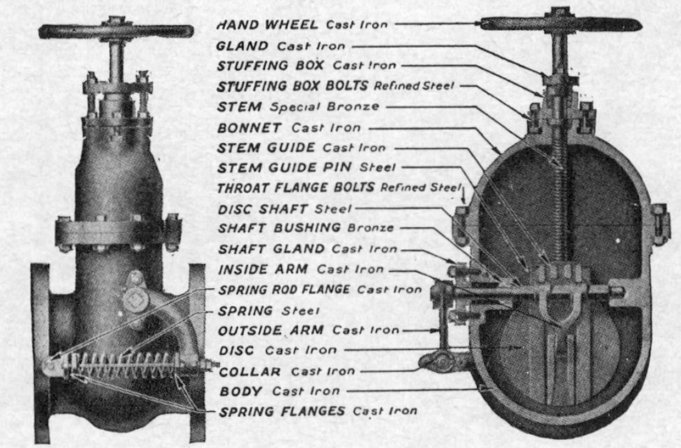

FIG. 14-24. Combined swing-gate and pressure-relief valve.

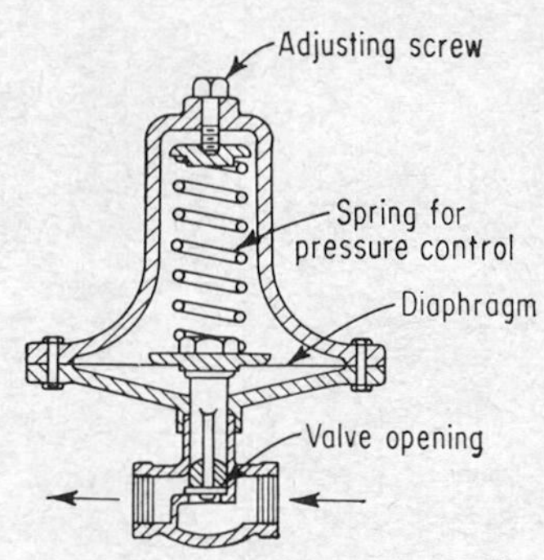

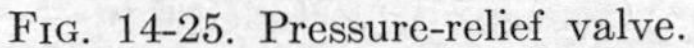

FIG. 14-25. Pressure-relief valve.

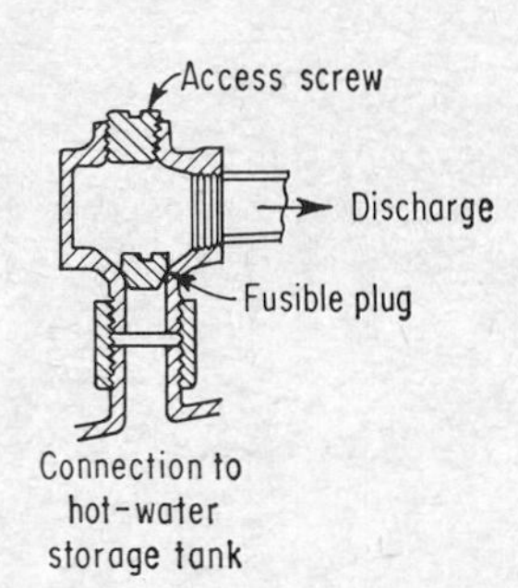

FIG. 14-26. Temperature–pressure-relief valve.

14-24, and 14-25. The control disk shown is held in place by an adjustable connection in the spring. When the pressure becomes too high, the disk is forced off its seat and the pressure in the tank or pipe is relieved. The valve should be designed with a free opening equal to or greater than the cross-sectional area of the pipe in which the pressure is to be relieved.

14-11. Temperature–Pressure-relief Valves.[1] Temperature–pressure-relief valves open to relieve hot-water or steam pressure when the temperature has reached a predetermined value. Temperature–pressure-

[1] See also *Domestic Eng.*, March, 1956, p. 202.

relief valves are shown in Figs. 8-9, 8-10, and 14-26. Such valves are more sensitive than pressure-relief safety valves and are preferred for hot-water-heater protection in some plumbing codes. The valves are designed to discharge water sufficiently rapidly to prevent the water temperature from rising above a predetermined limit. If the temperature cannot rise, the pressure is controlled. Hence, the selection of the size of the valve is based on the heat-unit (Btu) input of the device to be protected.

FIG. 14-27. Air-relief valve.

14-12. Air-relief Valves. Air-relief valves are shown in Figs. 14-27 and 14-28. These valves are used on water-supply systems in which air may become entrained to such an extent as to cause trouble. Water-supply pipes should be laid on such a slope that air will rise through the pipes to the valve, which is placed at the highest point on the system. The valve operates as follows: When the valve is filled with water, the float presses the valve shut and no water can escape. A bubble of air, rising through the air-relief pipe, will enter the valve chamber and displace some water. The float will drop, opening the valve to release air. Water will follow immediately into the valve chamber, raising the float and closing the valve again.

14-13. Air-inlet Valves, Vacuum Breakers, and Backflow Preventers. Air-inlet valves are used to admit air to a pipe or closed vessel to relieve vacuum.Types of air-inlet and air-relief valves are shown in Figs. 14-27 and 14-28. The type of valve shown in Fig. 14-27 can be used on hot-water storage tanks and in other locations where there is no flow of liquid under pressure past the valve.

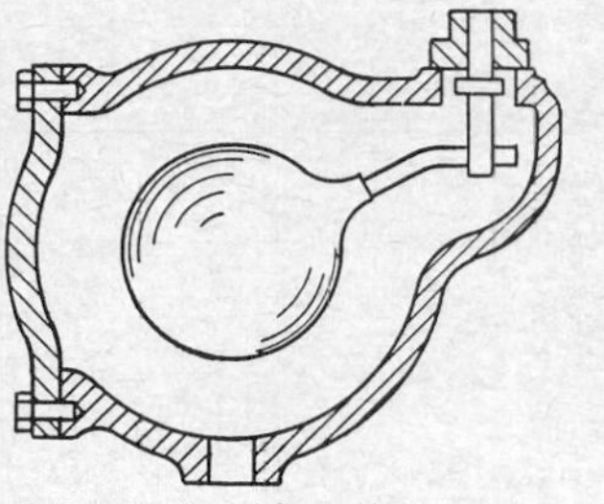

FIG. 14-28. Air-relief valve.

Vacuum breakers or backflow preventers[1] are required by most codes to prevent siphonage in water-supply pipes. Although the terms backflow preventer and vacuum breaker are not synonymous, they are sometimes so used in trade publications. A check valve is a backflow preventer. The vacuum breaker admits air to prevent the creation of a vacuum. The "Plumbing Manual"[2] states:

> **Backflow Preventers.** When any supply pipe is installed with a fixture or receptacle in such a manner that an approved air gap is not provided, an approved backflow preventer should be installed in the supply fitting or connection on the outlet side of the control valve.

[1] See also W. M. Dillon, How to Prevent Backflow in Water Lines, *Domestic Eng.*, October, 1955, p. 118.

[2] "Plumbing Manual," Report BMS66, National Bureau of Standards, 1940.

Some forms of vacuum breaker are illustrated in Fig. 14-29. One satisfactory device, shown in Fig. 14-30, has a central zone between two check valves in which the pressure is always less than the pressure on the drinking-water side of the device. There is a relief opening from the central zone to the atmosphere that discharges to the atmosphere at specified minimum rates without permitting the pressure in the intermediate zone to exceed the pressure on the inlet side by more than 0.5 psi.[1]

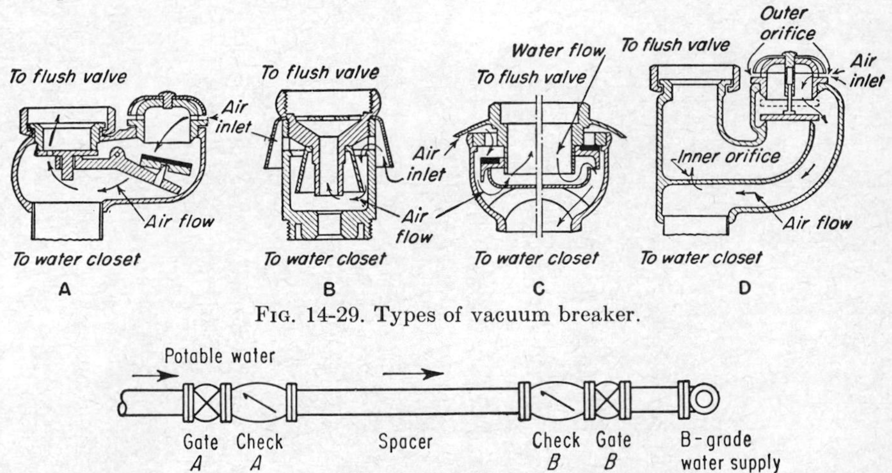

FIG. 14-29. Types of vacuum breaker.

FIG. 14-30. An approved interconnection between two water supplies in which the normal direction of flow is from the potable to the questionable supply. See also *J. Am. Water Works Assoc.*, March, 1933, p. 432.

Federal Specifications WW-P-541a, in Sec. E-38f, allow moving parts in the backflow preventer, but prohibit the use of springs or flexible parts. The specifications state:

The device shall prevent a reduction of pressure in the pipe greater than 1 in. of water when the outlet end of the flush pipe is closed or submerged in water and a vacuum of 15 in. of mercury is applied on the supply side. The critical level[2] shall in no case be below the outlet connection, and when the critical level is above that point it shall be shown by a horizontal line not less than 1/4 in. long accompanied by the appropriate symbols *C-L* or *C/L* clearly cast or stamped on the body of the device. When not indicated by the prescribed mark, the critical level shall be considered as being at the outlet end of the device.

The critical level of backflow preventers when installed shall be located at least 4 in. above the flood level[3] of the fixture except that where existing supplies

[1] See also *Water Works & Sewerage*, April, 1944, p. 127.

[2] The level to which a backflow preventer must be immersed in water in an open vessel before backflow will take place under a vacuum of 15 in. of mercury on the supply side of the device.

[3] See Glossary in Appendix I for definition.

which do not permit an elevation of 4 in. must be accommodated, the elevation of the critical level may be placed not less than 2 in. above the flood level of the fixture.

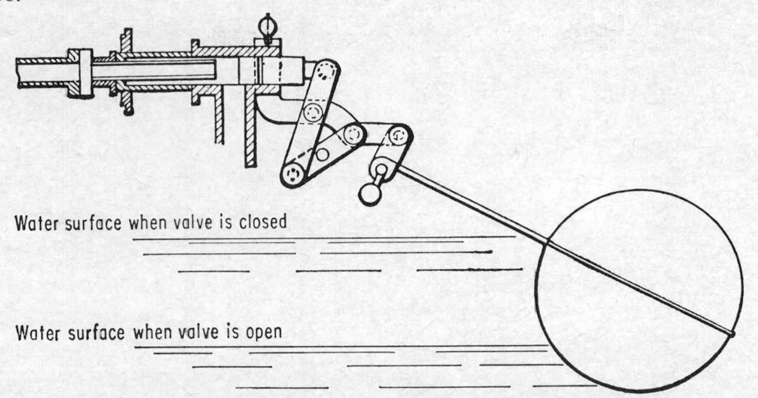

FIG. 14-31. Detail of float-controlled valve used on water-closet flush tanks.

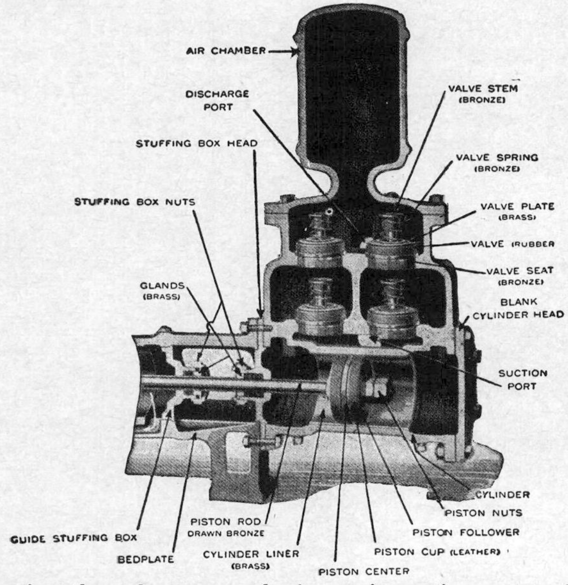

FIG. 14-32. Section through water end of a reciprocating pump showing valves, piston, packing glands, etc.

14-14. Float-controlled Valves. Float-controlled valves are in common use in automatic apparatus. They are used on flush tanks, on storage tanks, and in places where it is desired to maintain a constant level or a maximum height of water in a storage tank. Noiseless operation is desired. Principal causes of noise in such controls are the impingement of the high-velocity jet with entrained air into water already in the tank, and cavitation resulting from the high velocity of flow as the valve closes. Both of these can be overcome by submerging the jet, as at D in Fig. 15-7 and as discussed in Sec. 18-29. A common type of float-controlled valve used in a water-closet flush tank is shown in Figs. 14-31 and 15-7. Many other types and modifications are used; there is scarcely a complete plumbing installation that does not contain one or more float-controlled valves.

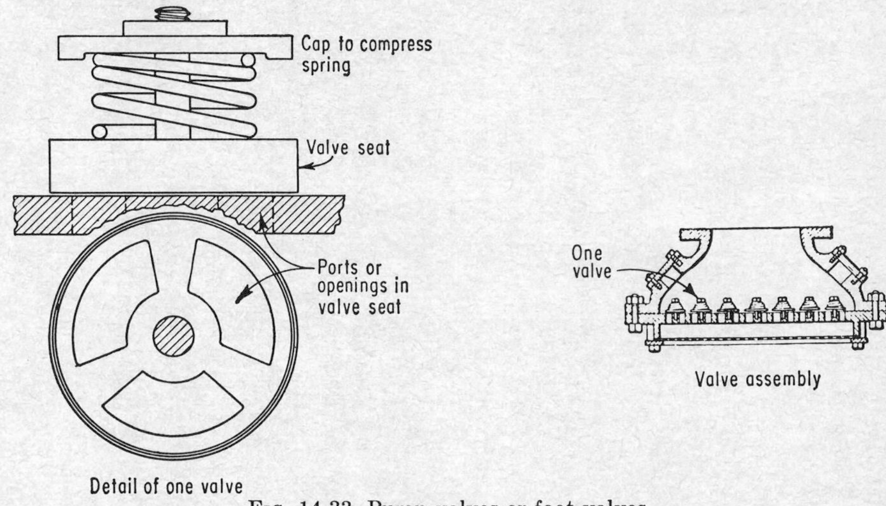

FIG. 14-33. Pump valves or foot valves

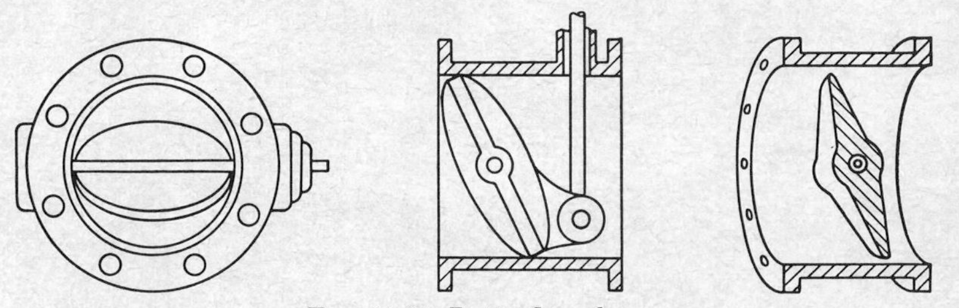

FIG. 14-34. Butterfly valve.

14-15. Valves for Reciprocating Pumps. The most common type of valve used in the suction and discharge chambers of reciprocating pumps is shown in Figs. 14-32 and 14-33. It acts as a check valve, allowing water to flow in one direction only. Pressure beneath the valve lifts the leather disk off of the valve seat, uncovering the port and permitting the passage of water. On the release of the pressure beneath the valve the reverse tendency of the water, aided by the spring, forces the disk back onto the seat.

14-16. Foot Valves. A foot valve used for the suction line of a pump is shown in Fig. 14-33. This consists of a large number of small valves, similar to the valve described in Sec. 14-15, placed in the bottom of the footpiece. Another type of foot valve consists of a single hinged flap in place of the many small valves shown in the figure.

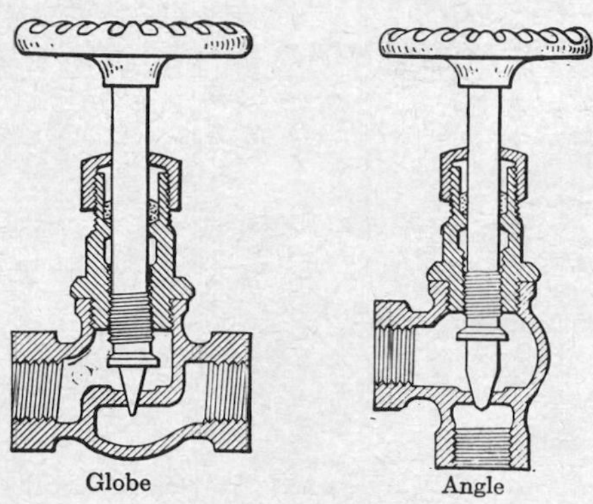

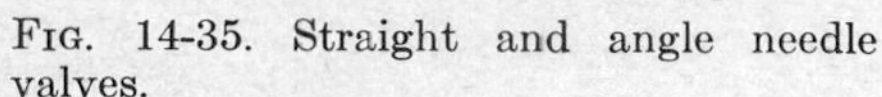

FIG. 14-35. Straight and angle needle valves.

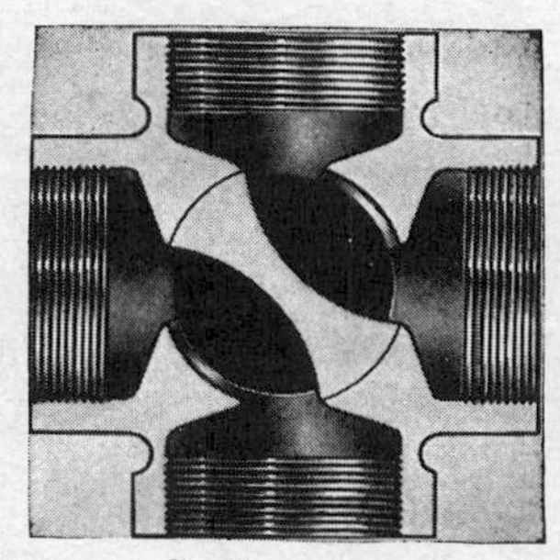

FIG. 14-36. Section through a four-way valve.

14-17. Butterfly Valves. A butterfly valve is illustrated in Fig. 14-34. It consists of a disk, somewhat like the damper in a stovepipe, that is shaped to fit the inside of the pipe as far as the disk will turn. When the valve is closed, the disk fits against a ground-faced seat, but unfortunately it is difficult to make and maintain a tight fit for such a valve. Butterfly valves are used in automatic apparatus, in throttle valves, and as quick-closing valves.

14-18. Needle Valves. Needle valves, such as illustrated in Fig. 14-35, are used in the handling of gases or where it is desired to obtain a fine adjustment of the opening of the port.

14-19. Three-way and Four-way Valves. Three-way and four-way valves are used in special mechanisms where it is desired to divert the flow from one incoming channel to any one of two or three possible outlet channels or where it is desired to connect an incoming and an outgoing channel immediately after a different pair of incoming and outgoing channels have been disconnected. One type of four-way valve is shown in Fig. 14-36.

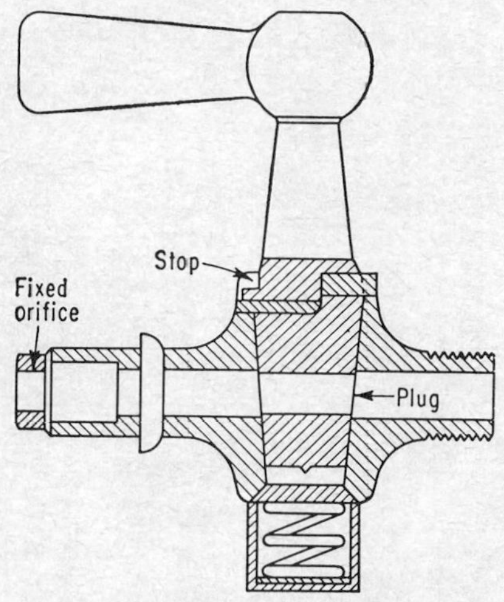

FIG. 14-37. Typical fixed-orifice plug valve for illuminating gas.

14-20. Gas Valves. Types of gas valve for illuminating gas are shown in Fig. 14-37 and for liquid petroleum gas in Fig. 9-5. Valves with special features are required for liquefied petroleum gas;[1] e.g., they have upset left-hand threads so that they cannot be interchanged with normal right-hand threaded valves. The purpose is to prevent interconnection of petroleum gas and other gases.

14-21. Fuller Faucets. The fuller faucet, shown in Fig. 14-38, is not in common use. The parts for the shutting off of water supply consist of a soft rubber ball marked A, which fits into a cup-shaped seat and is forced onto the seat by the turning of the faucet handle. The ball makes a tight joint, but it does not wear well, is affected by heat and by corrosive liquids, and is not easily replaced. The device has the advantage that the faucet can be opened wide by a one-fourth turn of the valve handle, but the quick closing of the valve may result in serious water hammer in the plumbing pipes.

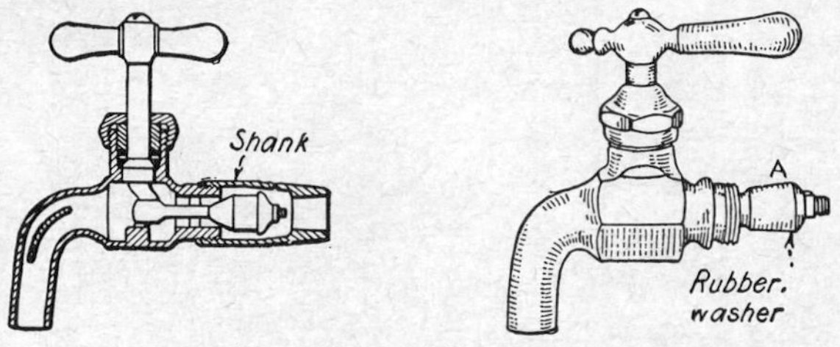

Fig. 14-38. Fuller faucet.

14-22. Compression Faucets. A compression faucet is an adaptation of a globe valve. It is commonly used on bathtub, lavatory, and some other plumbing fixtures. Its construction is illustrated in Fig. 14-6. The faucets are equipped with a soft gasket or packing, shown in the figure. This requires occasional renewal and may give trouble if exposed to hot water under high pressure. There are two general types of these faucets and many modifications. In one type the barrel is threaded to receive a correspondingly threaded plunger that carries the washer to the seat. In the second type the threaded stem actuates a squared plug which moves up and down inside a barrel squared to receive it. Compression faucets are open to many of the same objections as the globe valve, particularly with regard to tortuous passages, small waterway, and relatively high loss of head. The faucets require a number of turns of the handle or wheel to open or close them. The effort is sometimes bothersome to the user, but the slow motion is a preventive of excessive water hammer. Rates of discharge from compression faucets are shown in Table 14-1.

14-23. Self-closing Faucets. Self-closing faucets are used in public buildings and in institutions to conserve water. Their use is undesirable on lavatories in sleeping rooms in hotels without direct connection to toilets because the lavatories may be used as urinals and for other objec-

[1] See also *Domestic Eng.*, January, 1956, p. 135.

tionable purposes, and the self-closing faucet forces the legitimate user to place water in the lavatory for personal washing purposes.

A type of self-closing faucet in common use is illustrated in Fig. 14-1. It is of the compression type, but the valve disk is held on the seat by a strong spring that is compressed by a lever action or by means of a screw thread with a steep pitch. Although helpful for their purpose, which is to prevent water waste, quick-closing valves are not infallible, since they are more or less easily fastened open. Because they usually close quickly when operating properly, they nearly always produce water hammer. One type of self-closing faucet delivers a predetermined volume of water before it shuts off automatically. It has the advantages of leaving the user's hands free while the faucet is running and of eliminating water hammer as it slowly closes.

14-24. Mixing Faucets and Temperature-control Faucets. Mixing faucets are used to supply from one nozzle or opening water of any desired temperature between the limits of the temperatures of the cold-water supply and the hot-water supply. A type of faucet is available that will supply water at the same predetermined temperature whenever the faucet is turned on.

Three methods are in use for temperature control in mixing faucets: manual, pressure, and thermostatic. Manually controlled mixing faucets, illustrated in Figs. 14-2 and 14-3, are sometimes called *combination faucets.* The cold water and the hot water have a common outlet, the mixture of waters occurring in the nozzle of the faucet. Temperature is controlled by manipulating each valve handle as required. Such faucets are usually satisfactory on bathtubs, kitchen sinks, lavatories, and laundry trays, particularly when equipped with a long swinging nozzle that permits the discharge of water at various parts of the fixture. The faucets may be objectionable in some installations where there are marked differences of pressure between the two water supplies, cold and hot. Under such conditions the waters may not mix and nothing but cold or hot water will issue from the faucet, since some of the higher-pressure water flows through the base of the mixing faucet into the lower-pressure water supply.

Manually controlled mixing faucets may be objectionable on shower heads, particularly when there is a long pipe between the mixing faucet and the shower head. After the bather has adjusted the temperature of the water to his liking, a sudden change in pressure in one of the supply pipes may chill or scald him. Manual adjustment of the temperature at the shower head may be difficult because of the lag in temperature change after a movement of the hot-water or cold-water handle due to the effect of the length of pipe from the mixing faucet to the shower head.

No valve should be placed on the pipe or shower head downstream from the mixing faucet unless a special type of mixing faucet is used that will prevent flow from one supply pipe into the other. Permissible connections are illustrated in Figs. 14-8 and 14-9. If an ordinary stop valve were placed on the shower head or shower-head supply pipe and were closed when both ordinary supply faucets were open, water of one temperature at a high pressure would flow into the supply pipes of the water of lower pressure and different temperature. The difficulty might be overcome by placing a check valve on each supply pipe.

A pressure-controlled mixing valve is so constructed that a change in pressure on one side of a movable balanced piston inside the mixing valve will cause the piston to move and increase the flow from the low-pressure supply. Although the readjustment to balance pressure may not maintain a close temperature regulation, a moderately satisfactory temperature is maintained and the danger from sudden chill or scalding is avoided.

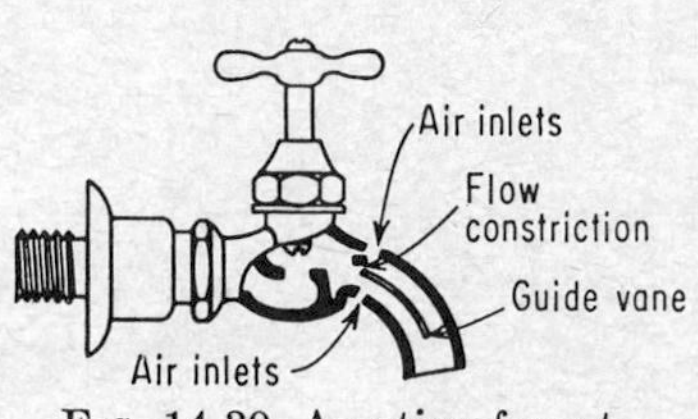

FIG. 14-39. Aerating faucet.

A thermostatically controlled mixing valve, as shown in Fig. 14-9, is capable of maintaining a closely controlled temperature. Such valves are useful in hospitals and in industries and laboratories where sudden or wide fluctuations of water temperature cannot be tolerated.

14-25. Miscellaneous Faucets. An aerating faucet that aspirates air into the stream of water discharging from the nozzle is shown in Fig. 14-39. The diameter of the stream is increased, its velocity is decreased, and splashing is diminished. A "versatile" sink faucet[1] can direct its stream at almost any desired angle, will shut off automatically if the jet spills from the sink, and can be reversed for use as a drinking fountain, and the downward stream can be aerated, but if it is turned upward the air is automatically cut off.

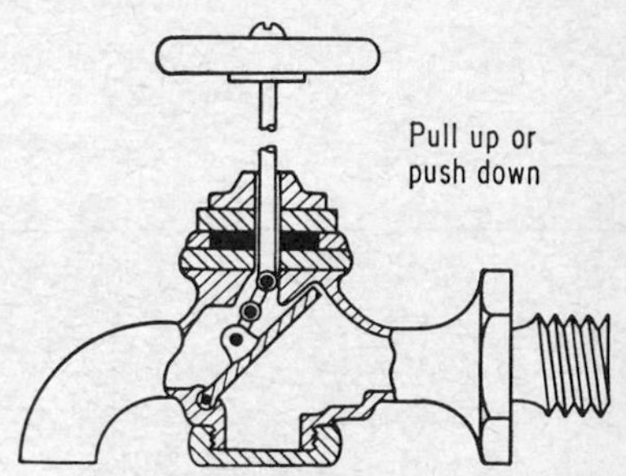

FIG. 14-40. Automatic tub or shower-bath nozzle.

Where a bathtub and overhead shower are connected to the same supply pipes, it is desirable to install a device, somewhat as shown in Fig. 14-40, that will automatically fall into such a position as to direct flow into the tub when the shower is turned off and will require manually lifting the knob on the top of the device while water is running into the bathtub in order to direct flow to the shower. The device consists of a butterfly valve that is held in position to supply the shower by the pres-

[1] Chase Brass and Copper Co.

sure of the water supplying the shower. The device serves to avoid the unexpected drenching that might otherwise be experienced by the user of the bathtub water-supply nozzle.

14-26. Pet Cocks. The pet cocks, air cocks, or faucets, shown in Fig. 14-41, are used to drain pipelines, as a valve, or on appurtenances; to test containers for the presence of liquid or gas under pressure behind the pet

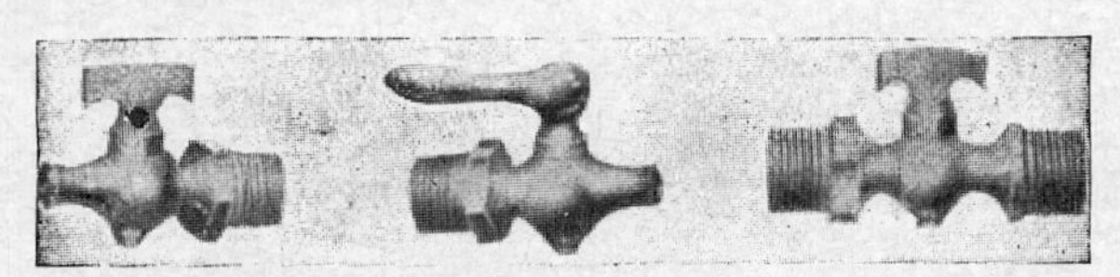

Fig. 14-41. Pet cocks, air cocks, or faucets.

cock; for connecting pressure gages; and for other purposes. They are seldom threaded for pipe larger than $\frac{3}{8}$ in.

14-27. Stop-and-waste Valves. The stop-and-waste valve, shown in Fig. 14-42, is used on water-supply lines and so operates that when the water supply is shut off, the water remaining in the pipe on the non-pressure side of the valve will drain out through the waste, which is opened when the pressure side of the valve is shut off. A stop-and-waste valve should be installed on the house end of all service pipes where the service pipe comes through the basement wall of the building, or if a meter is installed, the valve should be placed close to the meter and on the street side of it. The valve must be accessible and neither underground nor where it can become submerged. This type of valve is essential to the protection of water pipes from freezing when heat and water are turned off of a building in cold weather.

Fig. 14-42. Stop-and-waste valve.

14-28. Water-closet Flushometer Valves. Water closets can be flushed by admitting water to the closet bowl directly from the water-supply pipes. The installation of such valves is discussed in Secs. 15-24 and 15-27. To avoid the use of an excessive amount of water a valve should be used that will remain open just long enough to deliver the required amount of water and will then close automatically. The valve should be designed so that it cannot be held open.

Many types of water-closet flushometer valves are manufactured. A section through one type of valve manufactured by the Imperial Brass Company is shown in Fig. 14-43. It operates as follows: The flush valve has two main compartments or chambers, the upper one being bounded by the inside cover 3 and the rubber diaphragm 23, while the lower chamber is separated by this rubber diaphragm and is under direct water pressure at all times. The flush valve becomes active by merely a slight movement of the oscillating lever handle 18, which actuates the operating stem 13, thereby raising relief-valve stem 4. The raising of this stem opens hole 6, permitting water to escape from the upper to the lower chamber and thence into the closet bowl or urinal. The lowering of the pressure in the upper chamber results in the lifting of diaphragm 23, resulting in the opening of the main outlet from the supply pipe to the fixture, the water following the course marked by the arrow.

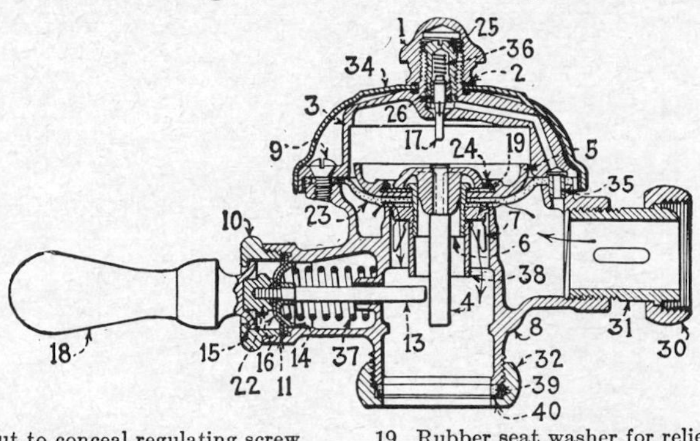

1. Cap nut to conceal regulating screw
2. Fibre washers (two required)
3. Inside cover
4. Relief-valve stem complete (Nos. 28 and 29)
5. Weight for diaphragm
6. Guide for relief-valve stem
7. Refill disk lock nut
8. Valve body
9. Screws for inside cover (five required)
10. Bonnet nut for handle
11. Brass friction washer for bonnet nut
13. Operating stem
14. Cup washer for operating stem
15. Bearing washer
16. Flexible rubber washer for operating stem.
17. Plunger for regulating screw
18. Handle, complete (porcelain or metal)
19. Rubber seat washer for relief valve
22. Cap nut for operating stem
23. Rubber diaphragm
24. Wire retaining ring
25. Regulating screw (body only)
26. Hexagon sleeve for regulating screw
30. Coupling nut for shut off
31. Coupling nipple for shut off
32. Coupling nut for flush connection
34. Shield or outside cover
35. By-pass tube
36. Spring for regulating screw
37. Spring for operating stem
38. Baffle tube
39. Rubber washer for flush connection
40. Brass friction washer for flush connection
41. Regulating screw, complete (Nos. 17, 25, 26, 36)

FIG. 14-43. Section through a flushometer valve. (*Imperial Brass Co.*)

In the meantime, the pressure in the upper chamber is being restored by the water flowing through the bypass channel 35, thus forcing diaphragm 23 downward and cutting off the flow of the water to the fixture.

The amount of flush is regulated by raising or lowering stem 17, which controls the amount of the opening of diaphragm 23.

WATER METERS AND GAS METERS

14-29. The Policy of Installing Water Meters. A meter is a device that measures the amount of water used and on which a charge for its use can be based. Some points for and against metering of water are summarized in Table 14-2. It is a peculiar fact that similar arguments are not raised against the installation of electric or gas meters. Where flat rates of charges are depended on for distributing the costs of water and

TABLE 14-2. SOME POINTS FOR AND AGAINST METERING

	For		*Against*
1.	It is only just that the consumer should pay in proportion to the amount he uses.	1.	Limited use of water may result in unhygienic conditions and disease.
2.	Waste is diminished, resulting in financial saving to all.	2.	Meters cost money to buy, to install, to maintain, and to read, partly defeating their own purpose.
3.	The poor actually pay more through taxes where meters are not used.	3.	Pressure losses through meters are appreciable and add to pumping cost.
4.	Loads on purification plants and pumps, etc., are minimized.	4.	The poor suffer more than the rich through water charges.
5.	Waste surveys are easier to conduct; leaks are more easily found.	5.	Consumer resents stand-by charge made when meter shows no water used.
6.	Most privately owned waterworks are 100 per cent metered.*	6.	The use on gardens is diminished, affecting economy and appearance of community.
7.	The careful consumer benefits; the careless consumer is penalized.	7.	Meter money might better be spent on waterworks improvement.
		8.	Waste can be more economically checked by inspection than by meter.

* It was illegal in Nevada for a private company to sell water by meter. (*See Water Works Eng.*, Feb. 26, 1941, p. 241.)

meters are not used, the small consumer may pay more than his proportionate share of the cost of water. Every householder should, therefore, insist on the installation of a meter for his own protection unless assured that the flat rates are favorable to him. Losses of pressure through and the capacities of water meters are given in Tables 14-3 to 14-5 and in Fig. 14-44. The loss of pressure through velocity meters is negligible, but since they are suitable only for high rates of flow, they cannot be used for domestic service.

14-30. Types of Water Meter. Water meters are known as *displacement meters* or as *velocity meters*. There are a number of types of displace-

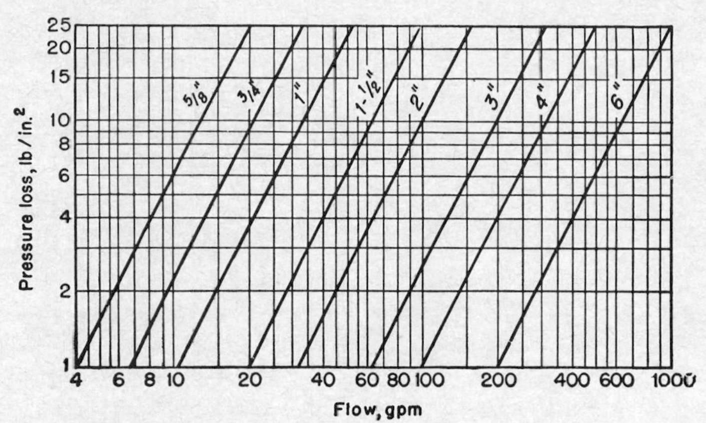

FIG. 14-44. Pressure losses in disk-type water meters.

Table 14-3. Dimensions and Capacities of Standard Water Meters
(American Water Works Association Standards)

Size, in.	Length, in.	Tail-piece length, in.	Maximum indication of initial dial, cu ft	Minimum capacity of register 100,000 cu ft	Normal test flow, gpm*	Mini-mum test flow, gpm	Capacity at 25 psi loss in pressure, gpm
			Disk meters				
5/8	7 1/2	2 3/8	1	1	1–20	1/4	20
3/4	9	2 1/2	10	10	2–34	1/2	34
1	10 3/4	2 5/8	10	10	3–53	3/4	53
1 1/2	12 5/8	2 7/8	10	10	5–100	1 1/2	100
2	15 1/4	3	10	100	8–160	2	160
3	24	...	10	100	16–315	4	315
4	29	...	100	1,000	28–500	7	500
6	36 1/2	...	100	1,000	48–1,000	12	1,000
			Compound meters				
1 1/2	....	2 7/8	10	100	2–100	1/2	100
2	....	3	10	100	2–160	1/2	160
3	....	...	10	100	4–315	1	315
4	....	...	100	100	6–500	1 1/2	500
6	....	...	100	1,000	10–1,000	3	1,000
8	....	...	1,000	1,000	16–1,600	4	1,600
10	....	...	1,000	1,000	32–2,300	8	2,300
12	....	...	1,000	1,000	32–3,100	14	3,100
			Current meters				
1 1/2	15 1/4 †	2 7/8	10	100	12–100	5	100
2	19	3	10	100	16–175	7	175
3	24	...	10	100	24–400	10	400
4	29 1/4	...	100	100	40–700	15	700
6	36 3/4	...	100	1,000	80–1,600	30	1,600
8	48 3/4	...	1,000	1,000	144–2,800	50	2,800
10	60	...	1,000	1,000	224–4,375	75	4,375
12	70	...	1,000	10,000	320–6,400	100	6,400

* Also "National Plumbing Code," Sec. D.7.3.

† The figures in this column represent maximum lengths. Minimum lengths are as follows: 1 1/2 = 13 in.; 2 = 15 1/4 in.; 3 = 20 in.; 4 = 22 in.; 6 = 42 in.; 8 = 26 3/4 in.; 10 = 30 in.; 12 = 36 in.

ment meters on the market that are known by the motion of the moving part or parts, as *reciprocating*, *rotary*, *oscillating*, and *nutating disk* meters. A displacement meter measures the quantity of flow by recording the number of times a container of known volume has been filled and emptied. A velocity meter measures the velocity of flow past a cross section of known area. The product of the velocity and the area will give the rate of flow at the instant of observation. An integrating device is necessary to record the total flow. Disk and piston meters are examples of displacement meters. Turbine and venturi meters are classed as velocity meters. Displacement meters are suitable only for low flows, and velocity meters only for high rates of flow. Capacities and dimensions of various types of meter are shown in Table 14-3.

TABLE 14-4. PERFORMANCE OF WATER METERS

Flow, gpm	Meters of various makes*	Niagara discount meters, all new†				From "Housing Code," p. 73‡		
	Head loss, ft of water					Size, in.	Normal test flow, gpm	Minimum test flow, gpm
	¾	¾	1	1¼	1½			
5	.....	2.0	1.1	0.7	0.2	⅝	1–20	¼
10	5–11	8.2	3.2	1.3	0.3	¾	2–34	½
15	7–25	17.2	6.3	2.4	0.7	1	3–53	¾
20	9–41	29.8	10.5	3.6	1.7	1½	5–100	1½
25	14–68	46.5	16.0	5.2	2.6	2	8–160	2
30	21–92	....	22.7	7.5	4.0	3	16–315	4
35	.....	....	29.5	10.2	5.5	4	28–500	7
40	.....	....	39.0	12.9	7.0	6	48–1,000	12
45	.....	....	....	16.0	9.0			
50	.....	....	....	19.4	10.8			
60	.....	....	....	27.2	15.0			
70	.....	....	....	37.0	21.0			
80	.....	....	....	....	27.0			

* Hartford Tests, *Eng. News-Record*, Dec. 12, 1918.
† Interpolated from test by V. R. Fleming at the University of Illinois.
‡ "*Registration.* The registration on the meter dial shall indicate the quantity recorded to be not less than 98 per cent nor more than 102 per cent of the water actually passed through the meter while it is being tested at rates of flow within the specified limits herein under normal test flow limits: There shall not be less than 90 per cent of the actual flow recorded when a test is made at the rate of flow set forth under 'minimum test flow.'" ("The Uniform Plumbing Code for Housing," Housing and Home Finance Agency, February, 1948.)

Standard specifications prepared by the American Water Works Association cover the following types of cold-water meter:

Cold-water meter type	*Date of publication in Journal*
Displacement	December, 1941
Current	April, 1946
Compound	April, 1946
Fire-service	February, 1947

Meters with moving parts (a classification which includes all turbine and disk meters) are sometimes protected by some form of screen, usually called a *fish screen*, to prevent the entrance of large objects into the meter.

All meters should be self-cleansing to prevent the accumulation of grit and other detritus from clogging the meter and wearing the moving parts. They should be constructed so that in case of freezing, some inexpensive and easily replaceable part will break, relieving the strain on other parts. Some meters are made partly of glass and of fragile cast iron for this purpose.

14-31. Capacity and Accuracy of Water Meters.[1] There seems little to influence the choice of any of the meters on the market in so far as accuracy and durability are concerned. However, meters of different makes show considerable variation in friction losses when meters are compared. A small meter of one make may work as efficiently as a larger or more costly meter of a different make. In general, within the field of their adaptation, meters can be relied on for accuracies greater than 99 per cent, although errors of 30 to 40 per cent have been found under unusual circumstances. In selecting the sizes of meters the data in Tables 14-3 and 14-4 can be used as a guide. It is usually better to select undersize rather than oversize meters because in the latter, although the life may be longer, the accuracy may be low and the loss of head high. Approximate values of pressure losses in various types of meter are given in Tables 14-3 to 14-5 and in Fig. 14-44.

14-32. The Disk Meter. A sectional view of a disk meter, given in Fig. 14-45, shows the common type used for residences. Water passing through the chamber in which the disk is located causes the dial to oscillate or nutate about its central spherical bearing with a spiral motion. The movement of the disk measures the filling and emptying of the disk chamber. The movements are transferred to the train of gears shown in the upper part of the meter, and the number of oscillations of the disk is recorded at the top of the meter in terms of the volume of water that has passed through the meter.

Two types of meter dial are shown in Fig. 14-46. In the circular-reading register each digit in the number expressing the amount of flow is shown on a different indicator. For example, one indicator shows the digit in tens of thousands, the next in thousands, the next in hundreds, then tens, and finally in units. The reading on the indicator in the figure is 19,517. Other arrangements of indicators may be used. The "straight-reading" dial shows the total flow directly in one place on the dial. Either type is satisfactory, but the former is simpler in construction and is more widely used.

All the moving parts of the disk meter are of metal except the disk and its spherical bearing, which are made of a composition not unlike hard rubber. The material is seriously affected by hot water, and where there

[1] For results of tests on water meters see G. E. Arnold, *J. Am. Water Works Assoc.*, 1954, p. 750.

is danger of hot water backing through the meter, a check valve should be installed as a protection. Meters made entirely of metal are available for measuring hot water, as explained in Sec. 14-38.

The accuracy of disk meters is high, even for very low rates of flow, and they maintain this accuracy through years of service. When in error

FIG. 14-45. Section of a disk, or nutating, meter. (*Neptune Meter Co.*)

Straight reading register Circular reading register

FIG. 14-46. Meter indicator dials. (*Neptune Meter Co.*)

they usually register less than the actual flow and almost all accidents of service tend to cause too low a record, except clogging of the displacement chamber.

14-33. Turbine Meters. A turbine meter is illustrated in Fig. 14-47. This is a velocity meter in which the velocity of the water flowing through the meter is measured by the speed of revolution of the turbine wheel. The volume of water passing through the meter is measured by the number of revolutions of the turbine wheel. This is recorded, in terms of water volume, on the meter dial. All the parts of the meter are of metal, making the meters suitable for use with hot water and, when properly constructed, for some corrosive liquids. The accuracy of turbine meters is satisfactory, but they have both upper and lower limits of usefulness.

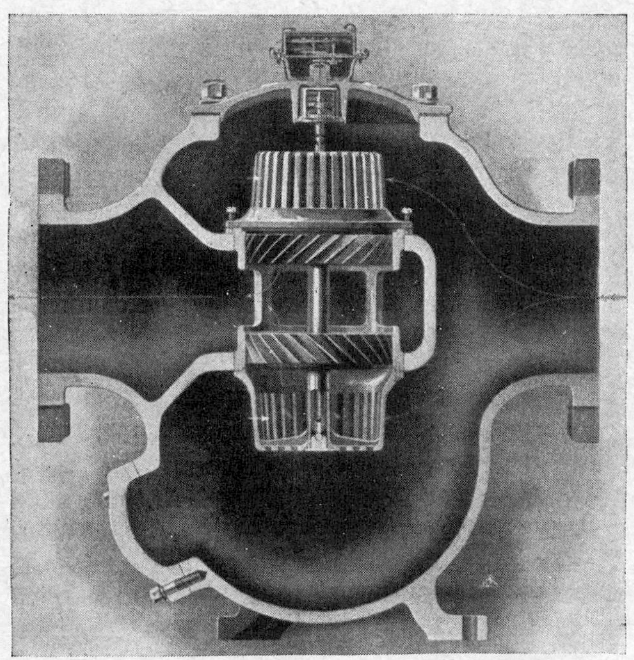

FIG. 14-47. Section of a turbine meter. (*Neptune Meter Co.*)

14-34. Compound Meters. To overcome the limited range of turbine and of large disk meters at low flows, a compound meter has been designed that combines the function of the turbine and the disk meter. The former measures the high flow, and the latter the low flow. When one part of the meter is recording, the other is idle. At the change-over from one to the other meter the accuracy is at its lowest.

14-35. Venturi Meters. A venturi meter, of the type shown in Fig. 2-3, operates without moving parts and hence is the simplest type of meter in use so far as its construction is concerned. It is a velocity meter, and it is suitable for measuring only high rates of flow. Rates of flow below its capacity limit are not accurately measured. It is not, therefore, suitable for use in measuring the low intermittent demand of most consumers, and its installation by the plumber will seldom be called for.

The difference between the pressure in the approach pipe and at the throat of the constriction is a direct measure of the rate of flow at the moment of observation, and this can be indicated on a gage. A somewhat complicated and expensive integrating mechanism is necessary to show the total flow through such a constriction or meter.

14-36. Elbow, Orifice, and Improvised Meters. An obstruction to the smooth flow of water in a pipe will cause turbulence and head loss that can be measured by observing the pressures in the pipe above and below the obstruction. The difference between the two pressures increases as the rate of flow increases. The relation between the two can be determined by calibration, a curve can be drawn to show the relation, and the rate of flow can be determined at any instant by referring the observed pressure loss to this calibration curve. An elbow or an orifice plate inserted in a pipeline constitutes an inexpensive velocity meter. The pressure gage or manometer should be placed one or two pipe diameters above the bend or elbow and the orifice and five or six or more lengths below the device to avoid the turbulence caused by the obstruction to flow. Such devices are used to show only the instantaneous rate of flow. They will not show the accumulated volume of flow that has passed through the device during a long period of time. An elbow meter is shown in Fig. 2-4.

14-37. Fire-service Meters. Special forms of meter are required for fire service because of the high rates and intermittent, irregular, or infrequent flow through the meter. Standard fire-service meters are of the compound type with the main meter of the velocity or proportional type and a small displacement meter for measuring the small flows. The capacities of fire-service meters, with a maximum head loss of 4 psi, are given in Table 14-5.

TABLE 14-5. CAPACITIES OF FIRE-SERVICE METERS
(Head loss of 4 psi)

Diameter of meter, in.	3	4	6	8	10	12
Flow, gpm	400	700	1,600	2,800	4,400	6,400

14-38. Hot-water Meters. Standard cold-water meters are unsuitable for use with hot water. The rubber and other compounds used in disks in cold-water meters are unsuited for use at temperatures above about 80°F. All-metal meters, with specifications otherwise similar to those for cold-water meters, are used for measuring hot water. In using hot-water meters it is essential that the pressure losses through the meter and the water temperature will not permit the formation of water vapor with resulting noise, inaccuracy, and probable damage to the meter.

14-39. Gas Meters. The dry type of meter illustrated in Fig. 9-4 is used almost exclusively in domestic plumbing installations. The capacity of the meter can be regulated by the internal adjustment that changes the stroke of the bellows. In view of the use of leather bellows and rings, the leather may harden with age and affect the performance of the meter. To avoid this difficulty the meter should be oiled at reasonable intervals.

TABLE 14-6. SIZES OF GAS METERS*

B-type meters			A-type meters		
Limits of flow, cu ft per hr	Pipe connections, in.	Cu ft per hr†	Size	Pipe connections, in.	Cu ft per hr†
3/100	3/8	100	6	1 1/4	600
5/150	1/2	150	11	2	1,100
10/300	3/4	300	25	2 1/2	2,500
20/450	1	450	40	3	4,000
30/600	1 1/4	600	60	4	6,000
60/1,250	1 1/2	1,250	170	8	17,000
100/1,850	2	1,850			
200/3,200	2 1/2	3,200			
500/9,000	6	9,000			

* See also Standards for Gas Service, *Natl. Bur. Standards Circ.* 405, 1934.

† On the basis of a pressure of 3 in. of water and a pressure loss of 1/2 in. through the meter.

If the gas passing through the meter contains tar, it is likely to coat the rings, rendering the meter worthless. Sizes and capacities of meters are shown in Table 14-6. The loss of pressure through the meter when delivering its rated capacity should be in the neighborhood of 0.2 in. of water and never more than 0.5 in. The installation of gas meters is discussed in Sec. 9-7.

CHAPTER 15

BATHROOMS, AND BATHROOM AND TOILET-ROOM FIXTURES

15-1. The Bathroom or Toilet Room.[1] The proper arrangement of the fixtures in a bathroom; heating, ventilation, and illumination; materials for floors, walls, and ceilings; etc., are as important as the type of fixture, the capacity of pipes, and other conditions in plumbing design. The best fixtures available connected to adequately designed and properly roughed-in drainage pipes but installed in a poorly ventilated and a poorly lighted room cannot be expected to give satisfaction to the user. Only a few principles of design and conditions of practice in heating, ventilating, and illuminating plumbing and bathroom installations will be given here. For a more complete discussion of the fundamentals and details, reference should be made to standard works on heating and ventilating.

15-2. Illumination. Illumination from an outside window or skylight is essential to the proper natural lighting of a bathroom. Light should not be cut off from the window or skylight by trees, buildings, walls, or other obstructions. Where it is not possible to obtain such natural illumination, artificial illumination must be depended upon. The amount of artificial illumination in a private bathroom is usually a matter of taste and exceeds the needs. A light placed high and near the center of the room is usually sufficient when the walls, ceiling, and floor of the room are light-colored or white. Additional lights are often placed on wall brackets, and usually one or two near the mirror over the lavatory. It is desirable, as a matter of safety, to place electric switches and equipment beyond the reach of a person touching an electric ground because bathroom pipes furnish such excellent grounding for electric current that a personal contact may be dangerous.

15-3. Heating. Heating may be by steam, hot water, or warm air. Gas, oil, or other open-flame heaters are undesirable in any room, but particularly so in a bathroom, because of its small size and impervious floor and wall construction, unless special provision has been made for

[1] See also How to Plan Bathrooms, series in *Am. Builder*, beginning November, 1955, p. 86.

ventilation. In the ordinary unventilated bathroom in a residence the presence of an open-flame heater with both door and window shut may result in death. For the same reason that ventilation is desirable, the required amount of heat is small.

In order to keep the required amount of heat small, the bathroom should be located in a protected part of the building with only one outside wall, the wall space being used for window, radiator, or built-in closet.

15-4. Ventilation. The ventilation of a bathroom in a dwelling usually depends on the door, the window, or a skylight. In some houses the bathroom door and window may be so substantially built and closely fitted as to afford no ventilation, and violent opening or closing of the door may weaken the seal of traps. Since the cleanliness of bathrooms and the absence of undesirable odors are dependent, to a great extent, on illumination and ventilation, some form of ventilation other than the door and window is desirable in homes and is essential in public toilet rooms. Air purifiers, deodorizers, and other proprietary devices are available to remedy undesirable conditions, but none can replace adequate illumination and ventilation.[1]

Plumbing codes may require that the toilet room be ventilated so that the air will be changed at least six times per hour. The change of air may be caused by forced or natural draft. Forced draft is created by fans, blowers, or special heating apparatus. Where dependence is on forced ventilation, some authorities recommend the use of a fan with a capacity of 35 cfm and with a maximum noise limitation of 50 decibels (db). There should be louvers in the door and high wall or ceiling louvers which close automatically when the fan is not operating. The louvers should open into the outside air or into a ventilating duct leading thereto. Natural drafts, which are commonly used in small buildings, are created by the use of flues and large air-duct openings.

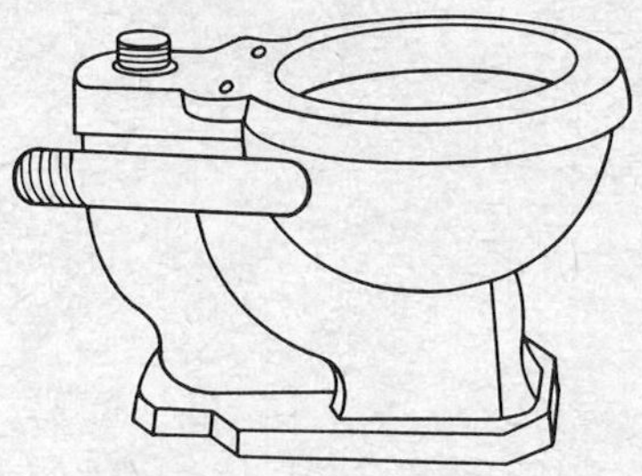

FIG. 15-1. Water closet with spud for local vent.

15-5. Local Vents. Bathrooms, toilet rooms, laundries, and similar rooms must be provided with fresh air. Foul air must be removed from them. The removal of foul air is the principal purpose of a local vent. It may open into the room through a register or other protected opening in the wall near the fixture to be vented, it may be connected into the fixture through a spud constructed in the fixture for the purpose as shown in Fig. 15-1, or it may be joined to the discharge pipe from the fixture, but above the trap, somewhat as shown in Fig. 15-2. Local-vent installa-

[1] See also *Domestic Eng.*, December, 1956, p. 128.

tions and connections are shown in Fig. 15-3. The upper end of the local vent is best carried through the roof and terminated under provisions similar to those for plumbing stacks stated in Sec. 12-11. Such provisions are not necessary, however, under all conditions. The local vent may terminate on the outside wall of a building, in an areaway, in a ventilating well surrounded by a building or buildings, in a ventilating flue, or other suitable condition. However and wherever the vent is terminated, provision must be made to avoid the return of odors and sounds to the interior of the building.

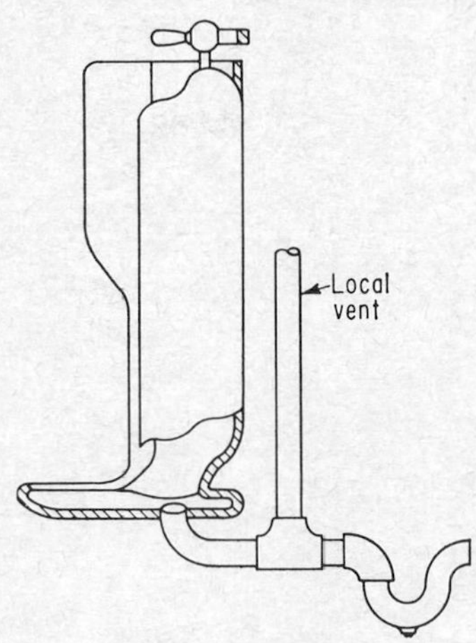

Fig. 15-2. Urinal with local vent.

If in a chimney flue, the local-vent terminal must be above all other openings in the flue to avoid the return of odors into the building. The vent pipe opening should be flush with the vertical face of the flue to avoid clogging the flue or the vent by accumulation of soot.

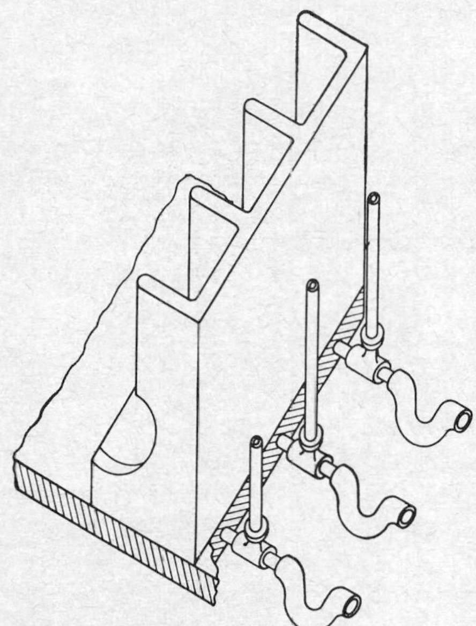

Fig. 15-3. Battery of urinal stalls with local vents.

The local vent pipe should be as short and direct as possible, with few changes in direction, each made with 45° or long-radius bends, sloping steeply upward from lower to upper ends. These provisions are made to prevent the accumulation of solids at bends in the pipe. Sheet-copper or galvanized-iron pipe or tubing should be used unless the local vent is exposed to corrosive gases, as in a chimney flue. Materials suitable for plumbing vent pipes should then be used. A local vent pipe from a single fixture should be 2 or preferably 2½ in. in diameter. Vents from fixtures in the same room can be joined into a single main vent whose cross-sectional area is equal to or only slightly smaller than the sum of the areas of the joining vents. It is not permissible to join local vents from different rooms, as sounds, odors, air-borne diseases, insects, and other indesirable features may be transmitted between rooms. A proprietary device is available which, it is claimed, will prevent the transmission of sounds in local vents without impeding the passage of air.[1] Difficulties are best avoided by separating local-vent terminals.

Natural ventilation cannot be depended on to accomplish the purpose of a local vent. Forced ventilation is usually impracticable and unavailable. The required draft can be induced by heating the local vent pipe. This can be done by placing it in a pipe run with a hot-water or a steam pipe, by terminating it in a chimney flue, by a gas flame burning under proper precautions in the bottom of the vent pipe or the flue or pipe run containing the vent pipe, or by other means.

15-6. Building Code Requirements. Few plumbing codes adequately cover the requirements for ventilation, which are more commonly classed under housing. In some governmental jurisdictions, either state or municipal, where the plumbing code is silent on such points, the building code, the housing code, or some other law, regulation, or ordinance will cover the point. Requirements of the "New York Building Code of 1941" and of the building code of the American Standards Association, ASA A 53.1–1946, have points of similarity. Some features of the New York code are:

Sec. C26-262.0. Every . . . [toilet] . . . room containing one or more water closets or urinals . . . shall be ventilated in at least one of the following ways:

1. *Windows opening to outer air.* By one or more windows, opening to a street or to a yard or court of lawful dimensions on the same lot or plot. Such window or windows shall have a clear area between stop beads of at least ten per cent of the floor area. At least 50 per cent of the required area shall be clear ventilating area but every window shall be at least 3 ft square in area and at least 1 ft in width.

2. *Windows opening on shafts or courts.* *a.* By a window of the size specified in paragraph 1 of this section opening on a vent shaft that extends to and through the roof, or into a court of lawful dimensions which has a cross-sectional area of at least ⅕ sq ft for every foot of height, but at least 9 sq ft and, unless such shaft opens to the outer air at the top, there shall be a net area of fixed louver openings in the skylight equal to the required shaft area.

3. *Individual vent flues or ducts.* *a.* By an individual vent flue or duct extending independently or any other flue or duct to and above the roof and having a cross-sectional area of at least 1 sq ft for one or two water closets or urinal fixtures and ⅓ sq ft additional for each additional water closet or urinal fixture.

b. Vent flues or ducts passing through two or more successive floors or through one or more floors and the roof shall run in shaft or shafts constructed as prescribed in Secs. C26.638.0 through C26.647.0.

c. Each flue or duct shall be equipped with an automatic closing damper where such flue or duct enters the shaft enclosure, and each flue or duct shall be equipped

[1] Manufactured by Industrial Acoustics Co., New York City.

at its upper termination with a wind-blown ventilator cap. Such damper or cap shall be designed in accordance with the rules of the board.

d. When two or more such flues or ducts are enclosed in a single shaft, each shall be covered with fire-retarding materials as prescribed by rules of the board.

4. *Skylights.* By a skylight in the ceiling, having a glazed surface at least 3 sq ft and arranged so as to provide fixed ventilating openings of at least 1½ sq ft to the outer air above the roof of the structure or into a court or yard of lawful dimensions, for one or two water-closet or urinal fixtures, and 1 sq ft additional for each additional water closet or urinal fixture.

5. *Mechanical exhaust ventilation. a.* By some approved system of mechanical exhaust ventilation of sufficient capacity to exhaust at least 40 cu ft of air per min per water closet and per urinal for public toilet rooms, and at least 25 cu ft per min per private bathroom.

b. Separate exhaust flues shall be provided for every 250 ft of height of structures, and such flues shall be of approved construction.

6. Interior bathrooms and water-closet compartments shall have fixed openings from adjacent rooms or corridors or from approved sources, ample to provide sufficient inflow of air to make exhaust ventilation effective.

7. It shall be unlawful to use pipe shafts as ventilating shafts.

The requirements of the "Wisconsin State Plumbing Code," issued by the State Board of Health in 1925, covering bathroom construction are as applicable today as when promulgated. The following has been abstracted from the Wisconsin code:

Sec. 5255. Artificial Light. Every toilet room, except in a private apartment, shall be artificially lighted during the entire period that the building is occupied, wherever and whenever adequate natural light is not available, so that all parts of the room are easily visible.

Sec. 5256. Size. Every toilet room shall have at least 10 sq ft of floor area, and at least 100 cu ft of air space, for each water closet and each urinal.

Sec. 5257. Floor. The floor and base of every toilet room shall be constructed of material (other than wood) which does not readily absorb moisture and which can be easily cleaned; except that wood floors may be used

1. In private apartments

2. If approved in writing by the Industrial Commission or the State Board of Health or their authorized agents, in existing buildings where there is an existing wood floor in good condition and where such toilet will be used by not more than five persons; provided further that such room must have an outside window or skylight

Sec. 5258. Walls and Ceilings. The walls and ceilings of every toilet room shall be completely covered with smooth cement or gypsum plaster, glazed brick or tile, galvanized or enameled metal, or other smooth, nonabsorbent material. Wood may be used if well covered with two coats of body paint and one coat of enamel paint or spar varnish. But wood or like material shall not be used for partitions between toilet rooms, nor for partitions that separate a toilet room from any room used by the opposite sex. All such partitions shall be as nearly soundproof as possible.

In large rooms a hose connection and floor drain should be provided.

Sec. 5259. Partitions between Fixtures. Adjoining water closets shall be separated by partitions. Each individual urinal or urinal trough shall be provided with a partition at each end and at the back, to give privacy.

Where individual urinals are arranged in batteries, a partition shall be placed at each end and at the back of the battery. A space of 6 to 12 in. shall be left between the floor and the bottom of each partition. The top of the partition shall be between 5½ and 6 ft above the floor. Doors with the top 5½ ft above the floor and the bottom 6 to 12 in. above the floor shall be provided for all water-closet compartments.

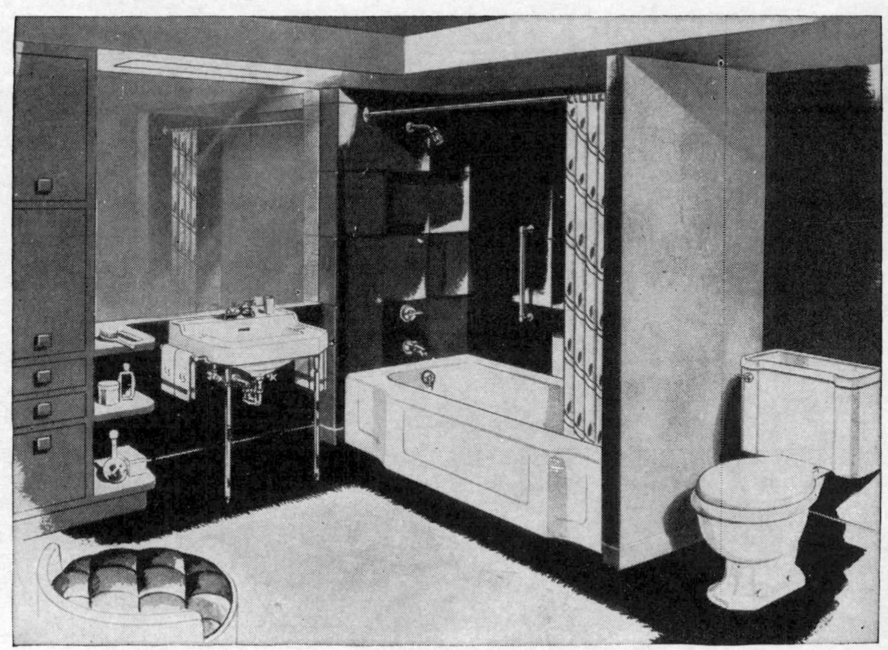

FIG. 15-4. A modern bathroom. (*Courtesy of Kohler of Kohler.*)

Water-closet compartments shall be not less than 30 in. in width, and shall be sufficiently deep to permit the door to swing past the fixture when opened.

15-7. Bathroom Interiors. Toilet- and bathroom floors and walls for a height of at least 3 or 4 ft, in private dwellings as well as in public buildings, should be of an impervious material that is easily cleaned. Among the satisfactory materials used are vitrified tile for the floor and enameled brick or glass for the wall, as specified above in Sec. 5258 of the Wisconsin code. Although white is desirable for the walls, floor, and ceiling, its monotony can be broken by the use of color and designs that do not detract from the illumination of the room. An interior view of a modern bathroom in which such materials have been used is shown in Fig. 15-4.

Other satisfactory materials used for floors include terrazzo, asphalt, concrete, or hardwood well laid and impregnated with a waterproof coat of linseed oil. Painted plastered walls above a 10-in. floor or baseboard have proved satisfactory for bathrooms in a residence, but they do not present an appearance of elegance or of present-day style, nor do they afford the ease in cleaning resulting from the use of impervious and smooth, nonabsorbent materials. In public toilet rooms slate can be used to form the walls of toilet compartments.

15-8. Industrial Washrooms and Locker Rooms.[1] Toilet rooms, washrooms, and locker rooms are necessary in industrial plants for the maintenance of proper hygiene. Washrooms should be centrally located, requiring no more than one flight of stairs or 200 ft of travel from the farthest workplace. Group washing fixtures are economical in use of floor space and of workers' time. The "National Plumbing Code" (Sec. 7.17) allows 18 in. of sink edge, either circular or straight, as the equivalent for one lavatory. From 8 to 10 users can be accommodated simultaneously at a 54-in.-diameter circular fountain. Precast stone and marble are most commonly used materials for such lavatories. A check list of equipment and of items to be considered in washroom and locker-room design includes location, illumination, ventilation, floor surface, washing equipment, soap dispensers, towels or warm-air driers, showers, toilet paper and holders, plumbing fixtures, cuspidors, lockers, waste baskets, toilet-article dispensers, rest rooms, bulletin boards, and other equipment.

15-9. Access Panels. An access panel, as the name implies, is a covered opening giving access to concealed plumbing such as valves, traps, and cleanouts that may require servicing. Although installed for a useful purpose, the panels may be adroitly concealed or worked into the decorative scheme of the bathroom. The "National Plumbing Code" (Sec. 7.6.2) requires an access panel or utility space so arranged as to make slip-joint connections accessible.

15-10. Bathroom Fixtures. The fixtures most commonly included in the bathroom of an American dwelling of moderate cost include a bathtub, a lavatory, and a water closet. The bathtub may have a shower-bath attachment. As the cost of the bathroom equipment increases, the elegance and number of fixtures may increase. Additional fixtures may include a sitz bath or bidet, a foot bath, a baby bath, or a shower-bath cabinet. It is sometimes desirable to place the water closet in a different room from the bathroom. This arrangement permits the use of the water closet when the bathroom is occupied, and it increases the availability of both.

A suggested check list of bathroom accessories is given in Table 15-1.

[1] See also *Natl. Safety News*, March, 1955, p. 63.

The arrangement of the fixtures within a residential bathroom, largely a matter of personal taste to be determined by the owner and the architect, should be based on the needs for heating, ventilating, lighting, and convenience.[1] No bathroom fixture should be placed against an outside wall, and economy in roughing-in will demand attention in the location of fixtures. The greatest economy can be approached by the use of the least amount of piping and fittings. The residential bathroom equipment and the rounghing-in shown in Fig. 11-3 represent an inexpensive and satisfactory arrangement. It is generally possible, but sometimes expensive, for an ingenious plumber to provide for any desired location or arrangement of fixtures. However, to minimize cost, short pipes, few fittings, and standard stock that is common and available locally should be used.

The number of bathroom fixtures specified by American Standards Association are stated in Tables 1-2 and 15-2.

TABLE 15-1. CHECK LIST OF BATHROOM FIXTURES

Clothes hanger
Electric outlets*
Face-tissue dispenser
Grab holds near bathtub and possibly in shower
Lingerie dryer†
Medicine cabinet
Shelves for linen and other storage
Soap cups or holders‡
Toilet-paper holder placed conveniently
Toothbrush holder placed conveniently
Towel cabinet
Towel racks†
Tumbler holders, one or two at lavatory and at dental lavatory
Vanity dresser and mirror

* One near lavatory for electric razor. Electric outlets should be placed to be inaccessible to a user in a bathtub or in a shower. Some codes require electric switches to be outide the bathroom.

† This may be a rack made up of stainless-steel, nickel-plated, or polished brass pipes through which hot water flows.

‡ Soap should be held above the bottom of its holder so that the soap touches at only a few points and air can circulate beneath it. The holder must be well drained. See *Air Conditioning, Heating, and Ventilating*, August, 1956, p. 124.

TABLE 15-2. MINIMUM NUMBER OF BATHROOM FIXTURES REQUIRED

Number of persons	1–9	10–24	25–49	50–100	Over 100
Toilets*	1	2	3	5	†
Urinals*	1	1	One for each 40 men		
Number of persons at one time	15	30	50	Over 50	
Washing facilities*	1	2	3	‡	

* ASA Z 4.1.

† One for each additional 30 persons.

‡ One for each additional 25 persons.

15-11. Plumbing-fixture Types and Standards. Plumbing fixtures may be as varied in form and color as the imagination of designers and manufacturers can make them, but there are basic sanitary, hygienic, and convenience requirements to which each fixture must conform.

[1] For bathroom layouts in mental hospitals see L. Blendermann, *Air Conditioning, Heating, and Ventilating*, February, 1957, p. 138.

Cleanliness and sanitation are paramount considerations in the selection of a fixture. For example, integral fixtures with the fixture and its back in one piece, as shown in Figs. 16-3*a*, 16-25, and others, are preferable to two-piece fixtures, such as a lavatory bowl with a separate top. Multiple-piece fixtures are conducive to the accumulation of dirt and the harboring of insects. For the sake of simplicity and economy, standard specifications have been prepared for plumbing fixtures, as listed in Table II-1.

15-12. Requirements for Plumbing Fixtures. The principal requirements of all plumbing fixtures are that they shall be sanitary, economical, and pleasing to the user. Sanitation is obtained by the use of nonabsorbent, easily cleaned materials; by the proper location and maintenance of the fixture; and by the use of traps, the seals of which will not be broken. The Housing Code[1] states (p. 25):

> **6.2.3. Protection of Water Supply.** All plumbing fixtures shall be supplied with water through an air gap as prescribed in ASA A 40.4–1942 [see Sec. 1-15 of this text]. Where such installations are not feasible they shall be provided with a backflow preventer as described, to be installed in accordance with the provisions of ASA A 40.6–1943 [see Sec. 14-13 of this text].

All fixtures other than water closets, pedestal urinals, bidets, and fixtures of similar type should be provided with a strong metallic strainer with outlet areas not less than that of the interior of the trap or waste pipe. The overflow pipe from a fixture should be connected to the house or inlet side of the trap, and the overflow should be so arranged as to be easily cleaned. If the overflow is erroneously connected to the drain pipe below the trap, a bypass is created for sewer air and gases to enter the room. The Housing Code[1] states (p. 24):

> **6.1.3. Overflows.** When any fixture is provided with an overflow, the waste shall be so arranged that the standing water in the fixture cannot rise in the overflow when the stopper is closed or remain in the overflow when the fixture is empty.

Economy is obtained through the production of durable fixtures at a low cost. The fixtures must be easily installed, and the maintenance cost must be minimum. Satisfaction to the user is obtained through pleasing appearance, graceful lines, attractive colors, and comfort in use. The architect and the plumber must consider what might be termed the engineering or practical features of fixtures, as well as the above-mentioned features that might be termed the *esthetics* of the fixtures. The engineering features include the rate of supply to and of discharge from the fixture, the type and number of fixtures required, the location of the

[1] "The Uniform Plumbing Code for Housing," Housing and Home Finance Agency, February, 1948.

fixture, together with other features pertinent only to certain installations and not of general importance.

15-13. Standard Dimensions. Various authorities have prepared standards for plumbing fixtures, as indicated in Table II-1. The "Plumbing Manual" states (p. 45):

> Federal Specification WW-P-541a is the accepted standard for plumbing fixtures in so far as it covers the material, kind, quality, size, and style of the different fixtures used.

Some standard dimensions from these specifications are given in Appendix II. In practice it may be necessary to obtain roughing-in dimensions from the manufacturer of any particular fixture that is to be installed. Some uniformity is followed in practice, however, which makes possible the drawing of plans and the layout of piping without detailed roughing-in dimensions.

15-14. Undersirable Types of Fixture. Wood should form no part of any plumbing fixture and should not be located near enough that it can be wet by the use of the fixture. Wood absorbs moisture, gives off odors when wet, and rots under conditions of alternate wetting and drying. Wood is an acceptab e material where properly protected from moisture. Wood washtrays or sinks fastened to the building should not be installed. Copper-lined or sheet-metal-lined wood fixtures are undesirable because of vermin and filth that may lodge between the wood and the metal. Pan and valve plunger, offset washout, and other water closets having invisible seals or unventilated space or walls not thoroughly washed at each flush should not be used. A water closet that makes possible the siphonage of its contents, under any conditions, into the water-supply pipes is prohibited. This does not preclude the use of water closets installed with a vacuum breaker or backflow preventer. Range closets are suitable only for temporary use, as in a ball ground, fair ground, etc. Long-hopper closets or similar appliances should be installed only in unheated places. No dry closet or chemical closet should be installed in a dwelling or other building when a common sewer is available.

15-15. Materials for Fixtures. The materials used for plumbing fixtures include vitrified porcelain, solid porcelain, enameled cast iron, black cast iron, tinned copper, galvanized iron, glass, soapstone, plastics, and marble. The first three materials named are used most extensively. They include semivitreous ware and terra cotta. Smooth, continuous surfaces that will show dirt and spots, yet are easily cleaned, and are free from cracks and crevices are desirable. A list of standard specifications is given in Table II-1, and abstracts from Federal Specifications covering materials for plumbing fixtures and from manufacturers' specifications for vitreous ware are included in Appendix II.

Vitrified porcelain is china, but the trade name *china* is frequently applied to unvitrified glazed ware made from baked clays. Vitrified ware is the most expensive and the most attractive in appearance of the three commonly used materials. It cannot be used for heavy fixtures, such as bathtubs and urinals, because the high temperature required for vitrification (2600°F) causes warping in the kiln. It is successfully used for lavatories, drinking fountains, and toilets. Solid porcelain is glazed only on the outside. It is not vitrified, and as a result, it will craze in time. Enameled cast iron is satisfactory except that it is affected by acids; even lemon and other citrus fruit juices will stain it. Enameled-iron fixtures should be made in one piece because cracks are difficult to clean. A standard test for vitreous-china fixtures was adopted on Oct. 1, 1926, by a manufacturers' committee cooperating with the Division of Simplified Practice of the U.S. Department of Commerce. Details of this test, together with definitions of trade terms, are given in those specifications. Other specifications for vitreous china are to be found in Commercial Standards CS 20–36 and in Federal Specifications WW–P–541a.

The protection of plumbing fixtures against damage during transportation and installation is necessary. Paper coverings pasted on the surfaces are common. A vinyl plastic coating applied with a brush and peeled off as a flexible film has been found useful and economical.[1]

TABLE 15-3. APPROXIMATE TIME IN SECONDS REQUIRED TO FILL A FIXTURE

Diameter of supply pipe, in.	Rate of discharge, gpm*			Water-closet flush tank, capacity 4 gal			Bathtub 4 ft long, capacity 50 gal			Wash basin, capacity 1 gal			Laundry tray, capacity 15 gal		
	H† = $5L$	$H = L$	$H = L/2$	$H = 5L$	$H = L$	$H = L/2$	$H = 5L$	$H = L$	$H = L/2$	$H = 5L$	$H = L$	$H = L/2$	$H = 5L$	$H = L$	$H = L/2$
3/8	8.8	4.0	2.8	27	60	85	342	750	1,071	7	15	22	105	225	330
1/2	14.0	6.3	4.4	17	38	55	215	480	747	4	9½	14	60	143	210
3/4	38.5	17.2	12.2	6½	14	20	80	174	246	1½	3½	5	23	52	75
1	79	35.3	25.0	3	7	9½	38	85	120	¾	1¾	2½	12	25	37½
1¼	138	61.7	43.7	1¾	4	5½	22	49	69	½	1	1½	7½	14½	22½

* It is assumed that the pressure is great enough to deliver this quantity through the faucet used on the fixture (see Table 14-1).

† H = head loss per unit length. L = length of pipe.

15-16. Supply and Discharge Rates. Information is given in Tables 3-4 to 3-6, and others on the rates of flow recommended for plumbing fixtures. Table 15-3 shows approximate units of time necessary to fill fixtures through pipes of different sizes, and Tables 3-4 and 7-2 contain recommendations for pipe sizes to fixtures. Sizes of traps, branch wastes,

[1] See *Architectural Record*, vol. 117, pp. 209 and 215, 1955.

and drainage pipes for fixtures and the rates of discharge to be expected from them are given in Tables 11-1 and 11-2.

15-17. Number of Fixtures Required. The number of fixtures needed in a residence is determined by the owner. He decides that he wishes one, two, or more bathrooms, and he decides how each bathroom is to be equipped. In schools, factories, public comfort stations, office buildings, hotels, and other buildings of a public or semipublic nature, the number of fixtures to be installed is usually left to the designer. Only special experience or knowledge of the experience of others can aid the judgment. The figures given in Table 1-2 show the number of fixtures recommended by the Housing Code.[1]

15-18. Number of Fixtures in Use Simultaneously. The ideal to be attained in selecting the number of fixtures in a public building is to

TABLE 15-4. NUMBER OF PLUMBING FIXTURES REQUIRED FOR MILITARY (MALE) PERSONNEL

Type of fixture	Number of fixtures													
	Number of persons													
	1	5	10	20	40	60	100	120	160	200	250	300	400	500
Showers:														
Arctic climate	1	1	1	1	1	2	3	4	4	6	7	8	11	14
Temperate climate	1	1	1	1	2	3	4	5	6	8	10	12	16	20
Tropical climate	1	1	1	1	2	4	6	7	9	11	14	16	22	27
Water closet	1	1	2	2	2	3	6	7	10	13	14	18	26	30
Lavatory	1	1	2	3	6	8	14	16	22	27	34	40	54	67
Urinal	1	1	1	1	2	2	3	4	4	5	5	9	10	13

supply that number, and no more, that will be in use simultaneously, with no persons waiting, at the moment of peak demand. The impracticability of attaining such an ideal seems evident. Studies made by the Hoover Committee[2] and others on the simultaneous use of fixtures may be summarized through the information given in Fig. 3-1. The probable maximum rate of discharge from any number of fixtures can be determined from this figure when the sum of the rates of discharge from all fixtures is known.

15-19. Water Closets. Dimensions of small toilet rooms are shown in Fig. II-68. Desirable features in a water closet include hygienic protection, cleanliness, comfort for the user, adequate flushing, silence in opera-

[1] *Ibid.*

[2] "Recommended Minimum Requirements for Plumbing in Dwellings and Similar Buildings," National Bureau of Standards, July 3, 1923.

tion, smooth passages, absence of mechanical parts except for flush control, and freedom from odors. The number of water closets and other fixtures is shown in Tables 1-2, 15-2, and 15-4 and is specified by ASA Z 4.1.

15-20. Types of Water Closet. Types of water closet include washdown, washout, siphon, siphon-jet, pneumatic siphon, and hopper. Some of these are shown in Figs. 15-5, II-65, II-66, and II-69 to II-71. The washdown and the washout closets depend on the force of the flushing water to remove solids from the pan on which they are deposited. The force of the water may be expended in doing this, so that the trap is not scoured and the fixture is spattered and fouled. The washdown closets shown in Fig. 15-5*A* and *C* are preferable to the washout closet shown at *D*, because solids fall into deeper water, are more easily flushed out, and the fixture tends to clear itself better. Siphonic-action closets avoid some of the objections to washdown and washout closets.

Siphon closets are divided into four types: siphon action, reverse trap, siphon jet, and pneumatic siphon. Of the types listed the siphon action is the least expensive. The reverse trap has the advantage over the siphon action in that it contains less water and a deeper seal, giving quicker and more positive action. When the fixture is flushed, it siphons itself because the downstream leg of the siphon is longer than the upstream leg. The energy of the siphon jet is directed into the trap. The siphon-jet closet has the advantages of the reverse trap and, in addition has a stronger action in self-cleansing. The flush in siphon-jet closets is strong because the energy of the jet is added to the downstream velocity of flow and to the self-siphonage of the fixture. Splashing sounds are subdued in siphon-jet closets as compared with washdown and washout fixtures.

A dry-hopper closet contains water only in the trap. A pedestal-hopper closet contains water in the bowl as well as in the trap. Short-hopper and pedestal-hopper closets have the trap above the floor. The former are more desirable in order that the water in the trap can be observed and the area of surface exposed to fouling above the trap is less. The long-hopper closet finds its use mainly in unheated compartments. In general, hopper closets are undesirable because of the area of dry surface exposed to fouling above the water in the trap.

Siphon or siphon-jet closets, with the trap above the floor in plain sight, are the most satisfactory type of water closet in use. Some are shown at *E* in Fig. 15-5 and in Figs. II-65, II-66, and II-70.

The Housing Code states (p. 25):

6.4.1. Types of Water Closet. Water-closet bowls shall be siphon jet, reverse trap, washdown, or blowout type with floor outlet or siphon jet or blowout type with wall outlet. Water-closet bowls and traps shall be made in one piece and shall be provided with integral flushing rims constructed so as to flush the entire

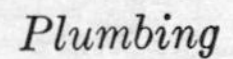

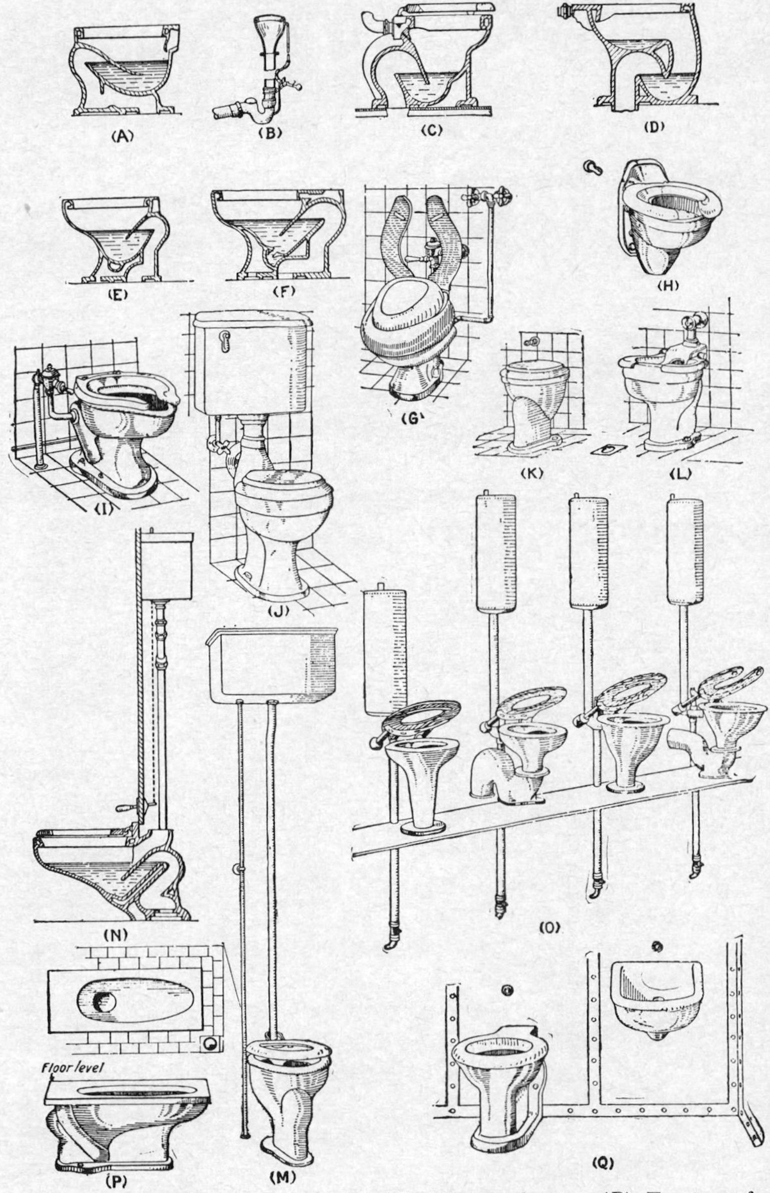

FIG. 15-5. Types of water closet. (*A*) Washdown closet. (*B*) Frostproof closet. (*C*) Washdown closet. (*D*) Washout closet. (*E*) Siphon-action closet. (*F*) Siphon jet closet. (*G*) Combination bedpan washer and water closet. (*H*) Wall-hung closet and flush valve handle. (*I*, *K*, and *L*) Closets showing flushometer valve handle. (*J*) Closet with low-down flush tank. (*M* and *N*) Closets with high-up flush tanks. (*O*) Automatic, seat-actuated pneumatic tanks for flushing. Two of the closets shown are frostproof hopper type. (*P*) Squat closet. Top is flush with the floor. (*Q*) Fixtures for penal institutions.

interior of the bowl. Water-closet bowls for public use shall be of elongated type with open front seat.

The "National Plumbing Code" (Sec. 7.6.1) prohibits the use of pan, valve, plunger, offset, washout, latrine, frostproof, and other water closets having an invisible seal or any unventilated space or having walls which are not thoroughly washed at each discharge and any water closet which might permit siphonage of the contents of the bowl back into the flush tank.

15-21. Water-closet Seats. Water-closet seats are so important a detail in proper plumbing that standard specifications have been prepared, such as those listed in Table II-1. Some abstracts from Federal Specifications are included in Fig. II-67. Hygiene and comfort are of first importance.

Comfort is provided through the style of the seat, its height above the floor, and the material of which it is made. It is not possible to suit all persons, but in general, a height of about 15 in. above the floor is satisfactory. Some persons prefer a seat sloping downward toward the rear. A seat that has the appearance of a chair is a satisfactory innovation in a residential toilet room, provided that all absorbent parts of the seat are protected from splashing and provided further that all parts are accessible for cleaning. Water-closet seats made of a material that (1) will not absorb moisture, (2) is a poor conductor of heat, (3) is easily cleaned, and (4) has no visible joints or cracks on the surface fulfill requirements in so far as materials are concerned. Seats made of varnished hardwood riveted, bolted, or glued together are too frequently used mainly because of a low first cost. They seldom fulfill the requirements listed above. They are being displaced in general use by more comfortable and durable types.

The seat should be securely fastened to the bowl, it should rest solidly on the rim, and a cover should be provided. The shapes and dimensions of various seats are shown in Figs. 15-6 and II-67. The spring seat shown at *O* in Fig. 15-5 and described in Sec. 15-28 is used in public comfort stations, correctional institutions, and locations where the user cannot be depended on to flush the toilet. The selection of any particular type of seat for a home is dependent on the choice of the householder. Health authorities emphasize the importance of selecting the open-front seat. An elongated, open-front seat is required by the "National Plumbing Code" in Secs. 7.7.1 and 7.7.6 for public toilets. The projecting-lip bowl is also essential to cleanliness, especially in public toilets. The depth and shape of the bowl and the location of the standing water should be such that the user is not splashed and urine is not spilled over the edge of the bowl.

The seat should be located so that there is room for the user between walls and fixtures. If the water closet is placed in a corner, the center of the seat should be 12 to 14 in. or more away from any wall, or if it is placed near another fixture, a plumb line through the nearest part of the adjacent fixture should be 12 to 14 in. or more away from the center of the seat. In public comfort stations the privacy of the user should be secured by placing the water-closet fixture in a separate compartment. It is not always feasible to place a door on the compartment because of abuses which sometimes occur in public toilets. The use of range closets is undesirable from the viewpoints of comfort and sanitation.

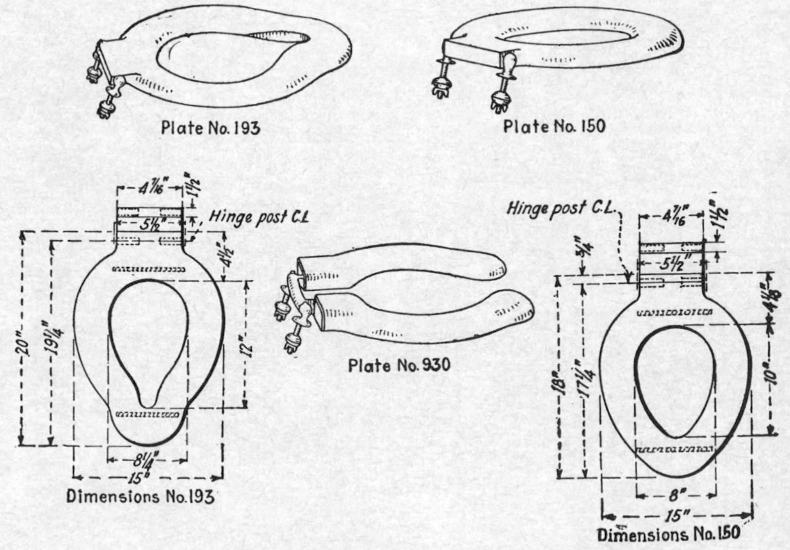

FIG. 15-6. Types of toilet seat. (*Church Seats.*)

15-22. Discharge Passages in Water Closets. The discharge passages in a water closet should be sufficiently large to pass a $2\frac{7}{8}$-in. ball, the depth of the seal of the trap should not be less than 3 or more than 4 in., and the amount of water in the trap should be in the neighborhood of 3 qt.

15-23. Frostproof Water Closets. A frostproof water closet, as shown at *B* in Fig. 15-5, is a long-hopper closet with its water-seal trap placed beneath the frost line. The sides of the long, funnel-shaped hopper exposed to soil and fecal matter become foul and in cold weather may become ice encrusted. The "Plumbing Manual" states (p. 14):

Sec. 703. Frostproof Closets. Frostproof closets may be installed only in compartments that have no direct access to a building used for human habitation or occupancy. The soil pipe between the hopper and the trap shall be of lead or cast iron enameled on the inside. The waste tube from the valve shall not be connected to the soil pipe or sewer.

The closet should be so arranged as to be adequately flushed, and the water-supply pipes and the trap should be adequately protected from freezing by placing them below the frost line. The hopper pipe should not, however, be more than 6 ft in length. Access to the trap should be provided by a manhole.

Where the closet is located at the upper end of a soil line, the soil line should be vented with a pipe not less than 2 in. in diameter carried above the roof of the closet and terminating 12 ft or more away from or 3 ft or more above any opening into a building.

15-24. The Flushing of Water Closets. Water closets are flushed by high-up or low-down tanks, direct-flushing valves, or pneumatic tanks. The Plumbing Committee of the American Plubic Health Association Year Book (1938–1939) recommended the following conditions as requisite for a satisfactory water-closet flushing tank:

1. Bottom of flush tank higher than rim of water closet
2. Tank overflow large enough to carry away readily full flow of water through the supply to the fixture when the valve is wide open and the pressure is maximum possible
3. The overflow drains into suitable fixture, thereby removing possible temptation to plug the overflow
4. Discharge opening and all other openings in water supply to tank not less than 1 in. above highest possible level of water surface in tank
5. No part of water-supply pipe, valves, or fittings submerged in water within tank
6. Free access of air into flush tank above water surface without passing through water closet
7. Tank so constructed and covered that foreign materials cannot accidentally be introduced into tank
8. Provision made for adequately hushing sound of water as it fills tank
9. Cover fastened on tank in such a way as to require use of tools to remove it

Note that desired conditions 5 and 9 in the preceding list are not commonly attained in practice.

Noise should be avoided in water-closet flushing. The silent closet has been designed, but few have been installed. Noises in flushing are discussed in Sec. 18-29.

15-25. Flushing Tanks. Because of the need for supplying water at a rate of 30 to 50 gpm for a short period the supply pipes from the tanks to the water closet should be of generous size, as described in Sec. 15-24. The use of so-called "high-up" tanks is so infrequent now that they may be considered as practically obsolete. Low-down tanks are most commonly used in household installation to avoid the large water-supply pipes required by flushometer valves. The use of flushometer valves is confined mainly to public buildings and comfort stations.

A close-coupled water closet may have the flush tank and seat in one unit with the tank only slightly above the seat or bowl. The "National Plumbing Code" (Sec. 7.7.4) states that in such a fixture the flush-valve seat shall be at least 1 in. above the rim of the bowl, so that the flush valve will close even if the closet trapway is clogged or any closet with flush valve seat below the rim of the bowl shall be so constructed that in case of trap stoppage water will not flow continuously over the rim of the bowl. A water closet with flush tank and bowl cast in one unit is shown in Fig. II-71.

15-26. Gravity Flush Tanks. Low-down gravity flush tanks for water closets are usually rectangular in plan, 5 to 6 in. wide, 18 to 20 in. long, and 14 to 15 in. high, with a capacity of about 4 to 6 gal. Materials used in them include vitrified porcelain, solid porcelain, and enameled cast iron, with a removable cover of the same material. A flat-top cover can be conveniently used as a bathroom shelf. Metal tanks with 12-oz[1] or heavier sheet copper may be lower in cost but are otherwise less satisfactory. Six-ounce sheet-copper lining has been used. The flush tank may be fastened to the wall behind the water closet to be flushed, with the bottom of the tank 6 in. or more above the toilet bowl. In some fixtures the tank is supported on the back of the toilet bowl, as in Fig. II-70, or it may be cast as an integral part of the unit, as shown in Fig. II-71.

Both high-up and low-down tanks should have a capacity of 5 to 8 gal, a flushometer should be able to deliver water at a rate of 40 to 50 gpm under normal minimum pressure, and a pneumatic tank should have a capacity of about 6 gal. Flush pipes for high-up tanks should not be less than 1½ in. in diameter and for low-down tanks not less than 2 in. The connection between the flush pipe and the water-closet bowl may be made by means of a slip joint.

Low-down flush tanks are preferred to high-up tanks, are more pleasing in appearance, and can be protected against back siphonage. The possibility of back siphonage from the toilet bowl into high-up tanks is more remote than in a low-down tank because of the height of the flush pipe.

The mechanism of a gravity flush tank is shown in Fig. 15-7, and of a float-controlled valve in Fig. 14-31. The mechanism operates as follows: When lever *F* is raised, the rubber ball *H* is lifted off of its seat and it is held at *G* by flotation. Water rushes out of the tank and ball *I* drops, opening the supply valve *B*. When the water level reaches *G*, ball *H* begins to drop and is suddenly pulled onto its seat by suction, thus cutting

[1] Weight per square foot.

off the flush. Valve *B* is supplying water to refill the tank through tube *D*. This tube is bent down to discharge water into the tank below the water level in the tank in order to maintain a silent flow. Water is also supplied from valve *B*, through pipe *A*, to the closet bowl in order to renew the seal in the trap. When float *I* reaches its original position, valve *B* closes and the tank is ready for the next flush.

All float-controlled or automatically controlled valves that discharge into fixtures, traps, or elsewhere should, where possible, be so installed that leakage from the valve is visible.

15-27. Flushometer Installations. Flushometer valves take water from the water-supply pipes at city pressure and deliver it directly into the closet bowl, thus doing away with noisy tanks and standing water and giving a short flush at a high rate. The amount of water delivered at one flush can be adjusted automatically, and with a properly designed bowl, flushometer valves will operate with less water than would be required by a flush tank. The operation of a flushometer valve is described in Sec. 14-28.

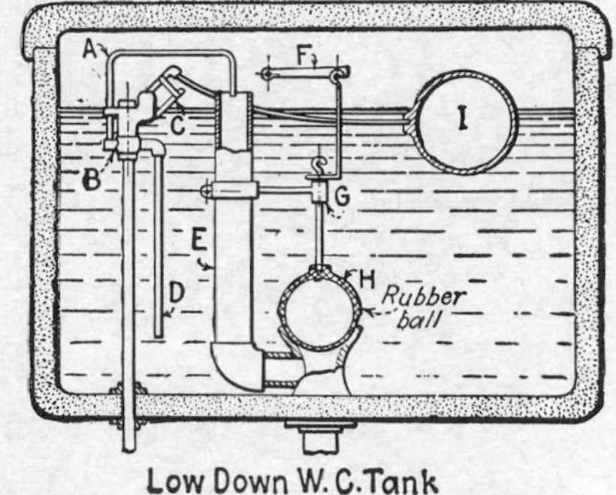

FIG. 15-7. Working parts of a water-closet flush tank.

Flushometer valves or similar devices should be connected to a water supply that will maintain a pressure of not less than 5 psi when it is flushing. A water-supply pressure between 35 and 50 psi is desirable. The valves should be constructed so that they cannot be held open for continuous discharge, and they should fulfill the conditions stated below without requiring regulation if the static pressure varies between 5 and 75 psi. The quantity of water discharged at one flush should be within the following limits:

Fixture	*Gallons*
Water closets and slop sinks, each	3–5
Pedestal or siphon-jet urinals, each	2–2½
For each 20 in. of urinal trough	2–2½
Flush rim or individual stall urinal, each	¾–2

Noise in flushometer valves is commonly caused by excessive pressure in the water supply. The noise can be diminished by an adjustment which forms a part of the valve. If this is not effective, a pressure-reducing valve or a perforated disk can be inserted in the supply pipe.

In the installation of flushometer valves care should be taken that water cannot back out of the fixture into the water-supply pipes when the pressure in the water-supply pipes is low or the water-supply pipes are drained. This can best be done by installing the fixture so that the end of the pipe cannot be submerged, the flushometer valve is above the highest possible water level in the fixture, and a backflow preventer or vacuum breaker is installed as described in Sec. 14-13.

An obstacle to the use of flushometer valves, which excludes them from use in most dwellings, is the large size of piping required from the water main to the valve. This size of pipe, as shown in Table 7-1, should not be less than 1¼ in., and it should preferably be larger. The difficulty can be overcome by the use of a pneumatic storage[1] and flush tank in conjunction with the flushometer valve or valves. A 20-gal tank is sufficient for one water closet, and a 30-gal tank should care for the flushing of six water closets. An installation is shown at *O* in Fig. 15-5.

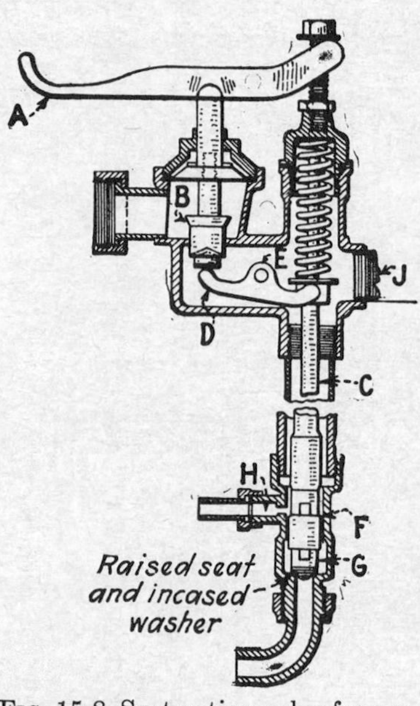

FIG. 15-8. Seat-action valve for use with pneumatic tank on frostproof water closets.

15-28. Pneumatic Flush Tanks. A pneumatic flush tank is shown at *O* in Fig. 15-5, and the details of a pneumatic flush valve are shown in Fig. 15-8. When the water-closet seat, attached to the arm *A*, is pressed down, valve *B* is pressed on its seat and shaft *C* is raised by arm *D* acting at fulcrum *E*. This raises sleeve valve *F* to uncover the inlet port *G* and cover waste port *H*. Water from the supply line enters the pneumatic tank *J*, compressing the air therein. When the water-closet seat is raised, sleeve valve *F* drops back to cover the inlet port *G* and at the same time flush valve *B* is raised from its seat and the waste port *H* is uncovered. A strong flush of water passes into the toilet bowl. Excess water is drained from the tank and valve through the waste valve *H*.

The use of pneumatic flush tanks is restricted mainly to locations in which (1) their unsightliness is not objectionable or in which they can be placed accessibly on the outside of the wall or inside of the wall against which the fixture to be flushed is placed, (2) the user of the water closet cannot be depended on to flush it, and (3) there is protection against freezing of the water in the flush tank.

15-29. Rate, Time, and Quantity of Flush. The rate of supply of water from any flush tank or flush valve to the bowl of a water closet or

[1] See also *Domestic Eng.*, June, 1956, p. 26.

other self-siphoning fixture with a resealing trap must be adjusted to the size and shape of the fixture and its passages so that the fixture is first filled with water and is then quickly and completely siphoned in order that no light, floating particles will be stranded in the fixture. A wash-down flush from a flushing rim, occurring simultaneously with the self-siphonage of the fixture, is desirable. All this action can be obtained by

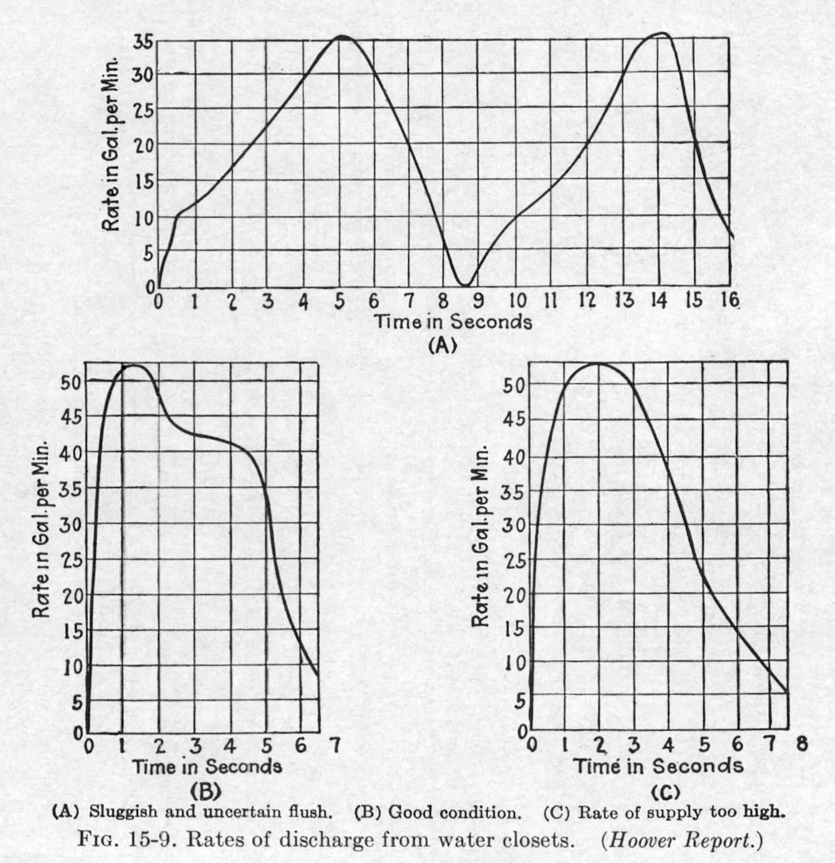

(A) Sluggish and uncertain flush. (B) Good condition. (C) Rate of supply too high.

FIG. 15-9. Rates of discharge from water closets. (*Hoover Report.*)

careful design and is a mark of a satisfactory fixture. The relative timing of the filling of the fixture and the occurrence of the self-siphonage must be properly adjusted. The Hoover Report states (p. 80):

If the rate of supply is too low, a complete break in the siphon action occurs and in extreme cases the minimum point approaches zero while the bowl is refilling following the first siphon action. This action is shown graphically in Fig. 15-9. The result is a sluggish and uncertain flush. If the rate of supply is too high, the siphon action becomes continuous with the bowl partly filled until the end of the flush.

Various conditions of flush are shown graphically in Fig. 15-9. In studying this figure it will be noted in *A* that the curve reaches two maxima, the first at about 45 gpm and the second at 26½ gpm. This is due to a double self-siphonage action that does not show in *B*. The curve in *C* illustrates a desirable rate-time discharge curve with proper siphoning action in the water-closet discharge.

It is stated further in the Hoover Report (pp. 81, 82):

> . . . A mean value of approximately 30 gpm is a rate of supply that, in general, will prove satisfactory. . . . The quantity of water required depends on the duration of the flush. The time varies from 6 to 10 sec. . . . This . . . indicates a range of from 3 to 5 gal as a serviceable quantity. No doubt there are many closets on the market which a smaller quantity would serve satisfactorily under certain conditions. The approximate mean values are, therefore, 4 gal supplied at the average rate of 30 gpm for 8 sec.

The flushing water, from whatever source, should be injected into the closet bowl from a flushing rim and through larger openings in the back of the bowl, under the seat, as shown in Fig. 15-5 at *A* and from *C* to *F*, inclusive. The flush should be distributed so as to wash down the walls of the bowl. The flush should come with sufficient rapidity to cause self-siphonage of the trap once and once only. The self-siphoning of the trap is essential to the carrying away of solid matter. A continuously running, nonsiphoning bowl will not cleanse itself. The siphonage is sometimes accelerated and the cleansing increased by the use of a jet at the bottom of the trap, as shown in Fig. 15-5 at *E*, *F*, and *N*.

15-30. Installation of Water Closets. Types of connection between water closets and the plumbing drainage pipes are shown in Fig. 15-10.

The weights of the fixture and the user are supported on the floor and not on the drainage pipe. The fixture and the drainage pipe are, therefore, seldom subjected to the same shocks, vibrations, and movements of the building, so that special care must be taken in making a joint that will stay tight. The flexible and vibrationproof lead bend is wiped into a brass ferrule which is calked into the soil pipe. The connection is made gastight and watertight by a metal ring and gasket[1] or a solder joint, as shown in the figure. A joint depending for its flexibility on a gasket of asbestos, rubber, or similar material is shown in Fig. 15-10 at (*b*). The use of putty in making the joint should be prohibited.

The "National Plumbing Code" (Sec. 2.27) permits the use of 3-in. lead bends and stubs or similar connections provided the inlet is dressed or expanded to receive a 4-in. floor flange; 3-in. iron bends or similar connections provided a 4- by 3-in. floor flange is used to receive the fixture horn; and 4- by 3-in. reducing bends. The Code states (Sec.

[1] See also *ibid.*, p. 70.

3.2.5) that flanges are to be not less than the following: brass, $\frac{1}{8}$ in. thick; cast iron or malleable iron, $\frac{1}{4}$ in. and not less than 2-in. calking depth; for hard lead, 1 lb 9 oz and not less than 7.75 per cent antimony by weight. Flanges are to be soldered to lead bends and calked, soldered, or screwed to other metal. Closet screws and bolts are to be of brass.

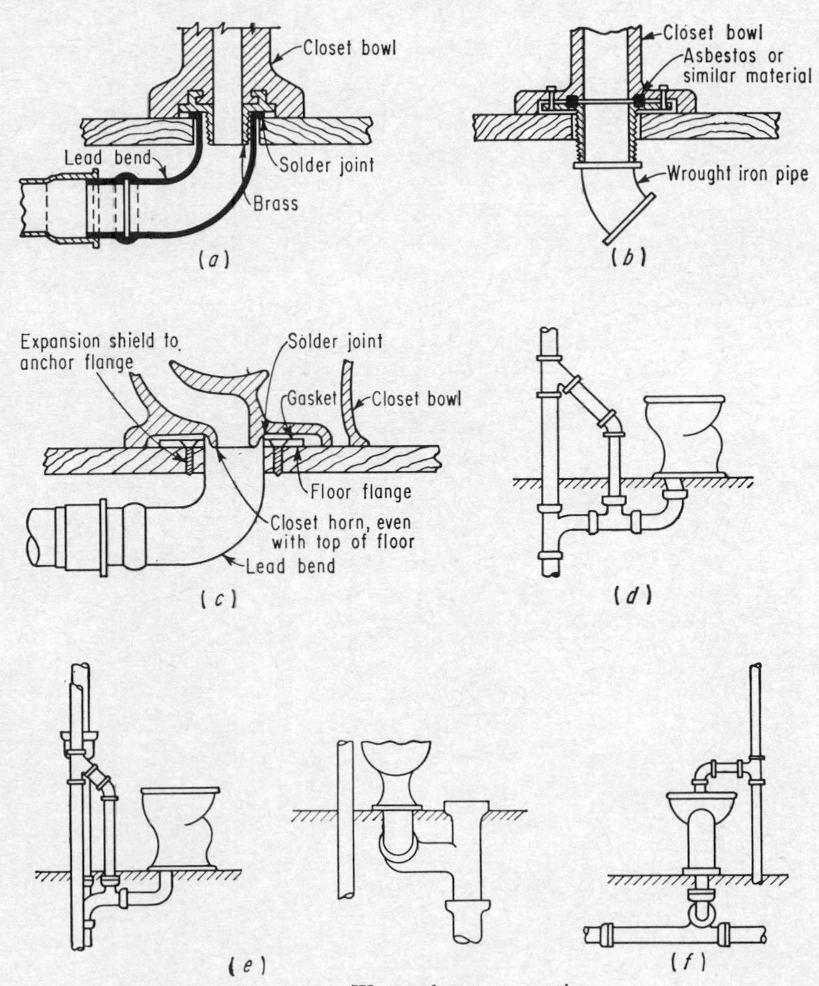

FIG. 15-10. Water-closet connections.

Wall-hung closets, such as are shown in Figs. 15-5 at *H* and II-69, with a similar connection to the floor-supported closets but with the weight of the fixture supported on the wall have the advantage of leaving the floor beneath the fixture accessible for cleaning. Wall-hung fixtures may, however, be slightly more expensive in first cost, and the maintenance of a gastight joint more difficult.

A vent pipe should not be connected to the closet bowl, other than to a precast spud, as the joint cannot be maintained gastight owing to shocks and vibration.

The use of a side-inlet closet bend is permitted by the "National Plumbing Code" which states (Sec. 12.6.4): "only in cases where the fixture is vented and in no case shall the inlet be used to vent a bathroom group without being washed by a fixture."

The Housing Code states, on p. 19:

4.3.10. Water Closet, Pedestal Urinal, and Trap Standard Service Sink. The connection between drainage pipes and water closets, floor-outlet service sinks, pedestal urinals, and earthenware trap standards, shall be made by means of brass, hard lead, or iron flanges, calked, soldered, or screwed to the drainage pipe. The connection shall be bolted to the earthenware by means of red-brass bolts with an approved gasket or washer or setting compound between the earthenware and the connection. The floor flange shall be set on an approved firm base.

CHAPTER 16

BATHING, KITCHEN, WASHING, AND OTHER FIXTURES

16-1. Waste Outlets and Overflows. The control of the outlet and the overflow are features of plumbing fixtures requiring attention in design and in selection. Overflow limitations to prevent cross connections are discussed in Sec. 15-12. Three types of waste outlet and overflow are shown in Fig. 16-1 and Figs. II-100 to II-103 of Appendix II. These are the plug-and-chain, the pop-up, and the standing waste.

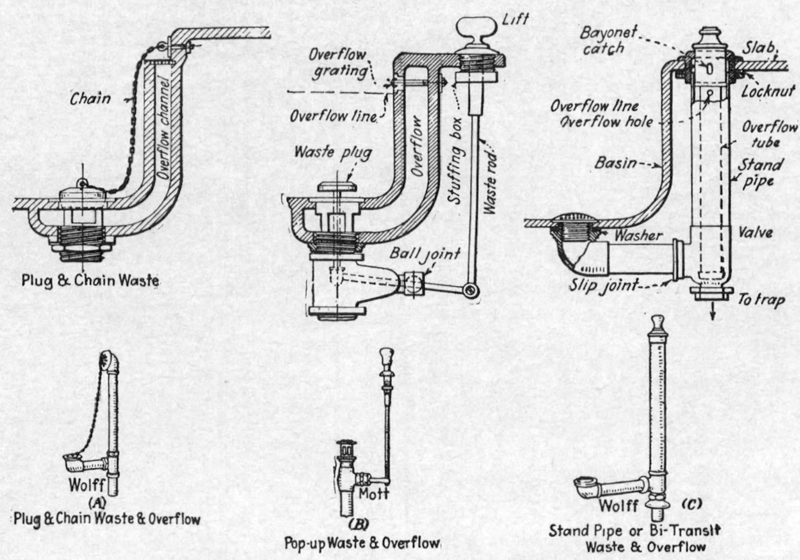

FIG. 16-1. Wastes and overflows for plumbing fixtures.

The plug-and-chain is inexpensive, certain in action, and usually watertight. It has objectionable features, however, including the possibility of failure, loss, or breakage of the chain followed by loss of the plug and the difficulty of cleaning the chain, which may carry foul water and particles from one user to the next.

The standing waste and overflow or bitransit waste is illustrated in Fig. 16-1*C*. In this type the standpipe is placed in position within an enclosure and is raised and lowered by hand. Although the standpipe is easily removable for cleaning, the fact that the soiled parts are hidden is conducive to neglect of cleaning, with a result that foul matter is carried from one user to the next and foul odors may escape into the room. Such objectionable conditions have been found that the use of this type of fixture is widely condemned.

The pop-up waste, shown at *B* in Fig. 16-1 and in Fig. 16-2, with exposed overflow opening provides protection against contamination of water in the fixture by contact with surfaces exposed to previously used water and when closed offers no obstruction to the use of the fixture.

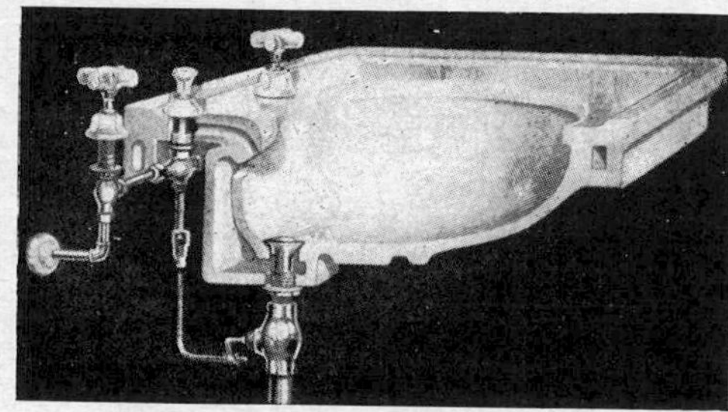

FIG. 16-2. Section through a lavatory showing a pop-up waste. (*The Sanitary Earthenware Specialty Co.*)

The device is not ideal because of possible difficulties with the moving mechanical parts that require adjustment and the poor fit of the waste plug in badly made fixtures.

An overflow pipe is provided and required in all fixtures as a safety measure to prevent the flooding of the building if the water is turned on too long. This overflow should be protected with a strainer in the same manner as is specified for the waste pipe. Among the common faults met with in overflow pipes is that they are too small to carry water away as fast as it will come in and the infrequent use of the overflow results in fouling of the interior of the pipe, which is conducive to clogging or leakiness or both without quick discovery. Great care should be taken to connect an overflow pipe to the drain pipe from its own fixture and to see that this connection is *above the trap;* that is, the overflow connection must be between the trap and the fixture.

16-2. Lavoratories and Wash Basins. Materials best suited for lavatories are vitrified porcelain, china, glazed earthenware, and enameled iron. Metal is also used and when properly designed is satisfactory. Its use is confined mainly to factories, theaters, jails, and other places where rough usage is anticipated or in situations subject to vibration, as in railway cars. The lavatory should be made of one piece, including the bowl and the table surrounding it. A lavatory consisting of a separate basin and table will not stand rough usage, and the joint between the table and the basin is difficult to maintain and will result in leakage and insanitary conditions.

The desirable features to be noted in the selection of a lavatory or wash basin are ease in cleaning, absence of cabinetwork, durability, economy, and attractive appearance. Various types of lavatory are shown in Figs. 16-3 and II-76 to II-82. The bowl is usually oval in shape with a steep slope to the bottom and with the waste in the center or toward the rear. The bowl has a capacity of about 1 gal up to the overflow. The waste opening should have as large a capacity as the waste pipe and should be protected by a heavy strainer. Strainers usually consist of two metal bars about $\frac{1}{8}$ to $\frac{3}{16}$ in. thick, crossing at right angles. The net open area between the bars should slightly exceed the cross-sectional area of the waste pipe. The basin should empty quickly, acting as a flush tank for its own trap and waste pipe. It should be self-cleansing, should not contain invisible fouling surfaces, and should be simple in appearance, operation, and maintenance.

Faucets for supplying water to lavatories are discussed in Chap. 14. In the selection of faucets and wastes for lavatories it is desirable that the fixture be kept as free as possible from obstruction. The faucets should, therefore, not project far over the basin. The air gap required between the faucet and the surface of water in the basin is discussed in Sec. 1-15. It is standard practice when both hot and cold water are available to place the cold-water faucet on the right of the user as he faces the fixture. It is convenient to place valves on each supply pipe to the fixture so that the water can be shut off when the faucets need repairs.

The number of fixtures required in washrooms[1] is shown in Tables 1-2, 15-2, and 15-4. Some authorities recommend a rule of thumb of one lavatory for each 10 users.

16-3. Setting of Lavatories. A lavatory may be set at any convenient point, against an inside wall, in a corner, or in a compartment. If it is placed in a compartment or closet, ventilation should be provided. The top of the lavatory should be placed about 30 in. above the floor. Where legs are used to support the lavatory, they should be made adjustable in order to fit the roughing-in.

16-4. Traps and Waste Pipes. The trap and the waste pipe should be simple and attractive in appearance with an outlet not less than $1\frac{1}{4}$ in. in diameter[2] and should be placed as close as possible to the fixture, within the limits stated in Sec. 11-43. The interior surfaces of the trap should be smooth and continuous to avoid clogging and for ease in cleaning.

[1] See also Yardstick for Washrooms, *Factory Management and Maintenance*, March, 1955, p. 90.

[2] "National Plumbing Code," Sec. 7.10.1.

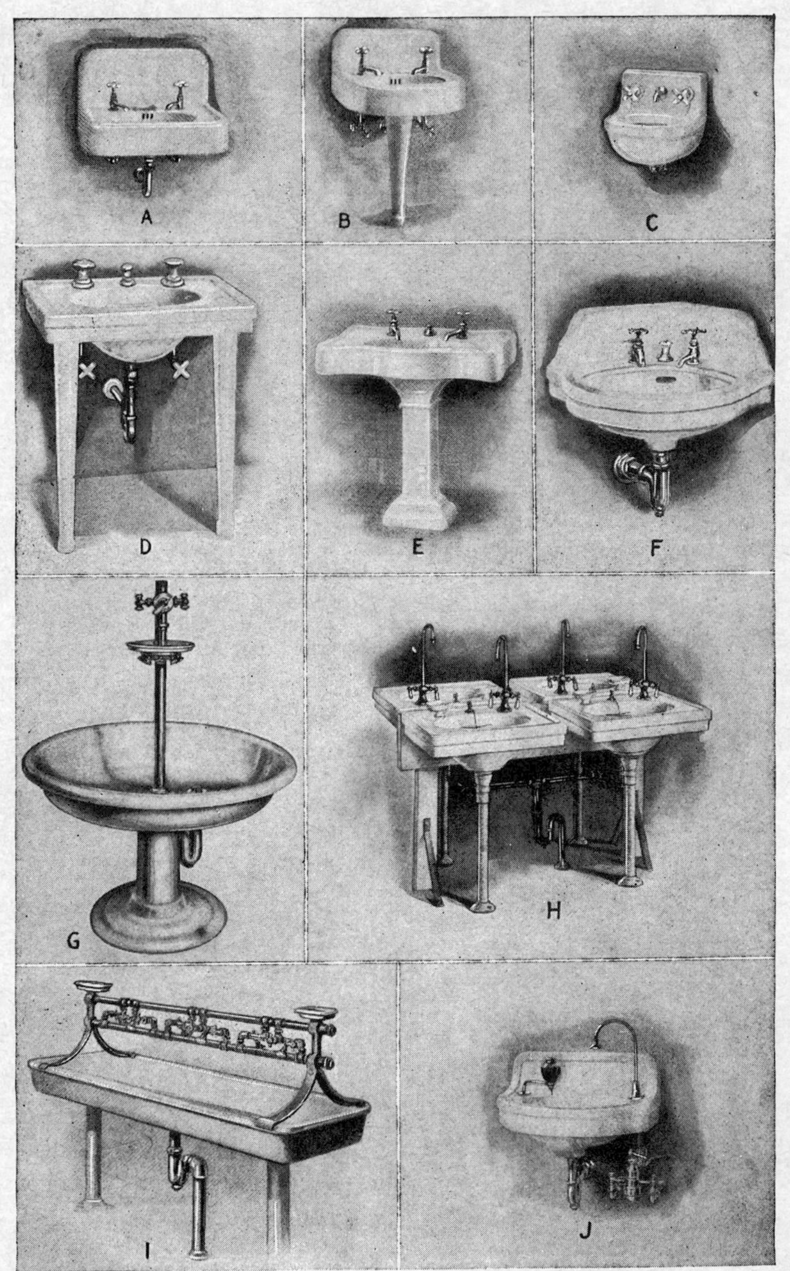

FIG. 16-3. Types of lavatory. (*A*) Wall type. (*Kohler.*) (*B*) Single leg. (*Kohler.*) (*C*) Dental lavatory. (*Mott.*) (*D*) Two legs. (*Sanitary Earthenware Specialty Company.*) (*E*) Pedestal. (*Richmond Radiator Company.*) (*F*) Corner type. (*Sanitary Earthenware Specialty Company.*) (*G*) Circular lavatory for industrial use. (*Ebinger Sanitary Manufacturing Company.*) (*H*) Factory type. (*Mott.*) (*I*) Barracks type, one trap. (*Chicago Potteries.*) (*J*) Surgical lavatory, knee-action faucet. (*Mott.*)

Enamel surfaces are especially suitable on water-closet and slop-sink traps. Wherever possible the waste pipe should pass through the wall at the back of the fixture rather than through the floor beneath it. All pipes should be exposed or easily accessible. Cabinetwork or other obstructions to complete exposure or accessibility are undesirable, except as allowed in Sec. 16-16, because of the moisture of condensation or leakage that may produce uncleanly conditions.

The connection between the trap and waste pipe from the basin should be made by means of a slip joint to allow for the slight movements and vibrations that are inevitable to the use of the basin. The trap should be supported by the waste pipe from the wall. Types of trap used are discussed in Chap. 11. Sizes of waste pipes are given in Sec. 11-7, and sizes of supply pipes in Chap. 7. Where a number of lavatories is to be set in battery, it is permissible and is considered good practice to permit them to discharge into a common waste pipe with one trap for a maximum of three lavatories, as shown in Fig. 11-6.

An undesirable feature of the piping arrangement shown is the length of untrapped waste pipe exposed to foul the air of a room.

16-5. Bathtubs. Among the requisite conditions for a satisfactory bathtub are that it shall be safe to use, sanitary, comfortable, easy to get in and out of, pleasing in appearance, watertight, easy to install, easy to maintain, and inexpensive.

Sanitation is provided through a shape that is self-draining and that does not have sharp angles or recesses in which dirt may collect. Safety is protected by avoiding steep, slippery slopes on the bottom and by providing handholds to assist the user. Handholds are not, however, necessarily a part of the bathtub.

Frequent cleaning, particularly after use, is necessary to preserve a clean surface. Materials that are satisfactory in this respect and that are easily cleaned include enameled iron, earthenware, and marble. Wooden tubs, metal-lined wooden tubs, and unlined or painted iron, copper, or zinc tubs are difficult to keep clean and should not be used. Their use is prohibited by some plumbing codes.

The tub is best made white or light-colored without decorative protuberances or recesses either inside or outside. The tub should be built into the wall and floor so that dirt cannot gain access to the built-in parts. If it is not convenient to build the bathtub into the building, then the tub should stand away from the wall and should be supported on legs above the floor so that access is possible for cleaning these areas.

Three common types of bathtubs are known as the *built-in tub*, the *bath on base*, and the *bath on legs*. These are illustrated in Figs. 16-4 and II-96. Tubs are known also by their shape as *ordinary*, *French*, and *Roman*, as shown in Fig. 16-5.

The built-in bath will occupy less room, since it can be placed against the wall, in a corner, or in a recess, as shown in Figs. 16-6 and II-96. The connection between the tub and the wall or the tub and the floor should be made watertight, somewhat as shown in Fig. 16-7.

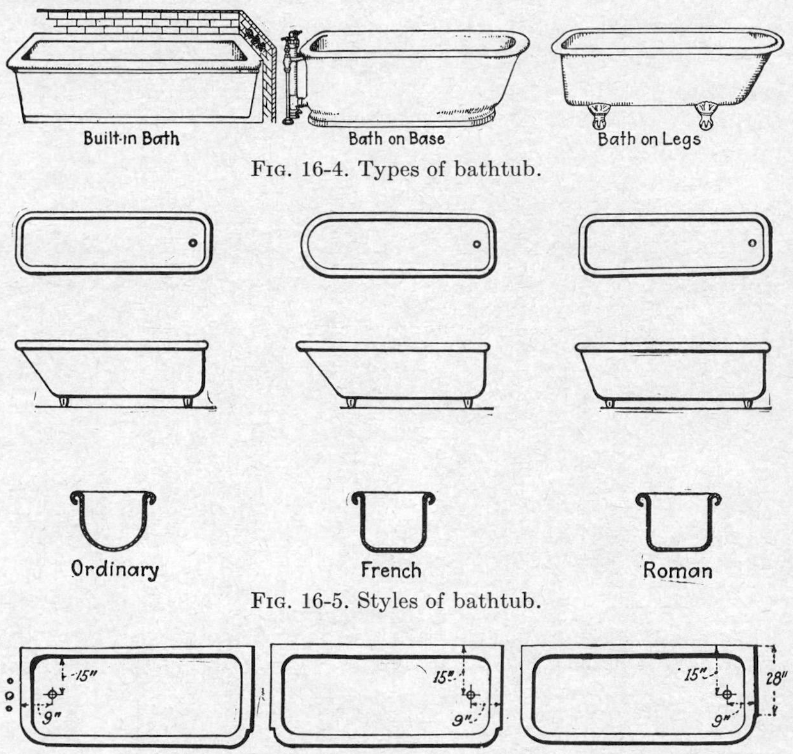

FIG. 16-4. Types of bathtub.

FIG. 16-5. Styles of bathtub.

FIG. 16-6. Diagram of locations of built-in bathtubs with approximate dimensions.

16-6. Water Supply to Bathtubs. Water is admitted to the bathtub in various ways. A common method is the use of a combination faucet, as shown in Fig. 14-3, or of separate faucets. The advantage secured from the use of a combination faucet lies primarily in the control of temperature of the incoming water. The placing of the valve handles outside the tub is desirable since it leaves the tub unobstructed. Under no condition should it be possible to submerge the nozzle of the faucet, an overrim supply, as shown in Fig. 16-8, being desirable to assure the impossibility of submergence.

FIG. 16-7. Details of connection between wall and built-in bathtub.

16-7. Bathtub Wastes and Overflows. Water can be drained from a bathtub at any convenient point on the bottom, but the best point is usually nearest the water-supply faucets. The bottom of the tub should slope toward the outlet at a slope of about ⅛ in. per ft. The advantage of having the outlet near the inlet is that obstructions to the use of the tub are concentrated at one point and drops from a leaky faucet do not flow the length of the fixture to reach an outlet. Comfort for the user is aided by placing the outlet near one end rather than in the middle or in the center at one side of the tub. An approved and a disapproved waste are shown in Fig. 16-8.

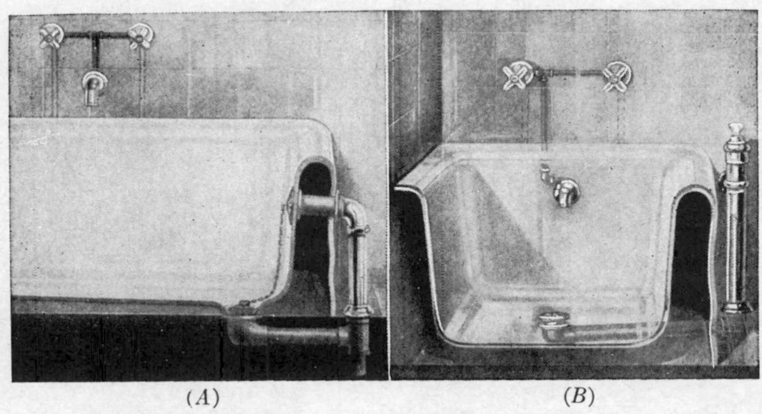

(*A*) (*B*)

Fig. 16-8. Bathtubs with supply-valve handles outside of the tub. (*J. L. Mott Co.*) (*A*) Nozzle outside of tub. Plug-and-chain waste. (*B*) Nozzle inside of tub Standing waste. Nozzle below rim. These are undesirable features.

Overflows and wastes for bathtubs should conform to the general requirements given in Sec. 16-1.

16-8. Sizes of Bathtubs. The most comfortable size of bathtub will permit the user to lie at full length in it. This calls for an over-all length of 8 ft or more to allow for the slope of the ends and the faucets. Such lengths are available but are not commonly used. Practically any size of bathtub is available, from an infant's tub up to lengths of 8 to 10 ft. Inside widths are in the neighborhood of 24 in. in the clear and a maximum depth of about 18 in. The edge of the tub should not be more than about 18 to 24 in. above the floor, except in hospitals. Some over-all dimensions of bathtubs are shown in Figs. II-95 and II-96.

16-9. Special Tubs. Tubs for bathing various parts of the body and for special purposes are sometimes included in the equipment of a bathroom. Principles similar to those applicable to the selection of a bathtub are applicable to the selection of special tubs.[1] A bidet or sitz bath is

[1] See also *Air Conditioning, Heating, and Ventilating*, May, 1956, p. 116.

illustrated in Fig. 16-9; a sitz bath or foot bath, in Fig. 16-10; a baby-bathing outfit, in Fig. 16-11, and a combination tub, in Fig. 16-12.

16-10. Hospital Tubs and Fixtures. Bathtubs are not recommended for general use in public institutions for many reasons. These include the

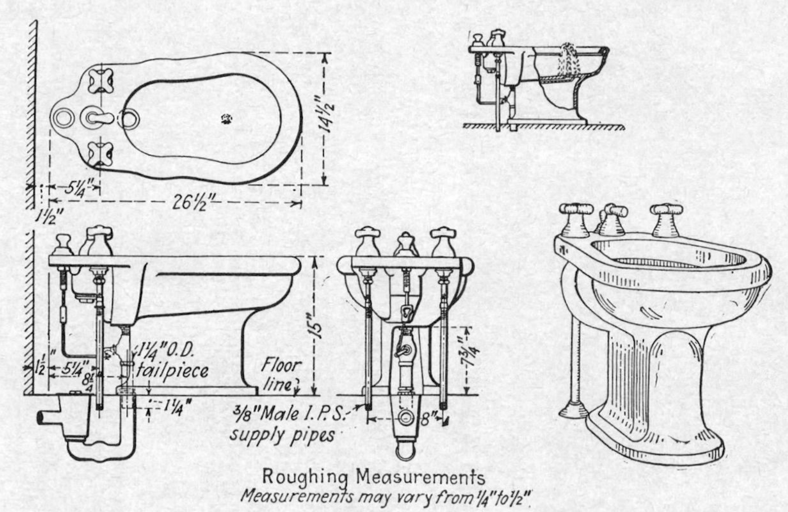

Fig. 16-9. Bidet. (*Thos. Maddock's Sons Co.*)

Fig. 16-10. A sitz, seat, or foot bath. (*J. L. Mott Co.*)

danger of contagion through uncleanliness, the time necessary for filling and emptying, abuse, and in hospitals—particularly for the insane—the danger of injury to the patient when bathing without an attendant. In private hospitals or for patients with special nurses, the bathtub may be used. The tubs may be either fixed or movable. The fixed tub should

FIG. 16-11. Mott standard baby bath with Leonard thermostatic temperature-control valve.

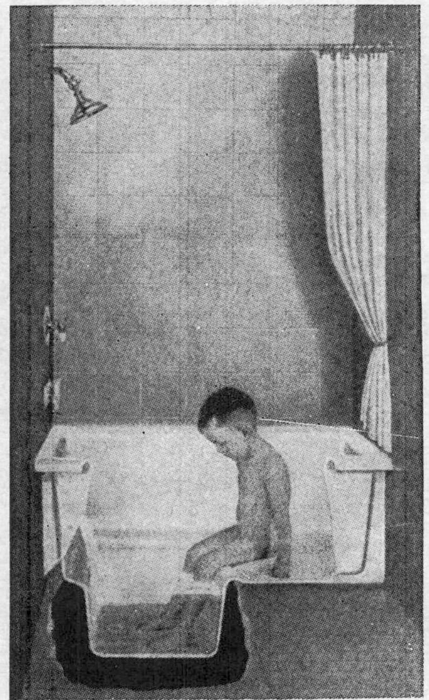

FIG. 16-12. Combination bathing tub, shower bath, and baby bath. (*Wheeling Sanitary Manufacturing Co.*)

fulfill all the requirements already described for ordinary tubs, and in addition it should be so located in the bathroom that the attendant can get all around it. The inlet valve and waste control are sometimes so arranged as to be controllable only by the attendant, either by a key or out of the reach of the patient in the tub. The movable tub is usually

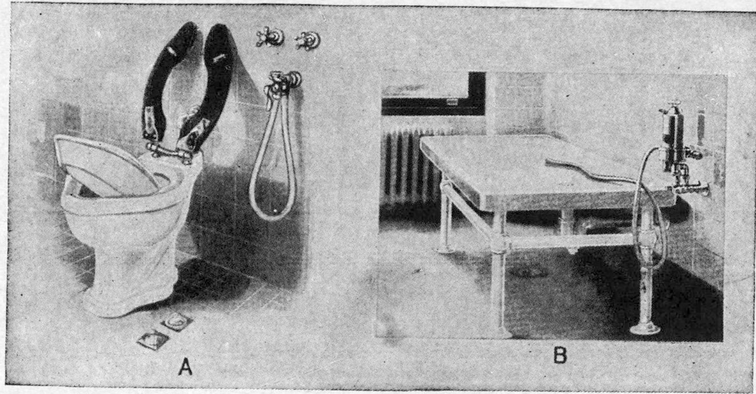

FIG. 16-13. Special types of hospital fixture. (*J. L. Mott Co.*) (*A*) "*Lombard-Duplex*" siphon-jet bedpan closet with extended front and back lip and integral bedpan lugs, top inlet, No. 8 birch-stained mahogany open-front and back-seat connection to wall; slow-closing, foot-action flush valve concealed in floor; regulating stop valve concealed in floor; No. 31 floor flange; D 2560 combination compression ½-in supply valves with five-ball metal handles; supply nozzle with hook, rubber hose, and hand spray with grip-control valve. (*B*) Marble irrigation slab with waste strainer, wrought-iron pipe frame finished white enamel, B-10 *Leonard* thermostatic mixing valve, check valves, loose-key stop valves, ½-in. supply pipes to wall, hose connection, 5-ft rubber hose, mercury thermometer, and 1½-in. half S trap with nipple to wall.

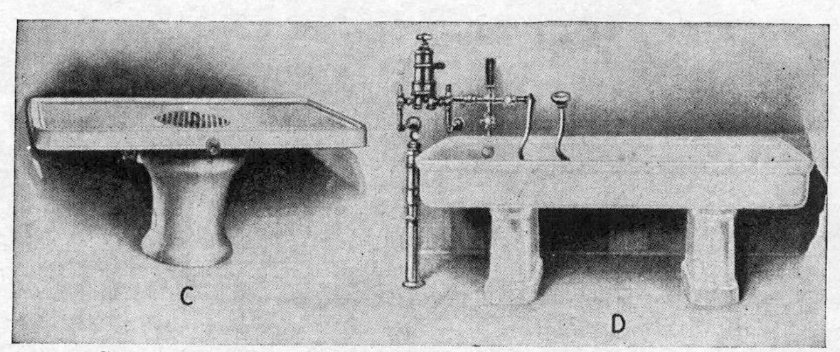

FIG. 16-14. Special types of hospital fixture. (*J. L. Mott Co.*) (*C*) "*Delano*" enameled-iron autopsy table with revolving top, spring catch, enameled inside flushing rim with special full-S trap, enameled-iron pedestal, crossbar strainer, compression control valves for hot and cold water with connections to flushing rim. (*D*) "*Hahneman*" Durecast roll-rim receiving bath, Durecast pedestals, standing waste, B-5 *Leonard* thermostatic mixing valve, check valves, loose-key stop valves, ½-in. supply valves to wall, mixed-water supply pipe, mercury thermometer, bell supply with control valve, hose connection with control valve, rubber hose, rubber-bound spray, and wall hook.

made of cast iron, enameled on the inside and painted on the outside. The tub is supported on rubber-tired wheels.and is provided with a long handle for pulling it about. It is filled from special faucets, not attached to the tub, placed over a floor drain into which the tub is emptied through a large outlet pipe controlled by a quick-opening gate valve. Types of special fixture used in hospitals are shown in Figs. 16-13 and 16-14.

16-11. Shower Baths. Among the advantages of shower baths are low cost of installation, small water consumption, immediate availability of

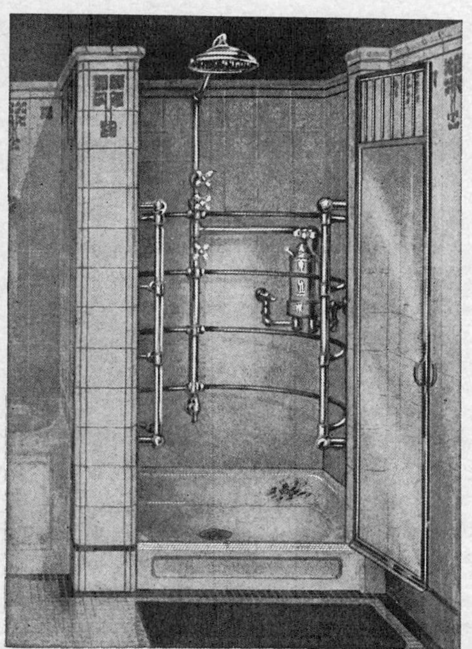

FIG. 16-15. Combined shower-bath and needle-bath compartment with thermostatic control valve. (*J. L. Mott Co.*)

use without the loss of time in filling and emptying a tub, cleanliness in that clean water is sprayed continuously over the body, and the sensation of stimulation that results from its proper use.

Showers are arranged in combination with a bathtub, as discussed in Sec. 14-25; in a separate compartment, as discussed in Sec. 16-12; or in batteries in open rooms for use in gymnasiums and similar places where a large number of persons will use them simultaneously. The number of shower baths recommended is shown in Tables 1-2 and 15-4.

16-12. Shower-bath Compartments. Shower baths are installed in special compartments, as illustrated in Figs. 16-15, II-68, and II-98. The walls of the compartment may be of glass, soapstone, slate, marble, or other easily cleaned and nonabsorbent material. The dimensions of slate shower stalls are shown in Fig. II-98 and Table II-85. The walls should extend all around the shower except at the entrance, which should be closed with a nonabsorbent curtain or door. The floor should be made of tile, marble, cement, enameled iron, or similar material, and it should slope gently toward the drain. The slope should be very slight, not over ¼ in. per ft, because of the danger of slipping on too steep a slope. In some baths a removable slatted wooden covering is placed on the floor to avoid the chill of the masonry or metallic floor and to remove the

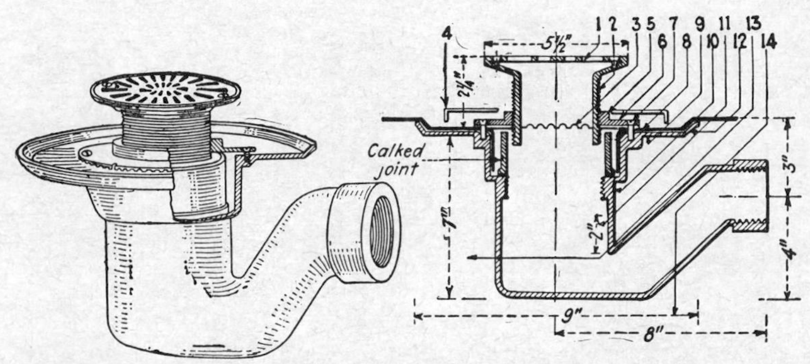

Fig. 16-16. Shower-bath floor drain. Patented Dec. 29, 1923. (*Greenwood Manufacturing Co.*) (1) Brass bar strainer NP diameter 5 in. and secured to brass extension tailpiece with removable brass screws. (2) Brass tailpiece 2½ in. long and threaded 3 in. iron pipe size. (3) Shows the grooves on the underside of the seepage plate through which the seepage drains back into the trap. (5) Seepage plate, tapped 3 in., connects the tailpiece to trap body with a screwed joint, the seepage plate being placed over the lead pan and connected to trap body by four brass screws. (6) Recess formed in the upper portion of trap into which the lead pan is secured to trap body by a calked joint. (7) Brass screws that secure the seepage plate to trap. (8) Lead pan which is permanently secured to trap by a calked joint. (9) Cast anchor which securely holds the trap body into the cement and forms a basin for the seepage waste. (10) Threads in the upper portion of trap for securing iron plug for the rough test. The plug also prevents dirt entering the trap and waste line during building construction. (11) Depth of water seal in 2-in. trap, 2 in. Depth of water seal in 3-in. trap, 3 in. (12) Ring forming basin of trap anchor. (13) Height of iron body 2-in. trap, 7 in. Height of iron body 3-in. trap, 10½ in. (14) Outside diameter of flange, 9 in.

danger of slipping. Valves for the control of the water supply should be within reach of the bather when under the shower, except in special cases in hospitals and public institutions. A floor drain for a shower-bath compartment is illustrated in Fig. 16-16.

The "National Plumbing Code" (Sec. 7.11.1) requires that "all shower compartments except those built directly on the ground, or those having metal enameled receptors shall have a lead or copper shower pan or . . ." its equivalent. The pan must turn up at least 2 in. on all sides above the finished floor level. The shower receptacle waste outlets are to be not less than 2 in. and have a removable strainer. The compartment is to have not less than 1,024-sq in. floor area with no dimension less than 30 in.

16-13. Temperature Control for Shower Baths. The control of temperature in a shower bath is difficult and offers an objection to its use.[1]

The simplest and least expensive arrangement is to lead cold water and hot water through separate pipes to the shower, at which there is a

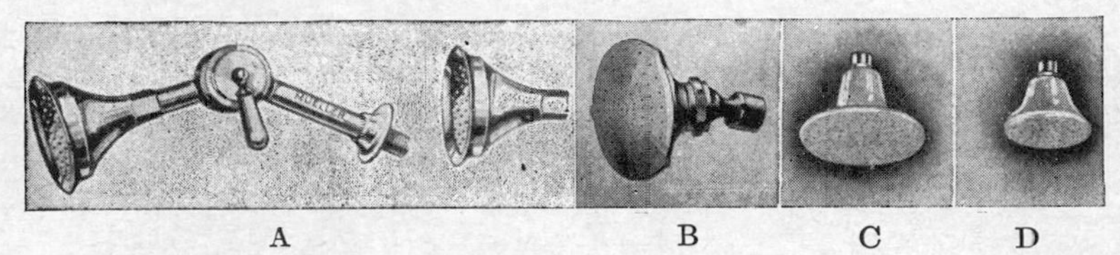

FIG. 16-17. Types of shower head. (*A*) Swinging joint, brass. (*Mueller.*) (*B*) Fixed spray, brass. (*Mueller.*) (*C*) Ball joint, brass. (*J. B. Wise Co.*) (*D*) Fixed spray, porcelain. (*Chicago Potteries Company.*)

tee leading the mixed waters to the shower head. Temperature is adjusted by manual control of the supply valves. Temperature control under such conditions is sensitive to pressure variations due to the use of water elsewhere in the building. In a battery of showers, as in a gymnasium, the temperature of the water can be controlled manually in a supply tank, an attendant watching the temperature of the water in the tank.

Mixing valves to replace the tee are so designed that the cold-water port is always opened before and closed after the hot-water port, thus minimizing the danger of scalding. Automatic mixing valves are available with which, by the turn of a single handle, as shown in Fig. 14-9, the temperature of the water can be changed. Another type, which is shown in the same figure, will deliver water at any desired temperature and is controlled by a thermostat.

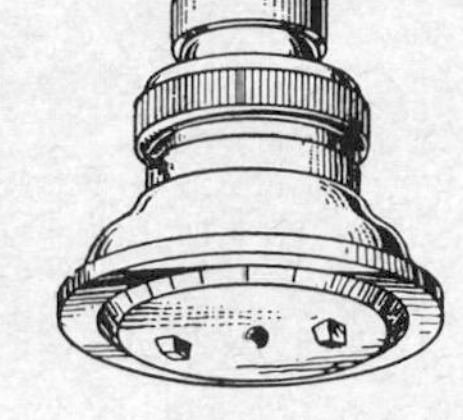

FIG. 16-18. Adjustable-spray ball-and-socket shower head. (*Speakman.*)

16-14. Shower Heads and Sprays. Various types of shower head or spray are illustrated in Figs. 16-17 and 16-18. The holes in a shower head should be large enough and of sufficient total area to prevent the building up of pressure above 20 to 25 psi in the spray when delivering the rated flow of water. A pressure higher than 5 to 10 psi will be uncomfortable to the bather, and a pressure much below 2 to 3 psi may prevent a proper distribution of water in a large shower head. An aerating head which aspirates air into the water somewhat diminishes the impact of the stream. A shower head 4 in. in diameter with 70 holes

[1] See also *ibid.*, March, 1956, p. 142.

about $\frac{1}{32}$ in. in diameter each is amply large for all ordinary requirements. Heads 6 to 8 in. in diameter are in use, and even larger ones are available.

The rate of supply of water to a shower bath should be about 15 to 35 gpm. The rate of flow from a shower head will fluctuate with the pressure, but shower heads are available that will give a constant rate of flow for any pressure between 15 and 125 psi. Recommended sizes of supply pipes to shower baths are given in Table 3-5 and of waste pipes in Table 11-1.

When installed to throw the spray vertically downward and when all the holes in the shower head are on the same level, the shower head will sometimes become air-bound and water will drip from it for some time after it has been turned off. The condition can be overcome by placing an air inlet in the shower head to admit air above the trapped water.

16-15. The Kitchen. Plumbing fixtures that may be found in a kitchen include a sink, a dishwasher, sometimes a washtray or combination fixture, and possibly a waste-food grinder. Where the kitchen is to be used also as a laundry, it may be desirable to provide a floor drain. Special sinks for the preparation of vegetables and for other purposes may be installed in hotels and restaurants.

16-16. Types of Kitchen Sink. Various types of kitchen sink are illustrated in Fig. 16-19 with dimensions in Figs. II-83 to II-90.

It is to be noted that a corrugated table or drainboard made of impervious material is placed on one or both sides of the sink. A back or splash plate of the same material is essential in a complete kitchen-sink installation where the sink is placed against the wall. Where the drainboard is to be placed on one side of the sink only, it should preferably be placed on the right of the user. Neither the sink nor the drainboard should be placed in a corner if another convenient location can be found. A drainboard in a corner is undesirable because it forces the dish wiper to step or reach around the dishwasher to reach the drainboard.

A hose is sometimes attached to the water-supply pipe to a kitchen sink to be used in washing dishes. To avoid a cross connection this hose must be installed so that it cannot hang submerged in the fixture as shown in Figs. 16-13 and 16-14, or a vacuum breaker must be installed on it.

Cabinet sinks are now more popular and are to be found in most new sink installations. Their construction of nonabsorbent metals, good ventilating provisions, and wide openings have overcome some of the earlier objections to enclosure of the supply and waste pipes under the sink.

The sinks shown in Fig. 16-19 can be obtained in almost any size up to 24 by 120 in. or even larger. The sink dimensions given in Appendix II are shown on the drawings. Most sinks are manufactured with a depth of about 6 in. A convenient size for a single-family residence is about 20 by 30 in.

16-17. Location of Kitchen Sink. Special attention should be paid to the location of the kitchen sink because of the amount of time spent at it by the users. It should be placed beneath a window or where the light

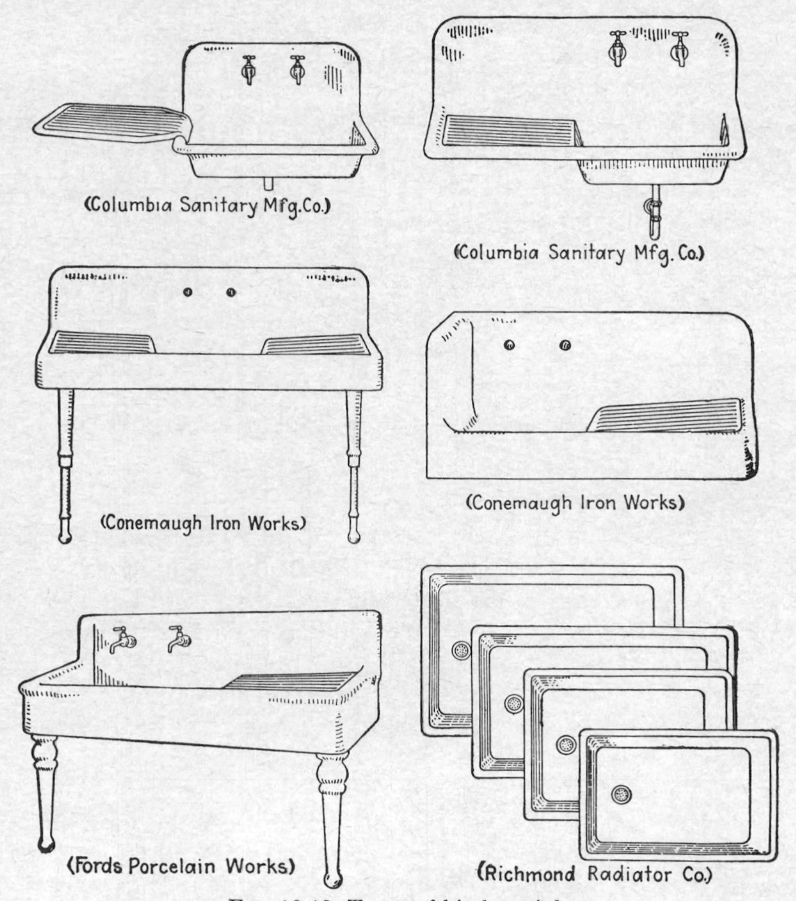

Fig. 16-19. Types of kitchen sink.

from a window will fall directly on it. This desirable location may force the placing of some piping in an outside wall. Where this cannot be avoided, special care should be taken to insulate *all* supply pipes and drainage pipes in cold climates. Artificial illumination should also be considered. It is not satisfactory to illuminate a kitchen with a single central light, since the shadow of the user will fall upon the work in the sink when the sink is placed against the wall. In hotels and institutions kitchen sinks are sometimes placed in the center of the room where they are accessible from either side.

16-18. Height of Sink above the Floor. The height of the kitchen sink above the floor is an important consideration for the comfort of the user but difficult of proper attainment because of diversity in the height of users. A common fault is placing the sink too low. The sink should be at such a height that anyone washing dishes therein need not bend over. It is more comfortable to raise the hands slightly than to bend over for a long time. A short person can stand on a board to reach the high sink, but a tall person must stoop to reach a low sink. For these reasons there is more danger of getting the sink too low than too high. In general, the distance from the floor to the bottom of the sink should not be less than 30 in. and preferably higher.

> The proper height of a sink is the height at which the individual woman can work most comfortably and efficiently. There can be no *standard* heights for sinks until there is a *standard* height and build for all women.[1]

The table or apron on the side of the sink, as shown in Fig. 16-19, should be not less than 36 in. above the floor and preferably higher. Where legs or brackets are used to support the sink, they should be made adjustable in order to fit the roughing-in work.

16-19. Materials for Kitchen Sinks. The principal materials in present-day use for kitchen sinks are enameled iron, earthenware, cast iron, slate, soapstone, copper, and metal lined with zinc. Enameled iron is the more desirable material, particularly for household use, because of its attractive appearance. Materials other than enameled iron are sometimes used in hotels, restaurants, and institutions because of their durability and lower first cost and their ability to resist stains. They are usually more difficult to keep clean, however. Soapstone, slate, earthenware, or lead-lined metal sinks are used almost exclusively in chemical laboratories because of their greater resistance to corrosion.

16-20. Water Supply to Kitchen Sinks. Water may be supplied to a kitchen sink through separate faucets or through a mixing faucet with a short nozzle or a long swinging nozzle, as shown in Figs. 14-2 and 14-5. The custom of providing a threaded end on the cold-water supply nozzle is falling into disuse.

16-21. Drainage Pipes for Kitchen Sinks. The waste from any sink should be protected by a perforated strainer the top of which is flush with the bottom of the sink, thus giving a smooth bottom surface. This strainer should be removable for cleaning and to give access to the trap. It should fit snugly into its recess and should be fastened there securely. The holes in the strainer should be about $\frac{1}{8}$ to $\frac{1}{4}$ in. in diameter, and

[1] From an advertisement of the U.S. Sanitary Manufacturing Co.

there should be sufficient area of holes to assure the quick drainage of the sink and flushing of the drain pipe. The total area of the holes should be slightly greater than the cross-sectional area of the waste pipe. This will require about 150 ⅛-, 70 3/16-, or 35 ¼-in. holes. The strainer plate may be made of brass or enameled iron. Sizes of waste pipes from kitchen sinks are given in Table 11-1.

The trap should be placed as close to the sink as possible to avoid exposing soiled pipe surfaces to the air of the room and to avoid the blows

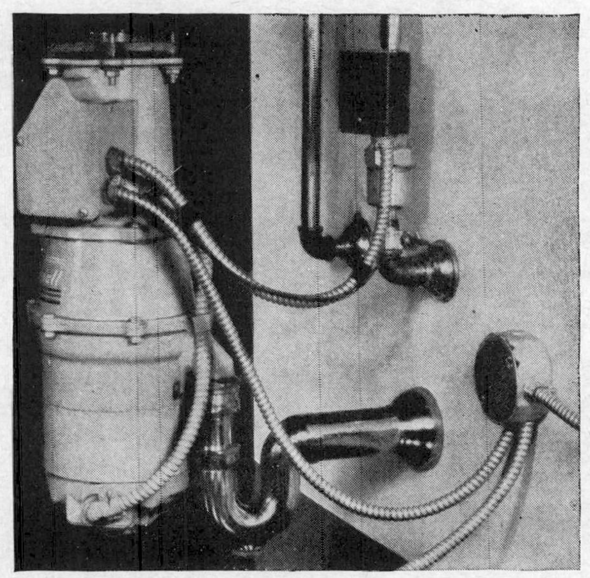

Fig. 16-20. Electric waste-food grinder. (*General Electric Co.*)

to which a low-hung trap is likely to be subjected. The drain pipe should pass through the wall well above the floor and behind the sink for the same reason. The connection between the sink waste and the trap should be made with a slip joint, unless a lead trap is used, to allow for vibration and movement of the fixture.

16-22. Waste-food Grinders. Devices to grind waste food and to dispose of it through the plumbing drainage pipes are to be found in many kitchens. Details of one of these devices, placed beneath a kitchen sink, are shown in Figs. 16-20 and 16-21. Sizes of water-supply piping and of drainage pipes should be the same as for kitchen sinks without such devices, and no special precautions need be taken to prevent clogging of the waste pipe. Such devices have functioned satisfactorily for years. The "National Plumbing Code"[1] permits the installation of waste-food grinders in a residence without a grease trap.

[1] Sec. 6.11.1.

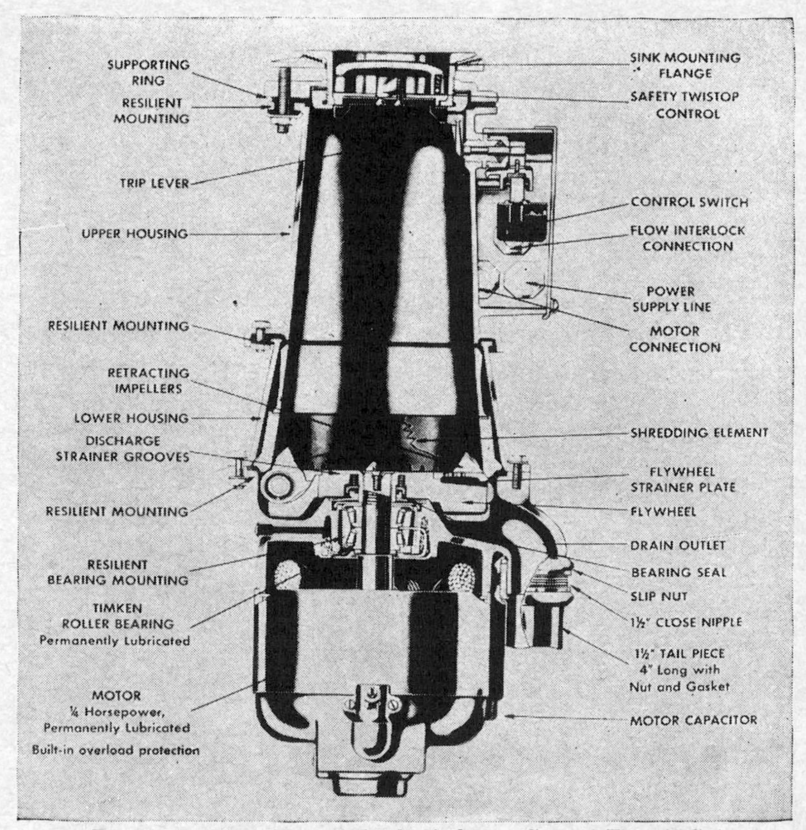

Fig. 16-21. Electric waste-food grinder. (*General Electric Co.*)

16-23. Pantry Sinks. A pantry sink is illustrated in Fig. 16-22, and a high gooseneck faucet used in such sinks is shown in Fig. 14-7. Such sinks and faucets are used principally for the cleaning of silverware, cut glass, and fine china and for drawing water for table use. They are frequently made of copper, in which breakable articles are less subject to fracture than in a sink of enamel iron, soapstone, porcelain, or similar hard material. As a rule, too much wood is used in the construction of the sink to assure immaculate cleanliness, and it is difficult to keep the joint between the metal and the wood free from filth. High gooseneck faucets are generally used in these sinks so that a pitcher can be placed under them for filling.

Fig. 16-22. A copper pantry sink encased in wood. (*J. L. Mott Co.*)

The drain pipe should be $1\frac{1}{4}$ in. in diameter, and each water-supply pipe not less than $\frac{3}{8}$ in. Otherwise the principles and considerations applicable to kitchen sinks are applicable to the selection and installation of pantry sinks.

16-24. Dishwashers. In roughing-in for an automatic dishwasher the plumber must provide for water supply and drainage and should avoid

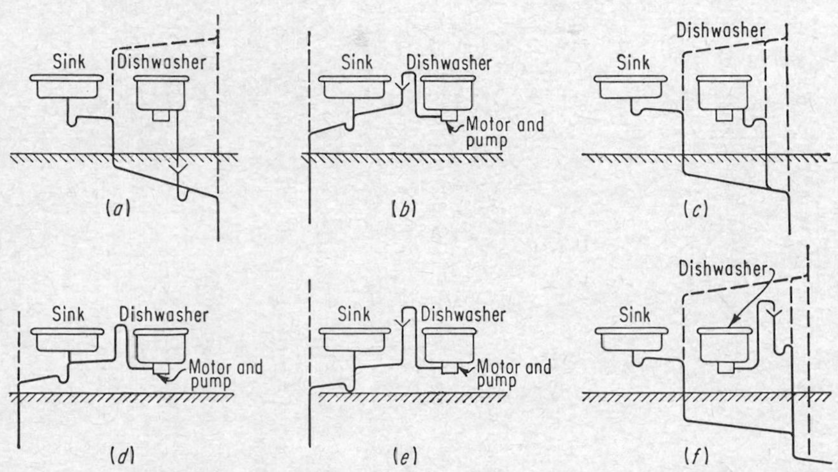

FIG. 16-23. Diagrams of dishwasher drainage connections. (*a*) Although this gives sanitary protection to the dishwasher, there is a hidden-sink problem in the basement. (*b*) This is superior to (*a*), but the hidden-sink problem remains. (*c*) This is probably the most widely used connection. However, complete sanitary protection to the dishwasher is not certain. If the system should not function properly, backflow might occur and drain water back up into the dishwasher and sink. (*d*) An objection to this setup is that a simple siphon exists that might bring sink wastes into the dishwasher. Probability of contamination from backflow is less than in (*c*), however, because of added height above the floor line. (*e*) This system affords complete protection of the dishwasher and produces no additional problem in the house system. (*f*) This system gives the same protection as (*e*) with "the dishwasher discharging into a separate trap through a *broken connection* located above the overflow rim of the sink. (*e*) and (*f*) are the most sanitary methods of connection to the drain system. (*e*) does have the added advantage that less structural change to the house is required than in (*f*)." (*Plumbing and Heating J., August,* 1954, *p.* 64.)

cross connections. Some suggested hookups are shown in Fig. 16-23. Working details of an automatic dishwasher[1] are shown in Fig. 16-24. The sizes of water-supply pipes and drainage pipes should be in accordance with the recommendations for kitchen sinks.

A fault sometimes found with automatic dishwashers is the existence of a cross connection which is not easy to recognize. The device should

[1] A hookup and discussion of dishwasher sanitation are given in *Domestic Eng.*, November, 1956, p. 116.

be inspected thoroughly for such a defect in order that the plumbing code is not violated.

16-25. Combination Fixtures and Washing Machines. A fixture that includes a kitchen sink and a washtub combined in one unit is known as a *combination fixture.* Such a fixture is shown in Fig. 16-25. Both movable

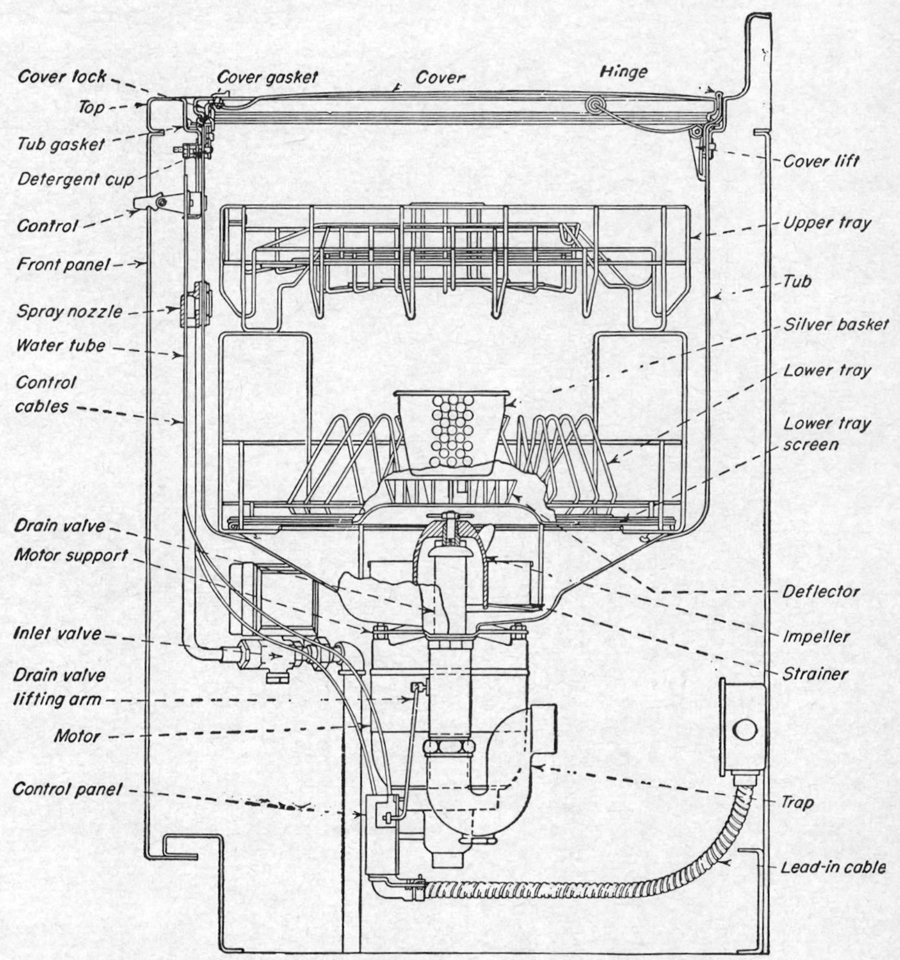

FIG. 16-24. Electric dishwasher. (*The General Electric Co.*)

and fixed washing machines are available. Water-supply and drainage pipes, of the sizes indicated in Tables 3-5, 7-1, and 11-1, should be provided for these machines. The waste pipe from movable washing machines is discharged into a floor drain of the type shown in Figs. 11-39 and 13-10 and discussed in Secs. 11-39, 13-14, and 16-30.

16-26. Slop, or Service, Sinks. Slop sinks, known also as *service sinks*, such as are shown in Figs. 16-26 and II-91, are used for drawing water for scrubbing and cleaning and to receive the contents of scrub buckets and vessels containing slops. They are advantageous in saving wear and tear on bathtubs and toilets. They are sometimes more

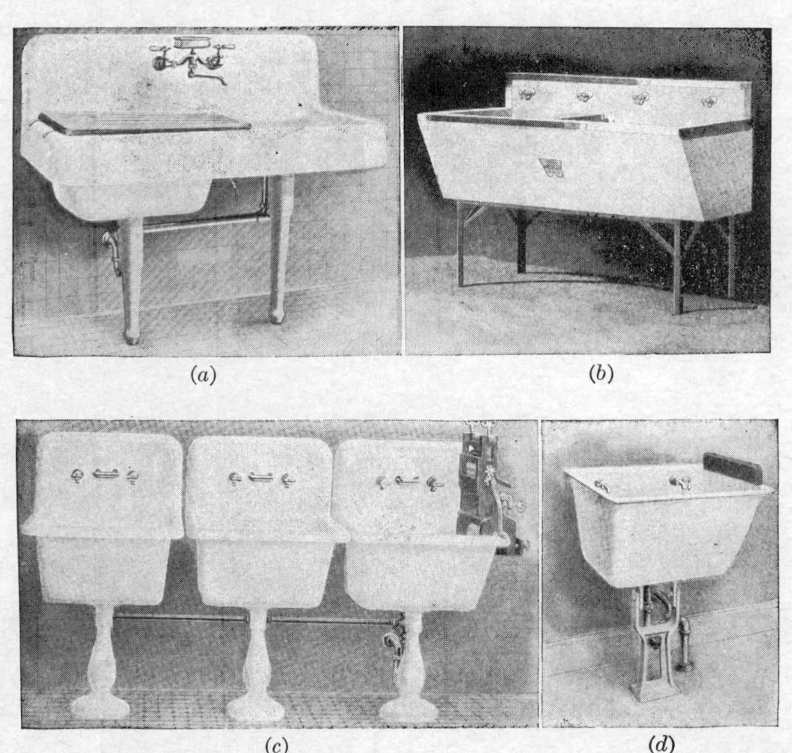

FIG. 16-25. Combination fixtures and laundry trays. (*a*) Laundry-tray and kitchen-sink combination, enameled iron. Tray overflows into sink. Adjustable legs. Swinging, combination faucet. (*Kohler.*) (*b*) Two-tray combination with integral back, stoneware. (*Chas. Wesley.*) (*c*) Triple combination of laundry trays, enameled iron. (*Kohler.*) (*d*) Single laundry tray with adjustable legs. Below-rim inlets are undesirable.

convenient than such fixtures, and they are indispensable in hospitals, hotels, and institutions.

The size of a slop sink should be about 20 to 24 in. square and 10 to 12 in. deep. The edge of the sink should be placed about 24 in. above the floor to minimize lifting and to leave room for the trap beneath. The amount of time spent at the sink is usually short, so that the effect of bending over the sink is not objectionable.

The water-supply and drainage equipment for such sinks is similar to that for kitchen sinks, and in addition provision is sometimes made for flushing as for a water closet or urinal. Flushed slop sinks of the type shown in Figs. 16-13, 16-26, and II-91 are particularly useful in hospitals. If a hose is to be used, as shown in Fig. 16-13*A*, a vacuum breaker should be used to prevent a cross connection, as explained in Sec. 1-15.

16-27. Laundry Trays or Washtubs. Laundry trays are plumbing fixtures used for the washing of clothing and other textiles. They are sometimes made of slate, earthenware, soapstone, enameled iron, or porcelain. The use of these materials will give satisfactory results,

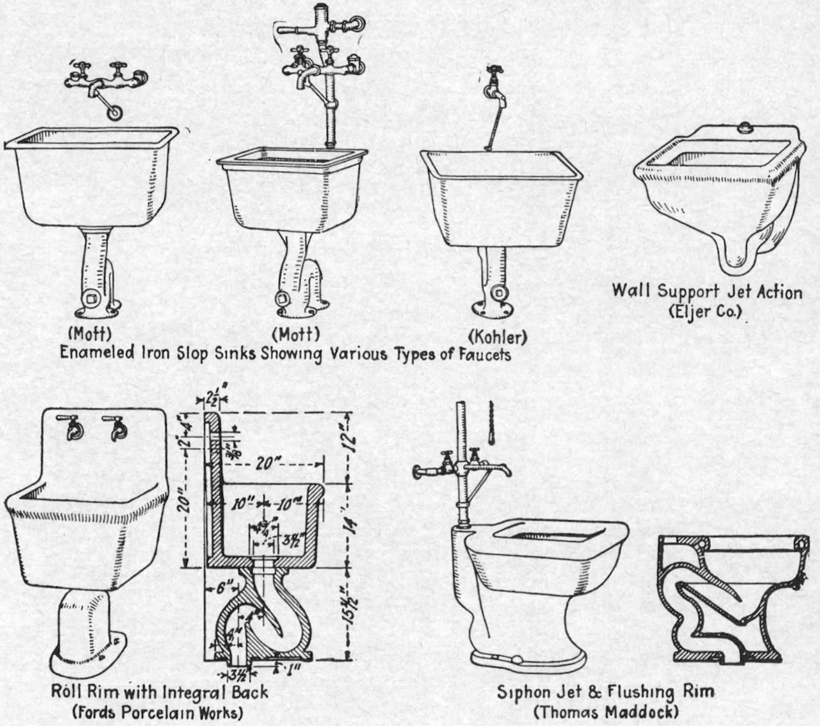

FIG. 16-26. Types of slop sink.

porcelain being the most expensive and infrequently used. Laundry trays are more generally standardized than most plumbing fixtures, and there is less variety from which to choose. Laundry-tray installations are shown in Figs. 16-25,[1] II-92, and II-93.

In the selection of a laundry tray, sanitation, cleanliness, watertightness, and ease in maintenance are assured by the choice of a tub of white or light-colored material made in one piece with curves instead of angles in the corners. The fixture should drain quickly and should act as a flush tank for the drain pipes. The dimensions of the inside of washtrays

[1] The single tray shown in the figure is undesirable because of the underrim inlets.

that have been found to be satisfactory and that are generally available are width, 20 in.; distance from front to back at top, 18 in., and at bottom, 12 in.; and depth, 14 in. The height of the tray above the floor should be between 30 and 36 in. Standard dimensions for slate trays are shown in Figs. II-92 and II-93 and in Table II-84. Laundry trays are usually supported on legs, and these should be adjustable in length to make the setting of the fixture easier. It is better to set the fixture too high than too low, for a short person can place a step in front of the tray if it is too high while a tall person must bend uncomfortably if it is too low.

Laundry trays should be located in well-illuminated, ventilated, and heated rooms as an inducement to thorough and cleanly work and to assure the comfort of the user. A cover is sometimes provided for laundry trays, which is used as a table when the tray is not in use for washing. The cover may be of wood, enameled iron, or other impervious materials. The use of wood is not recommended. The entire cover should be detachable or all parts should be readily accessible for cleaning purposes. Laundry trays are usually installed in sets of two or three, as shown in Fig. 16-25. Provision should be made for the attachment of a clothes wringer to the trays in such a manner as to give security without injury to the tray. This may be done by bolting or clamping a piece of wood to the tray and attaching the wringer to the wood.

Groups of two or three laundry trays may discharge through the same drain pipe and trap in the manner illustrated in Fig. 16-25. A large number of trays discharging through a single trap would leave an undesirable amount of untrapped waste pipe exposed to the air of the room. Economy is the purpose of using a single trap. An objection to its use is the flow of water through the common waste pipe from one tub to another in the unit when a full tub is emptied while another is empty.

A fixture that is popular in small apartments and for crowded installations or for light laundry work, such as for baby clothes, finery, etc., is the combination laundry tray and sink shown in Fig. 16-25*a*. In this installation the cover to the tray is used as a drain board to the sink. The single trap, which may be used for this combination fixture, should be placed near the sink in order to leave as small a surface as possible exposed to the fouler wastes. Where laundry trays are placed within the kitchen of apartment buildings or each apartment has independent laundry trays, it is possible that such trays may not be used for long periods of time. The waste pipe from the tray should therefore discharge through the trap under the kitchen sink. Where no sink or other frequently used fixture is available, the laundry trays should not be installed.

The satisfactory drainage and trapping of a combination fixture is difficult. If the units are separately trapped and the laundry tray is not frequently used, the trap seal may be lost by evaporation. If a single trap is used for the fixture, as shown in Fig. 16-25*a*, dirty water may flow from one into the other part of the fixture through the common waste pipe. The "Plumbing Manual" (p. 12) and the Housing Code (p. 20) both permit the use of the single trap, provided

> . . . that one fixture is not more than 6 in. deeper than the other and that the waste outlets are not more than 30 in. apart.

Water is usually admitted to the laundry tray through two faucets: one, on the right of the user, for cold water and the other for hot water. These are placed close up to the side of the tray, as shown in Fig. 16-25*b* and *c*. In a combination fixture or double laundry tray, a swinging nozzle may be used to swing over such compartments, as shown in Fig. 16-25*a*. The single tray with adjustable legs, shown at *d* in the figure, is objectionable because of below-the-rim inlets.

The bottom of the laundry tray should be flat with a slope of about ⅛ in. per ft toward the waste outlet. The outlet should be protected by a screen or two crossed bars, somewhat as described in Sec. 16-2. The net area of the openings should be greater than the cross-sectional area of the drain pipe. The waste is almost invariably closed with a plug and chain. Pop-up wastes would be unsatisfactory because of difficulties when the contents of the tub rest on the waste at the time it is to be opened. Overflows are not always used on laundry trays, but the combination unit shown in Fig. 16-25*a* includes an overflow for the tray. The sizes of water-supply pipes are given in Tables 3-5 and 7-1, types of faucet are discussed in Chap. 14, and the sizes of drain pipes are shown in Table 11-1.

16-28. Urinals. Many types of urinal are available, some of which are shown in Figs. 15-3 and 16-27, with sketches and dimensions in Table II-81 and in Figs. II-72 to II-75. Few urinals are completely satisfactory because of the need for protecting and draining the wall and floor from dripping and splashing. This is troublesome with all but wall-flushed, integral unit stall urinals, as shown in Fig. 16-27. Lip urinals should be provided with a flushing rim, as shown in Fig. 16-27 at *B*, *C*, *E*, and *G*. A wall flush unit is shown at *A*. The unit at *B* is undesirable unless adequate provision is made for draining, cleaning, and ventilation of the floor. In setting the bowl or trough of a urinal it should be no more than 20 to 25 in. above the floor, and in a battery of urinals the stalls should not be closer than 30 in. on centers. The number of urinals required in toilet rooms are shown in Tables 1-2, 15-2, and 15-4.

Materials selected for urinals should be durable and noncorrodible and should have a hard, glazed, nonabsorbent surface. Bare metallic surfaces corrode. The fixture should be constructed in one piece, without crack,

Size	18"	24"	Size	18"	24	Size	18"	24"
A	18"	24"	E	5"	5"	J	3½"	3½"
B	38"	38"	F	6"	6"	K	8½"	8½"
C	3¾"	3¾"	G	4½"	4½"	L	14"	14"
D	10"	10"	H	3"	3"	M	13"	13"

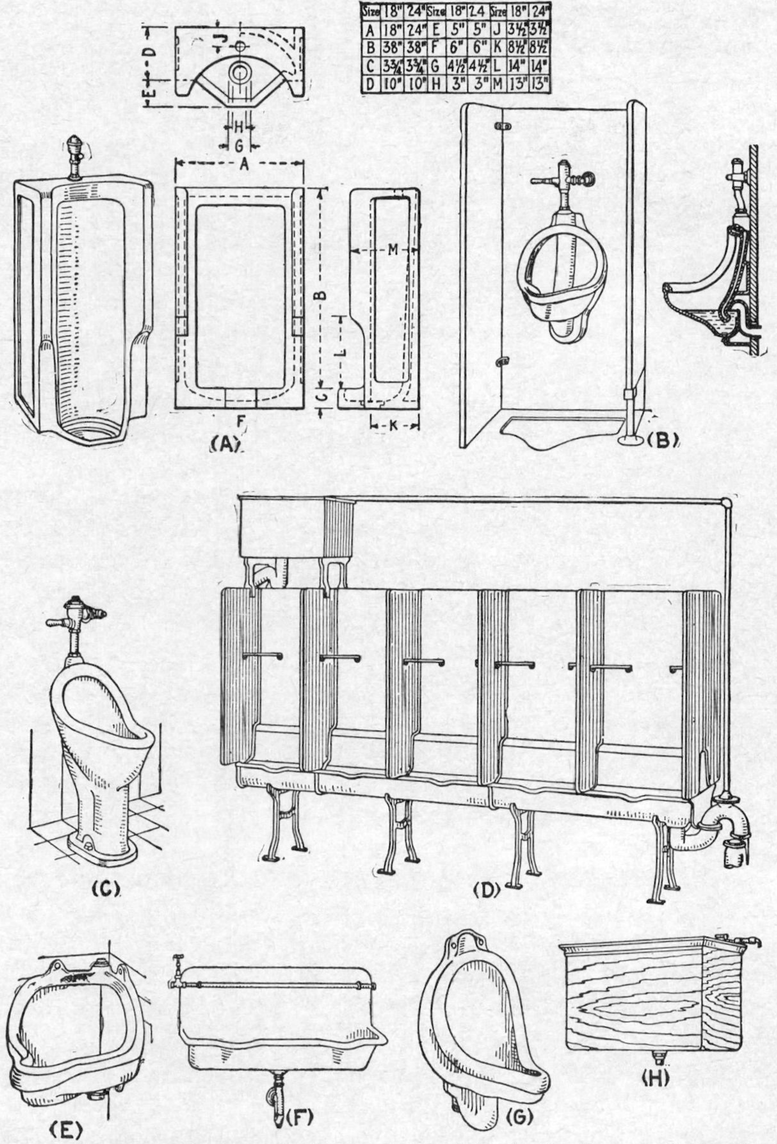

FIG. 16-27. Types of urinal. (*A*) Stall urinal. (*Woodbridge Ceramic Co.*) (*B*) Wall urinal showing two types of stall wall. (*Mott.*) (*C*) Pedestal urinal with flushometer valve and siphon jet. (*Trenton Potteries Co.*) (*D*) Trough urinal with stalls. (*Mott.*) (*E*) Corner urinal, wall type. (*Trenton Potteries Co.*) (*F*) Trough urinal. (*Kohler.*) (*G*) Wall urinal with integral trap. (*Canadian Pottery Co.*) (*H*) Automatic flushing tank. (*Becker Manufacturing Co.*)

joint, or recess. Vitreous ware, earthenware, and enameled iron fulfill these requirements. Enameled iron is not recommended because it does not stand the abuse such fixtures often receive, and it will crack, chip, or craze. Stall urinals are made of solid porcelain. Pedestal and wall-hung urinals may be made of vitreous china. Marble, slate, soapstone, etc., should not be used in the fixture, since they are slightly absorbent and the fixture cannot be constructed in one piece. These materials can be used for the walls of stalls separating urinals. Porcelain is not altogether satisfactory for urinals because it may chip or craze.

The side walls and back of urinals should be made of hard, durable, impervious material. For this purpose slate, soapstone, glass, marble, and any of the materials of which the urinal itself is properly made are satisfactory. The floor under the urinal should be covered with an impervious material. The "National Plumbing Code" states (Sec. 7.8.6) that wall and floor space to a point 1 ft in front of the urinal lip, 4 ft above the floor, and at least 1 ft each side of the urinal shall be lined with nonabsorbent material. The floor should be provided with a drain, and provision for flushing the floor should be made. There should be a local vent in which there is a draft, either forced or direct. There is probably no fixture that disseminates more odor than an improperly installed and maintained urinal. These fixtures are universally abused; urine is slopped over the sides and on the floor; cigars, cigarettes, and matches are thrown into the bowls; and the floor is expectorated on. It is essential, therefore, that care and attention be given to the selection and maintenance of urinals.

The outlet or waste from the fixture should be protected by a perforated-brass, nickel-plated brass, or enameled-iron strainer, or five or six holes about 1/4 in. in diameter may pass through the material of the fixture to serve as an outlet. Some urinals are equipped with an overflow waste cast into the fixture. The purpose of this is to prevent spilling from the fixture in case of stoppage of the drain pipe. Each urinal should be separately trapped. Cast iron or brass is preferable to lead, steel, or wrought iron for traps and drain pipes to avoid corrosion.

Sizes of water-supply and drainage pipes for urinals are shown in Tables 3-5, 7-1, and 11-1.

Some urinals are provided with traps included within the fixture. These are known as *integral traps*, one type of which is illustrated in the pedestal urinal shown in Fig. 16-27*C* and the wall urinal in Fig. 16-27*G*. These are designed to operate similarly to a water closet so that the trap will be emptied by self-siphonage and the seal will be restored by the last water of the flush and the refilling of the flush tank. A local vent installed just above the trap or immediately below the strainer is helpful in keeping down odors only where a draft is provided through the vent. It is desirable that the urinal be so designed that water is retained in the bowl or beneath the strainer to dilute urine falling into it. For this purpose and to assure thorough flushing, urinals are sometimes equipped with siphon or siphon-jet traps similar to those used for water closets.

The use of a trough urinal, such as is shown in Fig. 16-27*D*, with or without stalls, is limited by the "National Plumbing Code" (Sec. 7.8.3) to "places of temporary occupancy." They are to be figured on the basis of one urinal for each 18 in. of length, with a minimum length of 24 in. The flush tank should have a capacity of not less than 1½ gal for each 2 ft of trough length.

16-29. Flushing of Urinals. The flushing of urinals should be provided for either automatically or by some manual device such as a hand-pull chain on a flush tank or a flushometer valve. Manually operated flushing devices on urinals are not recommended because of the reluctance of the public to use them. The volume of each flush should not be less than 2 gal, and the capacity of the flush tank should be about 3 gal. In automatic types of flushing device the frequency of flush should depend on the frequency of use. The period between flushes can be adjusted by means of the supply valve to the flush tank. The supply rate is arranged to fill the tank once in 5 to 15 min, as desired.

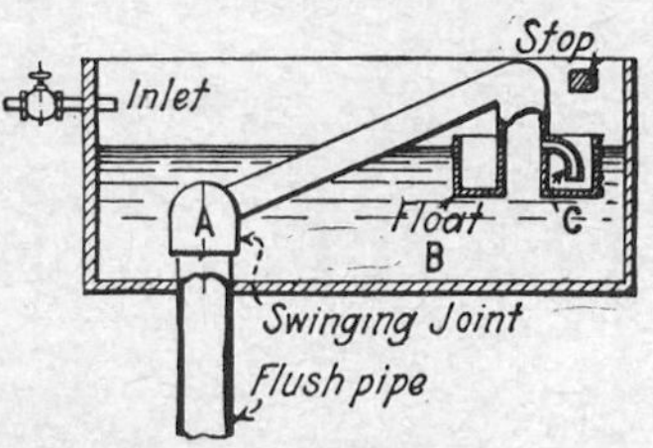

Fig. 16-28. Automatic urinal flush tank.

An automatic flush tank is illustrated in Fig. 16-28. It operates as follows: The supply valve is opened sufficiently to fill the tank in the desired time between flushes; the rising water level raises the float *B* and with it the arm of the flush pipe, which is pivoted about point *A*. When the float has raised to the flushing level of the tank, the arm stops rising and the float is filled with water and sinks. Water rushes down the flush pipe to the urinal, and siphonic action drains the water from the float through the small pipe *C*. When the float and tank are emptied of water, the cycle of operations has been completed and a new cycle is ready to begin.

The flush pipe from the tank to the urinal should be about ¾ in. in diameter. It may be made of brass or nickel-plated brass tubing. The urinal bowl should be provided with a flushing rim to distribute the water in the bowl. Stall urinals with floor outlets should be so equipped as to have the entire vertical wall, as well as the floor bowl of the fixture, thoroughly flushed. The flushing water is distributed over the walls of the fixture as a fan-shaped spray. Trough urinals are usually flushed by means of a perforated pipe placed along the back of the trough.

The Housing Code recommends (p. 27):

6.8.1. Flushing. Urinal tanks shall flush automatically or shall be provided with a substantial hand-operated mechanism. Flushing capacity of tank shall be adequate for the type of urinal used.

16-30. Floor Drains. Floor drains are considered to be plumbing fixtures, according to Sec. 6.15.2. of the Housing Code. They are used to drain off water that falls on the floor. They are useful principally on impervious floors, and in dwellings they are placed in cellars, bathrooms, laundries, kitchens, and other places where water might otherwise accumulate on the floor. Types of floor drain are shown in Figs. 11-39, 13-10, and II-104 to 107, and a floor-drain connection is shown in Fig. 11-5. Floor drains should be placed in a depression in the floor toward which the entire floor slopes on a grade between 1/8 and 1/4 in. per ft. The top of the strainer should form the bottom of the depression in the floor. It should be set in some impervious material, such as concrete, that adheres to the strainer and its support to form a watertight joint with the support. A crack around the drain that will admit water will harbor insects and retain dirt. The drains should be of heavy, durable construction, preferably of cast iron throughout or cast iron with brass or nickel-plated brass cover. The strainer should consist of a hinged or removable cover perforated with holes 1/4 to 1/16 in. in diameter and slightly greater in total area than the cross-sectional area of the drain pipe. The "National Plumbing Code" states (Sec. 7.15.1) that the open area of the strainer shall be at least two-thirds of the cross-sectional area of the drain line. The drain pipe should be 2 in. in diameter or larger.

The drain pipe on floor drains located in dry cellars or other locations subject to infrequent use and in basement rooms where food may be stored should discharge into drain pipes not likely to be used for the conveyance of sewage. Where the sanitary sewer or a combined sewer offers the only drainage outlet for putrefactive wastes such as those from laundry trays and floor washings, the floor drain can be connected to the sanitary sewer or combined sewer and an additional fixture, not discharging human excrement, should discharge into the floor-drain trap to assure the maintenance of seal. Floor drains receiving the discharge from frequently used fixtures or otherwise used with sufficient frequency to avoid evaporation of the trap seal should discharge into the sanitary or a combined sewer. Floor-drain openings should be located not more than about 30 to 40 ft apart with floor sloping toward the drain.

A floor drain discharging into a sanitary or combined sewer should be equipped with a deep-seal trap with large water capacity, and some clear-water drain, such as icebox drain, tank overflow, or hydrant drip should be discharged into it. The "National Plumbing Code" states (Sec. 7.15.1) that the trap should have a minimum depth of seal of 3 in. A backwater valve, as shown in Fig. 14-19, can be placed in the line to prevent water from the sewer from flowing backward through the drain. Floor drains located above the ground floor should discharge untrapped into drains on lower floors, as suggested for safe wastes, and finally into drains with a permanent water seal.

Floor drains on upper floors are sometimes equipped with gate valves and water-sealed traps and are connected to the plumbing pipes. It is expected that the valves will be opened only for brief periods of time. Such an arrangement is not recommended, however, because of the probability of neglect to close the valves, with resultant exposure of the premises to sewer gases.

16-31. Drinking Fountains.[1] The purpose of a drinking fountain is to supply potable water for drinking that has not been contaminated by the previous user of the fountain. Standard specifications for drinking fountains have been prepared by the American Standards Association (Z 4.2, 1942). Rates of flow into drinking fountains and recommended sizes of supply pipes[2] are given in Table 7-1.

In the selection of a fountain, its purpose should be kept in mind, that is, to supply water without permitting the lips of the drinker to touch the fountain and to prevent water from dripping back on to the nozzle. Many types of fountain are available, some of which are shown in Figs. 16-29 and II-99. Some fountains are equipped with a small trough containing water placed near the ground for the use of small animals.

The nozzles shown at *a*, *b*, and *c* in Fig. 16-29 do not fulfill requirements; the nozzles at *d*, *e*, and *f* approach desired conditions; and the nozzles at *g* and *h* attain it. An objection to the last type of nozzle or spray is the difficulty sometimes experienced in attempting to drink from it.

The fountain nozzle should be constructed so that a high-velocity spray cannot shoot or be shot out of the nozzle. This can be accomplished by making the outlet orifice larger than the inlet pipe so that the velocity is reduced. Slits cut in the side of the nozzle will discourage mischievously inclined persons from using the nozzle as a squirt gun.

The "National Plumbing Code" states, in Chap. C.3 concerning drinking fountain nozzles, as follows:

C.3.1. Minimum Elevation. All drinking fountain nozzles including those which may at times extend through a water surface with orifice not greater than 7/16 (0.440) in. diameter or 0.150 sq in. area shall be placed so that the lower edge of the nozzle orifice is at an elevation not less than 3/4 in. above the flood-level rim of the receptacle.

[1] See also L. Blendermann, Drinking Water Systems, *Air Conditioning, Heating, and Ventilating*, May, 1957, p. 101, and *Domestic Eng.*, January, 1956, p. 126.

[2] See also Water Demand Rates for Drinking Water Coolers, *Refrig. Eng.*, December, 1954, p. 37, and *Air Conditioning, Heating, and Ventilating*, January, 1955, p. 112.

C.3.2. The ¾-in. elevation shall also apply to nozzles with more than one orifice, provided that the sum of the area of all orifices shall not exceed the area of a circle 7/16 in. in diameter.

C.3.3. Special conditions and certain other materials related to drinking fountains shall meet requirements as set forth in American Standard A 40.2–1942 and ASA Z 4.2, respectively.

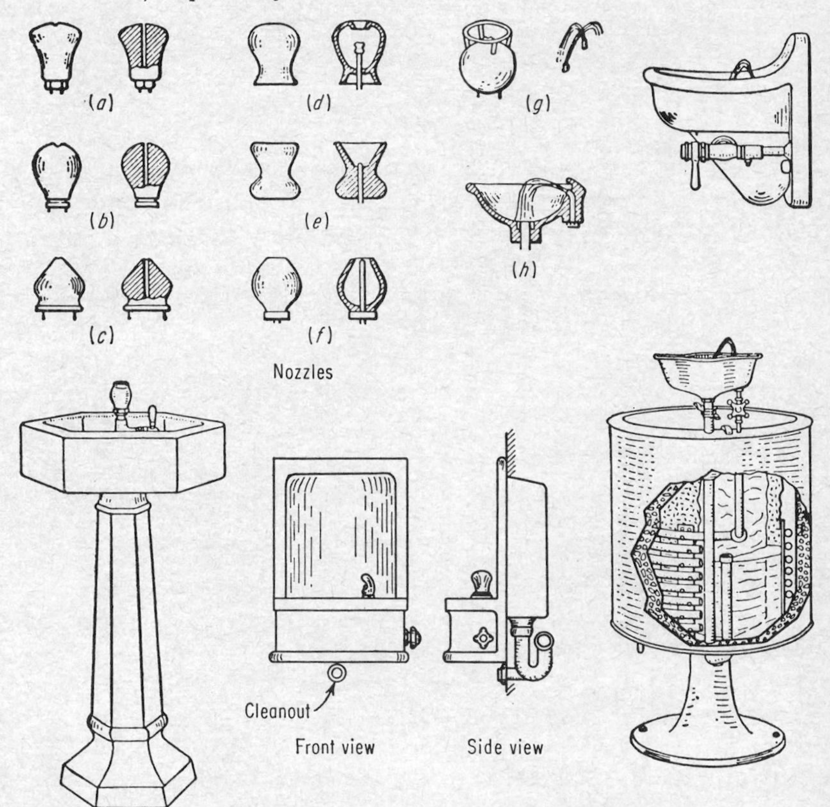

Fig. 16-29. Types of drinking fountain.

Some extracts from the "National Plumbing Code" standards for drinking fountains, which are the same as ASA Z 4.2–1942, are as follows:

C.8.1. The fountain shall be constructed of impervious material, such as vitreous china, porcelain, enameled cast iron, or stoneware.

C.8.2. The jet of the fountain should issue from a nozzle of nonoxidizing, impervious material set at an angle from the vertical such as to prevent the return of water in the jet to the orifice or orifices from whence the jet issues. The nozzle and every other opening in the water pipe or conductor leading to the nozzle should be above the edge of the bowl, so that such nozzle or opening cannot be flooded in case a drain from the bowl of the fountain becomes clogged.

C.8.3. The end of the nozzle should be protected by nonoxidizing guards to prevent the mouth and nose of the user from coming into contact with the nozzle. Guards should be so designed that the possibility of transmission of infection by touching the guards is reduced to a minimum.

C.8.4. The inclined jet of water issuing from the nozzle should not touch the guard, and thereby cause spattering.

C.8.5. The bowl of the fountain should be so designed and proportioned as to be free from corners which would be difficult to clean or which would collect dirt.

C.8.6. The bowl of the fountain should be so proportioned as to prevent unnecessary splashing at the point where the jet falls into the bowl.

C.8.7. The drain from the fountain should not have a direct physical connection with a waste pipe, unless the drain is trapped.

C.8.8. The water-supply pipe should be provided with an adjustable valve fitted with a loose key or an automatic valve permitting the regulation of the rate of flow of water to the fountain so that the valve manipulated by the users of the fountain will merely turn the water on or off.

C.8.9. The height of the fountain at the drinking level should be such as to be most convenient to persons using the fountain. The provision of several steplike elevations to the floor at fountains will permit children of various ages to utilize the fountain.

C.8.10. The waste opening and pipe should be of sufficient size to carry off the water promptly. The opening should be provided with a strainer.

The water-supply inlet to the drinking fountain should be adjusted by means of a special valve not easily found by the casual user so that only a gentle stream will issue from the fountain when the control valve is wide open. The control valve for the fountain, sometimes called a bubbler, should be in plain view and should be comfortably within reach of the user unless the fountain is to run continuously.

The nozzle may be of brass, which can be either plain or nickel-plated; enameled iron; porcelain; or other impervious materials. It should be placed about 36 to 40 in. above the floor or ground. The use of the fixed cup surrounding the nozzle, usually for the purpose of keeping the lips away from the nozzle, defeats its own purpose because the lips will come into contact with the cup. Cafeteria drinking fountains are supplied with a special type of nozzle, one type of which is shown in Fig. 14-7. The nozzle and control valve are so arranged that the patron can hold the glass and control the flow of water into it with one hand.

The waste pipe from the fountain should be trapped if it is discharged into any fixture or pipe containing other wastes. Where floor drains or other infrequently used fixtures exist, the fountain waste should be led into their traps.

16-32. Swimming Pools. The "Plumbing Manual" states (p. 46):

Standards for design, equipment, and operation of artificial swimming pools are included in the regulations of many state departments of health. These for the most part are based on the "Report of the Joint Committee on Bathing Places," Conference of State Sanitary Engineers and American Public Health Association. This Report is published by the U.S. Public Health Service as Supplement No. 139 to the Public Health Reports and is obtainable from the Superintendent of Documents, Washington, D.C.

16-33. Miscellaneous Fixtures. Among those fixtures that may be classed as "miscellaneous fixtures" are baptistries; fountains and ornamental ponds; aquaria; coffee urns in restaurants; soda-fountain and bar fixtures; ironers and laundry equipment attached to gas or water-supply pipes and to drainage systems; dental cuspidors, operating tables, and hospital equipment; chemical, bacteriological, and industrial laboratory equipment; and industrial machinery involving the use of gas, water, or other fluids. The list might be exhausting.

The principles of the design and installation of all fixtures are similar. The protection of health and hygiene are of primary importance. To attain this end the "Plumbing Manual" states (p. 14) that miscellaneous fixtures " . . . shall have supplies protected from backflow." The prevention of cross connections is another feature that should be guarded against by the plumber.

CHAPTER 17

WATER TREATMENT AND SEWAGE DISPOSAL

17-1. Methods of Water Purification. The safest and least expensive method to sterilize water is to boil it for at least 15 min. Since this method is uneconomical for the treatment of a public water supply and is not always convenient for household water supplies, some other method

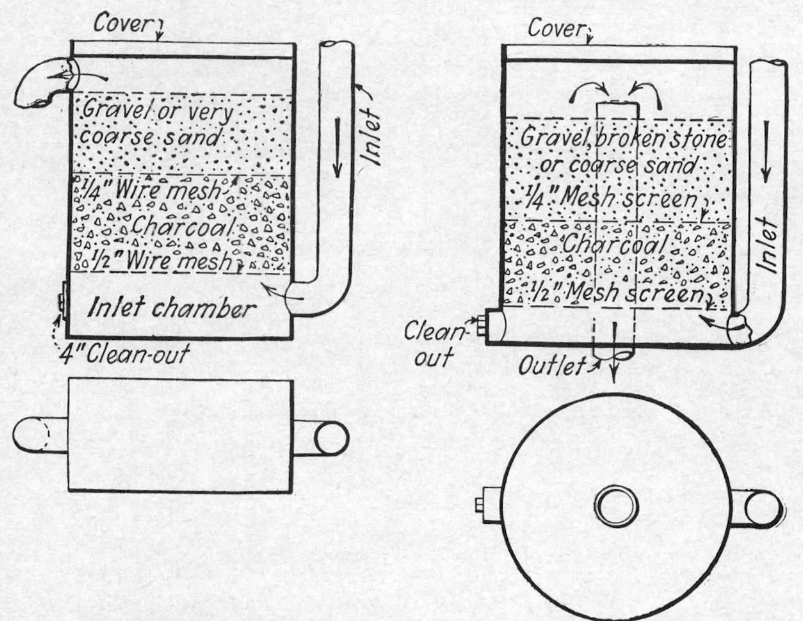

FIG. 17-1. Transparent view through charcoal rain-water filter. (*C. D. Puckett, Metal Worker, Apr.* 18, 1919, *p.* 491.)

that will deliver a continuous supply of potable water must be depended on. No method that is safe and certain for all conditions of small, intermittent private supplies is known. Filters containing 12-in. or deeper layers of gravel or coarse sand can be used to remove large suspended particles such as leaves, twigs, and sticks. Whatever is strained from the water, however, must ultimately be removed from the filter.

Devices that are used with some satisfaction for the removal of fine suspended matter and some bacteria include charcoal filters, porous earthenware, diatomite filters; ultraviolet-ray machines; ozonators; chlorine, bromine, or iodine tablets; and numerous proprietary filtering devices. Many of the devices and methods listed may be satisfactory. All are potentially dangerous because of the possibilities of abuse or neglect.

17-2. Household Filters. A charcoal filter for domestic use is illustrated in Fig. 17-1. This filter is especially applicable to use under a rain-water leader. The charcoal should be ground to a size of about ¼ to ½ in. Water will be too slow in passing through smaller pieces, and larger pieces may permit the passage of unfiltered water. A filter about 12 by 24 in. in plan and 30 in. deep, constructed as shown in Fig. 17-1, is satisfactory for a 4-in. rain leader. Upward filtration is better than downward filtration, since decomposing organic matter does not mat on the surface, the tendency to force it through the filter is less, and the filter is more easily cleaned. The filter is cleaned by removing the cover and pouring boiling water into it from the top. Large particles of matter can be removed from the filter through the hand hole at the bottom. Suspended matter and apparently some colloidal matter are removed from the water as it passes through the filter in contact with the charcoal, so that it issues from the filter clear and sparkling but not necessarily safely disinfected. The percentage of removal of bacteria by such a filter is low, and its value as a safeguard to health is inconsiderable. Small-sized charcoal filters are available that can be fastened to the end of a faucet. They are not recommended, because they become foul bacteria-breeding nests rather than bacteria removers.

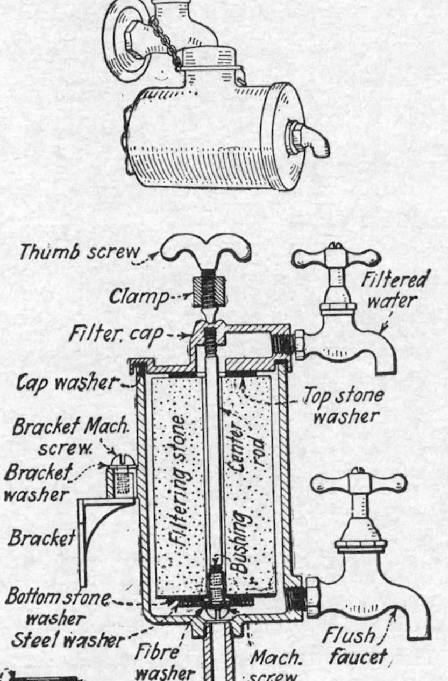

FIG. 17-2. Stoneware household filters. (*Roberts Filter Co.*)

The candle filter,[1] the stone filter shown in Fig. 17-2, and the Everpure filter shown in Fig. 17-3 are devices for household and small-quantity uses.

[1] See also *Domestic Eng.*, June, 1956, p. 164.

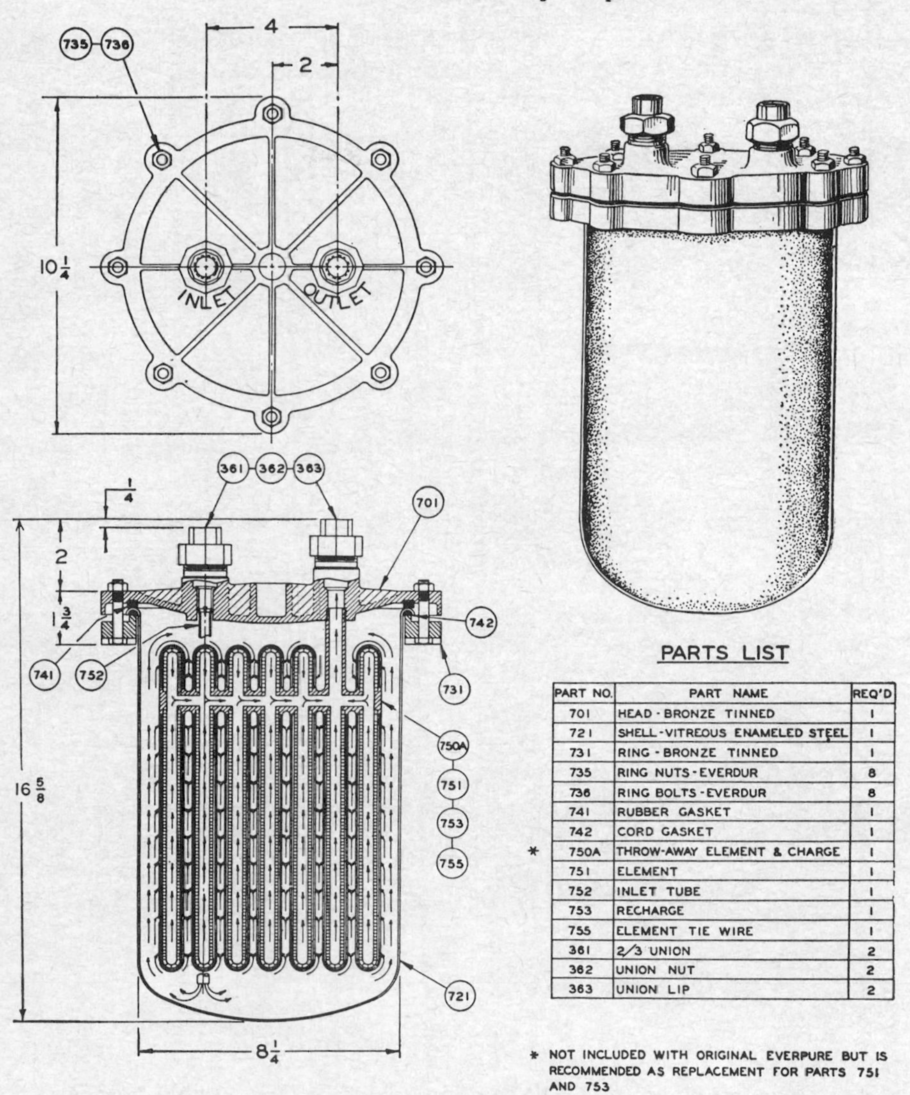

PARTS LIST

PART NO.	PART NAME	REQ'D
701	HEAD - BRONZE TINNED	1
721	SHELL - VITREOUS ENAMELED STEEL	1
731	RING - BRONZE TINNED	1
735	RING NUTS - EVERDUR	8
736	RING BOLTS - EVERDUR	8
741	RUBBER GASKET	1
742	CORD GASKET	1
* 750A	THROW-AWAY ELEMENT & CHARGE	1
751	ELEMENT	1
752	INLET TUBE	1
753	RECHARGE	1
755	ELEMENT TIE WIRE	1
361	2/3 UNION	2
362	UNION NUT	2
363	UNION LIP	2

* NOT INCLUDED WITH ORIGINAL EVERPURE BUT IS RECOMMENDED AS REPLACEMENT FOR PARTS 751 AND 753

FIG. 17-3. An Everpure filter.

17-3. Ultraviolet Ray. The ultraviolet-ray machine is an effective means for killing bacteria, and if clear water is passed through it, the water will be sterilized. The machine will have no effect in removing particles of suspended matter or in clarifying or decolorizing the water. The conditions necessary for the successful operation of the sterilizer are that all the water should flow in a thin, clear film close to the sterilizing ray, that the water should be well agitated (but not mixed with air), and that the exposure to the ray should be continuous.

Ultraviolet-ray machines are used for institutions and office buildings and particularly for sterilizing water for swimming pools. Their popularity lies in the absence of chemicals and the lack of effect on the taste or chemical quality of the water.

The amount of current consumed is dependent on the type of machine and whether one or a number of machines are used in series. A single machine can be expected to consume about $\frac{3}{8}$ kw in passing 120 gal of water in 1 hr. Two larger machines in series consume 1.54 kw in an hour while delivering 3,000 gal of water. The success, effectiveness, and durability of these machines have been demonstrated by their use.

17-4. Ozonizers. Machines for ozonizing water supplies have not been generally introduced and have found but little use. Not much is known about them at present, so that their installation would be little more than an experiment.

17-5. Sand Filters. The filtration of water through sand is accomplished in either gravity or pressure filters. The former are used almost exclusively for the treatment of public water supplies. The principal objection to their use for private water supplies is the necessity for pumping the filtered water after filtration because the water must flow by gravity from the bottom of the filter into the collecting basin, called the *clear-water basin.* Although gravity filter units are less expensive than pressure filters, the cost of the pump and the pumping usually renders their use uneconomical for small installations, particularly where all pressure necessary for distributing the water is supplied by the public waterworks. The rate of filtration through a gravity or pressure filter, where a high degree of purification is desired, should not exceed 2 gpm per square foot of filter surface. Gravity filters are usually of such size that they are constructed in place with concrete walls and bottom and with dimensions and capacities determined by the designer.

A pressure filter is a sand filter in a closed metal container resembling a steam boiler or a hot-water storage tank. Most pressure filters are proprietary devices built in sizes and capacities recommended by the Filter Standardization Committee of the American Society of Mechanical Engineers, December, 1916, as shown, together with other information, in Table 17-1.

A section through a pressure filter is shown in Fig. 17-4. It operates as follows: Untreated water passing into the filter is dosed with a solution of aluminum sulfate (alum) for the purpose of causing a gelatinous precipitate that will coagulate the suspended matter in the water. The dose is arranged to apply about 1 to 2 grains of alum per gallon of water treated, dependent on the turbidity, alkalinity, and other qualities of the water. The quantity of the dose of alum is usually regulated by an automatic apparatus. This water, which has been dosed with alum, enters the

TABLE 17-1. CAPACITIES OF PRESSURE FILTERS*

Inside diameter, in.	Capacity, gpm — Rate, gpm per sq ft: 2	3	4	5	Size of inlet and outlet pipe, in.	Inside diameter, in.	Capacity, gpm — Rate, gpm per sq ft: 2	3	4	5	Size of inlet and outlet pipe, in.
	Per cent bacterial removal: 97†	‡	95§	‖			Per cent bacterial removal: 97†	‡	95§	‖	
12	1½	2⅓	3	4	...	48	25	38	50	63	2½
14	2	3	4	5	...	54	32	48	64	80	2½
16	2¾	4¼	5½	7	...	60	40	59	80	99	2½
20	5	7	10	12	...	72	57	85	114	142	3
24	7	10	14	17	...	84	77	115	154	192	4
30	10	15	20	25	1½	96	100	150	200	250	4
36	15	21	30	36	1½	120	157	235	314	392	6
42	20	29	40	49	2						

* Filter Standardization Committee, *J. Am. Soc. Mech. Engrs.*, 1917, p. 425.
† Probable best possible results.
‡ Refiltration, swimming pools, etc.
§ Clarification, 95 per cent iron removal.
‖ Removal of suspended matter.

filter at the top and passes through the sand, leaving its suspended matter and some of its dissolved impurities on the surface of and in the sand bed. The water passes through the manifold and laterals of the collector system in the bottom of the filter and thence into the effluent pipe, which is connected to the distributing pipes of the building. The deposition of the impurities on and in the sand bed clogs the filter, requiring more pres-

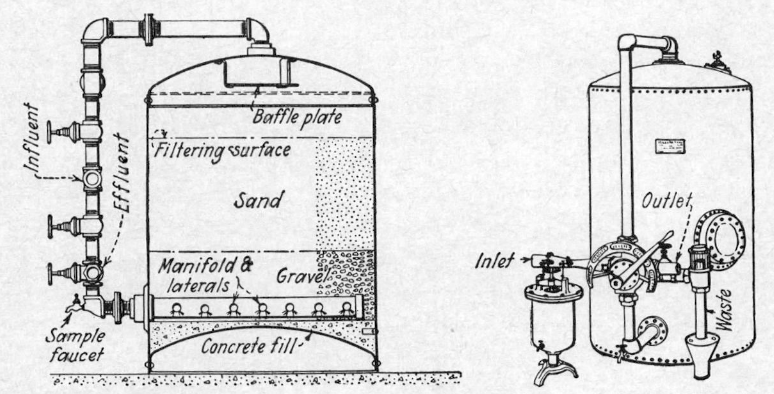

FIG. 17-4. Section and exterior view of a pressure filter. (*Roberts Filter Co.*)

sure to force the water through. When this pressure reaches about 10 to 15 ft of water, the filter should be washed. Regardless of the pressure loss through the filter, in operating without laboratory control the filter should be backwashed at least once a day to discourage the growth of organisms in it.

A filter is washed by stopping the normal process of filtration and forcing clean water upward through the filter in a direction opposite to that in normal operation. The dirty water from the filter is discharged to the sewer. The filter should be washed for from 5 to 8 min at a rate of about 15 gpm per square foot of sand surface. The manipulation of the valves for filtering and washing has been so simplified on some machines that only one handle need be turned when it is desired to wash the filter or to throw it back into service.

In washing, the pressure of the wash water should be about 15 psi. The filter should be washed until the wash water comes off clear and free from suspended matter. Washing must not be done too vigorously because of the danger of removing sand, too long because of the waste of water, or too slowly, for this will be of no value. In some installations it is proper to wash the filter with unfiltered water, but where high bacterial removal is desired, only filtered water should be used in washing. This may require the use of a storage tank that has a capacity of not less than 5 per cent of the amount of water filtered in 1 day. The filter should be washed at least daily. When it is properly maintained, the sand need never be removed. The loss of pressure through these filters varies between 6 in. when the filter is clean and 10 to 15 ft or more just before it is washed. The loss of head, or difference in pressure, between the inlet and outlet should not be allowed to exceed 10 to 15 ft, because of the packing of the sand layer and the danger of breaking through it, resulting in the delivery of unfiltered water.

Information concerning the dimensions, connections, care, and operation of any particular filter must be obtained from the manufacturer, since no standards for these factors have been adopted. The amount of attention necessary for their operation and their cost limit their use mainly to industrial plants, institutions, and estates.

17-6. Hardness in Water. Hard water is undesirable because of the amount of soap that must be dissolved before a lather can be produced.[1] The soap curd makes washing of clothing difficult and is objectionable in shampooing. Hard water is also undesirable in industrial supplies and in hot-water supplies because of the scale that it forms on heating surfaces and pipelines, the foaming that it causes in steam boilers, its injuriousness

[1] Approximately 13 oz of soap would be required to produce a lather in 100 gal of water with 100 ppm total hardness. Ppm means parts per million by weight. One grain per gallon is approximately 17.1 ppm.

to fine fabrics in laundries, its injuriousness to certain manufacturing processes such as dyeing and papermaking, and for other reasons. The softening of water is more essential for industrial uses than for domestic purposes. Any water containing more than 150 ppm of hardness can be softened economically.

Water is made hard primarily by the solution in it of carbonates and sulfates of calcium and magnesium. The chlorides and nitrates of calcium and magnesium are effective to an almost negligible degree in causing hardness. Total hardness is the sum of all the compounds in the water that cause hardness in it. Total hardness is expressed in various ways, the standard method in American waterworks practice being in milligrams per liter, by weight, in terms of calcium carbonate ($CaCO_3$).

TABLE 17-2. RELATION BETWEEN METHODS OF EXPRESSING HARDNESS OF WATER

	Milligrams per liter	Grains per gallon	Clark degrees	French degrees	German degrees
Milligrams per liter	1.0	0.058	0.07	0.1	0.056
Grains per gallon	17.1	1.0	1.20	1.71	0.96
Clark degrees	14.3	0.83	1.0	1.43	0.80
French degrees	10.0	0.583	0.7	1.0	0.56
German degrees	17.8	1.04	1.25	1.78	1.0

TABLE 17-3. RELATIVE HARDNESS OF WATER

Relative hardness	Extremely soft	Very soft	Soft	Moderately soft	Moderately hard	Hard	Very hard	Excessively hard	Almost too hard for use
Milligrams per liter carbonate of lime	15	30	45	90	110	130	170	230	250

The relation between various recognized methods for expressing hardness is shown in Table 17-2. The relative hardness of various waters is expressed approximately in Table 17-3.

A simple method for measuring the approximate hardness of a water is to determine the least quantity of a standard soap solution that, when shaken vigorously with a definite quantity of the water, will create a lather that will last for at least 5 min. The soap solution should be made according to the "Standard Methods of Water Analysis" of the American Public Health Association, and the standard procedure for making the test should be followed to obtain quantitative results. The procedure requires that 50 cu cm of the water to be tested shall be measured into a 250-cu-cm bottle. The soap solution should be added in small quantities of 0.2 to 0.3 cu cm at a time, the bottle being shaken vigorously after each addition. When the lather produced lasts at least 5 min, the amount of soap solution used is observed and the total hardness of the water expressed as milligrams per liter of calcium carbonate. This can be read from Table 17-4.

TABLE 17-4. TOTAL HARDNESS IN MILLIGRAMS PER LITER OF CALCIUM CARBONATE ($CaCO_3$) FOR EACH TENTH OF A MILLILITER OF SOAP SOLUTION WHEN 50 ML OF THE SAMPLE OF WATER ARE TESTED

Milliliters of soap solution	0.0	0.1	0.2	0.3	0.4	0.5	0.6	0.7	0.8	0.9
0.0	....	....	....	....	....	....	....	0.0	1.6	3.2
1.0	4.8	6.3	7.9	9.5	11.1	12.7	14.3	15.6	16.9	18.2
2.0	19.5	20.8	22.1	23.4	24.7	26.0	27.3	28.6	29.9	31.2
3.0	32.5	33.8	35.1	36.4	37.7	39.0	40.3	41.6	42.9	44.3
4.0	45.7	47.1	48.6	50.0	51.4	52.9	54.3	55.7	57.1	58.6
5.0	60.0	61.4	62.9	64.3	65.7	67.1	68.6	70.0	71.4	72.9
6.0	74.3	75.7	77.1	78.6	80.0	81.4	82.9	84.3	85.7	87.1
7.0	88.6	90.0	91.4	92.9	94.3	95.7	97.1	98.6	100.0	101.5

17-7. Methods of Softening Water. Drinking-water supplies are softened by the addition of slaked lime [$Ca(OH)_2$] or soda ash (Na_2CO_3) or both; by the zeolite or base-exchange process; by the use of synthetic organic detergents, the so-called *sulfonated-oil soaps,* and related substances; and by the addition of proprietary preparations usually containing organic compounds. The use of detergents, proprietary compounds, and boiling to soften water is confined mainly to domestic and industrial uses rather than to the softening of public water supplies.

The hardness produced by the carbonates of calcium and magnesium, known as *temporary hardness,* can be reduced by boiling. In the process of boiling, dissolved carbon dioxide is driven off, and the minerals, which are soluble in water only in the presence of dissolved carbon dioxide or carbonic acid gas, are precipitated. The precipitated minerals form the familiar scale on teakettles and the inside of hot-water pipes. The sulfates of calcium and magnesium cause what is known as *permanent hardness.* Some chemical must be added to throw them out of solution.

The amount of lime or sodium carbonate necessary is dependent upon the amount of calcium and magnesium dissolved in the water. An operator with some knowledge of chemistry is required to give constant attention to devices using lime or soda ash or both to soften water. Such softeners are usually operated similarly to a gravity water filter. The method is therefore not available for household use or for many institutional and industrial water supplies, and it is not of special interest to the plumber. Domestic, institutional, and industrial supplies can, however, be softened by zeolite water softeners that are available on the market and are in extensive and successful use in large and small installations.

Synthetic organic detergents, the so-called "sulfonated oil soaps" and related compounds, are proprietary products usually containing organic compounds that are used in the laundry for softening and in washing. The sulfonated compounds lather in hard water without producing a precipitate. They include sodium metaphosphate, trisodium phosphate,

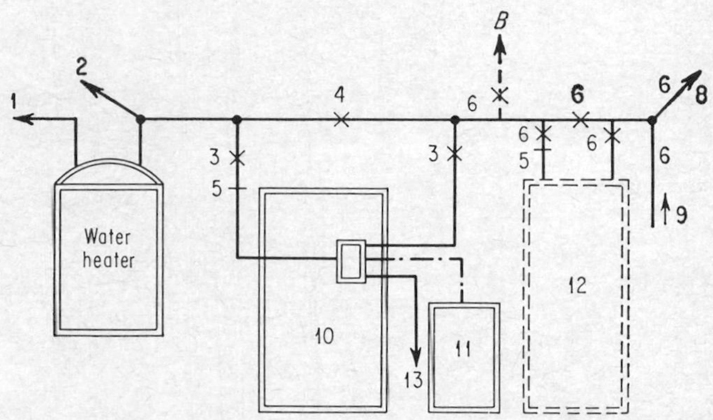

FIG. 17-5. Piping diagram for a water softener and iron removal plant. (*Based on W. D. Calhoun and Richard Hetherington, ABC's of Home Water Softeners, Plumbing and Heating J., August,* 1954, *p.* 107.)

1. Soft hot water. 2. Soft cold water. 3, 4, 5, 6. Valves. 8. Bypass. 9. Raw water. 10. Zeolite water softener. 11. Brine solution. 12. Iron removal. 13. Waste water from regeneration of softener.

sodium hexametaphosphate called Calgon, and sodium pyrophosphate. Calgon is a proprietary compound, highly hygroscopic, which is used in about a 25 per cent solution and is added in a concentration of about 1 ppm. These compounds are used also in the control of corrosion.

17-8. Zeolite Softeners. A zeolite is any chemical compound so imperfectly bound together that it will change its composition, dependent on the concentration of certain other chemicals in solution in its presence. The sodium zeolite, $NaAlSiO_4 \cdot 3H_2O$, known as *Permutit*, will exchange the sodium or magnesium for calcium when in solution in their presence and will take on the form $CaAlSiO_4 \cdot 3H_2O$. When the latter compound is placed in a solution containing a greater concentration of sodium, the sodium replaces the calcium and we have the original zeolite.

Zeolite as used in water softeners has the appearance of a coarse-grained sand with even-sized, hard, lustrous grains. It will absorb moisture from the atmosphere and must therefore be stored in dry places.

In practice, hard waters containing both temporary and permanent

TABLE 17-5. DIMENSIONS OF ZEOLITE WATER SOFTENERS

Grain capacity	Roughing-in dimensions, in.				
	A	B	C	D	E
45,000	47	12	40	15	37
60,000	49	14	46	19	45
90,000	58	14	46	19	45

hardness are passed through a layer of zeolite at a temperature not to exceed 100°F, so as not to injure the zeolite. Calcium and magnesium are absorbed by the zeolite and are exchanged for sodium, which goes off in solution in the water, but the presence of sodium in solution does not cause hardness. The water is partly filtered in passing through the softener and its hardness is reduced to *zero*. Upon the exhaustion of the zeolite the flow of hard water is cut off and a solution of sodium chloride (common salt) is passed into the softener and allowed to stand in contact with the zeolite. The sodium in the brine replaces the calcium and magnesium in the zeolite, which is thus restored, and the calcium and magnesium are discharged into the sewer with the wash water as calcium and magnesium chloride, together with such detritus as may be washed off the filter. The zeolite is now ready for use for softening more water, and the process can be carried on indefinitely without renewing the zeolite. In practice, it can be anticipated that about 5 per cent of zeolite will disintegrate and wash away annually. The only expense in connection with the operation of such a softener is the cost of a small amount of common salt.

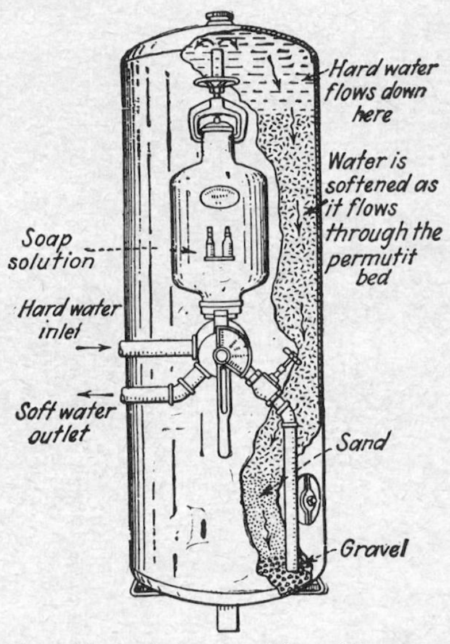

FIG. 17-6. A household zeolite water softener. (*Permutit Co.*)

A piping hookup for a softener installation[1] is shown in Fig. 17-5, and some approximate roughing-in dimensions are shown in Table 17-5. Factors to be considered in the installation of a softener include the following: (1) A bypass is needed around the installation. (2) Select piping material known to be satisfactory in the local water. (3) See that the pressure range is between 25 and 125 psi, with a pressure between 25 and

[1] See also *Domestic Eng.*, November, 1956, p. 94.

40 psi preferred. A sectional view of a household softener is shown in Fig. 17-6. The size of the softener depends on the amount of water to be treated and its hardness. A softener that is adequate for its purpose should operate for a week to 10 days without regeneration.

17-9. Zeolites. Some characteristics of zeolites are listed in Table 17-6. Precipitated synthetic zeolites with exchange values of 10,000 to 12,000 grains of calcium carbonate ($CaCO_3$) per cu ft are most commonly used in domestic and small water-supply services. With an efficient gel zeolite the salt consumption in regeneration may be as low as 0.3 to 0.5 lb per 1,000 grains of hardness removed; 0.6 lb is a safe figure to assume for cost estimates.

TABLE 17-6. CHARACTERISTICS OF ZEOLITES

Name or type	Description	Exchange value, grains of $CaCO_3$ per cu ft	Lb of salt per cu ft for regeneration
Glauconite, or greensand	Greenish-black, kidney-bean-shaped granules. A natural zeolite	2,700–3,800	1.4
Precipitated synthetic, or wet, process	Sand	6,000–20,000	4–8
Fused synthetic	Not widely used	4,000–8,000	4
Clay group	Granulated clay*	5,000–8,000	4

* Porous and consequently affected by turbidity. Not durable and hence not widely used.

An example of the computation of the necessary volume of zeolite in a softener is shown as follows:

Example. How many cubic feet of zeolite with an exchange value of 10,000 grains of hardness per cu ft will be required in a softener that is to handle 1,000 gal of water per day, containing 250 ppm of hardness, if the softener is to be regenerated weekly? How many pounds of salt will be required at each regeneration if 0.5 lb of salt is used per 1,000 grains of hardness removed?

Solution: First compute the total grains of hardness removed from the water between regenerations:

$$\frac{1{,}000 \times 8.3 \times 250 \times 7{,}000}{1{,}000{,}000} = 14{,}500 \text{ grains removed daily}$$

where

$$1{,}000 = \text{water used per day, gal}$$

$$8.3 = \text{weight of 1 gal of water, lb}$$

$$\frac{250}{1{,}000{,}000} = \text{grains of hardness in 1 lb of water}$$

$$7{,}000 = \text{grains per lb}$$

$$\frac{14{,}500 \times 7}{10{,}000} = 10.5 \text{ cu ft of zeolite required}$$

The salt requirement is

$$\frac{14{,}500 \times 7}{1{,}000 \times 2} = 51 \text{ lb of salt per regeneration}$$

17-10. Operation of Zeolite Water Softeners. The softener should never be operated after it is depleted without regeneration, since the softening material may be injured. The rate of softening, regardless of the hardness of the water, lies between 75 and 120 gal per hr per sq ft of surface of zeolite material, the thickness of the zeolite bed being about 2 or 3 ft. The thicker the bed, the greater is the permissible rate of softening, but a rate of over 120 gal per hr per sq ft of surface is undesirable because of the high velocity through the bed, which may prevent all the water from coming in contact with the zeolite grains for a sufficient length of time. The hardness of the water affects only the period between regeneration; it is not a factor in determining the rate of filtration, although, of course, a low rate should be used for hard waters to lengthen the period between regeneration.

It is seldom necessary or even desirable, on account of corrosive aggressiveness, to reduce hardness below 80 to 100 ppm. Since zeolites reduce hardness to zero, it may be both economical and desirable to bypass a portion of the hard water to mix with the softened water from the softener. The amount of bypassed water can be computed by proportion, as, for example, if in the preceding example it is desired to reduce the hardness of the water to 100 ppm:

The ratio of bypassed water to water passing through the softener would be 100:150.

The volume of zeolite required would be

$$\frac{10.5 \times 150}{250} = 6.3 \text{ cu ft}$$

The weight of salt required per regeneration would be

$$\frac{51 \times 150}{250} = 30.5 \text{ lb}$$

Further economy in the softening of water can be attained by delivering unsoftened water to flush tanks, garden hoses, and other services not requiring soft water. The original cost of the plumbing required for this separation of the services will be increased, but the saving in salt over a long period of time will probably render the system economical. In some installations only the hot-water supply is softened. It should be noted that hot water of zero hardness may be actively corrosive, particularly to steel pipes and tanks.

Where only one zeolite softening unit is used, it may be desirable to install a bypass to supply water during the time of 15 min required for regeneration and for emergencies.

The fact that regeneration is necessary is determined by a soap test. It is made according to particular directions, dependent on the quality of the soap solution used. In general, a known small quantity of soap solution (about 3 drops) is put into a bottle containing a small quantity (about 1 oz) of the water to be tested. The two are shaken together vigorously. If a froth or bead exists on the surface of the water for at least 5 min, it is not necessary to regenerate the softener. If the bead does not endure, the softener should be regenerated.

Steps in the regeneration of some domestic water softeners are as follows: (1) Shut off the supply of hard water to the softener. (2) Drain off the water in the softener. (3) Flood the softener with brine solution containing at least 10 per cent, by weight, of sodium chloride and allow a contact period of at least 5 min. (4) Draw off the brine and run it to waste. (5) Flush the softener with cold hard water, continuing to flush with running water until there is no salty or brackish taste to the water. (6) The softener is regenerated and is ready to be restored to service. Some softeners are regenerated automatically.

17-11. Removal of Dissolved Iron Compounds. Soluble compounds of iron are undesirable in public, in some industrial, and in most domestic water supplies because the iron compounds come out of solution and precipitate on plumbing fixtures and on textiles in the laundry, causing unsightly stains and occasionally tastes. Iron can sometimes be kept in solution by the addition of a chemical (polyphosphate) to the water, or it can more economically be precipitated by oxidation (aeration) followed by sedimentation and sand filtration to remove precipitated iron. The filter must be backwashed relatively frequently, the water used in washing going to the sewer. The removal of dissolved iron from the water supply of individual residences is not common, as the operation requires skill, attention, and special equipment.

SEWAGE DISPOSAL

17-12. Purpose and Methods. Sewage is the used water supply of a community, and from a household it generally constitutes the liquid waste and commonly contains human excrement. It is a putrid fluid, dangerous to health and repulsive to the senses. It must be disposed of in such a manner as to avoid danger to health or the creation of a nuisance. Methods for the treatment and disposal of sewage for isolated institutions and other relatively small establishments are limited to fewer processes than are available for larger municipal sewage-treatment plants. Methods of treatment for the individual residence are limited almost completely to the septic tank. Where sewage can be disposed of legally and without danger to the public health or other hazards into a large body of water, into the public sewer system, or onto the land, no treatment is required.

Methods of sewage disposal of special interest to the plumber include the cesspool, the septic tank, and the underground filter. Note that these methods of sewage treatment are biological. No chemical is used except in a few cases in which the final effluent may be disinfected with chloride or treated with lime before disposal.

17-13. Cesspools and Dry Wells.[1] A cesspool is a retainer or container for the temporary retention of sewage from which the sewage is withdrawn or seeps at a rate differing from the influent rate. A dry well is a cesspool constructed like a water well but in dry ground. The terms cesspool and dry well are sometimes used synonymously except that the dry well is properly a leaching cesspool constructed like a water well in dry ground. The use of cesspools and dry wells is looked upon with disfavor by health authorities, and both are a threat to the potability of underground water. The "National Plumbing Code" states:

B.12.1. Use. The use of cesspools for disposal of sewage and their installation will be accepted only if approval is obtained, before work is begun, from the Administrative Authority or the health department having jurisdiction.

B.12.2. Installation. Cesspool installations shall be considered only as a temporary expedient in those instances where connections to a public sewer system will be possible within a reasonable period of time.

B.12.3. Health Hazard. Because of the public-health hazard involved, extreme care shall be exercised in locating a cesspool. Under no circumstances shall the cesspool penetrate the ground-water stratum.

B.12.4. Construction. The construction of the cesspool shall comply with the requirements for seepage pits as given in Sec. B.10.

B.11.2. Size of Dry Well.[2] Large dry wells shall be constructed in general in accordance with Sec. B.10.

B.11.3. Small Dry Wells. For small dry wells handling limited quantities of water, the pit may consist of a 3-ft length of 18-in.-diameter vitrified-clay or cement pipe, filled with crushed rock or stone.

B.10.1. Seepage Pit. Seepage pits may be used either to supplement the subsurface disposal field or in lieu of such field where conditions favor the operation of seepage pits. . . .

B.10.2. Water Table. Care shall be taken to avoid extending the seepage pit into the ground-water table. Where the pit is used to receive the septic-tank effluent, the same limitations shall be placed on the location of the pit as on the cesspool (see Par. B.12.3).

[1] See also N. W. Wolpert, *Air Conditioning, Heating, and Ventilating*, April, 1956, p. 109.

[2] The "National Plumbing Code" apparently uses the term dry well to mean a seepage pit or a leaching cesspool.

B.10.3. Pit Lining. Except as provided in Par. B.10.6 the pit shall be lined with stone, brick, or concrete blocks laid up dry with open joints that are backed up with at least 3 in. of coarse gravel. The joints above the inlet shall be sealed with cement mortar. It is customary to draw in the upper section of the lining.

B.10.4. Pit Covers. A reinforced-concrete cover shall be provided, preferably to finished grade. If the cover is over 30 in. square, it shall have an access manhole.

B.10.5. Bottom of Pit. The bottom of the pit shall be filled with coarse gravel to a depth of 1 ft.

B.10.7. Size of Pit. The seepage pit shall be sized in accordance with provisions in Table 17-7.

TABLE 17-7. REQUIREMENTS FOR SEEPAGE PIT DESIGN
(From "National Plumbing Code," Table B.10.8)

Soil structure	*Effective absorption area required per bedroom, sq ft**
Coarse sand and gravel	20
Fine sand	30
Sandy loam or sand (*sic*) clay	50
Clay with considerable sand and gravel	80
Clay with small amount of sand and gravel	160

* In calculating absorption wall area of pit, gross diameter of excavation shall be used.

Cesspools may be either watertight or leaching. The former are used where no outlet is available. Sewage is retained in a tight cesspool until it is removed by pumps or buckets into a suitable receptacle or other means of disposal. Sewage seeps from a leaching cesspool into the surrounding ground. Cesspools are used properly only where no other satisfactory means of sewage disposal can be found. A leaching cesspool may require less attention in maintenance than a dry cesspool. Disadvantages of a cesspool may include restriction of the use of water in order not to overload the cesspool, cost and difficulty of maintenance, and pollution of the ground water, as applied to a leaching cesspool. The use of a leaching cesspool should be restricted to a small family, in a remote location, where there is loose, sandy soil and no danger of groundwater pollution.

There is no standard minimum safe distance between a water-supply well and a leaching cesspool or privy vault. The distance depends on local conditions. It may be as little as a few feet in tight clay, 50 ft in fine sand, and miles in cavernous limestone or fissured rock. As a rule of good practice the cesspool should be placed downhill from a well but even such a location will not assure a safe water supply when underground conditions are unfavorable. If strong salt brine poured into a cesspool increases the chloride content in a water well near by, there is underground flow between them.

Septic action which occurs in cesspools reduces the volume of solids to be removed from the cesspool. The combination of a tight cesspool overflowing into a leaching cesspool approaches the condition of a septic tank and underground disposal field described in Sec. 17-15 and minimizes the amount of solids to be disposed of separately.

The proper capacity of a leaching cesspool depends on the amount of water used by the family, on the absorbing characteristics of the ground,

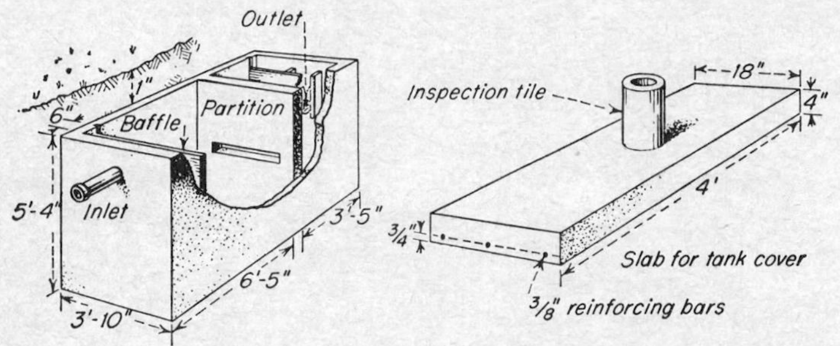

FIG. 17-7. Concrete septic tank. (*From Univ. Illinois Small Homes Council, Circ. G5.5*, 1947.)

and on other local conditions. A method of testing the absorptive characteristics of the ground is explained in Sec. 17-15. Where such a test cannot be made, a rule of thumb calls for 1 sq ft of percolating surface for each 5 gal per day discharged into the cesspool. In very porous conditions the percolating area may be diminished by as much as 50 per cent.

The proper capacity of a tight cesspool which receives only sewage and no roof-water or similar drainage may be fixed on a basis of 20 gal per day per user with a semiannual, monthly, or other frequency of cleaning.

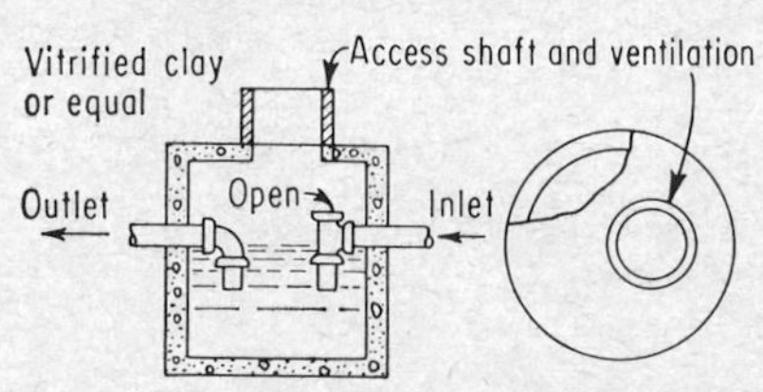

FIG. 17-8. Small, circular, concrete septic tank.

A tight cesspool may be constructed similarly to a septic tank, as described in Sec. 17-14. The walls and floor of a leaching cesspool may be built of brick or other materials, as specified in "National Plumbing Code," above. The top of the cesspool, tight or leaching, may be similar to the catch basin shown in Fig. 11-43 or to the septic tank in Fig. 17-7. The cesspool should be vented to the air to avoid pressures within it higher or lower than atmospheric. No wood should be used in the finished cesspool unless it will be permanently submerged because wood rots when it is alternately wet and dry.

Cesspools should be abandoned when sewers become available. An abandoned cesspool should be filled with any suitable material, as an unused, unfilled, and forgotten cesspool may become a murderous trap for which the owner is morally and legally liable.

17-14. Septic Tanks. A septic tank is a watertight container of sewage in which the period of retention is sufficiently long to permit some hydrolysis and gasification of the contents and from which the effluent flows at approximately the same rate at which the influent enters. A septic tank may be preferable to a cesspool for receiving the sewage from a residence

TABLE 17-8. MINIMUM CAPACITIES FOR SEPTIC TANKS SERVING AN INDIVIDUAL DWELLING
(From "National Plumbing Code," Table B.4.2)

Number of bedrooms	Maximum number of persons served	Nominal liquid cap. tank, gal*	Recommended inside dimensions							
			Length		Width		Liquid depth		Total depth	
			Ft	In.	Ft	In.	Ft	In.	Ft	In.
2†	4	500	6	0	3	0	4	0	5	0
3	6	600	7	0	3	0	4	0	5	0
4	8	750	7	6	3	6	4	0	5	0
5	10	900	8	6	3	6	4	6	5	6
6	12	1,100	8	6	4	6	4	6	5	6
7	14	1,300	10	0	4	6	4	6	5	6
8	16	1,500	10	0	4	6	4	6	5	6

* Liquid capacity is based on number of bedrooms in dwelling. Total volume in cubic feet includes air space above liquid level.

† Or less.

because, for the same service, it is smaller and less expensive to construct and operate, gives better results in the quality of the effluent, needs cleaning less frequently, and is more widely approved by health authorities because of the protection it affords to the neighboring ground water. Recommended capacities and dimensions of septic tanks are given in Table 17-8, and sections are shown in Figs. 17-7 to 17-9.

The "National Plumbing Code" specifications covering septic tanks are:

B.4.3. Multiple Compartments. In tank of more than one compartment, the inlet compartment shall have a capacity of not less than two-thirds of the total tank capacity.

B.4.4. Garbage Disposal. Where domestic garbage-disposal units are installed or contemplated, the capacity of the septic tank shall be at least 50 per cent greater than the requirements given in Table 17-8.

B.4.5. Length. Septic tanks shall be at least twice as long as they are wide.

B.4.6. Construction. Septic tanks shall be constructed of corrosion-resistant materials and be of permanent construction. The cover of the tank shall be

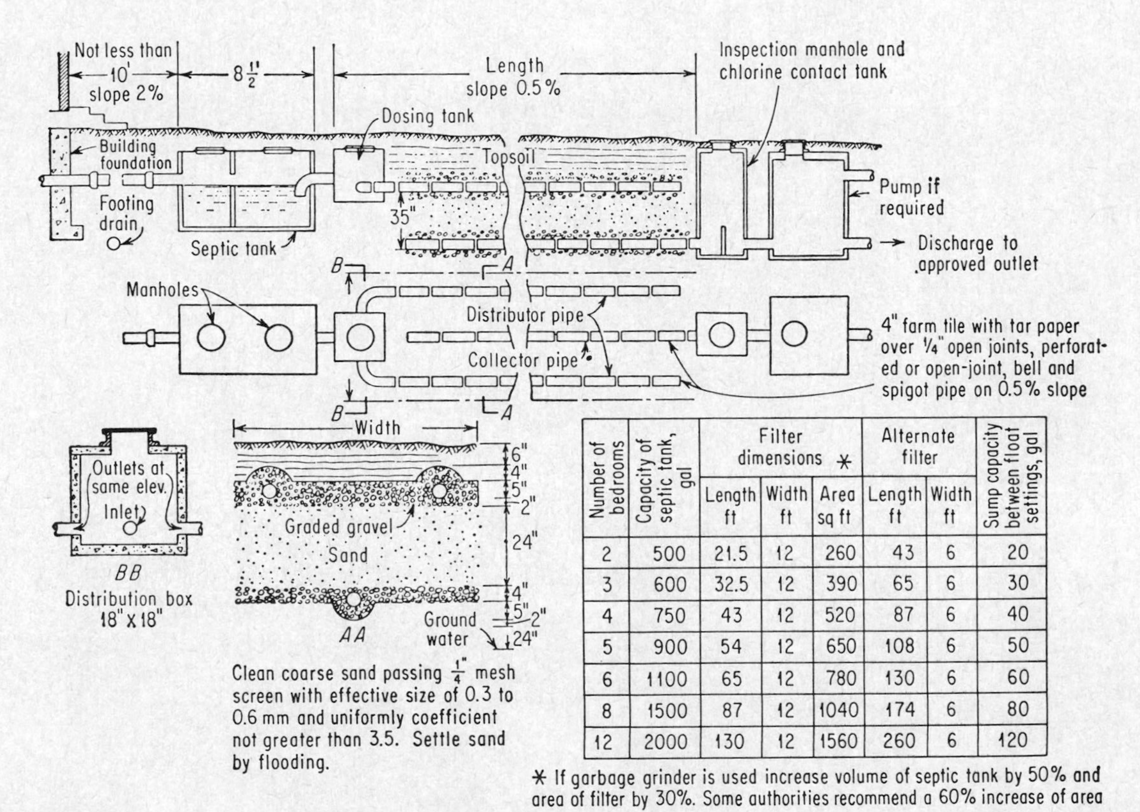

Number of bedrooms	Capacity of septic tank, gal	Filter dimensions *			Alternate filter		Sump capacity between float settings, gal
		Length ft	Width ft	Area sq ft	Length ft	Width ft	
2	500	21.5	12	260	43	6	20
3	600	32.5	12	390	65	6	30
4	750	43	12	520	87	6	40
5	900	54	12	650	108	6	50
6	1100	65	12	780	130	6	60
8	1500	87	12	1040	174	6	80
12	2000	130	12	1560	260	6	120

* If garbage grinder is used increase volume of septic tank by 50% and area of filter by 30%. Some authorities recommend a 60% increase of area

FIG. 17-9. Suggested layout for residential, subsurface disposal plant. *(From H. E. Babbitt and E. R. Baumann, "Sewerage and Sewage Treatment,"* 8th ed., *John Wiley & Sons, Inc., New York,* 1958. *Based on J. A. Salvato, Jr., Sewage and Ind. Wastes, August,* 1955, *p.* 909.)

designed for a dead load of not less than 150 psf and, if of concrete, should be reinforced and not less than 4 in. thick.

B.4.7. Manholes. The inlet compartment must be provided with one manhole. Other compartments may be provided with a manhole. Manholes shall be at least 20 in. square or 24 in. in diameter and provided with covers which can be sealed watertight. Manholes should be extended to grade. Where removable slab covers are provided, manholes are not required.

B.4.8. Baffles. If inlet and outlet baffles are used they shall extend the full width of the tank and be located 12 in. from the end walls. Such baffles shall extend at least 6 in. above the flow line. Inlet baffles shall extend 12 in. and outlet baffles 15 to 18 in. below the flow line.

B.4.9. Pipe Inlet and Outlet. In lieu of baffles, submerged pipe inlets and outlets may be installed consisting of a cast-iron sanitary tee with short section of pipe to the required depth as indicated in Sec. B.4.8.

B.4.10. Invert. The invert of the inlet pipe shall be located at least 3 in. above the invert of the outlet.

17-15. Underground Disposal of Effluent. A septic tank should not be allowed to discharge onto the surface of the ground or into a small body of surface water except under special circumstances. The effluent from the tank should be discharged into a porous tile or a seepage pit, or both, so placed that the liquid will seep into the upper layer of the surface soil as shown in Fig. 17-9. The length of the porous tile or the size of the seepage pit depends on the character and imperviousness of the ground. Some data on these requirements are given in Table 17-9 and in Fig. 17-9. A suggested layout for a rural residence disposal plan is shown in Fig. 17-9. The rate of filtration to be used is in the order of about 1 gal per sq ft of sand surface per day, using a medium-coarse building sand. Four-inch-diameter unglazed tile in an 18-in. trench can be used where the percolation coefficient is less than 0.24, 24-in. for a coefficient between 0.4 and 1.0, and 30-in. for a coefficient between 1.0 and 2.3. The *percolation coefficient* is the reciprocal of the rate, in gallons per day per square foot of trench bottom, at which sewage can be safely applied in the field. It can be computed from the expression

$$C = \frac{t + 6.2}{29}$$

where C = percolation coefficient

t = maximum time for water to fall 1 in. during percolation tests in saturated soil, min

Some coefficients given by Kiker[1] are coarse sand or fine gravel, 0.3; fine sand or light loam, 0.3 to 0.16; sandy clay or heavy loam, 0.6 to 1.3; medium clay, 1.3 to 2.0; and tight clay or rock, above 2.0.

[1] J. E. Kiker, Jr., *Florida Eng. Expt. Sta. Bull.* 23, 1948.

The method of conducting the percolating test is explained by Kiker, as follows:

1. Dig a hole about 1 ft square to the depth at which it is proposed to lay the tile drain.
2. Fill the hole with water and allow the water to seep away. When the water falls to within 6 to 8 in. of the bottom of the hole, observe the rate at which the water level drops.
3. Continue these observations until the soil is saturated and the water seeps away at a constant rate. Keep adding water until the rate becomes constant.
4. Compute the time required for the water to fall 1 in. after the soil becomes saturated. This is the standard percolation time t.

The reciprocal of the Kiker coefficient is the permissible rate of application of sewage in gallons per day per square foot of trench bottom area.

TABLE 17-9. SUBSURFACE ABSORPTION AREAS REQUIRED

Time for water to fall 1 in., min	From NPC*	From FHA†				
	Effective absorption area required in bottom of disposal trenches, sq ft per bedroom‡	Absorption area, sq ft per bedroom	Type of building	Length of 4 in. distributing pipe for drain, per person, ft — Absorption rate: Excellent	Moderate	Poor
2	50	50	Residences	20–30	30–50	50–100
3	60	60	Schools,	5–7½	7½–12½	12½–25
4	70	70	etc.			
5	80	80				
10	100	105				
15	130	125				
30	180	180				
60	240§	240				

* From "National Plumbing Code," Table B.6.2.

† From "Requirements for Individual Water-supply and Sewage-disposal Systems," Federal Housing Administration, revised Mar. 1, 1947. A minimum of 200 sq ft of effective absorption area (100 lin ft of 24-in. trench) shall be provided per living unit.

‡ A minimum of 150 sq ft shall be provided for each dwelling unit.

§ Where time for water to fall 1 in. is longer than 60 min, special design is required.

Some suggested lengths of 4-in. tile drains are given in Table 17-9. It is desirable, in laying the tile, that the slope be between 0.1 and 0.3 per cent to give both adequate distribution and adequate time for percolation of water through the open joints in the line. Where the length of underdrain is more than 200 to 300 ft, a dosing tank is recommended to assure filling the length of the tile. The backfill for the trench should be porous and should, preferably, not exceed about 30 in. in depth.

TABLE 17-10. MINIMUM STANDARDS FOR UNDERGROUND DISPOSAL-FIELD CONSTRUCTION
(From "National Plumbing Code," Table B.8.2)

Disposal-field construction	*Minimum standard*
Lines per field, minimum number	2
Individual lines, maximum length	100 ft
Trench bottom, minimum width	18 in.
Field tile, minimum diameter	4 in.
Field-tile lines, maximum slope	6 in. in 100 ft
Field trenches, minimum separation	6 ft
Effective absorption area, minimum per dwelling	See Table 17-9

The "National Plumbing Code" fixes minimum standards for disposal-field construction as shown in Table 17-10 and in the following sections:

B.9.2. Filter Material. The filter material shall cover the tile and extend the full width of the trench and shall be not less than 6 in. deep beneath the bottom of the tile. The filter material may be washed gravel, crushed stone, slag, or clean, bank-run gravel ranging in size from ½ to 2½ in. The filter material shall be covered by untreated paper or by a 2-in. layer of straw as the laying of the tile proceeds.

B.9.3. The size and minimum spacing requirements for disposal fields shall conform to those given in Table 17-11.

TABLE 17-11. SIZE AND SPACING OF TILES IN UNDERGROUND DISPOSAL FIELDS
(From "National Plumbing Code," Table B.9.4)

Width of trench at bottom, in.	Recommended depth of trench, in.	Spacing tile lines, ft*	Effective absorption area per lin ft of trench, sq ft
18	18–30	6.0	1.5
24	18–30	6.0	2.0
30	18–36	7.6	2.5
36	24–36	9.0	3.0

* A greater spacing is desirable where available area permits.

B.9.5. Absorption lines shall be constructed of tile laid with open joints. In the case of bell-and-spigot tile, it should be laid with ½-in. open joints, at 2-ft intervals with sufficient cement mortar at the bottom of the joint to assure an even flow line. In the case of agricultural tile, the sections shall be spaced not more than ¼ in., and the upper half of the joint shall be protected by asphalt-treated paper while the tile is being covered, unless the pipe is covered by at least 2 in. of gravel. Perforated clay tile or perforated bituminized-fiber pipe or asbestos-cement pipe may be used, provided that sufficient openings are available for distribution of the effluent into the trench area.

It is to be noted, as stated in Sec. 17-13, that a seepage pit can be used to supplement or in lieu of the subsurface disposal field.

17-16. Dosing Tanks or Intermittent Dosing. If the rate of flow from a septic tank is low, seepage may be concentrated at the upper end of a

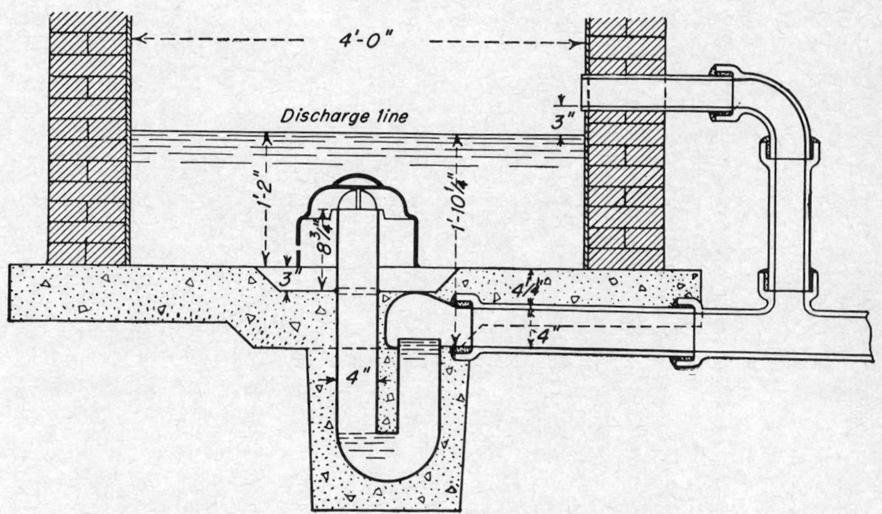

FIG. 17-10. Dosing siphon. (*Pacific Flush Tank Co.*)

long drainage line and the line may become clogged. The condition can be prevented by accumulating tank effluent in a dosing tank, such as is shown in Fig. 17-10. When the dosing tank discharges, the liquid flows rapidly enough to fill or partly fill the length of the drainage line, thus evenly distributing the seepage from the tile. The size of siphon and

TABLE 17-12. AUTOMATIC SIPHONS

Diameter of siphon, in.	Average rate of discharge, gpm	Min. diameter of tank at discharge, ft	Recommended min. capacity of tank, gal per hourly discharge
3	65	3	105
4	150*	3	187
5	325*	3	312
6	475*	4	416
8	1,350*	4	625

* From catalog of the Pacific Flush Tank Co.

dosing tank to be used can be approximated from the information in Table 17-12.

17-17. Care and Maintenance of Septic Tanks. The solid matters in sewage settle to the bottom and are decomposed by bacteria into gas and liquid. Little or no disinfecting chemical should be discharged into the tank because it may inhibit the bacterial action on which the work of the tank depends. In the process some of the solids are lifted to the surface by the rising gas to form scum on the tank surface. Bacterial action continues effectively in the scum as well as in the sludge in the bottom of the tank. The appearance of scum on the surface of a septic tank must not be interpreted as indicating an undesirable condition. When the liquid space between the scum and the sludge is about one-half of the total depth of the tank, cleaning is advisable.

To measure the depth of sludge in the tank lower the end of a suction hose, about 1½ or 2 in. in diameter, into the tank just below the scum. Start pumping slowly at the same time that the hose is slowly lowered without creating disturbance in the tank contents. The sudden appearance of black sludge in the discharge from the pump will indicate that the sludge level has been reached. If the scum layer is tough and strong, its thickness can be measured by pushing through it a flat disk, about 1 ft in diameter, to the center of which a rope has been tied. The disk is pushed through on edge and is pulled up gently, in a horizontal position, by the rope. When an appreciable difference is felt in the pull necessary to lift the disk, the bottom of the scum has been reached. It is usually not possible to conclude correctly concerning the need for cleaning a septic tank by viewing the top of the scum from an open manhole.

To clean the tank, stir the contents and pump or bail it out. If the sludge is too thick to be pumped out, water may be added as needed. The material removed may be discharged into a trench and buried, dumped into a city sewer, or, where conditions are suitable and health authorities permit, spread out on the land or dumped into a stream. It is to be noted that the gases formed in septic tanks may be poisonous, asphyxiating, or explosive. Care must be taken, therefore, not to enter a septic tank without assurance of adequate ventilation, and open flames should be kept away from the tank until it has been effectively ventilated. Neither the absence of an odor nor an open-flame test gives or assures safety. Asphyxiation, gas poisoning, or injuries from explosion may result. Safe and simple devices are available for the detection of dangerous gases or the absence of oxygen in enclosed spaces.[1]

[1] See also *Air Conditioning, Heating, and Ventilating*, June, 1956, p. 117.

CHAPTER 18

MAINTENANCE AND REPAIRS

18-1. Trouble Shooting. Maintenance and repair problems requiring the attention of the plumber include (1) stoppage of fixture outlets and of drain pipes; (2) burst or leaking pipes; (3) leaking faucets, flush valves, and other valves; (4) low water-supply pressures; (5) frozen pipes; (6) no hot water; (7) no cold water; (8) dirty water; (9) noises; (10) water in the basement; (11) bad odors; (12) closing or reopening a home; and others. To detect the causes and to make repairs may require knowledge, ingenuity, and equipment.

18-2. Stoppage of Lavatories and Similar Fixtures. First fill the fixture with water and surge vigorously with the pneumatic plunger, as shown in Fig. 18-1, to dislodge the stoppage and force it down the drain. Covering the overflow will help to concentrate the pressures produced on the drain where it is wanted. If this does not clear up the difficulty, drain the contents of the fixture through the cleanout plug in the bottom of the trap into a bucket on the floor. Remove the strainer and clean the drain pipe with a flexible wire or brush, as shown in Fig. 18-2. If the stoppage is not removed, it may be in either the branch or the main drainage pipes, and the cleanouts in them must be opened. The main drainage pipes can be cleaned by forcing through them a flexible wire or tape of the type shown in Fig. 18-2.

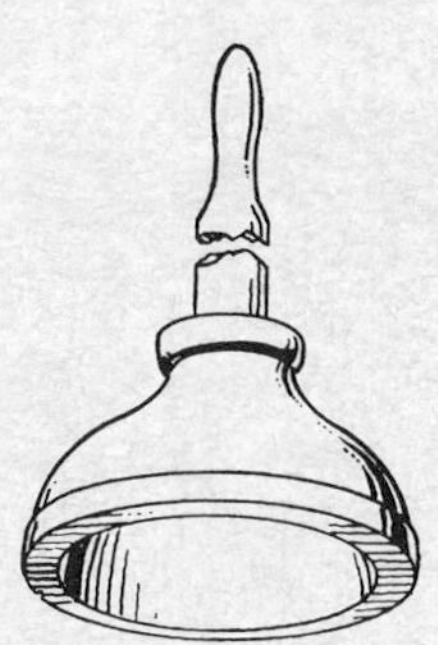

Fig. 18-1. Rubber suction and force cup.

18-3. Stoppage of a Water Closet. The pneumatic plunger should first be tried. If this is not successful, force a spiral wire of the type shown in Fig. 18-2 into the drainage pipe, twisting it as it advances. If stoppage is caused by paper or cloth, it may wrap around the wire so that it can be pulled out of the pipe. Cakes of soap, scrub brushes, and other hard objects may become lodged in the trap, necessitating the disconnection of the fixture at its base. Care must be taken in replacing the fixture that the seal between the fixture and the drain pipe is gastight.

18-4. Stoppage of Drain Pipes within the Building. Insert flexible, spiral wire, as shown in Fig. 18-2, into the clogged pipe through the cleanout provided, and attempt either to push the object along the pipe or to entangle it on the wire and pull it out. If the pipe is only partially closed and water will still flow down it, flush it with hot water and potash or washing soda, a proprietary drain-pipe solvent, or a proprietary substance that creates sufficient heat in the drain pipe to cause grease to go into solution. Lye or other strong alkali should be used cautiously, as soap may be formed with grease, increasing the difficulty of removing the stoppage.

18-5. A Cleanout Cover Sticks. Threaded cleanout covers are sometimes difficult to start. Too strong a pull on a wrench handle, especially if the length of the handle is increased by slipping a long piece of pipe

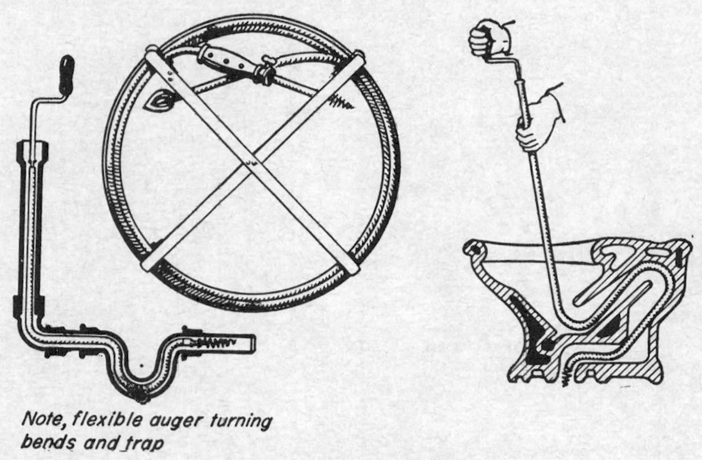

FIG. 18-2. Flexible wire for cleaning pipes.

over it, may break the drain pipe or a joint in it. If an ordinary pull on a well-fitting wrench is insufficient to start the cover, strike the wrench or the cover a few sharp blows at the same time that a steady pull is applied or apply heat to the outside of the pipe while the cover is kept cool and under a strain. If the cover is malleable, dry ice can be put on it.

18-6. Stoppage of the Building Sewer. Access to the sewer may be gained through the cleanout which is usually located near the basement wall or by excavating down to and breaking through the pipe. Stoppage of the building sewer is caused mainly by roots, grease, or hard objects. In any event bad stoppages must be pushed or washed into the public sewer or other outlet. Tools for cleaning sewers[1] are illustrated in Fig. 18-3.

18-7. Roots in Sewers. Tree roots, particularly those of the poplar, willow, and elm, will enter a sewer through minute holes in the joints and

[1] See also *Domestic Eng.*, October, 1956, p. 206.

growing in the sewer may fill it. Fungus growths occasionally cause trouble in sewers by forming a network of tendrils that catches floating objects and builds a barricade across the sewer. Difficulties from fungus growths are not common, but constant attention must be given to the removal of grit, grease, and roots. The addition of copper sulfate to sewage may aid the killing of tree roots, but the high dilution, intermittent application, and vigorous flushing between applications usually render the procedure valueless. Packing copper sulfate around the outside of pipe joints at the time the sewer is laid may be effective in discouraging roots for a few years. The placing of copper rings in pipe joints is a patented method that will stop the passage of roots through the joints.

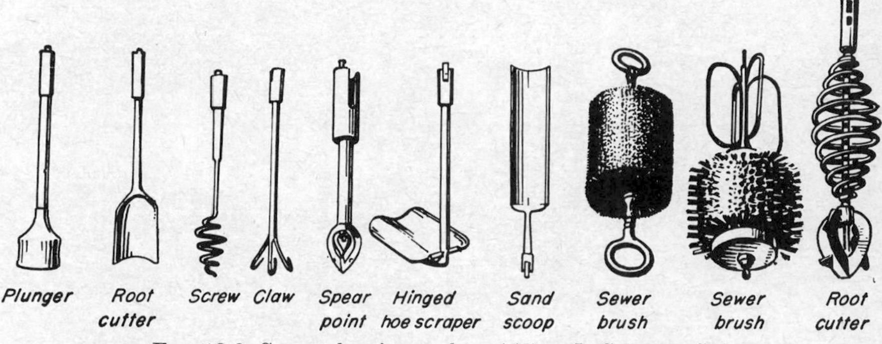

Fig. 18-3. Sewer-cleaning tools. (*Allan J. Coleman Co.*)

18-8. Location of Buried Sewers. If uncertainty exists concerning the exact location of a buried sewer one end of which is accessible, a flexible metal rod can be pushed into the sewer and the rod located electrically, as explained in Sec. 18-9 for the location of a buried metallic pipe. Sewers have been located by listening at one end of the pipe for the variation in intensity of sound produced by pounding on the ground surface or on a plank placed on the ground surface. The sewer should be found under the point where the maximum intensity of sound is produced.

18-9. Locating Buried Metallic Pipes. The unknown location of a buried metallic pipe can be found by methods similar to those described in Sec. 18-8 or possibly more easily by the use of electricity.[1]

One method works with the pipe as a part of a closed electric circuit, and in the other method an aerial picks up the magnetic field induced in the pipe by a radio transmitter or by previous excitation. In the former method, after the electric circuit has been completed by electrical con-

[1] See also Electronic Nose Finds Piping Leaks, *Plumbing, Heating, Air Conditioning*, March, 1953, p. 115.

tacts to two exposed points on the pipe, such as the point where it disappears through the building wall and a nearby fire hydrant or a neighbor's plumbing, the magnitude of the sound in earphones adjusted to receive electric currents and convert them to sound will increase as the pipe is approached and will fall silent over the pipe. In the latter method an electric dip needle points to the pipe. An illustration of the M scope, an example of the latter form of apparatus, is shown in Fig. 18-4.

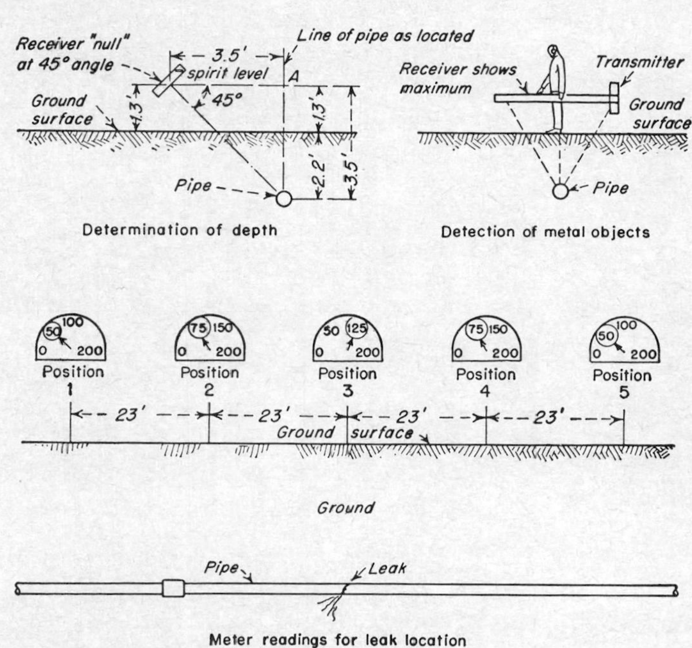

FIG. 18-4. Locating a buried pipe with an M scope. (*From Electronic Pipe Finding and Leak Locating, Water & Sewage Works, July,* 1946, *p.* R-101.)

18-10. Locating Leaks in Pipes. When the exact or approximate location of an underground pipe is known, a leak in a relatively short length of pipe, as in a water-service pipe, can be located by such direct methods of observation as the appearance of running water, melted snow, or green grass on a lawn or parking during a drought; the use of test rods thrust into the ground to find moisture; or sound-detecting instruments either transmitting the sound directly or amplifying it electrically. Instruments used include the aquaphone, water phone, dectaphone, sonograph, sonoscope, geophone, and Darley's leak locater. Procedures involving water hammer, following the hydraulic gradient, the addition of chemicals, volumetric displacement of water in the pipe, the addition of air or of coloring matter to the water in the pipe, and the use of radioactive tracers are all suitable to the purpose of locating a leak in a water pipe.

Unfortunately there is no correlation between the magnitude of a leak and the sound it makes. A small leak may be noisy; a large leak silent.

When direct observation fails to locate the leak, the first step may be the judicious manipulation of valves. If a leak can be controlled by a valve, the leak must be on the section of pipe controlled by that valve. Sometimes, where there are insufficient valves, the flow of water in a pipe can be stopped by freezing the water in the pipe as explained in Sec. 18-17. If the pipe is submerged, the addition of a tracer material to the water entering the pipe, such as common salt or fluorescein dye, may lead to the leak, or if air is pumped into the pipe, the rising bubbles will show the leak.

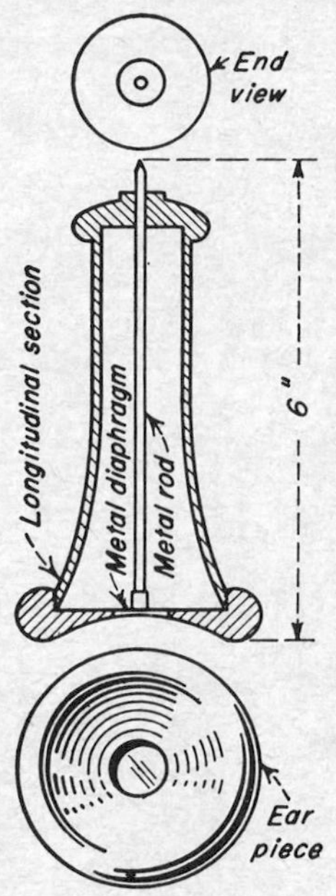

FIG. 18-5. Aquaphone.

The sound of a leak can sometimes be heard directly or with the aid of an instrument.[1] Devices used to detect or to amplify sound include a sounding rod, an aquaphone, a stethoscope, or an electronic sound amplifier. The proper part of the device is applied to the ear. Another portion is applied to the surface of the pipe and is moved along it toward the leak as the sound increases. Any convenient metal rod, plumbing tool such as a wrench or screwdriver, or an aquaphone, as shown in Fig. 18-5, can be used as a sounding rod. Electronic[2] devices are available that will amplify sound 10,000 times, excluding interfering sounds on a different pitch than the leak.

18-11. Repairing a Leaking Water Pipe. To repair a leaking water pipe under pressure when there is no shutoff valve it may be possible to unscrew the leaking portion of pipe and to replace it with a piece of suitable length on which a valve has been threaded. When threading the new piece of pipe into place, the plumber may stand behind a makeshift shield and put the pipe in place with the valve wide open. When the pipe is screwed in place, the valve is shut and repairs completed in the dry. In some cases a sleeve,[3] somewhat as shown in Fig. II-6 of Appendix II, can be used to stop a leak, whether or not the water has been shut off.

18-12. Leaks in Gas Pipes. Leaks in gas pipes can be located by observing the odor or by coating suspected points with soap solution or sputum. The appearance of bubbles will betray the leak. No open

[1] See also *Domestic Eng.*, April, 1957, p. 154.

[2] See also Electronic Nose Finds Pipe Leaks, *Plumbing, Heating, Air Conditioning*, March, 1953, and C. R. Fisher, Electronic Pipe Finding and Leak Locating, *J. Am. Water Works Assoc.*, December, 1946, p. 1330.

[3] See also *Air Conditioning, Heating, and Ventilating*, November, 1956, p. 134.

flame or acid should be used where an inflammable gas leak is suspected. Neither water nor air should be introduced into the gas line because of the danger of filling the gas pipes with water or creating an explosive mixture of gas and air. An inert gas such as nitrogen can, however, be introduced into pipe containing an inflammable gas and put under pressure to accelerate the leak. If the gas is not explosive, a small amount of Freon-12 can be placed in the pipe and searched for with a "Freon detector" in which the flame changes color when the leaking gas is found. Similarly ammonia can be mixed with the gas in the leaking pipe, and its appearance in the leak detected with the aid of a small amount of muriatic acid. The acid and the ammonia combine to produce a dense white cloud of ammonium chloride.

Gas leaks in pipes outside a building have been known to flow along the outside of the pipe for long distances, passing through a building wall with the pipe or entering a manhole or other enclosed space, with resulting fatalities. Where unusual difficulty is encountered in locating a gas leak, shut off the gas supply to the building, ventilate thoroughly, and subject the gas pipes to the air test described in Sec. 12-30.

18-13. Leaks in Water-closet Flush Valves. Either one of two conditions may occur when water runs continuously from a water-closet flush tank: (1) The float on the float valve is submerged more than it should be and water is escaping through the overflow. (2) The flush valve is not seated and water runs into the toilet bowl. In the first case examine the float to see that it is not waterlogged. If so, it should be replaced with a new float. If the inlet valve is leaking, a new seat or repacking may be called for. If both float and valve are in satisfactory condition, bend the bar holding the float so that the float will rise sooner and close the valve tightly. In the second case the rubber ball, the guide, or the flush valve seat needs adjustment or renewal.

18-14. Leaky Storage Tank. Leaks in wood tanks can sometimes be stopped without emptying the tank by throwing sawdust or oatmeal into the water near the leak. Small leaks in metal tanks can sometimes be controlled by the use of Smooth-on, a proprietary material which hardens, like cement, under water. If the leak is too large to be controlled, the tank can be emptied and the hole reamed out or enlarged sufficiently to pass a bolt through it with a washer on each end of the bolt. Tightening the nut on the end of the bolt compresses the washer and should stop the leak. Where the storage tank is closed, as in a hot-water storage tank, the bolt with gaskets can be lowered through a pipe opening in the top of the tank and pulled through the leaky hole by means of a hooked wire or other ingenious device.

18-15. Water-supply Pressure Is Low. If the difficulty lies in the public water supply and the condition is somewhat permanent, it will be necessary to install a pump to boost the pressure. If a pressure gage is placed on the water-service pipe immediately upstream from the meter and adequate pressure is shown in the service pipe when there is no flow in it, the trouble may be either in the service pipe or in the building. It is not due to inadequate pressure in the public water supply. Now, if the water is turned on through faucets in the building and the pressure drops abnormally, the service pipe is clogged or is inadequate in size. If the pressure does not drop on the gage and the difficulty persists, there is clogging or too small pipes in the building. The difficulty can be located by connecting a pressure gage at successive points on the piping in the building until the region of high pressure loss is located. The supply pipes in this region must be either cleaned or replaced with larger pipes.

18-16. Cleaning Water-service Pipes. Service pipes can be cleaned by either pushing or pulling through the pipe a scraper or cutting tool attached to a flexible metal tape or spring that is twisted by hand or by power as it passes through the pipe. A pneumatic tool is available for clearing clogged pipes which delivers a slug of compressed air at a pressure up to 60 to 80 psi, the air being compressed hydraulically.[1] In severe cases of stoppage the sudden release of a mixture of air and water at a pressure of over 1,000 psi has cleared the pipe without bursting it. In cleaning or replacing a service pipe it may be necessary to stop the flow in it by the insertion of a valve or by freezing.

18-17. Freezing to Stop Flow in Water Pipes. The flow of water through a pipe can be stopped temporarily where there is no valve by freezing the water in the pipe. To do this surround a short length of the pipe with dry ice wrapped in a cloth or wrap the pipe in an ice pack and pour on salt, ammonia, or other electrolyte to reduce the temperature. Water should not be flowing through the pipe at the time freezing is attempted. The section of pipe selected should be straight and long enough to allow expansion of the ice lengthwise in the pipe. The danger of bursting the pipe by this treatment is slight if the ice can expand, as explained in Sec. 18-18.

18-18. Thawing Frozen Water Pipes. Frozen water pipes burst because of the practically irresistible force of the expansion of water as it turns into ice. If the ice is allowed to expand as it forms, the pipe will not burst. This space for expansion may be provided lengthwise in the pipe. If the movement of the ice is obstructed, however, the pipe will burst. This may help to explain the occasional mystery of a burst frozen pipe in a warm kitchen. Pipes do not always burst at the spot where freezing occurs. Ice may form in the pipe at an exposed point. It will move along the pipe as it forms and, meeting an obstruction such as a valve, a bend, or a rough spot, will burst at that obstruction.

[1] See also *Air Conditioning, Heating, and Ventilating*, February, 1956, p. 115.

A frozen pipe can be thawed by applying cloths soaked in hot water, pouring hot water on the pipe, playing a jet of steam on it, injecting steam into it, applying an open flame to it, heating it by means of a charcoal fire, heating it by wrapping unslaked lime in a waterproof material around the pipe and pouring water into the lime, or by the use of electricity. All but the last method are messy and uncertain, and some are dangerous.

Thawing by electricity can be quick, inexpensive, and safe when properly done. When improperly done by untrained persons, it is dangerous. The procedure should be to connect one terminal of a welding machine to the pipe on each side of the frozen spot. The current and voltage are properly adjusted, and shortly after the current is turned on, the ice is melted and flow of water is resumed. The amount that the temperature is raised depends on the current I and the resistance R of the pipe and the length of time that the current is applied up to a maximum. A current of 100 amp will heat 1,000 ft of 1-in. pipe as rapidly as it will heat 10 ft of it, but it will take 100 times the voltage to do so. Some data concerning the factors involved are given in Table 18-1. These electrical factors can be formulated as

$$H = I^2Rt$$

where H = heat, joules
I = current, amp
R = resistance, ohms
t = time current is flowing, sec

Care must be taken to use the lowest amount of current suitable for the pipe material to avoid melting lead in pipe joints or other hazards of overheating and to use the lowest voltage necessary to avoid arcing, stray currents, and other hazards. The cost of the electricity used is in the order of a few cents.

18-19. Incorrect Temperature of Water. Cold water running from a faucet when it should be hot, or vice versa, may result from an interconnection between the two supplies, sometimes made for mixing purposes, and a stoppage or friction in one supply that forces flow into the other. The situation may be created by the restriction of flow in a shower head by a control valve downstream from a mixing valve. The remedy is to remove the stoppage, to increase the size of the lower-pressure supply pipes, to remove the offending mixing valve or faucet, or to revise the piping layout.

The trouble may be caused by the cold-water downcomer pipe in a hot-water storage tank. The downcomer may have come off or rusted off, or the vacuum-breaker or snifter hole in it may be so large as to permit cold water to pass into the hot-water supply without being heated. The remedy is to renew the downcomer pipe if it is the source of trouble. This can be determined best by inspection.

In any complicated system of piping and valves the layout should be diagrammed and studies made to determine the appropriate adjustments to avoid the mixing of the cold-water and hot-water supplies within the piping systems. Inadequate heating capacity may, of course, be the cause of inadequate hot-water supply.

TABLE 18-1. CONDITIONS AFFECTING THE THAWING OF FROZEN PIPES BY ELECTRICITY*

Diameter of pipe, in.	Pipe material	Resistance, ohms	Length, ft	Approx. volts†	Current, amp	Time, min	Size of leading wire to use	
							Amp	Size B & S
¾	Wrought iron	82.80	600	60	250	5	200	0
1	Wrought iron	......	600	60	300	10	225	00
1½	Wrought iron	......	600	60	350	10	275	000
2	Wrought iron	......	500	55	400	15	325	0000
3	Wrought iron	......	400	50	450	20		
4	Cast iron	684	400	50	500	60		
6	Cast iron	......	400	50	600	120		
8	Cast iron	......	300	40	600	240		
¾	Copper	9.35	400	40	500	30		
1	Copper	......	400	40	600	60		
1¼	Copper	......	300	35	600	60		
	Steel	63						
	Lead	123						

RELATIVE RESISTANCE OF PIPE METALS, EXPRESSED IN OHMS PER MIL FOOT OF A PIECE OF METAL 0.001 IN. IN DIAMETER AND 1 FT LONG

Metal....................	Copper	Steel	Wrought iron	Lead	Cast iron
Resistance, ohms..........	9.35	63.0	82.8	123.0	684

* From H. E. Babbitt and J. J. Doland, "Water Supply Engineering," 5th ed., McGraw-Hill Book Company, Inc., New York, 1955.

† The lowest possible voltage should be used.

18-20. Hot Water or Steam Comes from Cold-water Faucets. If hot water or steam issues from a cold-water faucet, a potentially dangerous condition exists. All water heaters should first be shut down. Some hot-water heater is out of control, and pressure is being relieved through the cold-water pipes. If this abnormal relief passage is blocked, there will be an explosion. If the hot water reaches the water meter, the meter will probably be injured by warping of the disk. The cause of the condition will probably be found in a stopped or frozen pipe, a valve closed that should be open, or other error in operation. If the system has functioned correctly at any time, there is probably nothing wrong with the piping layout; it is the equipment or the operation that has failed.

18-21. Air in Water Pipes. Air may enter plumbing from the public water supply as a result of the draining of the plumbing pipes, or it may have entered during a period of negative pressure siphoning air in through an open faucet or flush tank valve. Under any circumstances the unusual appearance of air in water pipes is indicative of an undesirable condition. The condition may be remedied by (1) inserting an air-relief valve at a high point in the piping system, (2) placing a frequently used faucet near the point of supply and concentrating the trouble there, (3) putting an aerating attachment on the faucet from which air is escaping, thus preventing splashing.

The condition must not be confused with dissolved air in water, discussed in Sec. 18-27.

18-22. Liming of Hot-water Heater. Liming of pipe coils or water backs occurs in hard-water regions. The condition manifests itself by low temperature of hot water, sluggish flow, and rumblings and crackling sounds, especially when water is drawn from the hot-water system. Liming can be prevented by such methods as (1) using only soft water in the hot-water system, (2) using soft water in the heater only, transferring the heat by means of a heat exchanger to the hard-water supply, and (3) adding softeners to the water supply. Methods of softening water are discussed in Sec. 17-7.

Lime can sometimes be removed from the system by circulating a solution of about 4 parts of water to 1 of muriatic acid through the pipes. The procedure is somewhat drastic, however, and damage may be done to the pipes by the acid. It may be better to remove the offending part of the system and soak it in the acid until improvement results or the part is replaced. Any attempt to scrape or to blow off the deposit is not likely to succeed.

18-23. Air Lock in Hot-water Pipe. An air lock may cause stoppage of circulation in a hot-water pipe. If no air valve or faucet is located at a high point in the hot-water pipe, proceed as follows: (1) Try to flush out the air lock by opening wide all faucets downstream from the high point, (2) study the piping system and by manipulation of valves and faucets reverse the direction of flow and flush out the air lock through a faucet, or (3) insert an air valve or a faucet at the highest point on the hot-water piping.

18-24. One of Two or More Hot-water Heaters Does Not Heat Water. When two or more heaters supply one or more storage tanks, the piping and valves may be complicated. If one of the heaters is not being used, cold water may be passing through the unused heater to mix with the hot-water supply. Check and readjust the valves, or insert a valve to control the condition.

18-25. Cold-water Pipes and Flush Tanks Drip or Sweat. The condition occurs when the temperature and humidity of the air are high and the temperature of the water supply is below the dew point. Remedies include wrapping or otherwise insulating the pipes or flush tank, as described in Sec. 7-20, running the cold-water pipe near a hot-water pipe, or running hot water into the flush tank.

18-26. Dirty Water or Red Water. If dirty water or red water is reported, the quality of the public supply or the source of the private supply should be checked. If the quality of the public water supply is unsatisfactory, the householder may treat it as described in Sec. 17-1 or 17-2 or by other means. If the public or private water supplies are of good quality, then corrosion or contamination is occurring within the building. Contamination may be entering at any point where the water surface is exposed to the air, as in a roof tank, an expansion tank, or a water-closet flush tank, or it may be the result of a cross connection. Discovery of the cause will suggest the remedy.

Corrosion is the most probable cause of red or discolored water. Hot water is more likely than cold water to be discolored by corrosion because of its greater corrosiveness. The remedy is to treat the water to prevent corrosion or to change plumbing materials to resist it.

The appearance of organic matter, sediment, or even living organisms in the water can be caused only by the quality of the water delivered to the building. Such difficulties sometimes arise in the distribution system after a high-quality filtered water has been put into it at the filtration plant. When such conditions appear, it is the responsibility of the waterworks authorities to overcome them.

18-27. Milky Water. The precipitation of dissolved oxygen from solution gives water a temporary milky appearance. The condition is harmless. It results from the solution of more air in the cold water under pressure than can remain in solution in the water exposed to the atmosphere at a warmer temperature. Hence, when water is drawn from a faucet its pressure is reduced and its temperature may be raised. Since the condition is harmless and disappears quickly, no remedy is necessary.

A milky appearance due to suspended solids can be remedied only by treatment of the water. This may consist of chemical precipitation, sand filtration, filtration through activated carbon, or combinations of these methods. A remedy may be required, as the condition is not necessarily harmless.

18-28. Prevention of Noises.[1] Causes of noises in plumbing systems are discussed in Sec. 7-17. Methods for the elimination of noise may be summarized as follows:

1. Hot-water heaters rumble or rattle. Clean clogged pipes, renew heating coil, avoid too rapid injection of cold water into the heater or storage tank, readjust the proportionate capacities of the heater and the storage tank to avoid overheating of water in either, readjust the rate of use of hot water by manipulation of a valve on the supply to the hot-water faucet, lower the temperature of the hot-water supply.

2. Hot-water pipes hiss or crackle. Clean clogged pipes, remove air binding by realigning and resupporting the pipe, insert rubber or felt between pipe and supports or surfaces in contact, tighten loose joints.

3. Pipes hiss or emit musical noises. The cause may be cavitation. If so, try to increase the internal pressure of water in pipe, eliminate sudden pressure drops in piping by increasing the size of pipe or equipment.

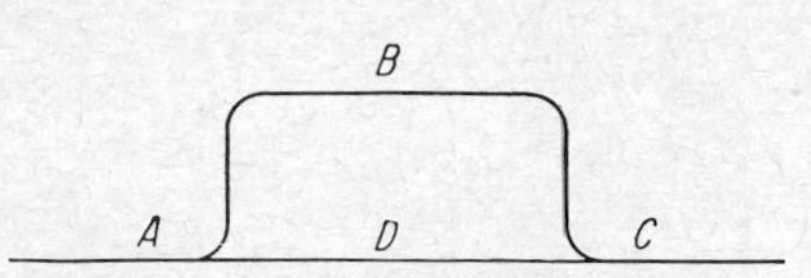

FIG. 18-6. A sound bypass.

4. Sound waves are liquid-borne. Allow the stream to pass through a short length of rubber hose, through an enclosed chamber partly filled with flexible, hollow, air- or gas-filled spheres, or through a chamber packed with steel wool.

5. Sound waves are borne by pipe walls. Connect a bypass pipe, as at B in Fig. 18-6, so that sound waves traveling from C to A along B and D arrive at A out of phase and nullify each other.

6. Pipes rattle or pound owing to movements in supports or between contact surfaces: (*a*) Insert silencing material such as rubber, felt, or steel spring between the pipe and its support or contact surface. (*b*) Tighten or renew loose connections or supports. (*c*) Lubricate the support or contact surface with heavy grease. (*d*) Avoid high temperatures or large changes in temperature of discharges into drainage pipes, or otherwise reduce the magnitude of changes of temperature in discharge pipes.

7. Water hammer is the cause. Remedies are discussed in Sec. 7-34.

8. Vibrations are being transmitted to a pipe from other pipes, supports, or other sources. Tighten supports to stop vibrations. Insert a flexible hose[2] in the pipeline.

[1] See also W. L. Rogers, Noise Production and Damping in Water Piping, *Heating, Piping, Air Conditioning,* January, 1956, p. 181; F. W. McGhan, *Plumbing and Heating J.,* January, 1952, p. 48; and W. L. Rogers, Experimental Approach to the Study of Noise and Noise Transmission in Piping Systems, *Trans. Am. Soc. Heating Ventilating Engrs.,* vol. 59, p. 347, 1953, and vol. 62, p. 39, 1956.

[2] See also *Domestic Eng.,* February, 1957, p. 148.

9. Valves or equipment hiss or sing. Replace noisy equipment, or increase the size of the valve and pipes to reduce velocity. Keep the velocity of cold water less than 10 fps and of hot water less than 4 fps.

11. Noise originating in large pipelines can be reduced by wrapping the pipe. This treatment is ineffective on relatively small household piping.[1]

18-29. Noises in Flushing. The noises in the operation of a water closet may be due to any one or a combination of causes, including the operation of the water-supply valve, the sucking sound due to siphonage, and the splashing of water. All can be suppressed to give almost silent operation, but subsequent loss of adjustment may restore the noise. Other factors entering into the production of the noise are small, high-velocity streams, rattling or singing of washers, and similar mechanical difficulties. Among the prolific causes of noise, and the producer of the most penetrating sounds sometimes heard throughout a building, is the slow closing of the water-supply valve which releases a small stream at a high velocity, resulting in a piercing whistle. The closing of a flushometer valve may result in a similar noise. Adjustment of the valves should overcome the difficulty.

The sucking noise made by the siphon and the splashing of water cannot be stopped completely in operation because it is usually inherent in the design of the fixture. It can, however, be suppressed to some extent by lowering the seat cover during the flush.

18-30. Odors. Odors from plumbing pipes result most frequently from the failure of a trap seal, breakage of a drainage pipe, the location of an opening in a drainage pipe too close to an opening into the building, and the exposure of too great a length of drain pipe between a fixture and its trap. Locating the source of sewage odors in occupied buildings is sometimes extremely difficult because of the subtle nature of odors and their peculiar travel habits. If the source of the odor is not immediately obvious, the plumber should suspect a fixture trap, floor drain, or other infrequently used trapped drainage connection. Other possibilities listed above should then be investigated. Resourcefulness, observation, and ingenuity must be combined sometimes to locate the sources of odors. A dead animal in the walls of the building may be causing the trouble. In one building a concealed opening left in a soil stack at the time of its installation was discovered by following the droppings of rats that had been using the opening as a path to and from the sewer.

18-31. Condensation of Water on Walls. The condensation of moisture on walls results when warm, humid air strikes walls whose temperature is below the dew point. This occurs in basements, kitchens, laun-

[1] See also C. M. Harris, "Handbook of Noise Control," McGraw-Hill Book Company, Inc., New York, 1957.

dries, bathrooms, and similar enclosures. Remedies include the installation of an electric dehumidifier or a dehumidifier using calcium chloride. The latter requires periodic attention to renew and possibly to dry the chemical. The former must have drainage to a floor drain or to a receptacle requiring periodic emptying. The blowing of warm air against the cold surfaces on which condensation is taking place or otherwise warming these surfaces or blowing the vapors out of the room before they condense on the cold walls should overcome the difficulty. Improving the ventilation of the room by diluting the humid air in it with cooler and drier air from the outside may be helpful.

18-32. Water in the Basement. Water in the basement may result from backflow from the public sewer flowing into the foundation drain and thence through the walls or floor of the building, from the breaking or the inadequate capacity of rain-water leaders or gutters, because of an inadequate yard or area drain, or from seepage of ground water through the building walls.

Remedies may be to install or to repair the backflow valve on the building foundation drain, to repair or to increase the capacity of the rain-water drainage facilities or the yard and area drains, to reroute the drainage channels, to waterproof the building walls, or to install foundation drains.

18-33. Soot-clogged Flues. Coal-heater, gas-heater, or other flues clogged with soot can be sprayed with a prepared flammable liquid[1] which, when fired, burns off the soot.

18-34. Closing a Building. When a building is to be closed for a long period of time, precautions must be taken against the evaporation of trap seals and against the freezing of water in pipes and fixtures. All traps should have the water in them replaced by a nonvolatile liquid such as kerosene oil. Salt can be placed in traps made of resistant materials. Unfortunately salt may penetrate some ceramic materials to stain the exterior and it may hasten the corrosion of some metals. It is, therefore, unwise to use salt to form an antifreeze solution unless it is known that no harm will be done. Other antifreeze solutions may be satisfactory provided their rate of evaporation is low. Supply pipes and storage tanks should be drained. The supply pipes are usually drained through the stop-and-waste valve located upstream from and adjacent to the meter in the basement. The operation of this valve is described in Sec. 14-27. The opening of faucets on upper floors will cause water to drain from the pipes through the stop-and-waste valve. Globe valves on horizontal lines may prevent complete drainage. However, sufficient air space may be left in the pipe to permit the formation and expansion of ice without damage to the pipes.

[1] See also *Domestic Eng.*, August, 1956, p. 220.

Water in hot-water storage tanks, pneumatic storage tanks, expansion tanks, floor tanks, and other receptacles must be drained from them at the points provided therefor. Care must be taken to see that special equipment, such as pumps, washing machines, and other water-holding devices, are drained. A few drops of water confined within the casing of a pump may freeze and crack the casing.

In reopening a building that has been closed, it is not sufficient to close all water outlets and to open the main water-supply valve in the basement. Such a procedure would entrap air in the pipes and closed tanks, with resulting water hammer as air escapes when faucets are opened. The main water-supply valve should be opened slowly, with faucets and drains throughout the building open. As the water rises in the building pipes and appears at the faucets, they should be closed progressively upward in the building. Air should be allowed to escape from closed hot-water storage tanks, the tanks of the water-softening equipment, and other receptacles in which air may be trapped, except air chambers provided for the prevention of water hammer.

18-35. Some Short Cuts in Repairs and Maintenance. Ingenuity and resoucefulness in repairs and maintenance may be exercised by (1) using a rubber ball, such as a fuller ball, to pack a valve with a poor seat; (2) coating pipes with shellac or varnish before painting to prevent subsequent peeling of paint from hot surfaces or from the penetration of tar on cast-iron pipes; (3) starting a screw stuck in metal by heating the head of the screw with a blowtorch or by holding a hot iron on the screw. If the head of the screw or the cover that is stuck is large enough, put dry ice on it but not on its surroundings. A screw stuck in wood can sometimes be started by heating it and allowing it to cool before starting. (4) A leak in a metallic ball float can sometimes be detected by steam escaping from the leak when the ball is gently heated. (5) To stop leaks in metallic tanks use some of the available antileak compounds used for automobile radiators, or if the tank is of wood or metal, use a substance, such as oatmeal, sawdust, or flaxseed, that will penetrate the leak and expand as it becomes water soaked.

APPENDIX I

GLOSSARY OF PLUMBING TERMS[1]

Air Conditioning. The process of treating air to control its temperature, humidity, cleanliness, and distribution to meet the requirements of the conditioned space.

Air Gap. The air gap in a water-supply system is the unobstructed vertical distance through the free atmosphere between the lowest opening from any pipe or faucet supplying water to a tank or plumbing fixture and the flood-level rim of the receptacle. See *A* in Fig. 1-5. Definition from ASA and NPC.

Air-lock. Air, vapor, or other gas entrapped between two liquid surfaces in a conduit or liquid container so that the flow of the liquid is impeded or stopped by the entrapped gas. Also called a *vapor lock.*

Alligator Wrench. A wrench with toothed, V-shaped jaws, fixed in position.

Backflow. The flow of water into a water-supply system from any source other than its regular source. NPC states:

> Backflow is the flow of water or other liquids, mixtures or substances into the distributing pipes of a potable supply of water from any source or sources other than its intended source.

Back-outlet Ell and Branch Ell. An ell with the outlet in the same place as run and on outside of curve. See also *heel outlet* in Fig. II-42.

Back Pressure. Air pressure in drainage pipes greater than atmospheric pressure.

Back Siphonage. Air pressure in plumbing pipes that is less than atmospheric pressure. NPC states:

> Back siphonage is the flowing back of used, contaminated, or polluted water from a plumbing fixture or vessel into a water-supply pipe due to a negative pressure in such pipe.

Back Vent or Back Venting. Sometimes synonymous with *vent* or *venting;* a pipe or system of pipes connected to the drainage pipes of a plumbing system for the purpose of supplying or removing air to relieve pressures above or below atmospheric. A back vent is that part of a vent line which connects directly with an individual trap beneath or behind a fixture and extends to a branch or main soil or waste pipe at any point higher than the fixture or fixture trap it serves (see Figs. 11-20 and I-1). This is sometimes called an *individual vent.*

Ball Cock. A faucet opened or closed by the fall or rise of a ball floating on the surface of water whose elevation is controlled wholly or in part by the faucet.

Ball Joint. A connection in which a ball is held in a cuplike shell that allows movement in any direction in the joint other than along the axis of the pipes that are joined.

[1] The "National Plumbing Code" is referred to herein as NPC, and the American Standards Association as ASA. Illustrations of some of these definitions are shown in Fig. I-1.

Bell or Hub. That portion of a pipe which, for a short distance, is sufficiently enlarged to receive the end of another pipe of the same diameter for the purpose of making a joint.

Bending Pin or Bending Iron. A tool used for straightening or expanding flexible pipe, especially lead pipe.

Bibb. Synonymous with *faucet, cock, tap, plug,* etc. The word *faucet* is preferred.

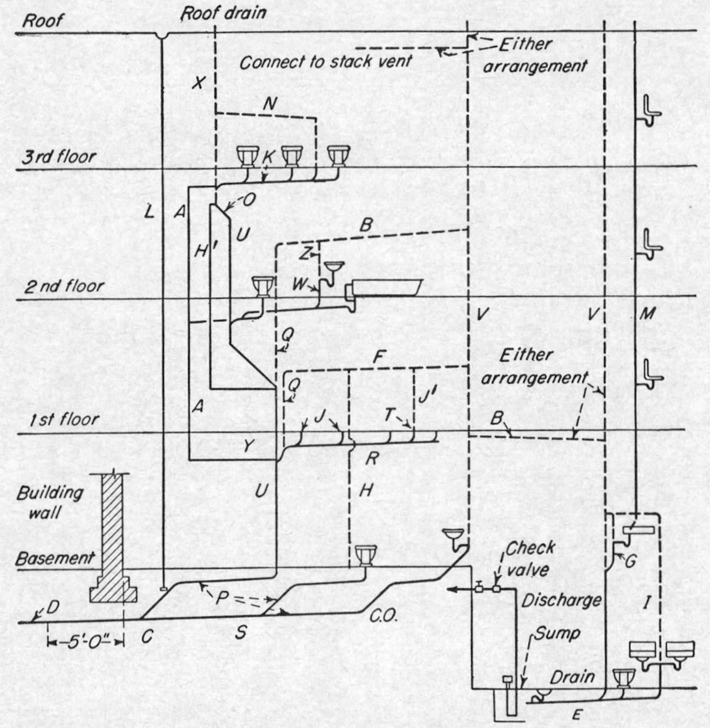

FIG. I-1. Illustration of definitions. (*A*) Branch interval. (*B*) Branch vent. (*C*) Building drain. (*D*) Building sewer. (*E*) Building subdrain. (*F*) Circuit vent. (*G*) Continuous waste and vent. (*H*) Dry vent. (*H'*) Double offset. (*I*) Dual vent (unit vent). (*J*) Fixture drain. (*J'*) Group vent. (*K*) Horizontal branch. (*L*) Leader. (*M*) Indirect waste. (*N*) Loop vent. (*O*) Offset. (*P*) Primary branch. (*Q*) Relief vent. (*R*) Return offset or jump-over. (*S*) Secondary branch. (*T*) Side vent. (*U*) Soil stack. (*V*) Vent stack. (*W*) Wet vent. (*X*) Stack vent. (*Y*) Yoke vent. (*Z*) Back vent. (*From "Plumbing Manual," Report BMS66, Subcommittee on Plumbing, Central Housing Committee, National Bureau of Standards,* 1940, *p.* 26.)

Bidet. A plumbing fixture used for washing the posterior of the body, especially the genitals (see Fig. 16-9).

Bitransit waste. A standing overflow (see Fig. 16-1).

Blank Flange. A flange without bolt holes. Sometimes also a *blind flange.*

Blind Flange. A flange without opening for the passage of water. It closes the end of a pipe (see Fig. II-9).

Blind Vent. A fixture vent terminating in a wall or in such a manner that only the appearance of a vent is provided.

Block Tin. Pure tin.

Blowoff. A controlled outlet from a pipeline used to discharge water, steam, vapor, sludge, or other fluid waste.

Bonnet. That portion of a gate valve into which the disk rises when the valve is opened (see Fig. 14-10).

Bossing Stick. A wooden tool for shaping lead for tank lining.

Box Union. A device for joining two threaded pipes that can be opened or separated without dismantling the pipes (see Figs. II-56 and II-57).

Branch. A pipe in a plumbing system into which no other branch pipes discharge. The branch pipe discharges into a main or submain. The NPC states that a branch discharges into a "riser or stack."

Branch Interval. A length of soil or waste stack corresponding, in general, to the height of one story, but not less than 8 ft, within which the horizontal branches from one floor or story of a building are connected to a stack (see Fig. I-1).

Branch Vent. "A vent connecting one or more individual vents into a vent stack or stack vent" (from NPC). Also a vent pipe to which two or more pipes venting fixtures are connected (see Fig. I-1).

Braze (verb). To solder with any alloy that is relatively infusible compared with common solder.

Building Drain. That part of the lowest horizontal piping of a plumbing drainage system which receives the discharge from soil, waste, and other drainage pipes within the building and conveys the wastes to the building sewer. Synonymous with *house drain* (see Fig. I-1).

Building Sewer. That part of the horizontal piping of a plumbing drainage system that extends from the end of the building drain to the public sewer or other outlet. It receives the discharge from the building drain (see Fig. I-1).

Building Storm Drain. NPC states:

> A building drain used for conveying rain water, surface water, ground water, subsurface water, condensate, cooling water, or other similar discharge to a storm sewer or a combined building sewer, extending to a point less than 3 ft outside of the building wall.

Building Storm Sewer. NPC states:

> The extension from the building storm drain to the public storm sewer, combined sewer, or other point of disposal.

Building Subdrain. That portion of a building drainage system which cannot drain by gravity into the building sewer (see Fig. I-1).

Building Trap. NPC states:

> A device, fitting, or assembly of fittings installed in the building drain to prevent circulation of air between the drainage system of the building and the building sewer.

See also *house trap* and *main trap* (see Fig. 11-1 at 76).

Bull-headed Tee. A tee in which the branch is larger than the run.

Burr. Roughness or extra metal protruding from the walls of a pipe, often as a result of improper finishing in the cutting of a pipe.

Bushing. A plug designed to be threaded into the end of a pipe. The plug is bored and tapped to receive a pipe of smaller diameter than that of the pipe into which the bushing is screwed.

Bypass. Any method which will permit water or other fluid to pass around a valve, fixture, appliance, connection, or pipe. Applied also to a connection between a drain pipe and a vent pipe that will provide a passage for sewer air to enter the building.

Bypass Vent. A vent stack parallel to a soil or waste stack with frequent connections at branch intervals between the two stacks (see Fig. 11-27).

Caliber. Internal diameter or bore.

Calking. Plugging an opening with oakum, lead, or other material that is pounded, tapped, or pushed into place. Also the material that is used to plug the opening.

Calking Recess. A counterbore or recess in the back of a flange into which lead can be calked for water pipes and similar connections.

Cap. A fitting into which the end of a pipe is screwed for the purpose of closing the end of the pipe (see Fig. II-52*b*).

Catch Basin. A receptable in which liquids are retained for a sufficient time to permit the deposition and retention of sediment in the basin (see Fig. 11-43).

Cesspool. A pit for the reception or detention of sewage. Sometimes called a *dry well*, especially when of relatively small diameter and large depth. Distinguished from a septic tank by the fact that water or sewage does not enter and leave the cesspool at the same time and rate.

Chain Tongs. A tool used to hold pipe from turning or to turn the pipe. It consists of a heavy bar with sharp teeth at one end. These teeth are held firmly impressed in the pipe by means of a chain wrapped around the pipe and attached to the bar (see Fig. 7-8).

Chase or Pipe Chase. A recess in a wall for the purpose of holding pipes and other conduits passing between floors in a building.

Check Valve. A valve that automatically closes to prevent the flow of fluid in a reverse direction (see Fig. 14-18).

Chipping Knife. A knife used for chipping or cutting lead.

Circuit Vent. NPC states:

> A branch vent that serves two or more traps and extends from in front of the last fixture connection of a horizontal branch to the vent stack.

See also *loop vent* (see Fig. I-1).

Close Nipple. A pipe fitting with outside threads used for connecting two pipes. The length of the threads and of the fitting are the shortest permissible by standard practice.

Close Return Bend. A short cast or malleable U-shaped pipe fitting with arms joined together, as shown in Figs. II-56 and II-57.

Closet Bolt. A bolt used for fastening the bowl of a water closet to the floor.

Closet Screw. A long screw with a detachable head, used for fastening a water-closet bowl to the floor.

Cock. See *faucet.*

Collar. A sleeve in back of a flange.

Combination Fixture. A fixture combining a kitchen sink and a laundry tray, with one or more units of each, into a single unit (see Fig. 16-25).

Common Vent. Same as *unit vent* and *dual vent* (see Fig. I-1).

Companion Flange. A flange drilled according to a standard that will fit the standard drilled holes in a flanged pipe or fitting (see Fig. II-9).

Compression Faucet or Valve. A faucet or valve in which the flow of water is shut off by means of a flat disk (either with or without packing) that is screwed down onto its seat (see Fig. 14-6).

Conductor. A vertical pipe to convey rain water, usually applied to such pipes when inside a building. See also *downspout* and *leader* (see Fig. I-1).

Continuous Vent and Continuous Waste-and-vent. A continuation of the drain to which the vent connects. A continuous waste-and-vent is further defined by the angle that the drain and vent at the point of connection make with the horizontal, for example, vertical continuous waste-and-vent, 45° continuous waste-and-vent, and flat (small-angle) continuous waste-and-vent (see Sec. 11-15 and Figs. 11-20 and I-1). NPC states:

> A continuous vent is a vertical vent that is a continuation of the drain to which it is connected.

Continuous Waste. A waste from two or more fixtures connected to a single trap (see Figs. 11-5 and 11-6). NPC states:

> A drain from a combination fixture or two or three fixtures in combination connected to a single trap.

Copper Bit. A tool used for soldering. Usually called a *soldering iron.*

Corporation Cock. A valve placed in a water-service pipe close to its connection with a water main. It is sometimes placed in the parking between the curb and sidewalk (see Fig. 14-15).

Coupling. A pipe fitting with inside threads only used for connecting two pipes (see Fig. II-55).

Cowl. A hood on the top of a vent pipe or soil stack (see Fig. 12-6).

Cross. A pipe fitting used for connecting four pipes at right angles (see Figs. II-51, II-58, and others).

Cross Connection. NPC states:

> Any physical connection or arrangement between two otherwise separate piping systems, one of which contains potable water and the other water of questionable safety, whereby water may flow from one system to the other, the direction of flow depending on the pressure differential between the two systems.

Sometimes called an *interconnection* but not to be confused with a *crossover.*

Crossover. A connection between two pipes in the same water-supply system or between two water-supply systems containing potable water.

Crossover Fitting. A small pipe fitting like a double offset or the letter U with ends turned out. Available only in small sizes and used to pass one pipe over another when they meet in the same plane (see Fig. II-56).

Crown. The crown of a trap. That part of the trap in which the direction of flow is changed from upward to downward (see Fig. 1-3).

Crown Vent. A vent pipe connected at the topmost point in the crown of a trap (see Fig. 11-25).

Crown Weir. The highest part of the inside portion of the bottom surface at the crown of a trap (see Fig. 1-3).

Cup Joint. A lead-pipe joint in which one end of the pipe is opened enough to receive the tapered end of the adjacent pipe.

Curb Box. A device usually consisting of a long piece of pipe or tubelike casing placed over a curb cock through which a key is inserted to permit the turning of the curb cock. Sometimes called *Buffalo box* (see Fig. 1-4).

Curb Cock. A valve placed in a water-service pipe usually at a point near the curb (see Figs. 1-4 and 14-16).

Dead End. NPC states:

> A branch leading from a soil, waste, or vent pipe, building drain, or building sewer, which is terminated at a developed distance of 2 ft or more by means of a plug or other closed fitting.

The extended portion of a pipe that is closed at the end opposite its connection to another pipe, pump, fixture, or other device.

Deep-seal Trap. A trap with a seal of 4 in. or more.

Developed Length. The length along the center line of the pipe and fittings.

Die. A tool for cutting pipe threads.

Dip of a Trap. The lowest portion of the inside top surface of the channel through a trap (see Fig. 1-3).

Direct Cross Connection. A continuous, enclosed interconnection or cross connection such that the flow of water from one system to the other can occur under the slightest pressure differential between the two piping systems.

Domestic Sewage. Same as *sanitary sewage.*

Dope. A compound with a pastelike consistency used on pipe threads when making a joint to lubricate the threads and assure a tight joint.

Double-bend Fitting. A pipe fitting shaped like the letter S (see Fig. II-31).

Double Offset. NPC states:

> Two changes in direction installed in succession or series in continuous pipe.

See *H'* in Fig. I-1.

Double Waste and Vent. See *unit vent.*

Downcomer. A pipe in which the flow is substantially downward.

Downspout. The vertical portion of a rain-water pipe. See also *leader* and *conductor.*

Drain. NPC states:

> Any pipe which carries waste water or water-borne wastes in a building drainage system.

Drainage Fitting. A cast-iron, threaded fitting used on drainage pipes. A distinctive feature is the shoulder against which the end of the connecting pipe rests so as to present a smooth and continuous interior surface. Sometimes called a *Durham fitting.*

Dresser. A tool used for straightening pipe, usually lead pipe and sheet lead.

Dresser Joint. See *Normandy joint.*

Drift. To drive a wooden plug through a soft-metal pipe, such as lead, to remove dents.

Drift Plug. A plug used in drifting.

Drop Ell. An ell with lugs in the sides by means of which it can be attached to a support (see Fig. II-56).

Drop Tee. A tee with lugs in the sides by means of which it can be attached to a support (see Fig. II-56).

Drum Trap. A trap consisting, substantially, of a cylinder with its axis vertical. The cylinder is larger in diameter than the inlet or outlet pipe, and it is usually about 4 in. in diameter with 1½-in. inlet and outlet pipes (see Fig. 11-39).

Dry Vent. A vent that does not carry water or water-borne wastes (see Fig. I-1 at *H*).

Dry Well. Usually synonymous with *cesspool.* Also a well constructed similar to a water well but intended to receive and dispose of sewage by dispersion into and absorption by surrounding underground materials.

Dual Vent. See *unit vent.*

Durham System. NPC states:

> Soil or waste systems where all piping is of threaded pipe, tubing, or other such rigid construction, using recessed, drainage fittings to correspond to the types of piping.

Dutchman. A lead nipple, not more than about 1 in. long, that is placed in a wiped joint to make up the desired length in joining two pipes which are too short.

Eccentric Fitting. A fitting in which the center line of the run is offset in the fitting.

Effective Opening. "The minimum cross-sectional area at the point of water-supply discharge measured or expressed in terms of . . . the diameter of a circle or if the opening is not circular the diameter of a circle of equivalent cross-sectional area." An ASA definition. Also NPC (see Fig. 1-5 and Table 1-4).

Elbow. A fitting joining two pipes at an angle (see Fig. II-56).

Ell. Same as *elbow*.

Escutcheon. A flange used on a pipe to cover a hole or opening in a floor or well through which the pipe passes (see Fig. 7-12).

Faucet. A valve on the end of a water pipe by means of which water can be drawn from or held within the pipe (see Fig. 14-1).

Female Thread. A thread on the inside of a pipe or fitting. Preferably called an *inside thread*.

Ferrule. A metallic sleeve, calked or otherwise, joined to an opening in a pipe into which a plug is screwed that can be removed for the purpose of cleaning or examining the interior of the pipe.

Finishing. Work done after the roughing-in.

Fittings. Parts of a pipeline other than straight pipe or valves, such as couplings, elbows, tees, unions, and reducers (see Fig. II-56).

Fixture. A receptacle attached to a plumbing system other than a trap in which water or wastes can be collected or retained for use and for ultimate discharge into the drainage pipes of a plumbing system.

Fixture Branch. "A pipe connecting several fixtures" (from NPC).

Fixture Drain. "The drain from a trap on a fixture to the junction of the drain with any other drain pipe" (from NPC).

Fixture Supply. "A water-supply pipe connecting the fixture branch" (from NPC).

Fixture Vent. A vent pipe leading from the drainage pipe from a fixture to the atmosphere or to another vent pipe.

Flange Union. A pair of flanges to be threaded onto the ends of pipes to be joined by bolts holding the flanges together (see Fig. II-56).

Flashing. A piece of sheet metal fitted under another piece of flat metal or other material, usually roofing, over the surface of which water is expected to run (see Fig. 12-5).

Flood Level or Flood-level Rim. "The top edge of the receptacle from which water overflows" (from NPC, see Fig. 1-5).

Flush Bushing. A bushing without a shoulder that fits flush into the fitting with which it is connected.

Flush valve. A special form of valve used in a flush tank for the purpose of controlling the flushing of a fixture or fixtures (see Fig. 15-7). Sometimes confused with *flushometer valve* or *flushometer*.

Flushometer or Flushometer Valve. NPC states:

> A device which discharges a predetermined quantity of water to fixtures for flushing purposes and is actuated by direct water pressure.

See Fig. 14-43 and Sec. 14-28.

Flux. Material used to aid in making solder flow and to prevent the oxidation of the materials to be soldered during the process of soldering.

Follower. Part of a threading tool that keeps the thread straight (see *J* in Fig. 7-8).

Fresh-air Inlet. A connection made to a house drain above the main trap leading to the outside air (see 79, 164, and 167 in Fig. 11-1).

Frostproof Closet. A toilet " . . . hopper that has no water in the bowl and has the trap and control for its water supply installed below the frost line" (from NPC, see Fig. 15-5 and Sec. 15-19).

Fuller Faucet. A faucet in which the flow of water is controlled by means of a rubber ball that is forced into the opening in the pipe (see Fig. 14-38).

Gasket. Packing of any material placed between two metal or similar surfaces that are to be drawn together in a watertight or gastight joint.

Gate Valve. A valve in which the flow of water is controlled by means of a circular disk fitting against and sliding on machine-smoothed faces, the motion of the disk being at right angles to the direction of flow. The disk is raised or lowered by turning a threaded stem connected to the handle of the valve. The opening of the valve is usually as large as the full bore of the pipe (see Fig. 14-10).

Globe Valve. A valve in which the flow of water is controlled by means of a circular disk which is forced against or withdrawn from an annular ring, known as the seat, which surrounds the opening through which water flows in the valve. The direction of movement of the valve is parallel to the direction of flow of water through the valve opening and normal to the axis of the pipe to which the valve is connected (see Fig. 14-12).

Gooseneck. A return-bend or small-size pipe one end of which is about 1 ft long and the other about 3 in. long. It is commonly used as a faucet for a pantry sink (see Fig. 14-7). Also the lead or similar flexible connection between a water-service pipe and a water main (see Figs. 1-4 and 3-2).

Ground Joint. A machined metal joint that fits tightly without gasket or packing.

Ground-key Valve. A valve or faucet through which the rate of flow of water is controlled by means of a circular plug or key that fits closely in a cylindrical or conical machine-ground seat. The axis of the plug is normal to the direction of the flow of water. The plug has a hole or passageway bored through it as a waterway (see Fig. 14-14).

Ground Water. Water that is standing in or flowing through the ground.

Group Vent. A branch vent that performs its functions for two or more traps (see Figs. 11-13 and I-1).

Hatchet Iron. A special form of soldering iron.

Header. A pipe of many outlets. The outlets are parallel and are frequently at 90° to the center line of the header. See also *manifold* (see Fig. 7-9).

Horizontal Branch. NPC states:

> A branch drain extending laterally from a soil or waste stack or a building drain, with or without vertical sections or branches, which receives the discharge from one or more fixture drains and conducts it to the soil or waste stack or to the building drain.

See Fig. I-1.

Horizontal Pipe. NPC states:

> A pipe that is installed in a horizontal position or that makes an angle of less than 45° with the horizontal.

House Drain. That part of the lowest horizontal piping of a drainage system which receives the discharge from soil, waste, and other drainage pipes within a building and conveys it to the building or house sewer (beginning 3 ft outside the building wall, according to NPC. See Fig. I-1). Same as *building drain.*

House Sewer. Same as *building sewer.*

House Slant. A tee or wye connection in a sewer for the purpose of connection to the house sewer.

House Trap. Same as *building trap* (see Fig. 11-1 at 76).

Hub. The enlarged end of a pipe made to provide a connection into which the end of the joining pipe fits (see Fig. II-1). Synonymous with *bell.*

Hydrant. A valve or faucet for drawing water from a pipe. The term is usually applied to an outside installation for supplying a relatively large quantity of water for sprinkling, watering, fire protection, and similar purposes (see Fig. III-19).

Hydronics. A coined word meaning the art or practice of heating and cooling with water.

Increaser. A coupling with one end larger than the other. Sometimes, more specifically, a pipe fitting to join the end of a small coupling with an inside thread to the end of a larger pipe with outside threads. The reverse of a *decreaser* (see Figs. II-44 and II-56).

Indirect Cross Connection. A potential cross connection such that the interconnection is not continuously enclosed and the completion of the cross connection depends on the occurrence of one or more abnormal conditions.

Indirect Vent. NPC states:

> A pipe installed to vent a fixture trap and which connects with the vent system above the fixture served or terminates in the open air.

See also *back vent* (see Fig. I-1).

Indirect Waste Pipe. NPC states:

> A pipe that does not connect directly with the drainage system but conveys liquid wastes by discharging into a plumbing fixture or receptacle which is directly connected to the drainage system.

See 24 and 116 in Fig. 11-1 and Fig. I-1.

Interceptor. NPC states:

> A device designed and installed so as to retain deleterious, hazardous, or undesirable matter from normal wastes and permit normal sewage or liquid wastes to discharge into the disposal terminal by gravity.

See Fig. 11-43.

Interconnection. A cross connection. Not to be confused with *crossover.*

Invert. The lowest portion of the inside of any horizontal pipe.

Inverted Joint. A fitting reversed in position, upside down, or turned in a contrary direction.

Joint Runner. An incombustible type of packing generally used for holding lead in a bell in the pouring of a joint (see Fig. 7-17).

Journeyman Plumber or Journeyman. A plumber who does plumbing work for another for hire.

Jumpover. See *return offset* (see Figs. I-1 and II-56).

Lap Weld. A weld in which two metallic surfaces are connected by lapping one over the top of the other. Frequently used in making small-size iron pipe from sheet metal.

Lateral. In plumbing, a secondary pipeline. In sewerage, a common sewer to which no other common sewer is tributary. It receives sewage from building sewers only.

Latrine. A water closet consisting of a continuous trough containing water. The trough extends under two or more adjacent seats.

Lavatory. A fixture designed for the washing of the hands and face. Sometimes called a *wash basin* (see Fig. 16-3).

Leaching Cesspool. A cesspool that is not watertight.

Lead Burning. Welding lead.

Lead Tacks. Pieces of lead that are soldered to lead pipe so that it can be attached to a support.

Lead Wool. Shredded lead. Used in packing lead joints.

Leader. A pipe to convey rain water from the roof to the building storm drain, combined building sewer, or other means of disposal. See also *conductor* and *downspout.* The leader is usually outside the building served (see Figs. 13-6, 13-7, 13-9, and I-1).

Length of Pipe. The length as measured along the center line.

Local Vent. A pipe or conduit to convey foul air from a plumbing fixture, room, or other space, to the outer air. NPC states:

> A local ventilating pipe is a pipe on the fixture side of the trap through which vapor or foul air is removed from the room or fixture.

See Figs. 15-1, 15-2, and 15-3.

Long Screw. A nipple 6 in. long with one thread 6 in. longer than the standard thread.

Loop or Circuit Vent. A vent pipe connected to a horizontal drainage pipe receiving the discharges from one or more otherwise unvented fixtures. The vent pipe rises above the overflow level or flood rim of the highest connected fixture connected to the vented drainage pipe, and the vent pipe is connected to a vent stack. If the vent pipe connects to the same vent stack or stack vent into which the vented fixtures discharge, the vent is a loop vent. If the vent pipe connects to some other vent stack or stack vent than that into which the vented fixtures discharge, it is a circuit vent. NPC states:

> A loop vent is the same as a circuit vent except that it loops back and connects with a stack vent instead of a vent stack.

See Fig. I-1.

Main Trap. See *building trap* (see Fig. 11-1 at 76).

Main Vent. A vent pipe to which branch and fixture vents are connected (see *H* in Fig. 11-4 and *B* and *F* in Fig. I-1).

Male Thread. A thread on the outside of a pipe or fitting. Preferably called an *outside thread.*

Malleable Iron. Cast iron that has been specially heat-treated to render it less brittle than ordinary cast iron.

Manhole. An opening constructed for the purpose of permitting a man to gain access to an enclosed space.

Manifold. A fitting or pipe with many outlets or connections relatively close together. Also a *header* (see Fig. 7-9).

Master Plumber. A person with knowledge of and experience in plumbing who employs journeymen or who conducts a plumbing business.

Matheson Joint. A bell-and-spigot joint in wrought pipe.

Needle Valve. A valve in which the opening, consisting of a small hole, is opened or closed by a long, needlelike spindle that is thrust into or is withdrawn from the hole (see Fig. 14-35).

Nipple. A short piece of pipe with outside threads used for connecting pipes or fittings in threaded joints (see Fig. II-55).

Nonsiphon Trap. A trap in which the diameter is not greater than about 4 in., the depth of seal is between about 3 and 4 in., and the volume of water held in the trap is not less than 1 qt. In general, a trap that is more difficult to siphon than the more commonly used P and S traps.

Normandy Joint. A joint in which the plain ends of two pipes are connected by a sleeve whose ends are made tight by rings of packing compressed between bolting rings and the sleeve. Modifications include *Dayton*, *Dresser*, and *Hammond joints*.

Nozzle. The outlet from a faucet or the end of a pipeline or hose so designed that the issuing stream of water is thrown in a shape or size different from the diameter of the pipe.

Oakum. Hemp or old hemp rope soaked in oil or other material to make it waterproof and to resist rotting.

Open Plumbing. Plumbing so that traps and drainage pipes and their surroundings beneath fixtures are ventilated, accessible, and open to inspection.

Open Return Bend. Similar to a *close return bend* except that the arms are separated (see Figs. II-56 and II-57).

Packing. A soft material used in making joints watertight or airtight by being squeezed or compressed in the joint.

Pedestal Urinal. A urinal supported on a single pedestal and not connected to a wall for support (see Figs. 16-27 and II-73).

Pet Cock. A ground-key faucet with an opening about ⅛ in. in diameter. Sometimes called an *air cock* (see Fig. 14-41).

Pilot Light. A small flame, used in gas-heating or gas-cooking devices, that burns constantly to ignite the main gas supply when it is turned on.

Pipe. From ASTM B 251-55T:

> Seamless tube conforming to the particular dimensions commonly known as "Standard Pipe Size."

This definition is for the purpose of distinguishing from the word *tube*, particulary in brass and copper conduits.

Pipe Stock. A die holder (see *J* in Fig. 7-8).

Pipe Tongs. A hand tool for gripping or turning pipes. See also *chain tongs*.

Pipe Wrench. A wrench with slightly curved, serrated jaws, designed to tighten the grip on the pipe as the handle is turned. Also called a *Stillson wrench* (see *M* in Fig. 7-8).

Plug. A pipe fitting with outside thread and projecting head, often square, that is used for closing the opening in another fitting. Also, sometimes, a *faucet*.

Plug Cock. Same as *ground-key faucet*.

Plumber. A person trained and experienced in the art of plumbing.

Plumber's Friend. A cup-shaped device of rubber on the end of a wood or metal handle used for forcing stoppages in pipes by the action of siphonage or compression. Also called a *pneumatic plunger* (see Fig. 18-1).

Plumber's Furnace. A gasoline-fired firepot or similar device for melting solder or heating soldering irons and similar service (see Fig. 7-8).

Plumber's Rasp. A coarse rasp for filing lead (see Fig. 7-8).

Plumber's Round Iron. A special form of soldering iron used for soldering seams in tanks.

Plumber's Soil. A mixture of lampblack and glue used in leadwork.

Plumbing. NPC states:

> Plumbing includes the practice, materials, and fixtures used in the installation, maintenance, extension, and alteration of all piping, fixtures, appliances, and all appurtenances in connection with any of the following: sanitary-drainage or storm-drainage facilities, the venting systems and the private water-supply systems, within or adjacent to any building, structure, or conveyance; and the practice and materials used in the installation, maintenance, extension, or alteration of the storm-water,

liquid waste, or sewerage and the water-supply system of any premises to their connection with any point of public disposal or other acceptable terminal.

Plumbing System. NPC states:

> The plumbing system includes the water supply and distribution pipes; plumbing fixtures and traps; soil, waste, and vent pipes; building drains and building sewers, including their respective connections, devices and appurtenances within the property lines of the premises; and water-treating and water-using equipment.

Pop Valve. A safety valve that is kept closed by the pressure of a spring against the valve. The valve is opened when the pressure of fluid against it overbalances the pressure of the spring.

Potable Water. NPC states:

> Water which is satisfactory for drinking, culinary, and domestic purposes and meets the requirements of the health authority having jurisdiction.

Pothook. A hook used for lifting the lead pot from the furnace.

Primary Branch. The primary branch of the building drain is the single sloping drain from the base of a stack to its junction with the main building drain or with another branch thereof (see Fig. I-1).

Private Sewer. A sewer that is privately owned.

Privy. An outhouse or structure used for the deposition of human excrement in a container or in a vault beneath the structure.

Privy Vault. A pit beneath a privy in which excrement collects.

Protected Waste Pipe. A waste pipe from a fixture that is not directly connected to a drain, soil, vent, or waste pipe.

Public Sewer. A sewer that is publicly owned or to which all abutters have equal rights of connection. A common sewer.

Raised-face Flange. A flange faced about $\frac{1}{32}$ in. higher inside the bolt circle.

Rake. The angle of the cutting edge of a tap or die.

Range Closet. A battery of seats placed close together or one continuous opening in a seat, all placed above a single water-bearing trough or receptacle designed to receive human excrement. Also called a *latrine.*

Ream. To cut the burr from the inside of a pipe or to increase the opening of a pipe or orifice by cutting with a circular motion.

Reamer. A tool used in reaming (see *K* in Fig. 7-8).

Reducer. A pipe fitting with inside threads, larger at one end than at the other. All such fittings having more than one size are reducers because of the custom of stating the larger size first. See also *increaser* (see Figs. II-51 and II-56).

Relief Vent. An auxiliary vent, supplementary to regular vent pipes, the primary purpose of which is to provide supplementary circulation of air between drainage and vent pipes (see Fig. I-1 and the dotted lines between 3 and 4 in Fig. 11-18). The NPC states:

> A relief vent is a vent the primary function of which is to provide circulation of air between drainage and vent systems.

Resealing Trap. A trap on a plumbing-fixture drain pipe designed so that the rate of flow at the end of a discharge from the fixture will seal the trap but will not cause self-siphonage.

Return Bend. An open return bend, usually with inside threads but applied also to a 180° bend in a pipe (see Figs. II-56 and II-57).

Return Offset. A return offset, or jumpover, is a double offset designed to return the slope of a pipe to its original line. See also *crossover fitting* (see Figs. I-1 and II-56).

Revent Pipe. NPC states:

> A revent pipe, sometimes called an individual vent, is that part of a vent pipeline which connects directly with an individual waste or group of wastes, underneath or back of the fixture, and extends either to the main or branch vent pipe.

The shower on the fourth floor in Fig. 11-1 illustrates a revented fixture.

Riser. A water-supply pipe that extends vertically one full story or more to convey water to branches or fixtures.

Roof Drain or Roof Gutter. A drain installed to receive water collected on a roof and to convey it to the leader, downspout, or conductor.

Roughing-in. NPC states:

> The installation of all parts of the plumbing system which can be completed prior to the installation of fixtures. This includes drainage, water-supply and vent piping, and all the necessary fixture supports.

The word may be a noun or a verb. It is sometimes used synonymously with *roughing.*

Run. That portion of a pipe or fitting continuing in a straight line in the direction of flow in the pipe to which it is connected. Sometimes an appreciable length of straight or approximately straight pipe.

Saddle Fitting. A fitting clamped to the outside of a pipe, the joint being made tight with a gasket (see Fig. 7-20).

Safe. A pan or other collector placed beneath a pipe or fixture to prevent or collect leakage or other unusual discharge of fluid from the pipe or fixture.

Safe Waste. The waste pipe from a safe.

Sand Trap or Sand Interceptor. A catch basin for the collection of sand or other gritty material (see Fig. 11-43).

Sanitary Sewage. Sewage containing human excrement and liquid household wastes. Also called *domestic sewage.*

Sanitary Sewer. A sewer intended to receive sanitary sewage with or without industrial wastes and without the admixture of surface-water, storm-water, or clear-water drainage.

Seal of a Trap. The vertical distance between the dip and the crown weir of a trap (see Fig. 1-3). Also the water in the trap between the dip and the crown weir.

Secondary Branch. Any branch in a building drain other than the primary branch (see Fig. I-1).

Self-siphonage. The breaking of the seal of a trap as a result of removing the water therefrom by the discharge of the fixture to which the trap is connected.

Separator. See *interceptor.*

Septic Tank. A watertight container of sewage in which the period of retention is sufficiently long to permit some hydrolysis and gasification of the contents and from which the effluent flows at approximately the same rate at which the influent enters (see Figs. 17-7, 17-8, and 17-9).

Service Box. See *curb box.*

Service Clamp. A saddlelike connection used on a water main for a service connection. Sometimes called a *pipe saddle.*

Service Ell. A 45 or 90° bend with an outside thread on one end and an inside thread on the other. Also called a *street ell* (see Fig. II-56).

Service Pipe. The pipe from the main in the street or other source of supply to the building served. Generally applied to water and gas pipes connected to the public distribution mains (see Fig. 1-4).

Service Tee. A tee with an outside thread on one end and an inside thread on the other end and on the branch (see Fig. II-56).

Sewage. The liquid wastes conducted away from residences, business buildings, or institutions, together with those from industrial establishments, and with such ground, surface, and storm water as may be present. More briefly sewage may be considered to be the used water supply of a community.

Sewer. A conduit whose purpose is to convey sewage.

Sewerage. *As a noun,* the works comprising a sewer system, pumping stations, treatment works, and all other works necessary to the collection, treatment, and disposal of sewage. *As an adjective,* having to do with the collection, treatment, or disposal of sewage.

Shave Hook. A leadworker's tool used for shaving or cutting lead.

Shoulder Nipple. A nipple somewhat larger than a close nipple. It has an unthreaded space of about ¼ in. between the outside threads.

Shrunk Joint. A joint made by shrinking a heated piece of pipe over the ends of two cool pipes.

Siamese Connection. A wye connection used on fire lines so that two lines of hose can be connected to a hydrant or to the same nozzle (see Fig. 10-2).

Side-outlet Ell. An ell with outlet at right angles to the plane of the run (see Fig. II-56).

Side Vent. "A vent connecting to the drain pipe through a fitting at an angle not greater than 45° to the vertical" (from NPC. See Fig. I-1).

Sill Cock. A faucet used on the outside of a building to which the garden hose is attached or is usually attached. It contains a hose thread rather than a pipe thread on the end of the faucet (see Fig. 7-7).

Sink. A shallow fixture, ordinarily with a flat bottom, that is commonly used in a kitchen or in connection with the preparation of food, for laboratory purposes, and for certain industrial processes. There are many types of special sink the purpose of which is indicated by the name prefixed before the word sink, such as *slop sink, vegetable sink,* etc.

Siphonage. A suction created by the flow of liquids in pipes. A pressure below atmospheric pressure.

Sitz Bath. A fixture for bathing parts of the body, especially the posterior (see Fig. 16-10).

Slant. A branch connection from a house sewer to a common sewer. See also *house slant.*

Sleeve. A cylindrical tube surrounding a pipe or shaft.

Slip Joint. A connection in which one pipe slides into another. The joint is made tight with a gasket, packing, or calking.

Slop Sink. A deeper fixture than an ordinary sink, intended for the receipt of slops. It is often equipped with an integral trap (see Fig. 16-26).

Socket. See *coupling.*

Socket Plug. A plug with a recess in the face into which a wrench will fit to turn the plug.

Soil Pipe. A pipe through which liquid wastes carrying human excrement can flow. Also a cast-iron pipe with bell-and-spigot ends used in plumbing to convey human excrement or liquid wastes.

Soil Stack. A vertical soil pipe conveying human excrement and liquid wastes (see Fig. I-1).

Soil Vent. That portion of a soil stack above the highest fixture-waste connection to it. Synonymous with *stack vent.*

Solder. *As a noun* "a metal or a metallic alloy used when melted to join metallic surfaces; especially an alloy of lead and tin." *As a verb:* "to join metallic surfaces with solder" (from NPC).

Soldering Iron. A piece of copper, rectangular in shape, about ½ in. thick and 2 in. long, pointed at one end. Used to hold heat as it is applied to the solder (see Fig. 7-8).

Spigot. The end of a pipe that fits into a bell. Also a word used synonymously with *faucet.*

Spud. A short connecting pipe between the meter and the supply pipe or a similar short piece of pipe.

Stack. A general term used for any vertical line of drainage or vent piping.

Stack Group. NPC states:

> A term applied to the location of fixtures in relation to the stack so that by means of proper fittings vents may be reduced to a minimum.

Stack Vent. The extension of a soil or waste stack above the highest horizontal drain connected to the stack (see Fig. I-1).

Stack Venting. Stack venting consists of a continuous venting of a single fixture connected directly into a soil or waste stack. Only the top fixtures or fixtures on a stack can be stack-vented. The fourth-floor fixtures at E in Fig. 11-7 are stack-vented.

NPC states:

> A method of venting a fixture or fixtures through the soil or waste stack.

Standpipe. A vertical pipe generally used for the storage of water. Also a pipe in which water stands unused ready for emergency or other irregular demand.

Stench Trap. A flap in a frame which opens to admit cellar drainage to a sewer and then closes to prevent sewer air from entering the building.

Stillson Wrench. Same as *pipe wrench* (see M in Fig. 7-8).

Stock. The tool that holds the dies in pipe threading and in the threading of bolts, etc. (see Fig. 7-8).

Stopcock. A small valve with a ground key (see Fig. 14-41).

Stop-and-waste Cock. A stopcock designed so that when the supply of water is shut off, a drain in the valve is opened through which water in the pipe downstream from the stopcock is drained to waste (see Fig. 14-42).

Storm Water. That portion of the rainfall or other precipitation which runs off over the ground surface or other catchment area and for such a short period following a storm as the flow exceeds the normal runoff.

Street Ell. Same as *service ell* (see Fig. II-56).

Sump. A pit or receptacle at a low elevation to which liquid wastes are drained.

Surface Water. That portion of the rainfall or other precipitation which runs off over the surface of the ground.

Swage. To increase or decrease in diameter by means of a special tool or process. The tool used in swaging.

Sweat Joint. A soldered joint heated by a flame instead of a soldering iron.

Sweating. The appearance of condensed moisture from the air on the surface of a cool pipe or fixture. The term is used also to indicate the soldering, welding, or brazing of metals.

Swedge. Same as *swage.*

Sweep Fitting. A fitting with a long-radius curve.

Swing Joint. A joint in a threaded pipeline permitting motion in the line in a plane normal to the direction of one part of the line (see Fig. 7-15).

Swivel Joint. Same as *swing joint.*

Tailpiece. A tee connection often used on the connection to a sink drain.

Tampion. A leadworker's tool of boxwood shaped like a toy top and used for swedging out the end of a lead pipe.

Tap. A tool used for cutting inside threads. Also to bore a hole in a pipe, tank, or other device. To cut threads on the inside of the hole.

Tapped Tee. A cast-iron bell-end tee with the branch tapped to receive a threaded pipe or fitting.

Tell Tale. An indicating device used to indicate the elevation of water surface in a tank.

Terneplate. Sheet iron or steel coated with an alloy of approximately 4 parts of lead and 1 part of tin.

Trap. A fitting so constructed with a water seal that when placed in a drainage pipe it will prevent the passage of air or gas through the drainage pipe but will not prevent the flow of liquids through it (see Fig. 1-3).

Trap Seal. See *seal.*

Tray. A fixture used in a laundry for washing. Commonly called a *tub* or a *washtub* (see Fig. 16-25).

Trimo Wrench. Same as *Stillson wrench.*

Tube. (To distinguish from *pipe,* particularly in copper or brass) "A hollow product of . . . any cross section (shape), having a continuous periphery." From ASTM B 251–55T.

Tucker Fitting. A cast-iron coupling one opening of which is threaded for screw pipe and the other opening of which has a hub to receive the spigot end of a pipe.

Union. A pipe fitting used for joining the ends of two pipes neither of which can be turned. There are two kinds of unions. See *box union* and *flange union* (see Figs. II-56 and II-57).

Union Coupling. A right- and left-handed threaded turnbuckle or sleeve nut used to join or draw two pipes together.

Union Vent. A dual vent or unit vent (see Fig. 11-26).

Unit Vent. One vent pipe which serves two traps (see Fig. 11-26).

Utility Vent. A vent in which the vent pipe rises well above the highest water level in the fixture vented and then turns down before connecting to the stack or main vent (see Fig. 11-28).

Vacuum. An air pressure less than atmospheric. Also *siphonage.*

Vacuum Breaker. A device to prevent the creation or formation of a vacuum (see Fig. 14-29).

Valve. A device used to control the flow of fluid or intended therefor.

Vapor Lock. Same as *air-lock.*

Vent. See *back vent.*

Vent System. NPC states:

> A pipe or pipes installed to provide a flow of air to or from a drainage system or to provide a circulation of air within such system to protect trap seals from siphonage and back pressure.

Vertical Pipe. A pipe or fitting that is installed in a vertical position and makes an angle of not more than 45° with the vertical.

Volumeter. A type of flushometer valve.

Washer. An annular ring threaded on the inside to be used as a lock nut. A smooth, flat annular ring to be placed under a nut or bolt head to fill a space or to protect the material under the bolt or nut. A flat annular ring of soft material used in valves and moving parts of equipment to prevent leakage.

Waste Pipe. NPC states:

A pipe which conveys only liquid wastes free from fecal matter.

Waste Stack. A vertical pipe used to convey liquid wastes not containing fecal matter and usually not containing any human excrement.

Waste Vent. That portion of a waste stack above the highest fixture-waste connection to it.

Water Back or Water Front. A small tank or other watertight container forming a portion of the lining of the firepot, or otherwise exposed to the fire, in a kitchen range or heating furnace (see Fig. 8-1).

Water Closet. A water-flushed plumbing fixture designed to receive human excrement directly from the user of the fixture. The term is sometimes used to indicate the room or compartment in which the fixture is located.

Weld. To join metals by heating or pounding while in contact until they flow together and adhere. The joint formed by welding.

Wet Vent. That portion of a vent pipe through which liquid wastes, other than water-closet wastes, flow (see Figs. 11-5 and I-1).

Wiped Joint. A joint between the ends of two pipes, usually lead pipes, formed manually by wiping a ball of solder about the ends of the two pipes (see Fig. 12-7).

Yarning Iron. A calking tool similar to a cold chisel except that the plane of the edge of the tool may be offset about 1 in. from the plane of the handle (see *C* in Fig. 7-8).

Yoke. The collar by which a lead trap is secured to its support. A pipe with two branches uniting them to form one stream. A vertical connection between a branch waste line or wet vent and a continuous-vent stack.

Yoke Vent. A vertical or 45° relief vent of the continuous waste-and-vent type formed by the extension of an upright wye-branch inlet of the horizontal branch to the stack. It becomes a dual yoke vent when two horizontal branches are thus vented by the same relief vent (see Figs. 11-15 and I-1). NPC states:

A pipe connecting upward from a soil or waste stack to a vent stack for the purpose of preventing pressure changes in the stacks.

APPENDIX II

EXTRACTS FROM, AND REFERENCES TO, STANDARD SPECIFICATIONS FOR PLUMBING MATERIALS, EQUIPMENT, AND PRACTICE[1]

Nomenclature and abbreviations: API, American Petroleum Institute; ASME, American Society of Mechanical Engineers; ASTM, American Society for Testing Materials; ASA, American Standards Association; AWWA, American Water Works Association; CS, commercial standards prepared under the National Bureau of Standards, and published by the U.S. Department of Commerce;[2] FS, specifications published by the Federal Specifications Board;[2] HI, Hydraulic Institute; MSS, Manufacturers Standardization Society of the Valve and Fittings industry;[3] NEMA, National Electrical Manufacturers Association; NPC, "National Plumbing Code";[4] SPR, Simplified Practice Recommendation, U.S. Department of Commerce.[1]

TABLE II-1. STANDARD SPECIFICATIONS

Item	*Standard*
Air gaps in plumbing systems	ASA A 40.6–1943
Aluminum, plates and sheets	FS QQ–A–561
Backflow preventers in plumbing (*known also as Vacuum breakers*)	FS WW–P–541a
Bolts	
steel	ASTM A 96
wrought iron	ASTM A 107
Brass, commercial:	
bars, plates, rods, shapes, sheets, and strips	FS QQ–B–611a
castings	FS QQ–B–621
fittings for flared copper tubes	ASA A 40.2–1936
leaded, sheet and strip	ASTM B 152–46T MSS SP–40–1946
sheets and strips	ASTM B 121–46T ASTM B 121–52 ASTM B 36–52
Bronze	
aluminum bars, plates, rods, shapes, sheets, and strips	FS QQ–B–666
manganese bars, forgings, plates, rods, and shapes	FS QQ–B–721a
phosphor bars, plates, rods, shapes, sheets, and strips	FS QQ–B–746
Burners, conversion, installation for house heating and water heating	ASA Z 21.8–1940
Bushings, with pipe threads	ASA B 16.14–1943
Caps, with pipe threads	ASA B 16.14–1943
Clay sewer pipe	*see Pipe, clay sewer*

[1] See also Sec. 4-2. For most recent information on standards see *Magazine of Standards* published by ASA.

[2] Usually obtainable through Superintendent of Documents, Washington 25, D.C.

[3] These standards may be subsequently submitted to and incorporated as ASA standards.

[4] Published by ASME.

TABLE II-1. STANDARD SPECIFICATIONS (*Continued*)

Item	*Standard*
Code	
national plumbing	ASA A 40.8–1955
requirements for light and ventilation	ASA A 53.1–1946
Compound, plumbing fixture setting	FS HH–C–536
Conductor, rain water	CS R 29–39
Conduit	
flexible, steel	FS WW′–C–566
rigid, steel, enameled	FS WW–C–571
rigid, steel, zinc-coated	FS WW–C–581a
Controls, furnace temperature limit, and fan	ASA Z 21.29–1941, R1947
Copper: bars, plates, rods, shapes, sheets, strips,	FS QQ–C–501a
tubing	*see Tubing, copper*
Copper-silicon alloy: bars, plates, rods, sheets, shapes, and strips	FS QQ–C–501a (1941)
Cork	
composition, gasket and sheet	FS HH–C–576
compressed, for thermal insulation	FS HH–C–561b
granulated, insulating	FS HH–C–571a
Couplings	
hose	*see Hose couplings*
pipe	*see Fittings*
Culverts, iron or steel, zinc-coated	FS QQ–C–806
Dishwashing machines	FS 00–M–31b
Drafting standards,	
American	ASA Z 14.1–1946 ASA Y 14.1–1947 to Y 14.6–1957
symbols for plumbing and revision	ASA Y 32.4–1955 ASA Z 32.2.2–1949 ASA Z 32.2.3–1949 ASA Z 32.2.4–1949
Drain tile	
clay	FS SS–T–310 ASTM C 4–24
Drain tile	
concrete	ASTM C 4–55 ASA A 6.1–1956
extra strength	ASTM C 200–44
Drainage fittings, cast brass, solder joint	ASA B 16.23–1955
Drinking fountains	ASA Z 4.2–1942
Electric code, National	ASA C 1–1956 National Board of Fire Underwriters, 70
Electric motors	FS CC–M–632a FS CC–M–641
Fire hose	*see Hose, fire*
Fittings	
brass, castings	ASTM B 62
for flared copper tubes	ASA A 40.2–1936
rough, screwed	CS 5–40
brass or bronze, screwed, 125 lb	ASA B 16.15–1947 FS WW–P–448a
brass or bronze, screwed, 250 lb	FS WW–P–461a
brass, bronze, or copper, solder joint	FS WW–P–456
bronze castings	ASTM B 61
bronze screwed, 125 lb	ASA B 16.15–1947 FS WW–P–460 1945 MSS SP–10
carbon steel and carbon molybdenum	ASTM A 234–44
cast-iron	
drainage	FS WW–P–491 ASA B 16.12–1942 ASA B 16.23–1953 MSS SP–8
flanged	ASA B 16a–1939 ASA B 16a–1–1943 ASA B 16b–1944 ASA B 16b2–1931
flared for copper water tubes	ASA A 40.2–1936
800-lb hydraulic pressure	ASA B 16b1–1931
long-turn sprinkler	ASA B 16g–1929
pipe	ASA B 16.12–1942 FS WW–P–501 ASA 21.1, 21.2, and 21.4 to 21.7, 1939

TABLE II-1. STANDARD SPECIFICATIONS (*Continued*)

Item	*Standard*
Fittings	
cast-iron	
pit-cast, for water pipe..	ASA 21.2–1939 ASTM A 44–41 AWWA May 12, 1908
screwed, 125 and 150 lb...	ASA B 16d–1941
soil..........	ASA 40.1–1935 ASTM A 74–42
malleable iron....	ASA B 16c–1939
galvanized....	ASTM A 93–52T FS WW–P–406 1944
screwed, 150 lb.	ASA B 16c–1939
pit cast-iron.....	ASA A 21.2–1939 ASTM A 44.41 AWWA May 12, 1908
screwed.........	FS WW–P–501a ASA B 16.12–1942
soil............	ASA A 40.1–1935 FS WW–P–401
soldered joint....	ASA A 40.3–1941
steel............	ASTM A 95, A105, and A181
butt-welded...	ASA B 16.9–1940
flanged........	ASA B 16c–1939
fusion welding.	ASTM A 126
socket welding.	ASA B 16.11–1946
welding.......	ASTM A 234
water tube, copper, flared...	ASA A 40.2–1936
soldered copper	ASA A 40.3–1941
wrought copper and wrought bronze, solder joint..........	ASA B 16.2–1951
Fixtures	
earthenware, vitreous glazed.	CS 111–43
enameled, cast-iron...........	CS 77–48
formed metal, porcelain enameled sanitary ware..........	CS 144–47
hospital.........	CS R 106–1941 SPR R 106–41
plumbing for land use...........	FS WW–P–542 FS WW–P–541b FS WW–P–541a 1947 FS WW–P–542
porcelain........	CS 4–29
sanitary, cast-iron enamel....	CS 77–40
steel enameled...	FS WW–P–542
Fixtures	
structural slate...	CS R 13–28
vitreous china....	CS 20–49 and CS 111–43
vitreous glazed...	CS 111–43
Flanges	
brass castings....	ASTM B 62
150 and 300 psi	ASA B 16.24–1953
bronze castings...	ASTM B 61
150 and 300 psi	ASA B 16.24–1953
cast-iron........	ASTM A 126 ASA B 16a–1939 B 1621–1943 FS WW–F–406a
class 250......	ASA B 16b–1944
25 lb..........	ASA B 16b2–1931
malleable iron....	ASTM A 197
for pipes and valves......	ASTM A 227–44T
standard......	FS WW–F–406a
steel............	ASTM A 95
forged........	ASA G 46.1–1947
fusion welding.	ASTM A 216
high temperature.......	ASTM A 182–46 ASTM A 105–46
pipe..........	ASA B 16e–1939 MSS SP–44–1955
valves and fittings......	ASA B 16e5–1943
welding.......	ASTM A 234
Flue linings, clay...	ASA A 62.4–1947
Gages	
pressure.........	FS GG–G–66
pressure and vacuum for air, ammonia, oil, steam, and water.........	FS GG–G–76
Gas appliances.....	*see also Water heaters*
central heating...	ASA Z 21.13–1943
connectors of flexible tubing and fittings........	ASA Z 21.32–1942 and R 1947
domestic, pressure regulators.	ASA Z 21.18–1934
ranges........	ASA Z 21.1–1956 Z 21.1.2–1956
semirigid tubing and fittings....	ASA Z 21.24–1941
thermostats.....	ASA Z 21.23–1940 and R 1947
Gas	
automatic pilots	ASA Z 21.20–1940 and R 1947

TABLE II-1. STANDARD SPECIFICATIONS (*Continued*)

Item	Standard
Gas	
cast-iron pipe	ASA A 21.3
control valves	ASA Z 21.21–1935 and R 1947
conversion burners	ASA Z 21.17–1940 ASA Z 21.8–1940
cooking equipment for hotels and restaurants	ASA Z 21.27–1940 ASA Z 21.26–1941
counter appliances	ASA Z 21.31 and R 1947
cylinders, compressed gas, seamless	FS RR–C–901
floor furnaces	CS 67–38
furnace temperature controls	ASA Z 21.29–1941
hair dryers	ASA Z 21.25–1937
ironers	ASA Z 21.7–1932 and R 1947
laundry stoves and hot plates	ASA Z 21.9–1940
ovens, baking and roasting	ASA Z 21.28–1941
ranges, domestic	ASA Z 21.1ES–1942 ASA Z 21.1–1956*ff* ASA Z 21.1.1a–1957
hotel and restaurant	ASA Z 21.4–1932
oven type	ASA Z 21.37–1948
shutoff valves	ASA Z 21.22–1935 R 1947
space heaters	ASA Z 21.11–1945
tubing, flexible	ASA Z 21.2–1938 R 1947
unit heaters	ASA Z 21.6–1940 and R 1947
valves	ASA Z 21.15–1944
water heaters	ASA Z 21.10–1945 ASA 21.10WS–1942 ASA Z 21.26–1941 and R 1947
Gaskets	
asbestos, corrugated	FS HH–G–71
metallic cloth	FS HH–G–76
metallic encased	FS HH–G–101
plumbing fixture setting	FS HH–G–116
rubber, molded, sheet, and strip	FS HH–G–156
Hangers, pipe	FS WW–H–171
Heaters	
electric, water storage, domestic	FS WW–H–196 ASA C 72.1–1949
garage, private	ASA A 21.4–1932
gas	*see Gas*
units, electric	FS WW–U–546
Hose	
couplings, for cotton and rubber-lined	FS WW–C–621a
coupling threads	ASA B 33.1–1935 ASA B 26–1925 and R 1947
fire, cotton and rubber-lined	ASA L 3.1–1941 ASTM D 296–38 NFPA* 194–1958 FS ZZ–H–451a
linen, unlined	FS JJ–H–571
garden and water	FS WW–C–623a
water, braided	FS ZZ–H–601
wrapped	FS ZZ–H–611
Hot plates and laundry stoves	ASA Z 21.9–1940 and R 1947
Illumination, building code requirements	ASA A 53.1–1946
Insulation, pipe, molded	ASTM C 168–1947
Iron	
malleable for fittings	MSS SP 22
sheets, galvanized	ASA G 8b1–1931 ASTM A 93–27
wrought, uncoated	ASTM G 23–1939 ASTM A 162–39
Iron and steel: sheet, black and zinc coated, galvanized	FS QQ–I–696 ASA G 8b.1 ASA G 8.8–1947 ASTM A 163–30 ASTM A 93–38T FS QQ–I–706a FS QQ–I–716
Joints	*see also type of joint wanted*
brazed	ASA B 31.1–1951
Lavatories	ASA Z 24.1–1935
Lead: calking	FS QQ–L–156 1934 CS 94–41
pig	FS QQ–L–171
sheet	FS QQ–L–201 1953

* National Fire Protection Association.

TABLE II-1. STANDARD SPECIFICATIONS (*Continued*)

Item	*Standard*
Leaded brass, sheet and strip	ASTM B 152-46T
Light and ventilation, building code requirements	ASA A 53.1-1946
Lock nuts with pipe threads	ASA B 16.14-1943
Low-water cutoff devices	ASA Z 21.36
Manhole frames and covers	ASA A 35.1-1941 AWWA *see J.* July, 1942
Markings, valves, fittings, etc	MSS SP 41-1953
Meters: cold water	AWWA, 1921
current, compound, and fire service	AWWA, 1923
Nipples, brass, copper, steel, and wrought iron	CS 5-40 and FS WW-N-351a
Nipples and bushings, soldering	ASTM B 43-46 FS WW-P-351
Packing, asbestos valve stem	FS HH-P-51
diaphragm	FS HH-P-61b
fiber, hard sheet	FS HH-P-91
hydraulic	FS HH-P-112
jute twisted	FS HH-P-117
leather	FS KK-L-177 FS KK-L-171a FS KK-L-231
metallic cloth	FS HH-P-32
metallic flexible	FS HH-P-126a
plastic	FS HH-P-131a
rod, braided	FS HH-P-34
rod, high pressure	FS HH-P-36
rope and wick	FS HH-P-41
rubber, cloth-insertion	FS HH-P-131a FS HH-P-151a
semimetallic	FS HH-P-166
spiral, gland, low-pressure	FS HH-P-71
Pipe, asbestos-cement	FS SS-P-351a
sewer	FS SS-P-331a
bituminized fiber	CS 116-44
brass	ASTM B 43-55 ASTM B 42-33 ASA H 27.1-1956 FS WW-P-351 1930
seamless, iron-pipe size	FS WW-P-351
Pipe	
cast-iron, and fittings	ASA A 21, 21.1, 21.2 and 21.4 to 21.7, 1939
cast-iron, for water bell-and-spigot, bolted joint	FS WW-P-421 ASA A 21.2-1953
cement-mortar lining	ASA A 21.4-1939
drainage, threaded, vent and waste	FS WW-P-356 ASA A 40.5-1943
cast-iron drainage fittings	ASA B 16.12-1942
cast-iron, flanged ends	AWWA Jan. 1, 1941
flanges and flanged fittings	ASA B 16, 16a, and 16b*ff*
gas and water, 50 psi	FS WW-P-00360 (GSA-FSS)
American Standard	ASME, Mar. 20, 1914
WSP for 25 lb	ASA B 16b2-1931
Class 250	ASA B 16b-1944
Class 125	ASA B 16a-1939
pit cast	ASTM A 44-41 ASA A 21.2-1953 AWWA May 12, 1908
screwed	ASA A 40.5 1943 FS WW-P-356
drainage fittings	ASA B 16.12-1953 FS WW-P-491a-1945
for 125 and 250 psi	ASA B 16d-1927 ASA B 16.4-1949 FS WW-P-501b (1945)
soil pipe and fittings	ASTM A 74-42 ASA A 40.1-1953 FS WW-P-401
cement lining	ASA A 21.4-1939
clay sewer	*See also Pipe, sewer, clay* FS SS-P-361a (1942)
coating, coal-tar enamel	AWWA 7 A.6-40
color bands for	ASA A 13.1-1956 ASA Z 53.1-1945

TABLE II-1. STANDARD SPECIFICATIONS (*Continued*)

Item	Standard
Pipe	
concrete, nonpressure	FS SS-P-371
pressure	FS SS-P-381; AWWA J. April, 1943
sewer	*see Pipe, sewer, concrete*
conductor, rain water	CS R 29-39
copper	ASTM B 42-52 and B43-33; ASA H 26.1-1949; FS WW-P-377 (1932)
copper and copper alloy	ASTM B 42; ASTM B 75; ASTM B 88; ASTM B 111
covering, mineral or rock	FS HH-P-386a
cork, molded	FS HH-P-381
magnesia, block, cement	FS HH-M-61
mineral wool	FS HH-M-371
electric fusion, welded	ASTM A 155
electric resistance, welded	ASTM A 135-44; ASA B 36.5-1945
fiber, bituminous	CS 116-44
fittings, bronze	FS WW-P-460; WW-P-471
cast-iron, drainage	FS WW-P-491a
screwed	FS WW-P-501c
iron and steel	FS WW-P-521c
flanges	*see Flanges*
iron, lap welded, seamless	ASA B 36.3-1939
open-hearth	ASTM A 253-55T; ASA B 36.23-1956
spiral	ASTM A 211-44
and steel pipe	ASA B 36.1
galvanized	ASTM A 93-52T; FS WW-P-406 (1944)
wrought	ASA B 36.10-1950
welded	ASA B 36.2-1956; ASTM A 12-55
lead	CS 95-41; CS 96-41; FS WW-P-325 (1944); Western Uniform Code 1948
malleable iron, screwed	ASA B 16.3-1951; CS 7-29
Pipe, malleable iron, screwed, fittings for 150 psi	ASA B 16.c-1939
manual for computation of strength and thickness	ASA A 21.1-1939
open-hearth iron	ASTM A 253-46
plugs	ASA B 16e-1936
pressure	ASA B 31.1-1942; ASA B 31.1a-1944; ASME April, 1947
sewer, bituminous fiber	CS 116-44
clay	ASTM C 13-44T; FS SS-P-361b
ceramic glazed	C 261-54; ASA 106.4-55
extra strength, ceramic glazed	ASTM C 278
perforated	C 211-1950
standard strength	ASTM C 13-54; ASA 106.3-1955
concrete	ASTM C 75, C 76, C 14-52; FS SS-P-371
vitrified clay	ASTM C 13-44T
soil and fittings cast-iron	*see Pipe, cast-iron, soil and fittings*
steel	ASA B 36.1; G 8.7-1945; ASTM A 120-46
black and white, hot-dipped galvanized	ASA B 36.20-1951; ASTM A 120-47
ferrous alloy	FS WW-P-403a
for bending	FS WW-P-404
for ordinary purposes	FS WW-P-406
flanges and fittings	ASA B 16.5-1953
high pressure and high service	ASTM A 106
mild	API 5L; ASTM A 53-55T; ASTM A 120-47; ASME Mar. 20, 1914; FS WW-P-406 (1944)

TABLE II-1. STANDARD SPECIFICATIONS (*Continued*)

Item	*Standard*
Pipe, steel	
open hearth	ASA B 36.23–1950 FS WW-P-406–1944
special uses	ASTM has 14 specifications
stainless	ASA B 36.19–1957
austenitic tubing for general use	ASA B 36.37 ASTM A 269–55
low-temperature service	ASA B 36.40 ASTM A 333–55T ASME SA 333 ASA B 36.41 ASTM A 334–55T ASME SA–334
sanitary tubing	ASA B 36.38
welded	ASA B 36.3–1942 ASA B 36.5–1945 ASTM A 106–42T ASTM A 135–44 ASTM A 53–44 and 55T ASTM B 36.1–45 ASTM A 120
austenitic	ASA B 36.26 ASTM A 312–55 A 333–57T
open hearth	ASA B 36.23 ASTM A 253–55T
ordinary uses	ASA B 36.20–1953 ASTM A 120–47
spiral	ASTM A 211–54 ASA B 36.16–1945
welded and seamless	ASA B 36.1–1956 ASME SA–53
zinc-coated	ASTM A 120–46 ASTM G 8.7–1947
wrought	ASA B 36.10–1950
wrought iron or steel up to 30 in.	AWWA 7A.4
threads	ASA B 2.1–1945 ASA B 33.1 ASME Dec. 22, 1886 FS GGG–P–351
tubing, nickel	ASTM B 161–41T ASTM B 167–41T ASTM B 165–41T
welded	ASA B 36.2–1939 ASTM A 72 FS WW–P–441b
Pipe, tubing	
wrought iron	ASA B 36.2–1950 ASA B 36.10–1939 ASTM A 72–52T FS WW–P–441b 1942
Piping systems, identification	ASA A13–1928 ASA A 13.1–1956
Plugs with pipe threads	ASA B 16.14–1943
Plumbing. FS covering many items are listed under WW and are available from Superintendent of Documents, Washington 25, D.C. See also "National Plumbing Code," ASA A 40.8–1955, published by ASME.	
Privy, sanitary	*see also Sanitary privy* ASA Z 4.3–1935
Pumps, centrifugal	HI
deep-well turbine	ASA B 58.1–1959
Range boilers and expansion tanks	CS RS–29
Sanitary privy	ASA Z 4.3–1935 Supplement 8 to U.S. Public Health Service Report *see also Privy, sanitary*
Sanitation in places of employment	ASA Z 4.1–1955
Sewer pipe	*see Pipe, sewer*
Sheet metal, wrought iron	ASA G 23–1939 ASTM A 162–39
galvanized	ASA G 8.8–1937
Shower stalls, metal partitions	CS R 101–29
Solder: brazing	FS QQ–S–551
flux, rosin-core	FS QQ–S–571b
metal	ASA H 11 ASTM B 32–46T
silver	FS QQ–S–561c
soft	ASTM B 32–49 FS QQ–S–571b (1947)
tin-lead	FS QQ–571a
Solder-joint fittings	ASA A 40.3–1941
Soldering paste, flux	FS O–F–506

TABLE II-1. STANDARD SPECIFICATIONS (*Continued*)

Item	*Standard*
Steel sheets, galvanized and black	ASA B 35.16–1945 ASTM A 211–44 ASA G 8.2–1947 ASTM A 163–39 ASTM A 93–46 FS QQ–I–716 (1942)
tinned, tin plate	FS QQ–I–706a
Steel, stainless, for valves, flanges, etc.	MSS SP–41–1953
Tanks, gravity and pressure	National Board of Fire Underwriters, Pamphlet 22
hot water, expansion	CS R 8–29
storage	CS R–25 CS R 8–29
Terneplate (roofing tin)	FS QQ–T–291 FS QQ–T–191
Threads: Acme screw threads	ASA B 1.5–1952
Stub Acme screw threads	ASA B 1.9–1953
nomenclature, definitions, and symbols	ASA B 1.7–1949 and R 1953
pipe	ASA B 33.1 FS GGG–P–351a ASA B 2.1–1945
ferrous plugs, bushings, lock nuts, and caps	ASA B 16.14–1943
screw threads, gages and gaging	ASA B 1.2–1951
Unified and American for nuts, screws, bolts, etc.	ASA B 1.1–1949
Tile drain, clay	FS SS–T–310
Tin, pig	FS QQ–T–371
Toilets, metal partitions	CS R 101–29
Traps: brass	CS R–21
lavatory and sink	CS R–21 SPR R21–46
lead	CS 96–41 FS WW–P–325 Lead Industries Assoc. 1940
Traps radiator, thermostatic	FS WW–T–696
Tubes, aluminum	FS WW–T–783a FS WW–T–788a FS WW–T–786a
Tubing: aluminum alloy	FS WW–T–756c, 783b, 785a, etc.
brass	FS WW–T–791–1931 ASTM B 135–52
copper, seamless	ASTM B 75–46aT ASTM B 88–55 ASA H 23.1–1956 ASA 26.1–1956 ASTM B 75–52 ASTM B 42–55 ASTM B 251–55T FS WW–T–797 and 799a
ferric, for general use	ASA B 36.36–1956 ASTM A 268–55 ASME SA 268
steel	FS WW–T–731c
water tube (K,L,M)	ASA H 23.1–1953 ASTM B 88–51 ASTM 38–51 FS WW–T–799a (1943) SPR 217–49
wrought steel	ASA B 36
Unions: brass or bronze	FS WW–U–516
malleable iron or steel, 250 lb.	FS WW–U–531
300 lb.	FS WW–U–536
Urinals	FS RR–U–691a
Vacuum breakers	*see Backflow preventers*
Valves, brass or bronze	ASTM B 61 and B 62 FS WW–V–51a and 54 (1946) FS WW–V–58 (1945)
bronze gate	MSS SP–37–1949, 1953 MSS SP–38–1955
cast iron	ASTM A 126 FS WW–V–58
ferrous flanged and welding end	ASA B 16.10–1939
gas, automatic, main	ASA Z 21.21–1935
control	ASA Z 21.22–1936

TABLE II-1. STANDARD SPECIFICATIONS (*Continued*)

Item	*Standard*
Valves	
globe and angle	FS WW–V–51
malleable iron	ASTM A197
radiator, 125 lb	FS WW–V–76b and 151
steel	ASTM A 95
fusion welding	ASTM A 216
welding fittings	ASTM A 234
Ventilation, building code requirements	ASA A 53.1–1946
Water-closet seats	CS 29–31
Water coolers, electric, drinking water	FS OO–C–566
Water heaters, electric	FS W–H–196 NEMA WH1–1949
household, American Standard	ASA C 72.1–1949
Water heaters	
electric, household noncorrosive	FS W–H–196b
gas, except sidearm	ASA Z 21.10.1–1956
sidearm	ASA Z 21.10.2–1956
instantaneous, steam-water	FS WW–H–191
Water-quality standards	Am. Public Health Assoc., 10th ed., 1955
Water-supply protection	ASA A 40.4–1942
Welding symbols and their use	ASA Z 32.1–1942
Zinc, slab, plate, and sheet	FS QQ–Z–301

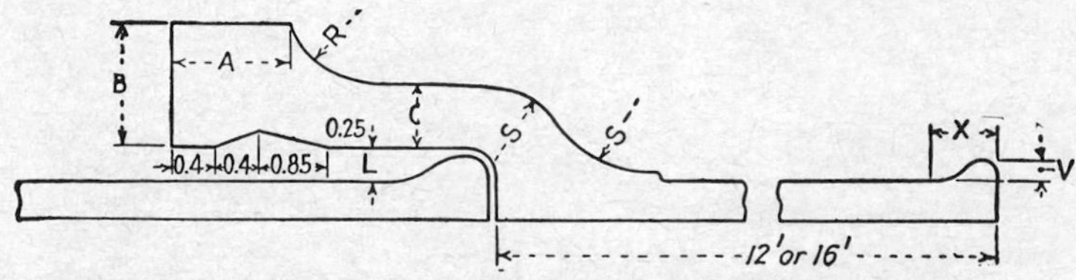

FIG. II-1. Section through a bell-and-spigot cast-iron water pipe.

Dimensions

Depth of bell = 4 in.

x = ¾ in. on sizes from 3 to 6 in. and 1 in. on sizes 8 to 36 in.

V = $\frac{3}{16}$ in. on sizes from 3 to 6 in. and ¼ in. on sizes 8 to 36 in.

A = 1½ in. on sizes 4 to 14 in.

B = 1.3 in. on size of 4 in., 1.4 in. on size of 6 in., 1.5 in. on size of 8 and 10 in. for Classes A and B, and 1.6 in. on size of 10 in. For Classes C and D and on size of 12 in. for Classes A and B.

C = 0.65 in. on size of 4 in., 0.7 in. on size of 6 in.; 0.75 in. on size of 10 in. for Classes A and B; 0.8 in. on size of 10 in. for Classes C and D and on size of 12 in. for Classes A and B; 0.85 in. on size of 12 in. for Classes A and B.

TABLE II-2. STANDARD DIMENSIONS OF CAST-IRON WATER PIPE
(Adopted by American Water Works Association, 1908, see Fig. II-1)

Bell and spigot

Nominal diam, in.	Classes	Actual outside diam, in.	Diameter of sockets, in.		Depth of sockets, in.		Dimensions, in.		
			Pipe	Special castings	Pipe	Special castings	*A*	*B*	*C*
4	A	4.80	5.60	5.70	3.50	4.00	1.5	1.30	0.65
4	B-C-D	5.00	5.80	5.70	3.50	4.00	1.5	1.30	0.65
6	A	6.90	7.70	7.80	3.50	4.00	1.5	1.40	0.70
6	B-C-D	7.10	7.90	7.80	3.50	4.00	1.5	1.40	0.70
8	A-B	9.05	9.85	10.00	4.00	4.00	1.5	1.50	0.75
8	C-D	9.30	10.10	10.00	4.00	4.00	1.5	1.50	0.75
10	A-B	11.10	11.90	12.10	4.00	4.00	1.5	1.50	0.75
10	C-D	11.40	12.20	12.10	4.00	4.00	1.5	1.60	0.80
12	A-B	13.20	14.00	14.20	4.00	4.00	1.5	1.60	0.80
12	C-D	13.50	14.30	14.20	4.00	4.00	1.5	1.70	0.85

Nominal diam, in.	Class *A* 100-ft head		Class *B* 200-ft head		Class *C* 300-ft head		Class *D* 400-ft head	
	Thickness, in.	Wt. of 12-ft length, lb	Thickness, in.	Wt. of 12-ft length, lb	Thickness, in.	Wt. of 12-ft length, lb	Thickness, in.	Wt. of 12-ft length, lb
3	0.39	175	0.42	194	0.45	205	0.48	216
4	0.42	240	0.45	260	0.48	280	0.52	300
6	0.44	370	0.48	400	0.51	430	0.55	460
8	0.46	515	0.51	570	0.56	625	0.60	670
10	0.50	685	0.57	765	0.62	850	0.68	920
12	0.54	870	0.62	985	0.68	1,100	0.75	1,200

Class E for 500-ft head; F for 600-ft head, G for 700-ft head, and H for 800-ft head.

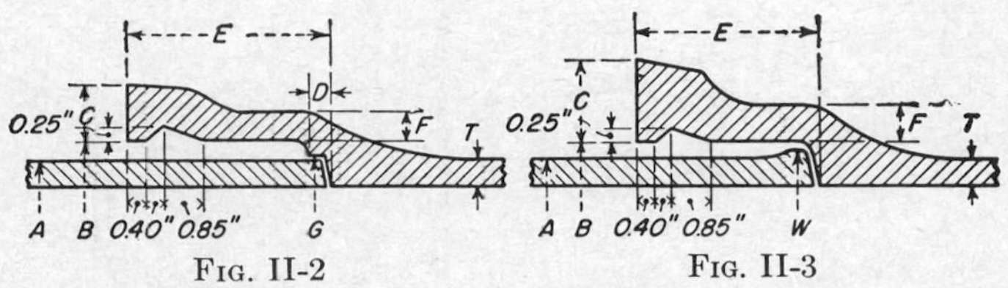

FIG. II-2 FIG. II-3

FIG. II-2. Cast-iron pipe, bell and spigot, type I. See Table II-3.
FIG. II-3. Cast-iron pipe, bell and spigot, type II. See Table II-3.

TABLE II-3. WEIGHTS AND DIMENSIONS OF CAST-IRON WATER PIPE
(Federal Specifications WW–P–421)

Type I. See Fig. II-2. Centrifugally cast in metal-contact molds in 12-ft and 18-ft lengths

Diam, in.	*A*, in.	*B*, in.	*E*, in.	*D*, in.	*G*, in.	For 150 psi			For 250 psi		
						T, in.	*C*, in.	*F*, in.	*T*, in.	*C*, in.	*F*, in.
4	4.80	5.60	3.30	0.30	4.93	0.34	1.06	0.48	0.38	1.06	0.48
6	6.90	7.70	3.88	0.38	7.18	0.37	1.13	0.52	0.43	1.13	0.52
8	9.05	9.85	4.38	0.38	9.31	0.42	1.18	0.57	0.50	1.18	0.57
10	11.10	11.90	4.38	0.38	11.43	0.47	1.23	0.63	0.57	1.23	0.63
12	13.20	14.00	4.38	0.38	13.43	0.50	1.28	0.69	0.62	1.28	0.69

Diam, in.	For 150 psi working press				For 250 psi working press			
	12-ft length		18-ft length		12-ft length		18-ft length	
	Wt. of pipe, lb	Wt. per ft with bell, lb	Wt. of pipe, lb	Wt. per ft with bell, lb	Wt. of pipe, lb	Wt. per ft with bell, lb	Wt. of pipe, lb	Wt. per ft with bell, lb
4	195	16.4	285	15.9	220	18.4	325	17.9
6	315	26.3	460	25.5	350	29.3	515	28.5
8	475	39.4	690	38.3	545	45.5	800	44.3
10	640	53.3	935	51.8	760	63.3	1,115	61.9
12	810	67.4	1,180	65.6	990	82.5	1,450	80.7

Type II. See Fig. II-3. Centrifugally cast in sand-lined molds in 16-, 16½-, and 20-ft lengths

Diam, in.	*A*, in.	*W*, in.	*B*, in.	*E*, in.	For 150 psi			For 250 psi		
					T, in.	*C*, in.	*F*, in.	*T*, in.	*C*, in.	*F*, in.
4	4.80	5.36	5.60	3.50	0.34	1.28	0.44	0.38	1.28	0.44
6	6.90	7.46	7.70	3.50	0.37	1.39	0.48	0.43	1.39	0.48
8	9.05	9.61	9.85	4.00	0.42	1.47	0.52	0.50	1.47	0.62
10	11.10	11.66	11.90	4.00	0.47	1.59	0.57	0.57	1.59	0.69
12	13.20	13.76	14.00	4.00	0.50	1.75	0.62	0.62	1.75	0.74

Diam, in.	For 150 psi working press				For 250 psi working press			
	16-ft length		16½-ft length*		16-ft length		16½-ft length*	
	Wt. of pipe, lb	Wt. per ft with bell, lb	Wt. of pipe, lb	Wt. per ft with bell, lb	Wt. of pipe, lb	Wt. per ft with bell, lb	Wt. of pipe, lb	Wt. per ft with bell, lb
4	255	16.1	265	16.0	290	18.1	300	18.0
6	410	25.7	425	25.6	460	28.7	475	28.6
8	615	38.6	635	38.5	715	44.6	735	44.5
10	835	52.2	860	52.1	995	62.3	1,025	62.2
12	1,055	66.1	1,090	65.9	1,300	81.1	1,335	81.1

* There are no 20-ft lengths smaller than 16 in. in diameter.

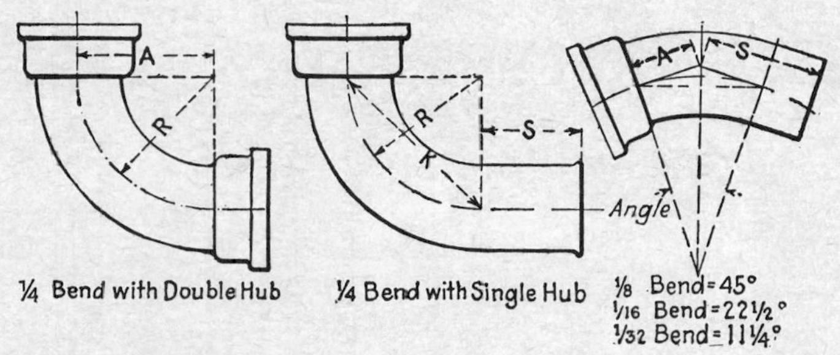

FIG. II-4. Standard specials. Bell-and-spigot cast-iron water pipe. Bends. See Table II-4.

TABLE II-4. STANDARD WEIGHTS AND DIMENSIONS OF BELL-AND-SPIGOT CAST-IRON BENDS OR CURVES

(Adopted by American Water Works Association, 1908. Dimensions in inches; weights in pounds; see Fig. II-4)

Diameter	One-fourth curve				One-eighth curve			One-sixteenth curve			One-thirty-second curve		
	A	*S*	Weight		*R*	*A*	Weight	*R*	*A*	Weight	*R*	*A*	Weight
			1 bell	2 bells			1 bell			1 bell			1 bell
4	16	24	82	94	24	9.94	66	48	9.55	66	120	11.82	66
6	16	24	130	140	24	9.94	105	48	9.55	105	120	11.82	104
8	16	26	200	211	24	9.94	150	48	9.55	150	120	11.82	150
10	16	28	278	280	24	9.94	202	48	9.55	202	120	11.82	192
12	16	28	366	366	24	9.94	265	48	9.55	265	120	11.82	250

All are Class D. In one-fourth bend $R = A$. $S = A + 6$ in. on one-eighth and one-sixteenth curve. $S = A$ on one-thirty-second curve.

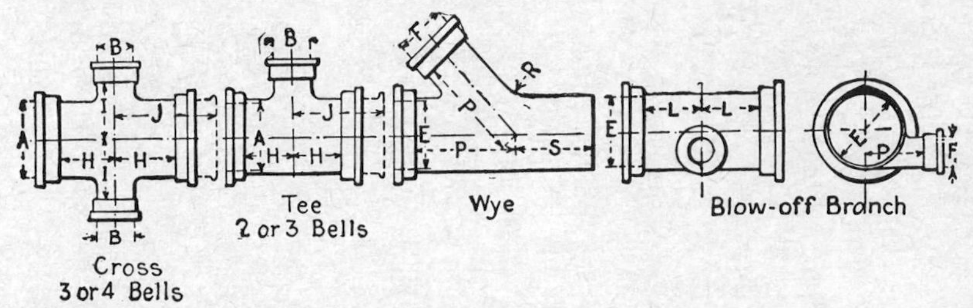

FIG. II-5. Standard specials. Bell-and-spigot cast-iron water pipe. Tees, crosses, wyes, and blowoff branches. See Table II-5.

TABLE II-5. STANDARD WEIGHTS AND DIMENSIONS OF BELL-AND-SPIGOT CAST-IRON TEES, CROSSES, WYES, AND BLOWOFF BRANCHES
(Adopted by American Water Works Association, 1908. Dimensions in inches; weights in pounds; see Fig. II-5)

All fittings, diameter		Tees and crosses						Wyes			Blowoff		
				Weight									
				Tees		Crosses							
Run	Branch	*H*	*J*	2 bells	3 bells	3 bells	4 bells	*P*	*S*	Weight	*L*	*P*	Weight
4	3	11	23	121	120	153	153						
4	4	11	23	125	128	164	166	10.5	11.5	103			
6	3	12	24	173	170	207	204						
6	4	12	24	185	183	223	221	12.0	12.0	159			
6	6	12	24	203	200	259	257	13.0	13.0	181			
8	4	13	25	262	255	301	294	14.0	14.0	221	12	7	227
8	6	13	25	278	270	333	325	15.0	14.0	253			
8	8	13	25	301	294	378	372	16.0	14.0	291			
10	4	14	26	356	338	395	377	....	....	...	12	8	286
10	6	14	26	371	352	424	406	17.0	15.5	348	12	8	300
10	8	14	26	389	371	461	443	18.0	15.5	392			
10	10	14	26	414	395	511	493	18.5	15.5	434			
12	4	15	27	473	445	514	486	....	....	...	12	10	365
12	6	15	27	486	458	540	512	19.0	15.5	490	12	10	379
12	8	15	27	502	474	573	545	21.5	15.5	553			
12	10	15	27	519	491	605	577	21.5	15.5	588			
12	12	15	27	540	512	651	623	21.5	15.5	632			

All are Class D. $H = I$. For wyes $R = 6$ in.

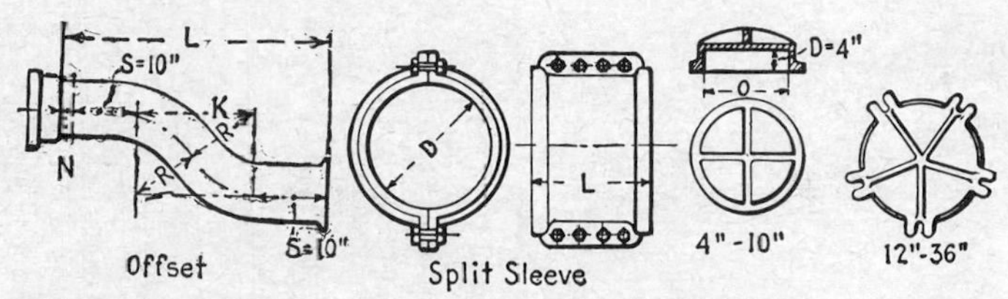

FIG. II-6. Standard specials. Bell-and-spigot cast-iron water pipe. Offsets, caps, split sleeves. See Table II-6.

TABLE II-6. WEIGHTS AND DIMENSIONS OF BELL-AND-SPIGOT CAST-IRON OFFSETS, CAPS, AND SPLIT SLEEVES

(Adopted by American Water Works Association, 1908. Dimensions in inches; weights in pounds; see Fig. II-6)

Diameter	Offset				Split sleeves							Caps	
								Bolts		Weight			
	R	*L*	*K*	Weight	*L*	*D*	Diameter of branch	Size	Number	Without branch	With branch	*D*	*O*
4	8	35.85	13.85	91	10	5.70	..	0.75	6	72	...	4.0	5.70
6	14	46.25	24.25	183	10	7.80	4	0.88	6	86	109	4.0	7.80
8	15	48.00	26.00	280	12	10.00	4	1.00	6	133	156	4.0	10.00
10	16	49.70	27.70	390	12	12.10	4	1.13	6	158	181	4.0	12.10
12	17	51.45	29.45	530	14	14.20	6	1.13	8	222	255	4.0	14.20

Ail are Class D. In offsets $S = 10$ in., and $N = 2$ in., all sizes.

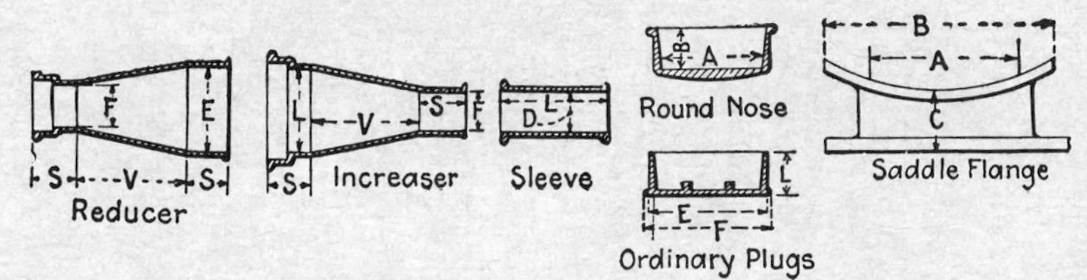

FIG. II-7. Standard specials. Bell-and-spigot cast-iron water pipe. Plugs, sleeves, reducers. See Table II-7.

TABLE II-7. STANDARD WEIGHTS AND DIMENSIONS OF BELL-AND-SPIGOT CAST-IRON PLUGS, SLEEVES, AND REDUCERS

(Adopted by American Water Works Association, 1908. Dimensions in inches; weights in pounds; see Fig. II-7)

Reducers					Sleeves				Plugs					
			Weights											
E	*F*	Spigot ends	Large end bell	Small end bell	Diameter	*D*	*L*	Weight	Diameter	*E*	*F*	*L*	Number of ribs	Weight
6	4	82	104	97	4	5.80	10	47	4	4.90	5.28	5.50	..	8
8	4	104	132	119	4	5.80	15	61						
8	6	121	150	143	6	7.90	10	68	6	7.00	7.35	5.50	..	14
10	4	131	162	146	6	7.90	15	87						
10	6	150	180	169	8	10.10	12	104	8	9.15	9.65	5.50	2	24
10	8	170	201	198	8	10.10	18	119						
12	4	163	201	179	10	12.20	14	123	10	11.20	11.70	6.00	2	38
12	6	181	218	202	10	12.20	18	176						
12	8	202	240	231	12	14.30	15	174	12	13.30	13.80	6.00	2	50
12	10	229	267	261	12	14.30	18	223						

V = 18 in. All are Class D.

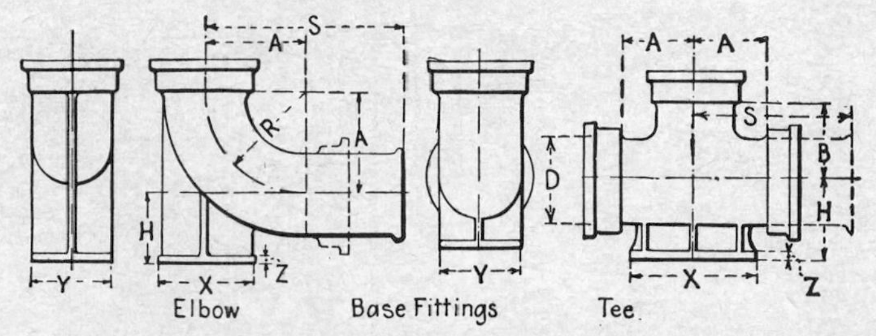

FIG. II-8. Standard specials. Bell-and-spigot cast-iron water pipe. Base elbows and tees. See Table II-8.

TABLE II-8. STANDARD WEIGHTS AND DIMENSIONS OF BELL-AND-SPIGOT CAST-IRON BASE ELBOWS AND TEES

(Adopted by American Water Works Association, 1908. Dimensions in inches; weights in pounds; see Fig. II-8)

Diameter	90° elbow, bell and spigot or two bells						Tees, three bells or two bells and one spigot							
	S	*H*	*X*	*Y*	*Z*	Weight one bell	*A* = *R*	*S*	*H*	*X*	*Y*	*Z*	Weight: Two bells	Weight: Three bells
4	24	5.50	9.0	5.0	0.78	107	11	23	5.5	9.0	5.0	0.68	136	139
6	24	6.50	11.0	7.5	0.83	171	12	24	6.5	11.0	7.5	0.72	223	220
8	26	7.50	13.5	9.5	0.90	260	13	25	7.5	13.5	9.5	0.77	333	326
10	28	9.00	16.0	11.5	1.02	366	14	26	9.0	16.0	11.5	0.86	464	445
12	28	10.00	19.0	13.5	1.13	498	15	27	10.0	19.0	13.5	0.93	613	585

For elbows $A = R = 16$ in. All are Class D.

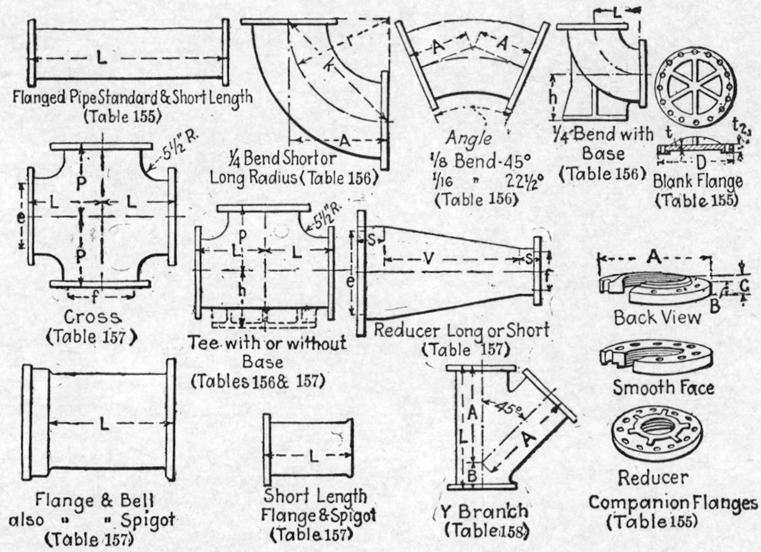

FIG. II-9. Standard specials with flanged ends for cast-iron water pipe. See Tables II-9 to II-12, inclusive. (*Selected from the American and the United States Cast-iron Pipe Companies' catalogs and adapted from the Standard Specifications of the American Water Works Association adopted May* 12, 1908.)

TABLE II-9. WEIGHTS AND DIMENSIONS OF CAST-IRON WATER PIPE, FLANGED
(For waterworks. Dimensions in inches; weights in pounds; see Fig. II-9)

Pipe diam	Flange diam	Flange thickness	Bolt circle diam	Number of bolts	Diam of bolts	Class A		Class B		Class C		Class D	
						Thickness	Wt. per length, 12 ft	Thickness	Wt. per length, 12 ft	Thickness	Wt. per length, 12 ft	Thickness	Wt. per length, 12 ft
3	7.5	¾	6.00	4	⅝	0.39	168	0.42	188	0.45	199	0.48	211
4	9.0	15/16	7.50	8	⅝	0.42	234	0.45	259	0.48	275	0.52	295
6	11.0	1	9.50	8	¾	0.44	358	0.48	398	0.51	421	0.55	451
8	13.5	1⅛	11.75	8	¾	0.46	498	0.51	549	0.56	614	0.60	654
10	16.0	13/16	14.25	12	⅞	0.50	672	0.57	759	0.62	840	0.68	916
12	19.0	1¼	17.00	12	⅞	0.54	876	0.62	998	0.68	1,109	0.75	1,216

TABLE II-10. WEIGHTS AND DIMENSIONS, CAST-IRON BASE TEES, AND PLAIN AND BASE BENDS, FLANGED

(For waterworks. Dimensions in inches; weights in pounds; see Fig. II-9)

Diameter	Weight of			90° plain and base bends		Base tees		
	1/4 bend	1/8 bend	1/16 bend	h*	Weight	$P = L$	h*	Weight
4	69	57	57	5.5	107	11	5.5	139
6	101	83	83	6.5	171	12	6.5	220
8	147	121	121	7.5	260	13	7.5	326
10	209	170	170	9.0	366	14	9.0	445
12	287	238	238	10.0	498	15	10.0	585

For one-fourth bends and base elbows $R = A = 16$ in. For one-eighth bend $R = 24$ in., $A = 9.94$ in. For one-sixteenth bend $R = 48$ in., $A = 9.55$ in.

* These are the same for straight or reducing fittings. Dimension on run controls.

Table II-11. Dimensions and Approximate Weights of Cast-iron Tees, Crosses, Reducers, Blank Flanges, and Short Lengths, Flanged; and Short Lengths, Flanged and Bell

(For waterworks. Dimensions in inches; weights in pounds; see Fig. II-9)

Tees and crosses					Reducers	
			Weight		Weight	
e	f	$L = P$	Tee	Cross	$V = 12$ $S = 2$	$V = 18$ $S = 4$
4	4	11.0	88	114	...	...
6	4	12.0	124	150	67	88
6	6	12.0	137	176	...	...
8	4	13.0	179	191	79	111
8	6	13.0	191	232	91	124
8	8	13.0	209	268	...	...
10	4	14.0	251	277	103	142
10	6	14.0	269	303	115	160
10	8	14.0	280	335	132	183
10	10	14.0	300	392	...	...
12	4	15.0	251	377	123	183
12	6	15.0	364	403	147	200
12	8	15.0	380	435	163	223
12	10	15.0	397	492	187	256
12	12	15.0	426	557		

Short lengths, weights Class D										
	Diam = 4 in.		Diam = 6 in.		Diam = 8 in.		Diam = 10 in.		Diam = 12 in.	
Length for all	Flange and spigot	Bell and flange	Flange and spigot	Bell and flange	Flange and spigot	Bell and flange	Flange and spigot	Bell and flange	Flange and spigot	Bell and flange
6	29	45	...	63	...	97	...	131	...	178
12	33	56	48	81	71	123	101	166	139	224
18	45	68	66	99	97	149	137	202	186	271
24	56	79	84	116	123	174	172	238	233	318
30	67	90	101	134	148	200	208	274	280	365
36	79	102	118	151	174	225	244	309	326	412
48	102	125	154	187	225	277	315	381	420	505
60	124	147	189	222	276	328	387	452	514	599
72	147	170	224	257	327	379	458	523	606	691

Blank flanges									
Diameter	D	t	t_2	Weight	Diameter	D	t	t_2	Weight
3	7.5	0.65	0.85	9	8	13.5	0.75	1.03	36
4	9.0	0.65	0.91	14	10	16.0	0.80	1.15	55
6	11.0	0.70	0.96	23	12	19.0	0.85	1.26	84

TABLE II-12. WEIGHTS AND DIMENSIONS OF FLANGED WYES
(Dimensions in inches; weights in pounds; see Fig. II-9)

Cast-iron water pipe, for waterworks

	American Cast Iron Pipe Co.				United States Cast Iron Pipe Co.				
Diameter	6	8	10	12	4	6	8	10	12
A	21.5	24.0	27.0	30.5	13.5	17.5	20.0	23.0	26.5
B	8.0	9.0	9.5	10.0	3.0	4.0	5.0	5.5	6.0
Weight	219	354	530	781	97	188			

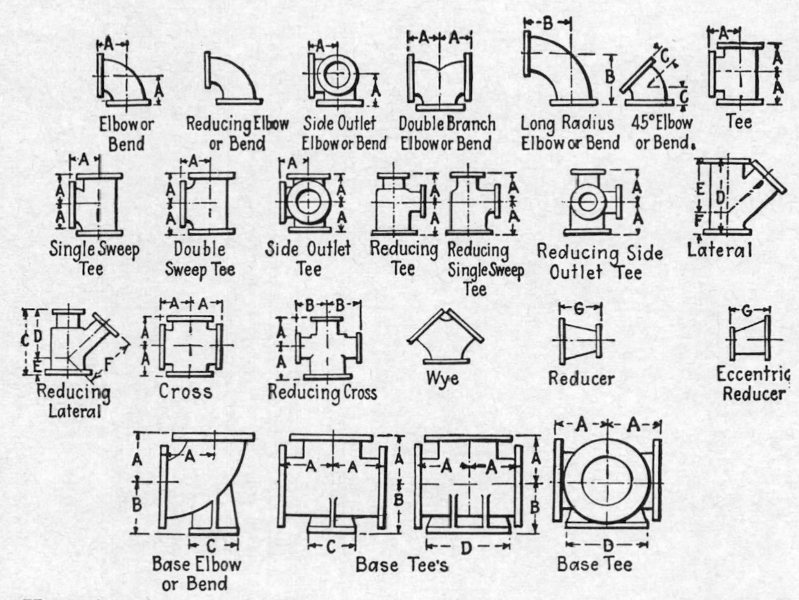

FIG. II-10. American standard fittings, flanged, for cast-iron pipe. See Table II-13.

TABLE II-13. STANDARD DIMENSIONS IN INCHES FOR FLANGED FITTINGS, AMERICAN STANDARD*
(For sketches, see Fig. II-10)

Size	1	1¼	1½	2	2½	3	3½	4	4½	5	6	7	8	9	10	12
Low-pressure† flanged fittings for close work																
AA	7	7½	8	9	10	11	12	13	14	15	16	17	18	20	22	24
A‡	3½	3¾	4	4½	5	5½	6	6½	7	7½	8	8½	9	10	11	12
B	5	5½	6	6½	7	7¾	8½	9	9½	10¼	11½	12¾	14	15¼	16½	19
C‡	1¾	2	2¼	2½	3	3	3½	4	4	4½	5	5½	5½	6	6½	7½
D	7½	8	9	10½	12	13	14½	15	15½	17	18	20½	22	24	25½	30
E	5¾	6¼	7	8	9½	10	11½	12	12½	13½	14½	16½	17½	19½	20½	24½
F	1¾	1¾	2	2½	2½	3	3	3	3	3½	3½	4	4½	4½	5	5½
G	.....	.....	.....	.....	.....	6	6½	7	7½	8	9	10	11	11½	12	14
Diameter of flanges	4¼	4⅝	5	6	7	7½	8½	9	9¼	10	11	12½	13½	15	16	19
Thickness of flanges	7/16	½	9/16	⅝	11/16	¾	13/16	15/16	15/16	15/16	1	1 1/16	1⅛	1⅛	1 3/16	1¼
Diameter of bolt hole circle in flanges	3	3⅜	3⅞	4¾	5½	6	7	7½	7¾	8½	9½	10¾	11¾	13¼	14¼	17
Number of bolts	4	4	4	4	4	4	4	8	8	8	8	8	8	12	12	12
Size of bolts§	7/16	7/16	½	⅝	⅝	⅝	⅝	⅝	¾	¾	¾	¾	¾	¾	⅞	⅞
Length of bolts	1¾	2	2	2¼	2½	2½	2¾	3	3	3	3¼	....	3½	....	3¾	3¾
Extra-heavy¶ flanged fittings																
AA‖	8	8½	9	10	11	12	13	14	15	16	17	18	20	21	23	26
A‡	4	4¼	4½	5	5½	6	6½	7	7½	8	8½	9	10	10½	11½	13
B	5	5½	6	6½	7	7¾	8½	9	9½	10¼	11½	12¾	14	15¼	16½	19
C‡	2	2½	2¾	3	3½	3½	4	4½	4½	5	5½	6	6	6½	7	8
D	8½	9½	11	11½	13	14	15½	16½	18	18½	21½	23½	25½	27½	29½	33½
E	6½	7¼	8½	9	10½	11	12½	13½	14½	15	17½	19	20½	22½	24	27½
F	2	2¼	2½	2½	2½	3	3	3	3½	3½	4	4½	5	5	5½	6
G	.....	.....	.....	.....	.....	6	6½	7	7½	8	9	10	11	11½	12	14
Diameter of flanges	4½	5	6	6½	7½	8¼	9	10	10½	11	12½	14	15	16¼	17½	20½
Thickness of flanges	11/16	¾	13/16	⅞	1	1⅛	1 3/16	1¼	1 5/16	1⅜	1 7/16	1½	1⅝	1¾	1⅞	2
Diameter of bolt hole circle in flanges	3¼	3¾	4½	5	5⅞	6⅝	7¼	7⅞	8½	9¼	10⅝	11⅞	13	14	15¼	17¾
Number of bolts	4	4	4	4	4	8	8	8	8	8	12	12	12	12	16	16
Size of bolts§	½	½	⅝	⅝	¾	¾	¾	¾	¾	¾	¾	⅞	⅞	1	1	1⅛
Length of bolts	2	2¼	2½	2½	3	3¼	3¼	3½	3½	3¾	3¾	4	4¼	4¾	5	5¼

Size, in.

	4		4½		5		6		7		8		9		10		12	
S = Standard low pressure *H* = Extra heavy	*S*	*H*	*S*	*H*	*S*	*H*	*S*	*H*	*S*	*H*	*S*	*H*	*S*	*H*	*S*	*H*	*S*	*H*
A, center to face ells and tees...	6½	7	7	7½	7½	8	8	8½	8½	9	9	10	10	10½	11	11½	12	13
B, center to face base flanges...	6½	7	6¾	7¼	7	7½	7½	8	8¼	8¾	8¾	9¼	9½	10	10	10½	10½	11
C, base flange, across flats of square or diameter of round..	6	6½	6	6½	7	7½	7	7½	7	7½	9	10	9	10	9	10	11	12½
D, anchorage flange, across flats of square................	9	10	9¼	10½	10	11	11	12½	12½	14	13½	15	15	16	16	17	19	19
Size of pipe support for round-base flange..................	2	2	2	2	2½	2½	2½	2½	2½	2½	4	4	4	4	4	4	6	6

* For American Standard see *Trans. ASME*, 1914.

† Low-pressure fittings for steam working pressures up to 25 psi and water working pressures up to 50 psi.

‡ Special angle fittings 1 to 45° use center-to-face dimensions of 45° elbow, and 46 to 90° use center-to-face dimensions of 90° elbow.

§ Bolt holes are drilled ⅛ in. larger than nominal diameter of bolts.

¶ For steam working pressures up to 250 psi.

‖ All extra-heavy flanges have a $\frac{1}{16}$-in. raised face inside of bolt holes. This raised face is included in face-to-face, center-to-face, and thickness-of-flanged dimensions.

NOTES: 1. All reducing fittings have the same center-to-face dimensions as straight-size fittings. The dimensions given in the tables are for "short-body" patterns. "Long-body" patterns are used when the outlets are larger than those given in the table, and they have, therefore, the same dimension as straight-size fitting.

2. The "long-body" pattern will always be used for fittings reducing on the run only, except double-sweep tees, on which the reduced end is always longer than the regular fittings. Dimensions on request to manufacturer.

3. Bull heads or tees having outlets larger than the run will be the same length center-to-face and face-to-face as a tee with all openings the size of the outlet.

4. Double-branch elbows, whether straight or ordinary, carry same center-to-face dimensions as regular elbows of largest straight size.

5. The dimensions of reducing fittings are regulated by the reduction of the outlet or branch.

6. Reducing elbows carry the same dimensions center-to-face as regular elbows of largest straight size.

7. All standard weight fittings and flanges have plain faces.

8. Bolt holes straddle the center line.

9. Bolt holes will not be spot-faced unless so ordered.

10. Square-head bolts with hexagonal nuts are recommended.

11. Hexagonal nuts on sizes 8-in. and smaller can be conveniently pulled up with open end wrenches of minimum design heads.

12. Where long radius fittings are specified it has reference only to elbows which are made in two center-to-face dimensions known as elbows and long-radius elbows, the latter being used only when so specified.

13. Side outlet elbows and side outlet tees, whether straight or reducing, carry the same dimensions, center-to-face and face-to-face, as regular tees having the same reductions.

14. Double-sweep tees are not made reducing on the run.

15. Standard tees, crosses, and laterals, reducing on run only, carry same dimensions face-to-face as largest straight size.

16. Tees, crosses, and laterals reducing on the outlet or branch use the same dimensions as straight sizes of the larger port.

17. Wyes are special and are made to suit conditions.

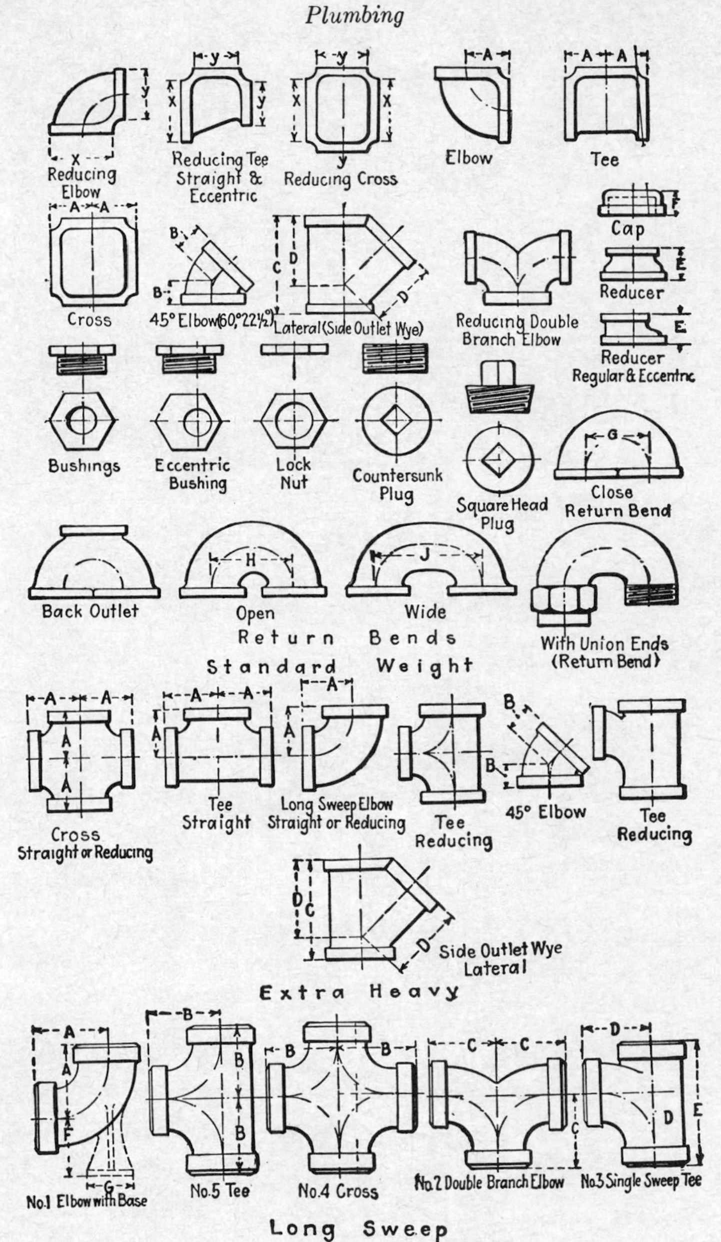

FIG. II-11. Threaded cast-iron fittings for water pipe. See Table II-14.

TABLE II-14. DIMENSIONS, IN INCHES, OF CAST-IRON THREADED FITTINGS FOR WATER SUPPLY
(Manufacturers' standard, Crane Co., see Fig. II-11)

Size	1/4	3/8	1/2	3/4	1	1 1/4	1 1/2	2	2 1/2	3	3 1/2	4	4 1/2	5	6	7	8	9	10	12
Standard weight																				
A	1 3/16	1 5/16	1 1/8	1 5/16	1 7/16	1 3/4	1 15/16	2 1/4	2 11/16	3 1/8	3 7/16	3 3/4	4 1/16	4 7/16	5 1/8	5 13/16	6 1/2	7 3/16	7 7/8	9 1/4
B	3/4	1 3/16	7/8	1	1 1/8	1 5/16	1 7/16	1 11/16	1 15/16	2 3/16	2 3/8	2 5/8	2 13/16	3 1/16	3 7/16	3 7/8	4 1/4	4 11/16	5 3/16	6
C	....	....	2 1/2	3	3 1/2	4 1/4	4 7/8	5 3/4	6 3/4	7 7/8	8 7/8	9 3/4	10 5/8	11 5/8	13 7/16	15 1/4	16 15/16	20 11/16	20 11/16	24 1/8
D	....	....	1 7/8	2 1/4	2 3/4	3 1/4	3 13/16	4 1/2	5 3/16	6 1/8	6 7/8	7 5/8	8 1/2	9 1/4	10 3/4	12 1/4	13 5/8	16 3/4	16 3/4	19 5/8
E	....	....	...	....	....	2 1/8	2 1/4	2 7/16	2 11/16	2 15/16	3 1/8	3 3/8	3 5/8	3 7/8	4 3/8	4 13/16	5 1/4	5 11/16	6 3/16	7 1/8
F	....	....	...	....	....	.....	.....	.....	.....	.....	.....	2 1/16	2 3/16	2 3/8	2 5/8	2 7/8	3 1/8	3 3/8	3 5/8	4 1/4
G	....	....	1 1/4	1 1/2	1 3/4	2 1/4	2 1/2	3 1/4	3 3/4	4 1/4										
H	....	....	...	1 7/8	2 3/8	3	3 1/2	4 1/2	5 1/2	6 1/2										
J	....	....	...	....	*	†	‡	§												
Extra heavy																				
A	....	....	...	....	2	2 1/4	2 9/16	3	3 1/2	4 1/8	4 11/16	5 1/8	5 1/2	6 1/8	7 1/4	8 1/8	9 1/8	......	11 3/8	13 3/8
B	....	....	...	....	1 3/8	1 1/2	1 5/8	1 15/16	2 1/4	2 1/2	2 9/16	2 3/4	3	3 5/16	3 3/4	4	4 1/4	......	4 7/8	5 1/2
C	....	....	...	....	....	4 7/8	5 3/4	6 3/4	7 7/8	8 7/8	.....	10 5/8	......	.....	15 1/4	16 15/16	18 1/8			
D	....	....	...	....	....	3 13/16	4 1/2	5 3/16	6 1/8	6 7/8	.....	8 1/2	......	.....	12 1/4	13 5/8	14			
Long-sweep fittings																				
A	....	....	...	2	2 1/2	3	3 1/2	4	4 1/2	5	5 3/4	6 1/2	7 3/4	7 1/2	9	10 1/2	11 3/4	13	14 1/2	17
B	....	....	...	2	2 1/2	3	3 1/2	4	4 1/2	5	5 3/4	6 1/2	7 3/4	7 1/2	9	10 1/2	11 3/4	Bush 10	14 1/2	17
C	....	....	...	....	2 1/2	3	3 1/2	4	5 1/4	6	6 5/8	6 1/2	7 3/4	9 1/4	10 5/8	10 1/2	11 3/4	Bush 10	14 1/2	17
D	....	....	...	2	2 1/2	3	3 1/2	4	4 3/4	5 1/2	5 13/16	7	7 3/4	8 1/2	9 1/4	10 1/2	11 3/4	10	14 1/2	17
E	....	....	...	3	3 7/8	4 1/2	5 1/4	6 1/8	7	8 1/8	8 13/16	10 1/2	11 3/8	12 1/4	14 1/4	15 1/2	17 3/8	16 1/2	21 1/8	24 3/4
F	....	....	...	....	....	.....	.....	.....	5 1/2	5 3/4	6 1/4	6 1/2	......	7	7 1/2					
G	....	....	...	....	....	.....	.....	.....	4 1/2	5	5	6	......	7	7					

Dimensions of reducing fittings not shown.
* 3, 4, 5, 6, and 8 in.
† 4 and 6 in.
‡ 4 7/8, 6, and 8 in.
§ 4 7/8, 6, 7, and 8 in.

Table II-15. Templates for Drilling Flanges
(American Standard for cast iron, dimensions in inches)

Standard and low pressure							Medium and extra heavy						
Size	Diameter of flange	Thickness of flange	Bolt circle	Number of bolts	Size of bolts	Length of bolts	Size	Diameter of flange	Thickness of flange	Bolt circle	Number of bolts	Size of bolts	Length of bolts
1	4	7/16	3	4	7/16	1 1/2	1	4 1/2	11/16	3 1/4	4	1/2	2
1 1/4	4 1/2	1/2	3 3/8	4	7/16	1 1/2	1 1/4	5	3/4	3 3/4	4	1/2	2 1/4
1 1/2	5	9/16	3 7/8	4	1/2	1 3/4	1 1/2	6	13/16	4 1/2	4	5/8	2 1/2
2	6	5/8	4 3/4	4	5/8	2	2	6 1/2	7/8	5	4	5/8	2 1/2
2 1/2	7	11/16	5 1/2	4	5/8	2 1/4	2 1/2	7 1/2	1	5 7/8	4	3/4	3
3	7 1/2	3/4	6	4	5/8	2 1/4	3	8 1/4	1 1/8	6 5/8	8	3/4	3 1/4
3 1/2	8 1/2	13/16	7	4	5/8	2 1/2	3 1/2	9	1 3/16	7 1/4	8	3/4	3 1/4
4	9	15/16	7 1/2	8	5/8	2 3/4	4	10	1 1/4	7 7/8	8	3/4	3 1/2
4 1/2	9 1/4	15/16	7 3/4	8	3/4	2 3/4	4 1/2	10 1/2	1 5/16	8 1/2	8	3/4	3 1/2
5	10	15/16	8 1/2	8	3/4	2 3/4	5	11	1 3/8	9 1/4	8	3/4	3 3/4
6	11	1	9 1/2	8	3/4	3	6	12 1/2	1 7/16	10 5/8	12	3/4	3 3/4
7	12 1/2	1 1/16	10 3/4	8	3/4	3	7	14	1 1/2	11 7/8	12	7/8	4
8	13 1/2	1 1/8	11 3/4	8	3/4	3 1/4	8	15	1 5/8	13	12	7/8	4 1/4
9	15	1 1/8	13 1/4	12	3/4	3 1/4	9	16 1/4	1 3/4	14	12	1	4 3/4
10	16	1 3/16	14 1/4	12	7/8	3 1/2	10	17 1/2	1 7/8	15 1/4	16	1	5
12	19	1 1/4	17	12	7/8	3 1/2	12	20 1/2	2	17 3/4	16	1 1/8	5 1/4

Both holes are drilled 1/8 in. larger than nominal diameter of bolts.

Table II-16. Templates for Drilling Brass Flanges
(Manufacturers' standard, dimensions in inches)

Heavy, for pressures up to 150 psi							Extra heavy, for pressures up to 250 psi						
Size	Diameter of flange	Thickness of flange	Bolt circle	Number of bolts	Size of bolts	Length of bolts	Size	Diameter of flange	Thickness of flange	Bolt circle	Number of bolts	Size of bolts	Length of bolts
1/4	2 1/2	9/32	1 11/16	4	3/8	1	1/4	3	3/8	2	4	7/16	1 1/4
3/8	2 1/2	9/32	1 11/16	4	3/8	1	3/8	3	3/8	2	4	7/16	1 1/4
1/2	3	5/16	2 1/8	4	3/8	1 1/4	1/2	3 1/2	13/32	2 3/8	4	7/16	1 1/2
3/4	3 1/2	11/32	2 1/2	4	3/8	1 1/4	3/4	4	7/16	2 7/8	4	1/2	1 1/2
1	4	3/8	3	4	7/16	1 1/4	1	4 1/2	1/2	3 1/4	4	1/2	1 3/4
1 1/4	4 1/2	13/32	3 3/8	4	7/16	1 1/2	1 1/4	5	11/32	3 3/4	4	1/2	1 3/4
1 1/2	5	7/16	3 7/8	4	1/2	1 1/2	1 1/2	6	9/16	4 1/2	4	5/8	2
2	6	1/2	4 3/4	4	5/8	1 3/4	2	6 1/2	5/8	5	4	5/8	2
2 1/2	7	9/16	5 1/2	4	5/8	2	2 1/2	7 1/2	11/16	5 7/8	4	3/4	2 1/4
3	7 1/2	5/8	6	4	5/8	2	3	8 1/4	3/4	6 5/8	8	3/4	2 1/2
3 1/2	8 1/2	11/16	7	8	5/8	2 1/4	3 1/2	9	13/16	7 1/4	8	3/4	2 3/4
4	9	11/16	7 1/2	8	5/8	2 1/4	4	10	7/8	7 7/8	8	3/4	2 3/4
4 1/2	9 1/4	23/32	7 3/4	8	3/4	2 1/2	4 1/2	10 1/2	7/8	8 1/2	8	3/4	2 3/4
5	10	3/4	8 1/2	8	3/4	2 1/2	5	11	15/16	9 1/4	8	3/4	3
6	11	13/16	9 1/2	8	3/4	2 3/4	6	12 1/2	1	10 5/8	12	3/4	3
7	12 1/2	7/8	10 3/4	8	3/4	2 3/4	7	14	1 1/16	11 7/8	12	7/8	3 1/4
8	13 1/2	15/16	11 3/4	8	3/4	3	8	15	1 1/8	13	12	7/8	3 1/2
9	15	15/16	13 1/4	12	3/4	3	9	16 1/4	1 1/8	14	12	1	3 1/2
10	16	1	14 1/4	12	7/8	3 1/4	10	17 1/2	1 3/16	15 1/4	16	1	3 3/4
12	19	1 1/16	17	12	7/8	3 1/4	12	20 1/2	1 1/4	17 3/4	16	1 1/8	4

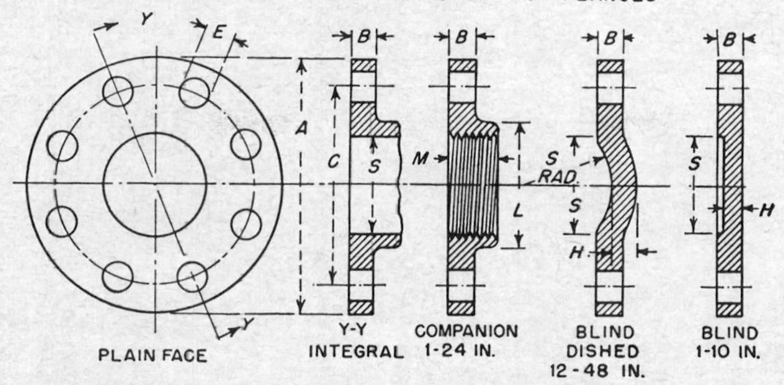

FIG. II-12. Flanges. See Tables II-15 to II-17, inclusive.

TABLE II-17. DIMENSIONS OF CLASS 125 CAST-IRON FLANGES
(Dimensions in inches, see Fig. II-12)

Iron pipe size	Flange diameter	Flange thickness, minimum	Wall thickness	Bolt circle diameter	Number of bolt holes	Bolt hole diameter	Threaded companion flanges		Length of bolts	Weights, lb	
							Hub diameter	Hub length		Companion flanges	Blind flanges
S	*A*	*B*	*H*	*C*		*E*	*L*	*M*			
1	4¼	7/16	3/8	3⅛	4	5/8	1 15/16	11/16	1¾	2	2
1¼	4⅝	½	7/16	3½	4	5/8	2 5/16	13/16	2	2	3
1½	5	9/16	½	3⅞	4	5/8	2 9/16	7/8	2	3	3
2	6	5/8	9/16	4¾	4	¾	3 1/16	1	2¼	5	5
2½	7	11/16	5/8	5½	4	¾	3 9/16	1⅛	2½	7	7
3	7½	¾	11/16	6	4	¾	4¼	1 3/16	2½	8	9
3½	8½	13/16	¾	7	8	¾	4 13/16	1¼	2¾	11	12
4	9	15/16	7/8	7½	8	¾	5 5/16	1 5/16	3	14	16
5	10	15/16	7/8	8½	8	7/8	6 7/16	1 7/16	3	17	20
6	11	1	15/16	9½	8	7/8	7 9/16	1 9/16	3¼	22	25
8	13½	1⅛	1 1/16	11¾	8	7/8	9 11/16	1¾	3½	31	42
10	16	1 3/16	1⅛	14¼	12	1	11 15/16	1 15/16	3¾	45	63
12	19	1¼	13/16	17	12	1	14 1/16	2 3/16	3¾	63	88

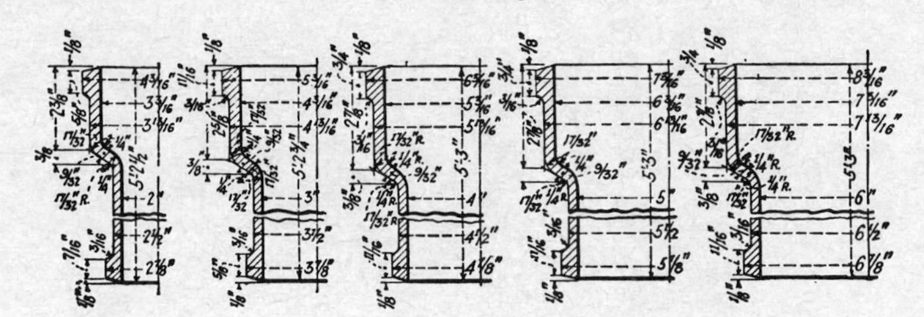

Fig. II-13. Section through extra-heavy cast-iron soil pipes with bell-and-spigot ends, showing dimensions. See Table II-18.

Table II-18. Weights of Bell-and-spigot Cast-iron Soil Pipe per 5-ft Length*
(Weights in pounds, see Fig. II-13)

Size, in.	2	3	4	5	6	7	8	10	12
Standard	17½	22½	32½	42	52	75	85	115	165
Medium	20	30	45	60	75	100	125	175	
Extra heavy	27½	47½	65	85	100	135	170	225	270

* Including hub or bell.

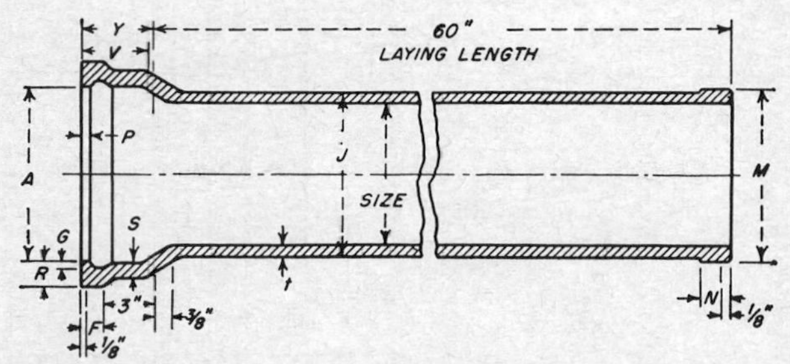

FIG. II-14. Cast-iron soil pipe. See Table II-19.

TABLE II-19. DIMENSIONS* OF HUBS AND SPIGOTS FOR CAST-IRON SOIL PIPE AND FITTINGS, IN INCHES
(Federal Specifications WW–P–401, see Fig. II-14)

Size	*A*	*M*	*J*	*t*	*S*	*R*	*F*	*N*	*G*	*V*	Distance from groove to end, *P*				Weights pipe, lb	
											Y	†	‡	§	Single hub	Double hub
2	3.06	2.75	2.38	0.12	0.18	0.37	0.63	0.56	0.13	2.28	2.50	0.25	0.38	0.31	25	26
3	4.19	3.88	3.50	0.18	0.25	0.43	0.69	0.63	0.13	2.56	2.75	0.25	0.38	0.31	45	47
4	5.19	4.88	4.50	0.18	0.25	0.43	0.75	0.63	0.13	2.81	3.00	0.25	0.38	0.31	60	63
5	6.19	5.88	5.50	0.18	0.25	0.43	0.75	0.69	0.13	2.81	3.00	0.25	0.38	0.31	75	78
6	7.19	6.88	6.50	0.18	0.25	0.43	0.75	0.69	0.13	2.81	3.00	0.25	0.38	0.31	95	100
8	9.50	9.00	8.63	0.25	0.34	0.59	1.06	1.00	0.19	3.38	3.50	0.31	0.50	0.44	150	157
10	11.63	11.13	10.75	0.31	0.40	0.65	1.06	1.00	0.19	3.31	3.50	0.31	0.50	0.44	275	225
12	13.75	13.13	12.75	0.31	0.40	0.65	1.31	1.25	0.19	4.13	4.25	0.38	0.63	0.56	270	285
15	17.00	16.25	15.88	0.37	0.46	0.71	1.31	1.25	0.25	4.06	4.25	0.38	0.63	0.56	375	395

* Tolerances: 6 in. and under, tolerance on $J = \pm 0.13$; on V and $Y = \pm 0.06$.
8 and 10 in., tolerance on $J = \pm 0.19$; on V and $Y = \pm 0.13$.
12 and 15 in., tolerance on $J = \pm 0.25$; on V and $Y = \pm 0.19$.

† Minimum, pipe and fittings.

‡ Maximum, pipe.

§ Maximum, fittings.

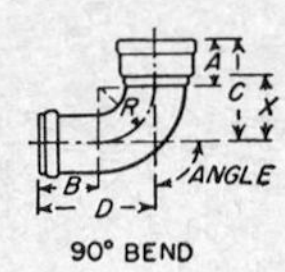

Fig. II-15

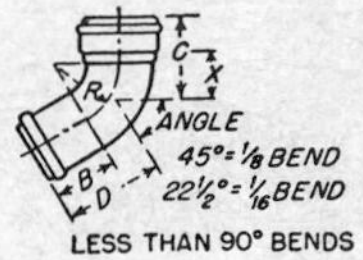

Fig. II-16

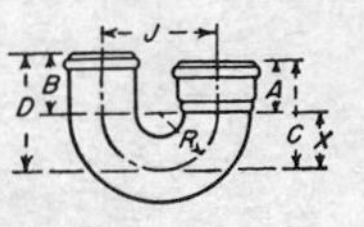

Fig. II-17

Table II-20. Weights and Dimensions of Large-sized 1/4, 1/8, and 1/16 Bends
(Federal Specifications WW–P–401)

1/4 bends, standard sweep (Fig. II-15)

Size	*A*	*B*	*C*	*D**	*R*	*X**	Weight, lb
8	4 1/8	5 1/2	10 1/8	11 1/2	6	6 5/8	51
10	4 1/8	5 1/2	11 1/8	12 1/2	7	7 5/8	78
12	5	7	13	15	8	8 3/4	111
15	5	7	14 1/2	16 1/2	9 1/2	10 1/4	169

1/4 bends, short sweep (Fig. II-15)

Size	*A*	*B*	*C*	*D**	*R*	*X**	Weight, lb
8	4 1/8	5 1/2	12 1/8	13 1/2	8	8 5/8	57
10	4 1/8	5 1/2	13 1/8	14 1/2	9	9 5/8	88
12	5	7	15	17	10	10 3/4	123
15	5	7	16 1/2	18 1/2	11 1/2	12 1/4	187

1/4 bends, long sweep (Fig. II-15)

Size	*A*	*B*	*C*	*D**	*R*	*X**	Weight, lb
8	4 1/8	5 1/2	15 1/8	16 1/2	11	11 5/8	67
10	4 1/8	5 1/2	16 1/8	17 1/2	12	12 5/8	103
12	5	7	18	20	13	13 3/4	141
15	5	7	19 1/2	21 1/2	14 1/2	15 1/4	212

1/8 bends (Fig. II-16)

Size	*A*	*B*	*C*	*D**	*R*	*X**	Weight, lb
8	4 1/8	5 1/2	6 5/8	8	6	3 1/8	41
10	4 1/8	5 1/2	7	8 3/8	7	3 1/2	61
12	5	7	8 5/16	10 5/16	8	4 1/16	87
15	5	7	8 15/16	10 15/16	9 1/2	4 11/16	129

1/16 bends (Fig. II-16)

Size	*A*	*B*	*C*	*D**	*R*	*X**	Weight
8	4 1/8	5 1/2	5 5/16	6 11/16	6	1 13/16	36
10	4 1/8	5 1/2	5 1/2	6 7/8	7	2	52
12	5	7	6 5/8	8 5/8	8	2 3/8	75
15	5	7	6 7/8	8 7/8	9 1/2	2 5/8	108

* Laying length.

TABLE II-21. WEIGHTS AND DIMENSIONS OF BENDS, CAST-IRON SOIL PIPE (Dimensions shown are for both manufacturers' standards and for Federal Specifications WW–P–401, except as noted. Dimensions in inches)

Dimension	Sweep	Size of fittings, in.					
		2*	2†	3‡	4‡	5‡	6‡
¼ bends = 90°							
A	All	3	2¾	3¼	3½	3½	3½
B	All	3	3	3½	4	4	4
C	Regular	6	5¾	6¾	7½	8	8½
	Short	8	7¾	8¾	9½	10	10½
	Long	11	10¾	11¾	12½	13	13½
D	Regular	6	6	7	8	8½	9
	Short	8	8	9	10	10½	11
	Long	11	11	12	13	13½	14
R	Regular§	3	3	3½	4	4½	5
	Short	5	5	5½	6	6½	7
	Long	8	8	8½	9	9½*–10†	10*–11†
X	Regular	3½	3¼	4	4½	5	5½
	Short	5½	5¼	6	6½	7	7½
	Long	8½	8¼	9	9½	10	10½
Weight, lb	Regular	6¾	5	10¼*	15	19	23½*
				10†			24†
	Short	8¼	6	12½*	17¾*	22½*	27½*
				13†	18†	23†	28†
	Long	10¼	8	15¾*	22	27½*	33½*
				16†		28†	34†

Dimension	Size of fittings, in.					
	2	2	3	4	5	6
⅕ bend = 72°, Mfg. Std. only						
C		5 3/16	5 13/16†	6 7/16	6¾	7⅛
D	MS	5 3/16	6 1/16	6 15/16	7¼	7⅝
X	only	2 11/16	3 1/16	3 7/16	3¾	4⅛
Weight, lb		6¼	9¾	13¾	17½	21½
⅙ bend = 60°, Mfg. Std. only						
C		4¾	5¼	5 13/16	6⅛	6⅜
D	MS	4¾	5½	6 5/16	6⅝	6⅞
X	only	2¼	2½	2 13/16	3⅛	3⅜
Weight, lb		6	9¼	13	16½	20
⅛ bend = 45°, both standards						
C	4¼	4	4 15/16*	5 3/16	5⅜	5 9/16
			4 11/16†			
D	4¼	4¼	4 15/16	5 11/16	5⅞	6 1/16
X	1¾	1½	1 15/16	2 3/16	2⅜	2 9/16
			8½*	12¼*	15¼*	18¼*
Weight, lb	5½	4	8†	12†	15†	18†
1/16 bend = 22½°, both standards						
C	3⅝	3⅜	3 15/16	4 5/16	4⅜	4½
D	3⅝	3⅝	4 3/16	4 13/16	4⅞	5
X	1⅛	⅞	1 3/16	1 5/16	1⅜	1½
Weight, lb	5	4	7¾	10¾	13¼	15¾
			8	11	13	16
Return or ½ bend, 180°*						
C		6	6¾	7½	8	8½
D	MS	6	7	8	8½	9
X	only	3¼	4	4½	5	5½
J		6	7	8	9	10
Weight, lb		8¾	14	20¼	26½	33½

* Manufacturers' standards only.
† Federal Specifications only.
‡ Both specifications, unless noted.
§ These figures for radius are applicable to all bends other than short-radius and long-radius bends.

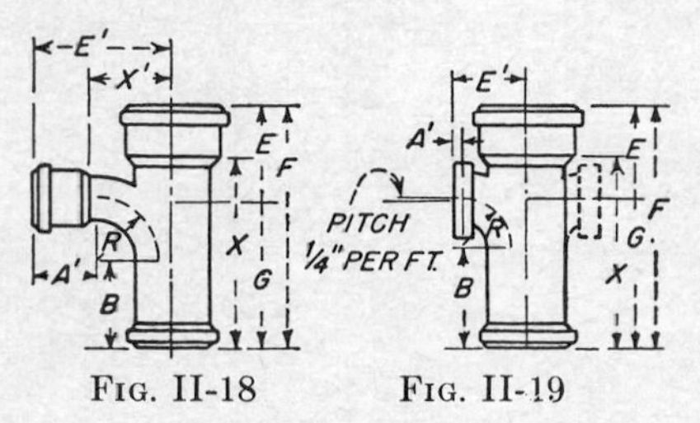

FIG. II-18 FIG. II-19

TABLE II-22. WEIGHTS AND DIMENSIONS OF CAST-IRON SOIL PIPE FITTINGS
(Federal Specifications WW–P–401. Dimensions in inches; weights in pounds*)

Sanitary tee branches, single (see Fig. II-18)

Size	A'	B	E	E'	F	G	R'	X†	X'†	Weight
2	2¾	3¾	4¼	5¼	10½	6¼	2½	8	2¾	8
3	3¼	4	5¼	6¾	12¾	7½	3½	10	4	16
4	3½	4	6	7½	14	8	4	11	4½	22
5	3½	4	6½	8	15	8½	4½	12	5	28
6	3½	4	7	8½	16	9	5	13	5½	34
8	4⅛	5¾	8¾	10⅛	20½	11¾	6	17	6⅝	72
10	4⅛	5¾	9¾	11⅛	22½	12¾	7	19	7⅝	108
12	5	7	11¾	13	26¾	15	8	22½	8¾	153
15	5	7	13¼	14½	29¾	16½	9½	25½	10¼	229
3 × 2	3	4	4¾	6½	11¾	7	3	9	4	14
4 × 2	3	4	5	7	12	7	3	9	4½	17
4 × 3	3¼	4	5½	7¼	13	7½	3½	10	4½	20
5 × 2	3	4	5	7½	12	7	3	9	5	20
5 × 3	3¼	4	5½	7¾	13	7½	3½	10	5	23
5 × 4	3½	4	6	8	14	8	4	11	5	26
6 × 2	3	4	5	8	12	7	3	9	5½	23
6 × 3	3¼	4	5½	8¼	13	7½	3½	10	5½	26
6 × 4	3½	4	6	8½	14	8	4	11	5½	29
6 × 5	3½	4	6½	8½	15	8½	4½	12	5½	32
8 × 2	3	5¾	5¾	9	14½	8¾	3	11	6½	43
8 × 3	3¼	5¾	6¼	9¼	15½	9¼	3½	12	6½	47
8 × 4	3½	5¾	6¾	9½	16½	9¾	4	13	6½	51
8 × 5	3½	5¾	7¼	9½	17½	10¼	4½	14	6½	55
8 × 6	3½	5¾	7¾	9½	18½	10¾	5	15	6½	57
10 × 4	3½	5¾	6¾	10½	16½	9¾	4	13	7½	70
10 × 5	3½	5¾	7¼	10½	17½	10¼	4½	14	7½	73
10 × 6	3½	5¼	7¾	10½	18½	10¾	5	15	7½	76
10 × 8	4⅛	5¾	8¾	11⅛	20½	11¾	6	17	7⅝	96
12 × 4	3½	7	7¾	11½	18¾	11	4	14½	8½	95
12 × 5	3½	7	8¼	11½	19¾	11½	4½	15½	8½	99
12 × 6	3½	7	8¾	11½	20¾	12	5	16½	8½	103
12 × 8	4⅛	7	9¾	12⅛	22¾	13	6	18½	8⅝	120
12 × 10	4⅛	7	10¾	12⅛	24¾	14	7	20½	8⅝	134
15 × 6	3½	7	8¾	13	20¾	12	5	16½	10	142
15 × 8	4⅛	7	9¾	13⅝	22¾	13	6	18½	10⅛	162
15 × 10	4⅛	7	10¾	13⅝	24¾	14	7	20½	10⅛	180
15 × 12	5	7	11¾	14½	26¾	15	8	22½	10¼	198

Sanitary tee branches, tapped, single and double (see Fig. II-19)

Size	A'	B	E	E'	F	G	R'	X†	Weight	
									Single	Double
2 × 2	1 3/16	4	4¼	3 1/16	10½	6¼	2¼	8	8	10
3 × 2	1 3/16	4¾	4¾	3 9/16	11¾	7	2¼	9	12	14
4 × 2	1 3/16	4¾	5	4 1/16	12	7	2¼	9	15	17
5 × 2	1 3/16	4¾	5	4 9/16	12	7	2¼	9	18	20
6 × 2	1 3/16	4¾	5	5 1/16	12	7	2¼	9	21	23

Size	B	E'	R'	Size	B	E'	R'
2	4½	2 13/16	1¾	5	5¼	4 5/16	1¾
3	5¼	3 5/16	1¾	6	5¼	4 13/16	1¾
4	5¼	3 13/16	1¾				

* For details of hubs and spigots see Table II-19.

† Laying lengths.

TABLE II-23. WEIGHTS AND DIMENSIONS OF CAST-IRON SOIL-PIPE FITTINGS
(Federal Specifications WW-P-401. Dimensions in inches; weights in pounds*)

Wye branches, single and double (Fig. II-20)

Size	*B*	*E*	*E'*	*F*	*G*	*X*	*X'*	Weight	
								Single	Double
2	3½	6½	6½	10½	4	8	4	8	11
3	4	8¼	8¼	13¼	5	10½	5½	17	23
4	4	9¾	9¾	15	5¼	12	6¾	24	32
5	4	11	11	16½	5½	13½	8	32	41
6	4	12¼	12¼	18	5¾	15	9¼	40	51
8	5½	15 5/16	15 5/16	23	7 11/16	19½	11 13/16	82	107
10	5½	18	18	26	8	22½	14½	133	168
12	7	21⅛	21⅛	31¼	10⅛	27	16⅞	186	236
15	7	25	25	35¾	10¾	31½	20¾	290	368
3 × 2	4	7 9/16	7½	11¾	4 3/16	9	5	14	18
4 × 2	4	8⅜	8¼	12	3⅝	9	5¾	17	21
4 × 3	4	9 1/16	9	13½	4 7/16	10½	6¼	20	26
5 × 2	4	8⅞	9	12	3⅛	9	6½	20	24
5 × 3	4	9⅝	9¾	13½	3⅞	10½	7	24	30
5 × 4	4	10 5/16	10½	15	4 11/16	12	7½	27	35
6 × 2	4	9 7/16	9¾	12	2 9/16	9	7¼	23	27
6 × 3	4	10⅛	10½	13½	3⅜	10½	7¾	27	33
6 × 4	4	10 13/16	11¼	15	4 3/16	12	8¼	31	39
6 × 5	4	11 9/16	11¾	16½	4 15/16	13½	8¾	35	45
8 × 2	5½	10⅞	11	14	3⅛	10½	8½	42	46
8 × 3	5½	11 9/16	11¾	15½	3 15/16	12	9	47	53
8 × 4	5½	12¼	12½	17	4¾	13½	9½	52	60
8 × 5	5½	13	13	18½	5½	15	10	57	66
8 × 6	5½	13 11/16	13½	20	6 5/16	16½	10½	63	73
10 × 4	5½	13 7/16	14⅛	17	3 9/16	13½	11⅛	74	82
10 × 5	5½	14 3/16	14⅝	18½	4 5/16	15	11⅝	80	89
10 × 6	5½	14⅞	15⅛	20	5⅛	16½	12⅛	86	97
10 × 8	5½	16½	16 15/16	23	6½	19½	13 7/16	110	135
12 × 4	7	15⅛	15 7/16	19¼	4⅛	15	12 7/16	97	105
12 × 5	7	15⅞	15 15/16	20¾	4⅞	16½	12 15/16	104	113
12 × 6	7	16 9/16	16 7/16	22¼	5 11/16	18	13 7/16	111	122
12 × 8	7	18 3/16	18¼	25¼	7 1/16	21	14¾	136	161
12 × 10	7	19 11/16	19 5/16	28¼	8 9/16	24	15 13/16	160	195
15 × 6	7	18¼	18¾	22¼	4	18	15¾	152	163
15 × 8	7	19⅞	20 9/16	25¼	5⅜	21	17 1/16	182	207
15 × 10	7	21⅜	21⅝	28¼	6⅞	24	18⅛	213	248
15 × 12	7	22 13/16	23 7/16	31¼	8 7/16	27	19 3/16	242	292

Wye branches, inverted, single and double (Fig. II-23)

Size	*E*	*E'*	*F*	*G*	*X*	*X'*	Weight	
							Single	Double
2	3¼	5⅞	12	8¾	9½	3⅞	9	12
3	4	7⅜	15¼	11¼	12½	4⅝	18	23
4	4½	8⅞	17	12½	14	5⅞	25	32
5	4¾	10⅛	18½	13¾	15½	7⅛	33	42
6	5	11⅜	20	15	17	8⅜	41	51
3 × 2	3¼	6⅝	13¾	10½	11	4⅛	15	18
4 × 2	3 1/16	7⅜	14	10 15/16	11	4⅞	18	21
4 × 3	3¾	8⅛	15½	11¾	12½	5⅜	22	27
5 × 2	2⅝	8 1/16	14	11⅜	11	5 9/16	22	25
5 × 3	3 5/16	8 13/16	15½	12 3/16	12½	6 1/16	25	31
5 × 4	4	9 9/16	17	13	14	6 9/16	29	37
6 × 2	2 3/16	8¾	14	11 13/16	11	6¼	25	29
6 × 3	2⅞	9½	15½	12⅝	12½	6¾	29	34
6 × 4	3 9/16	10¼	17	13 7/16	14	7¼	33	40
6 × 5	4¼	10¾	18½	14¼	15½	7¾	37	46

Wye-branch cleanout with screw plug on main (Fig. II-21)

Size	*E*	*E'*	*F*	*G*	*X*	IPS† tapping	Weight‡
2	5¼	6½	9¼	4	4	1½	9
3	6⅝	8¼	11⅝	5	5½	2½	15
4	7⅞	9¾	13⅛	5¼	6¾	3½	21
5	9⅛	11	14⅝	5½	8	4½	28
6	10⅜	12¼	16⅛	5¾	9¼	5½	37

Wye-branch cleanout with screw plug on branch (Fig. II-22)

Size	*E*	*E'*	*F*	*G*	*X*	IPS† tapping	Weight‡
2	6½	5¼	10½	4	8	1½	9
3	8¼	6⅝	13¼	5	10½	2½	15
4	9¾	7⅞	15	5¼	12	3½	21
5	11	9⅛	16½	5½	13½	4½	28
6	12¼	10⅜	18	5¾	15	5½	37

* For dimensions of hubs and spigots see Table II-19.
† Iron pipe size.
‡ Without plug.

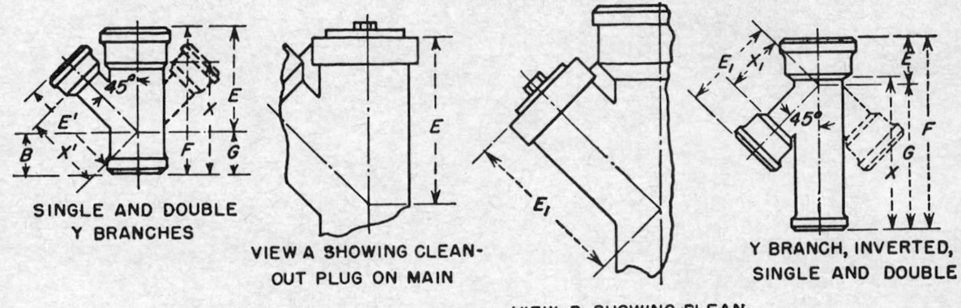

Fig. II-20 Fig. II-21 Fig. II-22 Fig. II-23

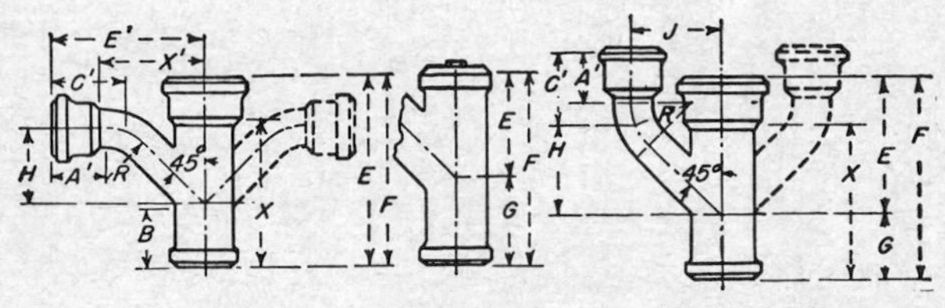

Fig. II-24 Fig. II-25 Fig. II-26

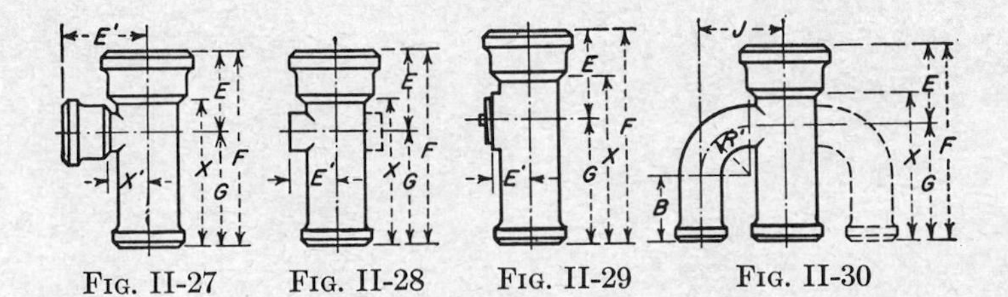

Fig. II-27 Fig. II-28 Fig. II-29 Fig. II-30

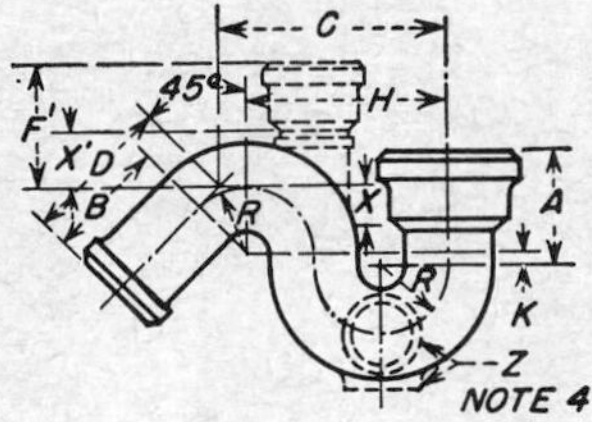

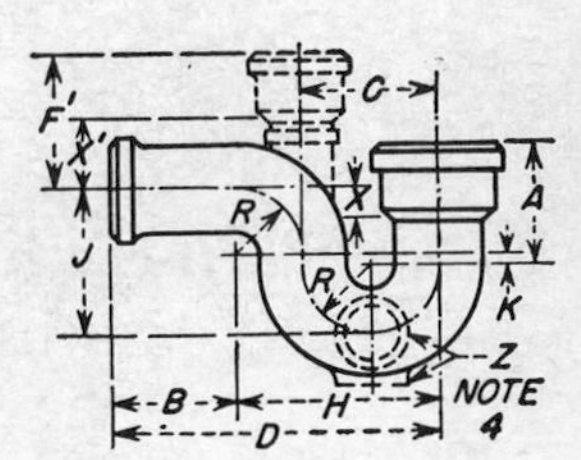

Fig. II-31 Fig. II-32 Fig. II-33

TABLE II-24. WEIGHTS AND DIMENSIONS OF CAST-IRON SOIL-PIPE FITTINGS
(Federal Specifications WW–P–401. Dimensions in inches; weights in pounds*)

Combination wye and 1/8 bend, single and double (Fig. II-24)

Size	A′	B, min.	C′	E	E′	F	G	H	R′	X†	X′†	Weight	
												Single	Double
2	2 3/4	3 1/2	4	6 1/2	7 3/8	10 1/2	4	3 3/8	3	8	4 7/8	10	15
3	3 1/4	4	4 11/16	8 1/4	9 3/4	13 1/4	5	5 1/16	3 1/2	10 1/2	7	20	29
4	3 1/2	4	5 3/16	9 3/4	12	15	5 1/4	6 13/16	4	12	9	29	43
5	3 1/2	4	5 3/8	11	14	16 1/2	5 1/2	8 5/8	4 1/2	13 1/2	11	38	56
6	3 1/2	4	5 9/16	12 1/4	15 7/8	18	5 3/4	10 5/16	5	15	12 7/8	50	75
3 × 2	3	4	4 1/4	7 9/16	8 1/4	11 3/4	4 3/16	4	3	9	5 3/4	15	20
4 × 2	3	4	4 1/4	8 5/16	8 3/4	12	3 11/16	4 1/2	3	9	6 1/4	18	23
4 × 3	3 1/4	4	4 11/16	9	10 1/4	13 1/2	4 1/2	5 9/16	3 1/2	10 1/2	7 1/2	24	33
5 × 2	3	4	4 1/4	8 5/8	9 1/4	12	3 3/8	5	3	9	6 3/4	21	26
5 × 3	3 1/4	4	4 11/16	9 1/2	10 3/4	13 1/2	4	6 1/16	3 1/2	10 1/2	8	27	36
5 × 4	3 1/2	4	5 3/16	10 1/4	12 1/2	15	4 3/4	7 5/16	4	12	9 1/2	33	47
6 × 2	3	4	4 1/4	9 5/16	9 3/4	12	2 11/16	5 1/2	3	9	7 1/4	24	29
6 × 3	3 1/4	4	4 11/16	10	11 1/4	13 1/2	3 1/2	6 9/16	3 1/2	10 1/2	8 1/2	30	39
6 × 4	3 1/2	4	5 3/16	10 3/4	13	15	4 1/4	7 13/16	4	12	10	36	50
6 × 5	3 1/2	4	5 3/8	11 7/16	14 1/2	16 1/2	5 1/16	9 1/8	4 1/2	13 1/2	11 1/2	42	60

Combination wye and 1/8 bend cleanout with screw plug on main (Fig. II-25)

Size	E	F	IPS‡ tapping	Weight§	Size	E	F	IPS‡ tapping	Weight§
2	5 1/4	9 1/4	1 1/2	9	5	9 1/8	14 5/8	4 1/2	34
3	6 5/8	11 5/8	2 1/2	17	6	10 3/8	16 1/8	5 1/2	47
4	7 7/8	13 1/8	3 1/2	26					

Wye branches, upright, single and double (Fig. II-26)

Size	A′	C′	E	F	G	H	J	R′	X	X′	Weight	
											Single	Double
2	2 3/4	4	6 1/2	10 1/2	4	4 1/2	4 1/2	3	8	6	10	15
3	3 1/4	4 11/16	8 1/4	13 1/4	5	5 1/2	5 1/2	3 1/2	10 1/2	7 7/16	20	30
4	3 1/2	5 3/16	9 3/4	15	5 1/4	6 1/2	6 1/2	4	12	8 11/16	28	42
5	3 1/2	5 3/8	11	16 1/2	5 1/2	7 1/2	7 1/2	4 1/2	13 1/2	9 7/8	37	56
6	3 1/2	5 9/16	12 1/4	18	5 3/4	8 1/2	8 1/2	5	15	11 1/16	47	70
3 × 2	3	4 1/4	7 9/16	11 3/4	4 3/16	5	5	3	9	6 3/4	16	20
4 × 2	3	4 1/4	8 5/16	12	3 11/16	5 1/2	5 1/2	3	9	7 1/4	19	24
4 × 3	3 1/4	4 11/16	9	13 1/2	4 1/2	6	6	3 1/2	10 1/2	7 15/16	23	34
5 × 2	3	4 1/4	8 5/8	12	3 3/8	6	6	3	9	7 3/4	22	27
5 × 3	3 1/4	4 11/16	9 1/2	13 1/2	4	6 1/2	6 1/2	3 1/2	10 1/2	8 7/16	27	38
5 × 4	3 1/2	5 3/16	10 1/4	15	4 3/4	7	7	4	12	9 3/16	32	46
6 × 2	3	4 1/4	9 5/16	12	2 11/16	6 1/2	6 1/2	3	9	8 1/4	25	30
6 × 3	3 1/4	4 11/16	10	13 1/2	3 1/2	7	7	3 1/2	10 1/2	8 15/16	30	41
6 × 4	3 1/2	5 3/16	10 3/4	15	4 1/4	7 1/2	7 1/2	4	12	9 11/16	35	49
6 × 5	3 1/2	5 3/8	11 7/16	16 1/2	5 1/16	8	8	4 1/2	13 1/2	10 3/8	40	60

* For details of hubs and spigots see Table II-19.

† Dimensions *X* and *X′* are laying lengths.

‡ Iron pipe size.

§ Without plug.

TABLE II-25. WEIGHTS AND DIMENSIONS OF CAST-IRON SOIL-PIPE FITTINGS (Federal Specifications WW–P–401. Dimensions in inches; weights in pounds)

Tee branches, single (Fig. II-27)

Size	E	E'	F	G	X'	X'†	Weight
2	4¼	4¼	10½	6¼	8	1¾	8
3	5¼	5¼	12¾	7½	10	2½	15
4	6	6	14	8	11	3	21
5	6½	6½	15	8½	12	3½	26
6	7	7	16	9	13	4	32
8	8¾	8¾	20½	11¾	17	5¼	67
10	9¾	9¾	22½	12¾	19	6¼	105
12	11¾	11¾	26¾	15	22½	7½	150
15	13¼	13¾	29¾	16½	25½	9	225
3 × 2	4¾	5	11¾	7	9	2½	13
4 × 2	5	5½	12	7	9	3	16
4 × 3	5½	5¾	13	7½	10	3	19
5 × 2	5	6	12	7	9	3½	19
5 × 3	5½	6¼	13	7½	10	3½	22
5 × 4	6	6½	14	8	11	3½	24
6 × 2	5	6½	12	7	9	4	22
6 × 3	5½	6¾	13	7½	10	4	25
6 × 4	6	7	14	8	11	4	27
6 × 5	6½	7	15	8½	12	4	30
8 × 2	5¾	7¾	14½	8¾	11	5¼	43
8 × 3	6¼	8	15½	9¼	12	5¼	47
8 × 4	6¾	8¼	16½	9¾	13	5¼	50
8 × 5	7¼	8¼	17½	10¼	14	5¼	53
8 × 6	7¾	8¼	18½	10¾	15	5¼	55
10 × 4	6¾	9¼	16½	9¾	13	6¼	70
10 × 5	7¼	9¼	17½	10¼	14	6¼	74
10 × 6	7¾	9¼	18½	10¾	15	6¼	78
10 × 8	8¾	9¾	20½	11¾	17	6¼	93
12 × 4	7½	10¼	18¾	11¼	14½	7¼	93
12 × 5	8	10¼	19¾	11¾	15½	7¼	97
12 × 6	8½	10¼	20¾	12¼	16½	7¼	100
12 × 8	9¾	10¾	22¾	13	18½	7¼	117
12 × 10	10¾	10¾	24¾	14	20½	7¼	130
15 × 6	8½	11¾	20¾	12¼	16½	8¾	140
15 × 8	9¾	12¼	22¾	13	18½	8¾	160
15 × 10	10¾	12¼	24¾	14	20½	8¾	177
15 × 12	11¾	13¼	26¾	15	22½	9	195

Tapped tee branches, single and double* (Fig. II-28)

Size‡	E	E'	F	G	X†	Weight	
						Single	Double
2	4¼	2	10½	6¼	8	7	8
3	4¾	2½	11¾	7	9	12	13
4	5	3	12	7	9	15	17
5	5	3½	12	7	9	18	19
6	5	4	12	7	9	20	22

Tee branches, cleanout with screw plug (Fig. II-29)

Size	E	E'	F	G	X	IPS§ tapping	Weight
2	4¼	2	10½	6¼	8	1½	7
3	5¼	2½	12¾	7½	10	2½	13
4	6	3	14	8	11	3½	17
5	6½	3½	15	8½	12	4	22
6	7	4	16	9	13	4	28

Vent branches, single and double (Fig. II-30)

Size	B	E	F	G	J	R'	X	Weight	
								Single	Double
3¼	4¼	4¼	10½	6¼	4½	3	8	9	12
4	5¼	5¼	12¾	7½	5½	3½	10	18	23
4	6	6	14	8	6½	4	11	25	32
4	6½	6½	15	8½	7½	4½	12	32	41
4	7	7	16	9	8½	5	13	41	51
4	4¾	4¾	11¾	7	5	3	9	14	17
4	5	5	12	7	5½	3	9	18	21
4	5½	5½	13	7½	6	3½	10	21	26
4	5	5	12	7	6	3	9	21	24
4	5½	5½	13	7½	6½	3½	10	24	29
4	6	6	14	8	7	4	11	28	35
4	5	5	12	7	6½	3	9	22	25
4	5½	5½	13	7½	7	3½	10	27	32
4	6	6	14	8	7½	4	11	31	38
4	8½	6½	15	8½	8	4½	12	36	45

* Intended for venting and cleanout purposes only, and branch openings are not intended for use as waste inlets.

† Laying length.

‡ Dimensions and weights given apply to branches tapped 1¼, 1½, and 2 in. IPS.

§ Iron pipe size.

TABLE II-26. WEIGHTS AND DIMENSIONS OF CAST-IRON SOIL-PIPE FITTINGS
(Federal Specifications WW–P–401. Dimensions in inches; weights in pounds)

S traps with or without vent and cleanout (Fig. II-31)

Size	A	B	F'	J*	K
2×2	3	$3\frac{1}{2}$	10	8	..
3×2	$4\frac{1}{2}$	$4\frac{1}{2}$	12	10	$\frac{1}{2}$
3×3	$4\frac{1}{2}$	$4\frac{1}{2}$	$12\frac{1}{4}$	10	$\frac{1}{2}$
4×2	$5\frac{1}{2}$	$5\frac{1}{2}$	14	12	$\frac{1}{2}$
4×3	$5\frac{1}{2}$	$5\frac{1}{2}$	$14\frac{1}{4}$	12	$\frac{1}{2}$
4×4	$5\frac{1}{2}$	$5\frac{1}{2}$	$14\frac{1}{2}$	12	$\frac{1}{2}$
5×4	$6\frac{1}{2}$	$6\frac{1}{2}$	$16\frac{1}{2}$	14	$\frac{1}{2}$
5×5	$6\frac{1}{2}$	$6\frac{1}{2}$	$16\frac{1}{2}$	14	$\frac{1}{2}$
6×4	$7\frac{1}{2}$	$7\frac{1}{2}$	$18\frac{1}{2}$	16	$\frac{1}{2}$
6×6	$7\frac{1}{2}$	$7\frac{1}{2}$	$18\frac{1}{2}$	16	$\frac{1}{2}$

Size	R	X*	X'*	Weight†	Weight‡
2×2	2	4	$7\frac{1}{2}$	9	12
3×2	$2\frac{1}{2}$	$5\frac{3}{4}$	$9\frac{1}{2}$	20	23
3×3	$2\frac{1}{2}$	$5\frac{3}{4}$	$9\frac{1}{2}$	20	24
4×2	3	$7\frac{1}{2}$	$11\frac{1}{2}$	30	33
4×3	3	$7\frac{1}{2}$	$11\frac{1}{2}$	30	34
4×4	3	$7\frac{1}{2}$	$11\frac{1}{2}$	30	36
5×4	$3\frac{1}{2}$	$9\frac{1}{2}$	$13\frac{1}{2}$	41	47
5×5	$3\frac{1}{2}$	$9\frac{1}{2}$	$13\frac{1}{2}$	41	48
6×4	4	$11\frac{1}{2}$	$15\frac{1}{2}$	54	62
6×6	4	$11\frac{1}{2}$	$15\frac{1}{2}$	54	63

$\frac{3}{4}$ S traps with or without vent and cleanout (Fig. II-32)

Size	A	B	C*	D*	F'	H
2×2	3	$3\frac{1}{2}$	$6\frac{13}{16}$	$4\frac{5}{16}$	$4\frac{1}{2}$	6
3×2	$4\frac{1}{2}$	$4\frac{1}{2}$	$8\frac{9}{16}$	$5\frac{9}{16}$	5	$7\frac{1}{2}$
3×3	$4\frac{1}{2}$	$4\frac{1}{2}$	$8\frac{9}{16}$	$5\frac{9}{16}$	$5\frac{1}{4}$	$7\frac{1}{2}$
4×2	$5\frac{1}{2}$	5	$10\frac{1}{4}$	$6\frac{1}{4}$	$5\frac{1}{2}$	9
4×3	$5\frac{1}{2}$	5	$10\frac{1}{4}$	$6\frac{1}{4}$	$5\frac{3}{4}$	9
4×4	$5\frac{1}{2}$	5	$10\frac{1}{4}$	$6\frac{1}{4}$	6	9
5×4	$6\frac{1}{2}$	5	$11\frac{15}{16}$	$6\frac{7}{16}$	$6\frac{1}{2}$	$10\frac{1}{2}$
5×5	$6\frac{1}{2}$	5	$11\frac{15}{16}$	$6\frac{7}{16}$	$6\frac{1}{2}$	$10\frac{1}{2}$
6×4	$7\frac{1}{2}$	5	$13\frac{11}{16}$	$6\frac{11}{16}$	7	12
6×6	$7\frac{1}{2}$	5	$13\frac{11}{16}$	$6\frac{11}{16}$	7	12

Size	K	R	X*	X'*	Weight†	Weight‡
2×2	...	2	$1\frac{1}{2}$	2	8	11
3×2	$\frac{1}{2}$	$2\frac{1}{2}$	$1\frac{1}{4}$	$2\frac{1}{2}$	18	21
3×3	$\frac{1}{2}$	$2\frac{1}{2}$	$1\frac{1}{4}$	$2\frac{1}{2}$	18	22
4×2	$\frac{1}{2}$	3	1	3	27	30
4×3	$\frac{1}{2}$	3	1	3	27	31
4×4	$\frac{1}{2}$	3	1	3	27	33
5×4	$\frac{1}{2}$	$3\frac{1}{2}$	$\frac{1}{2}$	$3\frac{1}{2}$	37	43
5×5	$\frac{1}{2}$	$3\frac{1}{2}$	$\frac{1}{2}$	$3\frac{1}{2}$	37	44
6×4	$\frac{1}{2}$	4	...	4	49	45
6×6	$\frac{1}{2}$	4	...	4	49	58

$\frac{1}{2}$ S traps with or without vent and cleanout (Fig. II-33)

Size	A	B	C	D	F'	H	J	K	R	X*	X'*	Weight†	Weight‡
2×2	3	$3\frac{1}{2}$	4	$9\frac{1}{2}$	$4\frac{1}{2}$	6	4	..	2	$1\frac{1}{2}$	2	8	11
3×2	$4\frac{1}{2}$	$4\frac{1}{2}$	5	12	5	$7\frac{1}{2}$	$5\frac{1}{2}$	$\frac{1}{2}$	$2\frac{1}{2}$	$1\frac{1}{4}$	$2\frac{1}{2}$	17	20
3×3	$4\frac{1}{2}$	$4\frac{1}{2}$	5	12	$5\frac{1}{4}$	$7\frac{1}{2}$	$5\frac{1}{2}$	$\frac{1}{2}$	$2\frac{1}{2}$	$1\frac{1}{4}$	$2\frac{1}{2}$	17	21
4×2	$5\frac{1}{2}$	5	6	14	$5\frac{1}{2}$	9	$6\frac{1}{2}$	$\frac{1}{2}$	3	1	3	25	28
4×3	$5\frac{1}{2}$	5	6	14	$5\frac{3}{4}$	9	$6\frac{1}{2}$	$\frac{1}{2}$	3	1	3	25	29
4×4	$5\frac{1}{2}$	5	6	14	6	9	$6\frac{1}{2}$	$\frac{1}{2}$	3	1	3	25	31
5×4	$6\frac{1}{2}$	5	7	$15\frac{1}{2}$	$6\frac{1}{2}$	$10\frac{1}{2}$	$7\frac{1}{2}$	$\frac{1}{2}$	$3\frac{1}{2}$	$\frac{1}{2}$	$3\frac{1}{2}$	34	40
5×5	$6\frac{1}{2}$	5	7	$15\frac{1}{2}$	$6\frac{1}{2}$	$10\frac{1}{2}$	$7\frac{1}{2}$	$\frac{1}{2}$	$3\frac{1}{2}$	$\frac{1}{2}$	$3\frac{1}{2}$	34	41
6×4	$7\frac{1}{2}$	5	8	17	7	12	$8\frac{1}{2}$	$\frac{1}{2}$	4	...	4	45	61
6×6	$7\frac{1}{2}$	5	8	17	7	12	$8\frac{1}{2}$	$\frac{1}{2}$	4	...	4	45	64
8×4	10	7	10	22	$8\frac{1}{4}$	15	11	1	5	$\frac{1}{2}$	$5\frac{1}{4}$	97	103
8×6	10	7	10	22	$8\frac{1}{4}$	15	11	1	5	$\frac{1}{2}$	$5\frac{1}{4}$	97	107
10×6	12	7	12	25	$9\frac{1}{4}$	18	13	1	6	$1\frac{1}{2}$	$6\frac{1}{4}$	166	175
12×6	$13\frac{1}{2}$	8	15	$30\frac{1}{2}$	$10\frac{1}{4}$	$22\frac{1}{2}$	15	...	$7\frac{1}{2}$	$1\frac{3}{4}$	$7\frac{1}{4}$	242	251
12×8	$13\frac{1}{2}$	8	15	$30\frac{1}{2}$	$10\frac{3}{4}$	$22\frac{1}{2}$	15	...	$7\frac{1}{2}$	$1\frac{3}{4}$	$7\frac{1}{4}$	242	264
15×8	$16\frac{3}{4}$	8	$18\frac{1}{2}$	$35\frac{3}{4}$	$12\frac{1}{4}$	$27\frac{3}{4}$	$18\frac{1}{2}$	...	$9\frac{1}{4}$	$3\frac{1}{4}$	$8\frac{3}{4}$	398	425

* Laying lengths.

† Without hub vent.

‡ With hub vent.

NOTE: Traps with tapped vent and cleanout shall have tappings of the sizes indicated in Table II-27.

TABLE II-27. DIMENSIONS OF TAPS FOR VENTS AND CLEANOUTS IN S TRAPS, ¾ S TRAPS, AND ½ S TRAPS*

Tapping for S† and ¾S‡ traps			Tapping for ½S§ traps					
Size	IPS‖ tapping at *Y*	IPS tapping at *Z*	Size	IPS tapping at *Y*	IPS tapping at *Z*	Size	IPS tapping at *Y*	IPS tapping at *Z*
2	1½	1¼	2	1½	1¼	8	..	3
3	2½	1½	3	2½	1½	10	..	3
4	3½	3	4	3½	3	12	..	3
5	4	3	5	4	3	15	..	3
6	4	3	6	4	3			

* All dimensions are in inches.
† See Fig. II-31.
‡ See Fig. II-32.
§ See Fig. II-33.
‖ Iron pipe size.

PITCH ¼ INCH PER FOOT

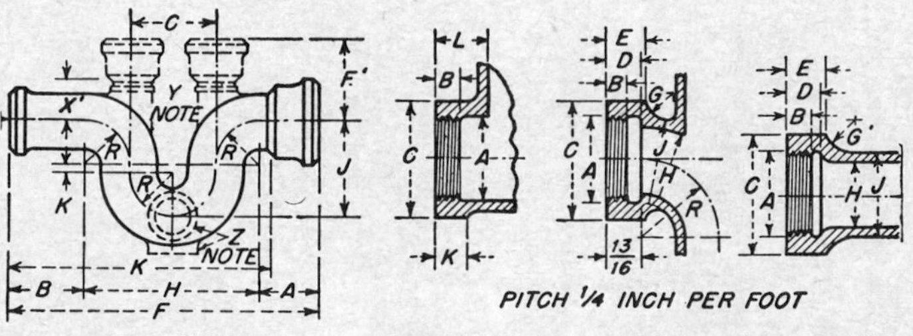

FIG. II-34 FIG. II-35 FIG. II-36 FIG. II-37

FIG. II-34. Running trap. See Table II-28.
FIG. II-35. Details of tapping boss, iron pipe size (IPS) 1¼ to 5½ in. See Table II-28.
FIG. II-36. Details of tapping boss, iron pipe size (IPS) 1¼ to 2 in. See Table II-28.
FIG. II-37. Details of tapping boss, iron pipe size (IPS) 1¼ to 4 in. See Table II-28.

Iron body ferrules with brass screw plug.

⅛ BEND OFFSETS

PLUG FOR HUB

6 IN. AND SMALLER

8 IN. AND LARGER

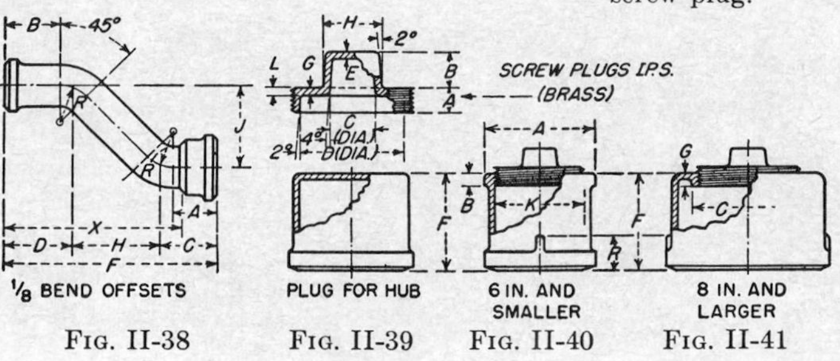

FIG. II-38 FIG. II-39 FIG. II-40 FIG. II-41

Table II-28. Weights and Dimensions of Cast-iron Soil-pipe Fittings (Federal Specifications WW-P-401. Dimensions in inches; weights in pounds)

Running trap with or without single or double vents and cleanout (Fig. II-34)

Size	*A*	*B*	*C*	*F*	*F′*	*H*	*J*	*K*	*R*	*X**	*X′**	Weight†	Weight‡	Weight§
2 × 2	3	3	4	14½	4½	8	4	..	2	12	2	9	12	15
3 × 2	3¼	4½	5	17¾	5	10	5½	½	2½	15	2½	19	22	25
3 × 3	3¼	4½	5	17¾	5¼	10	5½	½	2½	15	2½	19	23	27
4 × 2	3½	5	6	20½	5½	12	6½	½	3	17½	3	28	31	34
4 × 3	3½	5	6	20½	5¾	12	6½	½	3	17½	3	28	32	36
4 × 4	3½	5	6	20½	6	12	6½	½	3	17½	3	28	34	40
5 × 4	3½	5	7	22½	6½	14	7½	½	3½	19½	3½	37	43	49
5 × 5	3½	5	7	22½	6½	14	7½	½	3½	19½	3½	37	45	51
6 × 4	3½	5	8	24½	7	16	8½	½	4	21½	4	48	54	60
6 × 6	3½	5	8	24½	7	16	8½	½	4	21½	4	48	57	66
8 × 4	4⅛	7	10	31⅛	8¼	20	11	1	5	27⅝	5¼	103	109	115
8 × 6	4⅛	7	10	31⅛	8¼	20	11	1	5	27⅝	5¼	103	112	121
10 × 6	4⅛	7	12	35⅛	9¼	24	13	1	6	31⅝	6¼	175	184	193
12 × 6	5	8	15	43	10¼	30	15	...	7½	38¾	7¼	256	265	274
12 × 8	5	8	15	43	10¾	30	15	...	7½	38¾	7¼	256	278	300
15 × 8	5	8	18½	50	12¼	37	18½	...	9¼	45¾	8¾	414	436	463

Dimensions for tapping bosses (Figs. II-35 to II-37)

Size	*A*	*B*	*C*	*D*	*E*	*G*	*H*	*J*	*K*	*L*	*R*
1¼	1 15/16	7/16	2 11/16	¾	⅞	½	1½	1⅞	...	1	1¾
1½	1 15/16	7/16	2 11/16	¾	⅞	½	1⅞	1⅞	¾	1	1¾
2	2 7/16	7/16	3¼	¾	1 5/16	⅝	2	2⅜	...	1	2¼
2½	2 15/16	¾	3⅞	...	....	...	...	...	1		
3	3 9/16	¾	4⅝	1 3/16	1 5/16	1	3	3½	1		
3½	4 1/16	¾	5⅛	...	....	...	...	...	1⅛		
4	4 9/16	1 3/16	5¾	1¼	1 7/16	1⅛	4	4½	1⅛		
4½	5 1/16	⅞	6¼	...	....	...	...	...	1¼		
5½	6⅛	1 5/16	7 9/16	...	....	...	...	...	1⅜		

* Laying lengths.
† Without hub vent.
‡ With single hub vent.
§ With double hub vent.

TABLE II-29. WEIGHTS AND DIMENSIONS OF CAST-IRON SOIL-PIPE FITTINGS
(Federal Specifications WW–P–401. Dimensions in inches; weights in pounds)

1/8 bend offset. Diameter = 2 in.; A = 2 3/4; B = 3 1/2; C = 3 1/2; D = 4 1/4; R = 2; H = J

F	H	X	Weight
9 3/4	2	7 1/4	5
11 3/4	4	9 1/4	6
13 3/4	6	11 1/4	8
15 3/4	8	13 1/4	9
17 3/4	10	15 1/4	10
19 3/4	12	17 1/4	11
21 3/4	14	19 1/4	12
23 3/4	16	21 1/4	13
25 3/4	18	23 1/4	14
27 3/4	20	25 1/4	15
29 3/4	22	27 1/4	16
31 3/4	24	29 1/4	17

1/8 bend offset. Diameter = 3 in.; A = 3 1/4; B = 4; C = 4 1/4; D = 5; R = 2 1/2; H = J

F	H	X	Weight
11 1/4	2	8 1/2	10
13 1/4	4	10 1/2	12
15 1/4	6	12 1/2	14
17 1/4	8	14 1/2	16
19 1/4	10	16 1/2	18
21 1/4	12	18 1/2	20
23 1/4	14	20 1/2	22
25 1/4	16	22 1/2	24
27 1/4	18	24 1/2	26
29 1/4	20	26 1/2	27
31 1/4	22	28 1/2	29
33 1/4	24	30 1/2	31

1/8 bend offset. Diameter = 4 in.; A = 3 1/2; B = 4; C = 4 3/4; D = 5 1/4; R = 3; H = J

F	H	X	Weight
12	2	9	14
14	4	11	16
16	6	13	19
18	8	15	21
20	10	17	24
22	12	19	26
24	14	21	28
26	16	23	31
28	18	25	34
30	20	27	36
32	22	29	39
34	24	31	41

1/8 bend offset. Diameter = 5 in.; A = 3 1/2; B = 4 1/8; C = 4 15/16; D = 5 9/16; R = 3 1/2; H = J

F	H	X	Weight
12 1/2	2	9 1/2	17
14 1/2	4	11 1/2	21
16 1/2	6	13 1/2	24
18 1/2	8	15 1/2	27
20 1/2	10	17 1/2	30
22 1/2	12	19 1/2	33
24 1/2	14	21 1/2	36
26 1/2	16	23 1/2	39
28 1/2	18	25 1/2	42
30 1/2	20	27 1/2	45
32 1/2	22	29 1/2	48
34 1/2	24	31 1/2	51

1/8 bend offset. Diameter = 6 in.; A = 3 1/2; B = 4 1/8; C = 5 3/16; R = 4

H	J	F	X	Weight
2 3/8	2	13	10	21
4	4	15	12	25
6	6	17	14	28
8	8	19	16	32
10	10	21	18	36
12	12	23	20	39
14	14	25	22	43
16	16	27	24	46
18	18	29	26	50
20	20	31	28	54
22	22	33	30	57
24	24	35	32	61

Plug for hub (Fig. II-39)

Size	F	Weight
2	3 1/2	1 3/4
3	3 3/4	3
4	4	4 1/2
5	4	6
6	4	8
8	4 1/2	15
10	4 1/2	23
12	5 1/4	33
15	5 1/4	50

Screw plugs, IPS, § brass (Fig. II-39). B for 1 1/4 = 1/2; for 1 1/2, 2, and 2 1/2 = 3/4; all others = 1 in.

Size	A	C	D	E	G	H*	L†	Weight
1 1/4	1/2	3/4	1 5/16	5/32	1/8	1	5/32	1/4
1 1/2	5/8	3/4	1 1/2	3/16	1/8	1	3/16	3/8
2	5/8	1	2	3/16	1/8	1 1/4	3/16	1/2
2 1/2	3/4	15/16	2 3/8	3/16	5/32	1 1/4	1/4	3/4
3	3/4	1 5/16	2 15/16	3/16	5/32	1 5/8	1/4	1 1/4
3 1/2	3/4	1 1/4	3 7/16	1/4	3/16	1 5/8	1/4	1 1/2
4	7/8	1 5/8	3 15/16	1/4	3/16	2	5/16	2
4 1/2	7/8	1 9/16	4 7/16	5/16	7/32	2	5/16	2 1/2
5	1	1 15/16	4 15/16	5/16	7/32	2 3/8	3/8	3 1/2
5 1/2‡	1	1 15/16	5 7/16	5/16	7/32	2 3/8	3/8	4
6	1	1 7/8	5 15/16	3/8	1/4	2 3/8	3/8	4 3/4

Iron-body ferrules with brass screw plug (Figs. II-40 and II-41). C for 8 -in. size = 7 3/8 in.; for all larger sizes = 7 1/2 in

Size	A	B	F	IPS§ tapping	G	K	R	Weight
2	2 9/16	9/16	3 1/2	1 1/2	1/2	2	1 1/4	1 1/2
3	3 5/8	5/8	3 3/4	2 1/2	9/16	3	1 3/8	2 3/4
4	4 3/4	5/8	4 1/4	3 1/2	9/16	4	1 1/2	4
5	5 13/16	11/16	4 1/4	4 1/2	5/8	5	1 1/2	5
6	6 7/8	11/16	4 1/4	5 1/2	5/8	6 1/16	1 1/2	6
8	.....	....	4 1/2	6	3/4	8	1 7/8	12
10	.....	....	4 1/2	6	3/4	10	1 7/8	20
12	.....	....	5 1/4	6	3/4	12	2 3/8	28
15	.....	....	5 1/4	6	3/4	15	2 3/8	43

* Heads of plugs shall be hexagonal.

† When thread gage is screwed on tightly by hand, large end of gage shall be distance L ± 1 1/2 turns from surface of plug.

‡ Nonstandard pipe thread with same thread proportions as standard sizes.

§ Iron pipe size.

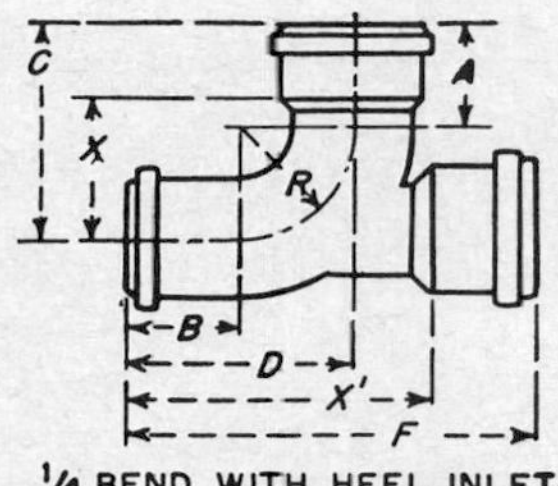

Fig. II-42

Fig. II-43

Fig. II-44

Table II-30. Weights and Dimensions of Cast-iron Soil-pipe Fittings
(Federal Specifications WW–P–401. Dimensions in inches; weights in pounds)

¼ bend, heel inlet (Fig. II-42)										Double hub (Fig. II-43)			
Size	A	B	C	D*	F	R	X*	X'*	Weight	Size	F	X*	Weight
3 × 2	3¼	3½	6¾	7	11½	3½	4	9	13	2	6	1	5
4 × 2	3½	4	7½	8	13	4	4½	10½	18	3	6½	1	8
4 × 3	3½	4	7½	8	13¼	4	4½	10½	19	4	7	1	11
5 × 2	3½	4	8	8½	14¼	4½	5	11¾	22	5	7	1	13
5 × 3	3½	4	8	8½	14½	4½	5	11¾	24	6	7	1	15
5 × 4	3½	4	8	8½	14¾	4½	5	11¾	25	8	8¼	1¼	35
6 × 2	3½	4	8½	9	15	5	5½	12½	27	10	8¼	1¼	50
6 × 3	3½	4	8½	9	15¼	5	5½	12½	29	12	10	1½	67
6 × 4	3½	4	8½	9	15½	5	5½	12½	30	15	10	1½	91

Reducer, spigot ends (Fig. II-43)

Size	B	F	X*	Weight	Size	B	F	X*	Weight
3 × 2	3¾	7¼	4¾	6	8 × 6	4½	9	6	21
4 × 2	4	7½	5	7	10 × 4	4½	9	6	25
4 × 3	4	7¾	5	9	10 × 5	4½	9	6	26
5 × 2	4	7½	5	8	10 × 6	4½	9	6	27
5 × 3	4	7¾	5	10	10 × 8	4½	9½	6	35
5 × 4	4	8	5	11	12 × 4	5¼	9½	6½	33
6 × 2	4	7½	5	9	12 × 5	5¼	9½	6½	34
6 × 3	4	7¾	5	11	12 × 6	5¼	9½	6½	35
6 × 4	4	8	5	12	12 × 8	5¼	10	6½	45
6 × 5	4	8	5	13	12 × 10	5¼	10	6½	51
8 × 2	4½	8½	6	17	15 × 6	5¼	9½	6½	44
8 × 3	4½	8¾	6	18	15 × 8	5¼	10	6½	55
8 × 4	4½	9	6	19	15 × 10	5¼	10	6½	60
8 × 5	4½	9	6	20	15 × 12	5¼	10¾	6½	67

Increaser, tapped (Fig. II-44)

Size	B	F	X*	Weight
1½ × 2	4	10½	8	7
2 × 3	4	11¾	9	9
2 × 4	4	12	9	11
2 × 5	4	12	9	12
2 × 6	4	12	9	14

Increaser, spigot and tapped

Size	B	F	X*	Weight
2 × 3	4	11¾	9	9
2 × 4	4	12	9	10
2 × 5	4	12	9	12
2 × 6	4	12	9	13
3 × 4	4	12	9	12
3 × 5	4	12	9	14
3 × 6	4	12	9	15
4 × 5	4	12	9	15
4 × 6	4	12	9	16
4 × 8	4	15½	12	32
5 × 6	4	12	9	18
5 × 8	4	15½	12	34
5 × 10	4	15½	12	44
6 × 8	4	15½	12	35
6 × 10	4	15½	12	45
6 × 12	4	16¼	12	58
8 × 10	5½	15½	12	55
8 × 12	5½	16¼	12	64
8 × 15	5½	16¼	12	82
10 × 12	5½	16¼	12	73
10 × 15	5½	16¼	12	88
12 × 15	7	16¼	12	97

* Laying lengths.

Table II-31. Weights and Dimensions of Extra-heavy Cast-iron Soil Pipe Bends, Bell and Spigot
(Manufacturers' standard)
(Dimensions in inches; weights in pounds; see Fig. II-45)

Fitting	Dimension	Size of fitting 2 in.	3 in.	4 in.	5 in.	6 in.
All fittings (except traps and offsets)	*A*	3	3 1/4	3 1/2	3 1/2	3 1/2
	B	3	3 1/2	4	4	4
	Regular *R*	3	3 1/2	4	4 1/2	5
	Short sweep *R*	5	5 1/2	6	6 1/2	7
	Long sweep *R*	8	8 1/2	9	9 1/2	10
All traps,* offsets† regular and one-eighth bends	*R*	2	2 1/2	3	3 1/2	4
	R	2	2 1/2	3	3 1/2	4
Return or one-half bend, 180°	*C*	6	6 3/4	7 1/2	8	8 1/2
	D	6	7	8	8 1/2	9
	X	3 1/4	4	4 1/2	5	5 1/2
	Wt	8 3/4	14	20 1/4	26 1/2	33 1/2
	J	6	7	8	9	10
One-fourth bend, regular 90°	*C*	6	6 3/4	7 1/2	8	8 1/2
	D	6	7	8	8 1/2	9
	X	3 1/2	4	4 1/2	5	5 1/2
	Wt	6 3/4	10 1/4	15	19	23 1/2
One-fourth bend short sweep 90°	*C*	8	8 3/4	9 1/2	10	10 1/2
	D	8	9	10	10 1/2	11
	X	5 1/2	6	6 1/2	7	7 1/2
	Wt	8 1/4	12 1/2	17 3/4	22 1/2	27 1/2

Fitting	Dimension	Size of fitting 2 in.	3 in.	4 in.	5 in.	6 in.
One-fourth bend, long sweep, 90°	*C*	11	11 3/4	12 1/2	13	13 1/2
	D	11	12	13	13 1/2	14
	X	8 1/2	9	9 1/2	10	10 1/2
	Wt	10 1/4	15 3/4	22	27 1/2	33 1/2
One-fifth bend, 72°	*C*	5 3/16	5 13/16	6 7/16	6 3/4	7 1/8
	D	5 3/16	6 1/16	6 15/16	7 1/4	7 5/8
	X	2 11/16	3 1/16	3 7/16	3 3/4	4 1/8
	Wt	6 1/4	9 3/4	13 3/4	17 1/2	21 1/2
One-sixth bend, 60°	*C*	4 3/4	5 1/4	5 13/16	6 1/8	6 3/8
	D	4 3/4	5 1/2	6 5/16	6 5/8	6 7/8
	X	2 1/4	2 1/2	2 13/16	3 1/8	3 3/8
	Wt	6	9 1/4	13	16 1/2	20
One-eighth bend, 45°	*C*	4 1/4	4 15/16	5 3/16	5 3/8	5 9/16
	D	4 1/4	4 15/16	5 11/16	5 7/8	6 1/16
	X	1 3/4	1 15/16	2 3/16	2 3/8	2 9/16
	Wt	5 1/2	8 1/2	12 1/4	15 1/4	18 1/4
One-sixteenth bend, 22 1/2°	*C*	3 5/8	3 15/16	4 5/16	4 3/8	4 1/2
	D	3 5/8	4 3/16	4 13/16	4 7/8	5
	X	1 1/8	1 3/16	1 5/16	1 3/8	1 1/2
	Wt	5	7 3/4	10 3/4	13 1/4	15 3/4

* The seal to be not less than 2 1/2 in.

† The angle of the bends of regular offsets shall not be more than 76°. The angle of the bends of one-eighth bend offsets shall be 45°. Wye-eighth bend offsets are produced by combining a regular full wye with a regular one-eighth bend, less the hub of the branch of the former and less the spigot end of the latter.

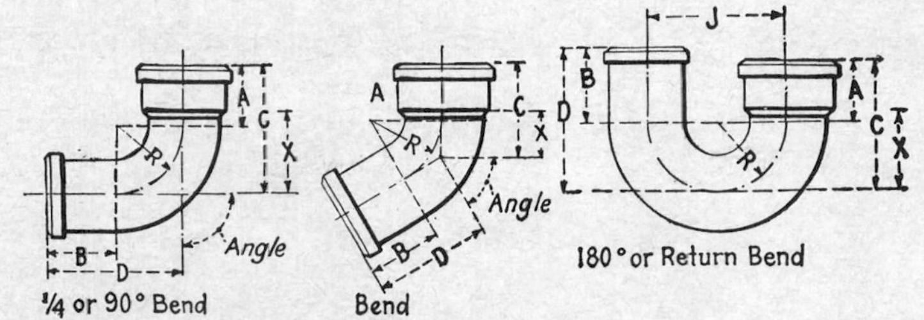

FIG. II-45. Details of cast-iron soil-pipe bends. Dimensions in Table II-31.

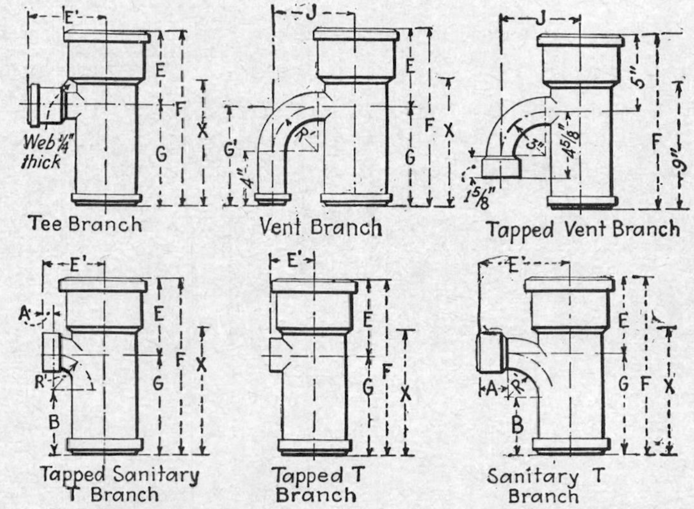

FIG. II-46. Details of cast-iron soil-pipe tee branches. Dimensions in Table II-32.

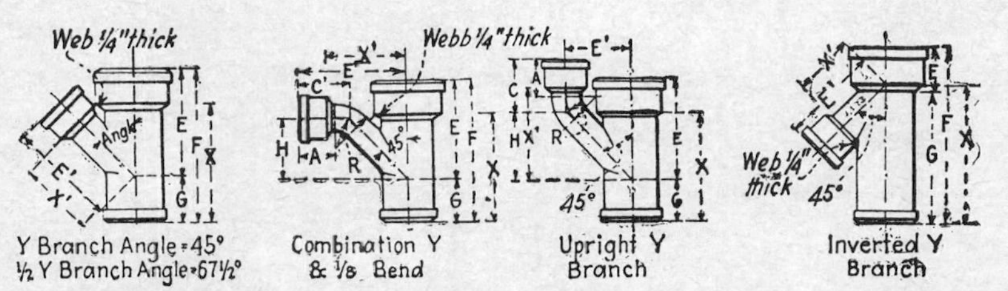

FIG. II-47. Details of cast-iron soil-pipe wye branch. Dimensions in Table II-33

TABLE II-32. WEIGHTS AND DIMENSIONS OF EXTRA-HEAVY CAST-IRON SOIL PIPE FITTINGS, TEE BRANCHES, BELL AND SPIGOT
(Dimensions in inches; weights in pounds; see Fig. II-46)
(Weights are ASTM A 74–18 standards. Dimensions are manufacturers' standards)

Size	A	E	E′	F	G	R′	X	Wt	E	E′	F	G	X	Wt
		Tapped sanitary tee branch							Tapped tee branch					
2 × 2	1/2	4 1/2	3	11 1/2	7	2 1/2	9	9	4 1/2	2	11 1/2	7	9	8 3/4
3 × 2	1/2	4 3/4	3 1/2	11 3/4	7	2 1/2	9	12	4 3/4	2 1/2	11 3/4	7	9	11 3/4
4 × 2	1/2	5	4	12	7	2 1/2	9	15 1/4	5	3	12	7	9	15
5 × 2	1/2	5	4 1/2	12	7	2 1/2	9	18	5	3 1/2	12	7	9	17 3/4
6 × 2	1/2	5	5	12	7	2 1/2	9	21 1/4	5	4	12	7	9	20 1/2
		Sanitary tee branch							Tee branch					
2	3	4 1/2	6	11 1/2	7	3	9	11	4 1/2	4 1/2	11 1/2	7	9	10 1/4
3	3 1/4	5 1/4	6 3/4	12 3/4	7 1/2	3 1/2	10	16 1/4	5 1/4	5 1/4	12 3/4	7 1/2	10	15 1/4
4	3 1/2	6	7 1/2	14	8	4	11	22 1/2	6	6	14	8	11	21
5	3 1/2	6 1/2	8	15	8 1/2	4 1/2	12	28	6 1/2	6 1/2	15	8 1/2	12	26 1/2
6	3 1/2	7	8 1/2	16	9	5	13	34 1/2	7	7	16	9	13	32 1/2
3 × 2	3	4 3/4	6 1/2	11 3/4	7	3	9	14	4 3/4	5	11 3/4	7	9	13 1/4
4 × 2	3	5	7	12	7	3	9	17 1/4	5	5	12	7	9	16 1/2
4 × 3	3 1/4	5 1/2	7 1/4	13	7 1/2	3 1/2	10	19 3/4	5 1/2	5 3/4	13	7 1/2	10	18 3/4
5 × 2	3	5	7 1/2	12	7	3	9	20	5	6	12	7	9	19 1/4
5 × 3	3 1/4	5 1/2	7 3/4	13	7 1/2	3 1/2	10	23	5 1/2	6 1/4	13	7 1/2	10	22
5 × 4	3 1/2	6	8	14	8	4	11	25 1/2	6	6 1/2	14	8	11	24 1/2
6 × 2	3	5	8	12	7	3	9	23	5	6 1/2	12	7	9	22 1/2
6 × 3	3 1/4	5 1/2	8 1/4	13	7 1/2	3 1/2	10	26	5 1/2	6 3/4	13	7 1/2	10	25
6 × 4	3 1/2	6	8 1/2	14	8	4	11	29	6	7	14	8	11	27 1/2
6 × 5	3 1/2	6 1/2	9 1/2	15	8 1/2	4 1/2	12	31 1/2	6 1/2	7	15	8 1/2	12	30

Vent branch							
Size	X	E	F	G and G'	J	R'	Wt
2	9	4½	11½	7	4½	3	11½
3	10	5¼	12¾	7½	5½	3½	7¾
4	11	6	14	8	6½	4	25
5	12	6½	15	8½	7½	4½	32½
6	13	7	16	9	8½	5	41
3 × 2	9	4¾	11¾	7	5	3	14¾
4 × 2	9	5	12	7	5½	3	17¾
4 × 3	10	5½	13	7½	6	3½	21½
5 × 2	9	5	12	7	6	3	21
5 × 3	10	5½	13	7½	6½	3½	24½
5 × 4	11	6	14	8	7	4	28½
6 × 2	9	5	12	7	6½	3	23½
6 × 3	10	5½	13	7½	7	3½	27½

Vent branch							
Size	X	E	F	G and G'	J	R'	Wt
6 × 4	11	6	14	8	7½	4	31½
6 × 5	12	6½	15	8½	8	4½	36

Size	Tapped vent branch		
	F	J	Wt
2 × 2	11½	4½	11¼
3 × 2	11¾	5	14¼
4 × 2	12	5½	17½
5 × 2	12	6	20¼
6 × 2	12	6½	23

TABLE II-33. WEIGHTS AND DIMENSIONS OF EXTRA-HEAVY CAST-IRON SOIL PIPE FITTINGS, BELL AND SPIGOT
(Dimensions in inches; weights in pounds; see Fig. II-47)
(Weights are ASTM A 74–18 standard. Dimensions are manufacturers' standard)

Size	E	E'	F	G	X	X'	Wt	E	E'	F	G	H	R	X	X'	Wt
	Wye branch							Combined wye and one-eighth bend								
2	6¾	6¾	11½	4¾	9	4¼	11	6¾	7¾	11½	4¾	3½	3	9	5¼	12
3	8¼	8¼	13¼	5	10½	5½	17	8¼	9¼	13¼	5	4 9/16	3½	10½	6½	18¾
4	9¾	9¾	15	5¼	12	6¾	24	9¾	10¾	15	5¼	5 9/16	4	12	7¾	27
5	11	11	16½	5½	13½	8	31½	11	12	16½	5½	6⅝	4½	13½	9	35
6	12¼	12¼	18	5¾	15	9¼	39½	12¼	13¼	18	5¾	7 11/16	5	15	10¼	44½
3 × 2	7 9/16	7½	11¾	4 3/16	9	5	14	7 9/16	8¼	11¾	4 3/16	4	3	9	5¾	15
4 × 2	8 5/16	8¼	12	3 11/16	9	5¾	17¼	8 5/16	8¾	12	3 11/16	4½	3	9	6¼	18¼
4 × 3	9	9	13½	14½	10½	6½	20½	9	9¾	13½	4½	5 1/16	3½	10½	7	22½
5 × 2	8 13/16	8 15/16	12	3⅜	9	6 7/16	20	8 13/16	9¼	12	3⅜	5	3	9	6¾	21
5 × 3	9½	9 11/16	13½	4	10½	6 15/16	23½	9½	10¼	13½	4	5 9/16	3½	10½	7½	25½
5 × 4	10¼	10 7/16	15	4⅜	12	7 7/16	27½	10¼	11¼	15	4¾	6 1/16	4	12	8¼	30½
6 × 2	9 5/16	9⅝	12	2 11/16	9	7⅛	23	9 5/16	9¾	12	2 11/16	5½	3	9	7¼	24
6 × 3	10	10⅜	13½	3½	10½	7⅝	27	10	10¾	13½	3½	6 1/16	3½	10½	8	29
6 × 4	10¾	11⅛	15	4¼	12	8⅛	31	10¾	11¾	15	4¼	6 9/16	4	12	8¾	34
6 × 5	11 7/16	11⅝	16½	5 1/16	13½	8⅝	35	11 7/16	12½	16½	5 1/16	7⅛	4½	13½	9½	39
	One-half wye branch							Upright wye branch								
2	5¼	5¼	11½	6¼	9	2¾	10¼	6¾	4½	11½	4¾	4½	3	9	6¼	12½
3	6¼	6¼	12¾	6½	10	3½	15½	8¼	5½	13¼	5	5½	3½	10½	7 7/16	19½
4	7¼	7¼	14	6¾	11	4¼	21½	9¾	6½	15	5¼	6½	4	12	8 11/16	28
5	8	8	15	7	12	5	27½	11	7½	16½	5½	7½	4½	13½	9⅞	36½
6	8¾	8¾	16	7¼	13	5¾	34	12¼	8½	18	5¾	8½	5	15	11 1/16	46

3 × 2	5 5/8	5 3/4	11 3/4	6 1/8	9	3 1/4	13 1/2	7 9/16	5	11 3/4	4 3/16	5	3	9	6 3/4	15 3/4
4 × 2	6 1/8	6 1/4	12	5 7/8	9	3 3/4	16 1/2	8 5/16	5 1/2	12	3 11/16	5 1/2	3	9	7 1/4	18 3/4
4 × 3	6 5/8	6 3/4	13	6 3/8	10	4	19	9	6	13 1/2	4 1/2	6	3 1/2	10 1/2	7 15/16	23
5 × 2	6 3/8	6 7/8	12	5 5/8	9	4 3/8	19 1/2	8 13/16	6	12	3 3/8	6	3	9	7 3/4	21 1/2
5 × 3	6 7/8	7 1/4	13	6 1/8	10	4 1/2	22	9 1/2	6 1/2	13 1/2	4	6 1/2	3 1/2	10 1/2	8 7/16	26 1/2
5 × 4	7 3/8	7 3/4	14	6 5/8	11	4 3/4	25	10 1/4	7	15	4 3/4	7	4	12	9 3/16	31 1/2
6 × 2	6 1/2	7 3/8	12	5 1/2	9	4 7/8	22 1/2	9 5/16	6 1/2	12	2 11/16	6 1/2	3	9	8 1/4	24 1/2
6 × 3	7 1/8	7 7/8	13	5 7/8	10	5 1/8	25 1/2	10	7	13 1/2	3 1/2	7	3 1/2	10 1/2	8 15/16	29 1/2
6 × 4	7 5/8	8 1/4	14	6 3/8	11	5 1/4	28 1/2	10 3/4	7 1/2	15	4 1/4	7 1/2	4	12	9 11/16	35
6 × 5	8 1/8	8 1/2	15	6 7/8	12	5 1/2	31	11 7/16	8	16 1/2	5 1/16	8	4 1/2	13 1/2	10 3/8	40

Inverted wye branch

2	3 1/2	5 7/8	13 1/2	10	11	3 3/8	11 1/2
3	4	7 3/8	15 1/4	11 1/4	12 1/2	4 5/8	18
4	4 1/2	8 7/8	17	12 1/2	14	5 7/8	25 1/2
5	4 3/4	10 1/8	18 1/2	13 3/4	15 1/2	7 1/8	33
6	5	11 3/8	20	15	17	8 3/8	41 1/2
3 × 2	3 1/4	6 5/8	13 3/4	10 1/2	11	4 1/8	15
4 × 2	3 1/16	7 3/8	14	10 15/16	11	4 7/8	18 3/4
4 × 3	3 3/4	8 1/8	15 1/2	11 3/4	12 1/2	5 3/8	22

Inverted wye branch

Size	*E*	*E′*	*F*	*G*	*X*	*X′*	*Wt*
5 × 2	2 5/8	8 1/16	14	11 3/8	11	5 9/16	22
5 × 3	3 5/16	8 13/16	15 1/2	12 3/16	12 1/2	6 1/16	25 1/2
5 × 4	4	9 9/16	17	13	14	6 9/16	29 1/2
6 × 2	2 3/16	8 3/4	14	11 13/16	11	6 1/4	25 1/2
6 × 3	2 7/8	9 1/2	15 1/2	12 5/8	12 1/2	6 3/4	29 1/2
6 × 4	3 9/16	10 1/4	17	13 7/16	14	7 1/4	33
6 × 5	4 1/4	10 3/4	18 1/2	14 1/4	15 1/2	7 3/4	37

For all fittings

Diameter of outlet	2	3	4	5	6
A for all fittings	3	3 1/4	3 1/2	3 1/2	3 1/2

Diameter of outlet	2	3	4	5	6
C for all fittings	4 1/4	4 11/16	5 3/16	5 3/8	5 9/16

TABLE II-34. WEIGHTS AND DIMENSIONS OF EXTRA-HEAVY CAST-IRON SOIL PIPE OFFSETS, BELL AND SPIGOT
(Dimensions in inches; weights in pounds; see Fig. II-48)
(Weights are ASTM A 74–18 standard. Dimensions are manufacturers' standard)

J	*F*	*H*	*X*	*Wt*	*F*	*H*	*X*	*Wt*	*F*	*H*	*X*	*Wt*	*F*	*H*	*X*	*Wt*
	2-in. straight offset				2-in. one-eighth bend offset				4-in. straight offset				4-in. one-eighth bend offset			
2	10½	2	8	7	10½	2	8	7	12	2	9	13¾	12	2	9	13¾
4	11	1	8½	8	12½	4	10	8¼	14	4	11	16¼	14	4	11	16¼
6	11½	1½	9	9	14½	6	12	9½	14	1½	11	18	16	6	13	18¾
8	12	2	9½	9¾	16½	8	14	10¾	14½	2	11½	19¾	18	8	15	21
10	12½	2½	10	10¾	18½	10	16	12	15	2½	12	21½	20	10	17	23½
12	13	3	10½	11¾	20½	12	18	13¼	15½	3	12½	23½	22	12	19	26
14	13½	3½	11	12¾	22½	14	20	14¾	16	3½	13	25	24	14	21	28½
16	14	4	11½	13¾	24½	16	22	16	16½	4	13½	27	26	16	23	31
18	14½	4½	12	14½	26½	18	24	17¼	17	4½	14	29	28	18	25	33½
20	15	5	12½	15½	28½	20	26	18½	17½	5	14½	30½	30	20	27	36
22	15½	5½	13	16½	30½	22	28	19¾	18	5½	15	32½	32	22	29	38½
24	16	6	13½	17½	32½	24	30	21	18½	6	15½	34	34	24	31	41
	3-in. straight offset				3-in. one-eighth bend offset				5-in. straight offset				5-in. one-eighth bend offset			
2	11¼	2	8½	10¼	11¼	2	8½	10¼	12½	2	9½	17¼	12½	2	9½	17¼
4	12¼	1	9½	12	13¼	4	10½	12	14½	4	11½	20½	14½	4	11½	20½
6	12¾	1½	10	13¼	15¼	6	12½	14	15	1½	12	22½	16½	6	13½	23½
8	13¼	2	10½	14¾	17¼	8	14½	15¾	15½	2	12½	25	18½	8	15½	26½
10	13¾	2½	11	16	19¼	10	16½	17¾	16	2½	13	27	20½	10	17½	29½
12	14¼	3	11½	17½	21¼	12	18½	19½	16½	3	13½	29½	22½	12	19½	32½
14	14¾	3½	12	18¾	23¼	14	20½	21½	17	3½	14	31½	24½	14	21½	35½
16	15¼	4	12½	20½	25¼	16	22½	23½	17½	4	14½	33½	26½	16	23½	38½
18	15¾	4½	13	21½	27¼	18	24½	25½	18	4½	15	36	28½	18	25½	41½
20	16¼	5	13½	23	29¼	20	26½	27	18½	5	15½	38	30½	20	27½	44½
22	16¾	5½	14	24½	31¼	22	28½	29	19	5½	16	40½	32½	22	29½	47½
24	17¼	6	14½	26	33¼	24	30½	31	19½	6	16½	42½	34½	24	31½	50½

	6-in. straight offset				6-in. one-eighth bend offset			
2	13	2⅜	10	21	13	2⅜	10	21
4	15	4	12	24½	15	4	12	24½
6	16	1½	13	28	17	6	14	28
8	16½	2	13½	30½	19	8	16	31½
10	17	2½	14	33	21	10	18	35½
12	17½	3	14½	35½	23	12	20	39
14	18	3½	15	38½	25	14	22	42½
16	18½	4	15½	41	27	16	24	46
18	19	4½	16	43½	29	18	26	49½
20	19½	5	16½	46½	31	20	28	53½
22	20	5½	17	49	33	22	30	57
24	20½	6	17½	51½	35	24	32	60½

For all offsets

Internal diameter	2	3	4	5	6	
A	3	3¼	3½	3½	3½	R = 2 in. M = 3¾ in. N = 4¾ in.
B	3	3½	4	4	4	

When J = 2 in., C = 3¾ in. and for all other offsets C = 4½ in. When J = 2 in., D = 4¾ in. and for all other offsets D = 5½ in.

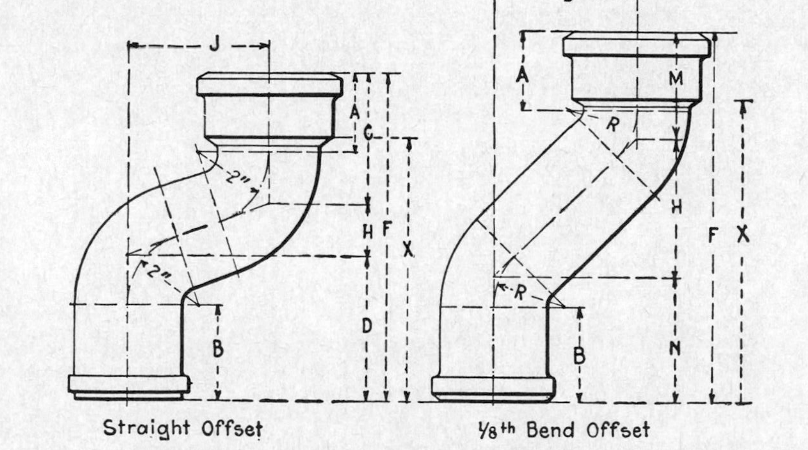

Fig. II-48. Details of cast-iron soil-pipe offsets. Dimensions in Table II-34.

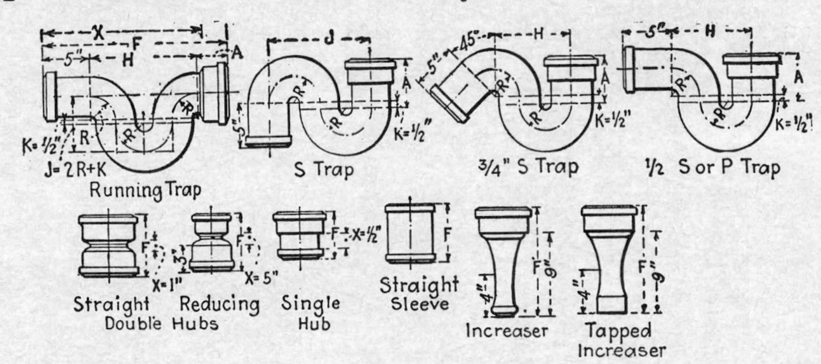

FIG. II-49. Details of cast-iron soil pipe fittings, traps, hubs, sleeves, increasers. Dimensions in Table II-35.

TABLE II-35. WEIGHTS AND DIMENSIONS OF EXTRA-HEAVY CAST-IRON SOIL PIPE FITTINGS, BELL AND SPIGOT

(Manufacturers' standard. Dimensions in inches; weights in pounds; see Fig. II-49)

Size	Double hub (straight)		Single hub		Straight sleeve	
	F	*Wt*	*F*	*Wt*	*F*	*Wt*
2	6	5¾	4	4¾	6	6¼
3	6½	8¼	5	6¾	6½	8½
4	7	10¾	5½	9¼	7	11¼
5	7	12¾	5½	10¾	7	13¼
6	7	14¾	5½	12½	7	15¼

Size	Double hub (reducing)		Increaser		Tapped increaser	
2 × 2					11½	7½
3 × 2	7½	6¼	13¾	8¾	11¾	7½
4 × 2	7½	7	12	10½	12	11

Size	Double hub (reducing)		Increaser		Tapped increaser	
	F	*Wt*	*F*	*Wt*	*F*	*Wt*
4 × 3	7¾	8½	12	12		
5 × 2	7½	8	12	12	12	12½
5 × 3	7¾	9¼	12	13½		
5 × 4	8	10¾	12	14¾		
6 × 2	7½	8¾	12	13½	12	14
6 × 3	7¾	10½	12	14¾		
6 × 4	8	11¾	12	16¼		
6 × 5	8	12¾	12	17½		

Traps

Size	*S* ¾ S ½ S	¾ S ½ S	Running *S* ¾ S ½ S	*S*	Weights				Running trap				
	A	*H*	*R*	*J*	*S*	¾ S	½ S	Running	*A*	*F*	*H*	*J*	*X*
2	3½	6	2	8	12	11¼	10¼	11¾	3	16	8	4½	13½
3	4½	7½	2½	10	20	18½	17¼	19	3¼	18¼	10	5½	15½
4	5½	9	3	12	30	27½	25½	27½	3½	20½	12	6½	17½
5	6½	10½	3½	14	40½	37½	34½	37	3½	22½	14	7½	19½
6	7½	12	4	16	53½	49	45	48	3½	24½	16	8½	21½

TABLE II-36. LAYING LENGTH OF EXTRA-HEAVY CAST-IRON SOIL PIPE FITTINGS, BELL AND SPIGOT AND THREADED WITH BELL AND SPIGOT
(Manufacturers' standard. Dimensions in inches)

Diameter of fittings on the run	Tee	Tapped tee	Sanitary tee	Wye	Tapped sanitary wye	One-half wye	Inverted wye	Tapped inverted wye	Wye and one-eighth bend	Upright wye	Vent branch	Tapped vent branch	Reducer	Increaser	Tapped increaser	Single hub	Double hub
2	9	9	9	9	9	9	11	11	9	9	9	9	5	9	9	½	1
3	10	9	10	10½	9	10	12½	11	10½	10½	10	9	5	9	9	½	1
4	11	9	11	12	9	11	14	11	12	12	11	9	5	9	9	½	1
5	12	9	12	13½	9	12	15½	11	13½	13½	12	9	5	9	9	½	1
6	13	9	13	15	9	13	17	11	15	15	13	9	5	9	9	½	1

Laying length of fittings reducing on branch are same as laying lengths of straight fittings.

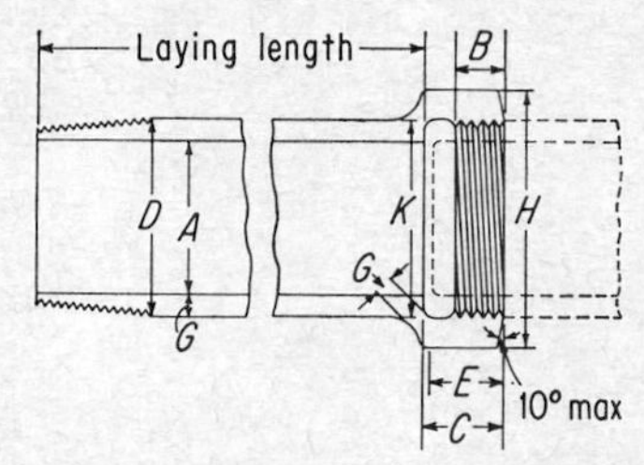

FIG. II-50. Threaded cast-iron pipe for drainage, vent, and waste services. See Table II-37. (*ASA* 40.5–1943.)

TABLE II-37. THREADED CAST-IRON PIPE FOR DRAINAGE, VENT, AND WASTE SERVICES*
(ASA 40.5–1943, see Fig. II-50)

Nominal pipe size	Pipe			Drainage hubs						Nominal weights	
	Nominal diameter, in.		Wall thickness	Thread length	Diameter of groove		End to shoulder†	Minimum band		Type A and barrel of type B per foot	Additional weight of hubs for type B
					Max	Min		Diam	Length		
	D	*A*	*G*	*B*	*K*	*K*	*C*	*H*	*E*		
1¼	1.66	1.23	0.187	0.42	1.73	1.66	0.71	2.39	0.71	3.033	0.60
1½	1.90	1.45	0.195	0.42	1.97	1.90	0.72	2.68	0.72	3.666	0.90
2	2.38	1.89	0.211	0.43	2.44	2.37	0.76	3.28	0.76	5.041	1.00
2½	2.88	2.32	0.241	0.68	2.97	2.87	1.14	3.86	1.14	7.032	1.35
3	3.50	2.90	0.263	0.76	3.60	3.50	1.20	4.62	1.20	9.410	2.80
4	4.50	3.83	0.294	0.84	4.60	4.50	1.30	5.79	1.30	13.751	3.48
5	5.56	4.81	0.328	0.93	5.66	5.56	1.41	7.65	1.41	19.069	5.00
6	6.63	5.76	0.378	0.95	6.72	6.62	1.51	8.28	1.51	26.223	6.60
8	8.63	7.63	0.438	1.06	8.72	8.62	1.71	10.63	1.71	39.820	10.0
10	10.75	9.75	0.438	1.21	10.85	10.75	1.92	13.12	1.93	50.234	
12	12.75	11.75	0.438	1.36	12.85	12.75	2.12	15.47	2.13	60.036	

* All dimensions are in inches except where otherwise stated.

† The length of thread *B* and the end to shoulder *C* shall not vary from the dimensions shown by more than ± the equivalent of the pitch of one thread.

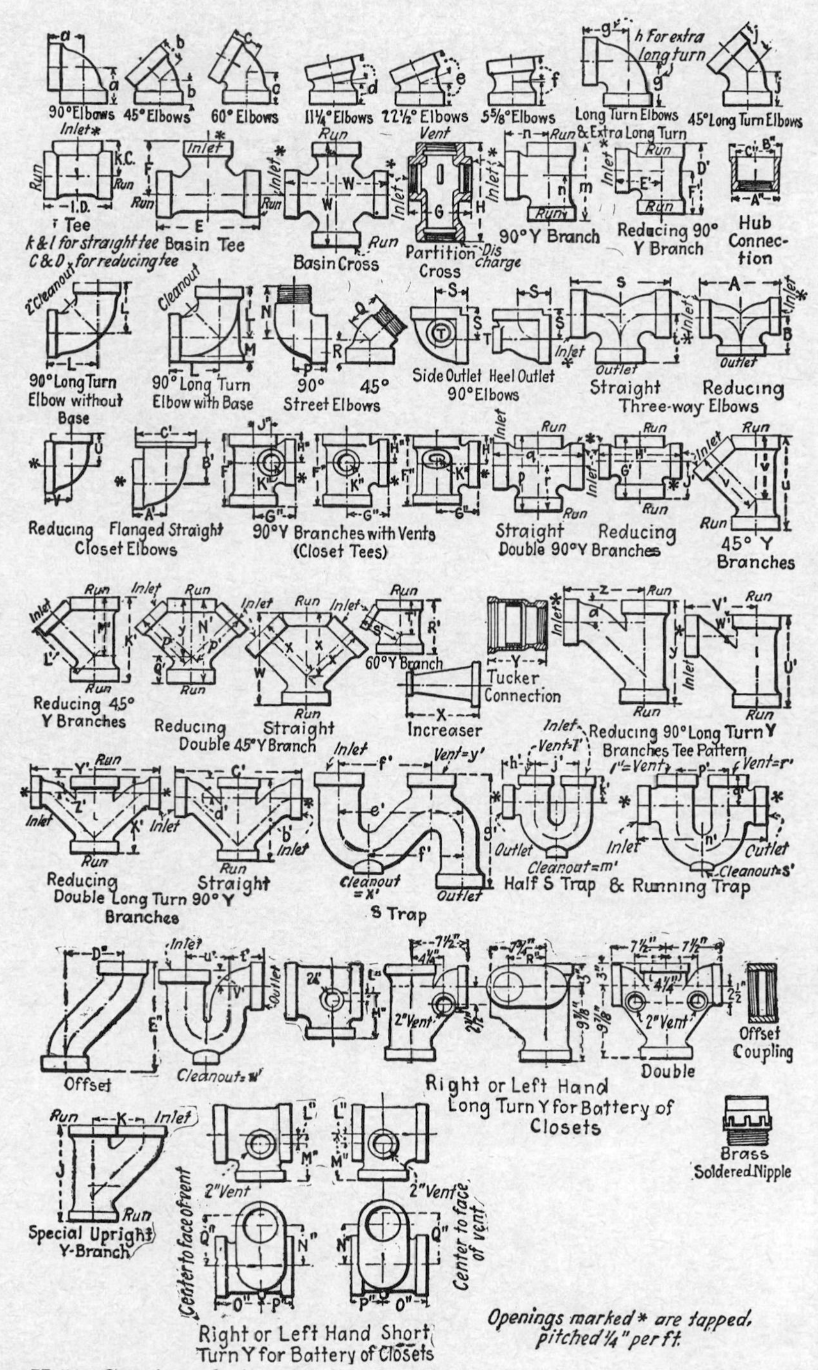

FIG. II-51. Cast-iron drainage fittings. Manufacturers' standard. Dimensions in Tables II-38 and II-39.

TABLE II-38. DIMENSIONS OF DRAINAGE FITTINGS

(Manufacturers' standard. Subject to slight variation and change; dimensions in inches; see Fig. II-51)

Dimension	1	1 1/4	1 1/2	2	2 1/2	3	4	5	6	7	8	10	12
a	.....	1 3/4	2 3/16	2 3/8	2 13/16	3 3/16	3 3/16	4 1/2	5 3/16	5 13/16	6 1/2	7 3/4	9
b	.....	1 5/16	1 7/16	1 3/4	2 1/16	2 3/8	2 3/4	3 3/16	3 1/2	3 7/8	4 3/16	4 7/8	5 1/2
c	.....	1 9/16	1 3/4	2	2 1/2	2 7/8	3 3/8	3 7/8	4 3/16	4 1/2	5 3/8	6 1/4	
d	.....	1	1 7/32	1 13/32	1 21/32	1 13/16	2 1/8	2 3/8	2 1/2	2 5/8	2 3/4	3	
e	.....	1 1/8	1 1/4	1 7/16	1 3/4	2	2 5/16	2 5/8	2 15/16	3 1/4	3 9/16	3 7/8	4 3/16
f	.....	1	1 5/16	1 1/2	.....	1 3/4	2	2 1/4	2 5/16	2 3/8	2 1/2	2 3/4	
g	.....	2 1/4	2 1/2	3 1/16	3 11/16	4 1/4	5 3/16	6 1/8	7 1/8	8 1/8	9	11	13
h	.....	3	3 1/2	4									
j	.....	1 3/4	1 7/8	2 1/4	2 5/8	2 7/8	3 7/16	4 1/8	4 7/8	5 1/2	6	7 1/2	8 3/4
k	.....	1 3/4	2 3/16	2 5/16	2 13/16	3 3/16	4	4 5/8	5 1/16	5 13/16	6 1/2	7 3/4	9
l	.....	3 1/2	4 3/8	4 5/8	5 5/8	6 3/8	8	9 1/4	10 3/8	11 5/8	13	15 1/2	18
m	3 1/4	3 3/4	4 1/4	5 3/16	6 5/16	7 1/4	8 3/4	10 5/16	11 15/16	13 5/8	15 1/16	20	24 1/4
n	1 15/16	2 1/4	2 1/2	3 1/16	3 11/16	4 1/4	5 3/16	6 1/8	7 1/8	8 1/8	9	11 3/8	13 3/4
p	3 1/4	3 3/4	4 1/4	5 3/16	6 5/16	7 1/4	8 3/4	10 5/16	11 15/16	13 5/8	15 1/16	20	
q	3 7/8	4 1/2	5	6 1/8	7 3/8	8 1/2	10 3/8	12 1/4	14 1/4	16 1/4	18	22 3/4	
r	1 15/16	2 1/4	2 1/2	3 1/16	3 11/16	4 1/4	5 3/16	6 1/8	7 1/8	8 1/8	9	11 3/8	
s	.....	4 1/2	5 1/4	6 1/4	7 3/8	8 5/8	10 3/8						
t	.....	2 1/4	2 5/8	3 1/8	3 11/16	4 5/16	5 3/16						
u	.....	5	5 1/2	6 7/16	7 7/8	9	10 7/8	12 15/16	14 3/4	16 11/16	18 13/16	20	24 1/4
v	.....	3 1/4	3 5/8	4 5/16	5 3/8	6 3/16	7 11/16	9 3/16	10 3/4	12 1/4	13 9/16	16 1/8	19 5/8
w	.....	5	5 1/2	6 7/16	7 7/8	9	10 7/8	12 15/16	14 3/4	16 11/16	18 13/16	20	
x	.....	3 1/4	3 5/8	4 5/16	5 3/8	6 3/16	7 11/16	9 3/16	10 3/4	12 1/4	13 9/16	16 1/8	
y	.....	4 3/4	5 3/8	7 1/16	8 1/4	9 13/16	13 3/4	15 3/4	18 11/16	21 5/8	24 9/16		
z	.....	3 5/8	4 1/8	5 7/16	6 1/4	7 1/2	9 7/8	12 1/4	14 9/16	16 7/8	19 5/16		
a'	.....	1 3/16	1 1/4	1 5/8	2	2 5/16	2 7/8	3 1/2	4 1/8	4 3/4	5 1/4		

b′	.....	4 3/4	5 3/8	7 1/16	8 1/4	9 13/16	13 3/4	15 3/4	18 11/16	21 5/8	24 9/16		
c′	.....	7 1/4	8 1/4	10 7/8	12 1/2	15	19 3/4	24 1/2	29 1/8	33 3/4	38 5/8		
d′	.....	1 3/16	1 1/4	1 5/8	2	2 5/16	2 7/8	3 1/2	4 1/8	4 3/4	5 1/4		
e′	.....	.....	....	8 1/4	......	10 3/4	14 1/8	16 3/4	20 1/4				
f′	.....	.....	....	6 1/4	......	8 1/16	10 5/8	12 9/16	15 3/16				
g′	.....	.....	....	8	......	11 1/4	13 3/4	16 7/8	19 3/4				
h′	.....	2 5/16	2 1/2	3 1/16	......	4 3/16	5 1/8	6 1/16	6 9/16	7 3/4	8 15/16		
j′	.....	3 1/8	3 3/8	4	......	5 3/8	7	8 3/8	10 1/8	11 1/4	12 3/8		
k′	.....	2	2 1/4	2 9/16	......	3 1/2	4 3/16	4 3/4	5 7/16	6 1/8	6 3/4		
l′	.....	1 1/4	1 1/2	2	......	3	4	4	4	5	6		
m′	.....	3/4	1	1	......	1 1/4	2	2	2	2	3		
n′	.....	7 3/4	8 3/8	10 1/8	......	13 1/4	17 1/4	20 1/2	23 1/4	26 3/4	30 1/4	36 1/4	41 7/8
p′	.....	3 1/8	3 3/8	4	......	5 3/8	7	8 3/8	10 1/8	11 1/4	12 3/8	14 3/4	17
q′	.....	2	2 1/4	2 9/16	......	3 1/2	4 3/16	4 3/4	5 7/16	6 1/8	6 3/4	8 1/4	9 1/2
r′	.....	1 1/4	1 1/2	2	......	3	4	4	4	5	6	6	6
s′	.....	3/4	1	1	......	1 1/4	2	2	2	2	3	3	3
t′	.....	1 15/16	2 1/8	2 9/16	......	3 3/8							
u′	.....	2	2 1/4	2 3/4	......	3 3/4							
v′	.....	3/4	7/8	7/8	......	1 3/16							
w′	.....	3/4	1	1	......	1 1/4							
x′	.....	.....	....	1	......	1 1/4	2	2	2				
y′	.....	.....	....	2	......	3	4	4	4				

TABLE II-39. DIMENSIONS OF DRAINAGE FITTINGS
(Manufacturers' standard. Subject to slight variation and change; dimensions in inches; see Fig. II-51)

Size	4 by 3	5 by 4	6 by 4	6 by 5
A	9 1/2	11 3/8	12 3/8	13 3/8
B	4 5/16	5 5/16	5 7/16	6 1/4

Size	1 1/2 by 1 1/4	2 by 1 1/2	3 by 2	4 by 2	4 by 3	5 by 2	5 by 3
C	2 5/16	2 1/4	3	3 9/16	3 3/4	4 3/16	4 3/8
D	4 5/8	4 1/8	5 1/2	6	7	6	6 7/8

Size	1 1/4	1 1/2	1 1/2 by 1 1/4	2	2 by 1 1/4
E	4 5/8	5 3/8	5 1/8	7	6 1/4
F	2 5/16	2 11/16	2 9/16	3 1/2	3 1/8

Size	2 by 1 1/2	2 1/2
E	6 1/2	8 1/2
F	3 1/4	4 1/4

Size	1 1/2 by 1 1/4
G	3 1/2
H	5 1/2

Size	2 by 1 1/2	2 1/2 by 1 1/2	2 1/2 by 2	3 by 2	3 by 3	4 by 2	4 by 4
J	5 7/8	5 15/16	7 1/8	9 11/16	9 11/16	7 1/8	11 3/4
K	2 15/16	3 5/16	3 5/8	4 5/8	4 5/8	4 9/16	5 13/16

Size	4
L	6 1/4
M	2 7/8

Size	1 1/4	1 1/2	2
N	2 5/8	2 15/16	3 5/16
P	1 5/8	1 13/16	2 1/8

Size	1 1/4	1 1/2	2
Q	1 7/8	2	2 7/16
R	1 3/16	1 1/4	1 3/4

Size	4
S	3 13/16
T	2

Size	4 by 5
U	4 7/16
V	3 13/16

Size	1 1/4	1 1/2	2	2 by 1 1/2
W	4 5/8	5 3/8	7	6 1/2

Size	2 by 1 1/2	2 1/2 by 2	3 by 2	4 by 2	4 by 3	5 by 2	5 by 3
X	9	9	9	9	9	9	9

Size	5 by 4	6 by 4	6 by 5	8 by 4	8 by 6
X	9	9	9	9	9

Size	2	3	4	5	6	8
Y	4 1/2	4 3/4	7	7	7	7

Size	4
A'	3 13/16
B'	4 5/8
C'	7

Size	1 1/2 by 1 1/4	2 by 1 1/2	2 1/2 by 1 1/2	2 1/2 by 2	3 by 1 1/2	3 by 2	4 by 1 1/2	4 by 2	4 by 2 1/2	4 by 3	3 by 4	5 by 1 1/2	5 by 2	5 by 3
D'	3 7/8	4 5/8	4 7/8	5 1/2	5 1/16	5 11/16	5 1/4	5 13/16	6 5/8	7 3/8	8 5/8	5 9/16	6 1/8	7 3/4
E'	2 1/2	2 15/16	3 1/16	3 5/16	3 5/16	3 5/8	3 13/16	4 1/16	4 7/16	4 3/4	5 3/16	4 3/8	4 5/8	5 1/4
F'	2 3/8	2 11/16	2 13/16	3 1/4	2 15/16	3 5/16	3	3 3/8	3 13/16	4 5/16	5 1/8	3 3/16	3 9/16	4 1/2

Size	5 by 4	6 by 2	6 by 3	6 by 4	6 by 5	8 by 3	8 by 4	8 by 6
D'	9 1/8	6 1/4	7 7/8	9 1/4	10 9/16	11 7/16	11 7/16	15 1/16
E'	5 11/16	5 1/8	5 3/4	6 3/16	6 5/8	7 1/4	7 9/16	8 5/16
F'	5 3/8	3 5/8	4 9/16	5 7/16	6 1/4	7 15/16	7 1/2	9 5/8

Size	1 1/2 by 1 1/4	2 by 1 1/2	2 1/2 by 2	3 by 1 1/2	3 by 2	4 by 2	5 by 3
G'	3 7/8	4 5/8	5 1/2	5 1/16	5 11/16	5 13/16	7 3/4
H'	5	5 7/8	6 5/8	6 5/8	7 1/4	8 1/8	10 1/2
J'	2 3/8	2 11/16	3 1/4	2 15/16	3 5/16	3 3/8	4 1/2

Size	5 by 4	6 by 2	6 by 4	6 by 5	8 by 6
G'	9 1/8	6 1/4	9 1/4	10 9/16	15 1/16
H'	11 3/8	10 1/4	12 3/8	13 1/4	16 5/8
J'	5 3/8	3 5/8	5 7/16	6 1/4	9 5/8

Size	2 by 1 1/2	2 1/2 by 1 1/2	2 1/2 by 2	3 by 1 1/2	3 by 2	3 by 2 1/2	4 by 1 1/2	4 by 2	4 by 3	5 by 2
K'	5 7/8	6 7/16	7 1/16	6 5/8	7 3/8	8	7 3/16	7 11/16	9 1/4	8 1/4
L'	4 1/8	4 5/8	5 1/8	5 1/16	5 3/8	5 13/16	6 1/16	6 5/16	7 3/16	7 3/16
M'	4 1/16	4 9/16	4 15/16	4 15/16	5 5/16	5 11/16	5 3/4	6	6 7/8	6 7/8

Size	5 by 3	5 by 4	6 by 2	6 by 3	6 by 4	6 by 5	7 by 3	7 by 4	7 by 5	7 by 6	8 by 4	8 by 6	10 by 4	10 by 6	12 by 6	14 by 6
K'	9 13/16	11 3/8	8 7/16	10	11 7/8	13	10	11 7/16	16 11/16	16 11/16	11 7/16	14 15/16	14	15 1/2	18	19 1/2
L'	7 7/8	8 1/2	8 1/16	8 3/4	9 3/8	10	9 5/8	10 1/4	10 15/16	11 9/16	11	12 3/8	12 7/8	14 1/4	15 3/4	17 3/4
M'	7 5/8	8 7/16	7 9/16	8 5/16	9 1/16	9 13/16	8 7/8	9 11/16	10 1/2	11 3/8	10 7/16	11 7/8	11 1/2	13 1/2	14 1/4	16 3/8

Size	2 by 1 1/2	2 1/2 by 1 1/2	3 by 1 1/2	3 by 2	4 by 2	4 by 3	5 by 2	5 by 3	5 by 4	6 by 2	6 by 4	8 by 4	8 by 6
N'	5 7/8	6 7/16	6 5/8	7 3/8	7 11/16	9 1/4	8 1/4	9 13/16	11 3/8	8 7/16	11 7/8	11 7/16	14 15/16
P'	4 1/8	4 5/8	5 1/16	5 3/8	6 5/16	7 3/16	7 3/16	7 7/8	8 1/2	8 1/16	9 3/8	11	12 3/8
Q'	4 1/16	4 9/16	4 15/16	5 5/16	6	6 7/8	6 7/8	7 5/8	8 7/16	7 9/16	9 1/16	10 7/16	11 7/8

Size	1 1/2 by 1 1/2	2 by 2
R'	4 5/8	5 1/2
S'	2 5/8	3 1/4
T'	2 5/8	3 1/4

Size	2 by 1 1/2	3 by 2	4 by 4	4 by 2	5 by 5	5 by 3	6 by 6	8 by 4
R'	4 15/16	6 7/16	9 3/8	6 11/16	11 1/8	8 1/2	13	11 7/16
S'	3 1/16	4 1/8	5 3/4	4 3/4	6 3/4	6	7 7/8	8 9/16
T'	2 15/16	3 15/16	5 3/4	4 5/16	6 3/4	5 9/16	7 7/8	7 3/8

Size	1 1/4 by 1	1 1/2 by 1	1 1/2 by 1 1/4	2 by 1 1/2	2 1/2 by 1 1/2	2 1/2 by 2	3 by 1 1/2
U'	4 1/4	4 1/2	5 1/8	5 3/4	5 3/4	7 3/8	5 15/16
V'	3 1/2	3 5/8	3 7/8	4 3/8	4 9/16	5 3/4	5
W'	1	1	1 3/16	1 5/16	1 5/16	1 5/8	1 5/16

Size	3 by 2	4 by 1 1/2	4 by 2	4 by 2 1/2	4 by 3	5 by 1 1/2	5 by 2	5 by 3	5 by 4	6 by 2	6 by 3	6 by 4	6 by 5	7 by 3	7 by 4	8 by 3
U'	7 9/16	6 1/16	7 11/16	8 5/8	10	6 5/16	7 3/4	10 1/4	13	7 15/16	10 3/8	13 1/16	16 1/16	10 7/8	13 5/16	10 5/8
V'	6 1/16	5 7/16	6 5/8	7 1/16	8 1/16	6	7 1/8	8 5/8	10 7/16	7 3/4	9 1/8	11	12 13/16	9 11/16	11 5/8	10 1/4
W'	1 5/8	1 5/16	1 5/8	2	2 3/8	1 3/8	1 5/8	2 3/8	2 7/8	1 5/8	2 5/16	2 7/8	3 1/2	2 5/8	2 7/8	2 5/16

Size	8 by 4	10 by 4	12 by 5
U'	13 3/8	14	17
Y'	12	13	15 1/2
W'	2 15/16	3	4

Size	1 1/2 by 1	1 1/2 by 1	1 1/2 by 1 1/4	2 by 1 1/2	2 1/2 by 1 1/2	3 by 1 1/2	3 by 2	4 by 2	4 by 3	5 by 4	6 by 2	6 by 4
X'	4 1/4	4 1/2	5 1/8	5 3/4	5 3/4	5 15/16	7 9/16	7 11/16	10	13	7 15/16	13 1/16
Y'	7	7 1/4	7 3/4	8 3/4	9 1/8	10	12 1/8	13 1/4	16 1/8	20 7/8	15 1/2	22
Z'	1	1	1 3/16	1 5/16	1 5/16	1 5/16	1 5/8	1 5/8	2 3/8	2 7/8	1 5/8	2 7/8

Size	6 by 5
X'	16 1/16
Y'	22 5/8
Z'	3 1/2

Size	2	3	4	5	6
A''	3 1/2	4 9/16	5 5/8	6 11/16	7 3/4
B''	3 13/16	4 15/16	6	7 1/16	8 3/16
C''	3	4 1/8	5 1/8	6	7 1/8

Size	2	2	2	2	3	3	3	3
D''	4	6	8	10	4	6	8	10
E''	7 1/2	9 1/2	11 1/2	13 1/2	8 3/4	10 3/4	12 3/4	14 3/4

Size	4	4	4	4	4	5	5	5	5	6	6	6	6
D''	4	6	8	10	12	6	8	10	12	6	8	10	12
E''	9 3/4	11 3/4	13 3/4	15 3/4	17 3/4	12 5/8	14 5/8	16 5/8	18 5/8	13 5/8	15 5/8	17 5/8	19 5/8

Size	*F''*	*G''*	*H''*	*J''*	*K''*
4 in.	8 3/4	5 3/16	3 9/16	2 5/16	2

TABLE II-39. DIMENSIONS OF DRAINAGE FITTINGS (*Continued*)

Tapping number	Size 4 by 4 in.							Size 5 by 4 in.						
	L''	*M''*	*N''*	*O''*	*P''*	*Q''*	*R''*	*L''*	*M''*	*N''*	*O''*	*P''*	*Q''*	*R''*
1	0	5¼	2⅝	5¼	3½	3	5¼	1¼	5¾	4⅞	5⅜	3¾	6¼	5¼
2	0	5¼	2¼	5¼	3½	3	4⅞	1¼	5¾	4½	5⅜	3¾	6¼	4⅞
3	0	5¼	1⅞	5¼	3½	3	4½	1¼	5¾	4⅛	5⅜	3¾	6¼	4½
4	0	5¼	1½	5¼	3½	3	4⅛	1¼	5¾	3¾	5⅜	3¾	6¼	4⅛
5	0	5¼	1⅛	5¼	3½	3	3¾	1¼	5¾	3⅜	5⅜	3¾	6¼	3¾
6	0	5¼	¾	5¼	3½	3	3⅜	1¼	5¾	3	5⅜	3¾	6¼	3⅜
7							3	1¼	5¾	2⅝	5⅜	3¾	6¼	3
8							2⅝	1¼	5¾	2¼	5⅜	3¾	6¼	2⅝
9							2¼	1¼	5¾	1⅞	5⅜	3¾	6¼	2¼
10							1⅞	1¼	5¾	1½	5⅜	3¾	6¼	1⅞
11							1½	1¼	5¾	1⅛	5⅜	3¾	6¼	1½
12							1⅛	1¼	5¾	¾	5⅜	3¾	6¼	1⅛

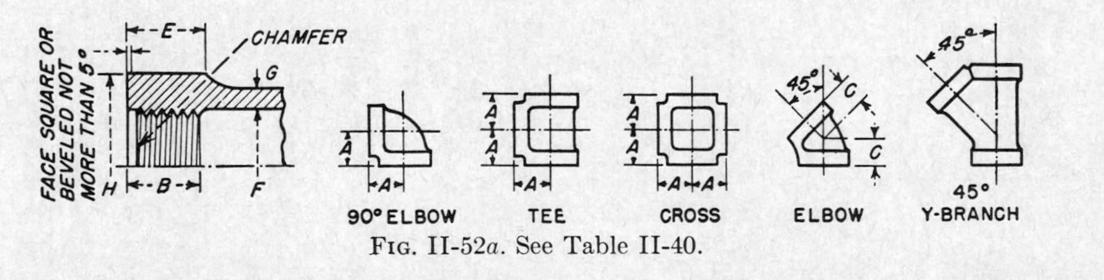

FIG. II-52*a*. See Table II-40.

TABLE II-40. CAST-IRON PIPE FITTINGS, SCREWED, 125 LB
(Federal Specifications WW-P-501b)

Nominal pipe size, in.	Detail dimensions (Fig. II-52*a*), in.*						Elbows, tees, crosses, and wye branches (Fig. II-52*a*)						
	Length of thread	Width of band	Inside diameter of fitting		Metal thickness	Outside diameter of band	Center to end, in.						
	B	*E*	*F*		*G*	*H*	*A*	*A*	*A*	*A*	*A*	*C*	
	Min.	Min.	Min.	Min.	Min.	Min.	90° elbow	90° pitched elbow†	90° *R* and *L* elbows‡	Tees	Crosses	45° elbows	45° wye branches
¼	0.32	0.38	0.54	0.58	0.11	0.93	0.81	....	....	0.81	....	0.73	
⅜	0.36	0.44	0.86	0.72	0.12	1.12	0.95	....	....	0.95	....	0.80	
½	0.43	0.50	0.84	0.90	0.13	1.34	1.12	1.12	....	1.12	....	0.88	
¾	0.50	0.56	1.05	1.11	0.15	1.63	1.31	1.31	1.31	1.31	1.31	0.98	*
1	0.58	0.62	1.32	1.38	0.17	1.95	1.50	1.50	1.50	1.50	1.50	1.12	*
1¼	0.67	0.69	1.66	1.73	0.18	2.39	1.75	1.75	1.75	1.75	1.75	1.29	*
1½	0.70	0.75	1.90	1.97	0.20	2.68	1.94	1.94	1.94	1.94	1.94	1.43	*
2	0.75	0.84	2.38	2.45	0.22	3.28	2.25	2.25	2.25	2.25	2.25	1.68	*
2½	0.92	0.94	2.88	2.98	0.24	3.86	2.70	....	....	2.70	2.70	1.95	*
3	0.98	1.00	3.50	3.60	0.26	4.62	3.08	....	....	3.08	3.08	2.17	*
3½	1.03	1.06	4.00	4.10	0.28	5.20	3.42	....	....	3.42	....	2.39	
4	1.08	1.12	4.50	4.60	0.31	5.79	3.79	....	....	3.79	3.79	2.61	*
5	1.18	1.18	5.56	5.66	0.38	7.05	4.50	....	....	4.50	....	3.05	
6	1.28	1.28	6.62	6.72	0.43	8.28	5.13	....	....	5.13	....	3.46	
8	1.47	1.47	8.62	8.72	0.55	10.63	6.56	....	....	6.56	....	4.28	

* Dimensions given conform to ASA B 16d–1941.

† Angle of center lines equals 90° plus angle for pitch of ¼ in. per ft.

‡ Right and left elbows shall have either four or more ribs or the letter L cast on the band at end with left-hand threads.

TABLE II-41. CAST-IRON PIPE FITTINGS, SCREWED, 125 LB[a]
(Federal Specifications WW–P–501b; see Fig. II-52*b*)

Caps

Nominal pipe size, in.	G, metal thickness, min. in.*†	P, height, min. in.†	F, inside diameter, min. in.
¼	0.11	‡	0.54
⅜	0.12	‡	0.68
½	0.13	‡	0.84
¾	0.15	‡	1.05
1	0.17	‡	1.32
1¼	0.18	‡	1.66
1½	0.20	‡	1.90
2	0.22	‡	2.38
2½	0.24	1.81	2.88
3	0.26	1.91	3.50
4	0.31	2.22	4.50
5	0.38	2.38	5.56
6	0.43	2.63	6.62
8	0.55	2.88	8.62

Return bends

Nominal pipe size, in.	R_1, in.	R_2, in.	R_3, in.**
½	1.25		
¾	1.50	1.88	
1††	1.75	2.50††	4.00‡‡
1¼††	2.25	3.00††	6.00‡‡
1½††	2.50	3.50††	6.00
2	3.25	4.50	6.00
2½	....	5.50	
3	....	6.50	

Reducers

Concentric: Nominal pipe size, in.	Concentric: M	Eccentric: Nominal pipe size, in.	Eccentric: Nominal pipe size, in.
¾ × ½	1.50	1¼ × 1	3½ × 1¼
1 × ¾	1.70	1¼ × ¾	3½ × 1
1 × ½	‖	1½ × 1¼	4 × 3½
1¼ × 1	2.13	1½ × 1	4 × 3
1½ × 1¼	2.25	1½ × ¾	4 × 2½
2 × 1½	2.32	2 × 1½	4 × 2
2 × 1¼	‖	2 × 1¼	4 × 1½
2 × 1	‖	2 × 1	4 × 1¼
2½ × 2	2.63	2 × ¾	4 × 1
3 × 2½	2.88	2½ × 2	5 × 4
3 × 2	‖	2½ × 1½	5 × 3½
4 × 3	3.38	2½ × 1¼	5 × 3
4 × 2½	‖	2½ × 1	5 × 2½
4 × 2	‖	3 × 2½	5 × 2
5 × 4	3.57	3 × 2	6 × 5
6 × 5	3.81	3 × 1½	6 × 4
6 × 4	‖	3 × 1¼	6 × 3½
8 × 6	5.25	3 × 1	6 × 3
		3½ × 3	6 × 2½
		3½ × 2½	6 × 2
		3½ × 2	8 × 6
		3½ × 1½	8 × 5

Reducing elbows

Nominal pipe size, in.	x, in.	z, in.	Nominal pipe size, in.	x, in.	z, in.
½ × ⅜	§	§	2½ × 1	§	§
¾ × ½¶	1.20	1.22	3 × 2½	2.83	2.99
1 × ¾¶	1.37	1.45	3 × 2	2.52	2.89
1 × ½	1.26	1.36	3 × 1½	§	§
1¼ × 1¶	1.58	1.67	3 × 1¼	§	§
1¼ × ¾	1.45	1.62	3½ × 3	§	§
1¼ × ½	§	§	4 × 3½	§	§
1½ × 1¼¶	1.82	1.88	4 × 3	3.30	3.60
1½ × 1	1.65	1.80	4 × 2½	§	§
1½ × ¾	1.52	1.75	4 × 2	2.74	3.41
2 × 1½¶	2.02	2.16	5 × 4	4.00	4.41
2 × 1¼	1.90	2.10	5 × 3	§	§
2 × 1	1.73	2.02	5 × 2½	§	§
2 × ¾	1.60	1.97	6 × 5	4.63	5.03
2½ × 2	2.39	2.60	6 × 4	4.13	4.94
2½ × 1½	2.16	2.51	6 × 3	§	§
2½ × 1¼	§	§	8 × 6	5.56	6.37

[a] See Table II-37.
* Tolerance, −10 per cent.
† G and P for 2½ in. and larger conform to ASA B 16d–1941. Cast caps 2 in. and smaller are not made to standard heights.
‡ P not standard in these sizes.
§ Dimensions conform to ASA B 16d–1941. Fittings not dimensioned are not made to standard center-to-end dimensions.
¶ Pitched elbows may be specified in these sizes. Angle of center lines equals 90° plus angle for pitch ¼ in. per ft.
‖ Dimensions conform to ASA B 16d–1941.
** Dimensions conform to ASA B 16d–1941 except those for the wide pattern.
†† Open-pattern R-and-L bends may be specified in these sizes. R-and-L return bends shall have two or more ribs or letter L cast on band with left-hand threads.
‡‡ Wide-pattern return bends, 1 and 1¼ in., shall be furnished with center dimensions R_3 as follows: 1 in., R_3 = 3.00; 1¼ in., R_3 = 4.00.

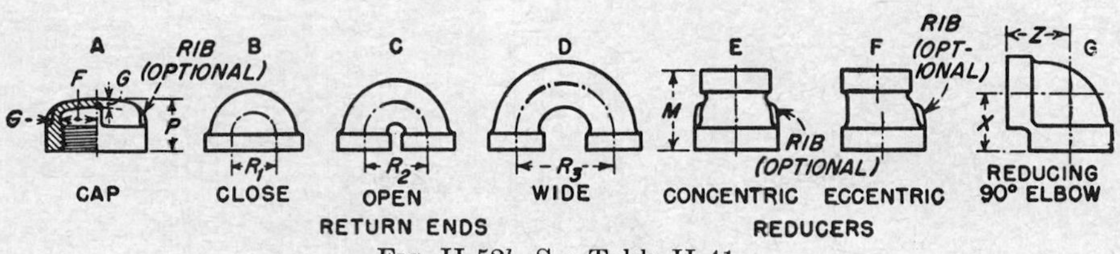

FIG. II-52*b*. See Table II-41.

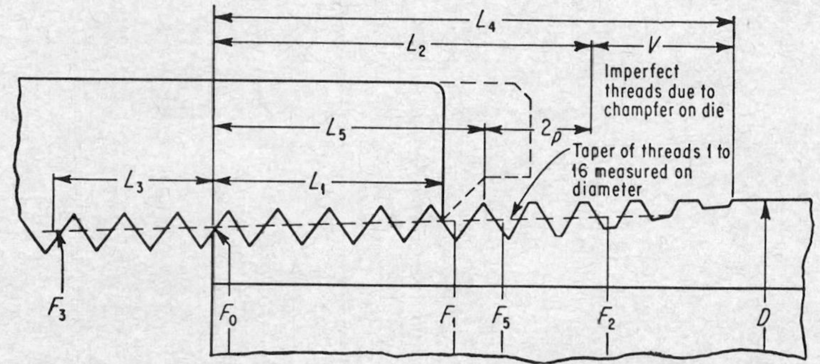

FIG. II-53. American Standard taper pipe threads. See Tables II-42 and II-43. (*ASA* B 2.1–1945.)

TABLE II-42. STANDARD PIPE THREADS*
(Federal Specification GGG–P–351)

Nominal size of pipe, in.	Number of threads per in.	Depth of thread h, in.		Taper thread dimensions, in. (see Fig. II-53)						Straight thread dimensions, in. (see Fig. II-53)				
		Taper pipe thread	Straight pipe thread	D	L_1	L_2	F_0 basic	F_1 basic	K	Major diameter,* basic	Minor diameter,* basic	Pitch diameter Max.	Pitch diameter Basic	Pitch diameter Min.
1/8	27	0.02963	0.405	0.405	0.180	0.26385	0.36351	0.37476	0.33388	0.40439	0.34513	0.37823	0.37476	0.37129
1/4	18	0.04444	0.540	0.540	0.200	0.40178	0.47739	0.48989	0.43294	0.53433	0.44544	0.49510	0.48989	0.48468
3/8	18	0.04444	0.675	0.675	0.240	0.40778	0.61201	0.62701	0.56757	0.67145	0.58257	0.63222	0.62701	0.62181
1/2	14	0.05714	0.840	0.840	0.320	0.53371	0.75843	0.77843	0.70129	0.83557	0.72129	0.78513	0.77843	0.77173
3/4	14	0.05714	1.050	1.050	0.339	0.54571	0.96768	0.98887	0.91054	1.04600	0.93172	0.99556	0.98887	0.98217
1	11½	0.06957	1.315	1.315	0.400	0.68278	1.21363	1.23863	1.14407	1.30819	1.16907	1.24678	1.23863	1.23048
1¼	11½	0.06957	1.660	1.660	0.420	0.70678	1.55713	1.58338	1.48757	1.65294	1.51382	1.59153	1.58338	1.57523
1½	11½	0.06957	1.900	1.900	0.420	0.72348	1.79609	1.82234	1.72652	1.89190	1.75277	1.83049	1.82234	1.81418
2	11½	0.06957	2.375	2.375	0.436	0.75652	2.26902	2.29627	2.19946	2.36583	2.22671	2.30442	2.29627	2.28812
2½	8	0.10000	.....	2.875	0.682	1.13750	2.71953	2.76216	2.61953	2.86216	2.66216	2.77388	2.76216	2.75044
3	8	0.10000	.....	3.500	0.766	1.20000	3.34062	3.38850	3.24063	3.48850	3.28850	3.40022	3.38850	3.37678
3½	8	0.10000	.....	4.000	0.821	1.25000	3.83750	3.88881	3.73750	3.98881	3.78881	3.90053	3.88881	3.87709
4	8	0.10000	.....	4.500	0.844	1.30000	4.33438	4.38712	4.23438	4.48713	4.28713	4.39884	4.38712	4.37541
5	8	0.10000	.....	5.563	0.937	1.40630	5.39073	5.44929	5.29073	5.54929	5.34929	5.46101	5.44929	5.43757
6	8	0.10000	.....	6.625	0.958	1.51250	6.44609	6.50597	6.34609	6.60597	6.40597	6.51769	6.50597	6.49425
8	8	0.10000	.....	8.625	1.063	1.71250	8.43359	18.50003	8.33359					
10	8	0.10000	.....	10.750	1.210	1.92500	10.54351	10.54531	10.44531					
12	8	0.10000	.....	12.750	1.360	2.12500	12.53281	12.61781	12.43281					

* $E_0 = D - (0.05 + 1.1)p$; $E_1 = E_0 + 0.0625L_1$; $L_2 = p(0.8D + 6.8)$; $h = 0.8p$.

TABLE II-43. PIPE THREADS—BASIC DIMENSIONS*
(American Standard taper threads. ASA B 2.1–1945)

Nominal pipe size, in.	Outside diam of pipe, in., D†	Threads per in., k†	Pitch of thread, p	Pitch diam of beginning of external thread	Hand-tight engagement			Effective thread external		
					Length, L_1‡		Diam, F_1*	Length, L_2‡		Diam, F_2
					In.	Threads		In.	Threads	
(1)	(2)	(3)	(4)	(5)	(6)	(7)	(8)	(9)	(10)	(11)
1/16	0.3125	27	0.03704	0.27118	0.160	4.32	0.28118	0.2611	7.06	0.28750
1/8	0.405	27	0.03704	0.36351	0.180	4.86	0.37476	0.2633	7.12	0.38060
1/4	0.540	18	0.05556	0.47739	0.200	3.60	0.48989	0.4018	7.23	0.60250
3/8	0.675	18	0.05556	0.61201	0.240	4.32	0.62701	0.4078	7.34	0.63750
1/2	0.840	14	0.07143	0.75843	0.320	4.48	0.77843	0.5337	7.47	0.79179
3/4	1.050	14	0.07143	0.96768	0.339	4.75	0.98887	0.5457	7.64	1.00179
1	1.315	11½	0.08698	1.21363	0.400	4.60	1.23863	0.6828	7.85	1.25630
1¼	1.660	11½	0.08698	1.55713	0.420	4.83	1.58333	0.7068	8.13	1.60120
1½	1.900	11½	0.08698	1.79609	0.420	4.83	1.82234	0.7235	8.32	1.84130
2	2.375	11½	0.08698	2.26902	0.436	5.01	2.29627	0.7565	8.70	2.31630
2½	2.875	8	0.12500	2.71953	0.682	5.46	2.76216	1.1375	9.10	2.79062
3	3.500	8	0.12500	3.34062	0.766	6.13	3.38850	1.2000	9.60	3.41562
3½	4.000	8	0.12500	3.38750	0.821	6.57	3.88881	1.2500	10.00	3.91562
4	4.500	8	0.12500	4.33438	0.844	6.75	4.38712	1.3000	10.40	4.41562
5	5.563	8	0.12500	5.39073	0.937	7.50	5.44929	1.4063	11.25	5.47062
6	6.625	8	0.12500	6.44609	0.958	7.66	6.50597	1.5125	12.10	6.54062
8	8.625	8	0.12500	8.43359	1.063	8.50	8.50003	1.7125	13.70	8.54062
10	10.750	8	0.12500	10.54531	1.210	9.68	10.62094	1.9250	15.40	10.66562
12	12.750	8	0.12500	12.53281	1.360	10.88	12.61781	2.1250	17.00	12.66562

* All dimensions are in inches (see Fig. II-53). The basic dimensions of American Standard pipe threads are given in inches to four or five decimal places. While this implies a greater degree of precision than is ordinarily attained, these dimensions are the basis of gage dimensions and are so expressed for the purpose of eliminating errors in computation.

† Refer to Fig. II-53 for these dimensions.

‡ Also length of thin ring gage and length from gaging notch to small end of plug gage.

TABLE II-44. PIPE THREADS*

Nominal pipe size	Threads per lin in.†		Engaged length (Fig. II-53)‡	Nominal pipe size	Threads per lin in.*		Engaged length (Fig. II-53)‡	Nominal pipe size	Threads per lin in.†		Engaged length (Fig. II-53)‡
	Std. wt.	Extra strong			Std. wt.	Extra strong			Std. wt.	Extra strong	
1/8	27	27	0.25	1 1/2	11 1/2	11 1/2	0.6875	5	8	8	1.25
1/4	18	18	0.375	2	11 1/2	11 1/2	0.75	6	8	8	1.3125
3/8	18	18	0.375	2 1/2	8	8	0.9375	7§	...	8§	1.3125§
1/2	14	14	0.500	3	8	8	1.0000	8	...	8	1.4375
3/4	14	14	0.5625	3 1/2	8	8	1.0625	9§	...	8§	1.4375§
1	11 1/2	11 1/2	0.6875	4	8	8	1.125	10	...	8	1.625
1 1/2	11 1/2	11 1/2	0.6875	4 1/2§	8§	8§	1.125§	12	...	8	1.75

* Dimensions are in inches unless otherwise stated.
† ASA B 2.1–1945.
‡ ASA B 2.1–1945 except as indicated.
§ Not standard.

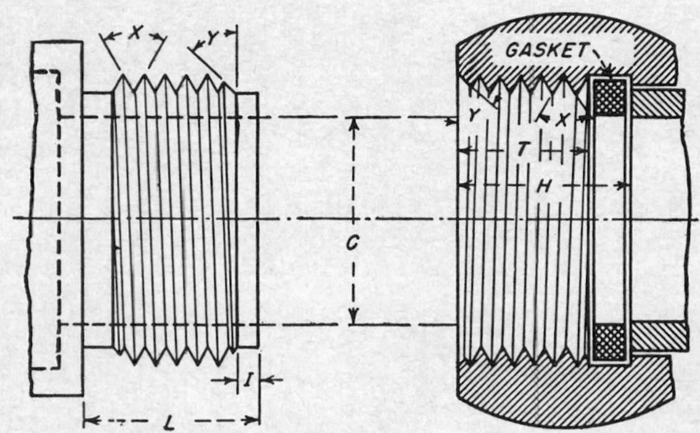

FIG. II-54. Hose-coupling screw threads. See Table II-45 and Fig. 10-3.

TABLE II-45. LENGTH OF THREAD FOR HOSE COUPLINGS AND NIPPLES
(See Fig. II-54)

Service and nominal size	Threads per inch	Length of nipple L	Depth of coupling H	Thread length for coupling T	Length of pilot I	Inside diameter of nipple C	Approx. number of threads in length T
Garden, ½, ⅝, ¾	11½	9/16	17/32	⅜	⅛	25/32	4¼
Chemical, ¾, 1	8	⅝	19/32	15/32	5/32	1 1/32	3¾
Fire, 1½	9	⅝	19/32	15/32	5/32	1 17/32	4¼
Other connections							
½	14	½	15/32	5/16	⅛	17/32	4¼
¾	14	9/16	17/32	⅜	⅛	25/32	5¼
1	11½	9/16	17/32	⅜	5/32	1 1/32	4¼
1¼	11½	⅝	19/32	15/32	5/32	1 9/32	5½
1½	11½	⅝	19/32	15/32	5/32	1 17/32	5½
2	11½	¾	23/32	19/32	3/16	2 1/32	6¾

All dimensions given in inches.
Thread and chamfer angles: X equals 60°; Y equals 35°.

TABLE II-46. PROPERTIES OF STANDARD WROUGHT-IRON[a] AND STEEL PIPE[b]
(Steel, ASTM A 120 and A 53; wrought-iron, ASTM A 72)

1	2	3	4	5	6	7	8	9	10	11	12	
Nominal size, in.	ASTM schedule[c]	Diameter		Wall thickness	Surface area, sq ft per lin ft		Section area, sq in.		Area of metal, sq in.	Volume, gal per lin ft	Weight plain end, lb per lin ft[d]	Working pressure, psi[e]
		OD	ID		OD	ID	OD	ID				
⅛	40(*s*)	0.405	0.269	0.068	0.106	0.0704	0.129	0.0568	0.0720	0.00295	0.244	314[f]
	80(*x*)	0.405	0.215	0.095	0.106	0.0563	0.129	0.0363	0.0925	0.00189	0.314	1,084[f]
¼	40(*s*)	0.540	0.364	0.088	0.141	0.0953	0.229	0.104	0.125	0.00541	0.424	649[f]
	80(*x*)	0.540	0.302	0.119	0.141	0.0791	0.229	0.0716	0.157	0.00372	0.535	1,353[f]
⅜	40(*s*)	0.675	0.493	0.091	0.177	0.129	0.358	0.191	0.167	0.00992	0.567	574[f]
	80(*x*)	0.675	0.423	0.126	0.177	0.111	0.358	0.140	0.217	0.00730	0.738	1,191[f]
½	40(*s*)	0.840	0.622	0.109	0.220	0.163	0.554	0.304	0.250	0.0158	0.850	697[f]
	80(*x*)	0.840	0.546	0.147	0.220	0.143	0.554	0.234	0.320	0.0122	1.09	1,266[f]
	XX	0.840	0.252	0.294	0.220	0.0660	0.554	0.0499	0.504	0.00259	1.71	3,824[f]
¾	40(*s*)	1.050	0.824	0.113	0.275	0.216	0.866	0.533	0.333	0.0277	1.13	604[f]
	80(*x*)	1.050	0.742	0.154	0.275	0.194	0.866	0.432	0.434	0.0225	1.47	1,078[f]
	XX	1.050	0.434	0.308	0.275	0.114	0.866	0.148	0.718	0.00768	2.44	3,134[f]
1	40(*s*)	1.315	1.049	0.133	0.344	0.275	1.36	0.864	0.494	0.0449	1.68	651[f]
	80(*x*)	1.315	0.957	0.179	0.344	0.251	1.36	0.719	0.639	0.0374	2.17	1,083[f]
	XX	1.315	0.599	0.358	0.344	0.157	1.36	0.282	1.08	0.0146	3.66	2,963[f]
1¼	40(*s*)	1.660	1.380	0.140	0.435	0.361	2.16	1.50	0.669	0.0777	2.27	440[f]
	80(*x*)	1.660	1.278	0.191	0.435	0.335	2.16	1.28	0.881	0.0666	3.00	805[f]
	XX	1.660	0.896	0.382	0.435	0.235	2.16	0.630	1.53	0.0328	5.21	2,318[f]
1½	40(*s*)	1.900	1.610	0.145	0.497	0.421	2.84	2.04	0.800	0.1058	2.72	417[f]
	80(*x*)	1.900	1.500	0.200	0.497	0.393	2.84	1.77	1.07	0.0918	3.65	756[f]
	XX	1.900	1.100	0.400	0.497	0.288	2.84	0.950	1.89	0.0494	6.41	2,122[f]
2	40(*s*)	2.375	2.067	0.154	0.622	0.541	4.43	3.36	1.07	0.174	3.65	376[f]
	80(*x*)	2.375	1.939	0.218	0.622	0.548	4.43	2.95	1.48	0.153	5.02	690[f]
	XX	2.375	1.503	0.436	0.622	0.393	4.43	1.77	2.66	0.0922	9.03	1,861[f]
2½	40(*s*)	2.875	2.469	0.203	0.753	0.646	6.49	4.79	1.70	0.249	5.79	505[f]
	80(*x*)	2.875	2.323	0.276	0.753	0.608	6.49	4.24	2.25	0.220	7.66	806[f]
	XX	2.875	1.771	0.552	0.753	0.364	6.49	2.46	4.03	0.128	13.7	2,048[f]
3	40(*s*)	3.500	3.068	0.216	0.916	0.803	9.62	7.39	2.23	0.384	7.57	454[f]
	80(*x*)	3.500	2.900	0.300	0.916	0.759	9.62	6.61	3.02	0.343	10.3	734[f]
	XX	3.500	2.300	0.600	0.916	0.602	9.62	4.15	5.47	0.216	18.5	1,829[f]
3½	40(*s*)	4.000	3.548	0.226	1.05	0.929	12.6	9.89	2.68	0.514	9.11	425[f]
	80(*x*)	4.000	3.364	0.318	1.05	0.881	12.6	8.89	3.68	0.462	12.5	692[f]
	XX[h]	4.000	2.728	0.636	1.05	0.714	12.6	5.85	6.72	0.304	22.9	1.699[f]
4	40(*s*)	4.500	4.026	0.237	1.18	1.05	15.9	12.7	3.17	0.661	10.8	403[g]
	80(*x*)	4.500	3.826	0.337	1.18	1.00	15.9	11.5	4.41	0.597	14.9	663[f]
	XX	4.500	3.152	0.674	1.18	0.825	15.9	7.80	8.10	0.405	27.5	1,602[f]
5	40(*s*)	5.563	5.047	0.258	1.46	1.32	24.3	20.0	4.30	1.04	14.6	498[g]
	80(*x*)	5.563	4.813	0.375	1.46	1.26	24.3	18.2	6.11	0.945	20.8	825[g]
	XX	5.563	4.063	0.750	1.46	1.06	24.3	13.0	11.3	0.673	38.6	1,951[g]
6	40(*s*)	6.625	6.065	0.280	1.73	1.59	34.5	28.9	5.58	1.50	18.0	467[g]
	80(*x*)	6.625	5.761	0.432	1.73	1.51	34.5	26.1	8.40	1.35	28.6	825[g]
	XX	6.625	4.897	0.864	1.73	1.28	34.5	18.8	15.6	0.978	53.1	1,912[g]
8	30(*s*)	8.625	8.071	0.277	2.26	2.11	58.4	51.2	7.26	2.66	24.7	351[g]
	40(*s*)	8.625	7.981	0.322	2.26	2.09	58.4	50.0	8.40	2.50	28.6	431[g]
	80(*x*)	8.625	7.625	0.500	2.26	2.00	58.4	45.7	12.8	2.37	43.4	753[g]
	XX	8.625	6.875	0.875	2.26	1.80	58.4	37.1	21.3	1.93	72.4	1,460[g]
10	(*s*)	10.750	10.192	0.279	2.81	2.67	90.8	81.6	9.18	4.24	31.2	285[g]
	30(*s*)	10.750	10.136	0.307	2.81	2.65	90.8	80.7	10.1	4.19	34.2	324[g]
	40(*s*)	10.750	10.020	0.365	2.81	2.62	90.8	78.9	11.9	4.10	40.5	405[g]
	60(*x*)	10.750	9.750	0.500	2.81	2.55	90.8	74.7	16.1	3.88	54.7	600[g]
12	30(*s*)	12.750	12.090	0.330	3.34	3.17	128	115	12.9	5.96	43.8	299[g]
	(*s*)	12.750	12.000	0.375	3.34	3.14	128	113	14.6	5.88	49.6	352[g]
	(*x*)	12.750	11.750	0.500	3.34	3.08	128	108	19.2	5.63	65.4	503[g]

[a] ASTM A 72 shows sizes of wrought iron approximately the same as steel except wall thicknesses are slightly greater.
[b] All dimensions are in inches unless otherwise stated.
[c] The numbers 30, 40, etc., refer to the ASTM schedule. The letter (*s*) refers to the former designation *standard weight*, the letter (*x*) refers to the former designation *extra strong*, and the letters *XX* refer to the former designation *double extra strong*.
[d] Threaded and coupled (T and C) pipe is slightly heavier.
[e] For working pressure for welded joints see Table II-47.
[f] Working pressure based on an allowable intensity of tensile stress of 6,225 psi at 250°F.
[g] Working pressure based on an allowable intensity of tensile stress of 8,400 psi at 250°F.
[h] 3½-in. double extra strong in ASTM specifications, but pipe of this size is still manufactured.

Standard-weight pipe is generally furnished with threaded ends in random lengths of 16 to 22 ft, although when ordered with plain ends, 5 per cent may be in lengths of 12 to 16 ft. Five per cent of the total number of lengths ordered may be jointers, which are two pieces coupled together. Extra-strong pipe is generally furnished with plain ends in random lengths of 12 to 22 ft, although 5 per cent may be in lengths of 6 to 12 ft.

TABLE II-47. ASTM SPECIFICATIONS FOR STEEL PIPE

Lap-welded, A 53 and A 120, available in sizes of 2 to 24 in. Allowable fiber stress* for A 53 is 9,000 psi and for A 120 is 8,400 psi.

Butt-welded, A 53 and A 120, available in sizes of 1/8 to 4 in. Allowable fiber stress for A 53 is 6,720 psi and for A 120 is 6,225 psi.

Seamless, A 53 Grades A and B, and A 120. The first two are available in sizes from 2 to 24 in. and A 120 in sizes of 2 to 12 in. Allowable fiber stress for Grade A is 12,000 psi, for Grade B is 15,000 psi, and for A 120 is 10,400 psi.

Electric resistance, A 53 Grades A and B, available in sizes of 2 to 24 in. Allowable fiber stress for Grade A is 10,200 psi, and for Grade B is 12,750 psi.

* The following formula is from "American Standard Code for Pressure Piping," American Society of Mechanical Engineers, revised November, 1952:

$$P = \frac{2S(t_m - C)}{D - 0.8(t_m - C)}$$

in which P = working pressure, psi
S = fiber stress, psi
t_m = 0.875 × wall thickness, in.
D = outside diameter, in.
C = joint factor. 0.05 for 1-in. and smaller or 0.065 for larger than 1-in.

† Under special conditions where applied under normal plumbing practice, fiber stresses for corresponding grades of seamless pipe can be used.

TABLE II-48. DIMENSIONS OF EXTRA-STRONG AND DOUBLE EXTRA-STRONG WROUGHT-IRON OR STEEL PIPE
(Manufacturers' standards)

Nominal size, in.	Actual diameter, in.		Thickness, in.	Circumference, in.		Transverse areas, sq in.			Length of pipe, ft, per sq ft of		Nominal weight, lb per ft of plain ends	Length, in ft of pipe containing 1 cu ft	Working pressure, psi
	External OD	Internal		External	Internal	External	Internal	Metal	External surface	Internal surface			
						Extra strong							
⅛*	0.405	0.215	0.095	1.272	0.675	0.129	0.036	0.093	9.431	17.766	0.314	3,966.4	2,340
¼*	0.540	0.302	0.119	1.696	0.949	0.229	0.072	0.157	7.703	12.648	0.535	2,010.3	2,200
⅜*	0.675	0.423	0.126	2.121	1.329	0.358	0.141	0.217	5.658	9.030	0.738	1,024.7	1,870
½*	0.840	0.546	0.147	2.639	1.715	0.554	0.244	0.320	4.547	6.995	1.087	615.0	1,750
¾*	1.050	0.742	0.154	3.299	2.331	0.866	0.433	0.433	3.637	5.147	1.473	333.0	1,468
1*	1.315	0.957	0.179	4.131	3.007	1.358	0.719	0.639	2.904	3.991	2.171	200.2	1,362
1¼*	1.660	1.278	0.191	5.215	4.015	2.164	1.283	0.881	2.301	2.988	2.996	112.3	1,150
1½*	1.900	1.500	0.200	5.969	4.712	2.835	1.727	1.068	2.010	2.546	3.631	81.5	1,054
2*	2.375	1.939	0.218	7.461	6.092	4.430	2.953	1.477	1.608	1.969	5.022	48.8	918
2½*	2.875	2.323	0.276	9.032	7.298	6.492	4.238	2.254	1.328	1.644	7.661	34.0	960
3*	3.500	2.900	0.300	10.996	9.111	9.621	6.605	3.016	1.091	1.317	10.252	21.8	856
3½*	4.000	3.364	0.318	12.566	10.568	12.566	8.888	3.678	0.954	1.135	12.505	16.2	993
4*	4.500	3.826	0.337	14.137	12.020	15.904	11.497	4.407	0.848	0.998	14.983	12.5	937
4½	5.000	4.290	0.355	15.708	13.477	19.635	14.455	5.180	0.763	0.890	17.611	10.0	888
5*	5.563	4.813	0.375	17.477	15.120	24.306	18.194	6.112	0.686	0.793	20.778	7.9	842
6*	6.625	5.761	0.432	20.813	18.099	34.472	26.067	8.405	0.576	0.663	28.573	5.5	815
7	7.625	6.652	0.500	23.955	20.813	45.664	34.472	11.192	0.500	0.576	38.048	4.18	820
8*	8.625	7.625	0.500	27.096	23.955	58.426	45.663	12.763	0.442	0.500	43.388	3.15	725
9	9.625	8.625	0.500	30.238	27.096	72.760	58.426	14.334	0.396	0.442	48.728	2.46	650
10*	10.750	9.750	0.500	33.772	30.631	90.763	74.662	16.101	0.355	0.391	54.735	1.93	582

Double extra strong													
½*	0.840	0.252	0.294	2.639	0.792	0.554	0.050	0.504	4.547	15.157	1.714	2,887.2	3,500
¾*	1.050	0.434	0.308	3.299	1.363	0.866	0.148	0.718	3.637	8.801	2.440	973.4	2,930
1*	1.315	0.599	0.358	4.131	1.882	1.358	0.282	1.076	2.904	6.376	3.659	511.0	2,720
1¼*	1.660	0.896	0.382	5.215	2.815	2.164	0.630	1.534	2.301	4.263	5.214	228.4	2,300
1½*	1.900	1.100	0.400	5.969	3.456	2.835	0.950	1.885	2.010	3.472	6.408	151.5	2,104
2*	2.375	1.503	0.436	7.461	4.722	4.430	1.774	2.656	1.608	2.541	9.029	81.2	1,835
2½*	2.875	1.771	0.552	9.032	5.564	6.492	2.464	4.028	1.328	2.156	13.695	58.5	1,920
3*	3.500	2.300	0.600	10.966	7.226	9.621	4.155	5.466	1.091	1.660	18.583	34.7	1,716
3½*	4.000	2.728	0.636	12.566	8.570	12.566	5.845	6.721	0.954	1.400	22.850	24.6	1,990
4*	4.500	3.152	0.674	14.137	9.902	15.904	7.803	8.101	0.848	1.211	27.541	18.5	1,871
4½	5.000	3.580	0.710	15.708	11.247	19.635	10.066	9.569	0.763	1.066	32.530	14.3	1,776
5*	5.563	4.063	0.750	17.477	12.764	24.306	12.966	11.340	0.686	0.940	38.552	11.1	1,682
6*	6.625	4.897	0.864	20.813	15.384	34.472	18.835	15.637	0.576	0.780	53.160	7.6	1,632
7	7.625	5.875	0.875	23.955	18.457	45.664	27.109	18.555	0.500	0.650	63.079	5.31	1,435
8*	8.625	6.875	0.875	27.096	21.598	58.426	37.122	21.304	0.442	0.555	72.424	3.88	1,268

Threads same as for standard pipe of same nominal size.
Bursting pressure computed as in Table II-47.
* Recommended Sept. 1, 1926, by Division of Simplified Practice, U.S. Department of Commerce.

Table II-49. Wrought-iron and Steel Pipe*
(Federal Specifications WW-P-441a and WW-P-406)

Diameter iron pipe size, in.	Outside diameter, in.	Standard weight									
		Wrought iron					Steel				
		Thickness, in.	Weight per ft lb		Test pressure, psi		Thickness, in.	Weight per ft lb		Test pressure, psi	
			Plain ends	Ends threaded†	Butt-welded	Lap-welded		Plain ends	Ends threaded†	Butt-welded	‡
⅛	0.405	0.070	0.244	0.245	700	.....	0.068	0.24	0.25	700	700
¼	0.540	0.090	0.424	0.425	700	.....	0.088	0.42	0.43	700	700
⅜	0.675	0.093	0.567	0.568	700	.....	0.091	0.57	0.57	700	700
½	0.840	0.111	0.850	0.852	700	.....	0.109	0.85	0.85	700	700
¾	1.050	0.115	1.130	1.134	700	.....	0.113	1.13	1.13	700	700
1	1.315	0.136	1.678	1.684	700	.....	0.133	1.68	1.68	700	700
1¼	1.660	0.143	2.272	2.281	700	1,000	0.140	2.27	2.28	800	1,000
1½	1.900	0.148	2.717	2.731	700	1,000	0.145	2.72	2.73	800	1,000
2	2.375	0.158	3.652	3.678	700	1,000	0.154	3.65	3.68	800	1,000
2½	2.875	0.208	5.793	5.819	.....	1,000	0.203	5.79	5.82	800	1,000
3	3.500	0.221	7.575	7.616	.....	1,000	0.216	7.58	7.62	800	1,000
3½	4.000	0.231	9.109	9.202	.....	1,000	0.226	9.11	9.20	.....	1,200
4	4.500	0.242	10.790	10.899	.....	1,000	0.237	10.79	10.89	.....	1,200
5	5.563	0.263	14.617	14.810	.....	1,000	0.258	14.62	14.81	.....	1,200
6	6.625	0.286	18.974	19.185	.....	1,000	0.280	18.97	19.19	.....	1,200
8	8.625	0.329	28.554	28.809	.....	1,000	0.322	28.55	28.81	.....	1,200
10	10.750	0.372	40.483	41.132	.....	900	0.365	40.48	41.13	.....	1,000
12	12.750	0.382	49.562	50.706	.....	800	0.375	49.56	50.71	.....	1,000
12§	12.750	0.336	43.773	45.000	.....	600	0.330	43.77	45.00	.....	1,000
		Extra strong									
⅛	0.405	0.098	0.314	.....	700	.....	0.095	0.31	.....	850	850
¼	0.540	0.122	0.535	.....	700	.....	0.119	0.54	.....	850	850
⅜	0.675	0.129	0.738	.....	700	.....	0.126	0.74	.....	850	850
½	0.840	0.151	1.087	.....	700	.....	0.147	1.09	.....	850	850
¾	1.050	0.157	1.473	.....	700	.....	0.154	1.47	.....	850	850
1	1.315	0.183	2.171	.....	700	.....	0.179	2.17	.....	850	850
1¼	1.660	0.195	2.996	.....	1,500	2,500	0.191	3.00	.....	1,100	1,500
1½	1.900	0.204	3.631	.....	1,500	2,500	0.200	3.63	.....	1,100	1,500
2	2.375	0.223	5.022	.....	1,500	2,500	0.218	5.02	.....	1,100	1,500
2½	2.875	0.282	7.661	.....	.....	2,000	0.276	7.66	.....	1,100	1,500
3	3.500	0.306	10.252	.....	.....	2,000	0.300	10.25	.....	1,100	1,500
3½	4.000	0.325	12.505	.....	.....	2,000	0.318	12.51	.....	.....	1,700
4	4.500	0.344	14.983	.....	.....	2,000	0.337	14.98	.....	.....	1,700
5	5.563	0.383	20.778	.....	.....	1,800	0.375	20.78	.....	.....	1,700
6	6.625	0.441	28.573	.....	.....	1,800	0.432	28.57	.....	.....	1,700
8	8.625	0.510	43.388	.....	.....	1,500	0.500	43.39	.....	.....	1,700
10	10.750	0.510	54.735	.....	.....	1,200	0.500	54.74	.....	.....	1,600
12	12.750	0.510	65.415	.....	.....	1,100	0.500	65.41	.....	.....	1,600
		Double extra strong									
½	0.840	0.307	1.714	.....	700	.....	0.294	1.71	.....	1,000	1,000
¾	1.050	0.318	2.440	.....	700	.....	0.308	2.44	.....	1,000	1,000
1	1.315	0.369	3.659	.....	700	.....	0.358	3.66	.....	1,000	1,000
1¼	1.660	0.393	5.214	.....	2,200	.....	0.382	5.21	.....	1,200	1,800
1½	1.900	0.411	6.408	.....	2,200	.....	0.400	6.41	.....	1,200	1,800
2	2.375	0.447	9.029	.....	2,200	2,800	0.436	9.03	.....	1,200	1,800
2½	2.875	0.567	13.695	.....	.....	2,800	0.552	13.70	.....	1,200	1,800
3	3.500	0.615	18.583	.....	.....	2,800	0.600	18.58	.....	1,200	1,800
3½‖	4.000	0.651	22.850	.....	.....	2,500	0.636	22.85	.....	.....	2,000
4	4.500	0.690	27.541	.....	.....	2,500	0.674	27.54	.....	.....	2,000
5	5.563	0.768	38.552	.....	.....	2,000	0.750	38.55	.....	.....	2,000
6	6.625	0.884	53.160	.....	.....	2,000	0.864	53.16	.....	.....	2,000
8	8.625	0.895	72.424	.....	.....	2,000	0.875	72.42	.....	.....	2,800

* Dimensions are in inches; weights are in pounds.
† Ends threaded and coupling on one end. Based on 20-ft lengths.
‡ Lap-welded pipe is not made in sizes of 1 in. and smaller; seamless pipe in these small sizes may be cold-drawn.
§ This is special-weight pipe.
‖ This is considered a special size.

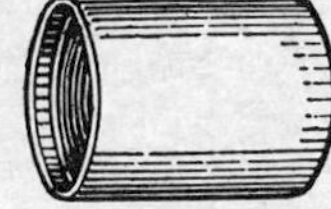

FIG. II-55. Couplings and nipples. See Table II-50.

TABLE II-50. DIMENSIONS OF COUPLINGS*

Nominal pipe size	Outside diameter					Length					Weight, lb		
	ASA B 2.1–1945		Not standard			ASA B 2.1–1945		Not standard			Not standard		
	Std.	*X* and *XX*†	Std.	*X*	*XX*	Std.	*X* and *XX*	Std.	*X*	*XX*	Std.	*X*	*XX*
1/8	0.563	0.563	19/32	....	....	13/16	1 1/16	15/16	.....	...	0.03		
1/4	0.719	0.719	3/4	....	....	1 3/16	1 5/8	1 1/32	.....	...	0.07		
3/8	0.875	0.875	29/32	0.95	....	1 3/16	1 5/8	1 5/32	1 1/2	...	0.11	0.23	
1/2	1.063	1.063	1 1/16	1.13	....	1 9/16	2 1/8	1 5/16	1 7/8	...	0.15	0.28	
3/4	1.313	1.313	1 11/32	1.44	1.66	1 5/8	2 1/8	1 9/16	2 1/8	1 7/8	0.25	0.50	0.70
1	1.576	1.576	1 5/8	1.63	1.90	2	2 5/8	1 13/16	2 3/8	2 5/8	0.42	0.56	1.12
1 1/4	1.900	2.054	1 31/32	2.07	2.22	2 1/16	2 3/4	2 1/16	2 5/8	2 7/8	0.60	0.90	1.50
1 1/2	2.200	2.200	2 5/16	2.31	2.44	2 1/16	2 3/4	2 5/16	2 7/8	3 1/8	0.81	1.35	1.88
2	2.750	2.875	2 23/32	2.81	3.19	2 1/8	2 7/8	2 9/16	3 1/8	3 3/8	1.18	1.80	3.55
2 1/2	3.250	3.375	3 5/16	3.31	3.62	3 1/8	4 1/8	2 7/8	3 1/2	3 5/8	1.70	2.40	4.50
3	4.000	4.000	3 15/16	4.00	....	3 1/4	4 1/4	3 1/16	3 11/16	...	2.45	3.46	
3 1/2	4.625	4.625	4 7/16	4.63	....	3 3/8	4 3/8	3 7/16	4 1/4	...	3.40	5.25	
4	5.000	5.200	4 15/16	5.13	....	3 1/2	4 1/2	3 7/16	4 1/4	...	3.50	6.80	
4 1/2	.....	......	5 17/32	....	....	.....	....	3 5/8	.....	...	4.70		
5	6.296	6.296	6 1/4	....	....	3 3/4	4 5/8	4 1/8	.....	...	8.50		
6	7.390	7.390	7 9/32	....	....	4	4 7/8	4 1/8	.....	...	9.70		
7	.....	......	8 9/32	....	....	.....	....	4 1/8	.....	...	11.10		
8	.....	9.625	9 1/4	....	....	.....	5 1/4	4 5/8	.....	...	13.60		
9	.....	......	10 5/16	....	....	.....	....	5 1/8	.....	...	17.40		
10	.....	11.750	11 5/8	....	....	.....	5 3/4	6 1/8	.....	...	31.10		
12	.....	14.000	13 7/8	....	....	.....	6 1/8	6 1/8	.....	...	44.20		

* Dimensions in inches unless otherwise stated.

† *X* = extra heavy; *XX* = double extra heavy.

Fig. II-56. Malleable-iron fittings for threaded pipe. Manufacturers' standard. See Tables II-51 and II-52.

TABLE II-51. DIMENSIONS, IN INCHES, OF MALLEABLE-IRON THREADED FITTINGS
(Manufacturers' standard; see Figs. II-56, II-57)

Dimension on figure	Ells, tees, crosses, wyes, etc. — Nominal internal diameter, in.															
	1/8	1/4	3/8	1/2	3/4	1	1 1/4	1 1/2	2	2 1/2	3	3 1/2	4	4 1/2	5	6
A	11/16	13/16	15/16	1 1/8	1 5/16	1 7/16	1 3/4	1 15/16	2 1/4	2 11/16	3 1/8	3 7/16	3 3/4	4 1/16	4 7/16	5 1/8
B	.....	3/4	13/16	7/8	1	1 1/8	1 5/16	1 7/16	1 11/16	1 15/16	2 3/16	2 3/8	2 5/8	2 13/16	3 1/16	3 7/16
C	.....	.....	2 1/8	2 1/2	2 7/8	3 7/16	4 1/16	4 1/2	5 7/16	6 1/4	7 1/4	.....	8 7/8			
D	.....	.....	1 7/16	1 11/16	2	2 7/16	2 15/16	3 5/16	4 1/16	4 11/16	5 9/16	.....	6 15/16			
E	.....	1	1 1/8	1 1/4	1 7/16	1 11/16	2 1/16	2 5/16	2 13/16	3 1/4	3 11/16	4	4 3/8			
F	17/32	5/8	3/4	7/8	1 1/16	1 3/16	1 1/4	1 5/16	1 7/16	1 5/8	1 3/4	1 15/16	2	.....	2 5/16	2 9/16
G	.....	1 1/16	1 3/16	1 5/16	1 1/2	1 11/16	1 15/16	2 1/8	2 1/2	2 7/8	3 3/16					
H	1 1/8	1 5/16	1 7/16	1 5/8	1 7/8	2 1/8	2 1/2	2 11/16	3 3/16	3 13/16	4 1/2	.....	5 11/16			
K	.....	15/16	1 1/16	1 3/16	1 5/16	1 1/2	1 11/16	1 7/8	2 1/4	.....	3	.....	3 3/4			
L	.....	5/8	11/16	13/16	15/16	1 1/16	1 1/4	1 3/8	1 11/16	.....	2 1/8	.....	2 1/2			
	Malleable-iron unions — Nominal size, in.															
	1/8	1/4	3/8	1/2	3/4	1	1 1/4	1 1/2	2	2 1/2	3	3 1/2	4	1 by 3/8	1 by 1/2	1 by 3/4
M	.....	13/16	15/16	1 1/8	1 5/16	1 7/16	1 3/4	1 15/16	2 1/4	2 11/16	3 1/8	.....	.....	1 3/16	1 1/4	1 5/16
N	.....	13/16	15/16	1 1/8	1 5/16	1 7/16	1 3/4	1 15/16	2 1/4	2 11/16	3 1/8	.....	.....	1 5/16	1 5/16	1 3/8
O	.....	3/4	13/16	7/8	1	1 1/8	1 5/16	1 7/16	1 11/16							
P	.....	1 13/16	2 1/16	2 5/16	2 5/8	3	3 7/16	3 13/16	4 3/8	4 7/8	5 5/8	.....	.....	2 5/8	2 11/16	2 13/16
Q	.....	2 7/16	2 3/4	3 1/16	3 1/2	3 15/16	4 7/16	4 3/4	5 3/8	6	6 15/16	.....	.....	3 9/16	3 5/8	3 3/4
R	.....	1 9/16	1 11/16	1 7/8	2 1/8	2 7/16	2 3/4	3 1/16	3 1/2							
S	.....	2 3/16	2 3/8	2 11/16	3	3 3/8	3 3/4	4	4 1/2							
T	1 1/2	1 5/8	1 3/4	1 7/8	2 1/8	2 3/8	2 5/8	2 15/16	3 1/4	3 9/16	3 15/16	4 5/16	4 5/8			
U	.....	2 1/4	2 7/16	2 11/16	3 1/16	3 5/16	3 11/16	4	4 5/16	4 11/16	5 5/16					
V	.....	3/4	13/16	7/8	15/16	1 1/16	1 3/16	1 5/16	1 1/2	1 5/8	1 13/16	2	2 1/4			
	Dimensions of return bends															
Center line to center line	close...	.....	.....	1	1 1/4	1 1/2	1 3/4	2 3/16	2 5/8							
	medium	.....	.....	1 1/4	1 1/2	1 7/8	2 1/4	2 1/2	3							
	open...	.....	.....	1 1/2	2	2 1/2	3	3 1/2	4	4 1/2	5					

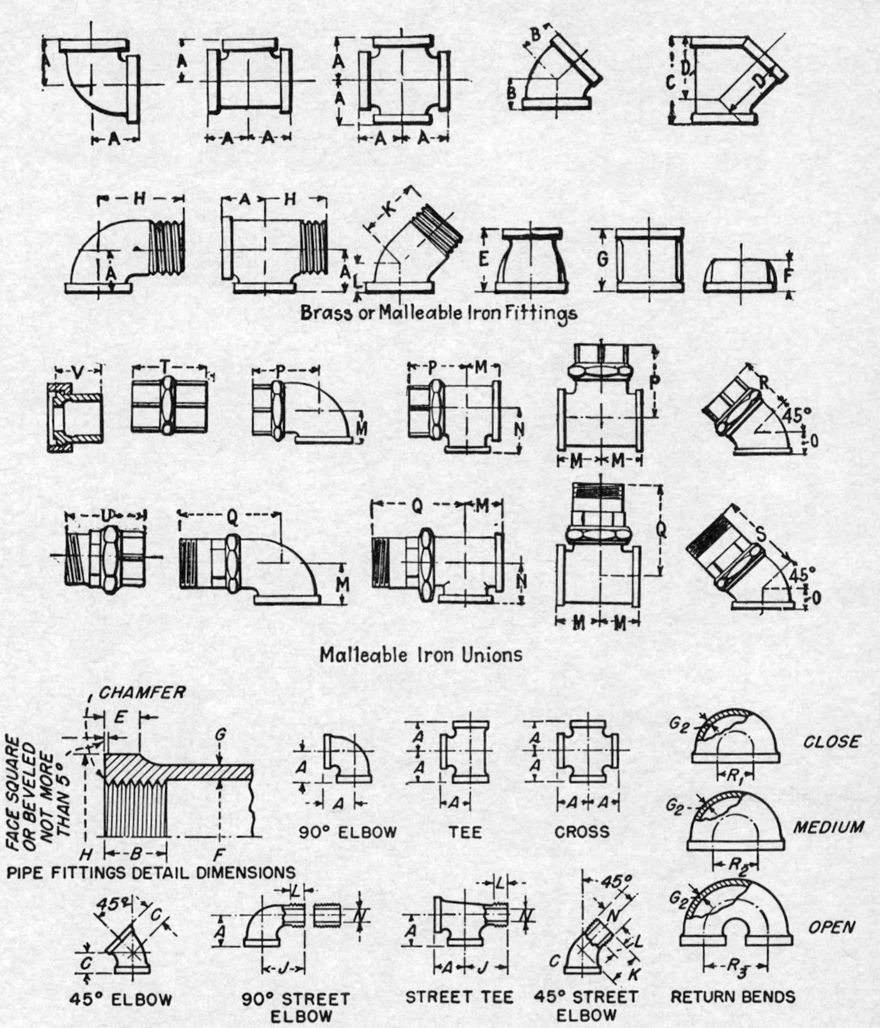

FIG. II-57. Malleable-iron pipe fittings. See Tables II-51 and II-52.

TABLE II-52. PIPE FITTINGS; MALLEABLE IRON, SCREWED
(Federal Specification WW-P-521a ¶; see Figs. II-56, II-57)

Nominal pipe size, in.	Pipe fittings; detail dimensions						Elbows, tees, crosses						
	B	E	F		G	H	Center to end, in.						
							A	C	J	J	K		
	Length of thread, min. in.	Width of band, min. in.	Inside diameter of fitting, in. Min.	Inside diameter of fitting, in. Max.	Metal thickness, min. in.	Outside diameter of band, min. in.	90° elbows, tees and crosses	45° elbows	90° street elbows	Street tees	45° street elbows	Length of external thread L	Port diameter N, max.
⅛	0.25	0.20	0.40	0.44	0.09	0.69	0.69	*	1.00	....	*	0.26	0.20
¼	0.32	0.22	0.54	0.58	0.10	0.84	0.81	0.73	1.19	1.19	0.94	0.40	0.26
⅜	0.36	0.23	0.68	0.72	0.10	1.02	0.95	0.80	1.44	1.44	1.03	0.41	0.37
½	0.43	0.25	0.84	0.90	0.11	1.20	1.12	0.88	1.63	1.63	1.15	0.53	0.51
¾	0.50	0.27	1.05	1.11	0.12	1.46	1.31	0.98	1.89	1.89	1.20	0.55	0.69
1	0.58	0.30	1.32	1.38	0.13	1.77	1.50	1.12	2.14	2.14	1.47	0.68	0.91
1¼	0.67	0.34	1.66	1.73	0.14	2.15	1.75	1.29	2.45	2.45	1.71	0.71	1.19
1½	0.70	0.37	1.90	1.97	0.16	2.43	1.94	1.43	2.69	2.69	1.88	0.72	1.39
2	0.75	0.42	2.38	2.45	0.17	2.96	2.25	1.68	3.26	3.26	2.22	0.76	1.79
2½	0.92	0.48	2.88	2.98	0.21	3.59	2.70	1.95	3.86	....	....	1.14	2.20
3	0.98	0.55	3.50	3.60	0.23	4.28	3.08	2.17	4.51	....	....	1.20	2.78
3½	1.03	0.60	4.00	4.10	0.25	4.84	3.42	2.39	....	....	....	....	....
4	1.08	0.66	4.50	4.60	0.26	5.40	3.79	2.61	5.69	....	....	1.30	3.70
5	1.18	0.78	5.56	5.66	0.30	6.58	4.50	3.05	....	....	....	....	....
6	1.28	0.90	6.62	6.72	0.34	7.77	5.13	3.46	....	....	....	....	....

Nominal pipe size, in.	Coupling	45° wye branch			Cap
	W	T	U	V	P
⅛	0.96	....	....	....	‡
¼	1.06	....	....	....	‡
⅜	1.16	0.50	1.43	1.93	‡
½	1.34§	0.61	1.71	2.32	0.87
¾	1.52§	0.72	2.05	2.77	0.97
1	1.67§	0.85	2.43	3.28	1.16
1¼	1.93§	1.02	2.92	3.94	1.28
1½	2.15§	1.10	3.28	4.38	1.33
2	2.53§	1.24	3.93	5.17	1.45
2½	2.88	1.52	4.73	6.25	1.70
3	3.18	1.71	5.55	7.26	1.80
3½	.....	....	....	....	1.90
4	3.69	2.01	6.97	8.98	2.08
5	.....	....	....	....	2.32
6	.....	....	....	....	2.55

Return bends		
Nominal pipe size, in.	G_2, in.	R_1, in.
½	0.12	1.00
¾	0.13	1.25
1	0.15	1.50
1¼	0.16	1.75
1½	0.18	2.19
2	0.20	2.62
2½	0.24	
3	0.27	
4	0.32†	
Nominal pipe size, in.	**R_2, in.**	**R_3, in.**
½	1.25	1.50
¾	1.50	2.00
1	1.88	2.50
1¼	2.25	3.00
1½	2.50	3.50
2	3.00	4.00
2½	....	4.50
3	....	5.00
4	....	6.00†

* One-eighth-inch 45° elbows are not made to standard dimensions.
† Four-inch size does not conform to ASA B 16c–1939.
‡ Not standard dimensions.
§ Right and left couplings only in these sizes.
¶ Dimensions conform to American Standard B 16c–1939.

TABLE II-53. SOME DIMENSIONS OF STEEL PIPE FOR UNDERGROUND WATER SERVICE*

Nominal diameter, in.	Outside diameter, in.	Wall thickness, in.	Weight per ft, lb	Working pressure, psi	Nominal diameter, in.	Outside diameter, in.	Wall thickness, in.	Weight per ft, lb	Working pressure, psi
4	4.500	0.125	5.84	690	14	14.000	0.188	27.65	340
4	4.500	0.337	14.98	1,870	14	14.000	0.500	72.09	890
6	6.625	0.188	12.89	710	18	18.000	0.188	35.67	260
6	6.625	0.432	28.57	1,630	18	18.000	0.500	93.45	690
8	8.625	0.188	16.90	550	20	20.000	0.188	39.67	240
8	8.625	0.500	43.39	1,450	20	20.000	0.500	104.13	630
10	10.750	0.188	21.15	440	24	24.000	0.188	47.68	200
10	10.750	0.500	54.74	1,160	24	24.000	0.500	125.49	520
12	12.750	0.188	25.15	370					
12	12.750	0.500	65.42	980					

* From "Standard Specifications American Water Works Association." See *J. Am. Water Works Assoc.*, p. 475, April, 1943.

There are eight or more thicknesses for each diameter between the greatest and the least thickness shown for each diameter in this table.

TABLE II-54. STANDARD GAGES AND WEIGHTS OF CORRUGATED STEEL PIPE*

Diameter, in.	Gage	Weight, lb per ft	End area, sq ft
8	16	7.3	0.35
10	16	9.0	0.55
12	16	10.5	0.79
15	16	12.9	1.23
18	16	15.3	1.77
24	16	25.2	4.14

* From Catalogue of American Rolling Mills.

TABLE II-55. COPPER TUBE*

Nominal pipe size	Type	Diameter		Wall thickness	Cross-sectional area, sq in.				Area of metal, sq in.	Volume, gal per lin ft	Weight, lb per lin ft†	Working pressure, psi‡
		OD	ID		OD	ID	OD	ID				
(1)	(2)	(3)	(4)	(5)	(6)	(7)	(8)	(9)	(10)	(11)	(12)	(13)
¼	*K*§	0.375	0.305	0.035	0.0982	0.0798	0.110	0.0730	0.0374	0.00379	0.145	918
	L§	0.375	0.315	0.030	0.0982	0.0825	0.110	0.0779	0.0324	0.00404	0.126	764
⅜	*K*	0.500	0.402	0.049	0.131	0.105	0.196	0.127	0.0695	0.00660	0.269	988
	L	0.500	0.430	0.035	0.131	0.113	0.196	0.145	0.0512	0.00753	0.198	677
½	*K*	0.625	0.527	0.049	0.164	0.138	0.306	0.218	0.0887	0.0113	0.344	779
	L	0.625	0.545	0.040	0.164	0.143	0.306	0.233	0.0735	0.0121	0.285	625
⅝	*K*	0.750	0.652	0.049	0.193	0.171	0.441	0.334	0.108	0.0174	0.418	643
	L	0.750	0.666	0.042	0.193	0.174	0.441	0.338	0.0934	0.0181	0.362	547
¾	*K*	0.875	0.745	0.065	0.229	0.195	0.601	0.436	0.165	0.0227	0.641	747
	L	0.875	0.785	0.045	0.229	0.206	0.601	0.480	0.117	0.0250	0.455	497
1	*K*	1.125	0.995	0.065	0.295	0.260	0.994	0.778	0.216	0.0405	0.839	574
	L	1.125	1.025	0.050	0.295	0.268	0.994	0.852	0.169	0.0442	0.655	432
1¼	*K*	1.375	1.245	0.065	0.360	0.326	1.48	1.22	0.268	0.0634	1.04	466
	L	1.375	1.265	0.055	0.360	0.331	1.48	1.26	0.228	0.0655	0.884	387
	M	1.375	1.291	0.042	0.360	0.338	1.48	1.31	0.176	0.0681	0.682	293
	DWV	1.375	1.295	0.040	0.360	0.339	1.48	1.32	0.163	0.0684	0.650	
1½	*K*	1.625	1.481	0.072	0.425	0.388	2.07	1.72	0.351	0.0894	1.36	421
	L	1.625	1.505	0.060	0.425	0.394	2.07	1.78	0.295	0.0925	1.14	359
	M	1.625	1.527	0.049	0.425	0.400	2.07	1.83	0.243	0.0950	0.940	289
	DWV	1.625	1.541	0.042	0.425	0.403	2.07	1.86	0.205	0.0969	0.809	
2	*K*	2.125	1.959	0.083	0.556	0.513	3.56	3.01	0.532	0.157	2.06	376
	L	2.125	1.985	0.070	0.556	0.520	3.56	3.10	0.452	0.161	1.75	316
	M	2.125	2.009	0.058	0.556	0.526	3.56	3.17	0.377	0.164	1.46	255
	DWV	2.125	2.041	0.042	0.556	0.534	3.56	3.27	0.288	0.142	1.07	
2½	*K*	2.265	2.435	0.095	0.687	0.638	5.41	4.66	0.755	0.242	2.93	352
	L	2.265	2.465	0.080	0.687	0.645	5.41	4.77	0.640	0.247	2.48	295
	M	2.265	2.495	0.065	0.687	0.653	5.41	4.89	0.523	0.254	2.03	234
3	*K*	3.125	2.907	0.109	0.818	0.761	7.67	6.64	1.03	0.345	4.00	343
	L	3.125	2.945	0.090	0.818	0.771	7.67	6.81	0.858	0.354	3.33	278
	M	3.125	2.981	0.072	0.818	0.780	7.67	6.98	0.691	0.362	2.68	230
	DWV	3.125	3.035	0.045	0.818	0.796	7.67	7.23	0.435	0.376	1.69	
3½	*K*	3.625	3.385	0.120	0.949	0.886	10.3	9.00	1.32	0.468	5.12	324
	L	3.625	3.425	0.100	0.949	0.897	10.3	9.21	1.11	0.478	4.29	268
	M	3.625	3.459	0.083	0.949	0.906	10.3	9.40	0.924	0.489	3.58	218
4	*K*	4.125	3.857	0.134	1.08	1.01	13.3	11.7	1.68	0.607	6.51	135
	L	4.125	3.905	0.110	1.08	1.02	13.3	12.0	1.39	0.623	5.38	256
	M	4.125	3.935	0.095	1.08	1.03	13.3	12.2	1.20	0.634	4.66	217
	DWV	4.125	4.009	0.058	1.08	1.05	13.3	12.6	0.67	0.656	2.87	
5	*K*	5.125	4.805	0.160	1.34	1.26	20.7	18.1	2.50	0.940	9.67	307
	L	5.125	4.875	0.125	1.34	1.28	20.7	18.7	1.96	0.971	7.61	234
	M	5.125	4.907	0.109	1.34	1.29	20.7	18.9	1.72	0.981	6.65	203
6	*K*	6.125	4.741	0.192	1.60	1.50	29.4	25.9	3.50	1.35	13.9	308
	I	6.125	5.845	0.140	1.60	1.53	29.4	26.8	2.63	1.39	10.2	221
	M	6.125	5.881	0.122	1.60	1.54	29.4	27.2	2.30	1.42	8.92	190
	DWV	6.125	5.959	0.083	1.60	1.56	29.4	27.9	1.58	1.55	6.10	
8	*K*	8.125	7.583	0.271	2.13	1.99	51.8	45.2	6.69	2.34	25.9	330
	L	8.125	7.725	0.200	2.13	2.02	51.8	46.9	4.98	2.43	19.3	239
	M	8.125	7.785	0.170	2.13	2.04	51.8	47.6	4.25	2.47	16.5	200
10	*K*	10.125	9.449	0.338	2.65	2.47	80.5	70.1	10.4	3.65	40.3	332
	L	10.125	9.625	0.250	2.65	2.52	80.5	72.8	7.76	3.79	30.1	241
	M	10.125	9.701	0.212	2.65	2.54	80.5	73.9	6.60	3.84	25.6	202
12	*K*	12.125	11.315	0.405	3.17	2.96	115	101	14.0	5.24	57.8	334
	L	12.125	11.565	0.280	3.17	3.03	115	105	10.4	5.45	40.4	225
	M	12.125	11.617	0.254	3.17	3.04	115	106	9.47	5.50	36.7	204

Table II-55 Copper Tube* (*Continued*)

* Dimensions are in inches unless otherwise noted.

† Weight per foot is based on tube without couplings.

‡ The following formula is from "American Standard Code for Pressure Piping," American Society of Mechanical Engineers, for plain end tubing, sweat joints, revised November, 1952.

$$P = \frac{2st_m}{D - 0.8t_m}$$

in which P = allowable pressure, psi
s = allowable fiber stress, 6,000 psi
t_m = minimum wall thickness, in.
D = maximum outside diameter, in.

§ Types *K* and *L* are furnished in both hard and soft metals. Type *M* is hard only. Standard lengths are 20 ft. Standard coils (1/4 to 1 1/2 in.) are 60 ft.

Note: Specifications ASA H 23.1-1956 and ASTM B 88-55 cover " . . . seamless copper tube especially designed for plumbing purposes, underground water services, drainage, etc., but also suitable for copper coil water heaters, fuel oil lines, gas lines, etc. For refrigeration and air conditioning the following is recommended: (*a*) when it is desired to use sweat fittings hard tube should be used and, (*b*) when it is necessary to use soft tube with sweat fittings, then rounding and sizing tools should be used.

"There shall be three types of copper water tube with principal uses as follows: *Type K* . . . for underground service, general plumbing and heating purposes involving severe service conditions and for gas, steam, and oil lines. *Type L* . . . for interior applications involving general plumbing and heating purposes. *Type M* . . . sizes 2 1/2 to 12 in. . . . for use with soldered fittings only, for interior applications involving general plumbing and heating purposes. Sizes 1 1/2 to 12 in. inclusive recommended for use with soldered fittings only, for waste, vent, and soil lines and for other interior, nonpressure applications.

"Dimensions, weights, and tolerances shall be as prescribed in . . . ASTM . . . B 251."

General requirements for wrought seamless copper and copper-alloy pipe and tube are given in ASTM B 251-55T. Information on other brass and copper tube is given in the following specifications, ASTM numbers:

Seamless copper tube, standard sizes	B 42
Seamless red brass pipe	B 43
Seamless copper tube, bright annealed	B 68 and B 75
Seamless copper water tube	B 88
Seamless brass tube	B 135

TABLE II-56. SEAMLESS COPPER PIPE* AND SEAMLESS BRASS PIPE†

Iron-pipe size, in.	Diameter		Thickness, in.	Threads per in.	Copper pipe weight per ft, lb	Copper pipe internal hydrostatic test pressure, min. psi	Brass pipe weight per ft		Brass pipe hydrostatic test pressure, min. psi
	Outside, in.	Inside, in.					Grade A, lb	Grades B and C, lb	
1/8	0.405	0.281	0.0629	27	0.259	1,000	0.253	0.246	1,000
1/4	0.540	0.375	0.0825	18	0.460	1,000	0.450	0.437	1,000
3/8	0.675	0.494	0.0905	18	0.643	1,000	0.630	0.612	1,000
1/2	0.840	0.625	0.1075	14	0.957	1,000	0.938	0.911	1,000
3/4	1.050	0.822	0.1140	14	1.30	1,000	1.27	1.24	1,000
1	1.315	1.062	0.1265	11 1/2	1.83	1,000	1.79	1.74	1,000
1 1/4	1.660	1.368	0.1460	11 1/2	2.69	1,000	2.63	2.56	1,000
1 1/2	1.900	1.600	0.1500	11 1/2	3.20	1,000	3.13	3.04	1,000
2	2.375	2.062	0.1565	11 1/2	4.23	900	4.14	4.02	1,000
2 1/2	2.875	2.500	0.1875	8	6.14	900	6.00	5.83	1,000
3	3.500	3.062	0.2190	8	8.75	900	8.56	8.31	1,000
3 1/2	4.000	3.500	0.2500	8	11.41	900	11.17	10.85	1,000
4	4.500	4.000	0.2500	8	12.94	800	12.66	12.29	900
5	5.563	5.062	0.2505	8	15.21	600	15.85	15.40	700
6	6.625	6.125	0.2500	8	19.41	500	18.99	18.44	600
8	8.625	8.000	0.3125	8	31.63	500	30.95	30.05	500
10	10.750	10.019	0.3655	8	46.22	400	45.20	43.91	500

Grade	Copper, per cent	Zinc	Lead, per cent max.	Iron, per cent max.
A	83–86	Remainder	0.06	0.05
B	65–68	Remainder	0.80	0.07
C	59–68	Remainder	0.80	0.07

* Federal Specifications WW–P–377, iron-pipe size.
† Federal Specifications WW–P–351, iron-pipe size.

TABLE II-57. BRASS AND COPPER TUBING*

Outside	Type A, 100 psi				Type B, 200 psi				Type C, 300 psi				Type D, 450 psi			
	Seamless brass,* and copper† for flanged fittings		Weight, lb per ft‡		Seamless brass, and copper for flanged		Weight, lb per ft		Seamless brass, and copper for flanged		Weight, lb per ft		Seamless brass, and copper for flanged		Weight, lb per ft	
	Inside diameter, in.	Thickness, in.	Brass	Copper	Inside	Thickness	Brass	Copper	Inside	Thickness	Brass	Copper	Inside	Thickness	Brass	Copper
0.405	...	...	...	...	...	...	...	...	...	...	...	...	0.281	0.062	0.253	0.269
0.540	...	...	...	...	...	...	...	...	...	...	...	...	0.410	0.065	0.368	0.376
0.675	...	...	...	...	...	...	...	...	...	...	...	...	0.545	0.065	0.472	0.483
0.840	...	...	...	...	...	...	...	...	...	...	...	...	0.710	0.065	0.600	0.613
1.050	...	...	...	...	...	...	...	...	...	...	...	...	0.920	0.065	0.763	0.780
1.315	...	...	...	...	...	...	...	...	1.185	0.065	0.968	0.989	1.183	0.066	0.982	1.004
1.660	...	...	...	...	...	...	...	...	1.530	0.065	1.235	1.262	1.492	0.084	1.577	1.612
1.900	...	...	...	...	1.770	0.065	1.421	1.452	1.768	0.066	1.442	1.474	1.709	0.096	2.053	2.109
2.375	...	...	...	...	2.245	0.065	1.789	1.828	2.209	0.083	2.266	2.316	2.135	0.120	3.224	3.295
2.875	2.745	0.065	2.176	2.224	2.739	0.068	2.274	2.324	2.675	0.100	3.306	3.379	2.585	0.145	4.716	4.820
3.500	3.370	0.065	2.660	2.719	3.334	0.083	3.378	3.453	3.256	1.122	4.910	5.018	3.146	0.177	7.006	7.162
4.000	3.870	0.065	3.047	3.115	3.810	0.095	4.420	4.517	3.720	0.140	6.436	6.580	3.596	0.202	9.139	9.342
4.500	4.370	0.065	3.434	3.510	4.286	0.107	5.600	5.724	4.186	0.157	8.123	8.303	4.044	0.228	11.603	11.861
5.563	5.427	0.068	4.451	4.550	5.299	0.132	8.540	8.730	5.175	0.194	12.408	12.683	5.001	0.281	17.178	18.073
6.625	6.463	0.081	6.315	6.455	6.309	0.158	12.172	12.442	6.163	0.231	17.596	17.985	5.955	0.335	25.102	25.658
8.625	8.415	0.105	10.657	10.893	8.215	0.205	20.563	21.018	8.023	0.301	29.848	30.509	7.753	0.436	42.534	43.476
10.750	10.488	0.131	16.573	16.939	10.238	0.256	32.004	32.713	10.000	0.375	46.349	47.375	9.662	0.544	66.142	67.607

TABLE II-57. BRASS AND COPPER TUBING (*Continued*)

* Dimensions are in inches unless otherwise stated.
† Seamless brass tubing. FS WW–T–791.
‡ Seamless copper tubing, for use with flanged fittings. FS WW–T–797.
§ For Type 1 brass. Types 2 and 3 are about 97.2 per cent of these weights (see Note 1 below).
NOTE 1. Grade 1 brass tubing is 83 to 86 per cent Cu; 0.06 per cent Pb; 0.05 per cent Fe; remainder, Zn.
Grade 2 brass tubing is 65 to 68 per cent Cu; 0.80 per cent Pb; 0.07 per cent Fe; remainder, Zn.
Grade 3 brass tubing is 58 to 68 per cent Cu; 0.80 per cent Pb; 0.07 per cent Fe; remainder, Zn.
Grade 1 is resistant to corrosion by salt water, salt air, or corrosive gases.
Grade 2 is intended for fresh-water service where corrosive conditions are more severe than ordinary.
Grade 3 is intended for ordinary fresh-water service.
NOTE 2. Standard straight lengths of pipe are 20 ft long. Standard coils (¼ to 1½ in.) are 60 ft long.

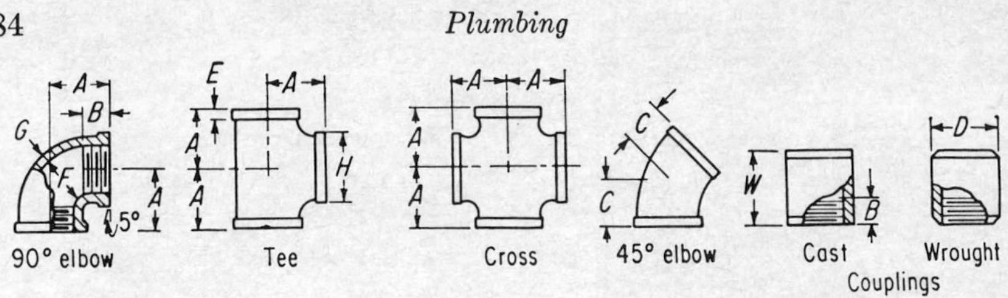

FIG. II-58. Brass and bronze fittings. See Tables II-58 and II-59.

TABLE II-58. DIMENSIONS OF BRASS OR BRONZE THREADED FITTINGS*
(From ASA B 16.15–1947, for 125 lb pressures)

Nominal pipe size	Center to end elbows, tees, crosses	Length of thread, min.	Center to end 45° elbows	Diameter wrought coupling	Band length, min.	Inside diameter of cast fitting F		Metal thickness, min.	Band diameter, min.	End to end straight coupling W	
						Max.	Min.			Cast	Wrought
	A	B	C	D	E			G	H		
1/8	0.54	0.25	0.42	9/16	0.14	0.44	0.41	0.08	0.67	0.80	0.83
1/4	0.71	0.32	0.56	11/16	0.16	0.58	0.54	0.08	0.81	0.97	1.03
3/8	0.82	0.36	0.63	27/32	0.17	0.72	0.68	0.09	1.00	1.05	1.11
1/2	1.01	0.43	0.78	1 1/16	0.19	0.90	0.84	0.09	1.17	1.29	1.36
3/4	1.18	0.50	0.89	1 5/16	0.23	1.11	1.05	0.10	1.42	1.43	1.50
1	1.43	0.58	1.06	.....	0.27	1.39	1.32	0.11	1.72	1.68	
1 1/4	1.69	0.67	1.22	.....	0.31	1.73	1.66	0.12	2.10	1.86	
1 1/2	1.84	0.70	1.30	.....	0.34	1.97	1.90	0.13	2.38	1.92	
2	2.12	0.75	1.45	.....	0.41	2.45	2.38	0.15	2.92	2.20	
2 1/2	2.70	0.92	1.95	.....	0.48	2.98	2.88	0.17	3.49	2.88	
3	3.08	0.98	2.17	.....	0.55	3.60	3.50	0.19	4.20	3.18	
4	3.79	1.08	2.61	.....	0.66	4.60	4.50	0.22	5.31	3.69	

* All dimensions are in inches unless otherwise stated. See Fig. II-58 for nomenclature.

TABLE II-59. FLANGED BRASS AND BRONZE FITTINGS, 90° ELBOWS, 45° ELBOWS, TEES, AND CROSSES
(Manufacturers' standard, Crane Co.; see Fig. II-58)

		1	1¼	1½	2	2½	3	3½	4	4½	5	6
Standard*	A	...	.....	....	4½	5	5½	6	6½	7	7½	8
	C	...	.....	....	2½	3	3	3½	4	4	4½	5
	Flange thickness	...	.....	....	½	9/16	⅝	11/16	11/16	23/32	¾	13/16
Extra heavy†	A	4	4¼	4½	5	5½	6	6½	7	7½	8	8½
	C	2	2½	2¾	3	3½	3½	4	4½	4½	5	5½
	Flange thickness	½	17/32	9/16	⅝	11/16	¾	13/16	⅞	⅞	15/16	1

* Pressures up to 150 psi.
† Pressures up to 250 psi.

Table II-60. Brass Calking Ferrules
("National Plumbing Code," Sec. 3.2.3. Also FS WW-P-448a)

Nominal pipe size, in.	Actual pipe size, in.	Length, in.	Weight	
			Lb	Oz
2	2¼	4½	1	0
3	3¼	4½	1	12
4	4¼	4½	2	8

Table II-62. Brass Soldering Nipples and Bushings
(Federal Specifications WW-P-448a and "National Plumbing Code," Sec. 3.2.4)

Soldering nipples shall be of brass pipe, iron-pipe size, or of heavy, cast red brass, not less than the following weights:

Diameter, in.	Weight, oz	Diameter, in.	Weight	
			Lb	Oz

Table II-63. Weight, Thickness, and

Trade name		3/8	1/2	5/8	3/4	7/8
Aqueduct	Weight*	½	⅝	¾	1	[illegible]
	Thickness†	0.072	0.072	0.070	0.077	[illegible]
	Pressure‡	75	60	45	40	[illegible]
Extra light	Weight	9/16	¾	1¼	1½	[illegible]
	Thickness	0.078	0.082	0.110	0.112	[illegible]
	Pressure	80	65	70	60	[illegible]
Light	Weight	¾	1	1¾	2	[illegible]
	Thickness	0.102	0.106	0.147	0.144	[illegible]
	Pressure	110	85	95	75	[illegible]
Medium	Weight	1	1¼	2	2¼	[illegible]
	Thickness	0.129	0.128	0.164	0.160	[illegible]
	Pressure	140	100	105	85	[illegible]
Strong	Weight	1½	1¾	2½	3	[illegible]
	Thickness	0.176	0.169	0.196	0.201	[illegible]
	Pressure	190	135	125	105	[illegible]
Extra strong	Weight	2	2½	3	3½	[illegible]
	Thickness	0.218	0.223	0.227	0.230	[illegible]
	Pressure	232	180	145	125	[illegible]
Double extra strong	Weight	2⅝	3	3½	4	[illegible]
	Thickness	0.265	0.254	0.256	0.259	[illegible]
	Pressure	280	200	164	140	[illegible]
Service and supply pipe: A or S up to 50 psi	Weight	.....	1½	2½	3	[illegible]
	OD, in.	.....	0.798	1.019	1.156	[illegible]
AA or XS up to 75 psi	Weight	.....	2	3	3½	[illegible]
	OD, in.	.....	0.876	1.082	1.212	[illegible]
AAA or XXS up to 100 psi	Weight	.....	3	3½	4¾	[illegible]
	OD, in.	.....	1.102	1.137	1.336	[illegible]
Waste pipe, min. recommended weight, D or XL§	ID	1¼	1½	2	2½	3
	Weight	2½	3½	4¾	5	6
	OD	1.486	1.776	2.284	2.75	3.25

* Pounds per foot length of pipe.
† Thickness of wall in inches.
‡ Computed working pressure, internal, psi = one-tenth bursting pressure on basis
§ From "Plumbing Manual," Report BMS66, p. 30, National Bureau of Standards.

TABLE II-64. LEAD SERVICE AND SUPPLY PIPES
(Water supply, soil, and waste. Federal Specification WW–P–325)

Classes and applications						Traps		
Class	Application	Working pressure max. psi	Nominal inside diameter, in.	Commercial designation East	Commercial designation West	Type	Nominal inside diameter, in.	Lengths
100	Service and supply	100	3/8 to 2, inclusive	*AAA*	*XXS*			
75		75		*AA*	*XS*	Full S	1 1/4, 1 1/2, 2	Short, long
50		50		*A*	*S*			
B	Soil and waste		1 1/4 to 6, inc.	*B*	*M*	Half S or P	1 1/4, 1 1/2, 2	Short, long
C			1 1/4 to 2, inc.	*C*	*L*			
D			1 1/4 to 6, inc.	*D*	*XL*			

Class 100					Class 75				
Inside diameter, nominal in.	Wall thickness, in.	Outside diameter, in.	Outside circumference, min. in.	Weight per ft., nominal lb	Inside diameter, nominal in.	Wall thickness, in.	Outside diameter, in.	Outside circumference, min. in.	Weight per ft, nominal lb
3/8	0.256	0.888	2 5/8	2.50	3/8	0.175	0.725	2 1/8	1.50
1/2	0.256	1.012	3 1/16	3.00	1/2	0.188	0.876	2 5/8	2.00
5/8	0.256	1.137	3 7/16	3.50	5/8	0.228	1.082	3 1/4	3.00
3/4	0.293	1.336	4 1/16	4.75	3/4	0.231	1.212	3 11/16	3.50
1	0.298	1.596	4 7/8	6.00	1	0.246	1.492	4 9/16	4.75
1 1/4	0.326	1.889	5 13/16	7.75	1 1/4	0.258	1.765	5 3/8	6.00
1 1/2	0.386	2.272	7	11.25	1 1/2	0.288	2.076	6 3/8	8.00
2	0.504	3.008	9 5/16	19.50	2	0.376	2.751	8 1/2	13.75
Class 50					Class B				
3/8	0.175	0.725	2 1/8	1.50	1 1/4	0.143	0.661	2 1/16	1.25
1/2	0.188	0.876	2 5/8	2.00	1 1/2	0.149	0.798	2 3/8	1.50
5/8	0.228	1.082	3 1/4	3.00	2	0.197	1.019	3 1/16	2.50
3/4	0.231	1.212	3 11/16	3.50	2 1/2	0.203	1.156	3 1/2	3.00
1	0.246	1.492	4 9/16	4.75	3	0.214	1.428	4 5/16	4.00
1 1/4	0.258	1.765	5 3/8	6.00	4	0.210	1.670	5 1/8	4.75
1 1/2	0.288	2.076	6 3/8	8.00	5	0.242	1.984	6 1/16	6.50
2	0.376	2.751	8 1/2	13.75	6	0.252	2.503	7 3/4	8.75
Class C					Class D				
1 1/4	0.139	1.528	4 11/16	3.00	1 1/4	0.118	1.486	4 1/2	2.50
1 1/2	0.165	1.830	5 5/8	4.25	1 1/2	0.138	1.776	5 7/16	3.50
2	0.177	2.354	7 1/4	6.00	2	0.142	2.284	7 1/16	4.75
					2 1/2	0.125	2.75	8 1/2	5.00
					3	0.125	3.25	10 1/16	6.00
					4	0.125	4.25	13 3/16	8.00
					5	0.125	5.25	16 3/8	10.00
					6	0.125	6.25	19 1/2	11.75

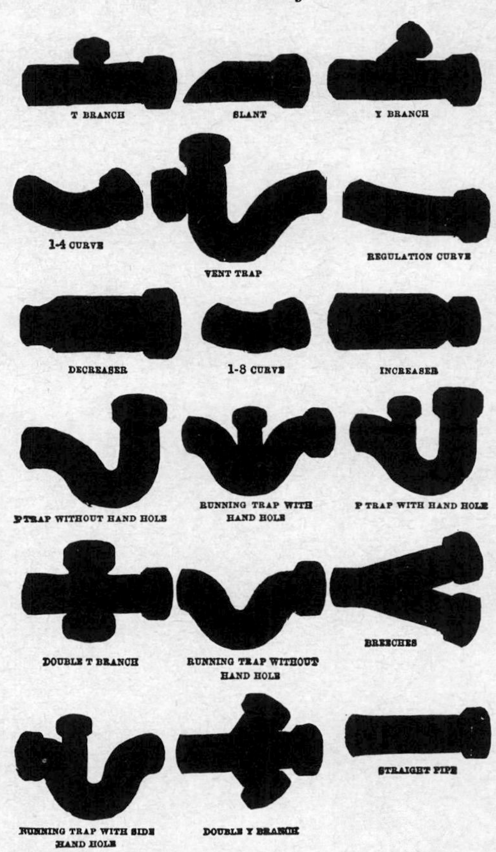

FIG. II-60. Vitrified-clay pipe and specials. See also ASTM C 12-54 for Installation of Clay Pipe. See Table II-65.

TABLE II-65. STRENGTHS AND DIMENSIONS OF VITRIFIED-CLAY SEWER PIPES*

Nominal pipe size	Standard strength							Safe supporting strength†	
	Outside diameter of barrel		Inside diameter of socket ½ in. above, base, min.	Depth of socket		Thickness of barrel		Standard strength	Extra strength
	Min.	Max.		Nominal	Min.	Nominal	Min.		
4	4⅞	5⅛	5¾	1¾	1½	½	7/16		
6	7 1/16	7 7/16	8 3/16	2¼	2	⅝	9/16	1,650	3,000
8	9¼	9¾	10½	2½	2¼	¾	11/16	1,950	3,000
10	11½	12	12¾	2⅝	2⅜	⅞	13/16	2,100	3,000
12	13¾	14 5/16	15⅛	2¾	2½	1	15/16	2,250	3,375

* Dimensions are in inches unless otherwise specified. Standard-strength specifications are from ASTM C 13-54. For extra strength see also C 200-55T; for standard-strength ceramic glazed sewer pipe see C 261; for extra-strength glazed-clay sewer pipe see C 278; for Recommended Practice for Installing Clay Sewer Pipe see C 12-54. There is no standard laying length; 2-ft lengths are specified, but longer lengths are available.

† In pounds per linear feet of pipe, the load being uniformly distributed over 180° of top of pipe.

TABLE II-66. STRENGTH AND DIMENSIONS OF CLAY SEWER PIPE
(Federal Specifications SS–P–361a)

Diameter, in.	Min. Avg. strength, lb per lin ft	Laying length: Nominal ft	Laying length: Limit of (−) variation in. per ft of length	Max. difference of length of two opposite sides	Outside diameter of barrel, in.: Min.	Outside diameter of barrel, in.: Max.	Inside diameter of socket at 1/2 in. above base, in.: Min.	Inside diameter of socket at 1/2 in. above base, in.: Max.	Depth of socket, in.: Nominal	Depth of socket, in.: Min.	Thickness of barrel, in.: Nominal	Thickness of barrel, in.: Min.	Thickness of socket at 1/2 in. from outer end, in.: Nominal	Thickness of socket at 1/2 in. from outer end, in.: Min.
Standard strength														
4	1,000	2, 3	1/4	5/16	4 7/8	5 1/8	5 3/4	6 1/8	1 3/4	1 1/2	1/2	7/16	7/16	3/8
6	1,000	2 1/2, 3	1/4	3/8	7 1/16	7 7/16	8 3/16	8 5/8	2 1/4	2	5/8	9/16	1/2	7/16
8	1,000	2 1/2, 3	1/4	7/16	9 1/4	9 3/4	10 1/2	11	2 1/2	2 1/4	3/4	11/16	9/16	1/2
10	1,100	2 1/2, 3	1/4	7/16	11 1/2	12	12 3/4	13 1/4	2 5/8	2 3/8	7/8	13/16	5/8	9/16
12	1,200	2 1/2, 3	1/4	7/16	13 3/4	14 5/16	15 1/8	15 3/4	2 3/4	2 1/2	1	15/16	3/4	11/16
15	1,400	3, 4	1/4	1/2	17 3/16	17 13/16	18 5/8	19 1/4	2 7/8	2 5/8	1 1/4	1 1/8	15/16	7/8
18	1,700	3, 4	1/4	1/2	20 5/8	21 7/16	22 1/4	23	3	2 3/4	1 1/2	1 3/8	1 1/8	1 1/16
21	2,000	3, 4	1/4	9/16	24 1/8	25	25 7/8	26 3/4	3 1/4	3	1 3/4	1 5/8	1 5/16	1 3/16
24	2,400	3, 4	3/8	9/16	27 1/2	28 1/2	29 3/8	30 3/8	3 3/8	3 1/8	2	1 7/8	1 1/2	1 3/8
Extra strength														
6	2,000	3	1/4	3/8	7 1/16	7 7/16	8 3/16	8 5/8	2 1/4	2	11/16	9/16	1/2	7/16
8	2,000	3	1/4	7/16	9 1/4	9 3/4	10 1/2	11	2 1/2	2 1/4	7/8	3/4	9/16	1/2
10	2,000	3	1/4	7/16	11 1/2	12	12 3/4	13 1/4	2 5/8	2 3/8	1	7/8	5/8	9/16
12	2,250	3	1/4	7/16	13 3/4	14 5/16	15 1/8	15 3/4	2 3/4	2 1/2	1 3/16	1 1/16	3/4	11/16
15	2,750	3, 4	1/4	1/2	17 3/16	17 13/16	18 5/8	19 1/4	2 7/8	2 5/8	1 1/2	1 3/8	15/16	7/8
18	3,300	3, 4	1/4	1/2	20 5/8	21 7/16	22 1/4	23	3	2 3/4	1 7/8	1 3/4	1 1/8	1 1/16
21	3,850	3, 4	1/4	9/16	24 1/8	25	25 7/8	26 3/4	3 1/4	3	2 1/4	2	1 5/16	1 3/16
24	4,400	3, 4	3/8	9/16	27 1/2	28 1/2	28 3/8	30 3/8	3 3/8	3 1/8	2 1/2	2 1/4	1 1/2	1 3/8

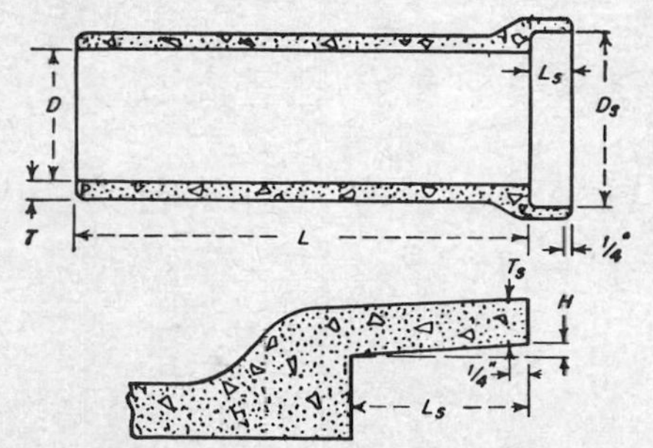

FIG. II-61. Concrete pipe. See Table II-67.

TABLE II-67. PROPERTIES OF UNREINFORCED CONCRETE SEWER PIPE, 4 TO 24 IN. IN DIAMETER
(ASTM Standards C 14–35)
(The letters in the column headings refer to Fig. II-61)

Inside diameter, in., D	Laying length, ft, L	Inside diameter at mouth of socket, in.,* D_s	Depth of socket, in.,† L_s	Thickness of barrel, in.,‡ T	Average crushing strength, lb per lin ft§		Limits of permissible variations‖				
					Three-edge bearing	Sand bearing	Length, in. per ft (−)‖	Internal diameter, in. Spigot (±)‖	Internal diameter, in. Socket (±)‖	Depth of socket, in. (−)‖	Thickness of barrel, in (−)‖
4	2, 2½, 3	6	1½	9/16	1,000	1,430	¼	⅛	⅛	⅛	1/16
6	2, 2½, 3, 4	8¼	2	⅝	1,000	1,430	¼	3/16	3/16	¼	1/16
8	2, 2½, 3, 4	10¾	2¼	¾	1,000	1,430	¼	¼	¼	¼	1/16
10	2, 2½, 3, 4	13	2½	⅞	1,100	1,570	¼	¼	¼	¼	1/16
12	2, 2½, 3, 4	15¼	2½	1	1,200	1,710	¼	¼	¼	¼	1/16
15	2, 2½, 3, 4	18¾	2½	1¼	1,370	1,960	¼	¼	¼	¼	3/32
18	2, 2½, 3, 4	22¼	2¾	1½	1,540	2,200	¼	¼	¼	¼	3/32
21	2, 2½, 3, 4	26	2¾	1¾	1,810	2,590	¼	5/16	5/16	¼	⅛
24	2, 2½, 3, 4	29	3	2⅛	2,150	3,070	⅜	5/16	5/16	¼	⅛

* When pipes are furnished having an increase in thickness over that given in the fifth column, the diameter of the socket shall be increased by an amount equal to twice the increase of thickness of the barrel.

† Minimum taper of socket. (H) is 1:20.

‡ Thickness of socket ¼ in. from its outer end shall be not less than three-fourths of the thickness of the barrel of the pipe.

§ Maximum absorption, 8 per cent.

‖ The minus (−) sign alone indicates that the plus variation is not limited; the (±) indicates variation in both excess and deficiency in dimension.

TABLE II-68. STRENGTH AND DIMENSIONS OF UNREINFORCED CONCRETE SEWER PIPE
(Federal Specifications SS–P–371)

Internal diameter, in.		Length, nominal ft	Shell thickness, min. in.	Socket depth, min. in.	Strength, average lb per lin ft
Nominal	Tolerance				
4	1/8	2½ or 3	½	1½	1,000
6	3/16	2½ or 3	5/8	2	1,100
8	¼	3 or 4	¾	2¼	1,300
10	¼	3 or 4	7/8	2½	1,400
12	¼	3 or 4	1	2½	1,500
15	¼	3 or 4	1¼	2½	1,750
18	¼	3 or 4	1½	2¾	2,000
21	5/16	3 or 4	1¾	2¾	2,200
24	5/16	3 or 4	2	3	2,400

TABLE II-69. ASBESTOS-CEMENT PIPE
(Federal Specifications SS–P–351)

Diameter, in.	Wall thickness, in. Class			Min. loads for flexure. Total applied load in lb			Min. applied loads for crushing tests, lb per ft Class		
	100	150	200	100	150	200	100	150	200
4	0.48	0.53	0.60	750	850	950	4,100	5,000	6,300
4½	0.49	0.55	0.64	950	1,050	1,250	3,900	4,800	6,400
5	0.50	0.57	0.68	1,150	1,350	1,650	3,600	4,700	6,500
6	0.51	0.62	0.75	1,650	2,100	2,600	3,200	4,600	6,700
8	0.52	0.69	0.91	2,900	4,000	5,400	2,500	4,400	7,400
10	0.59	0.85	1.10	.....	.....	.....	2,600	5,300	8,700
12	0.68	0.98	1.24	.....	.....	.....	2,900	5,900	9,300
18	0.97	1.39	1.87	.....	.....	.....	4,000	8,000	14,100
24	1.25	1.82	2.48	.....	.....	.....	5,000	10,300	18,600

TABLE II-70. STANDARD GAGES IN INCHES AND APPROXIMATE WEIGHTS OF SHEET METAL IN POUNDS PER SQUARE FOOT

Gage number	Birmingham (BWG) Stubbs iron wire	Brown & Sharpe (B & S) American	1884 British Imperial	American Steel and Wire Company	Trenton Iron Company	Stubbs steel wire	Old English, London	1893 U.S. plate iron and steel standard	Weights of metal, lb per sq ft United States standard*				
									Iron	Steel	Zinc	Copper	Lead
07†	.....	....	0.500	0.49	......	.....	....	0.500	20	20.4	18.00	22.63	29.4
05†	.....	....	0.432	0.43	0.45	.....	....	0.4375	17.50	17.85	15.75	19.85	25.75
03†	0.425	0.40964	0.372	0.3625	0.36	.....	0.425	0.375	15	15.3	13.5	17	22.05
0	0.340	0.32486	0.324	0.3065	0.305	.....	0.340	0.3125	12.50	12.75	11.25	14.18	18.38
1	0.300	0.2893	0.300	0.2830	0.285	0.227	0.300	0.28125	11.25	11.475	10.12	12.75	16.54
3	0.259	0.22942	0.252	0.2437	0.245	0.212	0.259	0.25	10.0	10.2	9.00	11.33	14.70
5	0.220	0.18194	0.212	0.2070	0.205	0.204	0.220	0.21875	8.75	8.925	7.875	9.91	12.87
7	0.180	0.14428	0.176	0.1770	0.175	0.199	0.180	0.1875	7.5	7.65	6.75	8.5	11.02
9	0.148	0.11443	0.144	0.1483	0.145	0.194	0.148	0.15625	6.25	6.375	5.625	7.08	9.19
10	0.134	0.10189	0.128	0.1350	0.130	0.191	0.134	0.140625	5.625	5.7375	5.0625	6.37	8.27
13	0.095	0.071961	0.092	0.0915	0.0925	0.182	0.095	0.09375	3.75	3.825	3.375	4.25	5.51
15	0.072	0.057068	0.072	0.0720	0.0700	0.178	0.072	0.0703125	2.8125	2.86875	2.53125	3.185	4.13
17	0.058	0.045257	0.056	0.0540	0.0525	0.172	0.058	0.05625	2.25	2.295	2.025	2.55	3.305
19	0.042	0.03589	0.040	0.0410	0.0400	0.164	0.040	0.04375	1.75	1.785	1.575	1.982	2.575
20	0.035	0.031961	0.036	0.0348	0.0350	0.161	0.035	0.0375	1.50	1.53	1.35	1.700	2.205
23	0.025	0.022571	0.024	0.0258	0.0250	0.153	0.027	0.028125	1.125	1.1475	1.012	1.256	1.654
25	0.020	0.0179	0.020	0.0204	0.0200	0.148	0.023	0.021875	0.875	0.8925	0.7875	0.991	1.287
26	0.018	0.01594	0.018	0.0181	0.0180	0.146	0.0205	0.01875	0.75	0.765	0.675	0.850	1.102
29	0.013	0.011257	0.0136	0.0150	0.0150	0.134	0.0155	0.0140625	0.5625	0.57375	0.50625	0.637	0.826
30	0.012	0.010025	0.0124	0.0140	0.014	0.127	0.01375	0.0125	0.5	0.51	0.45	0.566	0.735
33	0.008	0.00708	0.0100	0.0118	0.0110	0.112	0.01025	0.009375	0.375	0.3825	0.3375	0.425	0.551
35	0.005	0.005614	0.0084	0.0095	0.0095	0.108	0.009	0.0078125	0.3125	0.31875	0.28125	0.354	0.459
36	0.004	0.005	0.0076	0.0090	0.0090	0.106	0.0075	0.00703125	0.28125	0.286875	0.253125	0.3185	0.413
38	.....	0.003965	0.0060	0.008	0.0080	0.101	0.00575	0.00625	0.25	0.255	0.225	0.2835	0.3675
40	.....	0.003144	0.0048	0.007	0.0070	0.097	0.0045						

* A variation of 2½ per cent may be expected either way. Iron is assumed as 480 lb per cu ft; steel = 1.02 × iron; zinc = 0.9 × iron; copper = 1.133 × iron; and lead = 1.47 × iron.

† 07 means 0000000, 06 means 000000, etc.

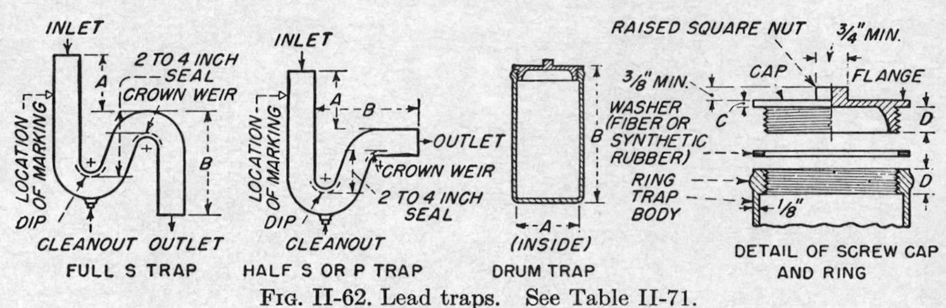

FIG. II-62. Lead traps. See Table II-71.

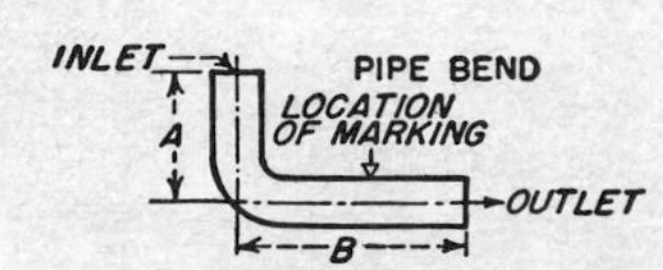

FIG. II-63. Lead-pipe bend. See Table II-71.

TABLE II-71. LEAD TRAPS AND PIPE BENDS*
(Federal Specifications WW–P–325)

Nominal inside diameter, in.	Weight per running foot	Short traps				Long traps			
		Dimensions		Total weight		Dimensions		Total weight	
Full S	Lb	*A*, in.	*B*, in.	Lb	Oz	*A*, in.	*B*, in.	Lb	Oz
1¼	2⅝	4½	7	4	9	4½	19½	7	6
1½	3⅛	4½	7	5	10	4½	19½	8	14
2	4⅛	4½	8	8	5	4½	19½	12	4
Half S or P									
1¼	2⅝	4½	7	4	0	4½	14	5	9
1½	3⅛	4½	7	4	15	4½	14	6	12
2	4⅛	4½	8	7	2	4½	14	9	3

Dimensions of drum traps				Pipe bends								
Trap sizes, *A* × *B*, in.	Screw cap and ring			1½-in. nominal inside diameter			3-in. nominal inside diameter			4-in. nominal inside diameter		
	C, in.	*D*, in.	Thread	Dimension, *A* × *B*, in.	Total weight Lb	Total weight Oz	Dimension, *A* × *B*, in.	Total weight Lb	Total weight Oz	Dimension, *A* × *B*, in.	Total weight Lb	Total weight Oz
4 × 8 4 × 9 4 × 10	5/32	9/16	3-in. straight pipe thread, 8 threads per in. free fit. Nominal pitch diameter, 3.388 in.	4 × 7	2	12	5½ × 10	7	3	5½ × 10	9	5
				4 × 12	4	1	5½ × 12	8	3	5½ × 12	10	10
				4 × 15	4	13	5½ × 15	9	12	5½ × 15	12	10
3 × 8 3 × 9 3 × 10	5/32	7/16	2-in. straight pipe thread, 11½ threads per in. free fit. Nominal pitch diameter, 2.296 in.	4 × 18	5	10	5½ × 18	11	4	5½ × 18	14	10
				4 × 20	6	2	5½ × 20	12	4	5½ × 20	16	10
				7 × 7	3	8	10 × 10	9	8	10 × 10	12	5
				7 × 12	4	13	10 × 12	10	8	10 × 12	13	10
				7 × 15	5	10	10 × 15	12	0	10 × 15	15	10
				7 × 18	6	6	10 × 18	13	8	10 × 18	17	10
				7 × 20	6	15	10 × 20	14	8	10 × 20	18	15

* See Figs. II-62 and II-63.

TABLE II-72. DIMENSIONS OF BRASS LAVATORY AND SINK TRAPS

Wrought traps					Cast bend traps					Wrought sink traps			
Size, in.	Gage	Style	Inlet, in.	Cleanout	Size, in.	Gage	Style	Inlet, in.	Cleanout	Size, in.	Gage	Style	Inlet, in.
1¼	20	P	1¼	No	1¼	20	P	1¼	No	1¼	17	P	1¼
1¼	20	P	1¼	Yes	1¼	20	P	1¼	Yes	1¼	17	P	1¼
1¼	20	S	1¼	No	1¼	20	S	1¼	No	1½	17	P	1¼ or 1½
1¼	20	S	1¼	Yes	1¼	20	S	1¼	Yes	1½	17	P	1¼ or 1½
1¼	20	S*	1¼	Yes	1¼	20	S*	1¼	Yes				
1¼	20	S*	1¼	No	1¼	20	S*	1¼	No	Wrought antisiphon traps (clean-sweep or ball pattern with cleanout)			
1¼	17	P	1¼	No	1¼	17	P	1¼	No				
1¼	17	P	1¼	Yes	1¼	17	P	1¼	Yes	1¼	20	P	1¼
1¼	17	S	1¼	No	1¼	17	S	1¼	No	1½	20	P	1¼ or 1½
1¼	17	S	1¼	Yes	1¼	17	S	1¼	Yes	1¼	20	S	1¼
1¼	17	S*	1¼	Yes	1¼	17	S*	1¼	Yes	1½	20	S	1¼ or 1½
1¼	17	S*	1¼	No	1¼	17	S*	1¼	No	1¼	17	P	1¼
1½	20	P	1¼ or 1½	No	1½	20	P	1¼ or 1½	No	1½	17	P	1¼ or 1½
1½	20	P	1¼ or 1½	Yes	1½	20	P	1¼ or 1½	Yes	1¼	17	S	1¼
1½	20	S	1¼ or 1½	No	1½	20	S	1¼ or 1½	No	1½	17	S	1¼ o r1½
1½	20	S	1¼ or 1½	Yes	1½	20	S	1¼ or 1½	Yes	Cast antisiphon traps (clean-sweep or ball pattern with cleanout)			
1½	20	S*	1¼ or 1½	Yes	1½	20	S*	1¼ or 1½	Yes				
1½	20	S*	1¼ or 1½	No	1½	20	S*	1¼ or 1½	No				
1½	17	P	1¼ or 1½	No	1½	17	P	1¼ or 1½	No	1¼	20	P	1¼
1½	17	P	1¼ or 1½	Yes	1½	17	P	1¼ or 1½	Yes	1½	20	P	1¼ or 1½
1½	17	S	1¼ or 1½	No	1½	17	S	1¼ or 1½	No	1¼	20	S	1¼
1½	17	S	1¼ or 1½	Yes	1½	17	S	1¼ or 1½	Yes	1½	20	S	1¼ or 1½
1½	17	S*	1¼ or 1½	Yes	1½	17	S*	1¼ or 1½	Yes	1¼	17	P	1¼
1½	17	S*	1¼ or 1½	No	1½	17	S*	1¼ or 1½	No	1½	17	P	1¼ or 1½
										1¼	17	S	1¼
										1½	17	S	1¼ or 1½

New York and Boston regulation traps					
1¼	P	1¼†	1¼		1¼-in. tubing outlet
1½	P	1½ or 1¼†	1½	Bag	1½- or 1¼-in. tubing outlet
2	P	2½ or 1¼†	1¼	offset	1¼-in. iron-pipe outlet
1½	P	1½ or 1¼‡	1½		1½- or 1¼-in. iron-pipe outlet

* 26 in. over-all.

† External. Antisiphon traps (S) to be furnished with either straight or offset outlet tube as required.

‡ Internal.

TABLE II-73. CAST-IRON GATE VALVES, SCREWED AND FLANGED
(Federal Specifications WW–V–58)*

Class	Size range, nominal pipe sizes				
	NS, in.	*NF*, in.	*OS*, in.	*OF*, in.	Types
A	2–6	2–12	2–6	2–12	I and II
B	2–4	2–12	2–4	2–12	I and II

Size, in.	Diameter of hand wheel, min. in.		Diameter of stem, min. in.		Thread length screwed-end valves, min. in.		Face-to-face dimensions of flanged-end valves, in.			
							Types I and II		Tolerances ±	
	Class		Class		Class		Class		Class	
	A	B	A	B	A	B	A	B	A	B
2	6	7	⅝	¾	0.75	1.00	7	8½	1/16	1/16
2½	7	7½	¾	¾	0.92	1.17	7½	9½	1/16	1/16
3	7	8	¾	⅞	0.98	1.23	8	11⅛	1/16	1/16
4	9	11	1	1⅛	1.08	1.33	9	12	1/16	1/16
5	10	12	1	1⅛	1.18	....	10	15	1/16	1/16
6	12	16	1⅛	1⅜	1.28	....	10½	15⅞	1/16	1/16
8	14	16	1¼	1½	....	....	11½	16½	1/16	1/16
10	16	18	1⅜	1½	....	....	13	18	1/16	1/16
12	18	20	1½	1⅝	....	....	14	19¾	⅛	⅛

* The following Federal Specifications are also applicable to gate valves: QQ–B–611; QQ–B–726; QQ–M–151; WW–F–406; GGG–P–351 (see also Table II-1).

Type I is wedge disk. Type II is double disk.

NS is nonrising stem, screwed ends.

NF is nonrising stem, flanged ends.

OS is outside screw and yoke, screwed ends.

OF is outside screw and yoke, flanged ends.

Class A is 125 psi steam pressure; 200 psi, for liquids and gases up to 150°F.

Class B is 250 psi steam pressure; 500 psi for liquids and gases up to 150°F.

TABLE II-74. GATE VALVES
(125 psi. Threaded and flanged. Federal Specifications WW–V–76a and WW–V–76b)

Construction		Sizes, in.	Types		Iron-body valves	
			WW–V–76a	WW–V–76b	Size of valve, in.	Face-to-face flanges I, II, III, in.
Bronze body	*A*	*T* ¼–3	I, II, III	I, II	¼	7
	A	*F* 1½–3	I, II, III	I, II	⅜	7½
	B	*T* ¼–3	I, II, III	I, II	½	8
	B	*F* 1½–3	I, II, III	I, II	¾	8½
Cast-iron body	*A*	*T* 2 –12	I, II, III	I, II	1	9
	A	*F* 2 –12	I, II, III	I, II	1¼	10
	C	*T* 2 –12	I, II, III	I, II	1½	10½
	C	*F* 2 –12	I, II, III	I, II	2	11½
		....	....	......	2½	13
		....	....	......	3	14

A = nonrising stem, interior screw
B = rising stem, interior screw
C = outside screw and yoke
T = threaded
F = flanged

NOTE: While some type II valves are made to these dimensions, they are not commercially standard for the industry. Valves are intended for use with steam pressures of less than 125 psi or for service with liquids or gases not exceeding 175 psi. Type I is single-wedge disk; type II is double disk, parallel seat; type III is double disk, taper seat.

TABLE II-75. FACE-TO-FACE DIMENSIONS OF FERROUS VALVES*,†

Nominal valve size	Class 125, cast iron						
	Flanged end, plain face				Distance face to face		
	Gate		Plug		Check	Check	Conical
	Wedge	Disk	Pattern	Regular			
1	....	....	3½	3½	....	....	7¼
1¼	....	....	....	6½			
1½	....	....	6½	6½	....	....	8¾
2	7	7	7	7½	8	8	10
2½	7½	7½	7½	8¼	8½	8½	10⅞
3	8	8	8	9	9½	9½	11¼
3½	8½	8½					
4	9	9	9	9	11½	11½	13⅞
5	10	10	10	11	13	13	
6	10½	10½	10½	13½	14	14	17¾
8	11½	11½	11½	...	19½	19½	21¾

* Extracted from American Standards ASA B 16.10–1957, with permission of the publisher, American Society of Mechanical Engineers, 29 West 39th Street, New York City.

† All dimensions are in inches.

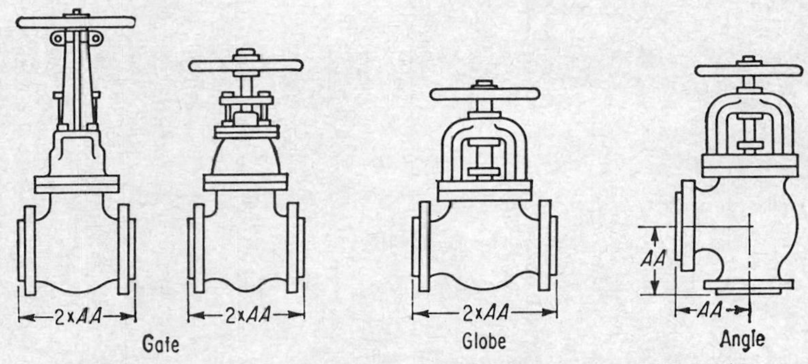

FIG. II-64. Flanged cast-iron and steel valves. See Table II-76.

TABLE II-76. DIMENSIONS OF FLANGED CAST-IRON AND STEEL VALVES*,†

Nominal pipe size	Contact surface to contact surface dimensions (2 × *AA*). See Fig. II-64									
	Cast iron					Steel				
	125	175†,‡	250†	800 Hy-draulic†	150†	300†	400†	600†	900†	1,500†
	1	2	3	4	5	6	7	8	9	10
	Globe and angle valves									
1	....	....	....	....	....	....	8½§	8½	10§	10
1¼	....	....	....	....	....	....	9§	9	11§	11
1½	....	....	....	....	....	7½	9½§	9½	12§	12
2	7	7¼	8½	11½	7	8½	11½§	11½	14½§	14½
2½	7½	8	9½	13	7½	9½	13§	13	16½§	16½
3	8	9¼	11⅛	14	8	11⅛	14§	14	15	18½
3½	8½	10	11⅞	....	8½	11⅞	.....	....	.....	21½
4	9	10½	12	17	9	12	16	17	18	26½
5	10	11½	15	....	10	15	18	20	22	27¾
6	10½	13	15⅞	22	10½	15⅞	19½	22	24	32¾
8	11½	14¼	16½	26	11½	16½	23½	26	29	39
10	13	16¾	18	31	13	18	26½	31	33	44½
12	14	17½	19¾	33	14	19¾	30	33	38	
	Wedge gate valves									
½										
¾	....	....	....	....	....	....	7½	7½		
1	....	....	....	....	....	....	8½	8½	10	10
1¼	....	....	....	....	....	....	9	9	11	11
1½	....	....	....	....	....	....	9½	9½	12	12
2	8	....	10½	....	8	10½	11½	11½	14½	14½
2½	8½	....	11½	....	8½	11½	13	13	16½	16½
3	9½	....	12½	....	9½	12½	14	14	15	18½
3½	10½	....	13¼	....	10½	13¼				
4	11½	....	14	....	11½	14	16	17	18	21½
5	13	....	15¾	....	14	15¾	18	20	22	26½
6	14	....	17½	....	16	17½	19½	22	24	27¾
8	19½	....	21	....	19½	22	23½	26	29	32¾

* Extracted from American Standards ASA B 16.10–1957, with permission of the publishers, American Society of Mechanical Engineers, 29 West 39th Street, New York City. Dimensions are in inches unless otherwise stated.

† These are pressure designations which refer to the primary service ratings in pounds per square inch of the connecting end flanges.

‡ The connecting end flanges of 175-lb valves are the same as those on 250-lb valves.

§ The contact surface to contact surface dimensions and connecting end flanges for 400-lb valves 3 in. and smaller are the same as for 600-lb valves, and the contact surface to contact surface dimensions and connecting end flanges for 900-lb valves 2½ in. in size and smaller are the same as for 1,500-lb valves.

NOTE: Female and groove-joint facings have bottom of groove in same plane as *flange edge*, and center to contact surface dimensions for these facings are reduced by the amount of the raised face.

TABLE II-77. THICKNESS OF MAGNESIA MOLDED PIPE COVERING, INCHES (Federal Specifications HH–M–61)

Nominal pipe size, in.	Thickness	
	Standard	Double standard
1½ and smaller	⅞	1¾
2–3½	1¹⁄₃₂	2¹⁄₁₆
4–6	1⅛	2¼
7–10	1¼	2½
12 and larger	1½	3

NOTE: Standard thickness of 1½ in. in single layer or double-standard thickness of 3 in. in two layers for all nominal pipe sizes of up to and including 10 in.

TABLE II-78. DIMENSIONS OF RANGE BOILERS*

Size, in.		Capacity, gal	Size, in.		Capacity, gal	Size, in.		Capacity, gal
Diameter †	Length ‡		Diameter †	Length ‡		Diameter †	Length ‡	
12	36	18	16	48	42	22	60	100
12	48	24	16	60	52	24	60	120
12	60	30	18	60	66	24	72	144
14	48	32	20	60	82	24	96	192
14	60	40						

* Recommendation No. 8, approved May 1, 1924, by the Division of Simplified Practice, U.S. Dept. of Commerce.

† Diameters refer to inside measurements.

‡ Length means length of sheet, not over-all length of boilers.

TABLE II-79. DIMENSIONS OF EXPANSION TANKS*

Size, in.		Capacity, gal	Size, in.		Capacity, gal	Size, in.		Capacity, gal
Diameter	Length		Diameter	Length		Diameter	Length	
12	20	10	14	30	20	16	36	32
12	30	15	16	30	26	16	48	42

* Recommendation No. 8, approved May 1, 1924, by the Division of Simplified Practice, U.S. Dept. of Commerce.

TABLE II-80. WEIGHTS AND CAPACITIES OF HOT-WATER STORAGE TANKS*

Galvanized and cold-welded		Weights, lb					Galvanized and cold-welded		Weights, lb					Copper: Standard for 100 psi, heavy for 150 psi	
		Galvanized		Cold-welded					Galvanized		Cold-welded				
Capacity, gal	Dimensions, in.	Standard, 150 psi	Extra heavy, 200 psi	Standard, 150 psi	Extra heavy, 200 psi	Double extra heavy, 250 psi	Capacity, gal	Dimensions, in.	Standard, 150 psi	Extra heavy, 200 psi	Standard, 150 psi	Extra heavy, 200 psi	Double extra heavy, 250 psi	Capacity, gal	Dimensions, in.
18	12 by 36†	50	...	53	...	...	48	14 by 72	106	...	...	...	...	30	12 by 60
21	12 by 42	58	...	62	...	...	52	16 by 60†	117	139	121	138	167	35	13 by 60
24	12 by 48†	60	72	65	81	109	53	18 by 48	124	...	130	...	...	40	14 by 60
24	14 by 36	61	67	66	83	...	63	16 by 72	140	166	...	...	...	50	16 by 60
27	12 by 54	68	84	72	90	...	66	18 by 60†	147	165	154	172	...	60	18 by 60
28	14 by 42	71	...	75	...	...	79	18 by 72	171	...	...	...	...	80	20 by 60
30	12 by 60†	73	87	79	94	127	82	20 by 60†	174	202	...	...	...	100	22 by 60
32	14 by 48†	75	...	82	...	...	98	20 by 72	199	...	...	...	...	120	24 by 65½
35	13 by 60	82	98	86	102	135	100	22 by 60†	202	229	...	...	...	125	24 by 69
36	12 by 72	89	...	...	...	...	120	24 by 60†	260	...	...	...	...	150	24 by 78½
36	14 by 54	84	...	89	...	...	144	24 by 72†	294	...	...	...	...	200	26 by 87
40	14 by 60†	89	105	92	110	144	168	24 by 84	325						
42	16 by 48†	98	115	102	120	...	192	24 by 96†	375						
47	16 by 54	110	...	114											

* Simplified Practice Recommendation No. 25, U.S. Department of Commerce, Dec. 31, 1924.

† Simplified Practice Recommendation No. 8, for Range Boilers, U.S. Department of Commerce, 1924.

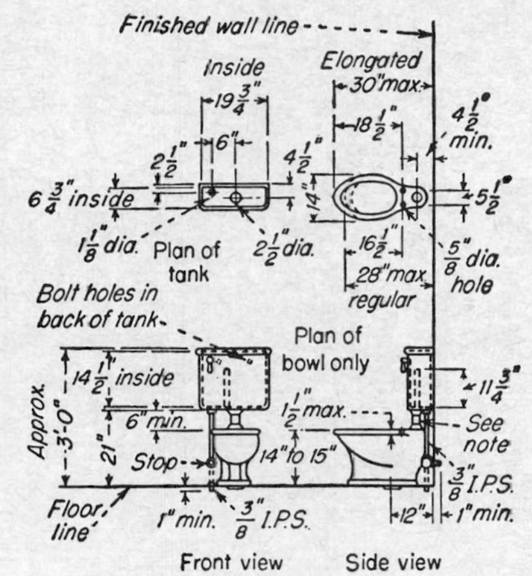

FIG. II-65. Siphon-jet water closet with low tank supply and floor outlet. NOTE: Flush pipe 2 in. OD or offset not over 1½ in. Washer joint at tank.

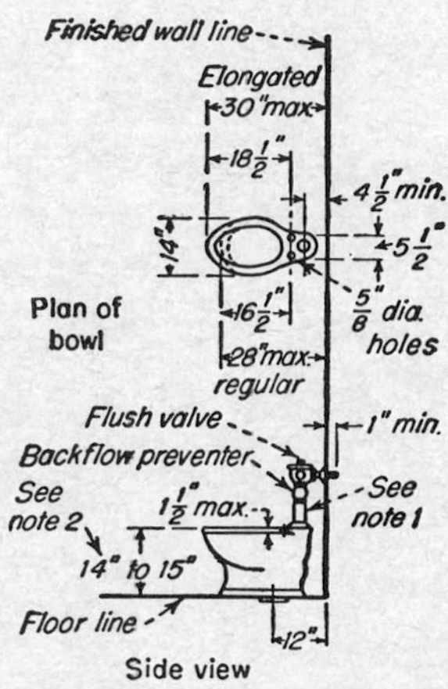

FIG. II-66. Siphon-jet water closet with flushometer valve supply and floor outlet. NOTE: Flush pipe 1½ in. OD, straight or offset not over 1½ in. Height of juvenile bowl 13 in.

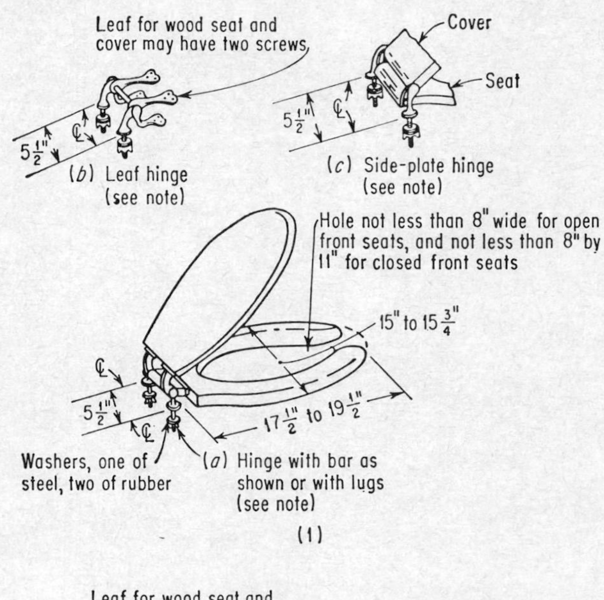

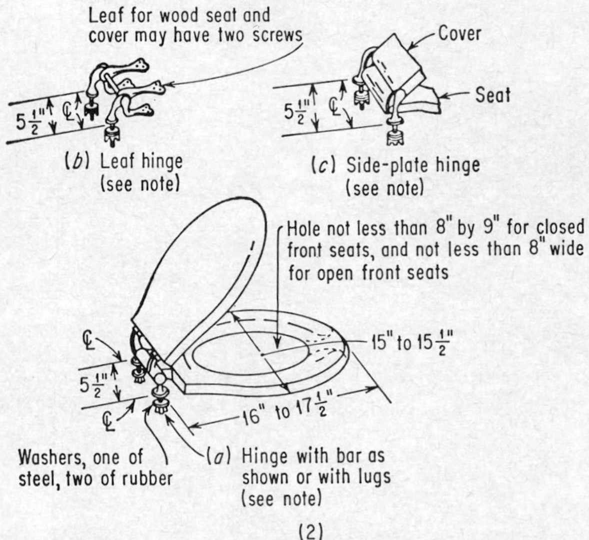

FIG. II-67. Water-closet seats. (1) Seat for elongated bowl. NOTE: Hinge with lugs or bar as shown at (*a*) for regular-weight composition seats, leaf hinge approximately as shown at (*b*) for lightweight composition seat and for square-back wood seat, and side plate hinge approximately as shown at (*c*) for extended-back wood seat. (2) Seat for regular bowl. NOTE: Hinge with lugs or bar approximately as shown at (*a*) for regular-weight composition seat and for square-back wood seat, and side-plate hinge approximately as shown at (*c*) for extended-back wood seat.

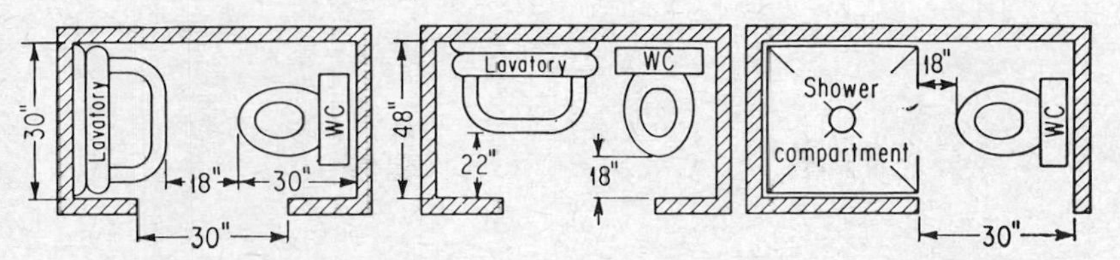

Fig. II-68. Water closet, lavatory, and shower compartments, showing minimum dimensions. See also Compartmentizing Is Here to Stay. *Am. Builder*, November, 1955, p. 86.

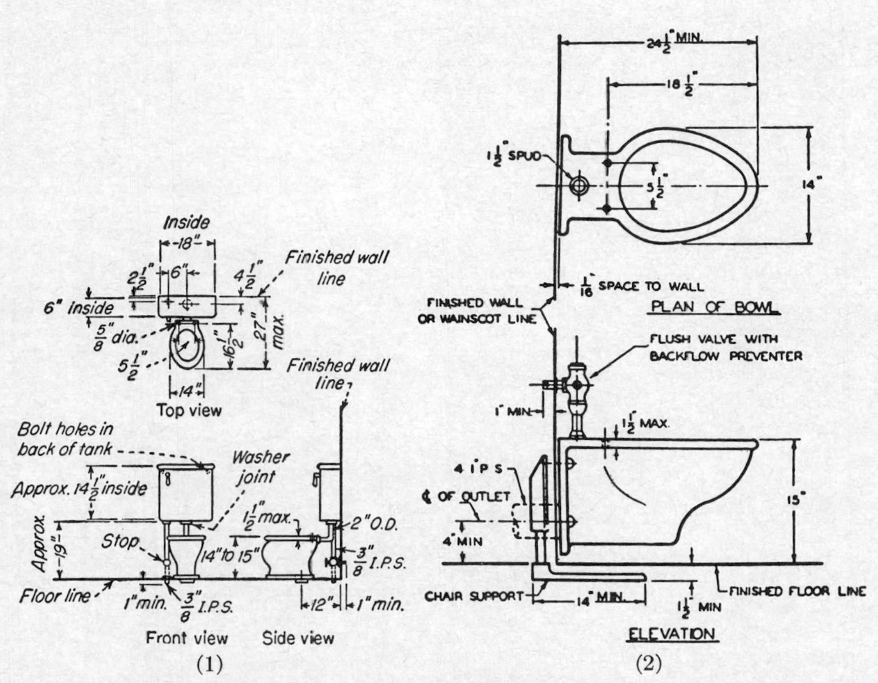

Fig. II-69. Water-closet installations. (1) Washdown water closet with tank. Note: Fixture may be as shown, may be with sealed flush connection, or may be close-coupled. (2) Siphon-jet wall-hanging water closet.

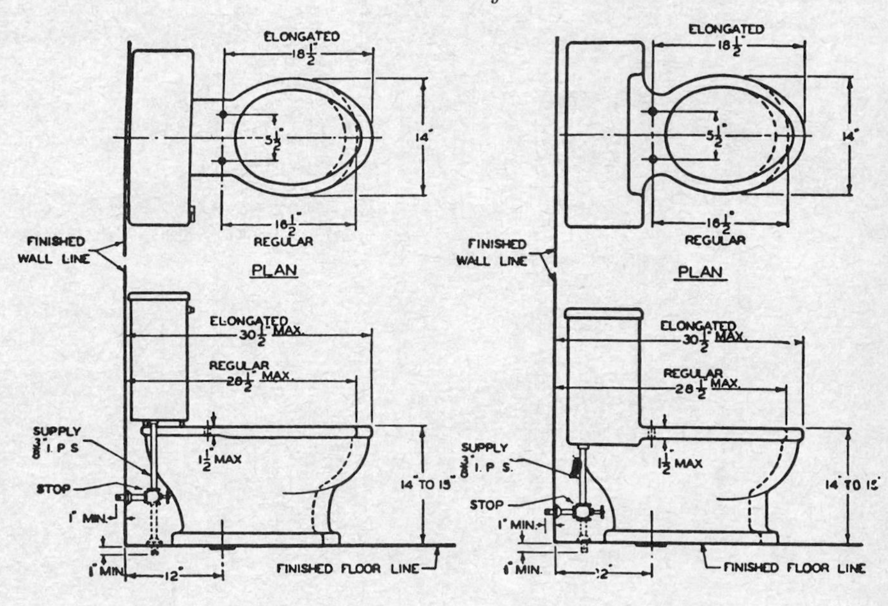

FIG. II-70. Sighon-jet close-coupled combination water closet.

FIG. II-71. Reverse-trap integral water closet and tank combination.

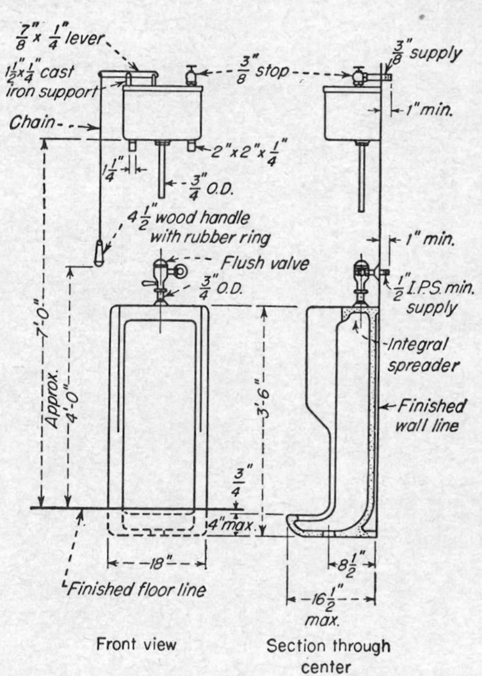

FIG. II-72. Stall urinal with high tank or flushometer-type valve.

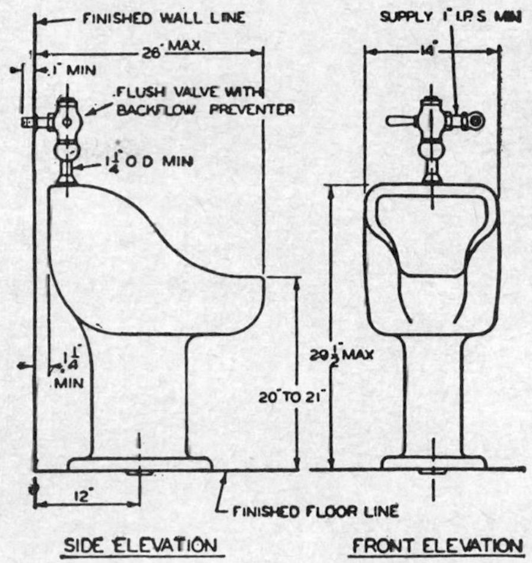

FIG. II-73. Pedestal urinal.

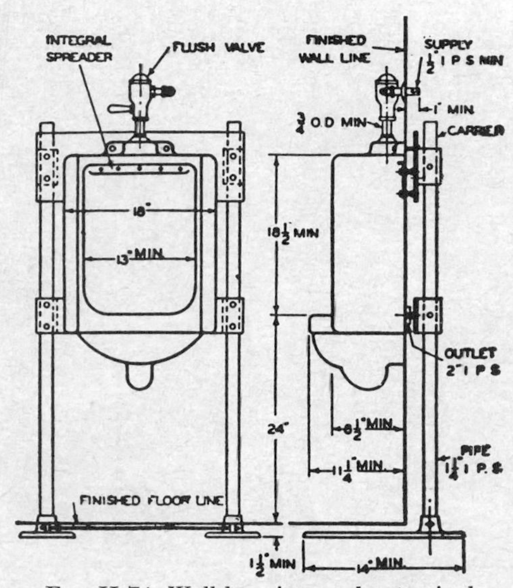

FIG. II-74. Wall-hanging washout urinal.

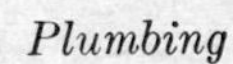

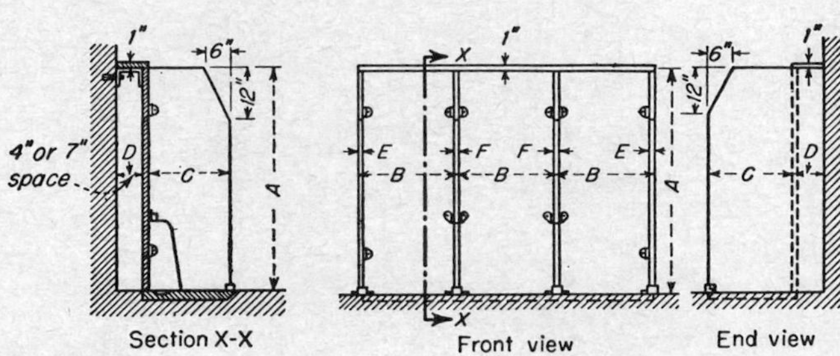

Fig. II-75. Urinals. See Table II-81. (*Division of Simplified Practice, U.S. Department of Commerce.*)

Table II-81. Urinals
(See Fig. II-75)

With partitions						Without partitions						Thickness of slate		
A		*B*		*C*		Height		Width inside		Inside depth of ends		*D*	*E*	*F*
Height		Width		Depth inside								Backs, in.	Ends, in.	Partitions, in.
Ft	In.	Ft	In.	Ft	In.	Ft	In.	Ft	In.	Ft	In.			
4	0	1	8	1	8			2	0					
4	6	2	0	1	2	4	6	4	0					
5	0	2	0	1	6	and		6	0					
5	6	2	0	1	8	5	0	8	0	1	8	1	1	1
								10	0					
								12	0					
								14	0					

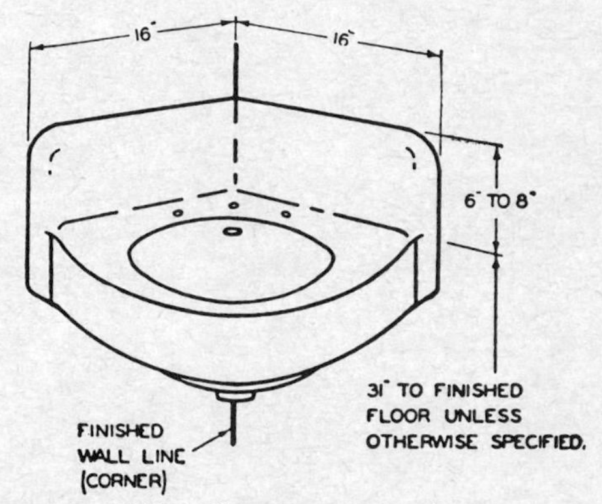

Fig. II-76. Cast-iron corner lavatory.

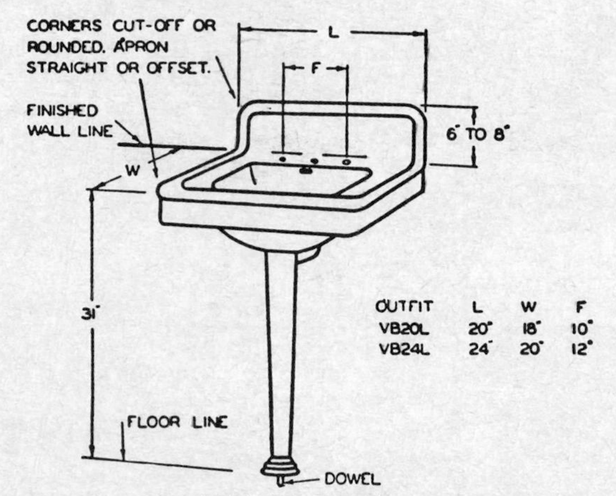

FIG. II-77. Vitreous lavatory.

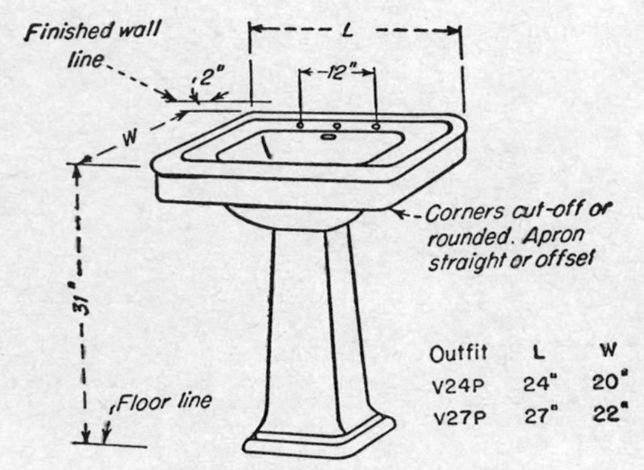

FIG. II-78. Vitreous lavatory.

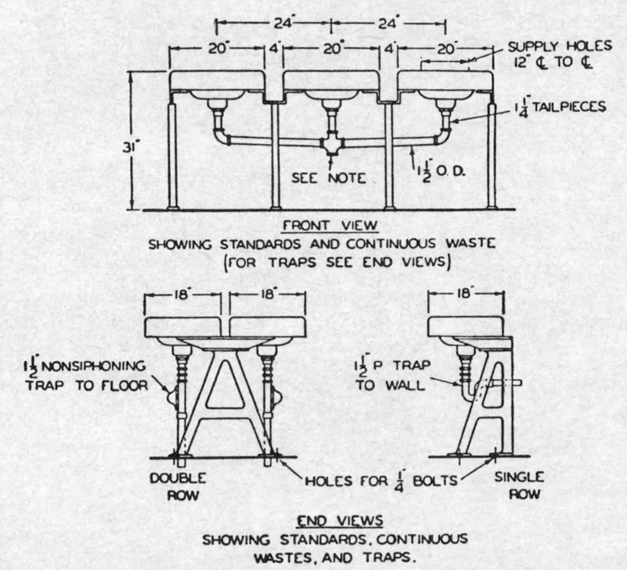

Fig. II-79. Cast-iron lavatories in battery. Note: For rows of three fixtures, slip-joint tee under an end fixture may be used in lieu of slip-joint cross as shown unless otherwise specified. Rows of two fixtures shall have slip-joint tee.

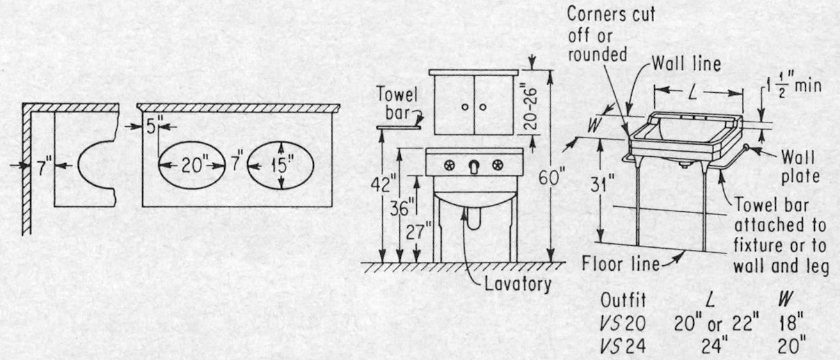

Fig. II-80. Lavatories. Recommended minimum dimensions. Vitreous lavatory with low shelf back.

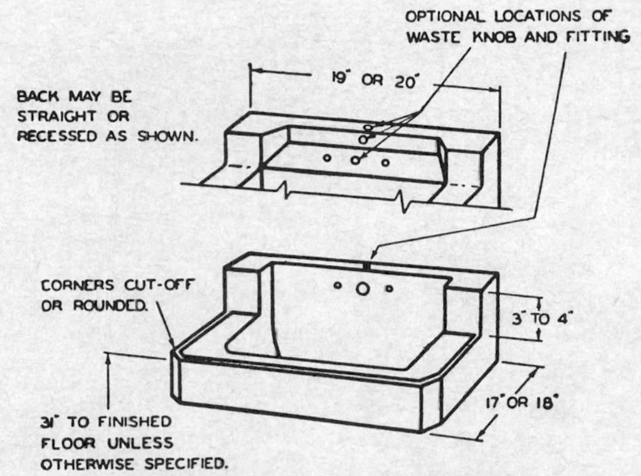

FIG. II-81. Cast-iron lavatory with shelf back.

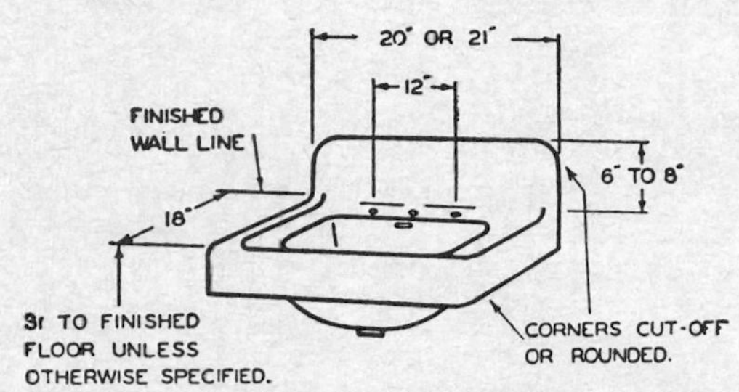

FIG. II-82. Cast-iron lavatory with regular back.

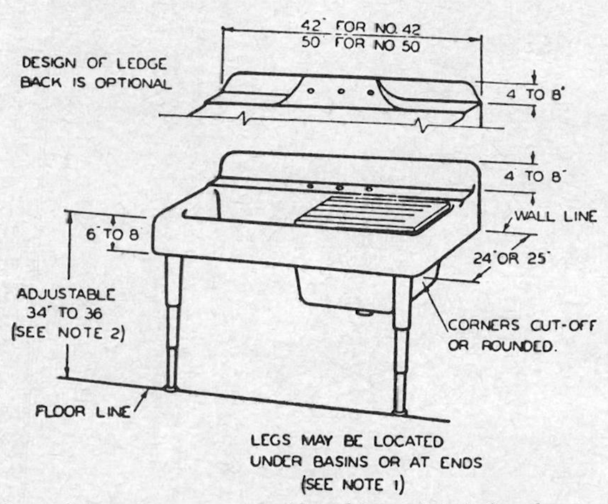

FIG. II-83. Sink and laundry-tray combinations with ledge back. NOTE: Fixture may have either straight or ledge back. Illustration shows fixture with right-hand tray and apron; fixtures with roll rim. Nos. M42 and M50 (not shown) shall have one leg only, located under tray. Install at 36-in. height unless otherwise specified.

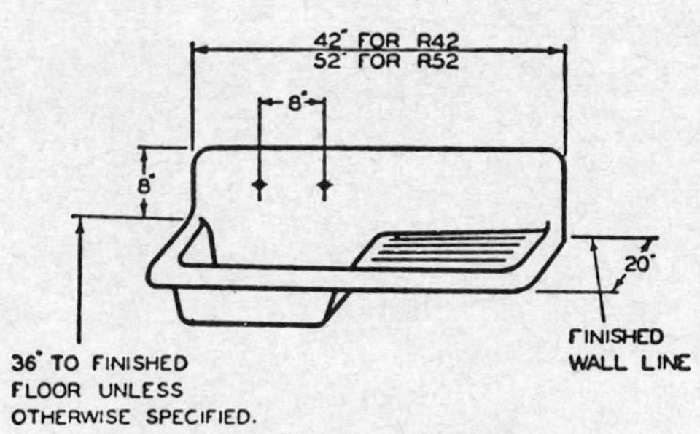

Fig. II-84. Cast-iron kitchen sink.

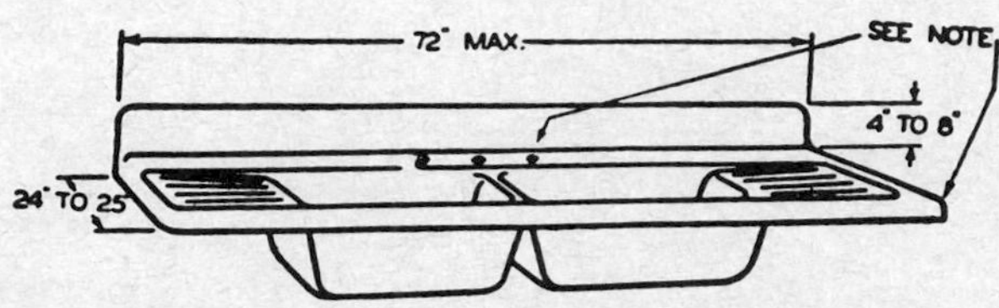

Fig. II-85. Cast-iron kitchen cabinet sink. Note: Holes for supply fittings shall be in ledge or in an angular panel. Corners may be rounded or cut off.

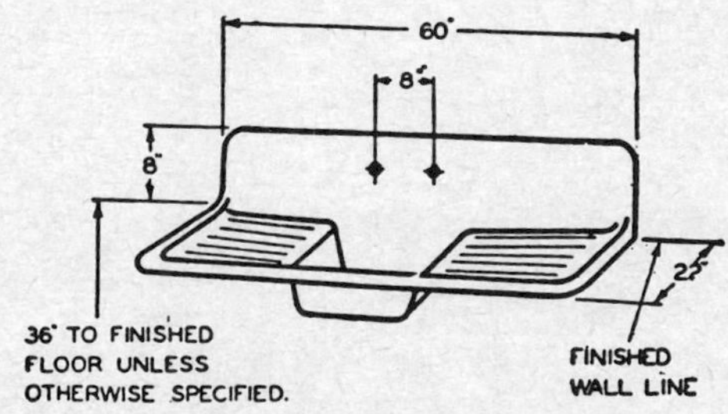

Fig. II-86. Cast-iron kitchen sink.

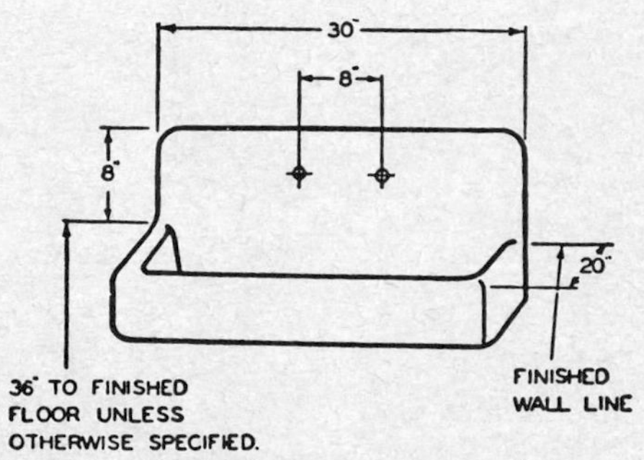

Fig. II-87. Cast-iron kitchen sink.

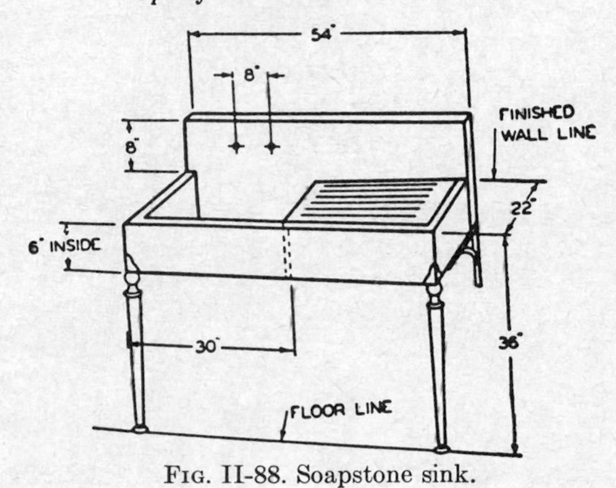

FIG. II-88. Soapstone sink.

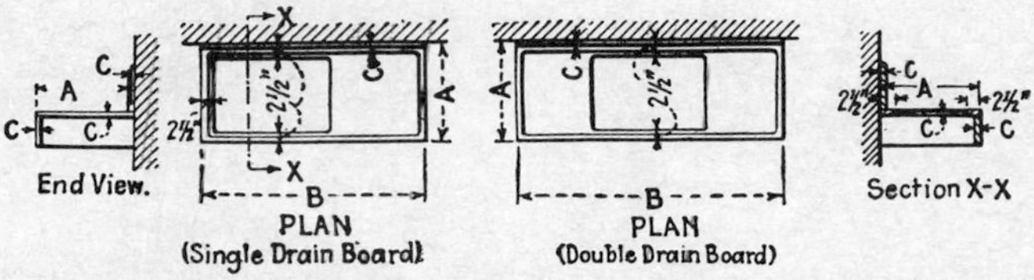

FIG. II-89. Slate sink tops. (See Table II-82.) (*From Simplified Practice Recommendation No.* 13, *approved Aug.* 1, 1924, *by Division of Simplified Practice, U.S. Department of Commerce.*)

TABLE II-82. SLATE SINK TOPS AND SLOP HOPPERS. SLOP HOPPERS ARE WITH OR WITHOUT INTEGRAL BACKS*
(See Fig. II-89)

Slate sink tops only			Slop hoppers—with or without integral backs				
A	*B*	*C*					
Width of tops, in.	Length of tops, in.	Thickness of slate, in.	Over-all width, in.	Over-all length, in.	Inside depth, in.	Style of front	Slate thickness, in.
18 to 22	up to 78	1¼	24	24	12 or 14	Sloping	1¼
24 to 30	above 78	1¼	24	30	12 or 14	Sloping	1¼
			24	36	12 or 14	Sloping	1¼

* From Simplified Practice Recommendation No. 13, approved Aug. 1, 1924, by Division of Simplified Practice, U.S. Department of Commerce.

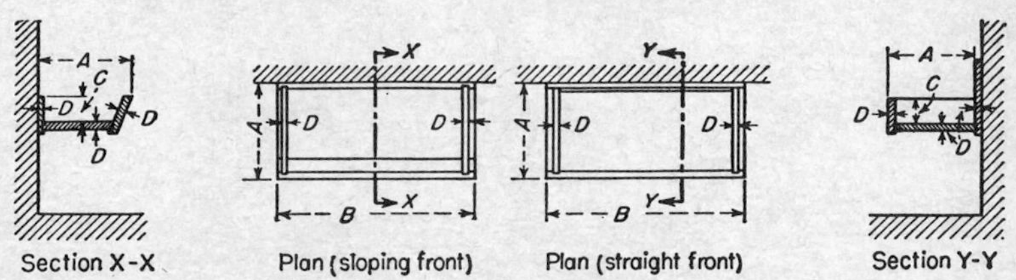

FIG. II-90. Sinks with or without integral backs. (See Table II-83.) (*From Simplified Practice Recommendations No. 13, approved Aug. 1, 1924, by Division of Simplified Practice, U.S. Department of Commerce.*)

TABLE II-83. SINKS WITH OR WITHOUT INTEGRAL BACKS*
(See Fig. II-90)

A Over-all width, in.	*B* Over-all length, in.	*C* Inside depth, in.	Style of front, straight or sloping	*D* Slate thickness, in.	*A* Over-all width, in.	*B* Over-all length, in.	*C* Inside depth, in.	Style of front, straight or sloping	*D* Slate thickness, in.
12	18	6	Yes	1¼	22	36	6	Yes	1¼
18	24	6	Yes	1¼	22	42	6	Yes	1¼
20	30	6	Yes	1¼	24	30	6	Yes	1¼
20	36	6	Yes	1¼	24	36	6	Yes	1¼
22	30	6	Yes	1¼	24	48	6	Yes	1¼

* From Simplified Practice Recommendation No. 13, approved Aug. 1, 1924, by Division of Simplified Practice, U.S. Department of Commerce.

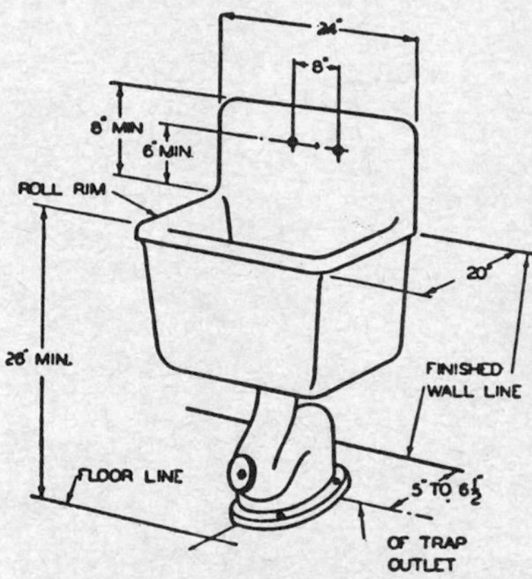

FIG. II-91. Cast-iron service sink.

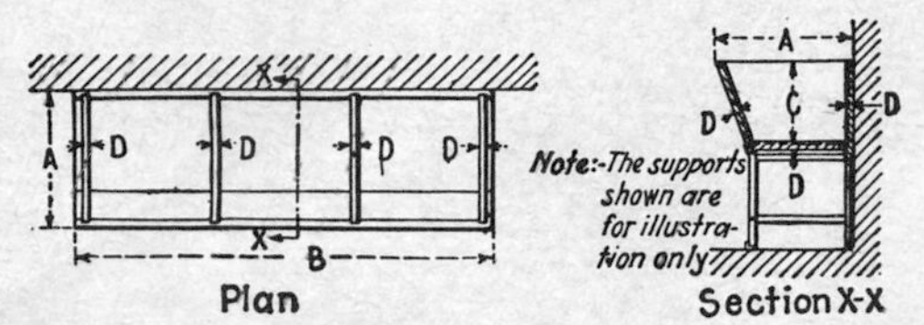

FIG. II-92. Slate laundry tubs with and without integral backs. See Table II-84. (*From Simplified Practice Recommendation No.* 13, *Approved Aug.* 1, 1924, *by Division of Simplified Practice, U.S. Department of Commerce.*)

TABLE II-84. SLATE LAUNDRY TUBS WITH OR WITHOUT INTEGRAL BACKS*
(See Fig. II-92)

Number of equal-sized compartments	*A*	*B*	*C*	*D*
	Compartment width, in.	Length, in.	Inside depth, in.	Slate thickness, in.
1	24	24, 30, and 36	12 or 14	1¼
2	24	48, 54, and 60	12 or 14	1¼
3	24	72, 78, and 84	12 or 14	1¼

* From Simplified Practice Recommendation No. 13, approved Aug. 1, 1924, by Division of Simplified Practice, U.S. Department of Commerce.

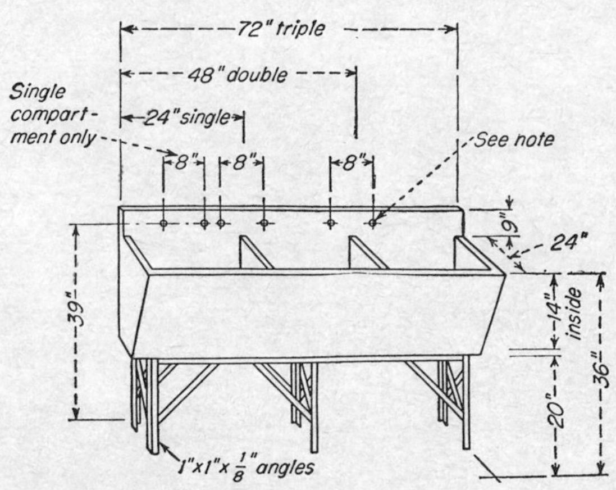

FIG. II-93. Soapstone laundry trays with high back. NOTE: Furnish single and double trays with two supply holes and triple trays with four supply holes.

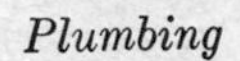

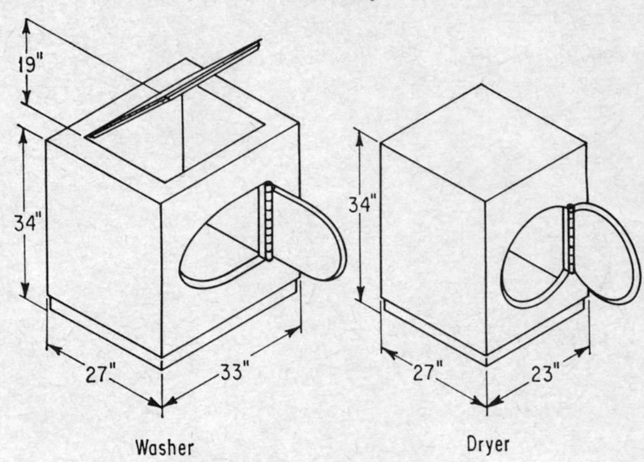

FIG. II-94. Washers and driers.

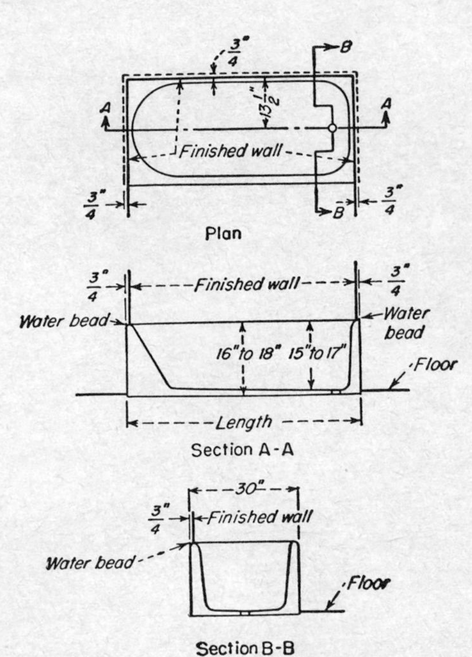

FIG. II-95. Recessed bathtubs, 54-, 60-, and 66-in. lengths. Drain at right or left end as specified.

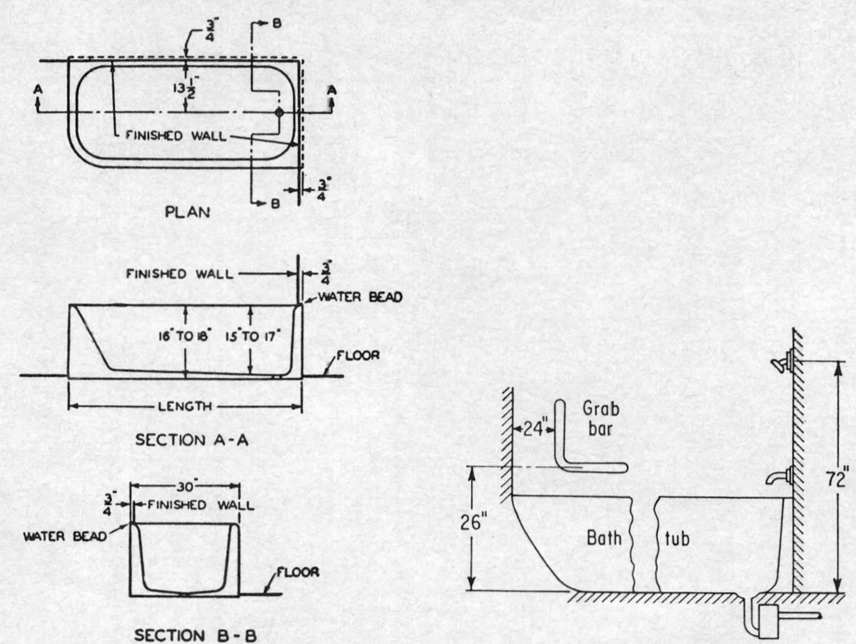

FIG. II-96. Corner bathtubs, 54-, 60-, and 66-in. lengths.

FIG. II-97. Bathtub grab-bar installation.

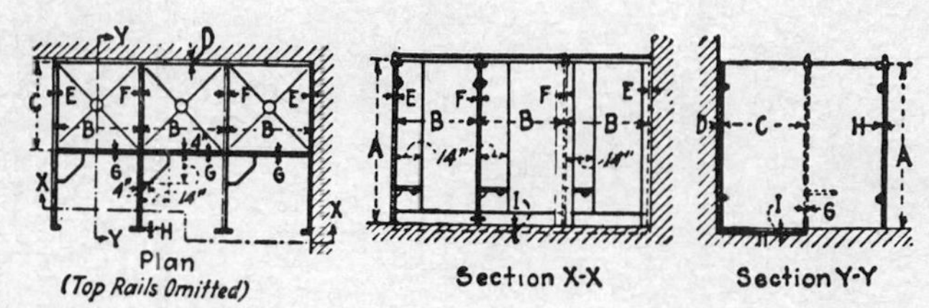

FIG. II-98. Shower stalls. See also Fig. II-68. (*From Simplified Practice Recommendation No.* 13, *approved Aug.* 1, 1924, *by Division of Simplified Practice, U.S. Department of Commerce.*)

TABLE II-85. SHQWER STALLS*
(See Fig. II-98)

A	*B*		*C*		Thickness of slate					
					D	*E*	*F*	*G*	*H*	*I*
Stall height	Stall width		Stall depth		Backs, in.	Ends, in.	Partitions, in.	Curbs, in.	Stiles, in.	Floor slabs, in.
	Ft	In.	Ft	In.						
6 ft 6 in. and 7 ft 0 in.	3	0	3	0	1	1	1	1¼	1¼	2
	3	0	3	6	1	1	1	1¼	1¼	2
	3	6	3	6	1	1	1	1¼	1¼	2

* From Simplified Practice Recommendation No. 13, approved Aug. 1, 1924, by Division of Simplified Practice, U.S. Department of Commerce.

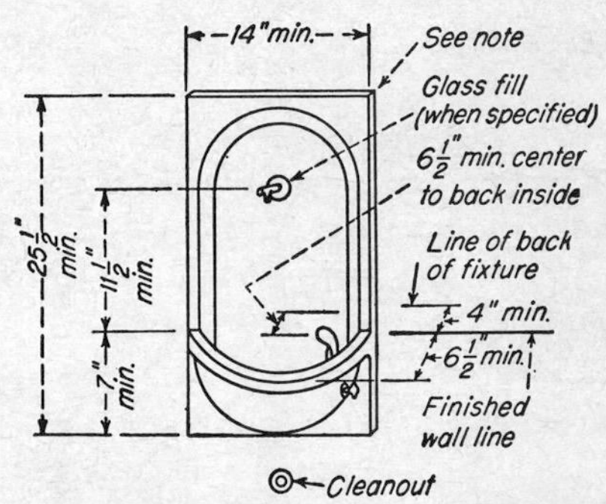

FIG. II-99. Semirecessed drinking fountain. NOTE: Top may be square (with or without mitered corners) or arched, and bottom shall be square unless otherwise specified.

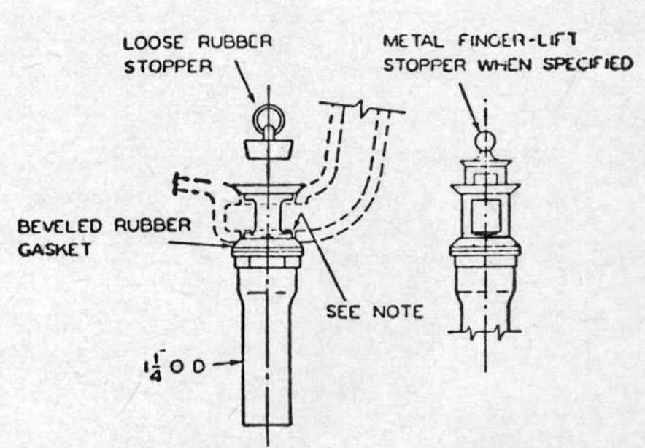

FIG. II-100. Lavatory drain plug (rubber stopper unless metal stopper is specified). NOTE: When the waste plug is assembled to the lavatory, the bottom of the overflow opening in the plug shall be below the bottom of the overflow channel to prevent trapping of water.

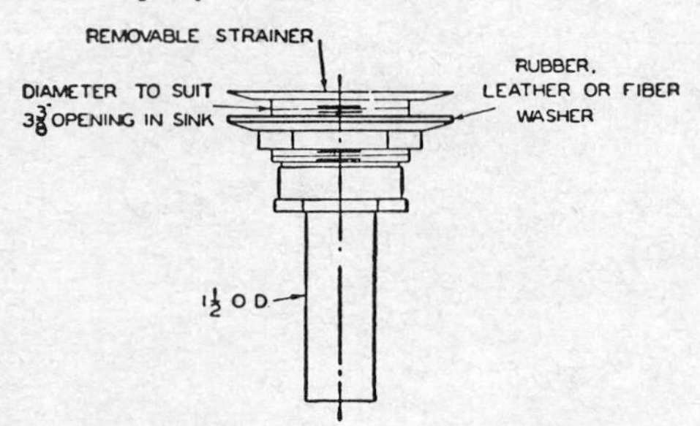

FIG. II-101. Sink-drain plug with flat strainer. NOTE: Minimum diameter of opening in sink, 3⅜ in.

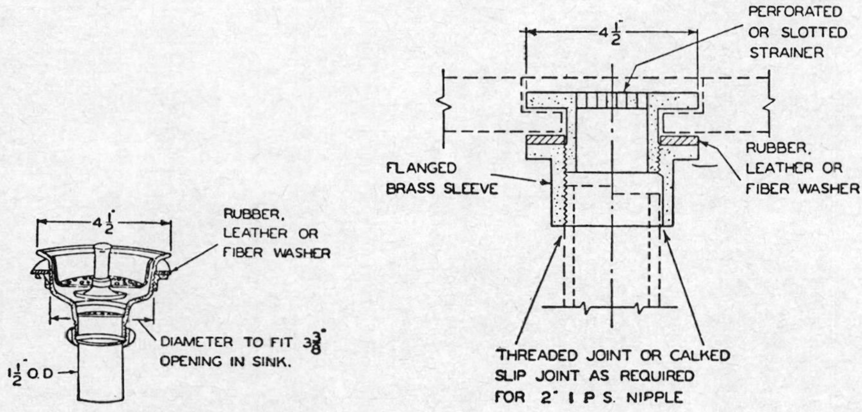

FIG. II-102. Sink-drain plug with cup strainer. NOTE: Minimum diameter of opening in sink, 3⅜ in.

FIG. II-103. Stall-urinal drain plug (waste strainer).

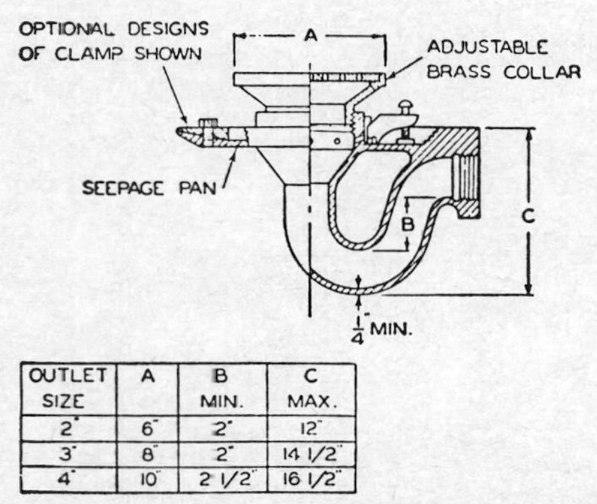

OUTLET SIZE	A	B MIN.	C MAX.
2"	6"	2"	12"
3"	8"	2"	14 1/2"
4"	10"	2 1/2"	16 1/2"

FIG. II-104. Adjustable floor drain with trap and side outlet.

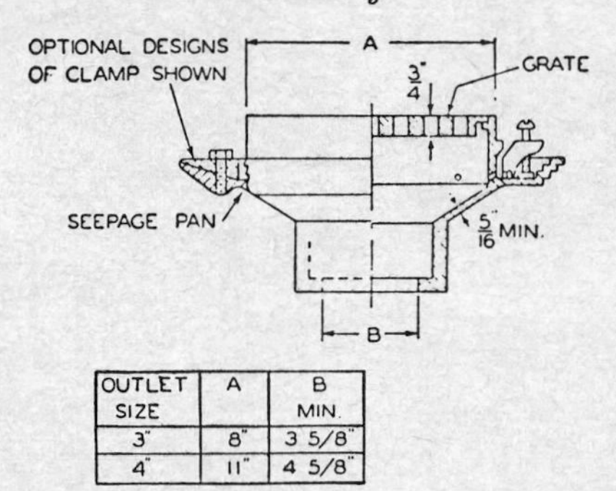

OUTLET SIZE	A	B MIN.
3"	8"	3 5/8"
4"	11"	4 5/8"

FIG. II-105. Floor or area drain with bottom outlet.

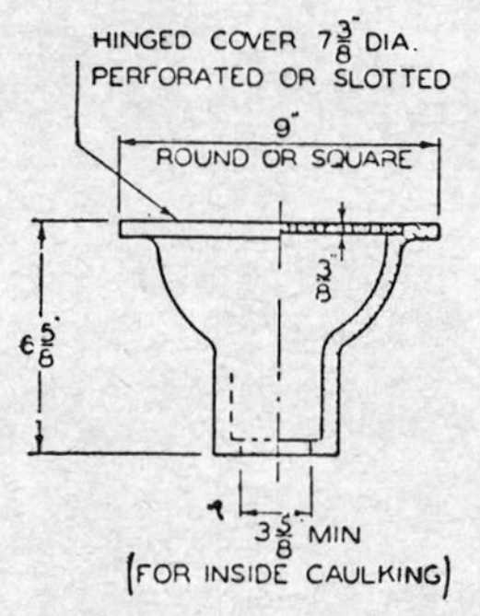

FIG. II-106. Area drain with bottom outlet.

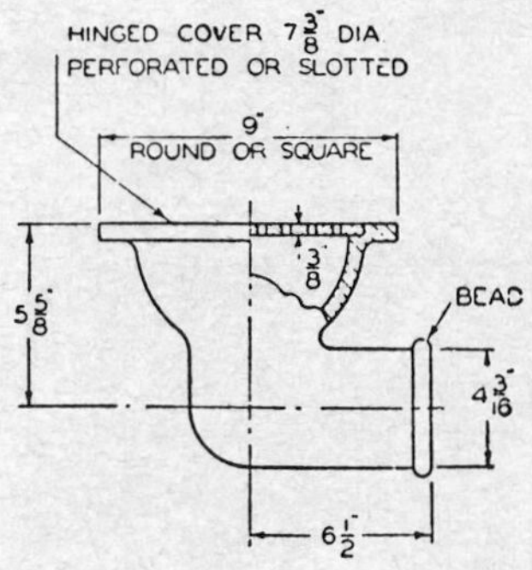

FIG. II-107. Area drain with side outlet.

APPENDIX III

INCORRECT PLUMBING FOR CORRECTION

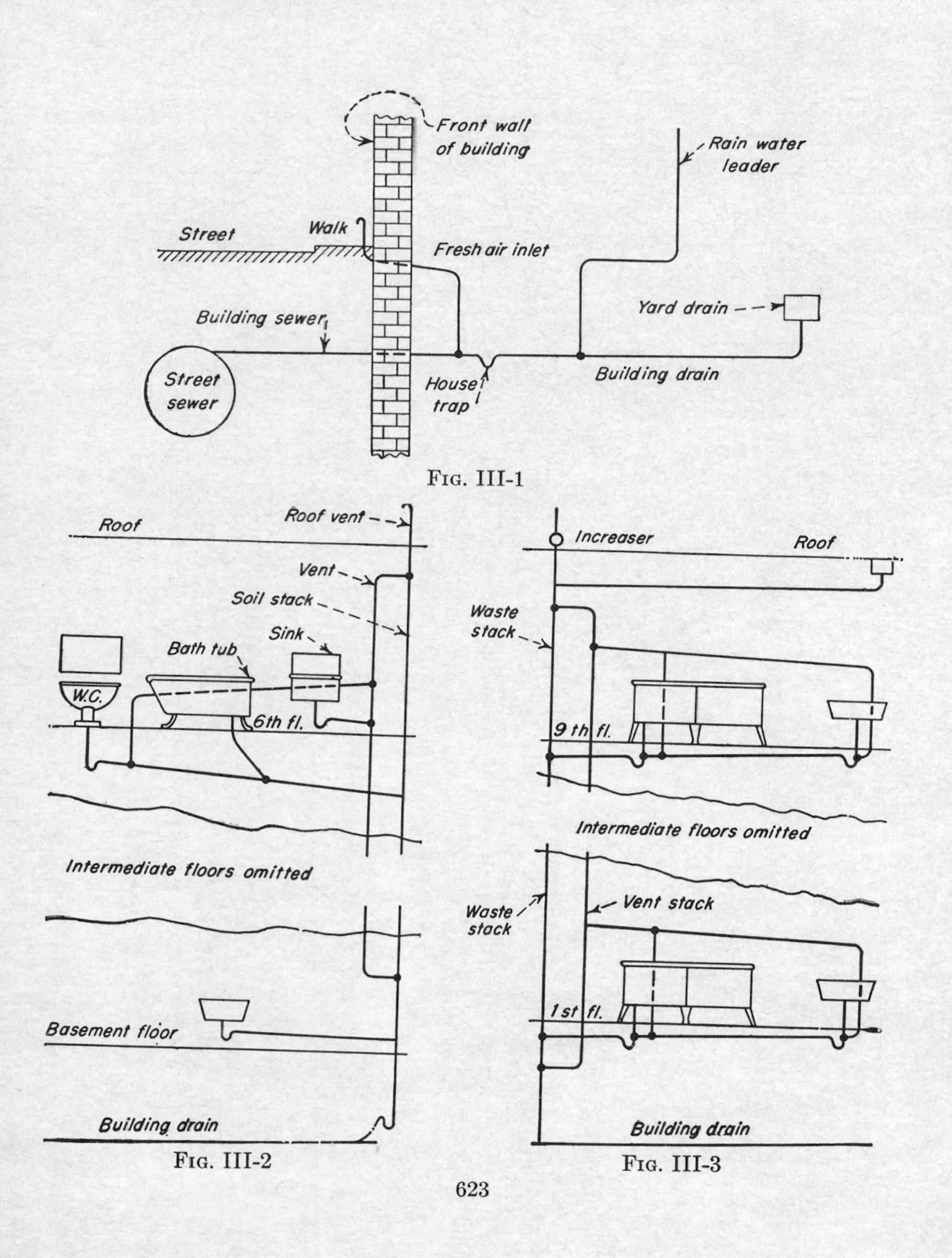

FIG. III-1

FIG. III-2

FIG. III-3

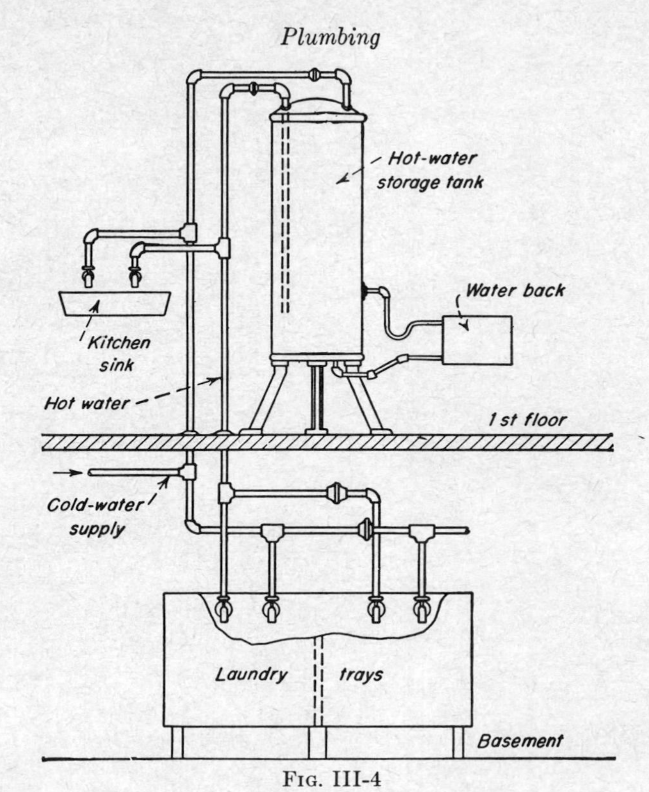
Hot-water storage tank
Water back
Kitchen sink
Hot water
1 st floor
Cold-water supply
Laundry trays
Basement

Fig. III-4

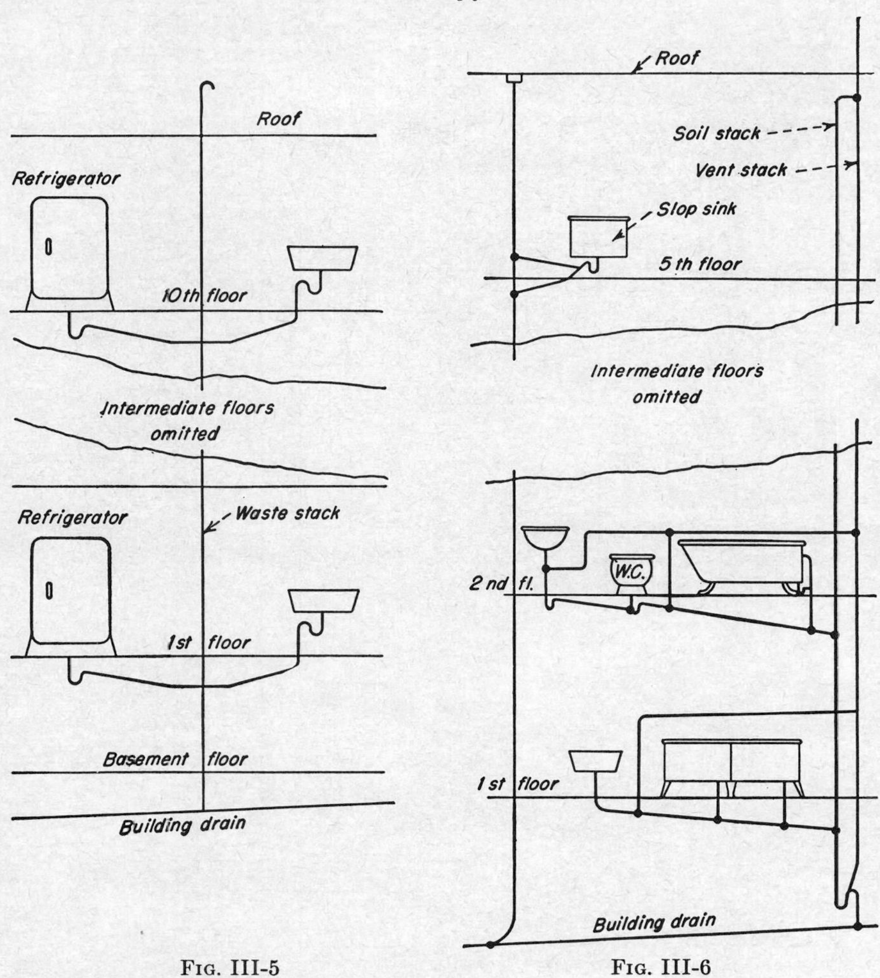

Fig. III-5

Fig. III-6

Fig. III-7. An interior view of a well-equipped bathroom. What is wrong here?

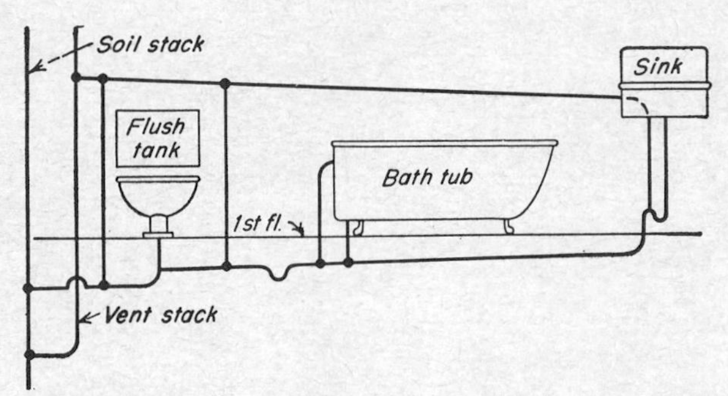

Fig. III-8

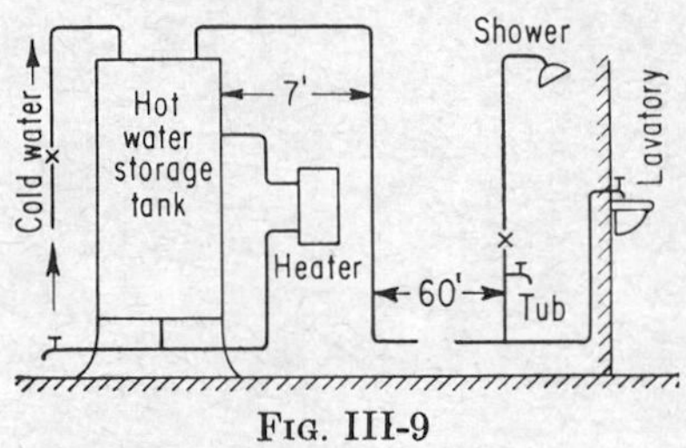

Fig. III-9

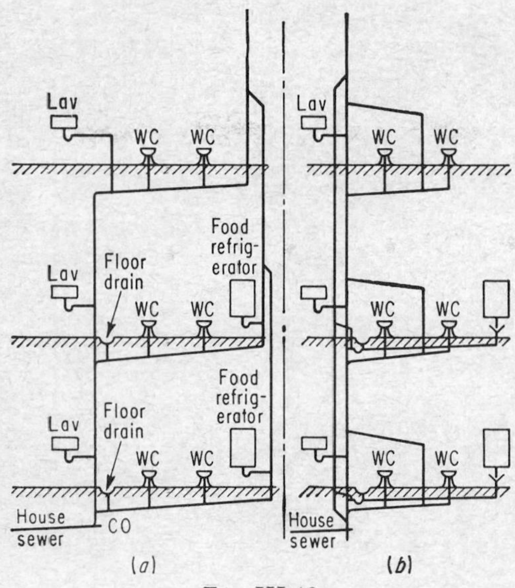

Fig. III-10

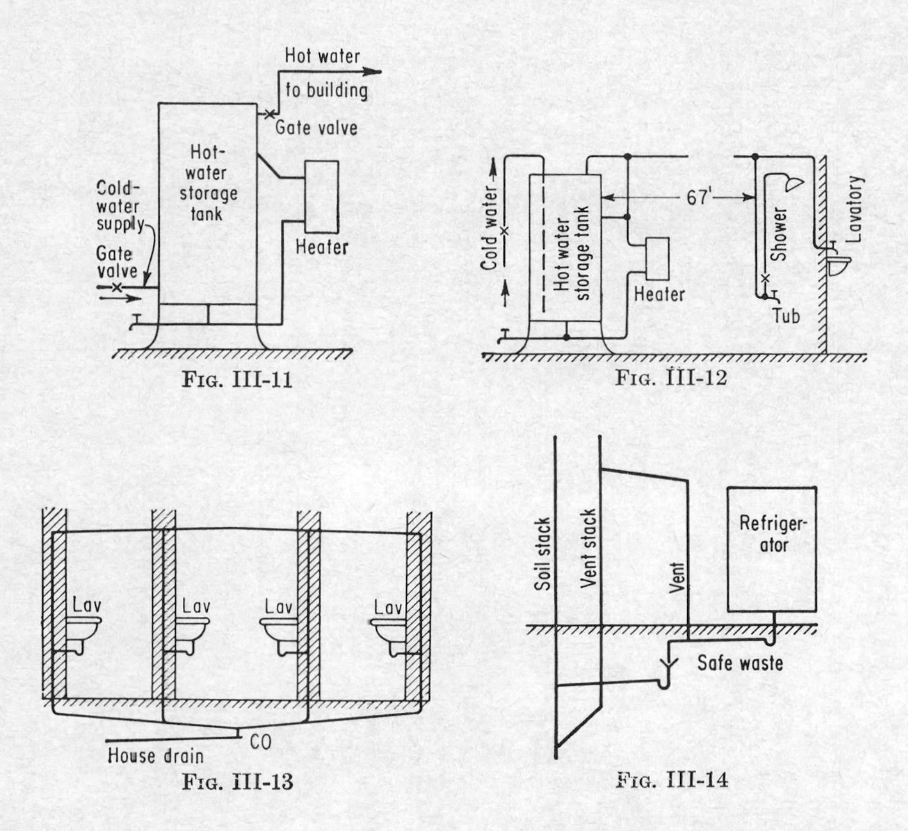

Fig. III-11

Fig. III-12

Fig. III-13

Fig. III-14

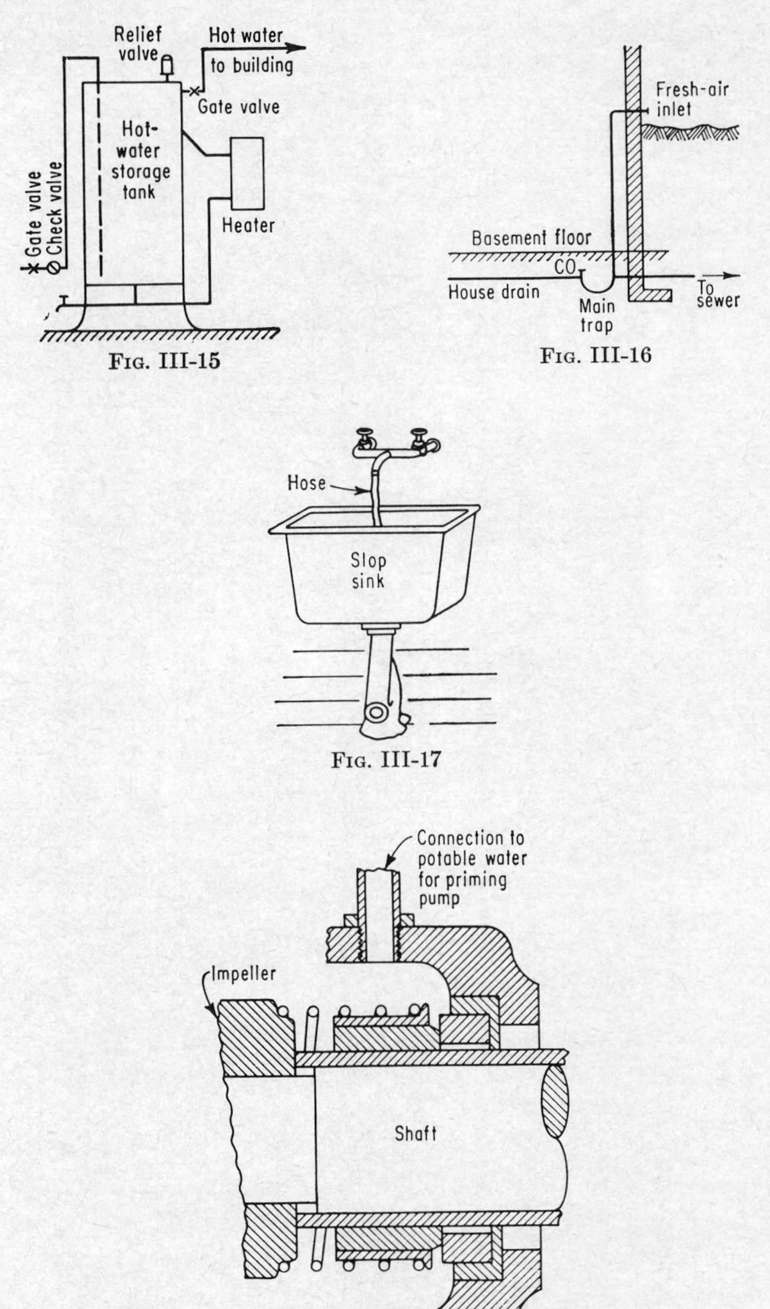

Fig. III-15

Fig. III-16

Fig. III-17

Fig. III-18. Connection of potable water supply for priming a centrifugal pump.

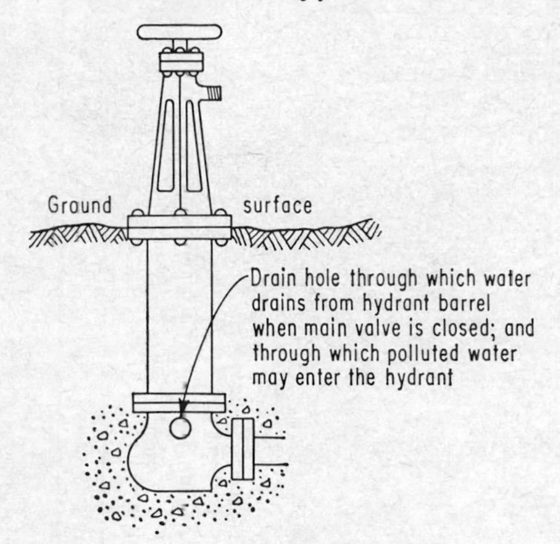

FIG. III-19

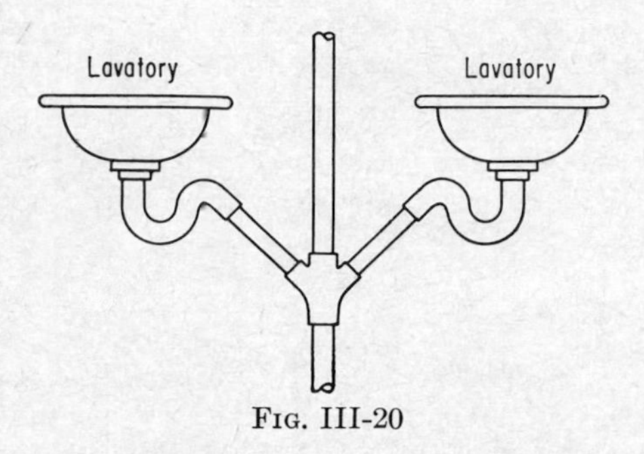

FIG. III-20

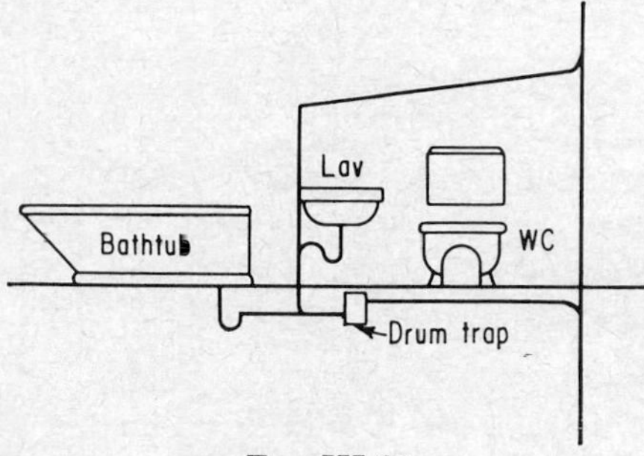

FIG. III-21

Explanation of Some of the Faults and Errors in Figs. III-1 to III-21

III-1. The fresh-air inlet should be connected to the building drain upstream from the house trap.

III-2. Vents should not connect at right angles. There can be no connections on "intermediate floors omitted," since there are insufficient stacks. No trap is shown on bathtub discharge. In a wet vent there should be no fixture between the water closet and the stack. The laundry tray is probably too far from the stack to be satisfactory without a vent.

III-3. Vents should not connect at right angles. Connection of roof drain to stack should be below connection of revent stack to stack vent.

III-4. Cold-water supply is shown at left in kitchen sink; hot- and cold-water supply inlets and outlets are reversed in storage tank.

III-5. Food containers should not drain directly, trapped or otherwise, into a soil pipe—in this case the house drain. Kitchen sink trap is too far from stack to remain unvented.

III-6. Vent from slop sink enters the stack vent too low; i.e., it is below the flood rim level of the slop sink. It should not be. Branch drain pipe on second floor is apparently double-trapped. Lavatory vent is apparently below flood rim of the lavatory. There is a bypass from the sewer through the bathtub overflow. No trap is shown on any fixtures on the first floor. The trap at the base of the stack above the building drain is unusual, unvented, unnecessary, and undesirable. It will cause strong back pressures in the drainage pipes on the lower floors of the building.

III-7. Inlets to both lavatories and to the bathtub are below the rim and are evidently not protected against back siphonage. No siphon breaker is shown on the bidet.

III-8. Vent should be above flood rim of sink. Trap on sink is crown-vented.

III-9. Cold water should enter at the bottom of the heater. Hot water should not be forced to flow down from the heater to the fixture. Correct conditions are shown in Fig. III-12.

III-10. (*a*) Venting is inadequate. Various fixtures will lose seals when others are discharged. (*b*) Shows corrections to erroneous installations in *a*.

III-11. It will be possible for water to drain from heater back into water-supply pipe in the event of low pressure, maybe resulting in explosion or other damage. The gate valve on the water-supply pipe, without pressure relief, invites trouble.

III-12. Nothing is wrong here. The installation corrects the condition shown in Fig. III-10.

III-13. No acceptable vent exists. The condition approaches a "utility vent," but the discharge of two fixtures simultaneously might result in self-siphonage.

III-14. The safe vent creates a bypass. In general, the venting of a safe waste into a plumbing system will cause a bypass.

III-15. There is nothing wrong here. It is a correction of the condition shown in Fig. III-11.

III-16. The vent is on the wrong side of the house trap and causes a bypass.

III-17. The hose hanging in the slop sink is a potential cross connection.

III-18. A cross connection is formed at the water connection to the sewage pump.

III-19. A cross connection is formed at the base of the hydrant. Sewage or polluted ground water may enter the hydrant through the drain hole.

III-20. The two wastes should discharge into the stack at a higher elevation with respect to the trap. As shown there will be self-siphonage.

III-21. The lavatory and the bathtub are double-trapped.

INDEX